3S Technology Applications in Meteorology

Spatial information technology and its integration, such as remote sensing, geographic information systems (GIS), and global navigation satellite systems (GNSS), known as 3S technology, have been extensively utilized in managing and monitoring natural disasters. This book illustrates the 3S integrated applications in the field of meteorology and promotes the role of 3S in developing precise and intelligent meteorology. It presents the principles of 3S technology and the methods for monitoring different meteorological disasters and hazards as well as their application progress. The case studies from the United States, Japan, China, and Europe were conducted to help all countries understand the 3S technology functions in handling and monitoring severe meteorological hazards.

FEATURES

- Presents integral observations from GNSS, GIS, and remote sensing in estimating and understanding meteorological changes
- Explains how to monitor and retrieve atmospheric parameter changes using GNSS and remote sensing
- Shows three-dimensional modelling and evaluations of meteorological variation processing based on GIS
- Helps meteorologists develop and use space-air-ground integrated observations for meteorological applications
- Illustrates the practices in monitoring meteorological hazards using space information techniques and case studies

This book is intended for academics, researchers, and postgraduate students who specialize in geomatics, atmospheric science, and meteorology, as well as scientists who work in remote sensing and meteorology, and professionals who deal with meteorological hazards.

3S Technology Applications in Meteorology

Observations, Methods, and Modelling

Edited by
Shuanggen Jin

CRC Press
Taylor & Francis Group
Boca Raton London New York

CRC Press is an imprint of the
Taylor & Francis Group, an **informa** business

Designed cover image: © Shuanggen Jin

First edition published 2024
by CRC Press
2385 NW Executive Center Drive, Suite 320, Boca Raton FL 33431

and by CRC Press
4 Park Square, Milton Park, Abingdon, Oxon, OX14 4RN

CRC Press is an imprint of Taylor & Francis Group, LLC

ISBN: 978-1-032-42513-9 (hbk)
ISBN: 978-1-032-42514-6 (pbk)
ISBN: 978-1-003-36311-8 (ebk)

DOI: 10.1201/9781003363118

Typeset in Times
by Apex CoVantage, LLC

Contents

Editor

Shuanggen Jin, PhD, is Vice-President and Professor at Henan Polytechnic University, Jiaozuo, China; Professor at Shanghai Astronomical Observatory, CAS, Shanghai, China; and Professor at Nanjing University of Information Science and Technology, Nanjing, China. He earned a BS in geodesy at Wuhan University, China, in 1999; and a PhD in geodesy at University of Chinese Academy of Sciences, China, in 2003. He has worked on satellite navigation, remote sensing, and space/planetary exploration with significant original contributions in the theory, methods, and applications of GNSS remote sensing. He has published over 500 peer-reviewed papers and 12 books/monographs with more than 12,000 citations and H-index >55, as well as more than 20 patents/software copyrights and over 100 invited talks. He has led over 30 projects from Europe's ESA, China-Germany (NSFC-DFG), and China's National Key R&D Program and NSFC. He has supervised over 30 PhD and 50 MSc candidates as well as 20 postdocs.

Professor Jin has been Chair of the IUGG Union Commission on Planetary Sciences, President of International Association of Planetary Sciences (IAPS), President of the International Association of CPGPS, Editor-in-Chief of *International Journal of Geosciences*, Editor-in-Chief of *Journal of Environmental and Earth Sciences*, Editor of *Geoscience Letters*, Associate Editor of *IEEE Transactions on Geoscience and Remote Sensing* and *Journal of Navigation*, Editorial Board member of *Remote Sensing*, *GPS Solutions*, and *Journal of Geodynamics*.

He received First Prize of Satellite Navigation and Positioning Progress Award, First Prize of China Overseas Chinese Contribution Award, 100-Talent Program of CAS, World Class Professor of Ministry of Education and Cultures, Indonesia, Chief Scientist of National Key R&D Program, China, Elsevier China Highly Cited Scholar, World's Top 2% Scientists, Fellow of the Electromagnetics Academy, Fellow of the International Union of Geodesy and Geophysics (IUGG), Fellow of the International Association of Geodesy (IAG), Fellow of the African Academy of Sciences, Member of the European Academy of Sciences, Member of the Russian Academy of Natural Sciences, Member of the Turkish Academy of Sciences, and Member of Academia Europaea.

Contributors

Mohamed Abdallah Ahmed Alriah
Nanjing University of Information Science
    and Technology
Nanjing, China

Andres Calabia
University of Alcala
Alcalá de Henares
Madrid, Spain
and
Henan Polytechnic University
Jiaozuo, China

Tingting Chen
College of Atmospheric Sciences
Lanzhou University
Lanzhou, China

Mohamed Darrag
Nanjing University of Information Science
    and Technology
Nanjing, China

Haiyong Ding
Nanjing University of Information Science
    and Technology
Nanjing, China

Jiajing Du
College of Atmospheric Sciences
Lanzhou University
Lanzhou, China

Zexia Duan
Nanjing University of Information Science
    and Technology
Nanjing, China

Ayman M. Elameen
Nanjing University of Information Science
    and Technology
Nanjing, China

Lei Fan
School of Geographical Sciences
Southwest University
Chongqing, China

Jinming Ge
College of Atmospheric Sciences
Lanzhou University
Lanzhou, China

Minmin Huang
Chuzhou University
Chuzhou, China

Weimin Huang
Faculty of Engineering and Applied
    Science
Memorial University
St. John's, Newfoundland and Labrador,
    Canada

Yan Jia
Nanjing University of Posts and
    Telecommunications
Nanjing, China

Shuanggen Jin
Henan Polytechnic University
Jiaozuo, China
and
Shanghai Astronomical Observatory
Chinese Academy of Science
Shanghai, China
and
Nanjing University of Information
    Science and Technology
Nanjing, China

Qinghao Li
College of Atmospheric Sciences
Lanzhou University
Lanzhou, China

Yize Li
College of Atmospheric Sciences
Lanzhou University
Lanzhou, China

Zheyu Liang
College of Atmospheric Sciences
Lanzhou University
Lanzhou, China

Chao Liu
Nanjing University of Information Science
and Technology
Nanjing, China

Xiao Liu
First Monitoring and Application Center
China Earthquake Administration
Tianjin, China

Yidong Lou
GNSS Research Center
Wuhan University
Wuhan, China

Usman Mazhar
Nanjing University of Information Science
and Technology
Nanjing, China
and
University of the Punjab
Lahore, Pakistan

Yuxing Song
Nanjing University of Information Science
and Technology
Nanjing, China

Lin Sun
Shandong University of Science
and Technology
Qingdao, China

Zhonghui Tan
National University of Defense
Technology
Changsha, China

Shiwen Teng
Ocean University of China
Qingdao, China

Ershen Wang
Shenyang Aerospace University
Shenyang, China

Mengya Wang
Nanjing University of Information Science
and Technology
Nanjing, China

Shuai Wang
State Grid Shanxi Electric Power
Research Institute
Taiyuan, China

Yong Wang
Tianjin Chengjian University
Tianjin, China

Jing Wei
University of Maryland
College Park, Maryland, USA

Jean-Pierre Wigneron
INRAE, Bordeaux Sciences Agro
Villenave-d'Ornon, France

Yang Xia
College of Atmospheric Sciences
Lanzhou University
Lanzhou, China

Zhiyu Xiao
Nanjing University of Posts and
Telecommunications
Nanjing, China

Yunjian Xie
Nanjing University of Information Science
and Technology
Nanjing, China

Qingyun Yan
Nanjing University of Information Science
and Technology
Nanjing, China

Yuanjian Yang
Nanjing University of Information Science
and Technology
Nanjing, China

Tengli Yu
Shenyang Aerospace University
Shenyang, China

Chi Zhang
College of Atmospheric Sciences
Lanzhou University
Lanzhou, China

Weixing Zhang
GNSS Research Center
Wuhan University
Wuhan, China

Zhenyi Zhang
GNSS Research Center
Wuhan University
Wuhan, China

Zhixuan Zhang
GNSS Research Center
Wuhan University
Wuhan, China

Wenjian Zheng
Maintenance and Test Center of CSG
EHV Power Transmission Company
Guangzhou, China

Shaohui Zhou
Nanjing University of Information Science
    and Technology
Nanjing, China

Yaozong Zhou
GNSS Research Center
Wuhan University
Wuhan, China

1 A Review of 3S Technology and Its Applications in Meteorology

Haiyong Ding and Shuanggen Jin

1.1 BACKGROUND

Meteorological disaster is the direct or indirect damages caused by the atmosphere to human life, economy, and national defense construction, which usually includes climate disasters and derived disasters (Guan et al., 2015; Ye, 2022). Climate disasters are directly caused by severe weather such as typhoons, rainstorms, blizzards, thunderstorms, hail, high temperatures, drought and other meteorological factors (Allan et al., 2006). The derived meteorological disasters are caused by meteorological factors. According to statistics, nearly 90% of natural disasters were related to the bad weather (United Nations, 2015). The derived meteorological disasters mainly include forest fires, debris flow and landslides. Traditional meteorological disaster monitoring mainly uses quantitative observation instruments. For example, anemometers, rain gauges and temperature measuring instruments are used to monitor high wind hazards, urban rainstorms and high temperatures, respectively. Traditional meteorological disaster monitoring methods have various deficiencies. For instance, mechanical rain gauges such as siphon type gauges are mainly used to monitor rainstorms (Li et al., 2010), while such instruments are cumbersome in operation and lack self-adjustment and self-adaptation ability. In addition, the monitoring of meteorological hazards requires not only a conventional meteorological observation network, but also a focus on the structural changes in the atmospheric boundary layer that affect human activities and the human habitat, as well as the impact of changes in the subsurface on meteorological conditions (Wang et al., 2009). However, such detection is difficult to achieve, only relying on traditional ground-based observation instruments.

3S technology specifically refers to the Global Navigation Satellite System (GNSS), Remote Sensing (RS), and Geographic Information System (GIS) (Zhang, 2020; Li, 2003). 3S technology, as the core technology of spatial informatics, can provide a full range of resource and environmental data and quickly obtain multi-platform, multi-temporal, multi-band, high-precision and high-resolution massive space-time information (Li, 2003; Li et al., 1998). The combination of 3S technology observations in meteorology can overcome the shortcomings of traditional meteorological instruments such as low accuracy and poor self-adaptive capability for observing certain meteorological parameters. For example, images acquired from satellite remote sensing can draw a map of water distribution in the atmosphere and rainfall distribution through light recognition equipment and infrared sensors, which can effectively improve the observation ability of rainfall in heavy rainstorms. GNSS Radio Occultation (GNSS RO) can achieve high vertical resolution and spatial coverage of the atmospheric parameter profiles, which can effectively solve the problems of poor stability and low vertical resolution of traditional observation means such as radar and reanalysis data and effectively improve the monitoring ability of meteorological disasters. 3S technology can not only enrich the means of meteorological observations and improve the ability of meteorological observations, but also collect, store and analyze meteorological data by GIS. For example, GIS can carry out statistical analysis of meteorological disaster data, assess meteorological disasters risks and improve the ability to defend against meteorological disasters. GIS can also evaluate the disaster

DOI: 10.1201/9781003363118-1

situation of meteorological disasters through the integration and analysis of information. Therefore, 3S technology has become an important means of meteorological observation, processing and analysis, and can improve the comprehensive management and assessment of meteorological disasters (Li, 2003; Li, 1998). With the development of 3S, more closely combined meteorological parameter observation and disaster monitoring and forecast will be constantly improved.

1.2   3S TECHNOLOGY AND DEVELOPMENT

1.2.1   BASIC CONCEPTS

The "3S" is the general name of Global Navigation Satellite System (GNSS), Geographic Information System (GIS), and Remote Sensing (RS), which is the integration of space technology, satellite navigation and positioning technology, computer technology, sensor technology and other disciplines (Li, 2003). The 3S technology system is used to quickly acquire spatial data using RS and GNSS and use GIS as the basic platform to store, manage and analyze spatial information.

The GNSS is a satellite-based radio timing and navigation system that provides high precision time and spatial location data for users in aviation, space, land and sea, online or offline (Lechner, 2000; Dow, 2009; Norman, 2012). The GIS refers to the representation of geospatial objects' nature, characteristics and motion state and all relevant and useful knowledge. GIS is a computer system that collects, stores, manages, analyzes, displays and applies geographic information (Goodchild, 2009). In a broad sense, RS refers to remote sensing technology for objects or natural phenomena without direct touch (Campbell & Wynne, 2011). In a narrow sense, RS is a modern technical science that uses various sensors (such as cameras and radars) to acquire surface information on various platforms at high altitudes and in outer space, and studies the shape, size, position and nature of ground objects and their relationship with the environment through data transmission and processing. GNSS and RS are respectively used to acquire point and surface spatial information or monitor its changes, while GIS is employed to store, analyze and process spatial data. Due to the obvious complementarity of the 3S technologies, one gradually realizes in practice that when 3S techniques are integrated in a unified platform, their respective advantages can be fully played.

Since the 1990s, 3S integration has attracted increasing attention and gradually developed into a new interdisciplinary discipline: geomatics (Gomarasca, 2009). But before that, the three went through independent and parallel development. RS obtains real-time, rapid geometric and physical qualitative or quantitative data on large areas of the landscape and environment. GNSS provides real-time or quasi-real-time target positioning information. GIS is a platform for storing, managing, analyzing and applying data from various sources. These three technologies have different characteristics. The 3S technology integrates the relevant parts of the three separated technologies to form a powerful and integrated system. It can provide users with accurate data and map information and realize the collection, processing and update of various spatial and environmental information quickly, accurately and reliably. As shown in Figure 1.1, the relationship of 3S technology is more like "one brain and two eyes".

The comprehensive application of 3S technology is one of the hot topics in current informatization applications. It is also a comprehensive informatization means to realize dynamic acquisition, editing and processing, storage management, analysis and mining of spatial information. The GNSS, RS and GIS are independent and complementary, which are widely used in many fields, such as meteorological disaster monitoring, intelligent transportation, flood monitoring, drought prevention and control, land management, landslide warning and ecological and environmental protection. The application of 3S technology in these fields can make good use of its advantages, enrich observation data, improve observation capacity, and achieve information integration, processing, analysis, prediction and display through GIS with improving the observation and governance level. The different needs of each application area have also contributed to and improved the continuous development of 3S technology.

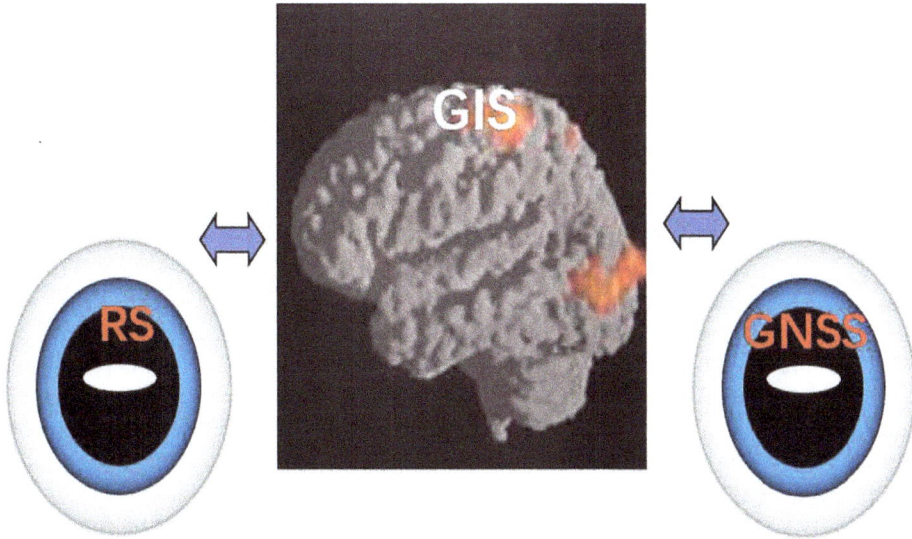

FIGURE 1.1 The relationship between GIS, RS and GNSS.

1.2.2 3S Technology Development

1.2.2.1 GNSS

Global Navigation Satellite System (GNSS) is a space-based radio navigation and positioning system that can provide users with all-weather three-dimensional coordinates, speed and time information at any place on the Earth's surface or near-Earth space (Wang, 2005; Jin, 2012). GNSS mainly includes four global navigation satellite systems, namely, China's BeiDou Navigation Satellite System (BDS), the United States' Global Positioning System (GPS), Russia's GLONASS satellite navigation system (GLONASS) and the European Union's GALILEO satellite navigation system (GALILEO).

GPS is the first global positioning system established and applied to navigation and positioning in the world. The US Department of Defense began to build GPS in 1973 and launched the first test satellite in 1978, and the entire GPS was completed in March 1994. The construction of GPS is divided into three stages. The first stage is the project demonstration and preliminary design stage. From 1973 to 1979, 4 experimental satellites were launched, and the ground receiver and tracking network were developed. The second stage is the comprehensive development and test stage as well as the networking stage, and the positioning accuracy of GPS was verified. The third stage is the practical networking stage, which began with the successful launch of the first GPS operational satellite in 1989 and ended with the complete completion of the GPS in 1994. The GPS satellite constellation is a combination of satellites in space, distributed in six orbits covering the entire Earth. The GPS has an orbital altitude of 20,200 km, an orbital inclination of 55 degrees, and an operating cycle of 11 hours and 58 minutes. GPS is one of the most significant achievements of space technology in the twentieth century. The emergence of GPS has expanded mapping and positioning technology from land and offshore to the entire ocean and outer space, from static to dynamic, from post-processing to real-time (quasi-real-time) with absolute and relative accuracy. The absolute and relative accuracy of GPS reaches the level of meter, centimeter and even sub-millimeter, which greatly broadens its application range and role in all walks of life.

The GPS consists of three main components: the space satellite component, the ground monitoring component and the user equipment. The ground monitoring part is composed of several tracking stations distributed all over the world. It is divided into master control station, monitoring stations, and ground antennas (Jin, 2012). The master control station is located at Falcon Air Force Base, Colorado. Its function is to calculate the correction parameters of satellite ephemeris and satellite clocks according to the GPS observation data of each monitoring station and send these data into the satellite through the ground antennas. At the same time, it also controls the satellite, issues instructions to the satellite and dispatches the standby satellite to replace the failed operational satellite. The master control station also performs a monitoring station. There are five monitoring stations. In addition to the main station, the other four are in Hawaii, the Ascension Islands, Diego Garcia and Kwajalein. The function of these stations is to receive satellite signals, monitor the working state of the satellite and provide satellite observation data for the master control station. Each of the five monitoring stations uses a GPS receiver to conduct integral Doppler observations and pseudo-distance measurements for each visible satellite and collect meteorological element data every six minutes. There are three injection stations, which are in the Ascension Islands, Diego Garcia and Kwajalein. The function of the injection station is to send the satellite ephemeris and clock corrections calculated by the master control station into the satellite. The station sends the ephemeris of 14 days three times a day each time. It also automatically sends signals to the master control station to report its operational status in minutes.

GLONASS is a satellite positioning system similar to the GPS, which was built by the Soviet Union in 1976. It went through several twists and turns, experienced the collapse of the Soviet Union, and is now managed by the Russian Space Agency. From 1982 to 1985, the Soviet Union successfully launched 3 simulated and 18 prototype satellites for testing. Due to the limited technology at that time, the average time of these satellites in orbit was only 14 months. In 1985, the GLONASS navigation system was officially under construction, and in 1985–1986, 6 real GLONASS satellites were launched. These satellites had improved frequency accuracy over the prototype, but the satellite life was still poor with an average of only 16 months. Since then, the Soviet Union has launched another 12 satellites with an average life span of 2 years. By 1987, GLONASS had launched a total of 30 satellites and realized 9 operational satellites in orbit. From 1988 to 2000, GLONASS launched 54 satellites and further improved the service life of the satellites. In 1996, the GLONASS space constellation was completed, and the system entered the phase of full operation and daily updates and maintenance.

The GLONASS also consists of three parts: satellite constellation, ground monitoring control station and user equipment (Hofmann-Wellenhof et al., 2007). The GLONASS satellite constellation consists of 24 satellites and one spare satellite evenly distributed in 3 nearly circular orbital planes, with 8 satellites in each orbital plane. As of 8 March 2015, GLONASS has 28 satellites with 24 satellites in operation, one in reserve, two in flight test, and one in inspection. The ground support system consists of a system control center, a central synchronizer, telemetry and remote-control stations (including laser tracking stations) and outfield navigation control equipment, all located in Russia. The system control center and central synchronization processor are located in Moscow, and the telemetry and remote-control stations are located in St. Petersburg, Ternopol, Yeniseysk and Komsomolskaya. GLONASS user equipment (receiver) can receive the navigation signal transmitted by the satellite, convert it into pseudo range and pseudo range change rate, and simultaneously extract and process the navigation message from the satellite signal. The receiver processor can process the these data and calculate the user's position, speed and time information. Unlike the GPS in the United States, GLONASS uses Frequency Division Multiple Access (FDMA) to distinguish satellites by carrier frequency (GPS is Code Division Multiple Access [CDMA], which distinguishes satellites by modulation code). The single point positioning accuracy of GLONASS is 16 m horizontally and 25 m vertically. To further improve GLONASS's positioning capability, Russia plans to spend four years updating the system, including improving ground monitoring and control station facilities and changing the frequency of the waves to further improve the positioning accuracy and system stability.

GALILEO Satellite Navigation System (GALILEO) is a global satellite navigation system developed and established by the European Union. The project was established in February 1999 and is jointly the responsibility of the European Commission and ESA. In 2011, the first two satellites of GALILEO were successfully launched. In 2012, the second batch of two satellites were successfully launched, indicating that GALILEO can initially play the function of accurate positioning on the ground. In December 2016, the GALILEO system was officially put into use. The satellite constellation component of the GALILEO system consists of three independent circular orbits, 30 GNSS Medium Earth Orbit (MEO) satellites (24 operational satellites and 6 spare satellites). Each orbital plane is evenly distributed with 10 satellites, and one serves as a spare satellite. The ground system built 3 control centers in Europe, 30–40 monitoring stations and 9 injection stations worldwide. The positioning principle is the same as GPS, and the navigation positioning accuracy is higher than other system at present, and can be combined with the existing GPS, BDS and GLONASS to achieve global navigation and positioning, so that the positioning accuracy is higher and the positioning time is faster. GALILEO provides open service, life safety, commercial, public concession, search and rescue and other basic services.

The BeiDou Navigation Satellite System (BDS) is a global navigation satellite system independently developed and operated by China. In the 1980s, China began to explore a navigation satellite system development path for its national conditions and formed a "three-step" development strategy to realize the goal of building BDS-1, BDS-2 and BDS-3 in three steps. The BDS-1 project construction was started formally in 1994. The successful launch of two navigation satellites in 2000 marked that China had established the first generation of an independent navigation satellite system, which provides China with positioning, timing, wide area difference and short message communication services. In 2004, the construction of the BDS-2 project was launched. Eight years later, the goal was achieved, and BDS began to provide services to users in Asia-Pacific. Construction of the BDS-3 began in 2009, and the basic system was completed in 2018. In 2020, the BDS-3 completed its global network and began to provide all-weather satellite navigation and positioning services, such as precise positioning, precise timing, satellite navigation and short message communication to global users.

The BDS consists of three parts: the space segment, the ground segment and the user segment. The space segment consists of several geostationary orbit satellites, inclined geosynchronous orbit satellites and medium earth orbit satellites. The ground segment includes several ground stations such as the master control station, time synchronization/injection station and monitoring station, as well as the operation and management facilities of the inter-satellite link. The user segment includes BeiDou navigation chips, modules, antennas and other terminal devices. The BDS has the following characteristics: firstly, the space segment of the BDS adopts a mixed constellation with three kinds of orbiting satellites. Compared with other global navigation satellite systems, it has more high-orbiting satellites and strong anti-occlusion capability, especially in low-latitude regions. Secondly, the BDS provides navigation signals with multiple frequency points, which can improve the service accuracy by combining multiple frequency signals. In addition, BeiDou can also integrate indoor and outdoor positioning technologies. A series of indoor and outdoor integrated navigation technologies such as BeiDou + inertial navigation system (Sun et al., 2016), BeiDou +ultra-broadband (Sun et al., 2020), BeiDou +WIFI, and BeiDou +5G are developing continuously. BeiDou's positioning accuracy is constantly improving, and its application scenarios are increasingly wide. Thirdly, the BDS innovatively integrates navigation and communication capabilities. BDS has five functions: real-time navigation, rapid positioning, precise timing, location reporting and short message communication services.

Since the BeiDou Navigation Satellite test system was formally offered in 2003, China has made great steady progress in theoretical research, application technology development and receiver manufacturing applications. It is widely used in many fields such as transportation, marine fishery, hydrological monitoring and management, meteorology, forest fire prevention and management, power dispatching and earthquake prevention and disaster reduction. It provides convenient services and efficiency for people and produces particularly significant social and economic benefits. The rapid development of the BeiDou Navigation Satellite System is complementary to the growing economic level and the improvement of China's comprehensive national strength. China will continue to promote the construction and application of the navigation satellite system, encourage more scientists, engineers and users to join this field, and actively promote the exchange and cooperation of new navigation and positioning technologies in China and abroad.

In addition to the four global navigation satellite systems mentioned in this section, GNSS also includes regional navigation systems and navigation-related enhancement systems, such as Quasi-Zenith Satellite System (QZSS) in Japan, Indian Regional Navigational Satellite System (IRNSS) in India, WAAS (Wide Area Augmentation System) in the United States and EGNOS (European Geostationary Navigation Overlay Service) in Europe. In addition, GNSS includes other satellite navigation systems under construction and to be built in the future. In the next 20 to 30 years, the development of a global, all-weather, high-precision, continuous, real-time global satellite navigation system will be further improved. The multi-navigation equipment or sensors are combined into a combined navigation system and a comprehensive navigation system are integrated with communication, navigation, command and other functions. Navigation equipment will achieve miniaturization, digitization, automation and unattended.

1.2.2.2   Remote Sensing

The basic principle of RS is shown in Figure 1.2. Based on the characteristics of different electromagnetic waves from different objects, sensors are used to detect the reflection and emission of electromagnetic waves to extract the information of these objects and realize remote recognition of objects (Khorram et al., 2012).

Satellite remote sensing can be divided into ultraviolet, infrared, multi-band, visible light and microwave remote sensing according to the electromagnetic wave segment. According to the sensor platform, it can be divided into ground remote sensing, aerial remote sensing and space remote sensing. RS has a relatively wide visual range and relatively fast updated information. It can take advantage of different electromagnetic wave characteristics of objects to obtain specific information of each object and identify objects with relatively far distances after sorting them out. Remote sensing observation can help get a large range of spatial data in a very short time and understand the dynamic change of the Earth. It can provide a large amount of information to monitor and understand the dynamic change process of the surface. With the continuous development of RS, it will play a more important role in mineral exploration, tidal flat monitoring, weather forecast, fire warnings and other aspects.

The development of remote sensing can be divided into four stages, which are the unrecorded ground remote sensing stage, the recorded ground remote sensing stage, the aerial remote sensing

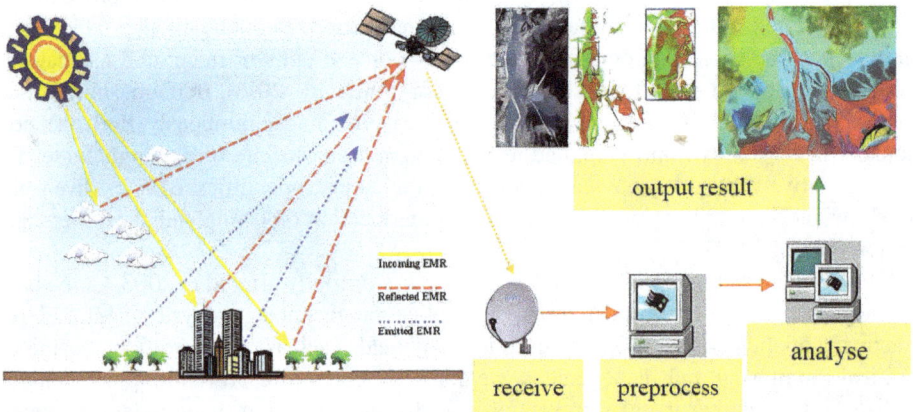

FIGURE 1.2   Basic principle of RS.

stage and the space remote sensing stage. Modern RS originated from the aerial film interpretation technology after the first photograph was obtained in 1858. With the development of technical means, especially after the successful launch of the world's first artificial satellite in 1957, RS has made a major breakthrough. Since then, the United States has launched Pioneer 2 and completed the mission to photograph the Earth's clouds. In 1960, the United States launched TIROS-1 and NOAA-1 solar synchronous satellites, which truly realized long-term exploration of the Earth from spacecraft. In 1961, the International Symposium on Remote Sensing of Environment was successfully held in the University of Michigan, USA. Since then, RS has developed rapidly as a new subject in the world. As an advanced and emerging space-based observation technology, RS has unique technical advantages when compared with traditional methods. First of all, RS has the characteristics of a large observation range, comprehensive and macroscopic, which provides favorable conditions for macroscopic study of various phenomena and their relationships. Secondly, the large amount of information in remote sensing images and the many technical means enable people to observe the Earth in a multi-faceted and all-weather capacity. In addition, remote sensing has the characteristics of fast information acquisition, short updating period and dynamic monitoring. These advantages of remote sensing prompt different countries to accelerate the development of remote sensing. In the late 1980s, France, Japan, China and India launched their remote-sensing satellites, and many countries have remote-sensing satellite programs. In the second half of 1999, the successful launch of the 1 m resolution commercial remote sensing satellite IKONOS marked the arrival of the high-resolution space remote sensing era. Then in 2000 and 2001, QuickBird with 0.61 m, OrbView-3/4 with 1 m resolution and other high-resolution commercial remote sensing satellites were launched successively, which greatly improved data selection from remote sensing images.

In recent years, with the continuous development and improvement of RS, the types of sensors carried on the remote sensing platform are constantly enriched, and the detection ability is constantly improved. The rapid development of radar interferometry, high resolution satellite remote sensing, hyperspectral remote sensing and other new technologies have promoted new applications of aerial remote sensing in many fields. At present, the RS will tend to be international cooperation, common development and common use in the world. At the same time, the spatial resolution and time resolution of the sensor will be further improved, and the multi-sensor integration will further improve the accuracy of data acquisition. In addition, the 3S integrated technology will continue to provide dynamic basic information and scientific decision-making for many industries.

1.2.2.3 GIS

Geographic Information Systems (GIS) are computer-based devices that use advanced computer technology to collect and apply data about geographic conditions on the surface of the earth. The information is further processed and analyzed to provide a more intuitive picture of the data and to provide the necessary information. With the "visualization" of GIS technology, we are able to grasp the changes in information through the feedback images. GIS is the integration of many professional disciplines, such as geospatial science and computer information technology. It is mainly composed of the computer hardware system, computer software system and spatial data. The hardware system is mainly used to collect, store and output geographic information data, and the software system is used to analyze data information. According to the content, GIS can be divided into two basic types: applied GIS and tool-type GIS. Applied GIS takes a certain profession, field or work as the main content, including thematic GIS and regional integrated GIS. Thematic GIS is the GIS with limited goals and specialty characteristics, serving a specific specialized purpose, such as forest dynamic monitoring information system and water resources management information systems. While regional information systems are mainly aimed at regional integrated research and comprehensive information services with different scales, such as national (Canada's National Geographic Information System), regional or provincial (Sweden's Stockholm Regional Information System), municipal and county level. Tool-type GIS provides the user with a package of tools, such as ArcGIS. It has spatial data input, storage, processing, analysis and other functions. According to the system structure, GIS can be divided into stand-alone GIS and network GIS. According to the data structure, it can be divided into vector data structure GIS, raster data structure GIS and mixed data structure GIS.

GIS technology was first started by Canadian researchers, who found that GIS technology needs computer processing technology to effectively improve its work efficiency and quality, and the effect of GIS technology has a significant correlation with the level of computer technology. In the early GIS technology, there were widespread problems such as insufficient functions and low speed of data information processing. In practical applications, GIS technology is closely combined with computer technology, RS and GNSS technology to realize real-time monitoring and analysis of GIS system information, providing users with possible or forthcoming situations. It also provides an important basis for users' follow-up work and decision-making. The electronic computer is developing towards miniaturization and intellectualization. It can process a large amount of data and information in a short period. At the same time, it can carry out intelligent analysis on a number of contents and reasonably predict the changes of relevant data and information in a certain stage in the future to provide decisions for the subsequent links. GIS technology has been applied in resource management, medical and health care, urban planning and design, disaster monitoring and other fields (Qin et al., 2015). In the future, it will be further developed towards the direction of intelligence, digitalization and high precision, and its application range will continue to expand.

There are four main areas of GIS technology development. The first one is the mobile GIS. In the narrow sense, mobile GIS refers to the GIS system that runs on mobile terminals and has the desktop GIS function. It does not interact with the server and is an off-line operation mode. The broad sense of mobile GIS is an integrated system which integrates GIS, GNSS, mobile communication, Internet service, multimedia technology and so on.

The second is three-dimensional GIS, which can only handle and manage two-dimensional graphics and attribute data. Three-dimensional GIS has a good advantage in replacing the traditional two-dimensional GIS, since it breaks the defects of two-dimensional GIS in the representation of spatial information and can accurately depict all parts and details of the city in the real three-dimensional space to promote the development of digitization, informatization and intelligence of the city.

The third is component GIS, which refers to the GIS provided by a group of components with some standard communication interface that allows cross-language applications based on component object platform. It can make GIS software more configurable, extensible and open, more flexible in use and more convenient in secondary development.

The last one is WebGIS. The main difference between WebGIS and GIS is that WebGIS integrates the functions of information browsing, uploading and downloading in the computer network to ensure that users can quickly query and analyze the contents of GIS data and information, further expanding the scope of information retrieval by users. Compared with the traditional GIS, the adaptability and application range of WebGIS have been further expanded with avoiding the tedious steps of information retrieval under the traditional GIS technology model. WebGIS is subdivided into two kinds: passive and active. The advantages of passive WebGIS lie in the high development rate, but its higher requirements on server performance and information retrieval time are relatively long, so the practicability is limited to a certain extent. The active technology does not use the server for information processing and retrieval. However, it sends the relevant program code to the client to achieve interaction with the customer, which is more efficient than the passive mode. Therefore, the application of active WebGIS has been increasing in recent years.

1.2.3 3S Integration

3S integration technology was proposed in the early 1990s and has been developed for more than 30 years. 3S technology integration is a new integrated technology based on RS, GIS and GNSS, which forms a whole by organically forming the relevant parts of RS, GIS and GNSS in three

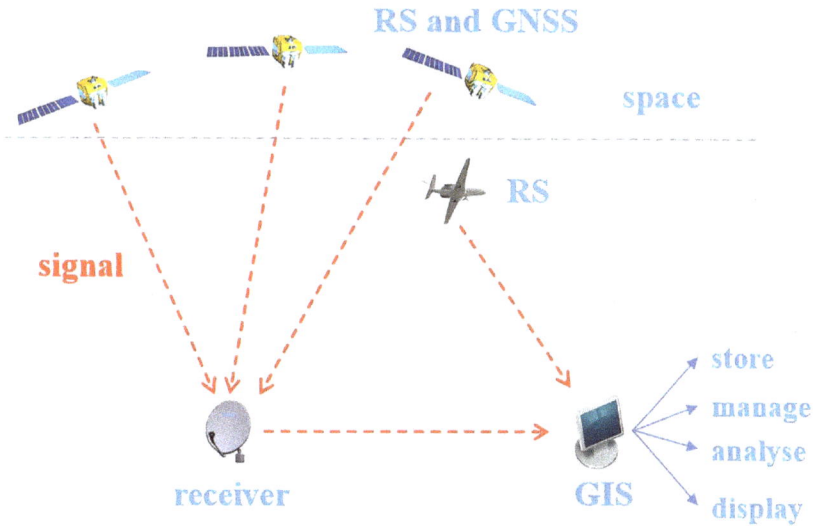

FIGURE 1.3 Brief principle of 3S technology.

independent technology fields and other high technology fields such as network technology and communication technology (Figure 1.3). 3S technology is one of the three major supporting technologies for the acquisition, storage, management, updating, analysis and application of spatial information in the current Earth observation system. It is an important technical means for the sustainable development of modern society, rational planning and utilization of resources, urban and rural planning and management, dynamic monitoring and prevention of natural disasters, etc. It is also one of the scientific methods for geological research towards quantification (Li, 1998; Li, 2003).

GNSS is mainly used to provide the spatial location of targets, including all kinds of sensors and delivery platforms such as vehicles, ships, aircraft and satellites in real time and quickly. RS is used to provide semantic or non-semantic information about targets and their environment in real time or quasi-real time and to discover various changes on the Earth's surface and update data for GIS in time. GIS, on the other hand, is a comprehensive processing, integration and dynamic access of spatial-temporal data from multiple sources management and dynamic access. As the basic platform of a new integrated system, GIS provides knowledge for intelligent data acquisition and analysis.

3S integration is the inevitable result of the development of GIS, GNSS and RS. After decades of development, 3S technology has been quite mature in terms of each technology. GNSS, GIS and RS are put forward as separate technologies, which constitute the three supporting technologies in the Earth observation system. But with the deepening of 3S technology research and application, people gradually realize that it is often difficult to meet the practical engineering application by using one of these technologies alone. Only by studying and applying the three technologies as a unified whole and making comprehensive use of the advantages of these technologies can provide comprehensive capabilities for Earth observation, information processing, analysis and simulation. GIS, RS and GNSS combine their advantages to compensate each other's shortcomings. The integrated application of 3S can be divided into the combination of GNSS and RS, the combination of RS and GIS, the combination of GNSS and RS, and the integrated application of the three techniques.

Combining RS with GNSS: Target positioning in remote sensing has always depended on ground control points. Supposed the remote sensing target positioning without ground control is to be realized in real time. In that case, the spatial position and sensor attitude of the instant acquired by remote sensing images need to be recorded synchronously by GNSS/INS. The pseudo-distance method is used for medium and low accuracy, and the phase difference method is used for high-precision positioning. GNSS dynamic phase difference has been used in aerial/aerospace photogrammetry for ground-free aerial triangulation, and is called GNSS photogrammetry. It can improve operation efficiency and save external workload. RS data volume is large and data accuracy is low, while GNSS has high data precision and low data volume, which can be organically combined to realize positioning, qualitative and quantitative earth observation.

Combining GIS with GNSS: Using the electronic map in GIS and the real-time differential positioning technology of GNSS receiver, various electronic navigation systems of GNSS + GIS can be formed, which are used in traffic, public security detection, vehicle and ship automatic driving. GNSS can be used as the data source of GIS to find the target, and GNSS data can also be used to update the GIS database, while GIS provides the technical means for GNSS to manage and analyze spatial data. At present, there are three integration methods of GIS and GNSS: one is GNSS single machine positioning + raster electronic map. The system can automatically calculate and display the best path according to the target position and the current position of vehicles and ships, guide the driver to reach the destination as soon as possible and give the driver a hint through multimedia means. The second way is GNSS single machine positioning + vector electronic map. This system is similar to the first one. The third method is GNSS differential positioning + vector/raster electronic map. The positioning accuracy can reach $\pm (1\text{--}3)$ m through the differential technology of two GNSS between fixed stations and mobile vehicles and vessels. At this time, the communication data link is needed, which can be one-way or two-way.

Combining RS with GIS: In this integrated mode, RS provides GIS with important data sources and data updating means, while GIS provides RS with technical means of spatial data management and analysis, which is used for the automatic extraction of semantic and non-semantic information. The combination of GIS and RS is the most widely used and mature technology. The key to the combination of the two lies in the software. The integration of GIS and RS can be used in global change monitoring, agricultural harvest area monitoring and yield estimation, automatic update of spatial data and so on. Remote sensing image processing and GIS are two separated systems using two separated databases, but file conversion tools are used to transfer files between different systems. Integrating remote sensing image processing and GIS into the same software system, a consistent user interface is used to process and display different types of data synchronously, but the tool library and database are separated. The same software system and database management system are used to realize the unified processing and management of remote sensing image and GIS spatial data.

The 3S integration is mainly to realize the dynamic management, analysis and application of multi-source information (multi-time, multi-scale and multi-type) in the same coordinate system. 3S integration is not a combination of equal structure, but a hierarchical organic combination. There are two main approaches: the integration approach centered on GIS (non-synchronous data processing) and the integration approach centered on GNSS/RS (synchronous data processing). The overall integration of 3S not only has the function of collecting, processing and updating data automatically and in real time, but also can analyze and apply data intelligently, provide scientific decision-making consultation for various applications and answer all kinds of complex questions that users may raise.

Due to the functional complementarity of RS, GIS and GNSS, various integration schemes were formed, which can give full play to their respective advantages and produce many new functions. The individual application of RS, GIS and GNSS can improve the accuracy, speed and efficiency of spatial data acquisition and processing, while the advantages of 3S integration are also manifested in the dynamic, flexible and automatic aspects. Dynamicity refers to the synchronization between data sources and the real world, the synchronization between different data sources, and the synchronization between data acquisition and data processing. Flexibility means that users can decide the corresponding data acquisition and data processing methods according to different application purposes and establish the connection and feedback mechanism between the two to complete the specified task most appropriately. Automation means the integrated system can automatically complete all links from data acquisition to data processing without manual intervention.

3S technology has been well applied in dynamic monitoring, crop yield estimation and other fields, thus opening new topics in the development of geography and other disciplines. Although 3S integration has been widely used, in the coming of the information age today, the development of digital Earth technology research and network information requires the combination of higher-level 3S technology and other high and new technologies, such as the combination of network technology, and distributed object technology, so as to form a multifunctional all-around integrated information system. However, there are still many problems that have not been solved, and the cooperation of more disciplines is needed.

3S integration is still a frontier in the field of spatial information science. Its development goal is "online connection, real-time processing". In order to realize 3S integration, it is necessary to explore the theory of 3S integration, improve the technical method of 3S integration and broaden the application range of 3S integration. 3S integration should solve the problems of data storage, data processing, data transmission and data visualization. From the perspective of RS, the appearance of a variety of high-resolution satellites makes the application of RS more and more extensive. The sensor technology is used to update GIS data to combine the two "S" in 3S more closely. The remote sensing image and GNSS can be directly linked to reflect the GIS signal in real time, and the three "S" are connected. 3S technology integration is an important part of spatial information science, with the concept of the "Digital Globe". Its importance is more and more prominent, and its application field is also expanding. In these increasing applications, higher requirements are put forward for the dynamic and real-time performance of 3S technology. To meet these demands, 3S technology integration must also be combined with communication technology and take full advantage of the current rapid development of communication technology to create a new era of geospatial information science.

1.3 CURRENT STATUS OF 3S METEOROLOGICAL APPLICATIONS

1.3.1 FLOOD MONITORING AND ASSESSMENT

Flooding is a natural disaster with high suddenness, high frequency and serious hazards. Floods not only damage the ecological environment, but also seriously threaten the safety of human life and property and stall the process of economic development (Aja et al., 2020; Ramkar, 2021). Therefore, a quick and effective analysis of flood simulation, risk zoning and risk assessment can provide a timely and effective indication of the damage caused by floods and help the government to carry out timely relief and formulate disaster prevention and mitigation policies to reduce the impact of flood disasters.

3S technology has been widely used in flood simulation and assessment due to its fast, convenient and efficient nature. The basic principle is that RS images are used to extract information efficiently and obtain information on changes in water bodies and the spatial analysis technology of GIS is used to simulate and analyze the flooded areas well as assess the risk of the affected area with dividing it into different levels of risk zones. In addition, the combination of RS imagery and GIS technology can simulate the entire process of flooding and post-disaster assessment as a complete flood monitoring system.

1.3.1.1 Flood Extraction from Multi-Source Remote Sensing

Because satellite remote sensing has a very good current situation and wide coverage, remote sensing data are widely used for flood information extraction and flood disaster monitoring. Zhou et al. (2021) extracted the flood range of the flood in Sri Lanka on 25 May 2017 using the data obtained by GF-3 and Sentinel-1 during and after the disaster. For Sentinel-1 data, it was first processed to form

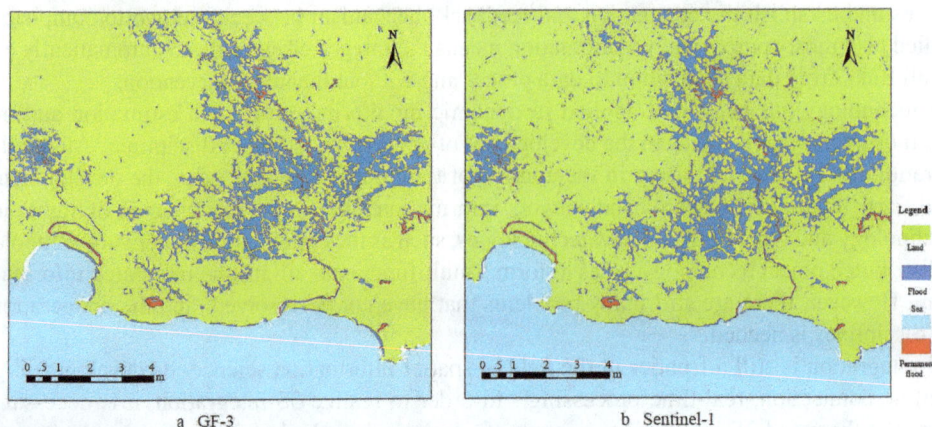

a GF-3   b Sentinel-1

FIGURE 1.4   Flood areas extracted from GF-3 and Sentinel-1 images.

Source: Zhou et al. (2021)

a binary map, then further generated a log histogram, and finally a suitable threshold was selected for flood extent extraction. For the GF-3 data, a threshold extracted from the minimum error method is used for binary segmentation to extract the flood range. The results are shown in Figure 1.4, where the official flooded areas are compared with the GF-3 flooded areas, which are much closer to the flooded areas with high accuracy. In addition, the combination of GF-3 and Sentinel-1 with a common calibration will give more information and make the data more reliable.

Wang and Zengzeng (2022) used Sentinel-1A data to monitor the catastrophic flood disasters occurred in the Poyang Lake area since the flood season in 2020 based on supervised classification and unsupervised classification. The range of water bodies before and after the disaster is extracted to show the extent of flooding. He et al. (2022) took Guangxi's "Xijiang Flood No. 1 in 2020" as an example to study the impact of floods on various regions in Guangxi. Due to the changes in light brightness before and after the flood (the power line damage and the collapse of buildings led to a significant decrease in light brightness), the NPP-VIIRS night light remote sensing data was used for radiation normalization, and the different method was used to extract the changed and the affected areas. The affected area and the degree of post-disaster recovery show that the affected area is mainly distributed in the urban built-up area.

1.3.1.2   Analysis of Flood Inundation Based on GIS

Using GIS to conduct inundation analysis to obtain the scope of the inundated area, and then mathematical analysis to calculate the height difference between the flood level and the DEM grid, the submerged water depth can be obtained. Overlay analysis in GIS can obtain the impact of floods on the watershed, and hydrological analysis can obtain information such as river network and water flow area. Zhang et al. (2021) studied the Jinpu New District of Dalian City using the DEM data and the observation data of flood level and tide level collected for many years. Based on the spatial analysis tool of ArcGIS, the active inundation analysis and calculation were carried out using the seed spreading method, and the scope of the inundated area was obtained, as shown in Figure 1.5(a). After the difference calculation between the flood level and the DEM grid, the submerged water depth distribution map is obtained, as shown in Figure 1.5(b).

1.3.1.3   Assessment of Flood Disaster Risk Based on GIS

Using GIS technology combined with the theory and method of natural disaster risk assessment, the flood disaster risk assessment is carried out from three aspects: the hazard of the hazard, the stability of the disaster-forming environment, and the vulnerability of the disaster-affected body. Flood

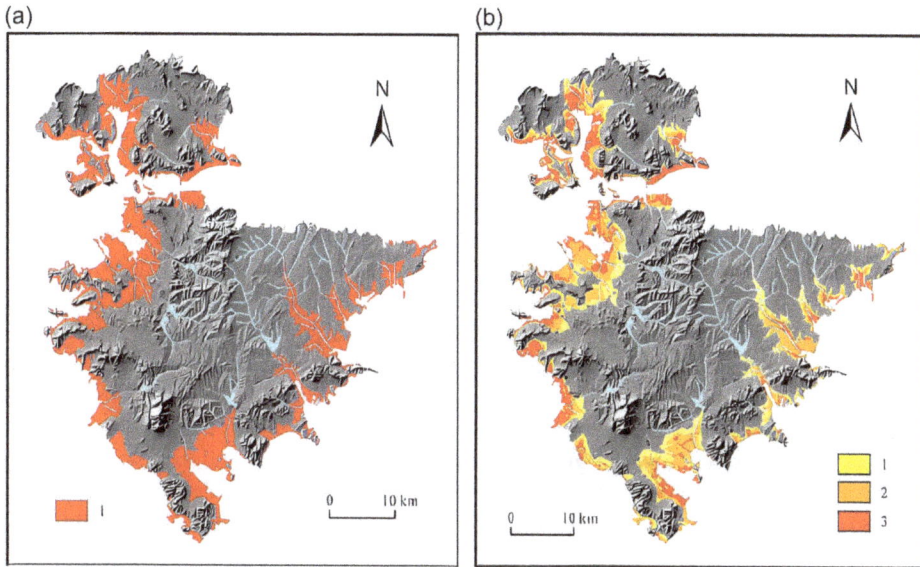

FIGURE 1.5 Maps of the flood-submerged area in Jinpu New District with flood-submerged area (a) and flood depth (b).

Source: Zhang et al. (2021)

risk assessment has two components: the flood risk impact assessment, which includes the disaster-causing factors and the disaster-forming environment factor, and the flood vulnerability impact assessment, which mainly refers to the hazard-bearing body. Sometimes the ability to prevent and mitigate disasters can also be included in the assessment of the impact degree of flood vulnerability. The Analytical Hierarchy Process (AHP) and the weighted comprehensive evaluation method are often used to evaluate the degree of influence of specific indicators in each part and then determine the weight (Ramkar, 2021; Seejata et al., 2018). Yi (2012) used GIS technology and the weighted comprehensive evaluation method for the urban area of Guilin City, Guangxi, and constructed a flood disaster risk assessment from the four aspects of disaster-causing factors, disaster-forming environment, disaster-bearing body and disaster prevention and mitigation capabilities. Then, the model was used to evaluate and analyze the flood disaster risk in the study area. Finally, the natural distance classification method was used to divide the flood disaster risk in Guilin into low-risk, medium-risk and higher-risk areas.

1.3.1.4 Flood Disaster Evaluation with Combining RS and GIS

Under the conditions of the flood submerged range with the help of high-precision DEM data, the elevation distribution of the water and land boundary (i.e., water surface) is obtained, and the water depth is calculated from the difference between the water surface elevation and the ground elevation. Dong et al. (2012) studied the flood disaster in Kouqian Town, Yongji County, Jilin Province, on 28 July 2010. Firstly, high-resolution remote sensing images were used to divide the grid according to study area's residential buildings to facilitate subsequent calculation of water depth. Then the remote sensed images were utilized to extract the normalized water index method to separate water bodies and non-water bodies. The segmentation threshold was determined by comparing with the standard false-color image, and then the submerged area of the flood was extracted. As for the water depth of the submerged area, GIS is used to process the elevation points of the flood submerged boundary first, and then the water surface elevation of each grid is obtained. Finally, the flood submerged water depth of each grid is calculated by the water depth GIS method to realize the complete evaluation of flood disasters from submerged range to water depth.

1.3.2   Drought Risk Mapping and Assessment

1.3.2.1   Drought Disaster

Meteorological disaster risk assessment requires the use of multidisciplinary theoretical knowledge such as meteorology, physical geography and disaster science, as well as certain methodologies such as analytic hierarchy process, risk index method and comprehensive weighting method. Based on GIS technology and disaster censuses, drought disaster risk analysis and assessment are made according to several elements such as the similarities and differences of the disaster-forming environment, intensity and frequency distribution of disaster-causing factors, and the vulnerability of the carrier. Among them, drought disaster refers to the social, economic and environmental conditions below a certain level caused by drought, which is the result of severe drought acting on fragile social, economic or sensitive ecological environment systems (Mishra et al., 2010). It is one of the most serious natural disasters in the world. Its frequency of occurrence, duration, the scope of influence and losses caused it to rank first among all natural disasters. With the rapid development of economy, population growth and the global climate change marked by climate warming, drought disasters tend to be further aggravated, causing incalculable damage to economic growth, social progress and ecological environment. Therefore, it is of great significance to scientifically assess the risk of drought disasters to mitigate disasters and improve economic and social benefits.

1.3.2.2   Drought Risk Assessment

Drought disaster risk assessment can characterize the form and degree of regional drought, which is the basis for formulating comprehensive disaster prevention and mitigation countermeasures. With the support of GIS software, a meteorological disaster database can be established, which includes basic geographic data, economic population data, meteorological data and disaster information. Through the comprehensive analysis of multiple factors such as the risk of hazard-causing factors, the sensitivity of the disaster-forming environment, the vulnerability of the hazard-bearing body, and the ability to prevent and mitigate disasters, several risk assessments factors such as the risk of the hazard-causing factor, the sensitivity of the disaster-forming environment, and the vulnerability of the hazard-bearing body are constructed. The evaluation model of nature, disaster prevention and resilience are established, and historical disaster data are combined to calibrate the model parameters. Finally, the disaster risk is evaluated, the division unit is determined, the disaster division level is divided, and the disaster division is carried out using the functions of GIS spatial overlay analysis, map spot merger and attribute database operation.

Chen et al. (2022) conducted the in-depth analysis of the causes of drought risk in the Loess Plateau, combining with the climate characteristics of the Loess Plateau and the occurrence of drought events. A natural disaster risk assessment system of "disaster stress-social vulnerability-exposure" was selected and corresponding remote sensing data and socioeconomic data were used as the data source of drought disaster risk. The drought disaster risk assessment model was constructed by Analytic Hierarchy Process (AHP). The spatial superposition analysis of the three index factors was carried out using GIS technology. Finally, the natural breakpoint method was used to analyze the drought risk classification and assessment.

Using the geographic data including county-level administrative division data of Qingyang City, digital elevation model (DEM) data, Palmer Drought Index (PDSI) data, vegetation cover index (NDVI) data, land surface temperature (LST) data, land cover data, soil moisture data, the constituent elements of the drought disaster risk assessment system can be determined. The Analytic Hierarchy Process (AHP) is used to calculate the weight of each index factor, and then GIS software is used to normalize the individual indicators contained in each element. The drought risk zoning can be constructed from the drought disaster data in Qingyang City. The technical framework is shown in Figure 1.6 (Chen et al., 2022).

Drought disaster risk assessment is jointly determined by multiple factors, and the dimension units of each factor are different. In order to eliminate the dimensional influence of the factors under

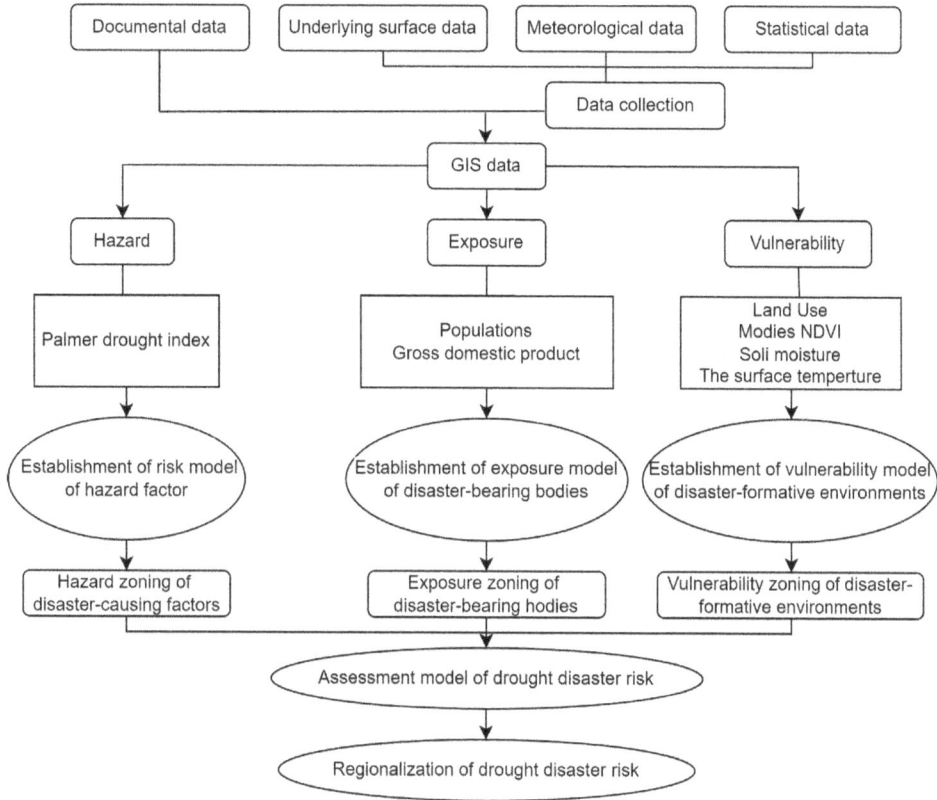

FIGURE 1.6 Framework of drought risk assessment in Qingyang City.

Source: Chen et al. (2022)

each index, normalization processing is required for each factor. According to the nature of indicators, it can be divided into positive indicators and negative indicators. Positive indicators are the reflected risk factor indicators. The larger the corresponding value, the higher the risk. The negative indicator is the larger for the corresponding value of the reflected risk factor indicator, the lower the risk is (Chen et al., 2022).

1.3.3 HIGH TEMPERATURE DISASTER RISK ASSESSMENT

1.3.3.1 High Temperature Disaster

High temperature disasters change the city's thermal environment, which will seriously impact the climate, air quality, hydrological conditions and urban soil of the whole city (Shan et al., 2022; Liu et al., 2015). At the same time, it will also change the distribution and activity of organisms in the city, resulting in a series of urban ecological problems. Research on high temperature disasters and putting forward countermeasures can effectively alleviate the impact of high temperature disasters on the ecological environment and improve the ecological environment, e.g., improving the abnormal weather caused by high temperature, reducing the frequency of urban disasters creating a suitable environment for urban organisms and restoring their growth environment.

High temperature disasters, especially persistent high temperature heat wave disasters, have a wide range of social impacts, which can endanger human health and cause diseases or death. At the same time, the formation of high temperature disasters is closely related to urban energy consumption and these factors complement each other. During the high temperature disaster in summer, the use of engineering measures to cool down consumes a lot of energy, and the energy consumption of electricity and water for urban production and living increases, which leads to the shortage of urban electricity consumption. When energy is used at the same time, a large amount of exhaust gas and heat will be emitted, which will aggravate the rise of urban temperature and aggravate the high temperature disaster in the city, forming a vicious circle. Therefore, through the study of high temperature disasters, strategies to deal with and mitigate high temperature disasters are proposed to reduce urban temperature and break this vicious circle.

The risk zoning of high temperature heat damage mainly includes four aspects: the hazard of hazard-causing factors, the sensitivity of disaster-forming environment, the vulnerability of hazard-bearing bodies, and the ability of disaster prevention and mitigation. The occurrence frequency of high temperature heat damage, topography, population density, local economy, etc., were selected as evaluation factors, and relevant indicators were established. The distribution of high temperature heat damage risk coefficient can be obtained by weighted synthesis and analytic hierarchy process.

1.3.3.2 High Temperature Disaster Risk Assessment

In order to provide references for reasonable responses to high temperature disasters, Fang et al. (2016) adopted a multi-factor weighted comprehensive evaluation method, using the meteorological observation data in Jiangsu, Zhejiang and Shanghai from 1961 to 2009 and the socioeconomic data in Jiangsu, Zhejiang and Shanghai from 2008 to 2010. Based on the comprehensive assessment of the disaster environment, disaster-bearing bodies and disaster resistance capabilities, the risk map of high temperature disasters in Jiangsu, Zhejiang and Shanghai was obtained. Using the daily maximum temperature data of 144 meteorological stations in Jiangsu, Zhejiang and Shanghai from 1961 to 2009, the daily maximum temperature, daily average temperature and duration of high temperature disasters are extracted. Based on the 2008–2010 Jiangsu, Zhejiang and Shanghai urban construction statistical yearbooks, economic yearbooks, statistical bulletins and other references, the socioeconomic data in Jiangsu, Zhejiang and Shanghai were obtained.

Traditional methods for analyzing disaster-causing factors include field investigation, searching for historical disaster data, building models, laboratory analysis and remote sensing acquisition. The risk of hazards refers to the abnormal degree of meteorological disasters, which is mainly determined by the scale and frequency of activities of hazards. The risk of disaster-causing factors is first determined based on the principle that the higher the level of high temperature disasters is, the greater the harm is, and the weight of the hazard-causing factors is determined (Fang et al., 2016).

Topography, land cover and water system are the main factors reflecting the high temperature disaster-forming environment. Among them, altitude has the most obvious impact on high temperature disasters, and the higher the altitude, the smaller the impact. Based on this principle, the terrain impact index of Jiangsu, Zhejiang and Shanghai is calculated. According to the sensitivity of land cover type to high temperature and the ordering principle of city, cultivated land, grassland, woodland and water area, assign values from high to low, and draw the impact index of different land use types in Jiangsu, Zhejiang and Shanghai. During the disaster period, the water system has the function of alleviating and mitigating the high temperature disaster, through the evaluation and analysis of the river network density and the flow buffer zone, the assignment of the water system impact index can be normalized (Fang et al., 2016).

The vulnerability of hazard-bearing bodies refers to the objects of meteorological disasters, including the material and cultural environment of human beings, which is the collection of various resources in human activities and the society in which they live. Combined with the socio-economic data of Jiangsu, Zhejiang and Shanghai, the vulnerability indicators of disaster-affected bodies are evaluated by selecting indicators such as population density, per capita GDP, proportion of cultivated land, and per capita energy consumption. The indicators of population density, per capita GDP, proportion of cultivated land, and per capita energy consumption are normalized and interpolated. The vulnerability index of high temperature disaster-affected bodies in Jiangsu, Zhejiang and Shanghai is calculated comprehensively using equal weight and natural breakpoint methods (Fang et al., 2016).

The grid calculator in the ArcGIS spatial analysis module was used to analyze the four factor layers of the high temperature disasters in Jiangsu, Zhejiang and Shanghai: the risk of hazards, the sensitivity of disaster-forming environments, the vulnerability of disaster-affected bodies, and the ability to resist disasters. Based on the superposition calculation, according to the grading method of high-risk areas, sub-high-risk areas, medium-risk areas, sub-low-risk areas and low-risk areas, the comprehensive risk zoning of high temperature disasters in Jiangsu, Zhejiang and Shanghai was obtained (Fang et al., 2016).

To assess the high temperature risk on 973 communities in Wuhan city, Shan et al. (2022) used the geography-weighted regression method using remote sensing data and geographic information data. A risk assessment model of high temperature disasters is established from disaster-causing danger, disaster-generating sensitivity and disaster-bearing vulnerability. The spatial distribution of high disaster-causing danger in the community is very consistent with its surface temperature. The spatial distribution of disaster-generating sensitivity in the community shows the spatial characteristics of the clustered distribution of high sensitivity areas.

1.3.3.3 High Temperature Disaster Risk Mapping

According to the high temperature characteristics of Fujian, combined with the analytic hierarchy process, expert scoring method and the spatial analysis function of GIS, the possibility, severity, sensitivity of the disaster-forming environment and the effectiveness of disaster prevention in Fujian were analyzed (Jin, 2017). Based on the evaluation of the regional differences in the degree of high-temperature disaster risk in Fujian, the risk level of high-temperature disasters in Fujian was divided, which provided the basis for the relevant departments to carry out urban planning, disaster management and formulate disaster prevention and mitigation measures.

Different data have been utilized to map a high temperature disaster risk zone, such as the daily temperature data from 67 meteorological stations in Fujian Province, the basic geographic information data, and social and economic data including the road area, GDP and per capita medical beds in each city at the end of 2009. The analytic hierarchy process was used to calculate the index weight coefficient. In fact, on the basis of establishing an orderly and hierarchical index system, the pros and cons of each index in the system were judged through pairwise comparisons between the indexes, and this evaluation result was used to synthesize and calculate the weight coefficient of each indicator. Through the analytic hierarchy process, the high temperature disaster risk assessment system is established by using three factors: the risk of disaster-causing factors, the sensitivity of the disaster-forming environment and the disaster-resistant ability. Comparing the scores to get the weight of each impact factor. The evaluation system is divided into target layer (A), index factor layer (B) and impact factor layer (C) from top to bottom. The index factor layer mainly considers the risk of disaster-causing factors, the sensitivity of disaster-forming environment and the ability to resist disasters.

Through the use of related functions in the spatial analysis and spatial statistics toolbox in ArcGIS software, based on remote sensing image data and basic geographic information data, the data are spatially calculated, integrated and superimposed, and visualized. The application of the GIS spatial analysis method provides technical support and guarantees for the analysis of the comprehensive risk of high temperature disasters, the spatial distribution characteristics of each index and the spatial heterogeneity of the factors affecting the built environment of the community. The ArcGIS software was used to perform inverse distance weighted interpolation, and the observation point data were interpolated into raster data. The grid overlay calculation is performed in ArcGIS, and natural breaks (Jenks) classification is used to obtain the distribution map of the high temperature risk level.

The degree of heat disaster risk is also affected by the effectiveness of local disaster control measures and post-disaster remedial measures. Disaster resilience refers to the ability of the disaster-affected area to resist and recover when and after being hit by a high temperature disaster. The ratio of per capita garden and green area, the number of medical beds per capita in hospitals and the per capita GDP are used to represent the effectiveness of existing disaster resistance capabilities. After normalizing these data, the weighted comprehensive evaluation method of AHP was used to obtain the weighting coefficients for the percentage of garden area per capita, number of hospital medical beds per capita and GDP per capita, respectively, and the disaster resistance index was calculated in ArcGIS software. The larger the index value, the more effective the control measures and the lower the risk of high temperature. The high temperature disaster resilience index of each city is divided into four grades: low, second-low, medium and high.

1.3.4 AGRICULTURAL FROST DAMAGE RISK MAPPING

1.3.4.1 Agricultural Frost Damage

Freezing injury is a kind of agricultural meteorological disaster that belongs to a low-temperature disaster. It refers to the sudden drop of the temperature near the plants to 0°C or below, causing the water seal in the crops to freeze, causing damage to the crops and affecting the plants' normal growth, leading to reduced production or crop failure. Freezing damage includes low temperature freezing, cold wave, strong cooling, frost, late spring cold and low temperature in autumn. The impact of meteorological disasters on agriculture has become an important factor restricting the steady and sustainable development of agriculture. Data show that the losses caused by agro-meteorological disasters such as droughts, frosts, floods and low temperature freezes account for more than 60% of agricultural disasters, and years of severe low temperature disasters can cause grain output losses of more than 10 billion kilograms. In addition, meteorological disasters are often accompanied by other secondary disasters, and the risk of disasters continues to increase. At the same time, disaster chains and disaster clusters will be formed, which will cause more serious and huge losses. Frost damage occurs in a wide range in my country, ranging from Heilongjiang in the north to Guangdong and Guangxi in the south.

The development of remote sensing technology has made it possible to monitor changes in large areas and long-term sequences. It can realize rapid extraction and accurate identification of crop information and obtain the required temperature information, spatial distribution of crops, planting area, growth status and production and other information. The spatial analysis ability and cartographic expression of GIS technology can establish the comprehensive risk index of freezing injury and the assessment model of freezing injury risk, and based on this, establish the freezing injury risk zoning. Remote sensing methods can be broadly divided into three types: minimum surface temperature retrieval methods, vegetation index difference methods and hyperspectral methods.

1.3.4.2 Agricultural Freezing Damage Monitoring

The vegetation index difference method is an index that can reflect the growth status of plants by combining data from different bands of hyperspectral data. After the crops are subjected to freezing injury and low temperature stress, the activity will decrease rapidly, and the vegetation index will also be reduced when the activity is reduced. The vegetation index and the degree of freezing injury show a significant positive correlation. Therefore, the severity of freezing injury can be judged by comparing the vegetation index before and after freezing injury of crops. The hyperspectral has the characteristics of high spectral resolution, strong band continuity, and a large amount of spectral information. It can monitor crop canopies and leaves and construct narrow-band spectral indices through hyperspectral to explore its impact on crops under low temperature stress.

3S technology has been utilized to assess the freezing damages on agriculture, and two case studies are given here. Wang et al. (2021) used the FY-3 satellite and the split window algorithm to invert the surface temperature, established a regression analysis with the ground minimum temperature measured by the meteorological station, and used the variational technology to correct the more accurate remote sensing ground minimum temperature, and then collected the late frost disaster indicators and winter wheat development period data to realize remote sensing monitoring of late frost. They developed a winter wheat late frost remote sensing monitoring system supported by GIS, and carried out remote sensing monitoring of late frost damage according to the freezing damage indicators of winter wheat at each period after jointing, produced a spatial distribution map of winter wheat occurrence, and calculated the affected areas of different regions and different levels of freezing damage, to realize remote sensing monitoring and evaluation of winter wheat. Li et al. (2015) selected the harmful extreme cold and frequency (danger) of winter wheat late frost, winter wheat planting area (exposure), irrigation-to-plow ratio and the number of machine wells per unit of irrigation area (vulnerability) to establish a risk assessment system, determine the weight of winter wheat late frost risk assessment indicators by analytic hierarchy process, and build a winter wheat late frost risk assessment model. Based on the climate data from 1984 to 2013, the risk assessment of winter wheat late frost in Henan Province was carried out. According to the risk assessment model of winter wheat late frost, ArcGIS was used for grid calculation, and the risk zoning map of winter wheat late frost in Henan Province was obtained (Li et al., 2015).

1.3.5 SNOW RISK MAPPING AND ASSESSMENT

1.3.5.1 Snow Mapping

Snow cover is an important part of land cover, an important source of water resources and one of the important elements in the global climate system. It can regulate river runoff and guarantee ecosystems' sustainable development. Its duration and coverage will affect the surface radiation and heat balance, the energy exchange of the earth-atmosphere system, etc. In addition, changes in snow cover also significantly affect the global and regional climate system, ecological environment and human production and life. Therefore, by obtaining or retrieving information about snow accumulation, the climate and ecological environment can be effectively adjusted, and the impact of snow accumulation on human production and life can be greatly reduced. Therefore, the snow cover is mainly elaborated from four directions: the calculation and inversion of snow cover depth, the identification and extraction of snow cover information, the temporal and spatial changes of snow cover, and the disaster risk analysis of snow cover.

Snow depth (SD) is one of the basic attributes of snow and an important parameter reflecting the distribution and change of snow. By obtaining continuous and uniform high-precision snow depth data, it can provide a scientific basis for research on climate change, water resource analysis and snow distribution. In addition, it can forecast, monitor, and warn of snowmelt flood disasters. The main methods of snow depth observation include ground observation and remote sensing data observation. However, ground observation mainly refers to meteorological stations or manual measurements. On the one hand, the obtained data is inefficient, scattered and poorly representative, and cannot meet the observation requirements of large-area snow depth information. On the other hand, it also has limited temporal and spatial resolution, high cost, low precision and other shortcomings; a single application of remote sensing data cannot ensure its accuracy. Therefore, using 3S data combined with measured data to calculate and invert snow depth has become increasingly widespread.

Currently GPS, InSAR and related extended technologies (such as GNSS-R, GNSS-IR and D-InSAR technology) are mainly used to retrieve snow depth. GNSS multipath information is obtained mainly through the signal-to-noise ratio at low altitude angles. Similarly, the multipath reflection information at low altitude angles has a significant impact on the signal-to-noise ratio. Therefore, the GNSS signal-to-noise ratio data at low altitude angles can be analyzed and processed, and then the surface environment parameters (snow depth) can be obtained. The band of the SAR satellite can penetrate the snow layer, so the SAR image is used to perform interferometric processing on the image data before and after the snowfall. The generated interference fringe pattern will also contain the phase information of the snow that can be used to retrieve the snow depth.

1.3.5.2 Snow Information Extraction

Snow information includes snow area, snow albedo, snow water equivalent and other information. Obtaining snow cover information can provide necessary and reliable reference materials for major research such as the hydrological cycle in cold regions, water resource management and snow

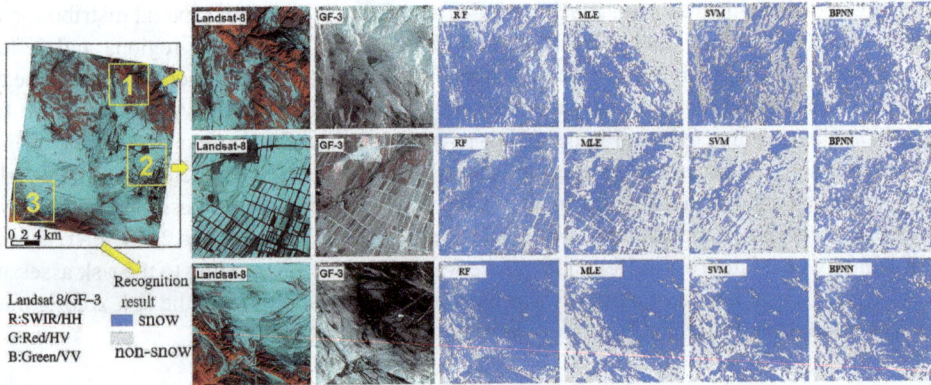

FIGURE 1.7   Details of Landsat-8, GF-3 and snow cover recognition.

Source: Ma et al. (2020)

disaster warning. Snow cover area (SCA) is one of the most important snow cover parameters. Rapid, real-time and accurate monitoring of changes in snow cover is of great significance for climate evolution simulation, water resource utilization management and disaster analysis and assessment. Snow albedo is one of the most important parameters of snow cover, and it has a significant impact on the snow hydrological process, snow mass-energy balance process and snowmelt runoff process. In addition, snow radiation information dominated by snow albedo can significantly affect the climate and hydrological cycle at different scales. Snow water equivalent (SWE), which is the liquid depth of snowmelt, is one of the main characteristics of snow cover and an important indicator for snowmelt runoff forecasting and water resource management. Therefore, obtaining snow water equivalent in time can provide a great reference value for snowmelt runoff forecast and water resource management.

Ma et al. (2020) extracted features for snow identification using domestic GF-3 data, five polarization decomposition methods (Pauli decomposition [a common coherent target decomposition method], H-A-α decomposition, Freeman decomposition, Yamaguchi decomposition and Anyang decomposition). The random forest method calculates the importance of each candidate feature; then selects the feature that contributes more to the recognition, constructs the feature optimization rule to generate the optimal feature set, and finally identifies the snow based on this feature set, forming a feature-based optimization method. The proposed method is compared with the three classifiers of the maximum likelihood method, support vector machine and BP neural network, as shown in Figure 1.7. It was found that the recognition accuracy was highest using the optimal feature set and the random forest method (the proposed method).

1.3.5.3 Snow Disaster Risk Assessment

Snow disasters are mainly caused by large-scale snow accumulation caused by heavy snowfall. The disasters seriously affect the environment and the survival and health of humans and livestock, and are likely to have a greater impact on transportation, communications, agriculture and electricity. Therefore, timely and effective disaster analysis of snow disasters can reduce their impact on the environment and human life and help the government and other relevant departments to make timely and effective disaster reduction and rescue measures.

Based on the principle of disaster risk assessment, GIS technology is used to select the influencing factors related to snow disasters, and the weights of the influencing factors are calculated. Finally, according to the weights, GIS is used for spatial superposition to obtain the snow disaster risk zoning map to further analyze the disaster. Xi (2020) extracted and processed snow data from 63 stations in Heilongjiang province from 1983 to 2015. He used methods such as trend analysis, spatial analysis in ArcGIS and Kriging interpolation to analyze the temporal and spatial variation characteristics of snow cover in Heilongjiang Province in the past 33 years. Combined with the theory of natural disaster risk, from the four aspects of the risk of disaster-causing factors, the sensitivity of disaster-forming environment, the vulnerability of disaster-affected bodies and the ability to reduce and prevent disasters, as well as the analytic hierarchy process and weighted comprehensive evaluation method, the snow disaster risk assessment index system in Heilongjiang Province was established and the risk was assessed for local snow disasters.

1.3.6 HAIL DISASTER RISK ASSESSMENT

1.3.6.1 Hail Disaster

Hail is a hard spherical, cone-shaped or irregularly shaped solid precipitation. It is a severe weather phenomenon caused by a strong convective weather system. Its impact range is small, and its time is short, but it is sudden and often accompanied by strong winds, development, rapid cooling and other paroxysmal weather. Hail occurs worldwide, and different countries have different degrees of hail disasters. Generally speaking, the geographical distribution characteristics are related to surface morphology and dimensional factors. Plains are less than mountainous areas, coastal areas are less than inland areas, and high-dimensional and low-dimensional areas are less than mid-latitude areas. However, due to its violent onset, often accompanied by thunderstorms or strong winds, it will cause great damage to agriculture, livestock, transportation, people's lives and property every year. Fujian Province, Tibet Autonomous Region and Inner Mongolia Autonomous Region are provinces with more frequent hailstorms and more serious hail disasters in China. According to statistics, the economic losses caused by hail disasters in China are as high as hundreds of millions or even billions of dollars every year. In China, hail mostly occurs in spring, summer and autumn, and April to July accounts for about 70% of the total. It is in the golden period of crop growth, so the occurrence of hail will cause a devastating blow to agriculture. In addition, hail will also cause losses in construction, communication, electric power, transportation and other industries. Therefore, it is very necessary to apply scientific detection equipment to warn of hail weather, implement artificial hail suppression operations and keep hail from growing, thereby reducing disaster losses.

Usually, a hail disaster lasts about ten minutes. Although the duration is not long, it will cause huge damage to vegetation and ecology. It will cause different damage to crops in different seasons. For example, autumn will cause sharp harvests. In spring, a large number of seedlings will die, and in summer, it will cause devastating damage to the growing crops. By processing the monitored data through satellite remote sensing technology, recording and analyzing the vegetation index before and after the hail disaster can monitor the damage's degree of size.

1.3.6.2 Hail Disaster Monitoring

Multi-temporal and multi-angle satellites can observe hail clouds and analyze the spectral characteristics, structural situation, range, boundary shape, color tone, shadow (specific to visible light cloud images) and texture features of hail clouds and other characteristics of hail clouds. By analyzing various features of satellite cloud images to obtain cloud types, horizontal scales, boundary shapes, relative heights and thicknesses, etc., hail clouds can be identified along with their spatial distribution and intensity.

GIS has accumulated a wealth of spatial data visualization and statistical analysis methods. It has the function of integrating various spatial data and performing powerful spatial analysis. It can provide powerful tool platform support for the identification of strong convective weather thunderstorms from radar data. The lightning location system can be used to identify hail clouds. During the formation and development of hail clouds, accompanied by strong lightning activities, the frequency of lightning increases sharply. Monitoring lightning activity using the Lightning Locating System allows the identification and location of hail clouds.

1.3.6.3   Hail Disaster Risk Assessment

3S technology has been used to assess the hail disaster risks. Peng et al. (2019) used fine-grained township hail frequency, historical disaster situation, DEM data, land and population density and other data in Chengde City to assess hail disasters in a targeted manner. A hail disaster risk assessment model is constructed from three aspects: disaster environment, disaster prevention and mitigation capabilities. The comprehensive weighted analysis method and the AHP are used to calculate the weight of each index, the evaluation index is gridded by GIS spatial analysis technology, and the hail disaster risk map of Chengde City was made with the grid as the basic evaluation unit. Yang et al. (2016) studied the hail disaster by using the hail data of 39 meteorological stations from 1981 to 2015 to fully analyze Tibet's climate background and economic environment. According to the disaster occurrence theory, the formation mechanism of regional hail distribution selects the factors of land-form, disaster frequency, population and social economy from three aspects: the disaster-causing environment, the possibility of disaster occurrence and the vulnerability of disaster-affected bodies. They used the cluster analysis method to establish mathematical models such as meteorological disaster disaster-pregnant background, disaster risk and hazard-bearing body vulnerability, and operated the attribute database and graphic database by MapInfo professional software to obtain various disaster background, disaster risk, and vulnerability evaluation layers. After layer stacking, plate merging and level division, the risk partitions of various meteorological disasters were obtained.

1.3.7   Wind Disaster Risks Assessment

1.3.7.1   Wind Disaster Assessment Method

At present, the evaluation method of wind disasters is mainly divided into indicator evaluation methods and statistical simulation methods. The evaluation of wind disaster indicators is mainly aimed at the macro range, such as the severity of a certain wind disaster to damage the affected areas or the general degree of destruction. The current research mainly adopts regression analysis, fuzzy comprehensive evaluation method and analytic hierarchy process.

The statistical simulation assessment of the wind disaster is mainly based on the intensity and frequency of wind disasters and the statistical simulation of the specific structure of the disaster, namely the frequency, intensity and specific characteristics and location distribution of the disaster body, are used to determine the threat of wind disasters. Corresponding statistical simulation is performed to evaluate wind disasters or risks. The statistical simulation method of typhoon disaster generally includes typhoon frequency, intensity, path simulation, wind field simulation, wind damage simulation of engineering structure and expected insurance loss simulation.

Index evaluation methods are commonly used, which are the mainstream method of typhoon disaster risk and loss assessment. There are many specific implementation methods. At present, the main problem of the statistical simulation method of wind disasters is that the complexity of each aspect of the simulation process and the choice of various parameters result in uncertainty and differences in the results. For example, in the meteorological module, the uncertainty and differences between the wind speed model and the near-ground wind farm parameters will cause the final estimated wind disaster loss to be more unstable.

The fuzzy comprehensive evaluation model is a comprehensive analysis method based on fuzzy transformation theory based on fuzzy reasoning, combining qualitative and quantitative, accurate and inaccurate. At present, it is widely used in the multi-index comprehensive evaluation. AHP is a relatively simple and feasible decision-making method. Its main advantage is that it can solve complex multi-objective problems. The AHP method is also a combination of qualitative and quantitative methods. It can quantify the qualitative factors, express the subjective judgment of people in mathematics, and test and reduce the subjective influence to a certain extent, making the evaluation more scientific. It can provide decision makers with a variety of decision-making methods, in the combination of quantitative and qualitative, according to the standard weight of each decision scheme.

1.3.7.2 Wind Disaster Risk Mapping Based on 3S Technology

Based on the annual extreme wind speed data, natural environment and social and economic factors in Hangzhou, Chen (2012) constructed a regional wind disaster risk assessment model by integrating disaster-causing factors, disaster-pregnant environment, disaster-bearing bodies and disaster prevention and mitigation capabilities. The appropriate disaster risk assessment indicators and the risk assessment indicators are selected with a greater correlation with the wind disaster, the spatial overlay analysis of each index is conducted using the spatial analysis technology and the risk status of the wind disaster are evaluated and mapped with the grid as the basic evaluation unit. The Hangzhou area is divided into five levels of risk: high, sub-high, medium, sub-low and low, as shown in Figure 1.8 (Chen, 2012).

Sun et al. (2021) used the powerful data function of ArcGIS to establish the Xingtai gale disaster risk database and construct the risk assessment model of gale disaster. Through comprehensive analysis and evaluation of the data, the risk map of the disaster-causing factors of the strong wind disaster, the disaster-pregnant environment sensitivity of the strong wind disaster, the vulnerability of the disaster-bearing body of the strong wind disaster, the disaster prevention and resilience of the strong wind disaster, and the strong wind disaster in Xingtai City were obtained.

Zhou et al. (2013) introduced the typhoon disaster risk index, comprehensively considered the risk, vulnerability, exposure and local disaster prevention and mitigation capabilities of typhoon disasters, and constructed a comprehensive risk assessment index system for typhoon disasters. Taking Zhejiang Province as an example, each evaluation index was quantified according to historical statistical

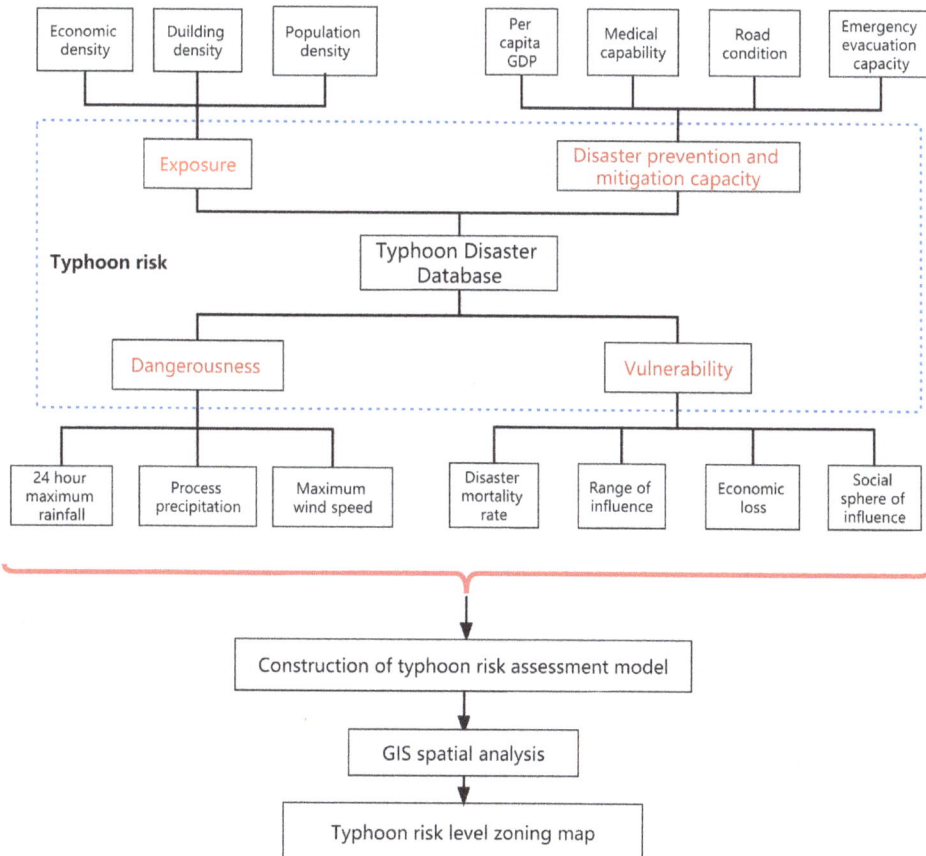

FIGURE 1.8 Flowchart of typhoon disaster risk assessment.

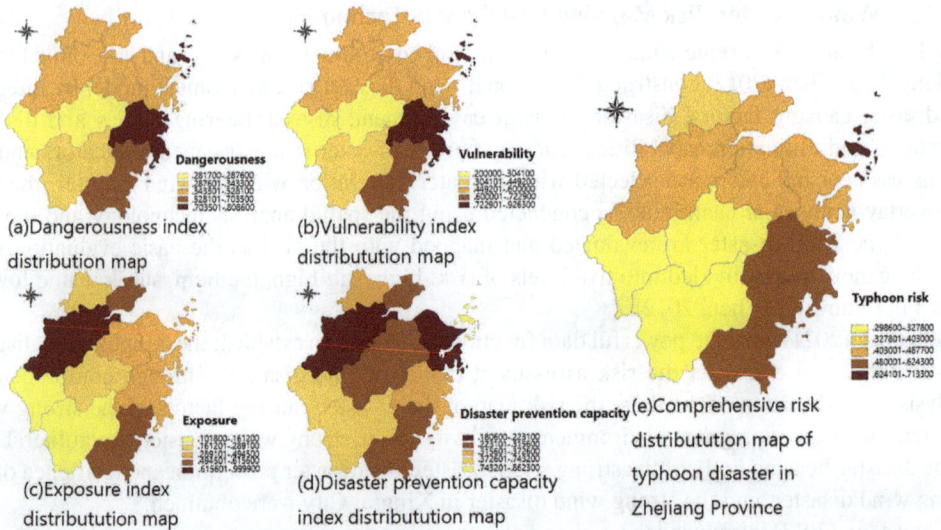

FIGURE 1.9 The risk index distribution map (a), vulnerability index distribution map (b), exposure index distribution map (c), disaster prevention and mitigation capacity index distribution map (d), and typhoon disaster comprehensive risk distribution map of various cities in Zhejiang Province (e) were obtained.

Source: Zhou et al. (2013)

data, and the spatial analysis function of GIS was used to overlay the evaluation indexes according to their weights. The risk index distribution map (Figure 1.9a), vulnerability index distribution map (Figure 1.9b), exposure index distribution map (Figure 1.9c), disaster prevention and mitigation capacity index distribution map (Figure 1.9d), and typhoon disaster comprehensive risk distribution map of various cities in Zhejiang Province (Figure 1.9e) were obtained (Zhou et al., 2013).

1.3.8 Lightning Risk Evaluation and Mapping

1.3.8.1 Lightning Risk Assessment Method

The formation and development of lightning disasters are subject to a variety of natural and socio-economic factors. According to its mechanism and change speed, the factors affecting the risk assessment of lightning disasters can be classified into four categories: lightning hazard, exposure, vulnerability and the ability of disaster prevention.

Lightning hazard is the dynamic factor causing lightning disasters. Meteorological disaster risk is a natural attribute, including disaster environment and disaster-causing factors. The loss caused by meteorological disasters depends on the risk of trigger factors. Hazard refers to the possibility of adverse events. Risk analysis is to study the possibility of adverse events from the perspective of risk inducing factors. The risk analysis of lightning disasters is to study the intensity and frequency of lightning in the area threatened by lightning. The intensity is expressed by the lightning intensity index, and the frequency is the probability, which can be expressed by lightning frequency. Here, the ground lightning frequency and intensity are mainly considered as the index factors to measure the lightning risk.

Exposure is divided into two elements: natural physical exposure and social physical exposure.

Vulnerability indicates the extent to which exposed objects in the affected area are affected by lightning disasters.

The ability of disaster prevention and mitigation mainly describes the level of social and economic development in the affected areas, reflecting the regional disaster bearing capacity and loss rate. Generally, it can be described by various statistical data, and has great regional differences and volatility in the time process.

The process of regional natural disaster risk formation, danger, exposure and vulnerability are indispensable. In the process of the formation of lightning disaster risk, in addition to the above natural disaster formation factors: danger, exposure and vulnerability, the role of disaster prevention and mitigation capacity in the risk of lightning disaster is relatively large. Therefore, when analyzing the risk of lightning disasters, it is necessary to consider the ability of disaster prevention and mitigation. Lightning disaster risk can be expressed as a multivariate function of dangerousness, exposure, vulnerability, disaster prevention and mitigation capacity.

The risk of lightning disaster is composed of four main factors: the risk of disaster-causing factors, the sensitivity of disaster-pregnant environment, the vulnerability of disaster-bearing bodies and the ability of disaster prevention and mitigation. Each factor is composed of several evaluation indexes. According to the theory of natural disaster risk and the formation mechanism of lightning disaster risk, the conceptual framework of lightning disaster risk zoning is established. According to the above conceptual framework of lightning disaster risk, a number of specific indicators are selected to evaluate the degree of lightning disaster risk.

1.3.8.2 Lightning Risk Mapping Based on 3S Technology

The "Regulations on the Prevention of Meteorological Disasters" implemented on 1 April 2010 stipulates:

> local people's governments at or above the county level should organize meteorological and other relevant departments to carry out meteorological disaster censuses on the types, frequency, intensity, and losses of meteorological disasters occurring in their administrative regions, establish a meteorological disaster database, conduct meteorological disaster risk assessment according to the types of meteorological disasters, and delineate meteorological disaster risk areas according to the distribution of meteorological disasters and the results of meteorological disaster risk assessment.

Lightning disaster is one of the ten most serious disasters announced by the United Nations. For a long time, lightning prevention work only focused on engineering protection, and lightning disaster risk mapping is lagging behind. Therefore, delineating the lightning disaster risk area based on the lightning disaster database is an urgent need to perform the administrative functions of the meteorological authorities. The lightning disaster risk zoning enriches the content of meteorological disaster comprehensive risk zoning and lays the necessary foundation for the preparation of meteorological disaster prevention planning.

Based on GIS technology, natural disaster risk assessment method and analytic hierarchy process, Lv et al. (2020) used lightning location monitoring data, geographic information data and socio-economic data in Jiangxi Province from 2010 to 2019 to carry out lightning disaster risk mapping from three aspects: disaster-causing factor, disaster-pregnant environment and disaster-bearing body, and formed lightning disaster risk zoning in Jiangxi Province. Based on the theory of natural disaster risk assessment, Cheng (2019) used lightning location data, geographic information data, socio-economic data and lightning disaster data to study the lightning disaster risk assessment and the risk of disaster-causing factors, the exposure of disaster-bearing bodies and the vulnerability of disaster-bearing bodies. The quantitative relationship between evaluation indicators and risk assessment was established, and the method of lightning disaster risk assessment in Henan Province was formed. At the same time, combined with GIS technology, the hazard distribution map of disaster-causing factors, the exposure distribution map of disaster-bearing bodies and the vulnerability distribution map of disaster-bearing bodies were formed. Finally, the comprehensive risk zoning map of lightning disaster in Henan Province was formed by superposition.

Based on the theory of natural disaster risk, Liu et al. (2019) selected 11 indicators to construct the lightning disaster risk index using the analytic hierarchy process and established the lightning disaster risk assessment model. By selecting the lightning disaster data, the grey correlation assessment of lightning disasters was carried out to verify the correctness of the zoning results. The assessment results were generally consistent with the distribution of lightning disaster risk zoning. The maps of the environmental sensitivity of lightning disaster, the risk of lightning disaster-causing factors, the vulnerability of lightning disaster-bearing bodies, the ability of lightning disaster prevention and mitigation, the comprehensive risk of lightning disasters, and the grey correlation degree of lightning disaster were obtained.

1.3.9   RISK MAPPING OF DENSE FOG

1.3.9.1   Dense Fog

Fog is a disastrous weather phenomenon in which a large number of water droplets or ice crystals are suspended in the air near the surface, and the horizontal visibility distance is reduced to less than 1 km. Fog seriously affects atmospheric visibility and can pose a certain threat to traffic, transportation and military activities. In recent years, the economic losses caused by fog-induced traffic accidents have increased with the city's development. According to statistics, about 1/4 of traffic accidents are caused by bad weather with low visibility, such as dense fog. The traffic accident rate on the expressway in dense fog weather is 10 times higher than that of normal. The pollutants carried by fog may induce various diseases and adversely affect human health. Therefore, the impact of fog on traffic and the human living environment has been widely recognized by all sectors of society.

In recent decades, many experts and scholars have made in-depth analysis, research and discussion on fog weather phenomenon through field observation and experiments on fog, summarized the long-term climate change characteristics and environmental impact of fog, studied fog prediction theory and methods, conducted numerical simulation research on fog and obtained many meaningful results.

The spatial and temporal distribution of fog days of different grades in different regions has strong locality, and the dense fog and heavy dense fog with low visibility and long duration are more harmful. The classification of fog days plays an important role in mastering the law of fog disaster and objectively evaluating the disaster. According to the general survey of fog meteorological data and disaster investigation, when the following three conditions are met, there are usually disasters (flight delays, traffic accidents, etc.). Therefore, the meteorological conditions for fog disasters are defined as follows: (1) More than half of the stations in the city have visibility < 1000 m (range); (2) Visibility ≤ 100 m (intensity) observed in some areas; (3) Duration ≥ 6 h (duration).

In the actual fog disaster risk assessment, the appropriate fog disaster classification standard can be selected according to the actual situation of the evaluated area to make the evaluation results more objective and accurate. The evaluation of fog disaster risk should consider the comprehensive effect of four factors, such as the risk of disaster-causing factors, the sensitivity of disaster-pregnant environment, the vulnerability of disaster-bearing bodies and the ability of disaster prevention and mitigation. The evaluation factors mainly include the location of the fog disaster, the scope of influence, the frequency of historical occurrence, the level of visibility. The environmental sensitivity of fog disasters refers to the effect of natural factors on the formation of fog disasters in a certain area without considering the weather background. Its evaluation indexes include the coastline index, river network density index and topography index. Vulnerability refers to the degree of vulnerability of a disaster-affected body to natural disaster events. The evaluation indicators mainly include road network density, population density and per capita GDP. The ability of disaster prevention and mitigation mainly includes the ability of human recovery and reconstruction of disaster-bearing bodies and the ability of government emergency response. The economic development level of the region and the construction of basic disaster prevention facilities are the important basis for the evaluation of disaster prevention and mitigation capacity.

1.3.9.2 Dense Fog Risk Evaluation

In the fog disaster risk assessment, some have tried to use the grey correlation method and achieved good results. Based on the observation data of foggy weather in the Beijing area in the past 20 years and the disaster survey data derived from foggy weather, Zhang et al. (2008) analyzed the disaster evaluation index and disaster classification of fog by using the grey correlation method, established the evaluation model and evaluated 22 fog disaster cases. The results showed that the grey correlation method had the characteristics of a small amount of calculation without requirement for the number of samples, and obeyed a certain distribution law. It is a simple and feasible method of fog disaster risk assessment combining qualitative analysis and quantitative estimation.

There are precedents to evaluate the risk of fog disaster by grid overlay analysis, such as Hu et al. (2010), using GIS software to divide a certain size of square grid units in the study area. The fog observation data are used to measure the fog disaster risk index in urban areas. The regular grid is used as the evaluation unit, and the road network density in the grid area is calculated grid by grid, which is used as the spatial vulnerability index of fog disasters. The population density in the grid is selected as the vulnerability index of fog disasters. The risk index of fog disaster is calculated according to the distribution ratio of 5:2:1. Ma et al. (2014) analyzed the temporal and spatial distribution characteristics of marine fog in Qingdao based on the climate data from 1978 to 2007. Marine fog is a meteorological disaster with a great impact on urban construction and social and economic activities in Qingdao. Combined with the fog disaster census data from 1984 to 2007, the index-weighted comprehensive model of disaster risk assessment was used to carry out the sea fog meteorological risk assessment with geographic information system (GIS).

1.3.10 FIRE RISK ASSESSMENT

1.3.10.1 Fire Risk

Due to the combined effects of climatic conditions, human activities, environmental factors, management and protection policies, the forest fire losses in different regions show significant regional differences (Bowman et al., 2009; Csiszar et al., 2006; Schroeder et al., 2008). In order to better prevent the occurrence of forest fires and effectively use resources, it is necessary to take different preventive policies for different regions and carry out forest fire risk mapping (Jaiswal et al., 2002; Dhar et al., 2023). Forest fire risk mapping refers to dividing the study area into different levels according to the risk of forest fires in the study area with the regional disaster-bearing capacity and socio-economic conditions (Eugenio et al., 2016). Forest fire risk mapping is used to estimate the possibility of forest fires and the potential losses caused by the vulnerability of forest systems to attacks, so as to quantitatively or qualitatively measure, predict, analyze and evaluate the potential losses caused by the uncertainty of forest fires in a certain period. Accurate fire risk classification can play a good early warning role, reduce the number of fires, protect forest resources, protect the life safety of firefighters and reduce the loss of personnel and resources caused by fires. According to the forest fire risk map, it can provide decision-making opinions for the forestry department in forest fire prevention, firefighting, fire prevention work construction and other aspects. Combined with qualitative and quantitative methods, the weight of each disaster-causing factor was determined by the analytic hierarchy process. According to the weight coefficient of different factors, the forest risk map model was established, and the risk zoning was carried out according to the model's results.

1.3.10.2 Fire Risk Evaluation

With the understanding of the fire process and mechanism, ones have successively studied fire risk models and risk map methods, including the Bayesian network, catastrophe progression method, cluster analysis method, cellular automata, optimal segmentation method, factor weighted overlay comprehensive analysis method, artificial neural network, entropy method, fuzzy comprehensive evaluation method, semi-parametric space model, logistic regression model and geographical weighted regression. Fire risk is the result of the comprehensive effect of hazard, exposure and vulnerability of bearing body, and regional fire prevention and disaster reduction capability.

The study on fire risk early warning has gone through three stages. In the 1960s and 1970s, meteorological factors were used as the main prediction parameters to predict fire risk. In the 1980s and 1990s, the development of remote sensing and geographic information system technology promoted the research of fire risk index, and the multi-factor fire risk prediction. Since the 1990s, the fire risk index has been calculated by taking into account the fuel conditions, climate conditions, terrain and other factors, making the use of the fire risk index faster and more convenient. Various indexes based on these factors, such as the Fire Weather Index (FWI), the Keetch-Byram Dry Index (KBDI) and the Forest Fire Danger Index (FFDI), have been widely used.

The development and application of satellite remote sensing technology have made contributions to the study of fire risk on combustible types and water content through remote sensing monitoring and inversion technology. For example, unsupervised classification methods are used to classify combustibles, greenness maps are used to quantify the moisture content of combustibles, vegetation index NDVI and VCI are used to monitor the state of combustibles, and remote sensing is also used to estimate drought and soil moisture for fire risk assessment (Kaufman et al., 1998). Many GIS spatial analysis methods are used for fire risk assessment. A multi-distance spatial clustering function was proposed by Ripley in 1976, belonging to the multi-distance spatial clustering analysis method. This function assumes that geographical things are uniformly distributed in space and counts the number of samples within the search circle according to a certain radius distance. By comparing the measured value and theoretical value of the average number of these samples and the ratio of sample density in the region, it finally determines whether the distribution characteristics of the actual observed geographical things are spatial aggregation, spatial divergence or spatial random. The ensemble empirical mode decomposition method decomposes the time series of a variable into the oscillation components and a nonlinear trend with different time scales such as inter-annual, inter-decadal and inter-year. Kernel Density Estimation can estimate the probability density value of geographical objects in their surrounding neighborhood by setting bandwidth without any prior density assumption.

Lin et al. (2013) used meteorological factors such as precipitation, maximum temperature, relative humidity, average wind speed, snow days and thunderstorm days, geographical factors such as elevation, slope and aspect, vegetation factors such as vegetation type and NDVI, and social factors such as traffic, population and residence as four risk factors for forest fire risk. The forest fire risk in Tibet was quantitatively evaluated by index normalization method, analytic hierarchy process and weighted comprehensive evaluation method. According to the fire risk level, the whole region was divided into five risk areas: low, lower, medium, higher and much higher. Using the analysis function of GIS software, the map of forest fire risk level in Tibet was compiled to provide a reference for improving the prediction level of forest fire risk.

Forest fire is the result of many natural factors and social factors. Its occurrence is closely related to climate, vegetation, human activities and terrain. According to the four risk factors of meteorology, geography, vegetation and society, the spatial distribution of forest fire risk factors was calculated, and the results of each risk factor were normalized to the range of 0–1 according to the normalization method.

Meteorological factors mainly include precipitation, temperature, sunshine duration, evaporation, wind and air humidity. Meteorological factors can affect fire occurrence and fire behavior by changing the fire environment. The area with the highest meteorological risk factors is located in the southeastern region, where the altitude is the lowest. The temperature is high, and the dry and wet seasons are distinct. The main precipitation is concentrated in the summer flood season, and the precipitation in the fire prevention period only accounts for less than 10% of the annual precipitation. The number of snow days is small, which is most conducive to forest fires.

Using high-resolution DEM data to derive the corresponding terrain data such as altitude, slope and aspect as the geographical factors of forest fire danger, different scores are given according to their different effects on forest fire. For forest fires, as the altitude increases, the temperature decreases, the vegetation distribution decreases and the possibility of forest fires decreases. The influence of slope on the occurrence and spread of forest fire was high in the middle and low on both sides. With the increase of slope, the surface runoff accelerated, the fuel on the ground was easy to dry and the fire risk was high. However, when the slope reached a steep slope, the distribution of trees decreased, and the risk of forest fire decreased. The slope direction directly affects the amount of solar radiation received by the ground, resulting in temperature differences in different slope directions. The southern slope receives higher solar radiation than the northern slope, and the air is drier and more likely to cause fire.

According to the provisions of the national forest fire danger zoning grade on the flammability of vegetation, different vegetation types are divided: the inflammable coniferous forest is given 1, the more inflammable broad-leaved forest is 0.7, the more refractory grassland is 0.3, the refractory plateau meadow is 0.1 and the non-combustible lake and desert are given 0. The spatial distribution of vegetation risk factors was calculated by normalized NDVI and vegetation type.

In addition to natural factors, human-induced fires are an important threat to forests. Therefore, forest fire risk factors formed by human factors must be considered. Considering the influence of traffic, population density and village density, the highway is analyzed by the 5 km buffer zone. The population density is obtained by dividing the agricultural population by the county area. Considering the low settlement of villages in Tibet, a buffer analysis of 3 km is used for villages. After standardizing the three indicators, the standardized spatial distribution of social risk factors is obtained.

The normalized meteorological, topographic, vegetation and social factors were given weights of 0.30, 0.15, 0.40 and 0.15, respectively. The comprehensive evaluation value of forest fire risk in Tibet was obtained by comprehensive calculation, and then the risk evaluation value in the range of 0–1 was obtained by standardization. According to the boundary of 0.2, it was divided into five levels: sub-low-risk areas, low-risk areas, medium-risk areas, sub-high-risk areas and high-risk areas, and the forest fire risk zoning map of Tibet was obtained.

1.4 OPPORTUNITIES AND PROSPECTIVE

The 3S technology has been widely used in meteorological disaster monitoring and assessment. RS technology is mainly used to obtain the data source of disaster monitoring at a large scale. GIS links spatial data and attributed data to perform spatial analysis, query and cartographic comprehensive management of geometric features in the study area. GNSS technology is mainly used to quickly obtain location information in real time. In the field of basic landslide data acquisition, 3S technology can play the key function in landslide geographic data acquisition and rapid update, establish a complete landslide catalog database as much as possible, and effectively express landslide mapping (Zhang et al., 2015). Also, 3S technology can obtain information such as elevation, vegetation coverage, surface humidity, water system distribution and topographic relief in a large area.

With the acceleration of urbanization, the demand for 3S technology is becoming increasingly urgent, and its application field is expanding. The construction of smart cities involves basic image map data and a large amount of spatial location data information. Based on the real-time location information provided by GNSS, the data transmission and processing capabilities of the system are further improved based on the existing path analysis of GIS, and the optimal path planning is provided in a short time. Reasonable diversion of traffic flow will help alleviate the current traffic congestion in large cities. Affected by human activities and climate change, natural disasters occur frequently, and emergency management of geological and meteorological disasters based on 3S technology is particularly critical. 3S technology has unique advantages in data acquisition and management, two-dimensional and three-dimensional visualization of spatial information, emergency monitoring and analysis, spatial data analysis and so on, which can help to better reduce the impact of disasters on people's lives and property.

Although 3S technology has made great achievements in the application of various fields, many problems also need to be solved in the practice process. (1) The inconsistency of data standards leads to low efficiency of 3S technology applications. Affected by different sensors of remote sensing satellites, the spatial resolution, radiation resolution and spectral resolution of satellite images from different sources are not uniform, and no better methods can be compatible with different data sources. In addition, raster data and vector data have different data structure characteristics. Integrating the advantages of raster and vector data to build a unified standard and compatible geographic information database with different data types is one of the goals of 3S technology integration. (2) Data accuracy needs to be further improved. At present, precise GNSS navigation and positioning for outdoor and indoor have not yet made a breakthrough, which makes it difficult to meet the needs of indoor high-precision positioning applications. Improving the accuracy of remote sensing data and GIS models will help apply 3S technology applications to more fields. (3) As a multidisciplinary interdisciplinary technology, 3S technology not only achieves internal integration, but also deeply integrates with other fields such as computer networks, artificial intelligence and big data. The construction of the 3S system requires not only the knowledge base of 3S technology, but also availability of computer software and hardware. Therefore, it is necessary to strengthen the cultivation of multidisciplinary high-quality compound talents and provide sufficient talent reserve for the development of 3S technology.

REFERENCES

Aja D., E. Elias, O. Obiahu. Flood risk zone mapping using rational model in a highly weathered Nitisols of Abakaliki Local Government Area, South-eastern Nigeria. Geology, Ecology, and Landscapes, 2020, 4(2): 131–139.

Allan C., A. Curtis, N. Mazur. Understanding the social impacts of floods in Southeastern Australia. Advances in Ecological Research, 2006, 39: 159–174.

Bowman D., Jr, K. Balch, P. Artaxo, et al. Fire in the earth system. Science, 2009, 324: 481–484.

Campbell J., R. H. Wynne. Introduction to remote sensing, Fifth Edition, The Guilford Press, 2011.

Chen S., A. Huo, D. Zhang, et al. Key technologies for drought disaster risk assessment in typical vulnerable areas of eastern Gansu Province. Agricultural Research in the Arid Areas. 2022, 40(2): 197–204.

Chen X. Division into districts on wind disaster risk of Hangzhou. Nanjing University of Information Science and Technology, 2012.

Cheng L. The application of analytic hierarchy process (AHP) and geographic information system (GIS) in lightning disaster risk-zoning in Henan province. Journal of Nanjing University of Information Science and Technology (Natural Science Edition), 2019, 11(2): 234–240.

Csiszar I., J. Morisette, L. Giglio. Validation of active fire detection from moderate-resolution satellite sensors: The MODIS example in Northern Eurasia. IEEE Transactions on Geoscience and Remote Sensing, 2006, 44(7): 1757–1764.

Dhar T., B. Bhatta, S. Aravindan. Forest fire occurrence, distribution and risk mapping using geoinformation technology: A case study in the sub-tropical forest of the Meghalay, India. Remote Sensing Applications: Society and Environment, 2023, 29: 1–19.

Dong S., L. Jiang, J. Zhang, et al. Research on flood vulnerability curves off rural dwellings based on "3S" technology. Journal of Catastrophology, 2012, 27(2): 34–39.

Dow J., R. E. Neilan, C. Rizos. The International GNSS Service in a changing landscape of Global Navigation Satellite System. Journal of Geodesy, 2009, 83: 191–198.

Eugenio F., A. dos Santos, N. Fiedler, et al. Applying GIS to develop a model for forest fire risk: A case study in Espirito Santo, Brazil. Journal of Environmental Management, 2016, 173: 65–71.

Fang X., W. Du, W. Quan, et al. Study on high temperature disaster risk regionalization in Jiangsu-Zhejiang-Shanghai region. Journal of Meteorology and Environment, 2016, 32(6): 109–115.

Gomarasca M. Basics of geomatics. Springer Dordrecht, 2009.

Goodchild M. F. Geographic information systems and science: Today and tomorrow. Annals of GIS, 2009, 15(1): 3–9.

Guan Y., F. Zheng, P. Zhang, et al. Spatial and temporal changes of meteorological disasters in China during 1950–2013. Nature Hazards, 2015, 75: 2607–2623.

He Y., X. Wang, C. Chai, et al. Flood damage assessment and visualization based on NPP-VIIRS nighttime light remote sensing. Journal of Natural Disasters, 2022, 31(3): 93–105.

Hofmann-Wellenhof B., H. Lichtenegger, E. Wasle. GNSS—Global navigation satellite systems: GPS, GLONASS, Galileo, and more. Springer Wien New York, 2007.

Hu H., Y. Xiong, S. Zhang. The risk assessment of the fog disaster based on vulnerability calculating related to the urban transportation network. Journal of Applied Meteorological Science, 2010, 21(6): 732–738.

Jaiswal R. K., S. Mukherjee, K. D. Raju, et al. Forest fire risk zone mapping from satellite imagery and GIS. International Journal of Applied Earth Observation and Geoinformation, 2002, 4(1): 1–10.

Jin S. Global navigation satellite systems—Signal, theory and applications. IntechOpen, 2012.

Jin X. The risk evaluation and regionalization of heat wave in Fujian Province within the background of risk society. Master Degree, Fujian Normal University, 2017.

Kaufman Y., R. G. Kleidman, M. D. King. SCAR-B fires in the tropics: Properties and remote sensing from EOS-MODIS. Journal of Geophysical Research Atmospheres, 1998, 103(D24): 31955–31968.

Khorram S., F. H. Koch, C. F. van Wiele, et al. Remote sensing, Springer, 2012.

Lechner W., S. Baumann. Global navigation satellite systems. Computers and Electronics in Agriculture, 2000, 25(1–2): 67–85.

Li D., Q. Li. The formation of geospatial information science. Advances in Earth Sciences, 1998, (4): 2–9.

Li D. Digital Earth and "3S" technology. China Surveying and Mapping, 2003, (2): 30–33.

Li H., Q. Li, X. Li, et al. Discussion on the algorithms of a new siphon rain gauge. Wseas Transactions on Circuits and Systems, 2010, 9(6): 389–398.

Li J., H. Zhang, S. Cao. Assessment and zonation of late frost injury of winter wheat in He'nan Province based on GIS. Journal of Arid Meteorology, 2015, 33(1): 45–51.

Lin Z., H. Lu, C. Luobo, et al. Risk assessment of forest fire disasters on the Tibetan plateau based on GIS. Resources Science, 2013, 35(11): 2318–2324.

Liu G., L. Zhang, B. He, et al. Temporal changes in extreme high temperature, heat waves and relevant disasters in Nanjing metropolitan region, China. Natural Hazards, 2015, 76: 1415–1430.

Liu X., L. You, H. Song, et al. Analysis and evaluation of lightning disaster risk regionalization based on GIS and AHP in inner Mongolia. Chinese Agricultural Science Bulletin, 2019, 35(20): 75–82.

Lv Z., Z. Yu, C. Wang. Risk zoning of regional lighting disaster in Jianxi province based on GIS technology. Meteorology and Disaster Reduction Research, 2020, 43(3): 228–233.

Ma T., P. Xiao, X. Zhang, et al. Recognition of snow cover based on features selection in GF-3 fully polarimetric data. Remote Sensing Technology and Application, 2020, 35(6): 1292–1302.

Ma Y., Y. Hao, Y. Wang. Characteristics of sea fog and risk assessment for fog disaster in Qingdao. Periodical of Ocean University of China, 2014, 44(11): 11–15 + 29.

Mishra A., V. P. Singh. A review of drought concepts. Journal of Hydrology, 2010, 391: 202–216.

Norman B. A brief history of Global Navigation Satellite Systems. The Journal of Navigation, 2012, 65(1): 1–14.

Peng J., D. Wang, Y. Zhao, et al. Hail disaster risk zoning in Chengde City based on GIS. Desert and Oasis Meteorology, 2019, 13(1): 105–109.

Qin D. Exploration of the application and development prospects of geographic information systems. Beijing Agriculture, 2015, 626(21): 179–180.

Ramkar P., S. M. Yadav. Flood risk index in data-scarce river basins using the AHP and GIS approach. Natural Hazards, 2021, 109: 1119–1140.

Schroeder W., E. Prins, L. Giglio, et al. Validation of GOES and MODIS active fire detection products using ASTER and ETM+ data. Remote Sensing of Environment, 2008, 112: 2711–2726.

Seejata K., A. Yodying, T. Wongthadam, et al. Assessment of flood hazard areas using Analytical Hierarchy Process over the Lower Yom Basin, Sukhothai Province. Procedia Engineering, 2018, 212: 340–347.

Shan Z., Y. An, L. Xu, et al. High-temperature disaster risk assessment for Urban Communities: A case study in Wuhan, China. International Journal of Environmental Research and Public Health, 2022, 19(1): 183.

Sun J., X. Zhang, B. Hou. Application of ABC-based BP neural network in integrated navigation system. Journal of Telemetry, Tracking and Command, 2016, 37(5): 40–48.

Sun J., Z. Zhao, D. Li, et al. Risk Zoning of Gale disaster in Xingtai City based on GIS in the past 35 years. Journal of Agricultural Catastropholgy, 2021, 11(6): 82–84 + 86.

Sun X., T. Li, X. Mao, et al. High precision Indoor/outdoor positioning system and positioning method based on Beidou UWB. Integrated Circuit Applications, 2020, 37(5): 118–119.

United Nations. The human cost of weather-related disasters 1995–2015. UN Report, 2015.

Wang F., Q. Wei, J. Chang, et al. Remote sensing monitoring technology of winter wheat late frost based on FY-3. Journal of Agricultural Catastropholgy, 2021, 11(2): 192–194.

Wang J. Application of GPS/GPRS/GIS integrated technology in vehicle positioning and monitoring. Wuhan University, 2005.

Wang L., L. Zengzeng, Remote sensing monitoring of Poyang Lake flood disaster in 2020 based on Sentinel-1A. Geospatial Information, 2022, 20(6): 3–46.

Wang Y., D. Zheng, Q. Li. Urban meteorological disaster. China Meteorological Press, 2009.

Xi W. Risk assessment and regionalization of snow disaster in Heilongjiang Province. Harbin Normal University, 2020.

Yang J., J. Lie, C. Yang. Risk assessment model of hail disaster in Tibet supported by GIS. Plateau and Mountain Meteorology Research, 2016, 36(2): 69–74.

Ye P. Remote sensing approaches for meteorological disaster monitoring: Recent achievements and new challenges. International Journal of Environmental Research and Public Health, 2022, 19, 3701.

Yi Y. Study on risk assessment of flood disaster in Guilin area based on GIS. Guangxi University, 2012.

Zhang C., M. Chen, R. Zheng. Landslide hazard risk assessment and zoning of Huadu District of Guangzhou based on "3S" technique and logistic regress-weighted SVM model. Journal of Ecology and Rural Environment, 2015, 31(6): 955–962.

Zhang J., J. Ni, S. Ma, et al. GIS-based analysis of flood submergence in Jinpu New District, Dalian City. Geology and Resources, 2021, 30(5): 590–594.

Zhang K. Review on geological disaster monitoring and early warning system based on "3S" technology in China. The Chinese Journal of Geological Hazard and Control, 2020, 31(6): 1–11.

Zhang S., D. Ding, Z. Fu, et al. Application of grey relational grade in fog disaster evaluation in Beijing Region. Journal of Catastrophology, 2008, 88(3): 54–56+61.

Zhou F., W. Zhang, L. Lei, et al. GF-3 and Sentinel-1 flood inundation information extraction. Geospatial Information, 2021, 19(6): 17–21.

Zhou Y., X. Cheng, J. Cai, et al. Study on comprehensive risk assessment of Typhoon disasters. China Public Security. Academy Edition, 2013, 30(1): 31–37.

2 Multi-GNSS Near-Real-Time Tropospheric Parameter Estimation and Meteorological Applications

Yidong Lou, Weixing Zhang, Yaozong Zhou, Zhenyi Zhang, and Zhixuan Zhang

2.1 INTRODUCTION

2.1.1 ROLES OF ATMOSPHERIC WATER VAPOR IN WEATHER SYSTEMS

The gas layers converged around the earth are called the earth's atmosphere, which, from top to bottom, can be divided into exosphere, thermosphere, mesosphere, stratosphere and troposphere according to temperature, as well as ionosphere and neutral atmosphere in terms of electron density as shown as Figure 2.1 (https://commons.wikimedia.org). The earth's atmosphere consists of gas molecules, electrons and ions as well as water, ice, dust and other liquid and solid particles. The gas molecules include nitrogen, oxygen and argon, accounting for about 99.96%, as well as carbon monoxide, carbon dioxide, ozone, methane, hydrogen, neon, helium and water vapor, accounting for about 0.04% (Zhang et al., 2011).

The atmospheric water vapor (AWV) is of the lesser contents, but one of the most active and important components in the earth's atmosphere. On the one hand, the AWV causes daily weather changes such as clouds, rain and lightning, and plays an important role in indicating extreme weather such as typhoons and rainstorms. On the other hand, the AWV is the most important greenhouse gas and the largest contributor to Earth's greenhouse effect (about 60%), and plays a vital role in Earth's ecosystem. In addition, the three-phase transformations of the gas, solid and liquid water are also very important for maintaining Earth's energy balance (Chung et al., 2014). The accurate knowledge of AWV is therefore crucial to human survival and development.

2.1.2 GNSS ADVANTAGES FOR AWV MONITORING

However, the accurate monitoring of AWV is very challenging work due to the high dynamic variation characteristics of AWV. The current monitoring techniques mainly include radiosonde, ground-based water vapor radiometer (WVR), satellite-based WVR, satellite-based infrared spectroradiometer, Global Navigation Satellite Systems Radio Occultation (GNSS-RO) and ground-based GNSS (Lou et al., 2022). Radiosonde generally restricts the number of balloon launchings to twice daily (00:00 and 12:00 UTC) at specific stations, considering the device cost and manpower consumption. Radiosonde measures the temperature, humidity, pressure and other parameters along the floating path, and is an important technical means for meteorology and climate research all over the world. However, influenced by the balloon ascending speed error and wind drift, the observation results are biased from the real (Jin and Wang, 2021). What's more, the 12 h sampling rate is inadequate to resolve the temporal variations of AWV, and the long-time observations usually include the inhomogeneity issue caused by radiosonde sensor changes (Zhang et al., 2019).

DOI: 10.1201/9781003363118-2

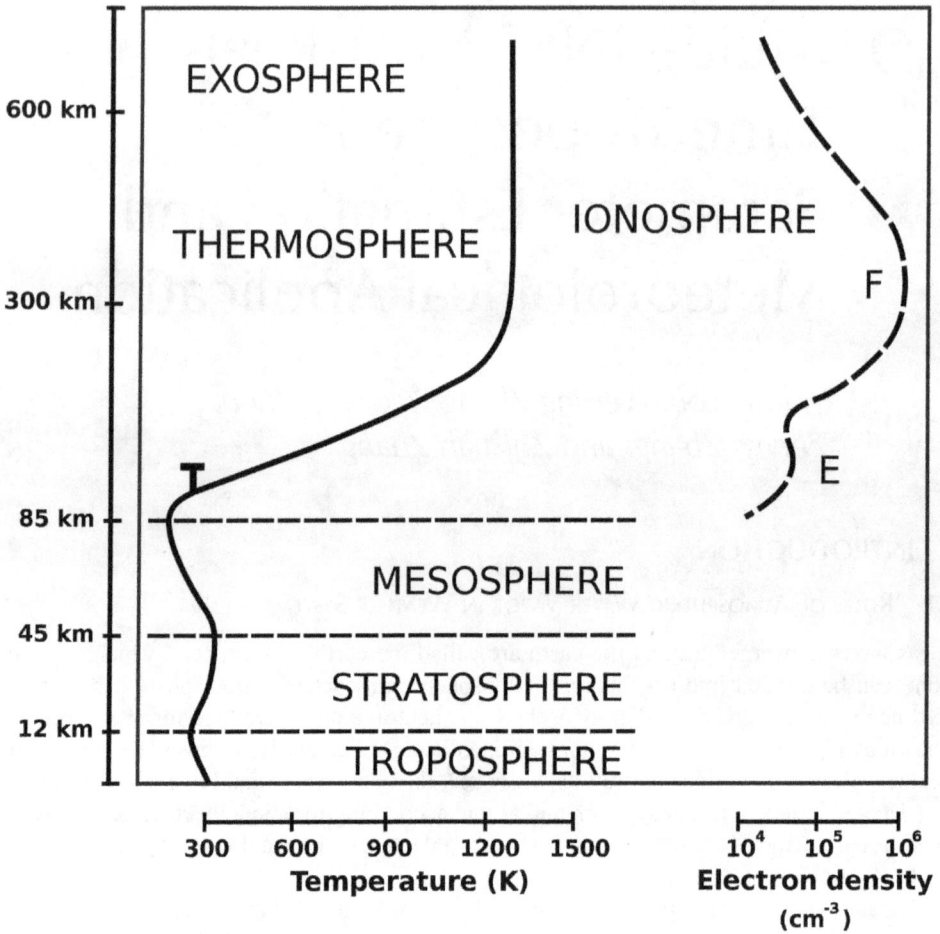

FIGURE 2.1   Structure of the earth's atmosphere.

Ground-based WVR upwardly measures the brightness temperature at several frequencies along the radiation signal path, and then converts the brightness temperature to precipitable water vapor (PWV) and liquid water using the frequency dependence and local meteorological observations (generally radiosonde data). However, the WVR instruments are relatively expensive and cannot be operated during moderate to heavy rain, and therefore the observation space coverage is poor (Bevis et al., 1992).

Satellite-based WVR downwardly measures the water vapor absorption lines from the hot backboard of the earth, and is more useful over the oceans than over land. Satellite-based WVR can provide good spatial but poor temporal coverage and has opposite characteristics to the ground-based units (Bevis et al., 1992). Satellite-based infrared spectroradiometry retrieves the AWV by the apparent reflectance of water vapor absorption. The most representative instrument of this channel is the moderate resolution imaging spectroradiometer (MODIS). Satellite-based infrared spectroradiometry can obtain the global coverage of AWV, but it is susceptible to the weather condition and cloud cover.

GNSS-RO retrieves the atmospheric vertical profile from the impacts of the atmosphere on the signal link between GNSS satellite and low Earth orbit (LEO) satellite. GNSS-RO has advantages of no calibration, all-weather, high precision, high vertical resolution, global coverage and so on. However, the GNSS-RO results have relatively poor performance in the low part of troposphere where most AWV is located.

The GNSS electromagnetic wave signals traveling through the neutral atmosphere are inevitably delayed and bent, yielding the hydrostatic and wet tropospheric delay. The wet tropospheric delay can be converted to PWV or slant water vapor (SWV) by using water vapor weighted mean temperature (Bevis et al., 1992), deriving the new research line of GNSS meteorology. Compared with other techniques, GNSS AWV monitoring has advantages of high precision, high time resolution, low cost and all-weather, attracting more and more attention from meteorologists and geodesists.

2.1.3 GNSS PWV Estimation and Applications

Since the proposition of GPS meteorology by Bevis et al. (1992), GNSS meteorology has been rapidly developed in the last 30 years. In the early stage, the GPS meteorology only focused on the PWV retrieval from post-processing double-difference network solution as an auxiliary product. With the rapid expansion of ground-based GPS networks, the low efficiency problem of the network solution became more and more serious. To overcome the low efficiency issue, the more flexible precise point positioning (PPP) technique based on the high-precision satellite orbit and clock correction product was proposed by Zumberge et al. (1997). After that, the GPS meteorology using the PPP technique has experienced great developments due to the advantages of flexibility and high efficiency. In 1998, the International GNSS Service (IGS) established the troposphere working group and operationally released the tropospheric products from the post-processed network or PPP solution, and the accuracy of the PWV derived from the IGS tropospheric products reaches about 1 mm.

However, the time delay of the tropospheric product reaches serval days or longer, and therefore cannot be applied in operational meteorological applications. Aiming at the time delay issue, researchers proposed the concept of near-real-time (NRT) PWV processing whereby the PWV is processed by using NRT or real-time observations and ultra-rapid or real-time satellite orbit and clock products. The accuracy of the NRT PWV is about 2 mm, and the time delays are decreased to less than 1 h and even 15 minutes, and, therefore, can be used for weather monitoring, forecasting and early warning, showing the brilliant prospects. Afterwards, IGS implemented the Real Time Pilot Project (RTPP) in 2007 and released IGS Real Time Service (RTS) in 2013 (https://igs.org). The real-time PWV processing is gradually catching the attention of meteorologists and geodesists due to shorter time delays, such as 60 s and 30 s, and acceptable accuracy of better than 3 mm.

After the successful launch of the first artificial satellite in 1957, GNSS has been developed rapidly for positioning, navigation and timing (PNT) applications. In 1995 and 1996, the United States and Russia established the Global Positioning System (GPS) and Global Navigation Satellite System (GLONASS), respectively (Li and Huang, 2005). In the twenty-first century, the development of GNSS entered the golden age, and different GNSS systems showed the trend of competition, cooperation and coexisting. China's BeiDou Navigation Satellite System (BDS) and the European Union Galileo Navigation Satellite System (Galileo) were constructed steadily, followed by GLONASS restart and modernization and GPS modernization.

China's BDS system construction has experienced three stages. The first stage completed the BDS-1 system by the end of 2000 and provided PNT services over China. The second stage completely established the BDS-2 system by the end of 2012 and provided services over the Asia-Pacific region. The last stage entirely constructed the BDS-3 system by 2020 and provided global PNT services. On July 31, 2020, the BDS-3 system was officially announced in service. The heterogeneous BDS-3 constellation includes 24 medium Earth orbit (MEO) satellites, 3 geostationary Earth orbit (GEO) satellites and 3 inclined geosynchronous orbit (IGSO) satellites (http://www.beidou.gov.cn). The development of Galileo also experienced three steps, namely in-orbit validation element (IOVE), in-orbit validation (IOV) and full operational capability (FOC). As of September 28, 2022, the number of available Galileo FOC satellites reaches 24 (https://igs.org/mgex/constellations/#galileo).

Around 2008, GLONASS initiated the restart and modernization plan. On the one hand, the global PNT capabilities of the GLONASS system were recovered by the launch of GLONASS-M satellites. On the other hand, the GLONASS-K satellites with code division multiple access (CDMA) signal were launched for the interoperability and compatibility with other GNSS systems. As of September 28, 2022, the GLONASS constellation contains 22 GLONASS-M satellites and 2 GLONASS-K satellites. GPS has undergone continuous satellite renewal, system maintenance and modernization. As of December 20, 2021, the GPS constellation includes 25 Block-II and 4 Block-III satellites (http://www.csno-tarc.cn).

At present, GNSS covers four GNSS systems, namely BDS, GPS, GLONASS and Galileo. All the GNSS systems compete, cooperate and interoperate with each other, and the number of visible GNSS satellites at user receiver is up to 40 to 50. The GNSS applications have penetrated into various fields of national economy and people's livelihood, such as national defense, military, aerospace, surveying, meteorology, transportation, communication, navigation, remote sensing and so on (Cao, 2016), generating great political, economic and social values. GNSS has become the key infrastructure in the national PNT system (Yang, 2016).

The multi-GNSS can provide much more abundant observations with respect to any single GNSS system that will be beneficial to the tropospheric products and PWV retrieval, especially for NRT and RT processing. Li et al. (2015b) analyzed the results of multi-GNSS PPP and RT PWV retrieval, and found the accuracy improvement of 25%–52% for multi-GNSS processing. Lu et al. (2017) and Dousa et al. (2018b) further pointed out the multi-GNSS zenith total delay (ZTD) accuracy improvement of 10%–22%. Hadas et al. (2020) showed an inner ZTD accuracy improvement of about 37% from the multi-GNSS estimation.

In recent years, with the rapid development of GNSS systems and continuous tracking networks, ground-based GNSS water vapor sounding has attracted more and more attention. The World Meteorological Organization (WMO) Global Climate Observing System Reference Upper-Air Network (GRUAN) has listed ground-based GNSS as the class I means of AWV monitoring (Seidel et al., 2009). The major international satellite navigation and meteorological organizations have successively set up working groups to carry out relevant research work. IGS established the troposphere working group in 1998 to carry out GNSS tropospheric parameter estimation and application research (https://igs.org/wg/troposphere). WMO GRUAN as well as Global Geodetic Observing System (GGOS) in Austria listed GNSS water vapor monitoring as one of the main observation tasks.

In 2005, the European Meteorological Association launched the EIG EUMETNET GNSS Water Vapor Programme (E-GVAP) (http://egvap.dmi.dk/), which is the most well-known NRT water vapor monitoring program in the world (Jones et al., 2020). After more than ten years of development, E-GVAP currently covers dozens of NRT ZTD analysis centers, and operationally processes the ZTD and horizontal gradient products over more than 3,500 globally distributed stations (mainly in Europe) for the extreme weather forecast and numerical weather assimilation applications (http://egvap.dmi.dk/). In 2019, the International Association of Geodesy (IAG) established the Inter-Commission Committee on "Geodesy for Climate Research" (ICCC) to carry out climate research based on space geodetic techniques such as GNSS.

In China, the China Meteorological Administration (CMA) has successively launched the projects of "GPS/MET Application Pre-Research", "Atmosphere and Ocean Space Monitoring and Early Warning Application Pilot Project Based on BDS", "National BDS Augmentation Service System (NBASS)–Meteorological Industry Data Processing Center" and other projects. Through more than 20 years of development, CMA has formed the norms, standards and processes for GNSS meteorological observation. In 2018, Wuhan University cooperated with E-GVAP and jointly built the first non-European and American E-GVAP analysis center in Wuhan University (WUHN) for NRT processing the tropospheric products of global and regional networks, realizing the product interaction and sharing with E-GVAP. In 2019, the WUHN analysis center was completed and operationally processed the NRT tropospheric products of more than 300 stations. In 2021, the WUHN analysis center was upgraded for multi-GNSS processing, deriving the multi-GNSS analysis center named WUHM (http://egvap.dmi.dk/).

2.2 NRT TROPOSPHERIC PARAMETER ESTIMATION AND PROCESSING

This section focuses on the NRT tropospheric parameter estimation. Sections 2.2.1, 2.2.2 and 2.2.3 introduce the basic theory of NRT GNSS data processing, including GNSS error sources, tropospheric models and tropospheric parameter estimation. Section 2.2.4 takes the operational WUHM analysis center as an example and describes the NRT tropospheric parameter processing method and flow as well as some results.

2.2.1 GNSS ERROR SOURCES

The GNSS observation impacts by various error sources. At the satellite end, the observation is mainly affected by the satellite orbit error, satellite clock error, satellite antenna phase center offset and variation, satellite hardware delay bias, relativistic effect and phase wind-up effect. In the propagation path, the error sources mainly include ionospheric delay error, tropospheric delay error and multipath effect. At the receiver and station, the receiver clock error, receiver antenna phase center offset and variation, receiver hardware delay bias, solid earth tide, ocean tide, polar tide, ocean load effect and atmospheric load effect are involved. In addition, the GNSS pseudorange and carrier phase observations also contain the measurement noises (Li and Huang, 2005).

The error sources can be handled mainly by three methods. The first method is by using a correction model or correction product whereby most errors in GNSS observation can be substantially eliminated. The second method is the linear combination of observations, such as ionosphere-free combination for eliminating the first-order ionospheric delay effect, and double-difference observations for eliminating the errors at both the satellite and receiver end and weaking the atmospheric delay error. The third method is parameter estimation, and it can be used for the highly dynamic tropospheric delay error, receiver clock error and so on. In addition, errors such as multipath effect and electromagnetic interference can be weakened by receiver hardware design or avoided by re-selecting station location (Li and Huang, 2005).

2.2.2 TROPOSPHERIC MODEL

2.2.2.1 Tropospheric Parametrization

The signals of GNSS satellites passing through neutral atmosphere (Figure 2.2) suffer the tropospheric slant total delay (STD), which is related to the elevation angle and the azimuth, and varies with time, which means that different GNSS observations are impacted by different STD. Therefore, how to deal with the STD in GNSS observations has become a difficulty in research.

Since the GNSS observation contains not only the tropospheric STD error, but also the ambiguity parameter and the errors related to both the satellite and receiver end, the STD error cannot be directly estimated due to the rank-deficit problem. Therefore, it is necessary to parameterize the tropospheric STD, establish the prior constraint relationship or model and reduce the number of tropospheric parameters to be estimated. Usually, the tropospheric STD is divided into isotropic and anisotropic delay wherein the isotropic part is modeled as zenith path delay and mapping function, and the anisotropic part is expressed the function of north–south and east–west horizontal gradient as (Landskron, 2017):

$$
\begin{cases}
SHD(e,\alpha) = MF_h(e) \cdot ZHD + MF_{gh}(e) \cdot \left[G_{nh} \cos(\alpha) + G_{eh} \sin(\alpha) \right] \\
SWD(e,\alpha) = MF_w(e) \cdot ZWD + MF_{gw}(e) \cdot \left[G_{nw} \cos(\alpha) + G_{ew} \sin(\alpha) \right] \\
STD = SHD + SWD
\end{cases}
\tag{2.1}
$$

where SHD and SWD are slant hydrostatic and wet delay. e and α denote the elevation angle and azimuth. MF_h and $MF_w(e)$ represent hydrostatic and wet mapping function, respectively. ZHD

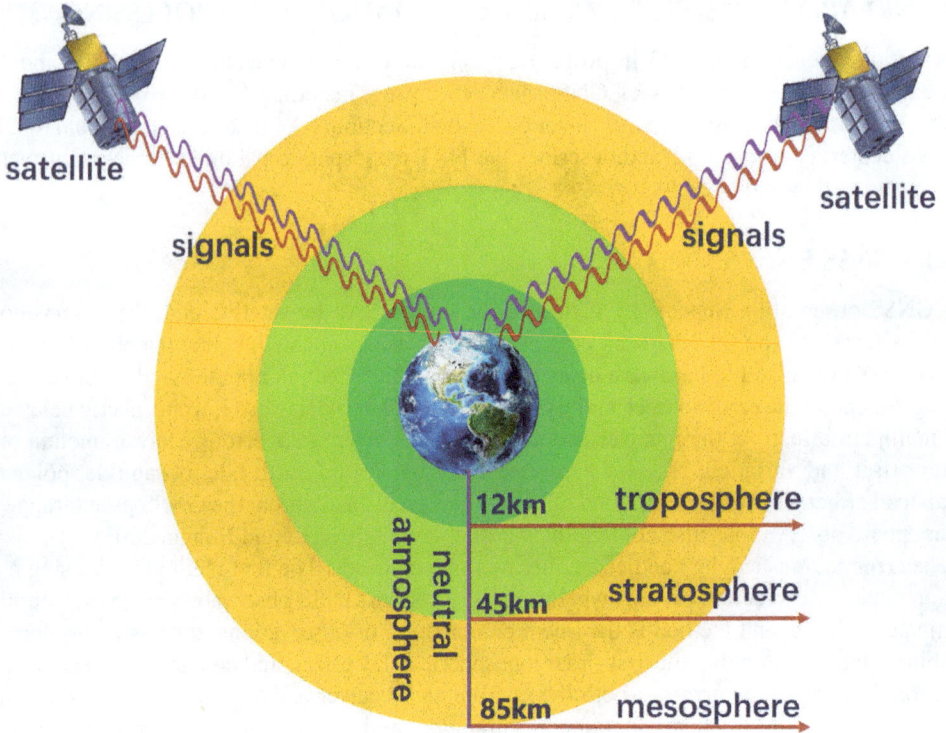

FIGURE 2.2   Diagram for GNSS signals traveling through neutral atmosphere.

and *ZWD* are zenith hydrostatic and wet delays. MF_{gh} and MF_{gw} stand for hydrostatic and wet gradient mapping functions. G_{nh} and G_{nw} are the hydrostatic and wet horizontal gradient in north–south direction. G_{eh} and G_{ew} are the hydrostatic and wet horizontal gradient in east–west direction. In GNSS data processing, the zenith path delay and mapping function models are commonly used to provide *a priori* isotropic delay, with the remaining tropospheric delay as a to-be-estimated parameter.

2.2.2.2 Zenith Path Delay Models

The zenith path delay models mainly contain two categories, namely the calculation model dependent on meteorological parameters and the correction model independent of meteorological parameters. When the meteorological observations are available, the calculation model, such as Hopfield, Saastamoinen, Black and Askne and Nordius, can accurately calculate the zenith path delay (Hopfield, 1971; Saastamoinen, 1972; Black, 1978; Askne and Nordius, 1987). In cases without meteorological observations, we can resort to the meteorological parameter models, such as the UNB3 model and GPT series model (Collins, 1999; Landskron and Böhm, 2018a). In addition, the correction model, which directly provides zenith path delay and does not depend on meteorological parameters, has also attracted attention, such as the SHAO series model, IGGtrop series model and TropGrid series model (Chen et al., 2020; Li et al., 2018; Schüler, 2014).

In June 2018, the European Centre for Medium-Range Weather Forecasts (ECMWF) released the fifth-generation re-analysis product (ERA5) (Hersbach et al., 2020). Focusing on the high spatial-temporal resolution advantages of ERA5 (0.25°×0.25° and 1 h), some scholars have further refined the zenith path delay model and established a series of high spatial-temporal resolution models, such as the HGPT series model, CPTw model and IGPT model (Mateus et al., 2021; Li et al., 2020; Li et al., 2021).

In GNSS data processing, the Saastamoinen (SAAS) and Askne and Nordius models are usually used for the *a priori* zenith path delay calculation aided by the GPT series model. The improved SAAS model can be expressed as (Davis et al., 1985):

$$\begin{cases} ZHD = 0.002277 \times \dfrac{P_S}{f(\varphi, h_S)} \\[3mm] ZWD = \dfrac{e_S}{f(\varphi, h_S)} \times \left(\dfrac{0.2789}{T_S} + 0.05 \right) \\[3mm] f(\varphi, h_S) = 1 - 0.00266 \cos(2\varphi) - 0.00028 h_S \end{cases} \tag{2.2}$$

where P_s (hPa), e_s (hPa) and T_s (K) are the site-wise pressure, water vapor pressure and temperature. φ (rad) and h_s (km) denotes the station latitude and height. Askne and Nordius (1987) further introduced the temperature lapse rate and water vapor lapse rate into the ZWD calculation and constructed a new ZWD calculation formula:

$$ZWD = \frac{10^{-6} \left(k_2' + k_3 / T_m \right) \cdot R_d}{(\lambda + 1) \cdot g_m} \cdot e_s \tag{2.3}$$

where k_2' and k_3 denote the atmospheric refraction index coefficients, which are 16.5221 K/hPa and 377600 K²/hPa, respectively. T_m (K) represents the water vapor weighted mean temperature. R_d stands for the dry gas constant and its value is 287.058 J/kg/K. g_m is the average gravity acceleration and takes value of 9.80665 m/s². λ denotes the water vapor lapse rate.

Landskron and Böhm (2018a) developed the highly comprehensive GPT3 model, which not only retains all the parameters from the GMF, GPT2 and GPT2w models, but also adds the horizontal gradient parameter. The model parameters contain mapping function coefficients (a_h and a_w), horizontal gradients (G_{nh}, G_{eh}, G_{nw} and G_{ew}), pressure (P), temperature (T), temperature lapse rate (α), water vapor weighted mean temperature (Tm), water vapor pressure (e), water vapor lapse rate (λ) and geoid undulation (undu). The GPT3 model takes into account the annual and semi-annual variations of the parameter, and the spatial resolution of GPT3 model is 1°×1°. In its usage, only the modified Julian date (MJD) as well as the station longitude, latitude and geodetic height are input for retrieving the target parameters, which can be widely used in space geodesy and meteorological research.

2.2.2.3 Mapping Function Model

The mapping function is the key to realize the conversion between zenith path delay and slant path delay, and is commonly expressed as the following continued fraction:

$$MF(e) = \frac{1 + \cfrac{a}{1 + \cfrac{b}{1 + c}}}{\sin(e) + \cfrac{a}{\sin(e) + \cfrac{b}{\sin(e) + c}}} \tag{2.4}$$

where a, b and c are the mapping function coefficients. The mapping function models mainly contain two categories, namely empirical model and discrete product. The well-known empirical mapping function models include NMF, GMF and GPT2 (Niell, 1996; Böhm et al., 2006a; Lagler et al., 2013). The commonly used discrete mapping function is VMF1, with spatial-temporal resolutions of 2.5°×2.0° and 6 h (Böhm et al., 2006b). The VMF1 products includes the post (ERA-Interim, EI), rapid (Operational, OP) and real-time (Forecast, FC) products, where the time delay of the rapid product is about 1 day and the real-time product can be real-time access (https://vmf.geo.tuwien.ac.at/).

On the basis of VMF1 and GPT2, Landskron and Böhm (2018a) established the latest-generation discrete and empirical mapping function of VMF3 and GPT3 by using the ERA-Interim reanalysis data from ECMWF. The VMF3 spatial-temporal resolutions are 1°×1° and 6 h. The GPT3 model consider the annual and semi-annual variations of a, b and c mapping function coefficients. Similar to VMF1, the VMF3 products also contain three modes of EI, OP and FC. The difference is that the VMF3 adopts a more rigorous modeling method and has higher spatial resolution, and therefore is more accurate than VMF1.

The mapping function carries out the conversion between zenith path delay and slant path delay, and the mapping function accuracy will impact the vertical (U) coordinate and ZTD estimations. Vey et al. (2006) analyzed the impacts of NMF and IMF mapping functions on GPS network solution, and found that the maximum ZTD difference reached 5 mm. Dousa et al. (2017) indicated that the U coordinate repeatability by using the VMF1 mapping function improved by about 10% from using the GMF model. Yuan et al. (2019) pointed out that, in real-time PPP water vapor retrieval, the ZTD RMS can reduce by about 1 mm by using the VMF1-FC mapping function to substitute the GPT2 and GPT2w models.

Zhou et al. (2021) evaluated the accuracy and GNSS PPP performance of GPT3, grid-wise VMF3-FC (VMF3-FC-G) and site-wise VMF3-FC (VMF3-FC-S) at 33 globally distributed IGS Multi-GNSS Experiment (MGEX) stations spanning 40 days in 2020, and the STD modeling accuracy and the coordinate repeatability difference distribution for the three mapping functions are shown in Table 2.1, Table 2.2 and Figure 2.3.

From Table 2.1, we can find that the STD modeling accuracy decreases with the elevation angle. The accuracy of VMF3-FC-S is better than VMF3-FC-G at all the elevation angles, indicating the superiority of the site-wise product. At 3°, 30° and 90° elevation angles, the improvements reach 3.2, 0.3 and 0.2 cm, respectively. As for the empirical GPT3 model, the modeling accuracy significantly decreases at all the elevation angles. At 3°, 30° and 90° elevation angles, the modeling accuracy

TABLE 2.1

Statistical STD Modeling Accuracy (cm) for the Three Mapping Functions

Model	3°	5°	7°	10°	15°	30°	70°	90°
GPT3	53.7	36.3	27.1	19.6	13.3	7.0	3.7	3.5
VMF3-FC-G	20.4	13.8	10.2	7.4	5.0	2.6	1.4	1.3
VMF3-FC-S	17.2	11.7	8.7	6.3	4.3	2.3	1.2	1.1

TABLE 2.2

PPP Coordinate Repeatability for the Three Mapping Functions

Models	BDS3/mm			GPS/mm		
	E	N	U	E	N	U
GPT3	2.8	2.0	5.7	1.5	1.4	4.7
VMF3-FC-G	2.8	2.0	5.2	1.6	1.4	4.2
VMF3-FC-S	2.8	2.0	5.2	1.5	1.4	4.2

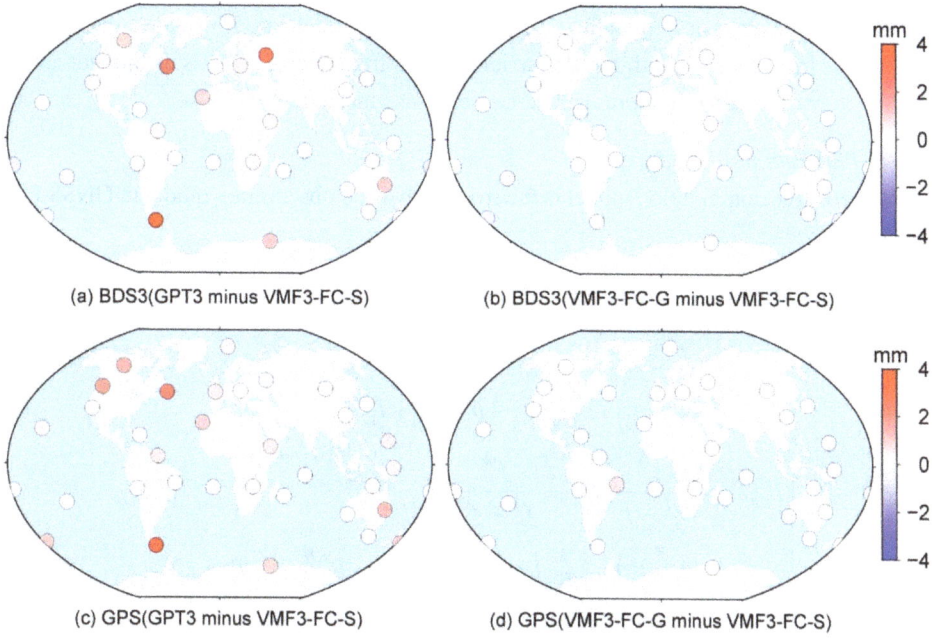

(a) BDS3(GPT3 minus VMF3-FC-S)   (b) BDS3(VMF3-FC-G minus VMF3-FC-S)

(c) GPS(GPT3 minus VMF3-FC-S)   (d) GPS(VMF3-FC-G minus VMF3-FC-S)

FIGURE 2.3 Distribution of the U coordinate repeatability differences.

decreases by 36.5, 4.7 and 2.4 cm compared with VMF3-FC-S, illustrating that the real-time mapping functions based on the forecast numerical weather model (NWM) are superior to those of the empirical model.

From Table 2.2, we can find that different mapping functions mainly impact the U coordinate repeatability. By adopting the VMF3-FC-G and VMF3-FC-S mapping functions to substitute GPT3, the statistical U coordinate repeatability improvements can reach 0.5 mm, while the U coordinate repeatability for the two VMF3-FC mapping functions is nearly identical. From Figure 2.3, we can further find that the U coordinate repeatability improvements are different at different stations. The stations located in North America (STJ3) and southern South America (RGDG) show very significant improvements, and the maximal improvement reaches about 4 mm. Based on the aforementioned modeling accuracy and PPP performance evaluations, we recommend using the two VMF3-FC mapping functions to substitute GPT3 in NRT or real-time GNSS data processing.

2.2.3 TROPOSPHERIC PARAMETER ESTIMATION

2.2.3.1 Function Model

Ignoring some errors, the GNSS pseudorange (P) and carrier phase (L) observation equations can be expressed as:

$$
\begin{cases}
P_{r,j}^s = \rho_r^s + c\left(t_r - t^s\right) + c\left(b_{r,j} - b_j^s\right) + I_{r,j}^s + T_r^s + e_{r,j}^s \\
L_{r,j}^s = \rho_r^s + c\left(t_r - t^s\right) + \lambda_j\left(B_{r,j} - B_j^s\right) + \lambda_j N_{r,j}^s - I_{r,j}^s + T_r^s + \varepsilon_{r,j}^s
\end{cases}
\tag{2.5}
$$

where s, r and j denote satellite, receiver and frequency, and ρ_r^s is the geometric distance from satellite to receiver. c is the speed of light in vacuum. t^s and t_r represent the satellite and receiver clock correction, respectively. $b_{r,j}$ and b_j^s stand for the pseudorange biases while $B_{r,j}$ and B_j^s are the uncalibrated phase delays at frequency j. $I_{r,j}^s$ is the ionospheric delay. T_r^s denotes the tropospheric STD. λ_j stands for the wavelength for the carrier phase at frequency j. $N_{r,j}^s$ is the integer ambiguity. $e_{r,j}^s$ and $\varepsilon_{r,j}^s$ represents the pseudorange and carrier phase measurement noise.

2.2.3.1.1 PPP Function Model

The ionosphere-free combination model is the frequently used observation model in GNSS PPP, and can be expressed as:

$$
\begin{cases}
P_{r,IF_{i,j}}^s = \dfrac{f_i^2}{f_i^2 - f_j^2} P_{r,i}^s - \dfrac{f_j^2}{f_i^2 - f_j^2} P_{r,j}^s \\[2mm]
\quad = \rho_r^s + c\left(t_r - t^s\right) + c\left(b_{r,IF_{i,j}} - b^{s,IF_{i,j}}\right) + T_r^s + e_{r,IF_{i,j}}^s \\[2mm]
L_{r,IF_{i,j}}^s = \dfrac{f_i^2}{f_i^2 - f_j^2} L_{r,i}^s - \dfrac{f_j^2}{f_i^2 - f_j^2} L_{r,j}^s \\[2mm]
\quad = \rho_r^s + c\left(t_r - t^s\right) + \lambda_{IF_{i,j}}\left(N_{r,IF_{i,j}}^s + B_{r,IF_{i,j}} - B^{s,IF_{i,j}}\right) + T_r^s + \varepsilon_{r,IF_{i,j}}^s
\end{cases}
\tag{2.6}
$$

where $P_{r,IF_{i,j}}^s$ and $L_{r,IF_{i,j}}^s$ denote the pseudorange and carrier phase ionosphere-free combination observations for i and j frequencies from satellite s to receiver r. f_i and f_j stand for the frequency values. $b_{r,IF_{i,j}}$, $b^{s,IF_{i,j}}$ and $e_{r,IF_{i,j}}^s$ are the receiver pseudorange bias, satellite pseudorange bias and pseudorange measurement noise after the ionosphere-free combination. $\lambda_{IF_{i,j}}$ and $N_{r,IF_{i,j}}^s$ are the wavelength and ambiguity for the ionosphere-free combination observation. $B_{r,IF_{i,j}}$, $B^{s,IF_{i,j}}$ and $\varepsilon_{r,IF_{i,j}}^s$ denote the receiver uncalibrated phase delays, satellite uncalibrated phase delays and phase measurement noise after the ionosphere-free combination.

In equation 2.6, the tropospheric STD T_r^s can be further written as:

$$
\begin{aligned}
T_r^s(e,a) = {} & MF_h(e) \cdot ZHD + MF_w(e) \cdot ZWD \\
& + MF_g(e) \cdot \left[G_n \cos(\alpha) + G_e \sin(\alpha)\right]
\end{aligned}
\tag{2.7}
$$

where the smoothly variable ZHD can be accurately corrected by using the *a priori* zenith path delay models. The rapidly changed ZWD cannot be greatly eliminated by the *a priori* models, and needs the additional estimation of the ΔZTD parameter for absorbing the un-modeled ZWD error. The horizontal gradient can also be corrected by the *a priori* horizontal gradient models such as GPT3 and GRAD (Landskron and Böhm, 2018b). However, considering the millimeter magnitude (generally smaller than 5 mm), the G_n and G_e horizontal gradient parameters are usually estimated in GNSS data processing.

In GNSS data processing, the satellite orbit and clock errors as well as the pseudorange bias at both the satellite and receiver end can be corrected by using the external products, and the rest of the parameters include the float ambiguity ($\bar{N}$) (including uncalibrated phase delay), inter-system bias (ISB) (if using multi-GNSS), station location (X_s, Y_s and Z_s), receiver clock correction (t_r) and tropospheric delay (ΔZTD, G_n and G_e). The to-be-estimated vector $\mathbf{X}$ can expressed as (Li et al., 2015c):

$$
X = \left(\bar{N}\ X_s\ Y_s\ Z_s\ t_r\ \Delta ZTD\ G_n\ G_e\ ISB\right)^T
\tag{2.8}
$$

After that, the uncalibrated phase delays can be corrected by external product or eliminated by double-differenced combination to restore the integer feature of the ambiguity parameter and can be fixed by using methods such as the least-squares ambiguity decorrelation adjustment (LAMBDA) method.

2.2.3.1.2 Double-Differenced Function Model

The double-differenced observation model is the fundamental function model in GNSS data processing. Ignoring some error sources, the pseudorange and carrier phase double-differenced observation equations for receivers k and ℓ and satellites i and j at first and second frequencies can be written as (Dach et al., 2015):

$$
\begin{cases}
P_{1k\ell}^{ij} = \rho_{k\ell}^{ij} + I_{k\ell}^{ij} + T_{k\ell}^{ij} \\[2mm]
P_{2k\ell}^{ij} = \rho_{k\ell}^{ij} + \dfrac{f_1^2}{f_2^2} I_{k\ell}^{ij} + T_{k\ell}^{ij} \\[2mm]
L_{1k\ell}^{ij} = \rho_{k\ell}^{ij} - I_{k\ell}^{ij} + T_{k\ell}^{ij} + \lambda_1 N_{1k\ell}^{ij} \\[2mm]
L_{2k\ell}^{ij} = \rho_{k\ell}^{ij} - \dfrac{f_1^2}{f_2^2} I_{k\ell}^{ij} + T_{k\ell}^{ij} + \lambda_2 N_{2k\ell}^{ij}
\end{cases}
\tag{2.9}
$$

where $P_{1k\ell}^{ij}$, $P_{2k\ell}^{ij}$, $L_{1k\ell}^{ij}$ and $L_{2k\ell}^{ij}$ are the pseudorange and carrier phase double-differenced observations at first and second frequencies, respectively. $\rho_{k\ell}^{ij}$, $I_{k\ell}^{ij}$ and $T_{k\ell}^{ij}$ denote the geometric distance, ionospheric delay and tropospheric delay after double-differenced combination. λ_1 and $N_{1k\ell}^{ij}$ are wavelength and ambiguity for the first frequency double-differenced combination observation, and λ_2 and are wavelength and ambiguity for the second frequency. The double-differenced combination observation can eliminate the impacts from satellite and receiver clock correction and weaken the atmospheric delay errors. In addition, the double-differenced ambiguity is with the integer characteristic and can be directly fixed by the LAMBDA method.

2.2.3.2 Random Model

The piece-wise constant assumes that the to-be-estimated parameter is constant in a Δt time interval, and considers the parameter time variations by adding a parameter in every Δt time interval. As for the tropospheric delay parameter estimation, the piece-wise constant can be expressed as:

$$
\begin{cases}
T(e, \alpha, t) = MF_h(e) \cdot ZHD + MF_w(e) \cdot (ZWD + \cdot ZTD_i) \\[1mm]
\qquad + MF_g(e) \cdot \left[G_{n,j} cos(\alpha) + G_{e,j} sin(\alpha) \right] \\[1mm]
t_{i-1} \leq t \leq t_i, t_i = t_{i-1} + \Delta t_{ZTD} \\[1mm]
t_{j-1} \leq t \leq t_j, t_j = t_{j-1} + \Delta t_{HTG}
\end{cases}
\tag{2.10}
$$

where t is the observation epoch. ΔZTD_i denotes the ith ΔZTD parameter. $G_{n,j}$ and $G_{e,j}$ are the jth G_n and G_e parameters. t_{i-1} and t_i are the lower and upper bound epochs for the ΔZTD_i parameter. Δt_{ZTD} stands for the segmentation time for ΔZTD estimation, and it generally takes 1 h or 2 h for post-processing and 15 min or 30 min for NRT processing. t_{j-1} and t_j are the lower and upper bound epochs for the $G_{n,j}$ and $G_{e,j}$ parameters. Δt_{HTG} represents the segmentation time for horizontal gradient estimation, which is generally taken as 6 h or 12 h for post-processing and 1 h or 2 h for NRT processing.

In GNSS data processing, the first-order discrete Gaussian Markov process can be used to describe the variation characteristics of tropospheric delay and satellite clock correction parameters. When the correlation time tends to infinity, the first-order discrete Gaussian Markov process degrades into the random walk process. The random walk process for the tropospheric delay estimation can be written as (Hadas et al., 2017):

$$E\left(\left|P_{t+\Delta t} - P_t\right|\right) = \varepsilon\sqrt{\Delta t} \tag{2.11}$$

where $P_{t+\Delta t}$ denotes the to-be-estimated tropospheric delay parameters (including ΔZTD, G_n and G_e) at $t + \Delta t$ epoch. P_t represents the parameters at t epoch. Δt is the time interval. ε stands for the random walk noise. In GNSS post-processing, the tropospheric parameters are generally estimated by using piece-wise constant and random walk between segments, and the random walk noises for the ΔZTD and G_n and G_e parameters are commonly set to be constants, such as $15\,mm/\sqrt{h}$ and $10\,mm/\sqrt{h}$, respectively. In GNSS dynamic or real-time data processing, the random walk mode is frequently used where the noise is determined according to the carrier (such as cellphone, train and unmanned aerial vehicle) moving speed and the tropospheric atmosphere state.

2.2.3.3 Parameter Estimation Method

The recursive least squares method is widely used in GNSS data processing. This method divides all to-be-estimated parameters to two types, namely time variant parameters (X) (e.g., receiver clock correction and tropospheric delay) and time invariant parameter (Y) (e.g., ambiguity), and the corresponding error equation can be expressed as:

$$V = AX + BY - L, P \tag{2.12}$$

where A is the coefficient matrix for time variant parameters (X), and B is the coefficient matrix for time invariant parameter (Y). L is the observation vector, and P denotes the weight matrix. Using the parameter elimination method to eliminate X, we can obtain the normal equation with respect to Y as:

$$\begin{bmatrix} A^T PA & A^T PB \\ B^T PA & B^T PB \end{bmatrix} = \begin{bmatrix} N_{11} & N_{12} \\ N_{21} & N_{22} \end{bmatrix} = \begin{bmatrix} A^T PL \\ B^T PL \end{bmatrix} \tag{2.13}$$

where N is the normal equation coefficient matrix. Setting $Z = N_{21}N_{11}^{-1}$, this equation can be transformed:

$$\begin{bmatrix} I & 0 \\ -Z & I \end{bmatrix}\begin{bmatrix} N_{11} & N_{12} \\ N_{21} & N_{22} \end{bmatrix} = \begin{bmatrix} N_{11} & N_{12} \\ 0 & \tilde{N}_{22} \end{bmatrix} = \begin{bmatrix} A^T PL \\ \tilde{B}^T PL \end{bmatrix} \tag{2.14}$$

where I is the unit matrix, and $\tilde{N}_{22} = B^T PB - B^T PAN_{11}^{-1}A^T PB$. Setting $J = AN_{11}^{-1}A^T P$, $\tilde{N}_{22}$ can be written as $\tilde{N}_{22} = B^T \left(I - J\right)^T P\left(I - J\right)B$. Setting $\tilde{B} = \left(I - J\right)B$, the new normal equation respect to Y can be expressed as:

$$\tilde{B}^T P\tilde{B}Y = \tilde{B}^T PL \tag{2.15}$$

After that, Y can be estimated by using equation (2.15), and X can be calculated as:

$$\tilde{X} = N_{11}^{-1}\left(A^T PL - N_{12}Y\right) \tag{2.16}$$

where $\tilde{X}$ is the estimated matrix for X. Through the classified recursive processing, the normal equation size can be reduced, and the large matrix inversion operation can be avoided.

2.2.4 Near-Real-Time Processing Results

2.2.4.1 NRT Processing Method

In GNSS data processing, the tropospheric delay, satellite and receiver clock corrections and station height component are strongly correlated as shown as Figure 2.4 (Nilsson et al., 2013), and the high-precision satellite clock correction product is therefore very important to improve the GNSS station height and tropospheric delay estimation. To date, the real-time satellite clock correction product is still hard to support the NRT tropospheric product estimation with high precision and reliability, and we generally need to additionally estimate the NRT satellite clock correction in the NRT data processing.

There are two data processing methods for NRT tropospheric parameter estimation. The first one is a one-step network solution method. This method simultaneously estimates the satellite clock correction and tropospheric parameters by double-differenced network solution. The second one is the two-step method. This method firstly estimates the satellite clock correction by double-differenced network solution, and then estimates the tropospheric delay by using the estimated NRT satellite clock correction product and PPP solution.

However, the data processing efficiency of the one step method will significantly decrease with the increase of the station number. In addition, the one step method can only estimate the relative ZTD and horizontal gradient product if the network size is smaller than 500 km (Duan et al., 1996). Therefore, we choose the two-step method for establishing the WUHM analysis centers.

2.2.4.2 NRT Processing Flow

We established the EGVAP WUHM analysis center based on the high precision GNSS data processing software PANDA. The WUHM center processes the tropospheric ZTD and horizontal gradient product at more than 300 stations in NRT mode. The flow chart for WUHM center is shown in Figure 2.5. The data and product preparations are necessary before NRT processing. The hourly

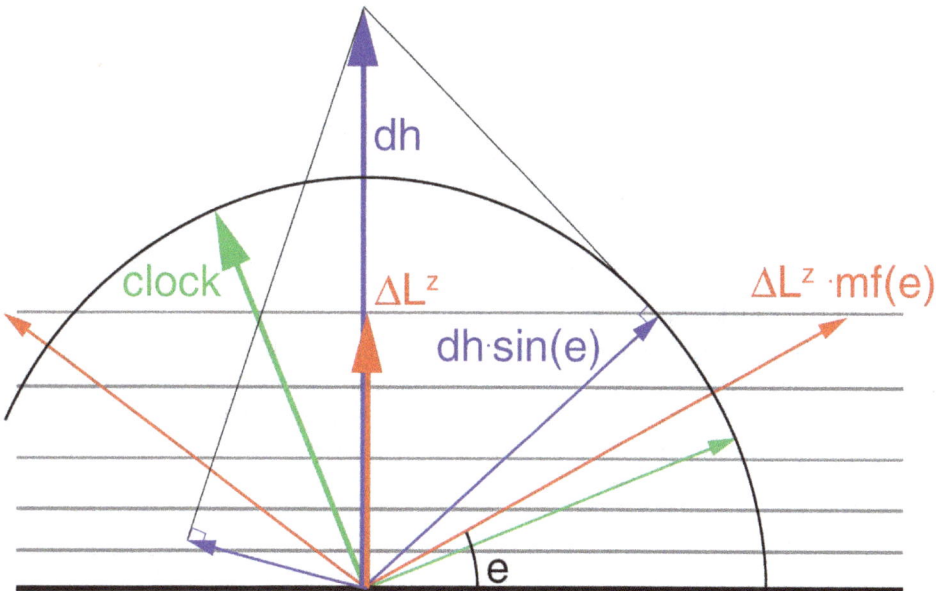

FIGURE 2.4 Different elevation dependence for tropospheric delay (red), clock (green) and height component (blue).

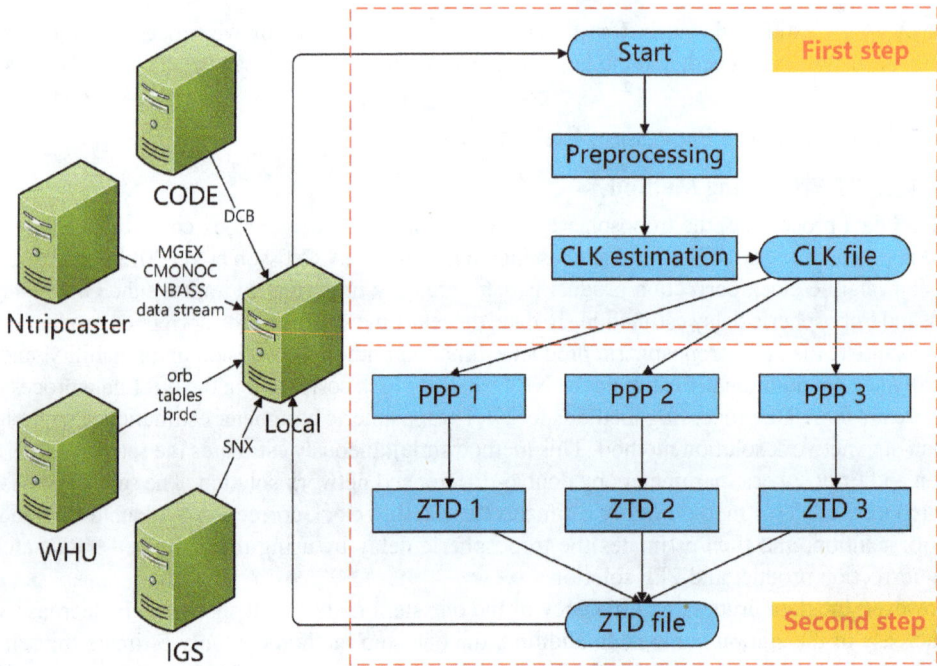

FIGURE 2.5   Flow chart for WUHM analysis center.

multi-GNSS RINEX data in the local server are derived from the data streams of the MGEX, CMONOC and NBASS networks. The table files, ultra-rapid satellite orbit products, broadcast ephemeris files and SINEX files are downloaded from IGS analysis centers.

The first step is the NRT satellite clock correction processing by the double-differenced network mode. The 6 h observation data from the globally distributed MGEX stations and nationally distributed NBASS stations and real-time satellite orbit product are used for the satellite clock correction estimation in 6 h sliding window length and hourly sliding mode, deriving the NRT satellite clock correction files.

The second step is the NRT tropospheric product processing by subnetwork mode. The 6 h observation data from MGEX, CMONOC and NBASS networks are divided into serval subnetworks, and then the PPP processing for every subnetwork is carried out by using the estimated satellite clock correction files in parallel processing mode, deriving the tropospheric parameter for every subnetwork. Finally, the tropospheric parameter files for these subnetworks are merged into one file. By adopting the two-step method, the processing time for more than 300 stations is less than 50 min.

2.2.4.3   Satellite Clock Correction

We took the post-processing satellite clock correction product from IGS Wuhan University analysis center as a reference and evaluated the accuracy of the estimated NRT satellite clock correction accuracy, and the RMS and STD results for GPS week 2195 are shown in Figure 2.6. We can find that the STDs for the four GNSS systems are all better than 0.1 ns, indicating the good performance of the estimated satellite clock correction.

2.2.4.4   Zenith Total Delay

We also evaluated the estimated NRT ZTD accuracy by taking the post-processing ZTD product as a reference, and the RMS distribution and histogram for GPS week 2195 are shown in Figure 2.7. We can find that the RMSs for all stations are generally smaller than 20 mm, and the statistical RMS for NBASS, CMONOC and MGEX are better than 10 mm.

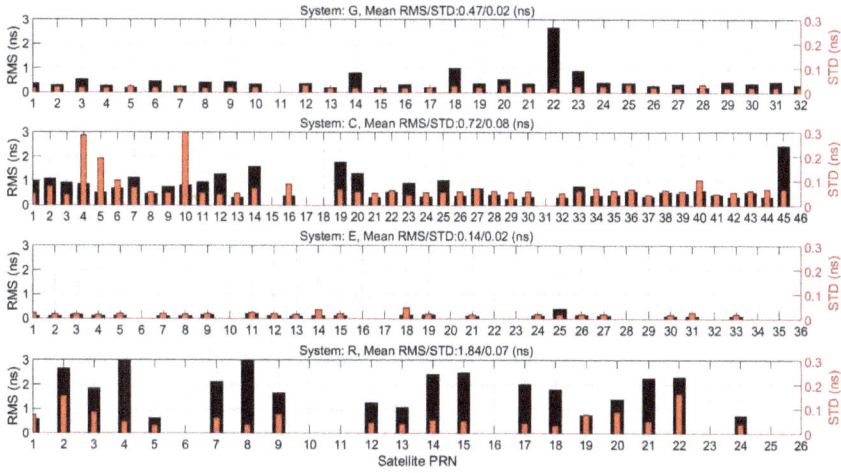

FIGURE 2.6   NRT satellite clock correction RMS and STD (GPS week: 2195).

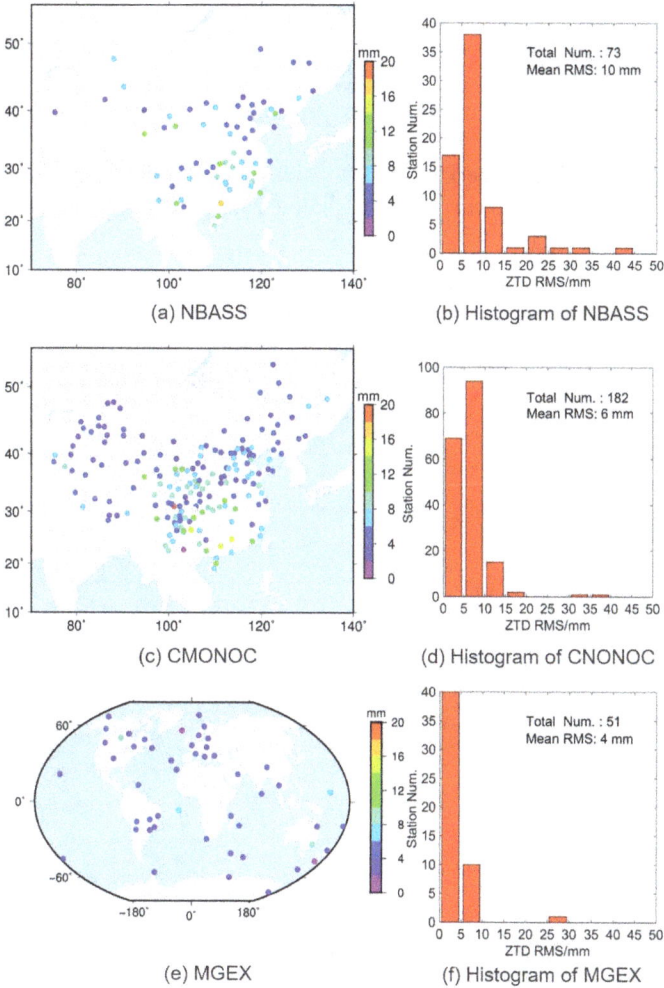

FIGURE 2.7   NRT ZTD RMS distribution and histogram (GPS week: 2195).

2.3 METEOROLOGICAL APPLICATIONS

2.3.1 Data Assimilation and Weather Forecasting

2.3.1.1 Status

In the area of weather forecasting, even the most advanced numerical weather prediction systems are currently unable to accurately predict the location, time and intensity of extreme precipitation, mainly due to the inadequate ability to describe the spatial and temporal variability of atmospheric humidity. Ducrocq et al. (2002) found the importance of the initial humidity field for improving forecasts of heavy precipitation events. Vedel and Huang (2004) and Poli et al. (2007) assimilated GPS-ZTD products to numerical forecast models using three-dimensional variational assimilation (3D-Var) and four-dimensional variational assimilation (4D-Var), respectively, and found improvements in precipitation forecasting. Lindskog et al. (2017) used 2.5 km horizontal resolution based on 3D-Var to assimilate GPS-ZTD data into the HARMONIE-AROME model and found that it was possible to improve effective forecasting, particularly for humidity, by up to one and a half days. Mahfouf et al. (2015) showed that although GNSS-ZTD accounts for a relatively small proportion of all assimilated observations, ZTD can systematically improve the accuracy of moisture initial fields and thus precipitation forecasts in short-term forecasts compared to other water vapor observations.

According to an assessment by the Met Office in the UK, ground-based GNSS ranked second in average contribution to numerical weather prediction (Figure 2.8) among all observational tools (Jones et al., 2020). With the development of GNSS systems and the construction of ground-based station networks, refined atmospheric water vapor products will be more beneficial to improve the quality of numerical weather forecasts (Singh et al., 2019). However, there are still problems with the assimilation of ground-based BeiDou/GNSS water vapor products for forecasting that need further research. For example, the current assimilation strategies and methods ignore the spatial

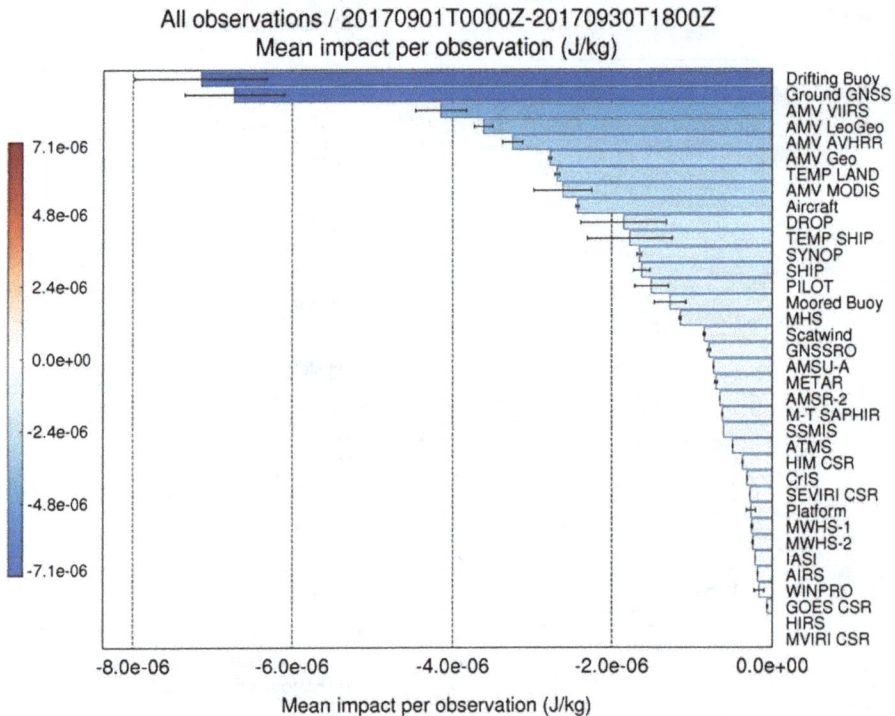

FIGURE 2.8 Contributions of different meteorology observations on weather forecasting.

correlation of GNSS water vapor products, and the forecast bias becomes larger after increasing the GNSS station density, which cannot truly exploit the advantages of high spatial and temporal resolution of ground-based GNSS. The current numerical assimilation forecasting studies usually carry out spatial and temporal sparse degradation correlation processing, and do not pay enough attention to the accuracy of BeiDou/GNSS water vapor products, which cannot fully exploit the advantages of high accuracy and high resolution of BeiDou/GNSS water vapor, and further optimization of the existing assimilation operators and methods is needed.

The most commonly used platform for numerical assimilation forecasting is the weather research and forecasting model (WRF), a next-generation mesoscale numerical weather prediction system designed for atmospheric research and operational forecasting applications, which began to be developed in the late 1990s by a collaborative effort between the National Center for Atmospheric Research (NCAR), the National Oceanic and Atmospheric Administration (NOAA) and the National Centers for Environmental Prediction (NCEP). The latest version of the WRF is version 4.4 (September 2022). With its support for parallel computing and system scalability, the WRF includes powerful features such as model forecasting and data assimilation, allowing researchers to use the WRF to carry out simulations based on real atmospheric conditions or simulated conditions. The WRF currently has a cumulative total of over 57,800 registered users in over 160 countries as of 2021 and is used extensively in real-time forecasting at NCEP and other national meteorological centers as well as laboratories, universities and companies. The Advanced Research WRF (ARW) modeling system has been in development for the past 20 years and is applicable to scientific research for model forecasting. It is in the public domain and is freely available for community use. The overall framework and processing flow of WRF-ARW is shown in Figure 2.9 (Skamarock et al., 2019), which mainly consists of four parts: WRF preprocessing system (WPS, Real), forecast model (WRF-ARW), WRF data assimilation system (WRFDA) and post-processing and visualization tools.

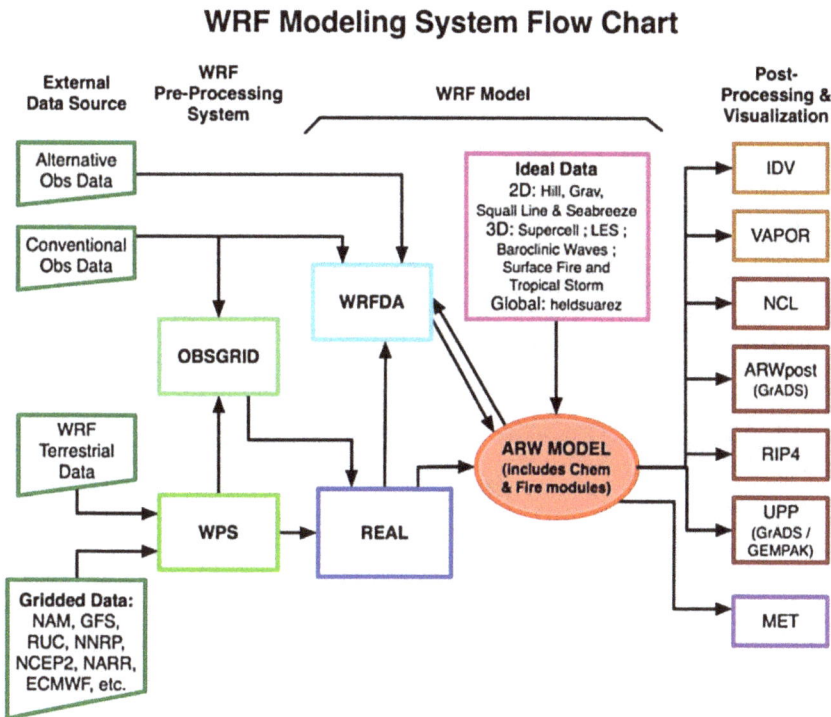

FIGURE 2.9   Flowchart for the WRF modeling system version 4.

2.3.1.2 Methods

Data assimilation is a technique that optimally combines observations with the initial background field using some criteria to obtain an improved and updated background field. The data assimilation system (WRFDA) module of the WRF model supports both 3D-Var and 4D-Var, where the difference is that the 4D-Var includes a numerical forecast model and considers the time difference between background and observations. Because the 4D-Var method is very resource demanding and complicated, the most widely used method in meteorological operations is still 3D-Var, which will be introduced in the following text. The core idea of 3D-Var is to assimilate the different observations into a representation of the analysis field (analysis, i.e., the updated background field) based on the criterion of minimizing the objective function of the deviation of the analysis field from the initial background field and the observations, which can be defined as:

$$\xi(x_a) = \frac{1}{2}(x_a - x_b)^T B^{-1}(x_a - x_b) + \frac{1}{2}(y_0 - H(x_a))^T R^{-1}(y_0 - H(x_a)) \tag{2.17}$$

where x_a is the analysis to be resolved, x_b is the initial background, y_0 represents the observations assimilated, H is the observing matrix, and B and R are initial background field error and observations error, respectively. The workflow of the WRFDA for GNSS ZTDs is given in Figure 2.10.

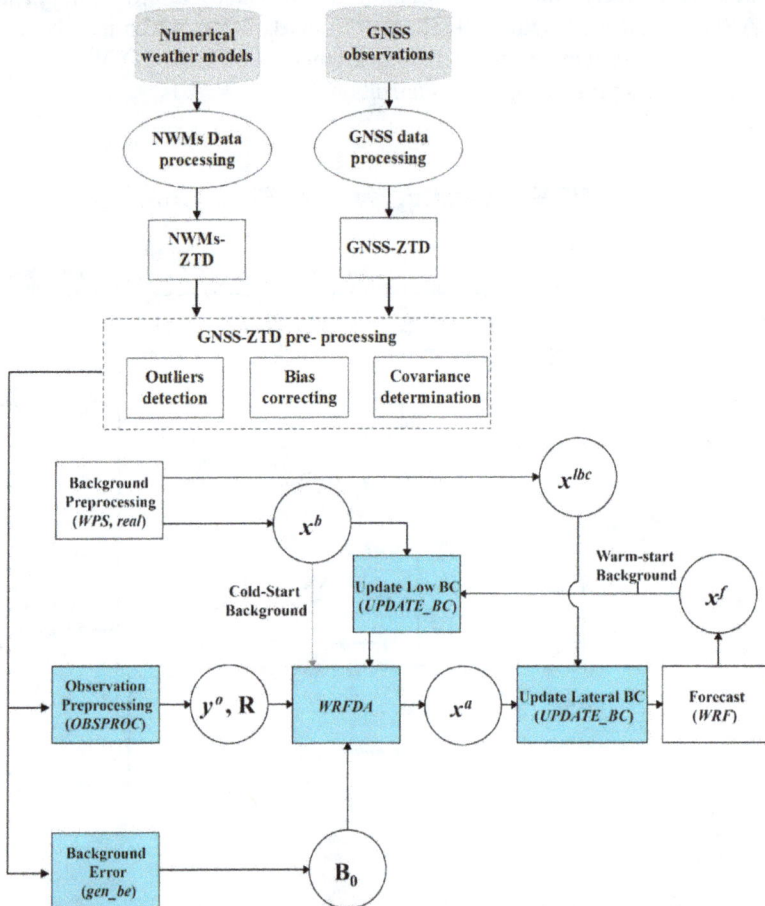

FIGURE 2.10   Workflow of WRFDA based on GNSS ZTDs and WRF forecasting.

GNSS ZTD can be assimilated only after passing quality control, whereby outliers and bias are detected and removed and covariance is determined to contain observation error information. It is common to use NWMs ZTDs to help quality control. Thereafter, quality assured GNSS ZTDs will be saved in LITTLE_R format, which is the system's intermediate observation data in the WRFDA preprocessing, and then extracted into a usable data assimilation format by the OBSPROC (observation preprocessor) module.

2.3.1.3 Results

GNSS ZTD assimilation has been studied in the past decades, with abundant assessment carried out and finer techniques developed. Boniface et al. (2009) studied heavy rain in the Mediterranean region and indicated that the assimilation of GPS ZTD can further improve the precipitation forecast when the atmosphere state can be described well with traditional observations assimilated (Figure 2.11). Rohm et al. (2019) found that GNSS data assimilation can significantly change moisture fields and rain, whereas GNSS ZTD always reduces the humidity field bias and improves the first 24-hour rain forecast. Singh et al. (2019) indicated that ZTD assimilation can benefit the lower to middle tropospheric moisture, upper air temperature, and middle and upper tropospheric wind, where errors reduced up to 4% compared to the model run without ZTD assimilation. Macpherson et al. (2008) found a mixed impact of ZTD assimilation on the precipitation forecast, with positive impacts for heavy rainfall events but damage for small threshold rainfall accumulation. GNSS ZTD has the advantage of high temporal and spatial resolution. However, some studies found that this may not help in WRF assimilation. Nykiel et al. (2016) found that assimilating GNSS ZTD too frequently can lead to incorrect results when the method 3D-Var is used and without upper atmosphere data assimilated together. Assimilating GNSS ZTD together with the upper air and surface data every 6 hours outperforms assimilating every hour or only assimilating ZTD. In general, most studies indicated positive impacts of GNSS ZTD assimilation but generally in certain cases and at certain areas and heights. Stable contribution of GNSS ZTD to assimilation needs further development of assimilation technique and improvement of GNSS ZTD accuracy.

2.3.2 PPP Augmentation

2.3.2.1 Status

Precise point positioning (PPP) has been widely used during the past decades with centimeter or decimeter level accuracy (Zumberge et al., 1997). However, PPP still suffers from some limitations in its real-time applications. Due to the high coupling between tropospheric delay

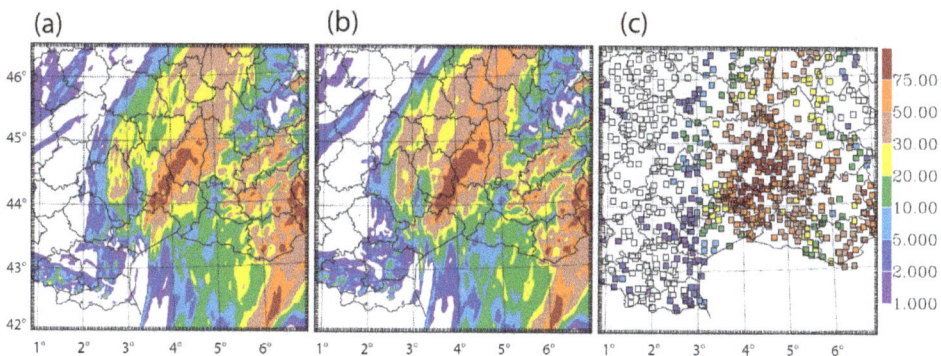

FIGURE 2.11 24-hour accumulated precipitation on November 22, Mediterranean region for (a) only traditional observations assimilated; (b) GPS ZTD also assimilated; (c) rain-gauge observations.

and the vertical position, PPP processing usually requires a long time to separate these two parameters for convergence, which significantly limits its real-time applications (De Oliveira et al., 2017). Therefore, the introduction of an accurate *a priori* ZTD helps to constrain the ZTD in PPP processing and thereby accelerates the convergence efficiently. The currently used *a priori* ZTD values are derived from empirical models, such as UNB3m model (Leandro et al., 2006) and GPT series models (Böhm et al., 2015), which can provide approximate tropospheric delay without external data as inputs. However, empirical models usually have difficulty in capturing the high variability in ZTD, limiting their accuracy in real-time applications. It is thus necessary to adopt external data sources to obtain precise *a priori* ZTD. With the development of the numerical weather model (NWM) and PPP-RTK technique, the precise ZTD relying on meteorological reanalysis products and measured atmospheric information have been used to interpolate ZTD for users, providing more accurate real-time ZTD than *a priori* empirical models. Dousa et al. (2018b) summarized that the standard empirical models have an accuracy limit of about 3.5 cm, NWM-derived ZTD is capable of achieving an accuracy of approximately 0.8–1.2 cm, and GNSS-derived ZTD estimated using real-time PPP can be accurate to within 0.5–0.9 cm.

In recent years, there have been some studies focusing on wide-area real-time tropospheric product establishment. For example, Zhang et al. (2018) utilized the CMONOC network and the IGGtrop empirical model (Li et al., 2015a) to generate a grid-based tropospheric product (GTP) based on the undifferenced and uncombined PPP (UU-PPP) technique. Lou et al. (2018) developed an inverse scale height model based on ERA-Interim reanalysis data and applied it to real-time ZTD solutions at GPS stations to generate real-time ZTD grid product (RtZTD) over China for PPP users. Zheng et al. (2018) established a real-time tropospheric grid point (RTGP) model in China with improved modeling the zenith wet delay. They adopted the decrease factor of water vapor pressure and the temperature lapse rate provided by the GPT2w empirical model, and introduced a modified height parameter for the uniformity of ZWD. It is noteworthy that in operational applications, precise real-time tropospheric products can be difficult to achieve. Near-real-time tropospheric products are sufficient for numerical assimilation forecasting applications, since we can generally assume that ZTD variations are small over a short period of time, which enables our NRT products to be adopted in real-time positioning and navigation as well. In practical terms, NRT products can assist in the rapid convergence of positioning by supporting the generation of enhanced ZTD model products.

2.3.2.2   Methods

The generation and application of real-time ZTD grid product proposed by Lou et al. (2018) can be briefly introduced as follows, with the overall flowchart shown in Figure 2.12.

Firstly, the meteorological reanalysis product (ERA-Interim) is used to estimate ZTD at a given location. The estimation method divides ZTD into two sections, including the ZTD section from the location to the top level and the hydrostatic section above the top level, and calculates each section separately. Secondly, the ZTD profiles derived from ERA-Interim are used to analyze the vertical variations, results of which show near exponentially decreasing trends with altitude. Based on this analysis, the inverse scale height model can be determined.

$$\text{ZTD} = \text{ZTD}_0 \cdot \exp\left(-\frac{1}{H} \cdot h\right) \tag{2.18}$$

where ZTD_0 is the zenith tropospheric delay at mean sea level (MSL), $1/H$ denotes the inverse scale height, and h is the target height (m) above MSL.

Thirdly, the temporal variations of $1/H$ are analyzed based on the ERA-Interim ZTD time series, which mainly contain annual and semi-annual periodical signals. Thus, the function as expressed in equation (2.19) is used to fit $1/H$ at each grid point.

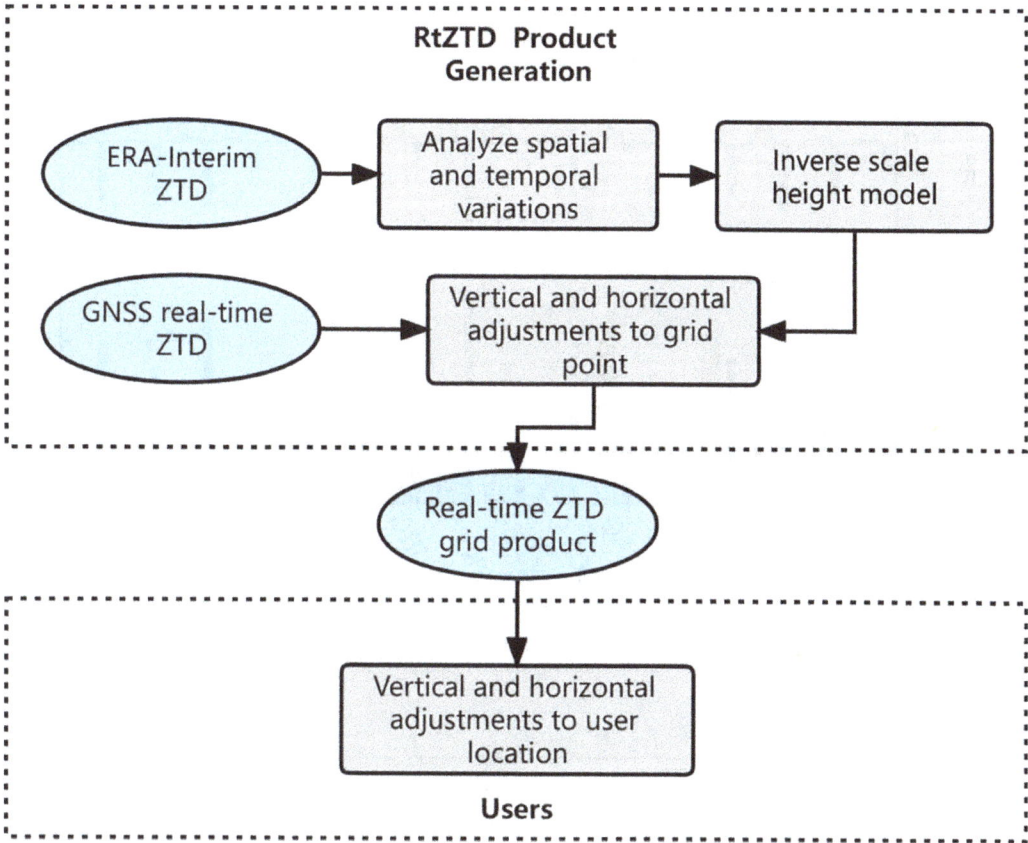

FIGURE 2.12 Real-time ZTD grid product generation and application flowchart.

$$1/H = a_0 + a_1 \cos(\frac{doy}{365.25} \cdot 2\pi) + a_2 \sin(\frac{doy}{365.25} \cdot 2\pi)$$
$$+ b_1 \cos(\frac{doy}{365.25} \cdot 4\pi) + b_2 \sin(\frac{doy}{365.25} \cdot 4\pi)$$

(2.19)

where doy is the day of the year, a_0 is the mean value, (a_1, a_2) and (b_1, b_2) are the annual and semi-annual amplitudes, respectively. Finally, the real-time ZTD grid product for each grid point is generated from the GPS-derived ZTD at stations within 1000 km. Based on this inverse scale height model, the GPS-derived ZTD is converted from the station height to the grid point height. Then the inverse distance weighted (IDW) method is applied to horizontally interpolate ZTD to the grid point. In practical application, the real-time ZTD grid product can be converted to the target height using the inverse scale height model, and then horizontally interpolated from the nearest four surrounding grid points to the target location.

2.3.2.3 Results

The real-time ZTD grid product (RtZTD) generated by Lou et al. (2018) can provide precise *a priori* ZTD for PPP users, which can help accelerate the PPP convergence. Taking post-processing ZTD as references, the real-time ZTD provided by RtZTD has a bias of 0.39 cm and an RMS of 1.56 cm, significantly better than the empirical models (Figure 2.13).

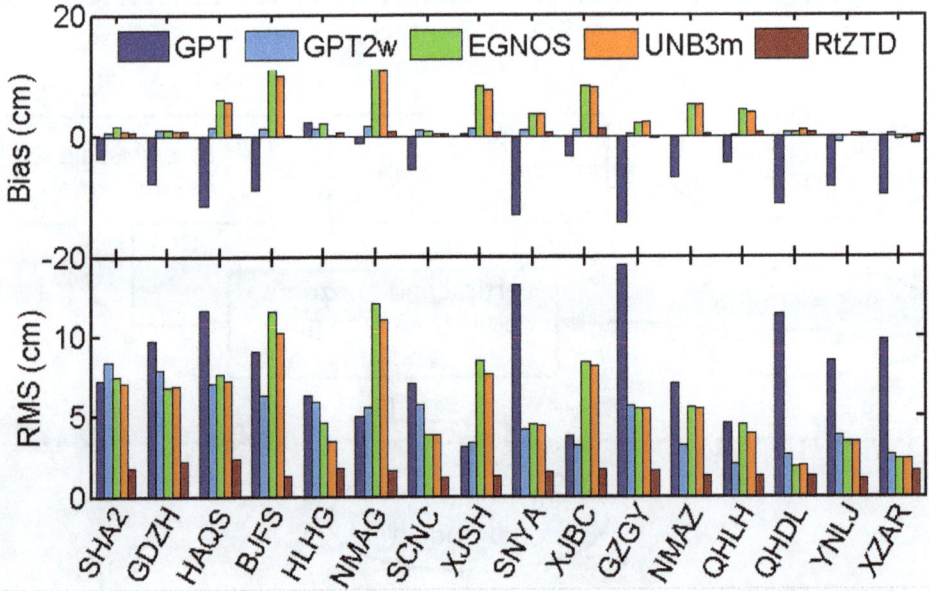

FIGURE 2.13 Accuracy comparison of the real-time ZTD grid product with empirical models.

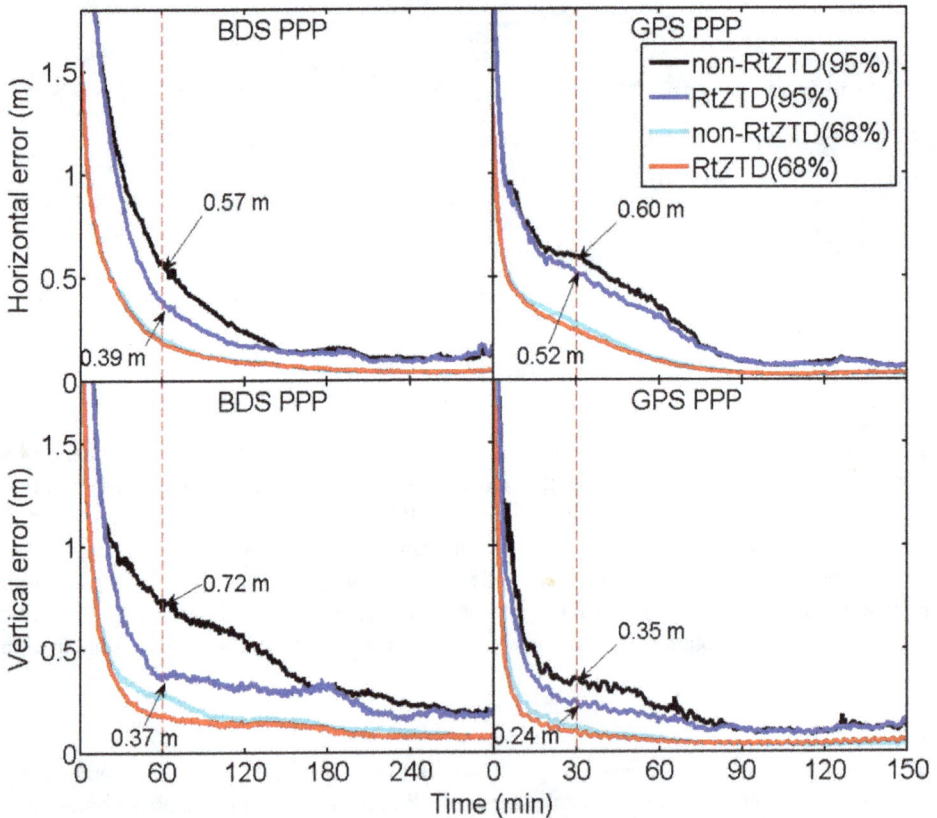

FIGURE 2.14 Convergence time for BDS/GPS PPP in horizontal component and vertical component.

TABLE 2.3
Convergence Time for Non-RtZTD PPP and RtZTD PPP (unit: min)

	Non-RtZTD PPP				RtZTD PPP			
	H (95%)	H (68%)	V (95%)	V (68%)	H (95%)	H (68%)	V (95%)	V (68%)
BDS	84.5	62	149.5	87.5	57.5	58	52	30.5
GPS	73.5	60.5	62.5	36	71.5	57.5	44.5	27

Besides, as Figure 2.14 and Table 2.3 indicate, the BDS PPP convergence time introducing RtZTD as *a priori* achieves approximately 32% (6%) and 65% (65%) improvements in the horizontal and vertical components in 95% (68%) situation, compared with the cases adopting *a priori* ZTD provided by Saastamoinen and GPT2w models. For GPS PPP, the convergence time also improves over 30% in the vertical direction.

2.4 CONCLUSIONS

In past decades, GNSS tropospheric parameter estimation has been through the rapid developments from post-processing to NRT processing as well as single system processing to multi-system processing. Compared with the single system post-processing, the multi-system NRT processing has the advantages of short time delay and acceptable accuracy, and therefore is very prospective in weather forecasting application and real-time PPP augmentation. In this chapter, we described the basic theory of multi-GNSS tropospheric parameter NRT estimation and summarized the current status of operational multi-GNSS NRT processing by taking the E-GVAP WUHM analysis center as an example. Then the applications of the NRT ZTD in data assimilation, weather forecasting and PPP augmentation were introduced. We validated that the operational multi-GNSS NRT tropospheric parameter processing is robust and effective for the weather forecasting and PPP augmentation applications.

REFERENCES

Askne J, Nordius H. 1987. Estimation of tropospheric delay for microwaves from surface weather data. Radio Science, 22(3), 379–386. DOI:10.1029/RS022i003p00379.

Bevis M, Businger S, Herring T A et al. 1992. GPS meteorology: remote sensing of atmospheric water vapor using the global positioning system. Journal of Geophysical Research: Atmospheres, 97(D14), 15787–15801. DOI:10.1029/92jd01517.

Black H D. 1978. An easily implemented algorithm for the tropospheric range correction. Journal of Geophysical Research: Solid Earth, 83(B4), 1825–1828. DOI:10.1029/JB083iB04p01825.

Böhm J, Möller G, Schindelegger M et al. 2015. Development of an improved empirical model for slant delays in the troposphere (GPT2w). GPS Solutions, 19(3), 433–441.

Böhm J, Niell A, Tregoning P et al. 2006a. Global Mapping Function (GMF): A new empirical mapping function based on numerical weather model data. Geophysical Research Letters, 33(7), 1–4. DOI:10.1029/2005GL025546.

Böhm J, Werl B, Schuh H. 2006b. Troposphere mapping functions for GPS and very long baseline interferometry from European Centre for Medium-range Weather Forecasts operational analysis data. Journal of Geophysical Research, 111(B2), B02406, 1–9. DOI:10.1029/2005jb003629.

Boniface K, Ducrocq V, Jaubert G et al. 2009. Impact of high-resolution data assimilation of GPS zenith delay on Mediterranean heavy rainfall forecasting. Annales Geophysicae, 27. DOI:10.5194/angeo-27-2739-2009.

Cao C. 2016. Introduction to BDS and GNSS systems. Beijing: Electronic Industry Press.

Chen J, Wang J, Wang A et al. 2020. SHAtropE-a regional gridded ZTD model for China and the surrounding areas. Remote Sensing, 2020(12), 165. DOI:10.3390/rs12010165.

Chung E S, Soden B, Sohn B J et al. 2014. Upper-tropospheric moistening in response to anthropogenic warming. Proceedings of the National Academy of Sciences, 111(32), 11636–11641. DOI:10.1073/pnas.1409659111.

Collins J P. 1999. Assessment and development of a tropospheric delay model for Aircraft users of the global positioning system. Department of Geodesy and Geomatics Engineering Technical Report No. 203, University of New Brunswick, Fredericton, New Brunswick, Canada, 1999, 174 pp.

Dach R, Lutz S, Walser P. 2015. Bernese GNSS software version 5.2. User manual, Astronomical Institute, University of Bern.

Davis J L, Herring T A, Shapiro I I et al. 1985. Geodesy by radio interferometry: Effects of atmospheric modeling errors on estimates of baseline length. Radio Science, 20(6), 1593–1607. DOI:10.1029/RS020i006p01593.

De Oliveira PS Jr, Morel L, Fund F et al. 2017. Modeling tropospheric wet delays with dense and sparse network configurations for PPP-RTK. GPS Solutions, 21(1), 237–250.

Dousa J, Vaclavovic P, Elias M. 2017. Tropospheric products of the second GOP European GNSS reprocessing (1996–2014). Atmospheric Measurement Techniques, 10(9), 3589–3607.

Dousa J, Václavovic P, Zhao L et al. 2018b. New adaptable all in-one strategy for estimating advanced tropospheric parameters and using real-time orbits and clocks. Remote Sensing, 10(2), 232.

Duan J, Bevis M, Fang P et al. 1996. GPS meteorology: Direct estimation of the absolute value of precipitable water. Journal of Applied Meteorology and Climatology, 35(6), 830–838.

Ducrocq V, Ricard D, Lafore J-P et al. 2002. Storm-scale numerical rainfall prediction for five precipitating events over France: On the importance of the initial humidity field. Weather and Forecasting, 17, 1236–1256. DOI:10.1175/1520-0434(2002)017<1236:Ssnrpf>2.0.Co;2.

Hadas T, Hobiger T, Hordyniec P et al. 2020. Considering different recent advancements in GNSS on real-time zenith troposphere estimates. GPS Solutions, 24(4), 1–14.

Hadas T, Teferle F N, Kazmierski K et al. 2017. Optimum stochastic modeling for GNSS tropospheric delay estimation in real-time. GPS Solutions, 21(3), 1069–1081. DOI:10.1007/s10291-016-0595-0.

Hersbach H, Bell B, Berrisford P. 2020. The ERA5 global reanalysis. Quarterly Journal of the Royal Meteorological Society, 146(730), 1999–2049. DOI:10.1002/qj.3803.

Hopfield H S. 1971. Tropospheric effect on electromagnetically measured range: Prediction from surface weather data. Radio Science, 6(3), 357–367. DOI:10.1029/RS006i003p00357.

Jin S, Wang X. 2021. Principles and applications of GNSS meteorology. Beijing: China Meteorological Press.

Jones J, Guerova G, Douša J et al. 2020. Advanced GNSS tropospheric products for monitoring severe weather events and climate. Springer.

Lagler K, Schindelegger M, Böhm J et al. 2013. GPT2: empirical slant delay model for radio space geodetic techniques. Geophysical Research Letters, 40(6), 1069–1073. DOI:10.1002/grl.50288.

Landskron D. 2017. Modeling tropospheric delays for space geodetic techniques. Doctoral Dissertation, Department of Geodesy and Geoinformation, TU Wien, Supervisor: Böhm J.

Landskron D, Böhm J. 2018a. VMF3/GPT3: refined discrete and empirical troposphere mapping functions. Journal of Geodesy, 92(4), 349–360. DOI:10.1007/s00190-017-1066-2.

Landskron D, Böhm J. 2018b. Refined discrete and empirical horizontal gradients in VLBI analysis. Journal of geodesy, 92(12), 1387–1399. DOI:10.1007/s00190-018-1127-1.

Leandro R, Santos M, Langley R. 2006. UNB neutral atmosphere models: development and performance. Proceedings of the 2006 National Technical Meeting of The Institute of Navigation, 564–573.

Li J, Zhang B, Yao Y et al. 2020. A refined regional model for estimating pressure, temperature, and water vapor pressure for geodetic applications in China. Remote Sensing, 12(11), 1713. DOI:10.3390/rs12111713.

Li T, Wang L, Chen R et al. 2021. Refining the empirical global pressure and temperature model with the ERA5 reanalysis and radiosonde data. Journal of Geodesy, 95(31), 1–17. DOI:10.1007/s00190-021-01478-9.

Li W, Yuan Y, Ou J et al. 2015a. New versions of the BDS/GNSS zenith tropospheric delay model IGGtrop. Journal of Geodesy, 89, 73–80.

Li W, Yuan Y, Ou J et al. 2018. IGGtrop_SH and IGGtrop_RH: Two improved empirical tropospheric delay models based on vertical reduction functions. IEEE Transactions on Geoence & Remote Sensing, 56(9), 5276–5288. DOI:10.1109/TGRS.2018.2812850.

Li X, Dick G, Lu C et al. 2015b. Multi-GNSS meteorology: real-time retrieving of atmospheric water vapor from BeiDou, Galileo, GLONASS, and GPS observations. IEEE Transactions on Geoscience and Remote Sensing, 53(12), 6385–6393. DOI:10.1109/TGRS.2015.2438395.

Li X, Zus F, Lu C et al. 2015c. Retrieving of atmospheric parameters from multi-GNSS in real-time: Validation with water vapor radiometer and numerical weather model. Journal of Geophysical Research: Atmospheres, 120(14), 7189–7204.

Li Z, Huang J. 2005. GPS measurement and data processing. Wuhan: Wuhan University Press.

Lindskog M, Ridal M, Thorsteinsson S, Ning T. (2017). Data assimilation of GNSS zenith total delays from a Nordic processing centre. Atmospheric Chemistry and Physics, 17, 13983–13998. DOI:10.5194/acp-17-13983-2017.

Lou Y, Huang J, Zhang W et al. 2018. A new zenith tropospheric delay grid product for real-time PPP applications over China. Sensors, 18(1), 65.

Lou Y, Yaozong Z, Weixing Z et al. 2022. Review on the high-accuracy and high-resolution processing of ground-based BDS/GNSS water vapor and its applications. Journal of Geomatics, 47(5), 1–11.

Lu C, Chen X, Liu G et al. 2017. Real-time tropospheric delays retrieved from multi-GNSS observations and IGS real-time product streams. Remote Sensing, 9(12), 1317.

Macpherson S R, Deblonde G, Aparicio J M et al. 2008. Impact of NOAA ground-based GPS observations on the Canadian Regional analysis and forecast system. Monthly Weather Review. 136, 2727–2746. DOI:10.1175/2007mwr2263.1.

Mahfouf J-F, Ahmed F, Moll P et al. 2015. Assimilation of zenith total delays in the AROME France convective scale model: a recent assessment. Tellus A: Dynamic Meteorology and Oceanography, 67, 26106. DOI:10.3402/tellusa.v67.26106.

Mateus P, Mendes V B, Plecha S M. 2021. HGPT2: an ERA5-based global model to estimate relative humidity. Remote Sensing, 13(11), 2179. DOI:10.3390/rs13112179.

Niell A E. 1996. Global mapping functions for the atmosphere delay at radio wavelengths. Journal of Geophysical Research: Solid Earth, 101(B2), 3227–3246. DOI:10.1029/95JB03048.

Nilsson T, Böhm J, Wijaya D D et al. 2013. Path delays in the neutral atmosphere. In: Böhm J, Schuh H (eds) Atmospheric effects in space geodesy. Springer atmospheric sciences. Berlin: Springer, pp 73–136. ISBN: 978-3-642-36931-5.

Nykiel G, Figurski M, Kroszczynski K et al. 2016. Assimilation of GNSS ZTD data from local dense GNSS networks in WRF model. In: COST ES1206" GNSS4SWEC" WG Meeting.

Poli P, Moll P, Rabier F et al. 2007. Forecast impact studies of zenith total delay data from European near real-time GPS stations in Météo France 4DVAR. Journal of Geophysical Research, 112. DOI:10.1029/2006JD007430.

Rohm W, Guzikowski J, Wilgan K, Kryza M. (2019). 4DVAR assimilation of GNSS zenith path delays and precipitable water into a numerical weather prediction model WRF. Atmospheric Measurement Techniques, 12, 345–361. DOI:10.5194/amt-12-345-2019.

Saastamoinen J. 1972. Atmospheric correction for the troposphere and stratosphere in radio ranging satellites. The Use of Artificial Satellites for Geodesy, Geophysical Monograph Series, 15, 247–251.

Schüler T. 2014. The TropGrid2 standard tropospheric correction model. GPS Solutions, 18(1), 123–131. DOI:10.1007/s10291-013-0316-x.

Seidel D J, Berger F H, Diamond H J et al. 2009. Reference upper-air observations for climate: Rationale, progress, and plans. Bulletin of the American Meteorological Society, 90(3), 361–369.

Singh R, Ojha SP, Puviarasan N et al. 2019. Impact of GNSS signal delay assimilation on short range weather forecasts over the Indian Region. Journal of Geophysical Research: Atmospheres, 124, 9855–9873. DOI:10.1029/2019JD030866.

Skamarock W C, Klemp J B, Dudhia J et al. 2019. A description of the advanced research WRF model version 4. NCAR Tech. Note NCAR/TN-556+STR, 145. DOI:10.5065/1dfh-6p97.

Vedel H, Huang X-Y. 2004. Impact of ground based GPS data on numerical weather prediction. Journal of the Meteorological Society of Japan. Ser. II, 82, 459–472. DOI:10.2151/jmsj.2004.459.

Vey S, Dietrich R, Fritsche M et al. 2006. Influence of mapping function parameters on global GPS network analyses: Comparisons between NMF and IMF. Geophysical Research Letters, 33(1), 1–4. DOI:10.1029/2005GL024361.

Yang Y. 2016. Concepts of comprehensive PNT and related key technologies. Acta Geodaetica et Cartographica Sinica, 45(5), 505–510.

Yuan Y, Holden L, Kealy A et al. 2019. Assessment of forecast Vienna Mapping Function 1 for real-time tropospheric delay modeling in GNSS. Journal of Geodesy, 93(9), 1501–1514. DOI:10.1007/s00190-019-01263-9.

Zhang H, Ding A, Lei W. 2011. Atmospheric refraction theory in space Geodesy. Beijing: Surveying and Mapping Press.

Zhang H, Yuan Y, Li W et al. 2018. A grid-based tropospheric product for China using a GNSS network. Journal of Geodesy, 92(7), 765–777.

Zhang W, Lou Y, Cao Y et al. 2019. Corrections of radiosonde-based precipitable water using ground-based GPS and applications on historical radiosonde data over China. Journal of Geophysical Research: Atmospheres, 124(6), 3208–3222.

Zheng F, Lou Y, Gu S et al. 2018. Modeling tropospheric wet delays with national GNSS reference network in China for BeiDou precise point positioning. Journal of Geodesy, 92(5), 545–560.

Zhou Y, Lou Y, Zhang W et al. 2021. On the accuracy and PPP performance evaluation of the latest generation of real-time tropospheric mapping function. Geomatics and Information Science of Wuhan University, 46(12), 1881–1888.

Zumberge J F, Heflin M B, Jefferson D C et al. 1997. Precise point positioning for the efficient and robust analysis of GPS data from large networks. Journal of Geophysical Research: Solid Earth, 102(B3), 5005–5017.

3 Spatial-Temporal Variation of GNSS ZTD and Its Responses to ENSO Events

Tengli Yu, Yong Wang, Shuanggen Jin,
Ershen Wang, and Xiao Liu

3.1 INTRODUCTION

In recent years, global meteorological disasters such as rainstorms, floods, high temperatures, droughts, and typhoons have occurred frequently, and the impact of extreme climate events on people's health and the sustainable development of society has become more serious. As the global monitoring and early warning service capacity of severe weather constantly improve, comprehensive urban and rural disaster prevention and mitigation have significantly improved. However, with global warming, the risk of natural disasters has further intensified, and extreme weather tends to be stronger, heavier, and more frequent. The causes are complex and difficult to predict accurately. The monitoring and early warning accuracy of meteorological disaster events need further improvement to ensure people's health and property safety. The heavy rain event in July 2021 affected many jurisdictions in Henan Province, China, resulting in significant economic losses and social impacts. The National Meteorological Center of the China Meteorological Administration indicated that this event is related to La Niña (the cold phase event of the El Niño-Southern Oscillation event) and that the GNSS meteorological elements had a good indication for this disastrous rainstorm event (Shi et al., 2022). China is located in the East Asian monsoon region. The El Niño-Southern Oscillation (ENSO) event affects the East Asian monsoon precipitable water vapor (PWV) transport by affecting the changes of atmospheric circulation, such as the Western Pacific subtropical high, which indirectly leads to the extreme climate in China (Chen et al., 2018). It is of great significance to use continuous zenith tropospheric delay (ZTD) series to monitor the evolution of ENSO events with interannual and interdecadal variations.

Global Navigation Satellite System (GNSS) technology has the advantages with continuous operation, low cost, high accuracy, and high spatial and temporal resolution (Wang and Liu, 2012; Wu et al., 2021). Using GNSS technology to remotely sense the environmental status of the earth's atmosphere and surface meteorology and conduct research and application of meteorological theories and methods, such as measuring atmospheric temperature, PWV content, soil moisture, and sea wind, which is called GNSS meteorology. It's a proven tool for weather and climate monitoring (Bevis et al., 1992, 1994). The standard product of GNSS meteorology is the ZTD. The GNSS signal is influenced by the refraction effect of the atmospheric medium when passing through the troposphere, thus causing a delay in signal transmission. The resulting path delay is called ZTD (Jin et al., 2011). GNSS ZTD has fewer sources of error because it avoids the subsequent inversion process (Zhou et al., 2020a). The researchers found that GNSS ZTD has the potential to monitor meteorological disasters and evaluated the application value of GNSS ZTD in identifying climate events such as rainstorms, drought, and heavy haze weather. GNSS ZTD is widely used in numerical weather forecasts and extreme climate monitoring (Giannaros et al., 2020; Guo et al., 2021; Wang et al., 2019; Yao et al., 2016, 2018; Zhao et al., 2021; Zhou et al., 2020a).

The ENSO event is manifested as a large-scale abnormal sea surface temperature variation in the eastern equatorial Pacific Ocean. It appears as a La Niña event in the cold phase and an El Niño event in the warm phase (Ren et al., 2012). The ENSO event not only directly causes catastrophic events such as droughts and floods in the tropical Pacific and its surrounding areas but also indirectly affects climate parameters in other regions in a "teleconnection" manner, such as temperature, pressure, precipitation, clouds, and radiation balance, which causes meteorological disasters (Ren et al., 2012). The influence mechanisms of El Niño events on extreme rainfall in different regions of China have been discussed by related scholars (Liu et al., 2018; Song et al., 2020). Experiments show that El Niño events and precipitation strongly correlate, leading to the phenomenon of "southern flooding and northern drought" in mainland China. According to the research, the response characteristics of climate to ENSO events differ in different regions of mainland China. El Niño and La Niña events cause roughly opposite environmental effects (Cao et al., 2013; Lei and Huang, 2018). Different types of ENSO events will lead to global temperature increase and annual average precipitation decrease globally in El Niño years. In La Niña years, the global temperature is low, and the yearly average precipitation increases (Wang and Gong, 1999). Since the 1990s, the frequency of new types of El Niño events has increased significantly. This new El Niño pattern is different from the traditional one. The anomalous warming area of the maximum sea surface temperature (SST) is not in the eastern equatorial Pacific but in the central equatorial Pacific. Researchers refer to this new type of El Niño event as the Central-Pacific type of El Niño (CP-El Niño) and the traditional El Niño event as the Eastern-Pacific type of El Niño (EP-El Niño) (Kao and Yu, 2009). This criterion also distinguishes La Niña events. ENSO events with different SST anomaly types have different atmospheric effects (Cao et al., 2013; Ren et al., 2012; Zhang et al., 2018).

Compared with precipitation, GNSS ZTD data has better temporal continuity and is suitable for capturing long-term climate change patterns. Therefore, GNSS meteorology can be applied to ENSO monitoring. Foster et al. (2000) first explored the impact of ENSO events on climate change using two GNSS stations in the tropical monsoon climate zone. Barindelli et al. (2018) and Zhao et al. (2020) proved the correlation between ENSO events and meteorological factors using GNSS observation data. Yao et al. (2013) demonstrated that GNSS ZTD has a significant semi-annual variation period, and the abnormal change of this cycle has responses to El Niño events. Wang et al. (2021) explored the impact of El Niño events on the climate of mainland China by using regional GNSS ZTD in China. The aforementioned research shows that GNSS ZTD can indicate the evolution of ENSO events, which provides theoretical support for applying GNSS ZTD to ENSO events monitoring. However, there are differences in the case selection in the foregoing studies, which lead to different results, and some regions have insufficient quantitative analysis. Therefore, the use of GNSS ZTD for monitoring and early warning of extreme climate events can systematically quantify the impact of climate anomalies in different regions of China on various ENSO events and provide technical reference for meteorological monitoring and forecasting departments and disaster prevention and control departments.

3.2   GNSS OBSERVATIONS AND ZTD CALCULATIONS

3.2.1   The Principle of GNSS ZTD Measurement

When the radio wave signal of the GNSS satellite passes through the atmosphere, it will be affected by the atmospheric medium's refraction effect, resulting in the signal transmission delay. The delay caused by crossing the ionosphere is called the ionospheric delay. The delay caused by crossing the troposphere and stratosphere is called the tropospheric delay. The ionospheric delay can be eliminated by using GNSS dual-frequency receivers. The tropospheric delay can be estimated by using model simulation. The ZTD includes the zenith hydrostatic delay (ZHD) and the zenith wet delay (ZWD). The ZHD is mainly related to meteorological elements such as pressure and temperature. The ZWD contains the primary information on atmospheric PWV. The ZHD content is more than

90%, but it is relatively stable, with minor changes in a short period. The ZWD accounts for a small proportion of the ZTD but contains almost all PWV information in the atmosphere. It is active, mainly for atmospheric PWV inversion. Assuming that GNSS satellite signals propagate along a straight line at the speed of light in a vacuum, the atmospheric delay can be defined as the difference between the length of the actual propagation path and the assumed straight-line length. Based on this definition, a GNSS ZTD simulation model can be constructed.

ZTD can be expressed by the increased length of the signal propagation path:

$$\Delta L = \int_L n(s)ds - G \tag{3.1}$$

where $n(s)$ represents the atmospheric refractive index at s on the signal propagation path L; G represents the path length between the satellite and the receiver (not affected by atmospheric refraction interference). The previous equation can be expressed as:

$$\Delta L = \int_L \left[n(s) - 1 \right] ds + (S - G) \tag{3.2}$$

where S represents the path length along L, $\int_L \left[n(s) - 1 \right] ds$ is affected by the slowing down of signal propagation, $(S - G)$ is affected by signal bending. When the ray points in the zenith direction, the ray is straight and $(S - G)$ is 0. The atmospheric refractive index N is often used to indicate $\left[n(s) - 1 \right]$, $N = 10^6 \times (n - 1)$.

The atmospheric refractive index N can be calculated by using the functional relationship between air temperature, air pressure, and water vapor pressure, which can be expressed as:

$$N = 77.6 \times \left(\frac{P}{T} \right) + 3.73 \times 10^5 \times \left(\frac{P_\omega}{T^2} \right) \tag{3.3}$$

where P is the total surface atmospheric pressure (hPa), T is the surface atmospheric temperature (K), P_ω is the water vapor pressure (hPa).

3.2.2 ZTD SOLUTION MODEL

Currently, the main models used to calculate ZTD are the Hopfield model (Vey et al., 2010), the Saastamoinen model (Vey et al., 2009), and the Black model (Suparta, 2013).

3.2.2.1 Hopfield Model

In the ideal gas state, the Hopfield model solution equation is as follows:

$$ZTD = ZHD + ZWD = \frac{K_d}{\sin\left(E^2 + 6.25\right)^{\frac{1}{2}}} + \frac{K_\omega}{\sin\left(E^2 + 2.25\right)^{\frac{1}{2}}} \tag{3.4}$$

$$ZHD = 155.2 \times 10^{-7} \times \frac{P_s}{T_s} \times \left(h_d - h_s \right) \tag{3.5}$$

$$K_\omega = 155.2 \times 10^{-7} \times \frac{4810}{T_s^2} \times \left(h_\omega - h_s \right) \tag{3.6}$$

$$h_d = 40136 + 148.72 \times (T_s - 273.16) \tag{3.7}$$

$$h_\omega = 11000 \tag{3.8}$$

where ZTD is the zenith tropospheric delay (m), ZHD is the zenith hydrostatic delay (m), ZWD is the zenith wet delay (m). T_s is the absolute temperature (K). P_s is the surface pressure (hPa). E is the satellite elevation angle (°). h_d is the effective height from the top of the troposphere to the geoidal surface (m); h_ω is the effective height of tropospheric moisture to the geoidal (m).

3.2.2.2 Saastamoinen Model

The formula for the Saastamoinen model is expressed as follows:

$$ZTD = \frac{0.002277}{\sin E} \times \left[P_s + \left(\frac{1255}{T_s} + 0.05 \right) \times e_s - \frac{B}{\tan^2 E} \right] \times W(\varphi \bullet H) + \delta R \tag{3.9}$$

where $W(\varphi \bullet H) = 1 + 0.0026 \times \cos 2\varphi + 0.00028 \times h_s$, φ is the latitude of the station (°), h_s is the altitude of the station (km), B is the list function of h_s, δR is the list function of E and h_s. After numerical fitting, the formula can be expressed as:

$$ZTD = \frac{0.002277}{\sin E'} \times \left[P_s + \left(\frac{1255}{T_s} + 0.05 \right) \times e_s - \frac{a}{\tan^2 E'} \right] \tag{3.10}$$

$$E' = E + \Delta E \tag{3.11}$$

$$\Delta E = \frac{16''}{T_s} \times \left(P_s + \frac{4810}{T_s} e_s \right) \times \cot E \tag{3.12}$$

$$a = 1.16 - 0.15 \times 10^{-3} \times h + 0.716 \times 10^{-3} \times h_s^2 \tag{3.13}$$

3.2.2.3 Black Model

The formula for the Black model is expressed as follows:

$$ZTD = K_d \times \left[\sqrt{1 - \left(\frac{\cos E}{1 + (1 - l_0) \times \frac{h_d}{r_s}} \right)^2} - b(E) \right] + K_\omega \times \left[\sqrt{1 - \left(\frac{\cos E}{1 + (1 - l_0) \times \frac{h_\omega}{r_s}} \right)^2} - b(E) \right] \tag{3.14}$$

$$l_0 = 0.833 + \left[0.076 + 0.00015 \times (T_s - 273.16) \right]^{-0.3E} \tag{3.15}$$

$$b(E) = 1.92 \times \left(E^2 + 0.6 \right)^{-1} \tag{3.16}$$

$$h_d = 148.98 \times (T_s - 3.96) \tag{3.17}$$

$$h_\omega = 13000 \tag{3.18}$$

$$K_d = 0.002312 \times (T_s - 3.96) \times \frac{P_s}{T_s^2} \tag{3.19}$$

$$K_\omega = 0.20 \tag{3.20}$$

In equation (3.16), $b(E)$ represents the path bending correction, and the other parameters have the same meaning as in the Hopfield model.

In high-precision GNSS data processing, the ZTD solution model is first selected according to the actual situation, and the model's value is calculated as an approximate value. Then the accurate value of tropospheric delay is estimated by rigorous adjustment calculation. It has been verified that the results of the three models are equivalent when the station elevation is less than 1 km, while the Hopfield model is not applicable when the station elevation is greater than 1 km (Wang et al., 2012).

The basic observation equation of the carrier phase in GNSS data processing is as follows (Zhou et al., 1999):

$$\varphi_i^j(t) = \frac{f}{c} \times \rho_i^j(t) + f \times [\delta t_i(t) - \delta t^i(t)] - N_i^j(t_0) + \frac{f}{c} \times [\Delta_{i.I_p}^j(t) + \Delta_{i.T}^j(t)] \tag{3.21}$$

where $\varphi_i^j(t)$ is the carrier phase from the satellites j to the observation station T_i at the observation epoch t_i. f is the carrier frequency. c is the propagation speed of the electromagnetic wave. $\rho_i^j(t)$ is the geometric distance from the satellites j to the observation station T_i at the observation epoch t_i. $[\delta t_i(t) - \delta t^i(t)]$ is the clock error of the receiver clock relative to the satellite clock at the observation epoch t_i. $N_i^j(t_0)$ is the carrier phase integer cycle. $\Delta_{i.I_p}^j(t)$ is the effect of the ionospheric delay at the observation epoch t_i. $\Delta_{i.T}^j(t)$ is the influence of the tropospheric delay at the observation epoch t_i, which can be obtained from the GNSS data processing results.

3.2.3 Mapping Function

During the GNSS data processing, the tropospheric delay in the zenith direction is estimated as an unknown for each station and projected onto each oblique direction by the hydrostatic mapping function and the wet delay mapping function. The model is as follows:

$$atdel(el) = dryzen \times drymap(el) + wetzen \times wetmap(el) \tag{3.22}$$

where el is the satellite elevation angle, $dryzen$ is the hydrostatic delay, $wetzen$ is the wet delay, $drymap$ is the hydrostatic delay mapping function, $wetmap$ is the wet delay mapping function.

The mapping function is a mathematical model related to the elevation angle of each delay. Mapping functions are approximately equal to the cosecant of elevation angles. However, the curvature of the earth and the propagation path of GNSS signals in the atmosphere make the mapping function and the cosecant theorem significantly deviate. The accuracy of the mapping functions affects not only the delayed solution accuracy but also the positioning accuracy. Mapping functions are divided into two main categories: the empirical mapping function model represented by NMF and GMF, which are used without introducing external data information; the other is the mapping function model based on the numerical weather model represented by VMF1 (Jiang, 2000). These three models are all continued fractions, which can meet the accuracy requirements even at an altitude angle of 3° (Niell et al., 2001):

$$
mf(e) = \frac{1 + \dfrac{a}{1 + \dfrac{b}{1 + c}}}{\sin(e) + \dfrac{a}{\sin(e) + \dfrac{b}{\sin(e) + c}}}
\tag{3.23}
$$

where e is the elevation angle, and the coefficients a, b, and c are constants far less than 1, which are related to factors such as surface temperature, station latitude and altitude. Currently, most mapping functions are constructed based on Eq. (3.23).

3.2.3.1 NMF Model

The projection function of the model consists of two components: the dry component mf_d and the wet component mf_w. The dry component is:

$$
mf_d(e) = \frac{1 + \dfrac{a_d}{1 + \dfrac{b_d}{1 + c_d}}}{\sin(e) + \dfrac{a_d}{\sin(e) + \dfrac{b_d}{\sin(e) + c_d}}} + \left[\frac{1}{\sin(e)} - \frac{1 + \dfrac{a_{ht}}{1 + \dfrac{b_{ht}}{1 + c_{ht}}}}{\sin(e) + \dfrac{a_{ht}}{\sin(e) + \dfrac{b_{ht}}{\sin(e) + c_{ht}}}} \right] \times \frac{H}{1000}
\tag{3.24}
$$

where, e is the altitude angle, $a_{ht} = 2.53 \times 10^{-5}$, $b_{ht} = 5.49 \times 10^{-3}$, $c_{ht} = 1.14 \times 10^{-3}$, and H is the positive height. When the station is between 15° and 75° latitude, the coefficients a_d, b_d, and c_d can be obtained by interpolation from Eq. (3.25):

$$
p_d(\varphi, t) = p_{avg}(\varphi_i) + [p_{avg}(\varphi_{i+1}) - p_{avg}(\varphi_i)] \times \frac{\varphi - \varphi_i}{\varphi_{i+1} - \varphi_i} +
$$
$$
\left\{ p_{amp}(\varphi_i) + [p_{amp}(\varphi_{i+1}) - p_{amp}(\varphi_i)] \times \frac{\varphi - \varphi_i}{\varphi_{i+1} - \varphi_i} \times \cos\left(2\pi \frac{t - 28}{365.25} \right) \right\}
\tag{3.25}
$$

where φ denotes the latitude of the station, φ_i represents the latitude corresponding to Table 3.1, t is DOY (day of year). The mean and amplitude of the coefficients are shown in Tables 3.1–3.2.

The wet projection function of NMF is calculated by Eq. (3.23); the projection coefficients a_d, b_d, and c_d can be obtained by using Eq. (3.26) and Table 3.3.

$$
p_w = p_{avg}(\varphi_i) + (p_{avg}(\varphi_{i+1}) - p_{avg}(\varphi_i)) \times \frac{\varphi - \varphi_i}{\varphi_{i+1} - \varphi_i}
\tag{3.26}
$$

TABLE 3.1

The Mean of the NMF Model Dry Component Projection Function Coefficient

Latitude(deg)	$a_d \times 10^{-3}$	$b_d \times 10^{-3}$	$c_d \times 10^{-3}$
≤15	1.2769934	2.9153695	62.620505
30	1.268323	2.9152299	62.837393
45	1.2465397	2.9288445	63.721774
60	1.2196.49	2.9022565	63.824265
≥75	1.2045996	2.9024912	64.258455

TABLE 3.2

The Amplitude of the NMF Model Dry Component Projection Function Coefficient

Latitude(deg)	$\Delta a_d \times 10^{-3}$	$\Delta b_d \times 10^{-3}$	$\Delta c_d \times 10^{-3}$
≤15	0	0	0
30	1.2709626	2.1414979	9.0128400
45	2.6523662	3.0160779	4.3497037
60	3.4000452	7.2562722	84.795348
≥75	4.1202191	11.723375	170.37206

TABLE 3.3

The Wet Delay Projection Function of the NMF Model

Latitude(deg)	$a_w \times 10^{-4}$	$b_w \times 10^{-3}$	$c_w \times 10^{-2}$
≤15	5.8021897	1.4275268	4.3472961
30	5.6794847	1.5138625	4.6729510
45	5.8118019	1.4572752	4.3908931
60	5.9727542	1.5007428	4.4626982
≥75	6.1641693	1.7599082	5.4736038

3.2.3.2 VMF1 Model

The form of the VMF1 model is similar to that of NMF. The coefficients a_h and a_w are obtained by interpolating the grid generated from the measured meteorological data with a spatial resolution of $2.5° \times 2°$ and a temporal resolution of 6 h. The coefficient b takes a fixed value of 0.0029. The coefficient c was divided into two cases considered for the southern and northern hemispheres and was calculated using Eq. (3.27), and the coefficients in the equation are shown in Table 3.4.

$$c = c_0 + \left[\left(\left(\cos \left(\frac{doy - 28}{365} \cdot 2\pi + \varphi \right) \right) + 1 \right) \cdot \frac{c_{11}}{2} + c_{10} \right] \cdot \left(1 - \cos(\varphi) \right) \qquad (3.27)$$

The accuracy of the VMF1 projection function is higher than that of the NMF model, but the VMF1 projection function is expressed by the grid form, which is inconvenient to use. It needs to be calculated using the measured meteorological parameters, and the formula is complicated.

TABLE 3.4

The Correlation Coefficient Used to Calculate the Dry Delay Projection Coefficient c

Hemisphere	c_0	c_{10}	c_{11}	A
Northern Hemisphere	0.062	0.000	0.006	0
Southern Hemisphere	0.062	0.001	0.006	π

3.2.3.3 GMF Model

b_d, b_w, c_d, and c_w in the GMF model follow the calculation of the VMF1 model. The annual mean and amplitude of the coefficients a_d and a_w are subjected to spherical harmonic expansion. The spherical harmonic coefficients are obtained by least squares fitting.

$$a = a_{avg} + a_{amp} \cdot \cos(\frac{doy - 28}{365} \cdot 2\pi)$$

$$a_i = \sum_{n=0}^{9} \sum_{m=0}^{n} P_{nm}(\sin\varphi) \cdot [A_{nm} \cdot \cos(m \cdot \varphi) + B_{nm} \cdot \sin(m \cdot \varphi)]$$

(3.28)

The accuracy of the GMF model is similar to that of VMF1 and does not require measured meteorological data. It is currently the most commonly used projection function model.

The China Mainland Tectonic Environment Monitoring Network (CMTEMN) has 262 GNSS continuous observation stations, including 31 Crustal Movement Observation Network of China (CMONOC) stations (Figure 3.1). The data of the CMONOC reference stations have been recorded

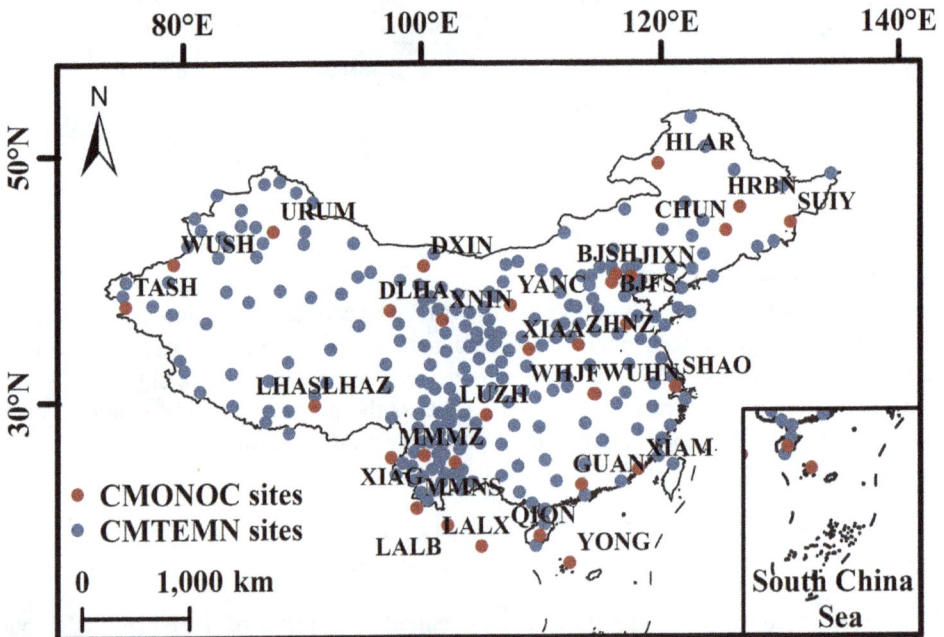

FIGURE 3.1 GNSS sites distribution from CMONOC and CMTEMN.

since 1999. The new 231 continuous stations of the CMTEMN have been producing data since mid-2010, and nearly 12 years of GNSS observations have been accumulated. The processing strategy is as follows: the ephemeris type is precision ephemeris, loose solution mode is selected, satellite cutoff height is 10°, and GMF mapping function is selected. The estimated time resolution is 1 hour, and the default horizontal gradient of GAMIT is used. The length of observation data accumulated at each site is different. The different times of network engineering and land-based network construction lead to the different lengths of the accumulated observations at each station. This study gets the longest time series of GNSS ZTD: 2008/1–2021/6, and the shortest time series: 2011/1–2021/6. We removed sites with large numbers of missing values, such as LALX, LALB, and YONG. For the hourly ZTD series with a small number of missing values, we filled the data using the neighbor mean interpolation method in SPSS software, and the GNSS ZTD series were denoised using wavelet transform (Yu et al., 2022).

3.3 GNSS ZTD SPATIAL-TEMPORAL VARIATION ANALYSIS

3.3.1 GNSS ZTD Spatial Distribution Characteristics

The EOF method was used to analyze the typical spatial distribution characteristics of GNSS ZTD in the Chinese mainland. EOF can decompose the original data set into patterns ordered by their temporal variances. The initial field of relevant variables is decomposed into several unrelated spatial functions and temporal coefficients without losing the original data information (Hannachi and Neill, 2001). The first few modes that pass the significance test in the EOF analysis results contain the primary variation information of the actual field. Regarding the EOF decomposition, the covariance matrix of the original data is constructed at first (Hannachi et al., 2007). The matrix form of the spatial and temporal grid data is:

$$X = \begin{bmatrix} x_{11} & x_{12} & \cdots & x_{1n} \\ x_{21} & x_{22} & \cdots & x_{2n} \\ \vdots & \vdots & \ddots & \vdots \\ x_{m1} & x_{m2} & \cdots & x_{mn} \end{bmatrix} \tag{3.29}$$

where $X(i, t)$ is the observation corresponding to position i ($i \in (1, m)$) and time t ($t \in (1, n)$), m is the spatial station, and n is the time series.

Construct covariance matrix C:

$$C_{m \times m} = \frac{1}{n} X \cdot X^T \tag{3.30}$$

The eigenroots $\left(\lambda_1, \lambda_2, \cdots, \lambda_m \right)$ and eigenvectors $V_{m \times m}$ of C are:

$$C_{m \times m} \times V_{m \times m} = V_{m \times m} \times E_{m \times m} \tag{3.31}$$

$$E = \begin{bmatrix} \lambda_1 & 0 & \cdots & 0 \\ 0 & \lambda_2 & \cdots & 0 \\ \vdots & \vdots & \ddots & \vdots \\ 0 & 0 & \cdots & \lambda_m \end{bmatrix} \tag{3.32}$$

where λ_1 is the eigenvector value corresponding to the first mode of the EOF. A higher λ value indicates that its corresponding modality is more essential and contributes more to the total variance.

The explanatory rate of the k modality to the total variance can be expressed as:

$$\rho_k = \frac{\lambda_k}{\sum_{i=1}^{m} \lambda_i} \times 100\% \tag{3.33}$$

The analysis results are tested for significance (North et al., 1982):

$$\Delta\lambda = \lambda \sqrt{\frac{2}{N^*}} \tag{3.34}$$

where λ represents the characteristic root. N^* represents the degree of freedom of data. λ is ranked in order, and its error range is analyzed. If the error ranges of the two λ before and after overlap, it does not pass the significance test.

We interpolated the EOF results of GNSS ZTD using the Kriging method to optimize the visualization of the ZTD spatial distribution. According to the analysis results, the first and second spatial modes passed the significance test (North et al., 1982). The first mode variance contribution is 87.4%, the second mode variance contribution is 4.04%, and the cumulative contribution of the two modes' variance is 91.44% (Figure 3.2[a–b]). The first mode spatial eigencoefficients are all positive values, reflecting the consistent increase and decrease of GNSS ZTD variation in mainland China, and show an increasing trend from northwest to southeast (Figure 3.2[a]). The second mode has both positive and negative spatial eigencoefficients, reflecting the differences in GNSS ZTD variations in different regions (Figure 3.2[b]). Related studies have shown significant differences in atmospheric humidity and precipitation distributions in different climate regions, and ENSO events have different effects on different climate regions (Cao et al., 2013; Lei et al., 2018; Ren et al., 2012). Therefore, the GNSS ZTD distribution characteristics were compared with the five climatic type divisions, and the results are in good agreement. The first mode shows that the GNSS ZTD spatial distribution forms a significant boundary at the junction of the temperate continental zone (TCZ), mountain plateau zone (MPZ), and temperate monsoon zone (TMZ), and tropical monsoon zone (TPMZ). It shows that the GNSS ZTD in TMZ, TPMZ, and subtropical monsoon zone (SMZ) is higher than in TCZ and MPZ. The GNSS ZTD distribution characteristics of the second mode form a significant dividing line at the junction of TMZ, MPZ, and SMZ. It shows the distribution characteristics that GNSS ZTD increases in the TMZ and decreases in other regions.

The TPMZ and SMZ have high temperatures and rainfall all year round. The TMZ is hot and rainy in summer and cold and dry in winter. The TCZ is far from the ocean and lacks humid air mass transport with low annual rainfall. The MPZ is rainy on the windward side of humid air currents with less precipitation on the leeward side and inside the plateau. Related studies show that the distribution characteristics of GNSS ZTD correlate with the distribution characteristics of meteorological factors such as atmospheric humidity and precipitation (Isioye et al., 2018; Wang et al., 2019). Therefore, the experimental analysis was combined with related research results. The

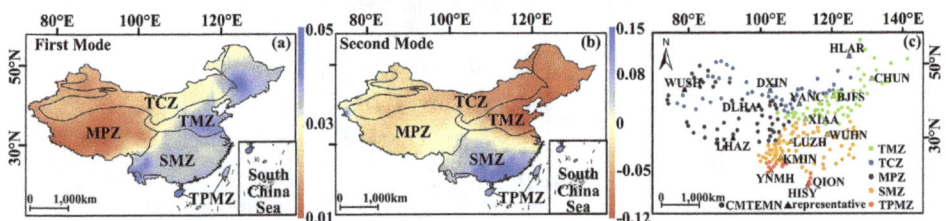

FIGURE 3.2   GNSS ZTD spatial characteristics and regional division results.

regional division of mainland China was carried out according to five major climate types to explore the response patterns of climate to ENSO events in different regions. The distribution of CMTEMN sites after division is shown in Figure 3.2(c). Fifteen representative sites with special notes (triangles) are the stations that show the results.

3.3.2 GNSS ZTD Time-Frequency Variation Characteristics

The GNSS ZTD time series has significant nonlinear characteristics. The fast Fourier transform (FFT) and wavelet transform (WT) methods are used to explore the change rules of GNSS ZTD time series from the perspective of the time domain and frequency domain, respectively. Discrete Fourier transform (DFT) establishes the corresponding relationship between the time and frequency domain characteristics and realizes the signal's mutual conversion in the time and frequency domains. Compared with the computationally intensive DFT, the FFT method accelerates the computational process. It can successfully capture the energy change in multi-dimensional data and significantly reduce the computational complexity of the DFT using unit complex roots as rotation factors. FFT can reflect the amplitude and phase characteristics of continuous signals in the frequency domain that cannot be reflected in the time domain (Zheng, 2015). The core equation is:

$$x(k) = \frac{1}{N} \sum_{j=1}^{N} X(j) W_N^{(j-1)(k-1)} \tag{3.35}$$

$$W_N = e^{-j\frac{2\pi}{N}} \tag{3.36}$$

where $x(k)$, $k = 1, 2, \cdots, N$ represents the distribution of the signal in the frequency domain. $X(j)$, $j = 1, 2, \cdots, N$ represents the distribution of the signal in the time domain.

The GNSS ZTD time series of mainland China were analyzed in three dimensions: year, month, and day. The oscillation characteristics of GNSS ZTD were analyzed in the frequency domain using the FFT with the Hanning window. The analysis results show no apparent regional difference in the frequency domain oscillation characteristics of GNSS ZTD in mainland China. Four uniformly distributed stations are selected to present the analysis results (Figure 3.3).

In Figure 3.3, cpy (cycle per year) represents the significant change cycle in the year dimension, cpm (cycles per month) represents the significant change cycle in the month dimension, and cpd (cycle per day) represents the significant change cycle in the day dimension. The horizontal axis represents the frequency, and its reciprocal is the corresponding period. The most significant period corresponds to the maximum amplitude. It can be seen from the annual scale analysis that GNSS ZTD has a significant annual period and semi-annual period. Excluding the interference of these two change periods, it is found that there is a more significant 9-month period (corresponding to frequency 1.1 cpy) and a seasonal period (corresponding to frequencies of 3 cpy and 4 cpy). It can be seen from the monthly scale analysis that GNSS ZTD has many significant change periods within 3 days–3 months (corresponding to the frequency interval of [0.3, 10]). From the figure of the daily scale analysis results, it can be concluded that there are significant daily and semi-daily wave variation cycles in GNSS ZTD. The GNSS ZTD wavelet coefficients corresponding to each change period will be extracted by the wavelet transform method to verify the accuracy of the significant change period analysis. The significant change period will be judged according to the size of its amplitude.

The fast Fourier transform provides a more accurate frequency domain location of signals, while the time domain location corresponding to frequency domain information can be provided by the wavelet transform (WT). Fourier analysis is the main support theory of traditional signal analysis, and its global variation characteristics lead to certain limitations of the analysis. The WT was

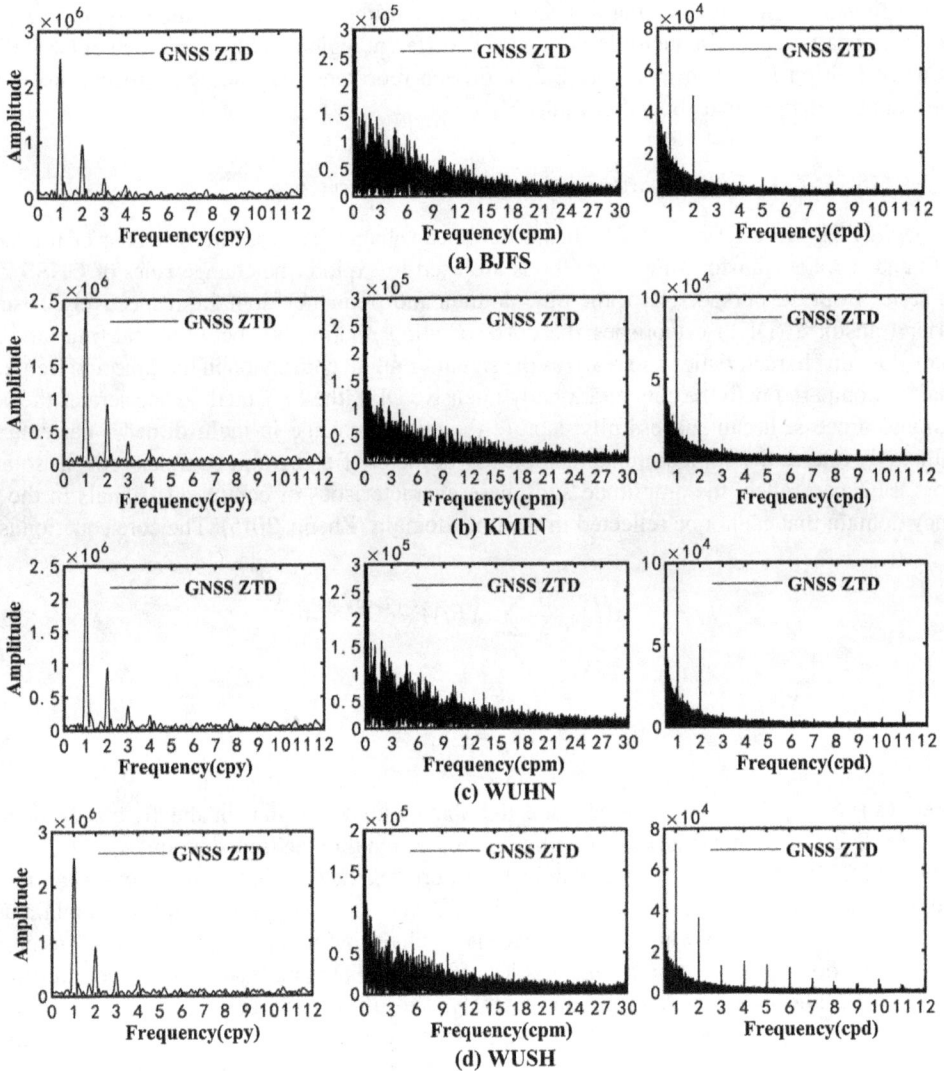

FIGURE 3.3 Significant change period of GNSS ZTD time series.

generated through the continuous improvement of the Fourier transform. It is another effective time-frequency analysis method after Fourier transform in signal processing, image processing, and many nonlinear scientific fields. It has a good "zoom" function, which refines the signal progressively on multiple scales using extension and translation operations to achieve the effect of time subdivision at high frequencies and frequency subdivision at low frequencies (Li et al., 2020). WT is widely used in signal processing, image processing, and many nonlinear scientific fields due to its time-frequency localization and multi-resolution analysis capabilities (Cai et al., 2019). The Db6 wavelet basis function with tightly supported standard orthogonal wavelets was chosen to carry out a time-frequency analysis of the GNSS ZTD time series (He et al., 2019). The principle of WT layering is to decompose the signal $f(t)$ (frequency band is $0–F$) into $J+1$ signal bands using $2, 4, 8, \cdots, 2^J$ as the scale, where $0 - F/2^J$ is the low-frequency term (AJ) and $F/2^n - F/2^{n-1}$ $(n = 1, 2, 3, \cdots, J)$ is the high-frequency term (D1, D2, . . ., DJ). The corresponding periods of each frequency band are shown in Table 3.5.

TABLE 3.5

The Corresponding Period of Wavelet Coefficients at Each Scale

Layers	D1	D2	D3	D4	D5	D6	D7	D8	D9
Period (h)	2–4	4–8	8–16	16–32	32–64	64–128	128–256	256–512	512–1024
Layers	D10	D11	D12	D13	D14	D15	D16	A16	
Period (h)	1024–2048	2048–4096	4096–8192	8192–16384	16384–32768	32768–65536	65536–131072	131072–∞	

The data time length of this study corresponds to the D16 layer period range. After the test, the GNSS ZTD time series was divided into 16 layers using the Db6 wavelet basis function to satisfy the stratification qualification criteria. D1–D16 layers are high-frequency terms, and the A16 layer is low-frequency terms (Figure 3.4).

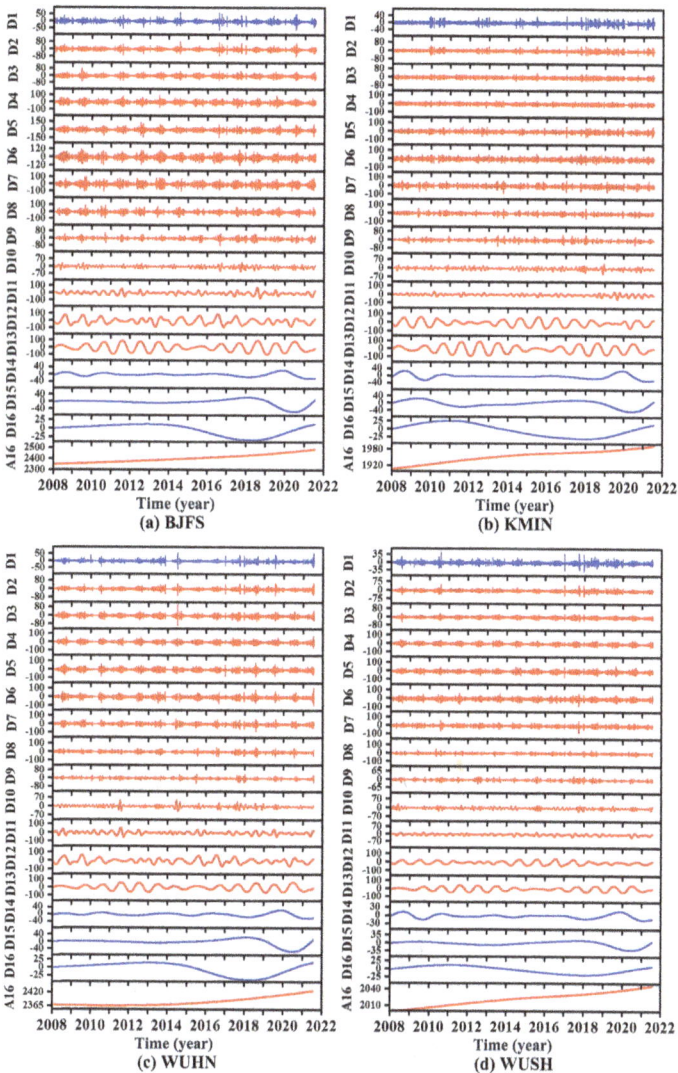

FIGURE 3.4 Wavelet coefficients for each scale of GNSS ZTD time series.

It can be seen from Figure 3.4 that the WT can divide the original time series into high-frequency terms of different period scales and can obtain the characteristics of the variation of any high-frequency term in the time domain. The significance of the period can be judged by the amplitude of wavelet coefficients in each layer. The analysis results of each site show that the wavelet coefficient amplitude thresholds of the D1, D14, D15, and D16 layers are significantly smaller than those of other layers. Therefore, the corresponding cycles of these layers are not significant and are not the dominant periods of GNSS ZTD timing variation. Among the wavelet coefficients with more significant amplitudes, D2–D4 layers correspond to diurnal and semi-diurnal periods, and D5–D10 layers correspond to cycles of variation within 3 days–3 months. This conclusion corresponds to the frequency interval (0.3, 10) where significant variation cycles exist in the FFT analysis results. The D11 layer corresponds to seasonal variation periods, layer D12 corresponds to semi-annual variation periods, and layer D13 corresponds to annual variation periods (containing significant 9-month variation periods). The threshold values of wavelet coefficients at each scale coincide with the results of the FFT analysis. It is concluded that the variation of the GNSS ZTD time series is mainly driven by the high-frequency term in layers D2–D13 and the trend term (A16). Wavelet coefficients in layers D1, D14, D15, and D16 are the anomalous variation time series of GNSS ZTD. The D1 layer corresponds to a smaller period. It is treated as noise because the amplitude of the normal wavelet component is larger when the orthogonal wavelets are processed for signals, in contrast to the uniform variation of the noise in the high-frequency part exactly. After WT decomposes the signal, the larger amplitude is the useful signal, and the smaller amplitude is generally the noise. Related research shows that the noise generally exists in D1 and D2 layers after the original signal is processed by WT (Peng and Chen, 2016). The D1 layer wavelet coefficients of GNSS ZTD are removed as the noise layer, and the other wavelet coefficients are reconstructed to denoise each station's original GNSS ZTD sequence in mainland China.

In summary, the spatial distribution of GNSS ZTD shows an increasing trend from northwest to southeast, and the regional pattern is more consistent with the five major climate-type divisions in mainland China. There are significant annual periods, 9-month periods, semi-annual periods, seasonal periods, diurnal and semi-diurnal waves, and other significant variation periods in the GNSS ZTD time series.

3.4 THE CORRELATION FEATURES BETWEEN GNSS ZTD ANOMALY AND ENSO EVENT

3.4.1 GNSS ZTD ANOMALY SEQUENCE

In this chapter, GNSS ZTD anomaly sequences were used to analyze the impact of ENSO events on climate anomalies in mainland China. The GNSS ZTD anomaly time series can be obtained by removing the trend signal and the significant periodic signal (Zhao et al., 2020). The original time resolution of GNSS ZTD is 1 hour, and its monthly average was taken to match the time resolution of the ENSO discriminant index and to explore the correlation characteristics between them. The GNSS ZTD monthly mean time series is decomposed into 7 layers using the WT method. According to the table of corresponding periods of wavelet coefficients at each scale (Table 3.5), it is known that the D1 layer corresponds to 2–4 monthly variation periods, the D2 layer corresponds to semi-annual periods, the D3 layer is an annual period, and A7 layer is a trend term. It can be concluded that the change of GNSS ZTD is mainly driven by the trend term (A7), the annual period term (D3), the semi-annual period term (D2), and the seasonal period term (D1) by comparing the amplitudes of wavelet coefficients at each scale. Therefore, GNSS ZTD anomaly sequences can be obtained by removing D1, D2, D3, and A7 and reconstructing the high-frequency terms in the D4–D7 layer. The results are shown in Figure 3.5. The black solid line is the original GNSS ZTD monthly mean sequence, the red line is the primary GNSS ZTD driver term, and the dark blue line is the reconstructed GNSS ZTD anomaly sequence.

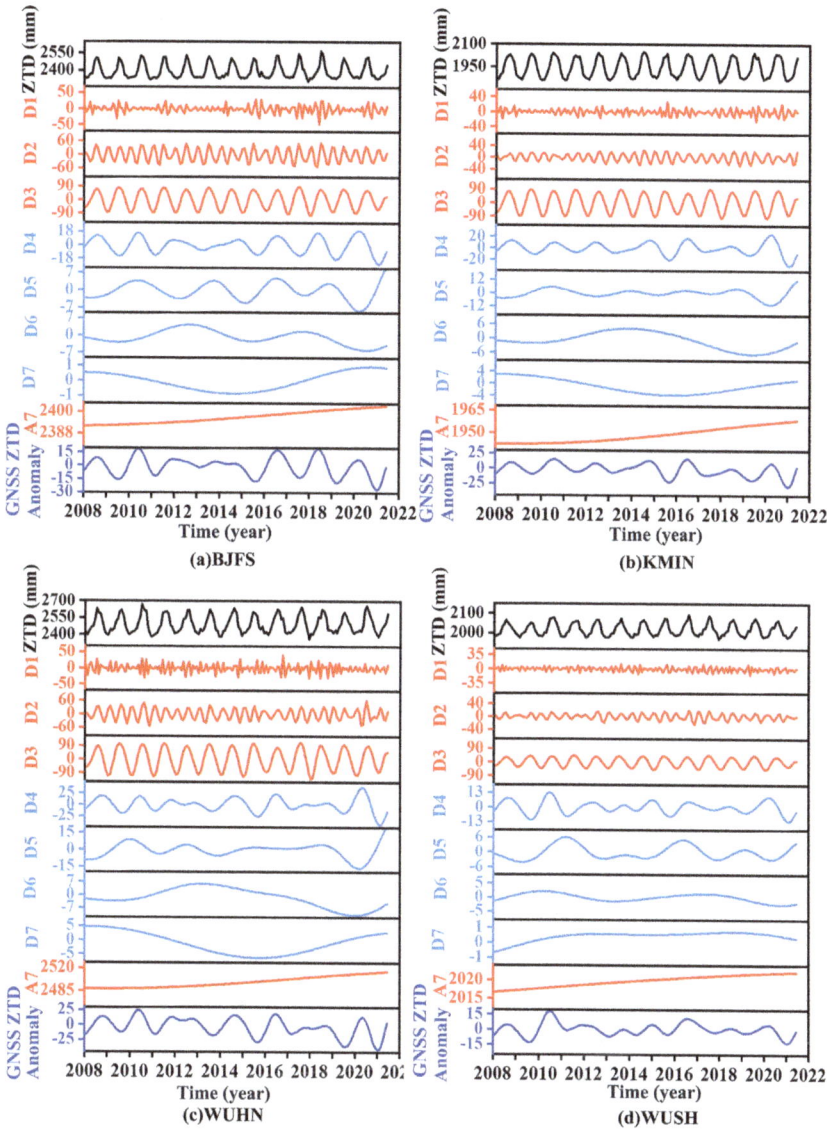

FIGURE 3.5 GNSS ZTD anomaly sequence reconstruction.

3.4.2 ENSO Events Discriminatory Index

ENSO events consist of two components: El Niño (mainly for the ocean) and the Southern Oscillation (mainly for the atmosphere). These two components are manifestations of the same phenomenon in different media, and the two phenomena constitute a cyclic system. There is no significant pattern in the evolution of ENSO events, and there are differences between each event. The formation mechanism of the ENSO phenomenon is a question that has been explored by related scholars. The ENSO event discriminatory index has been evolving with the continuous improvement of monitoring tools and monitoring areas. The Southern Oscillation Index (SOI) was the most commonly used in the early days (McBride and Nicholls, 1983), which discriminated against ENSO events mainly from the atmospheric perspective. Trenberth (1997) later verified that using the sea surface temperature (SST) index for the Niño 3.4 region of the eastern Pacific was more accurate. Figure 3.6

shows the key monitoring areas for ENSO events, where the Niño $1+2$ area (0–10°S, 90–80°W) is the smallest of the monitoring areas. It is located in the coastal region of South America, where El Niño was first detected. The Niño 3 region (5°S–5°N, 150–90°W) used to be the main observational area for monitoring and predicting ENSO events. However, Trenberth found that the key region for ENSO event sea-air interactions is further west. The Niño 4 region (5°S–5°N, 160°E–150°W) captures the SST anomaly in the central equatorial Pacific more accurately. Therefore, the Niño 3.4 region (5°S–5°N, 170–120°W) becomes the main observation area. The National Oceanic and Atmospheric Administration (NOAA) defines the 3-month sliding average of the Sea Surface

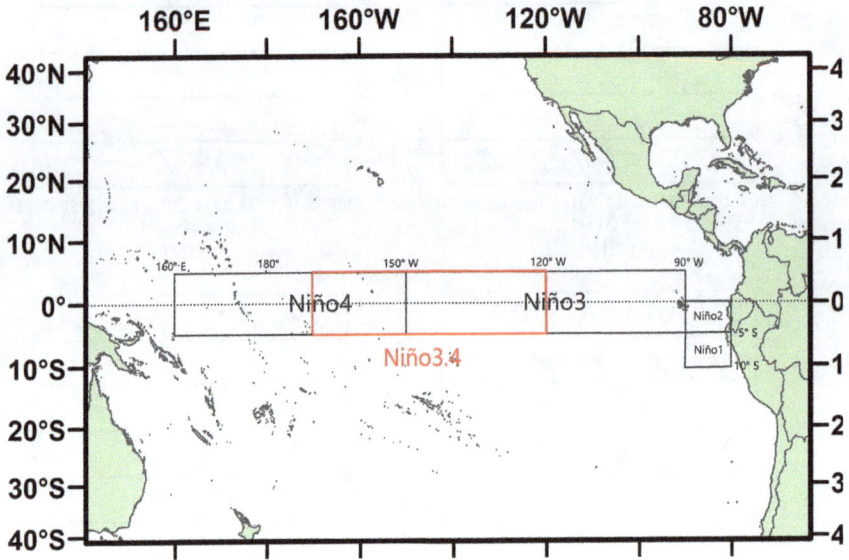

FIGURE 3.6   The sea surface temperature monitoring area.

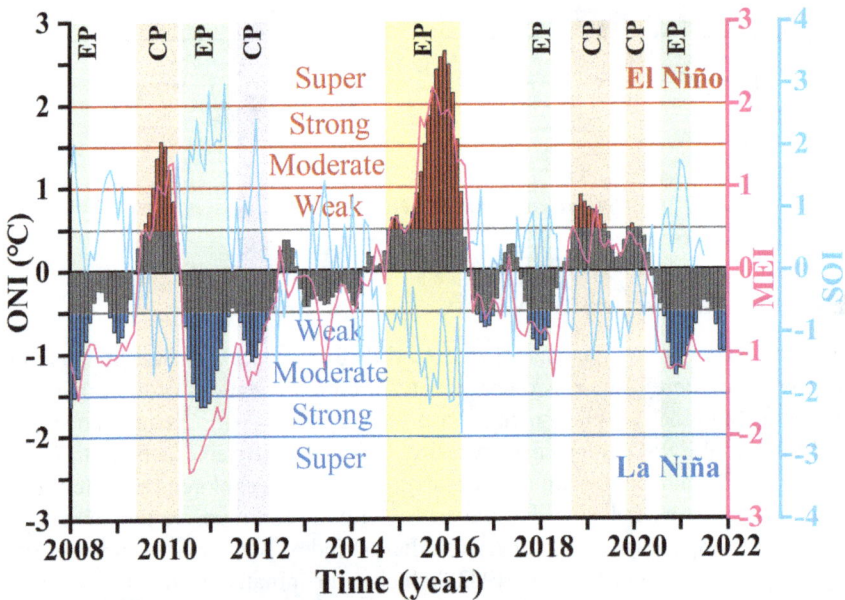

FIGURE 3.7   The time series of ENSO events discrimination index.

Temperature Anomaly (SSTA) in the Niño 3.4 region as the Oceanic Niño Index (ONI). Since the variation of Niño 3.4 SSTA and SOI are sometimes inconsistent, Wolter and Timlin proposed a Multivariate ENSO Index (MEI) (Wolter and Timlin, 1998). The index incorporates multiple atmospheric and oceanic meteorological elements to provide a more comprehensive characterization of ENSO events. A new version of MEI (MEI.v2) has now been created, which was obtained by principal component analysis of five variables: sea level pressure (SLP), sea surface temperature (SST), surface zonal winds (U), surface meridional winds (V), and outgoing longwave radiation (OLR). These five variables were obtained by principal component analysis. In this chapter, the three most commonly used ENSO event discriminant indices, SOI, ONI, and MEI, will be analyzed separately to screen out the index with the strongest correlation with GNSS ZTD in mainland China. NOAA provides SOI, ONI, and MEI data (https://psl.noaa.gov/enso/), all with a temporal resolution of 1 month (Figure 3.7). When SOI ≤ -0.5 for 5 consecutive months or more, an El Niño event was determined. When SOI ≥ 0.5 for 5 consecutive months or more, a La Niña event was determined. When ONI/MEI ≥ 0.5 for at least 5 consecutive months, it is judged as an El Niño event, and when ONI/MEI ≤ -0.5 for at least 5 consecutive months, it is judged as a La Niña event.

3.4.3 CONSTRUCTION OF GENERALIZED ADDITIVE MODELS

The time series of GNSS ZTD anomaly and ENSO event discriminant index have complex nonlinear variation characteristics. Therefore, the linear or nonlinear correlation characteristics between GNSS ZTD anomalies and various ENSO events in different regions of mainland China are determined based on generalized additive models (GAMs). The ENSO event discriminant index with the most significant correlation with the GNSS ZTD anomaly sequence was selected. GAMs can simultaneously fit explanatory variables with linear and complex nonlinear relationships with response variables into the model (Hastie and Tibshirani, 1986). Its core equation is:

$$g(u) = f_1(x_1) + f_2(x_2) + \cdots + f_i(x_i) + X_j\theta + \alpha \tag{3.37}$$

where u represents the expected value of the response variable, $g(u)$ represents the connection function, $x_1, x_2, \cdots, x_i$ represents the explanatory variable, and $f_i(x_i)$ is a smooth function of the linear or nonlinear relationship between the explanatory variable and the response variable.

The model does not require the analyst to pre-specify the form of the nonlinear relationship. It uses a smooth spline function to establish the relationship between the explanatory and response variables and can automatically select the appropriately segmented polynomial (He and Lin, 2017). The model was constructed using the mgcv package in R × 64 4.2.0 software (https://mirrors.bfsu.edu.cn/CRAN/). The three ENSO event discriminant indices ONI (°C), MEI, and SOI were used as explanatory variables. The GNSS ZTD anomaly in each region was used as the response variable to construct the GAMs. The degree of correlation between each explanatory variable and response variable was analyzed, and the analysis results are shown in Table 3.6. In the table, * * * indicates that the variable is significant at the 0.001 level. The effect graphs of ONI, MEI, and SOI on the variation of GNSS ZTD anomaly sequences in different regions were obtained (Figure 3.8).

In Table 3.6, the equivalent degree of freedom (edf) is the number of variables that are not restricted in their values when calculating a statistic. When the edf is 1, it means linear correlation; when the edf is greater than 1, it means nonlinear correlation. The F-value represents the test statistic set, and the larger the F-value indicates the more significant the relative importance of the influencing factors. The P-value represents the significance index, and the smaller the P-value indicates the more significant correlation. The adjusted coefficient of determination (R^2) is the ratio of the sum of squares of regression to the sum of squares of total deviations. The higher the R^2 and deviance explained indicates the better model fit and the stronger the correlation between the two.

TABLE 3.6

Analysis Results of GNSS ZTD Anomalies and ENSO Index GAMs Models in Each Region

Region	Explanatory variables	edf	F-value	P-value	Deviance explained (%)	R²
TPMZ	ONI	8.49	120.21	<2e-16 ***	49.11	0.4869
	MEI	8.23	123.59	<2e-16 ***	50.5	0.501
	SOI	8.56	50.39	<2e-16 ***	44.71	0.4426
SMZ	ONI	8.87	560.79	<2e-16 ***	46.43	0.4635
	MEI	8.9	824.8	<2e-16 ***	54.6	0.545
	SOI	8.89	490.87	<2e-16 ***	43.21	0.4313
TMZ	ONI	8.72	546.41	<2e-16 ***	47.97	0.4787
	MEI	8.96	615.4	<2e-16 ***	52	0.519
	SOI	8.87	468.69	<2e-16 ***	43.92	0.4382
TCZ	ONI	8.86	440.72	<2e-16 ***	51.6	0.514
	MEI	8.93	1508.3	<2e-16 ***	48.63	0.4854
	SOI	8.9	360.17	<2e-16 ***	45.63	0.4549
MPZ	ONI	8.43	325.87	<2e-16 ***	43.43	0.4331
	MEI	8.54	378.43	<2e-16 ***	43.89	0.4376
	SOI	8.66	230.78	<2e-16 ***	41.48	0.4135

In Figure 3.8, the blue shaded areas represent the upper and lower limits of the 95% confidence intervals. The solid line represents the smoothed fit curve of each explanatory variable to the GNSS ZTD anomaly series. The horizontal axis indicates the actual values of each explanatory variable, and the values in parentheses in the vertical axis indicate the edf. The analysis results show that all ENSO event discriminant indices significantly affect the GNSS ZTD anomaly variation in each region at the $P < 0.001$ level. This result indicates that ENSO events are statistically significant as explanatory variables for GNSS ZTD anomaly variation. ONI, MEI, and SOI are all non-linearly correlated with GNSS ZTD in each region (the edf greater than 1). The GAMs of MEI-GNSS ZTD anomaly for each region has the highest deviance explained and R^2, indicating that MEI is the best correlated with the GNSS ZTD anomaly in each area. Therefore, MEI is involved in further analytical studies as the best ENSO event discriminant index.

3.5 GNSS ZTD VARIATION AND RESPONSE TO ENSO

3.5.1 THE RESPONSE THRESHOLD OF GNSS ZTD ANOMALY TO ENSO

Quantifying the correlation between MEI and GNSS ZTD anomaly series is beneficial for a more effective analysis of the influence patterns of ENSO events on climate change in different regions of mainland China. Since the GNSS ZTD anomaly is non-linearly correlated with MEI, the correlation coefficients of the two are analyzed by moving the window correlation analysis (MWCA). The moving window size was first determined, and each independent window's local correlation coefficient was calculated. Eventually, a smoothed time series of correlation coefficients was generated (Zhao et al., 2020). The best common period between GNSS ZTD anomaly and MEI is used as a moving window to reduce the effect of temporal heterogeneity between MEI and GNSS ZTD anomaly sequences. The significant change periods of MEI and GNSS ZTD anomaly sequences were extracted separately using the FFT method, and a comparative analysis was carried out. The analysis results show no significant difference in each station's primary variation periods of GNSS

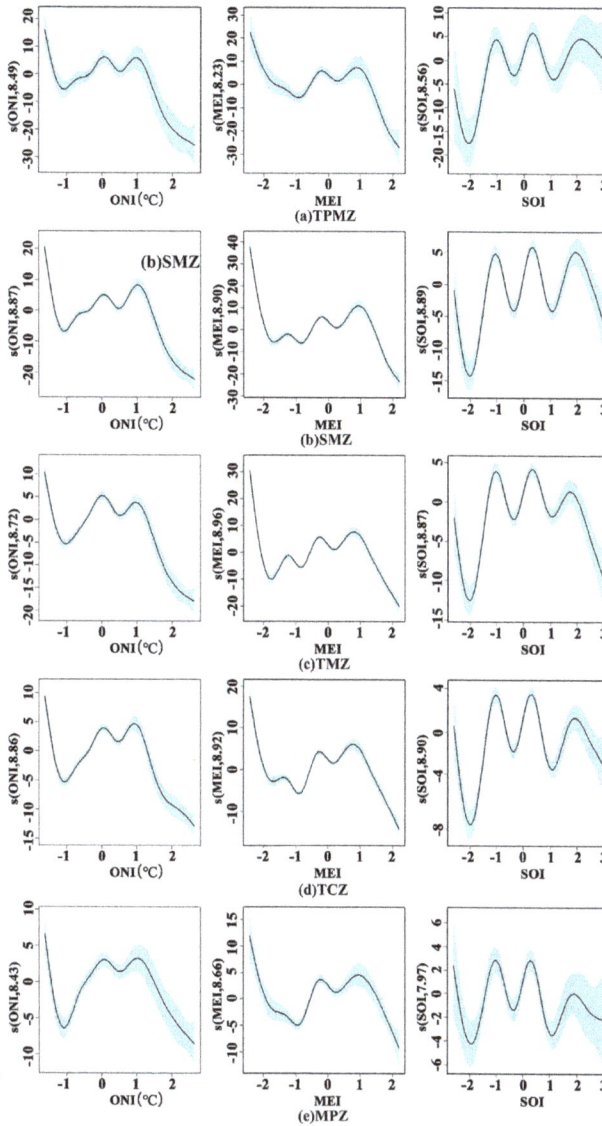

FIGURE 3.8 Correlation characteristics of ENSO discriminant index with GNSS ZTD anomaly in each region of mainland China.

ZTD anomaly sequences. Four evenly distributed stations were taken as examples, and the common period analysis results are displayed (Figure 3.9).

As can be seen from Figure 3.9, there are significant 90-month (corresponding to frequency 0.011 cpm), 40-month (corresponding to frequency 0.025 cpm), 24-month (corresponding to frequency 0.042 cpm), and 18-month (corresponding to frequency 0.056 cpm) significant variation periods of MEI. The GNSS ZTD anomaly sequences of each region also have the same variation period, the most significant of which is the 24-month variation period. The moving window correlation analysis of GNSS ZTD anomaly with MEI at each site was conducted to analyze the best common period using four common periods as the moving windows. The magnitudes of the average correlation coefficients of GNSS ZTD anomaly and MEI in each region under different moving windows were compared, and the analysis results are shown in Table 3.7.

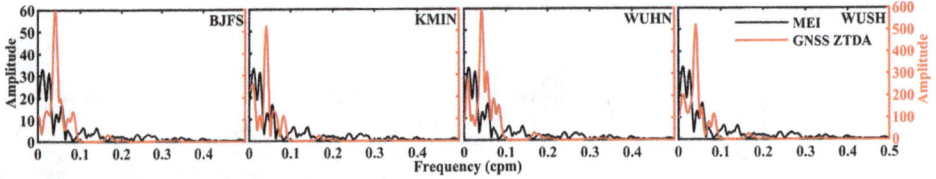

FIGURE 3.9   Common significant variation period of MEI and GNSS ZTD anomaly sequences.

TABLE 3.7
Comparison of the Correlation Coefficient between GNSS ZTD Anomaly and MEI under Different Moving Windows

Regions	90-Month	40-Month	24-Month	18-Month
TPMZ	0.234	0.332	0.463	0.577
SMZ	0.258	0.353	0.467	0.569
TMZ	0.266	0.356	0.452	0.559
TCZ	0.209	0.340	0.435	0.556
MPZ	0.234	0.349	0.451	0.556

The correlation coefficients in Table 3.7 all pass the 0.01 level of a significance test. The comparison results show that the correlation of each region under the 18-month moving window is better than that under other window lengths, and 18 months is the optimal common period. Therefore, the moving time window is set to 18 months. The results of sliding correlation analysis between GNSS ZTD anomaly series and MEI in each region are shown in Figure 3.10 (randomly selected uniformly distributed stations). Different types of ENSO events have different effects on the GNSS ZTD anomaly in various regions of mainland China. The EP-El Niño event positively affects the GNSS ZTD anomaly in tropical and subtropical monsoon zones. In the temperate monsoon, temperate continental, and mountain plateau zones, the GNSS ZTD anomaly has contrary responses to the development and recession years of the EP-El Niño event. The EP-El Niño event development year has a negative effect on the GNSS ZTD anomaly in these three regions. In contrast, the EP-El Niño recession year positively affects them. The presumed reason is that in the development year of the EP-El Niño event, the East Asian summer monsoon weakened, and the central monsoon rain belt in summer shifted southward. It leads to sufficient precipitable water vapor in the southern region and is prone to high temperature and drought in the northern region. In the decay year of the EP-El Niño event, the Western Pacific subtropical high (WPSH) is stronger and located southward. The westward shift of the WPSH transports precipitable water vapor from the Pacific Ocean to the southern and central-eastern regions of China. Under the influence of the East Asia-Pacific "remote correlation", the Asian west blocking high and east blocking high establish low-pressure troughs in the middle and high latitudes. It is conducive to continuously transporting Arctic Ocean precipitable water vapor to northwest China and north China. Therefore, the GNSS ZTD in China is constantly rising during this period (Wu et al., 2017; Zhai et al., 2016; Zhou et al., 2020b). The CP-El Niño event has a negative effect on the GNSS ZTD anomaly of temperate monsoon, temperate continental, tropical monsoon, and mountain plateau zones. It has a positive impact on the subtropical monsoon zone. The reason is the northward position of the WPSH during the CP-El Niño event, and a large amount of Pacific evaporative precipitable water vapor is transported to the subtropical monsoon zone (Chen et al., 2019; Liu et al., 2019; Yuan et al., 2012). During the EP-La Niña and CP-La Niña events, the GNSS ZTD anomaly in China showed positive responses to them. The presumed reason is that during the La Niña event, the SST in the equatorial eastern Pacific decreased, and

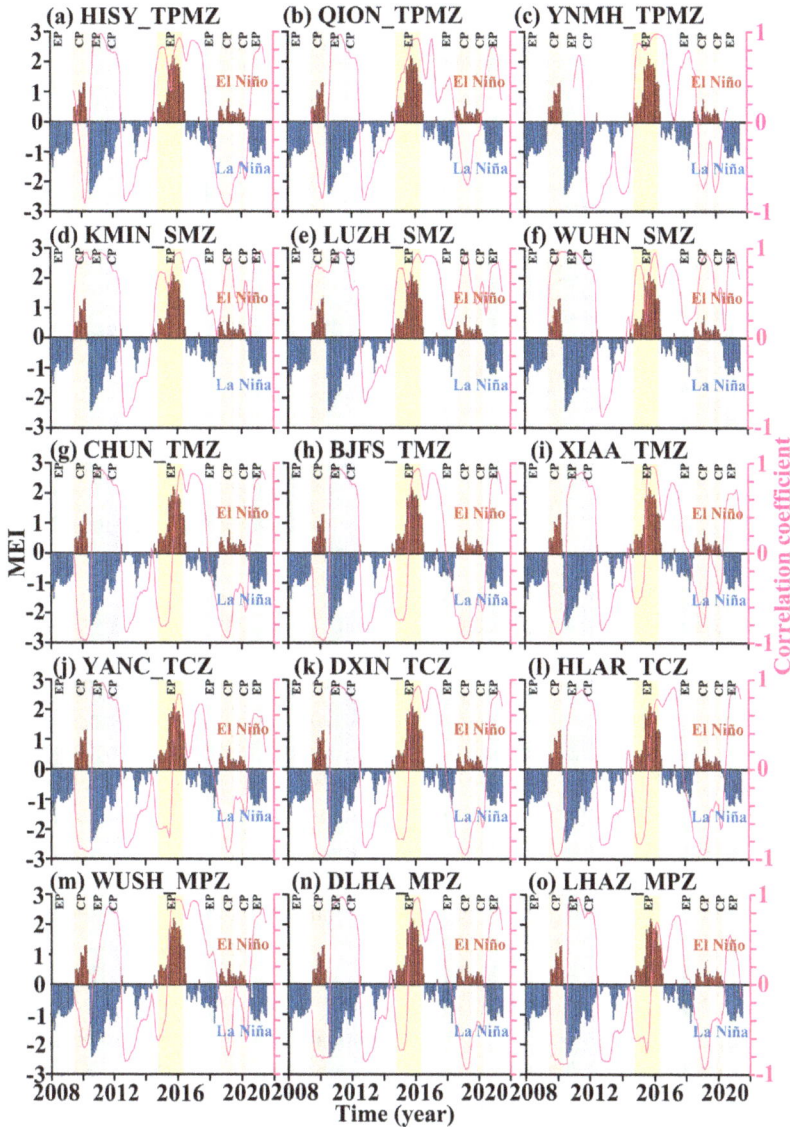

FIGURE 3.10 The moving window correlation analysis of GNSS ZTD anomaly and MEI in each region.

the current sea temperature in the western Pacific increased, resulting in a northward shift of the WPSH. The East Asian monsoon intensifies, and the major monsoon rain belts are northward, leading to abundant precipitable water vapor in temperate monsoon, temperate continental, and mountain plateau zones. Due to the strong cold air masses from Siberia, Mongolia is rapidly moving to the southern regions, so precipitable water vapor is continuously transported to southern China. The meeting of cold and warm air currents increases rain and snow in central China (Shi et al., 2022; Zhang, 2021). The precipitable water vapor and rain/snow content are closely related to GNSS ZTD.

This analysis shows that the GNSS ZTD anomaly in each region of mainland China has a more significant response to ENSO events. In this study, a linear fit of the correlation between MEI and GNSS ZTD anomaly series was used to quantify the effect of ENSO on GNSS ZTD anomaly in different regions of mainland China and to investigate the response threshold of GNSS ZTD anomaly to ENSO events. The correlation coefficients between MEI and GNSS ZTD anomaly

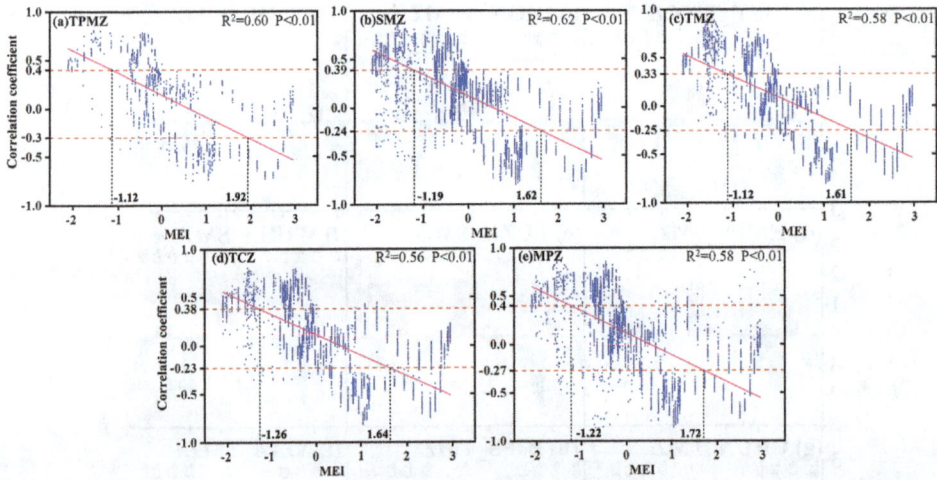

FIGURE 3.11   The MEI threshold for GNSS ZTD anomaly in each region response to ENSO events.

sequences were divided using the percentile method for quantitative analysis. The percentile method is a simple method commonly used to represent the distribution status of variables. The percentile method is commonly used to represent a variable's distribution status. The calculation process sorts the sample data from smallest to largest and calculates the corresponding cumulative percentile. At present, the 25-percentile and 75-percentile methods are widely used. The total sample size is divided into three parts, with the upper and lower percentiles being 75% and 25%, respectively. The middle 50% is the normal part (Zhao et al., 2020). Figure 3.11 shows the scatter plot of the correlation between MEI-GNSS ZTD anomaly and MEI. From the analysis results, it can be seen that different regions have different response thresholds to ENSO events. The MEI thresholds for GNSS ZTD anomaly in the tropical monsoon zone response to El Niño and La Niña events are -1.12 and 1.92, respectively. Therefore, the anomaly changes of GNSS ZTD in China's tropical monsoon zone when the MEI exceeds the range of (-1.12, 1.92) are influenced by El Niño and La Niña events. The MEI thresholds for GNSS ZTD anomaly in the subtropical monsoon zone response to El Niño and La Niña events are -1.12 and 1.61, respectively. When the MEI exceeds the range of (-1.12, 1.61), the anomaly changes of GNSS ZTD in the subtropical monsoon zone of China are influenced by ENSO events. Similarly, the MEI thresholds for the GNSS ZTD anomaly response to ENSO events are (-1.19, 1.62) for temperate monsoon climate regions. The MEI thresholds for the GNSS ZTD anomaly to ENSO events are (-1.26, 1.64) for temperate continental zones and (-1.22, 1.72) for mountain plateau zones.

3.5.2   The Response Pattern of GNSS ZTD Frequency Domain Variation to ENSO

In this chapter, the response of GNSS ZTD frequency domain oscillation characteristics to ENSO events is investigated to more comprehensively analyze the impact of ENSO events on GNSS ZTD in China. The ENSO event occurrence period and its corresponding normal climate period were intercepted from the complete time series, and the significant variation period amplitudes of GNSS ZTD were compared for each period. The change frequency (cpm) of each element was analyzed using FFT, and the change period (frequency) and amplitude were accurately extracted using the [pks, locs] function. Three normal climate periods and one ENSO event period were selected for comparative analysis to ensure the accuracy and reliability of the results. A total of nine ENSO events were included in this study period, and six ENSO events that occurred after 2010 were selected for analysis based on the completeness of GNSS ZTD data and the duration of ENSO events (Table 3.8).

TABLE 3.8
Screening Results of ENSO Event Periods and Normal Climate Periods

Event Type	Abnormal Climate	Normal Climate1	Normal Climate2	Normal Climate3
EP-El Niño	2015/5–2016/4	2012/5–2013/4	2013/5–2014/4	2016/5–2017/4
CP-El Niño-1	2018/9–2019/6	2012/9–2013/6	2013/9–2014/6	2016/9–2017/6
CP-El Niño-2	2019/11–2020/3	2012/11–2013/3	2013/11–2014/3	2016/11–2017/3
EP-El La Niña-1	2010/6–2011/5	2012/6–2013/5	2013/6–2014/5	2016/6–2017/5
EP-El La Niña-2	2020/8–2021/3	2012/8–2013/3	2013/8–2014/3	2016/8–2017/3
CP-El La Niña	2011/8–2012/3	2012/8–2013/3	2013/8–2014/3	2016/8–2017/3

3.5.2.1 The Effect of El Niño Events on the GNSS ZTD Significant Change Period

This study period includes three El Niño events, of which October 2014 to April 2016 is an EP-El Niño event period. Three normal climate periods and one EP-El Niño event period were selected for comparison and analysis. This EP-El Niño event lasts 19 months, and only one time series of normal climate corresponds to it in the studied time series range (2012/10–2014/4). We intercepted 12 months (2015/5–2016/4) forward and backwards from the peak time of this event to carry out the analysis to ensure the accuracy of the results. The significant variation periods of GNSS ZTD during four periods were analyzed using FFT to explore the differences and patterns during the EP-El Niño event compared with normal climate periods. Due to the intercept length of 12 months, the FFT results only show the GNSS ZTD significant variation periods within 9 months and less. These include the 9-month period (corresponding to a frequency of 0.11 cpm) and the significant variation period within 0.8–3 months (corresponding to a frequency interval of [0.3, 1.3]) as well as the daily and semi-diurnal waves (corresponding to frequencies of 1/30cpm and 1/60 cpm, respectively). The analysis results of three uniformly distributed stations in each climate type region were selected separately for presentation (Figures 3.12–3.16).

As seen from Figure 3.12, the amplitude of the 9-month significant variation period (corresponding to frequency 0.11 cpm) of GNSS ZTD in the TPMZ during the EP-El Niño has decreased to varying degrees compared to the three normal climate periods. The amplitude of the GNSS ZTD 0.9-month variation period for HISY and QION has increased to different degrees. The amplitude of the GNSS ZTD 2.5-month variation period at QION has increased. There is a significant increase in the amplitude of the 3-month and 1.5-month significant change periods for the GNSS ZTD at the YNMH site. The amplitude of the GNSS ZTD diurnal and semi-diurnal waves at the HISY site and the semi-diurnal wave amplitudes at the QION and YNMH sites decreased. The GNSS ZTD diurnal wave amplitude at the QION and YNMH sites increased.

From Figure 3.13, the amplitude of the 9-month significant variation period of GNSS ZTD in the subtropical monsoon climate region during the EP-El Niño event decreased to varying degrees compared to the three normal climate periods. The amplitudes of daily and semi-diurnal waves of GNSS ZTD at KMIN decreased, while those at LUZH and WUHN increased significantly.

According to Figure 3.14, the amplitude of the 9-month significant change period of GNSS ZTD in the TMZ during the EP-El Niño event is significantly lower than that of the three normal climate periods. In the frequency interval of (0.2, 1.4), the amplitude of GNSS ZTD significant variation during the EP-El Niño event has increased in different degrees compared with the normal time. During the EP-El Niño event, the 3-month period amplitudes of CHUN and BJFS are significantly higher, the 1.3-month period amplitude of XIAA is significantly higher, and the 0.8-month period amplitude of all three stations is enhanced. The diurnal wave

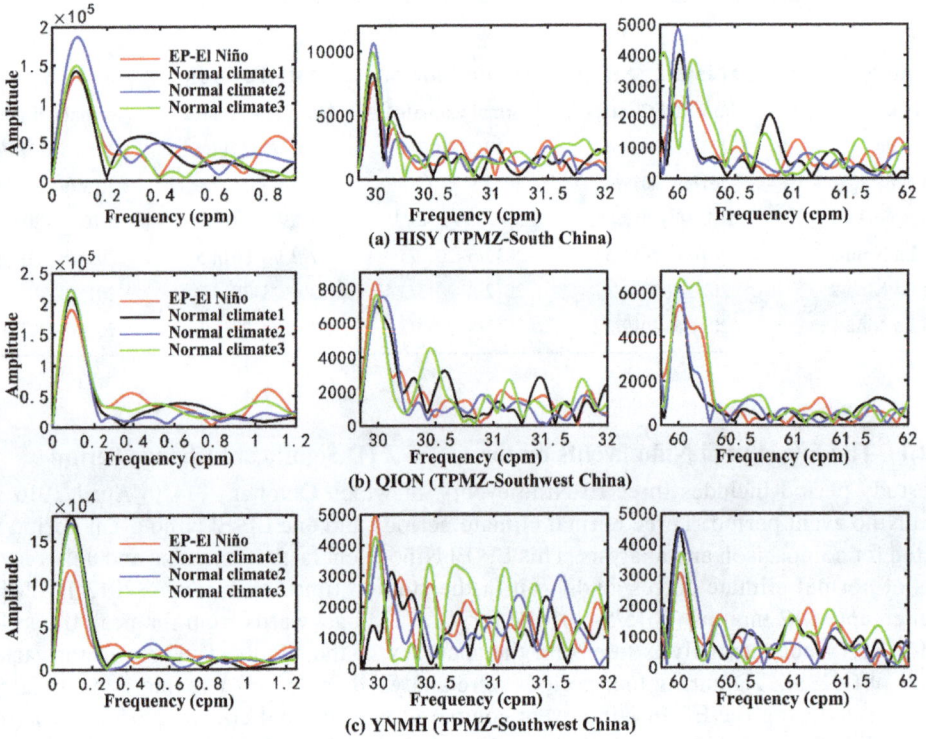

FIGURE 3.12 The effect of EP-El Niño on the GNSS ZTD significant change period in the TPMZ.

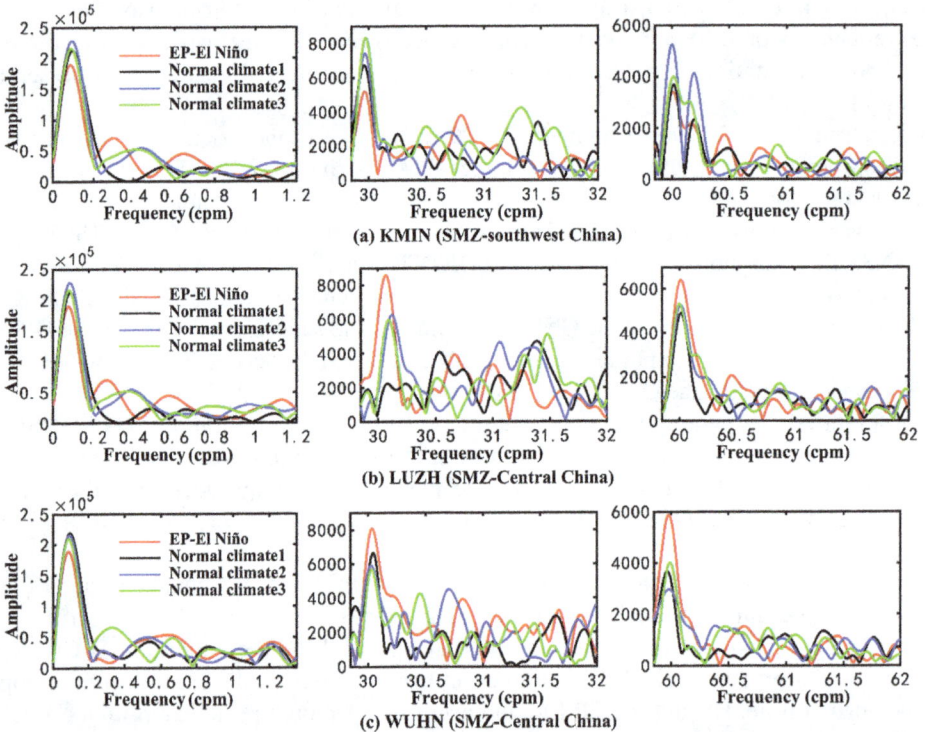

FIGURE 3.13 The effect of EP-El Niño on the GNSS ZTD significant change period in the SMZ.

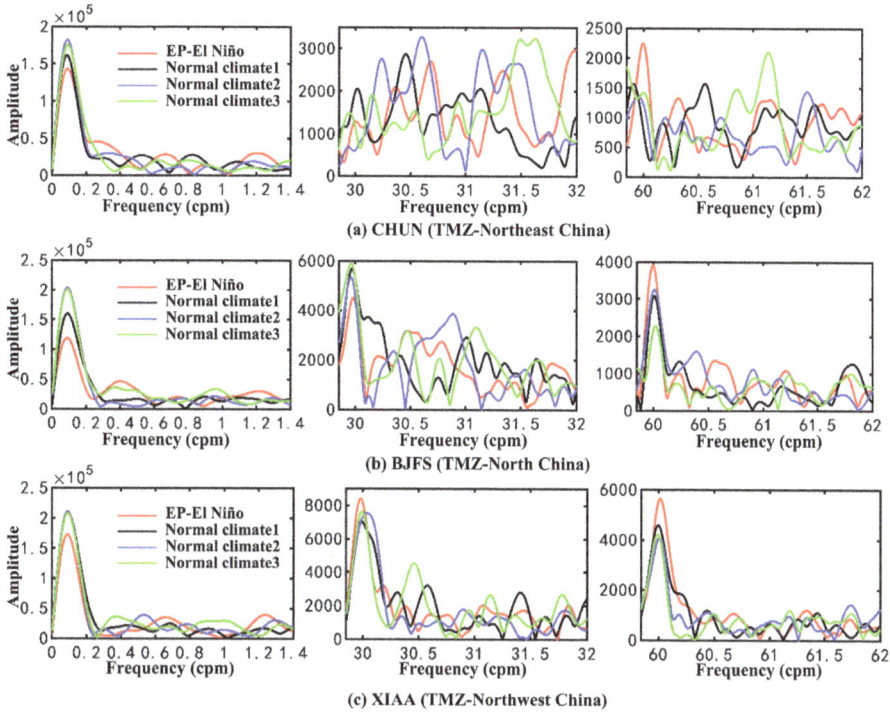

FIGURE 3.14 The effect of EP-El Niño on the GNSS ZTD significant change period in the TMZ.

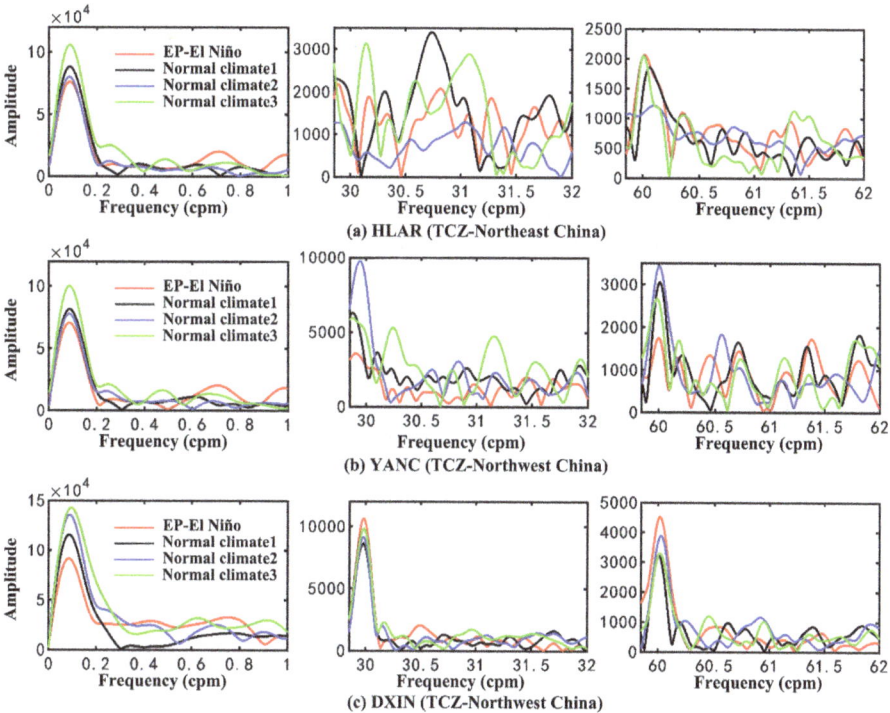

FIGURE 3.15 The effect of EP-El Niño on the GNSS ZTD significant change period in the TCZ.

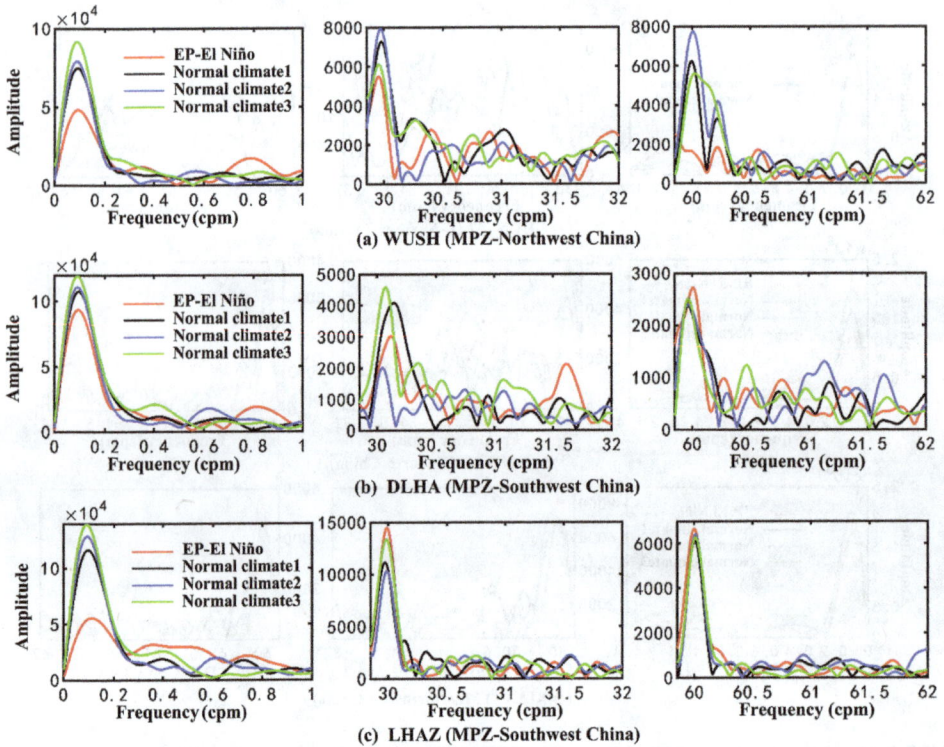

FIGURE 3.16 The effect of EP-El Niño on the GNSS ZTD significant change period in the MPZ.

amplitude decreases at BJFS and increases at XIAA, and the semi-diurnal wave amplitude increases at all three sites.

Figure 3.15 shows that the GNSS ZTD period variation patterns of individual stations in TCZ are relatively consistent. The amplitude of the significant variation period of the GNSS ZTD about 9 months during the EP-El Niño event is slightly lower than that of the three normal climate periods. In the frequency range of (0.6, 1), the amplitude of GNSS ZTD significant variation period during the EP-El Niño event has a more obvious increase compared with the normal climate period. The amplitudes of the 1.5-month and 1-month periods of the GNSS ZTD at each station are significantly higher. The amplitudes of the diurnal and semi-diurnal waves of the GNSS ZTD at the YANC station decrease, and the amplitudes of the diurnal and semi-diurnal waves of the GNSS ZTD at the DIXN station increase.

From Figure 3.16, the amplitude of the 9-month period of GNSS ZTD in the MPZ during the EP-El Niño event decreases to different degrees compared to the three normal climate periods. The amplitudes of the GNSS ZTD significant change period at the LHAZ site in the (0.3, 0.6) frequency interval are significantly higher. The 1.2-month period amplitudes of the GNSS ZTD at the WUSH and DLHA sites are significantly higher. The amplitudes of the diurnal and semi-diurnal periods of the GNSS ZTD at the LHAZ site and the semi-diurnal periods of the GNSS ZTD at the DLHA site increase to different degrees. The amplitudes of diurnal and semi-diurnal periods of GNSS ZTD at the WUSH site have decreased to various degrees. The results of all the analyzed sites in China are summarized as shown in Figure 3.17.

The analysis results of each station show that the amplitude of the 9-month significant variation period of GNSS ZTD in mainland China is reduced to different degrees under the influence of the EP-El Niño event. The amplitudes of the significant variation periods of 0.8, 1.2, 1.5, and 3 months have increased to different degrees. Only a few stations have no variation pattern, and these stations

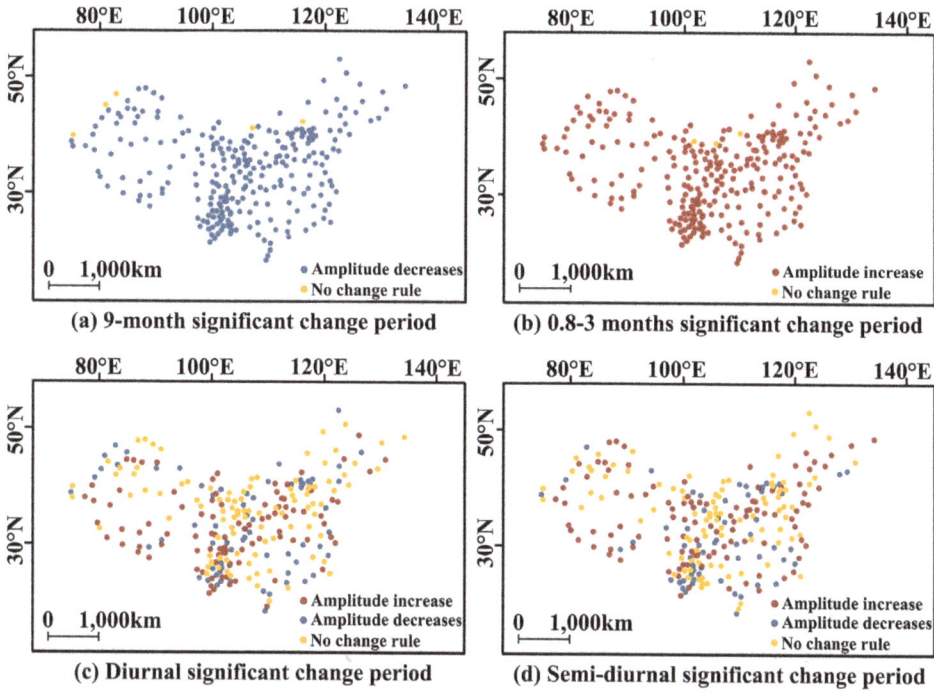

FIGURE 3.17   Response patterns of GNSS ZTD significant change period amplitudes to EP-El Niño events.

are located at higher latitudes in mainland China, where the GNSS ZTD (PWV) content is low, and the variation is not active. Moreover, during this EP-El Niño event, the northerly wind flow in the northern region was suppressed, which affected the normal circulation of precipitable water vapor in some regions (Liu et al., 2018). Some stations' GNSS ZTD diurnal and semi-diurnal periods respond to the EP-El Niño event, but there is no significant change pattern. Because the GNSS ZTD small-scale period is significantly affected by local climatic conditions and is prone to fluctuations and amplitude instability, the regional pattern of GNSS ZTD diurnal period and semi-diurnal period amplitude changes are not significant.

The study period contains two CP-El Niño events. For CP-El Niño-1, three normal climate periods, 2012/09–2013/06, 2013/09–2014/06, and 2016/09–2017/06, were screened for comparison and analysis. For the CP-El Niño-2, three normal climate periods, 2012/11–2013/03, 2013/11–2014/03, and 2016/11–2017/03, were selected for comparison with it. FFT analyzed the significant variation period of GNSS ZTD in each period, and the differences and patterns of GNSS ZTD time series oscillation characteristics between the CP-El Niño event and normal climate periods were investigated. The duration of the CP-El Niño-1 is 10 months. The FFT results can only show GNSS ZTD significant variation periods of 3 months and less and some linear trends of 9-month significant variation periods. The significant period of 3 months or less includes a 1.2–3 months significant period (corresponding to frequency interval [0.3, 0.8]) and diurnal and semi-diurnal periods (corresponding to frequencies of 1/30 cpm, 1/60 cpm, respectively). The duration of the CP-El Niño-2 is 5 months. The FFT results only show GNSS ZTD significant variation periods of 2 months and less, including 1.2–1.5 months significant variation periods (corresponding to the frequency interval of [0.6, 0.8]) as well as diurnal and semi-diurnal periods. The analysis results of the GNSS ZTD significant variation periods of some stations in different regions in response to the CP-El Niño-1 are presented in Figures 3.18–3.22.

Figure 3.18 shows that the amplitude of GNSS ZTD significant change period in the TPMZ during the CP-El Niño-1 event has different degrees of increase in the frequency interval of (0.4,

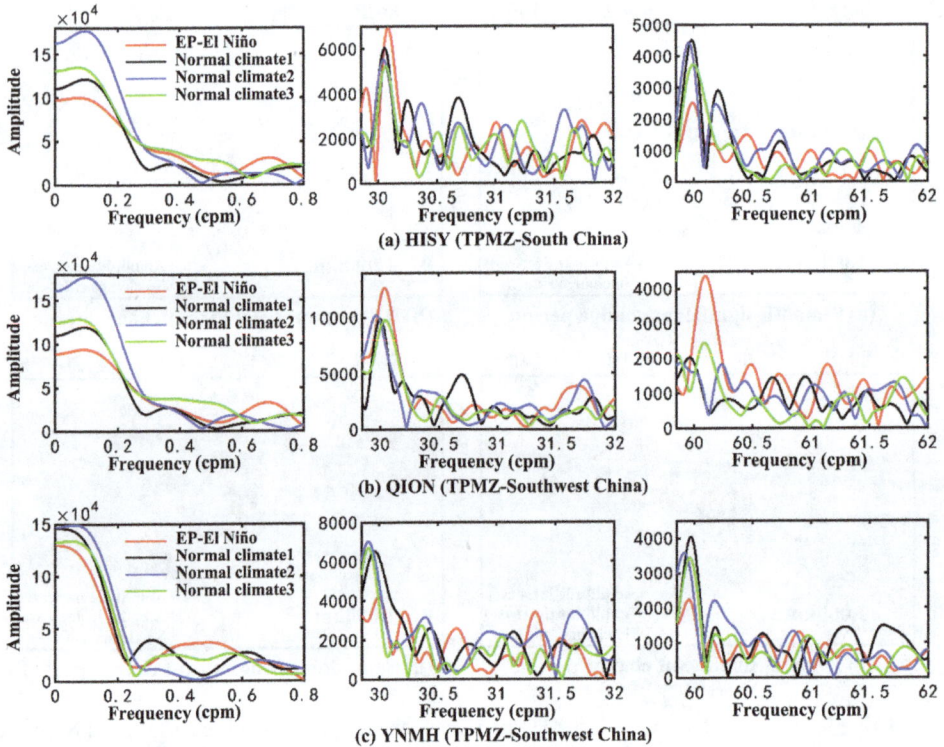

FIGURE 3.18 The effect of CP-El Niño-1 on the GNSS ZTD significant change period in the TPMZ.

0.7) compared with the normal climate. The 1.4 monthly cycles amplitude of GNSS ZTD at HISY and QION sites increases. The amplitude of GNSS ZTD diurnal waves at HISY and QION sites increases. The amplitude of the GNSS ZTD diurnal wave at the YNMH site decreases. The amplitude of the GNSS ZTD semi-diurnal wave at the QION site increases while that of all other sites decreases. Due to the limitation of time series length, the 9-month significant variation period of GNSS ZTD cannot be shown completely. However, the trend lines of the 9-month significant variation period can be judged. The amplitude of the GNSS ZTD 9-month significant variation period in the TPMZ decreases during CP-El Niño-1.

According to Figure 3.19, the amplitude of GNSS ZTD significant change period in the SMZ during the CP-El Niño-1 event has different degrees of increase in the frequency interval of (0.4, 0.7) compared with the normal climate. The 2-month period amplitude of GNSS ZTD at KMIN and WUHN sites increased. The 2.5-month period amplitude of GNSS ZTD at LUZH and KMIN sites increased. The 1.4-month period amplitude of the GNSS ZTD at the LUZH site increases. The diurnal period amplitude of the GNSS ZTD at the WUHN site decreases, while there is no significant change at other sites. The semi-diurnal period amplitude decreases at all three sites. The amplitude of the 9-monthly significant variation period at each site decreases as judged by the partial trend line of the 9-monthly period of GNSS ZTD.

As seen in Figure 3.20, the amplitude in the (0.3, 0.7) frequency interval of the GNSS ZTD in the TMZ during the occurrence of CP-El Niño-1 has increased to different degrees compared to the normal climate. The 2-month period amplitude of the GNSS ZTD at the CHUN and BJFS sites is significantly increased. The 3-month period amplitude of the GNSS ZTD at the XIAA site is significantly raised. The daily period amplitude of GNSS ZTD increases at the CHUN site. The diurnal and semi-diurnal periods amplitude of GNSS ZTD decreases to different degrees at BJFS and

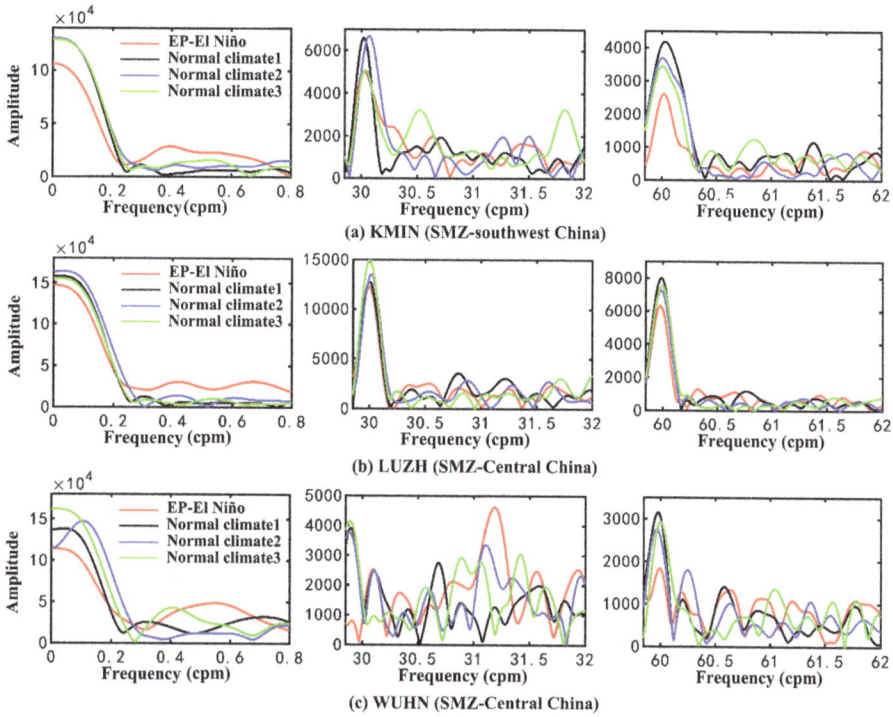

FIGURE 3.19 The effect of CP-El Niño-1 on the GNSS ZTD significant change period in the SMZ.

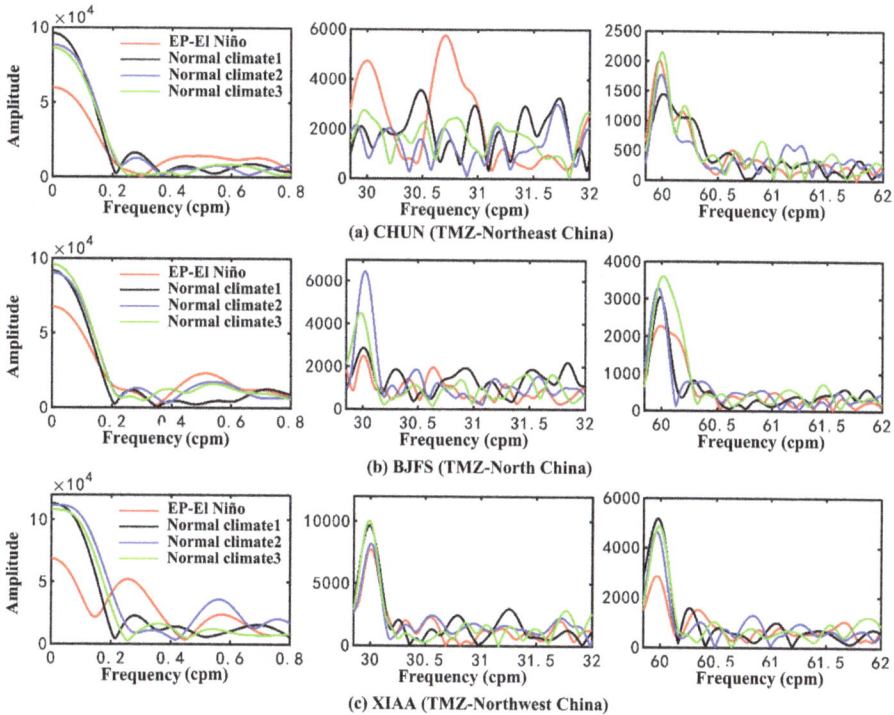

FIGURE 3.20 The effect of CP-El Niño-1 on the GNSS ZTD significant change period in the TMZ.

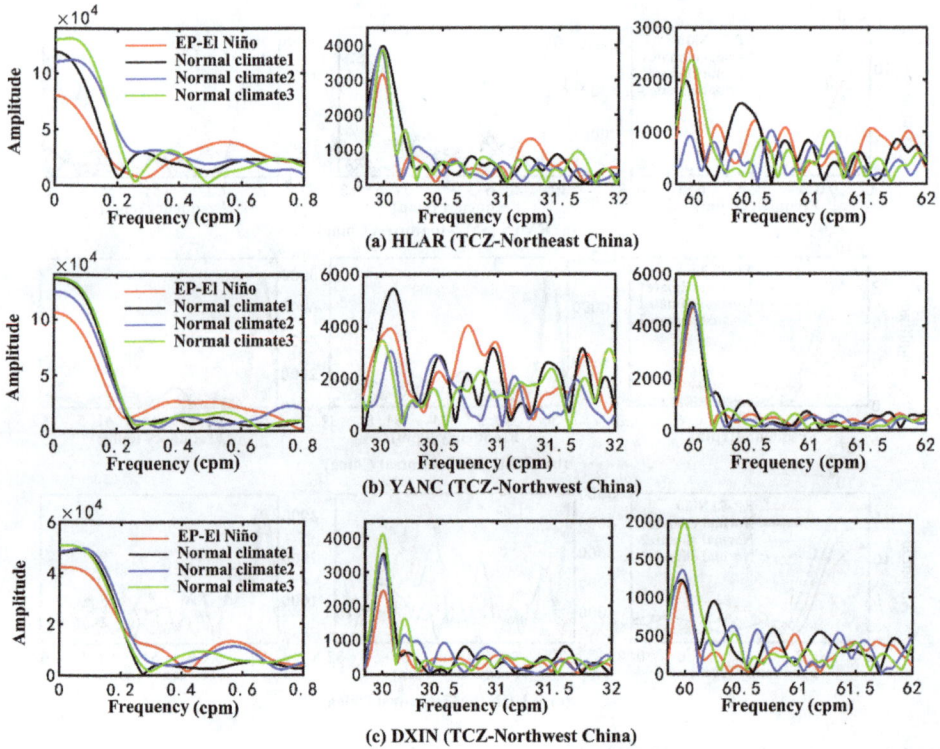

FIGURE 3.21   The effect of CP-El Niño-1 on the GNSS ZTD significant change period in the TCZ.

XIAA sites. The 9-monthly significant change period amplitude of the GNSS ZTD in this region decreases as judged by the partial trend line of the 9-monthly significant change period.

As shown in Figure 3.21, the amplitude in the (0.3, 0.7) frequency interval of the GNSS ZTD in the TCZ during the occurrence of CP-El Niño-1 has increased to different degrees compared to the normal climate period. The 1.8-month period amplitude of the GNSS ZTD at all three sites increases. The daily period amplitude of the GNSS ZTD at the DXIN and HLAR sites decreases. The GNSS ZTD semi-diurnal period amplitude at the HLAR site increases, the GNSS ZTD semi-diurnal period amplitude at the YANC and DXIN sites does not change significantly. The 9-monthly significant change period amplitude of the GNSS ZTD in this region decreases as judged by the partial trend line of the 9-monthly significant change period.

Figure 3.22 shows that the GNSS ZTD significant change period amplitude in the (0.3, 0.7) frequency interval during the CP-El Niño-1 has increased to different degrees than the normal period in the MPZ. The 3-month and 1.7-month period amplitudes of GNSS ZTD at the WUSH site are significantly higher. The 1.4-month period amplitude at the DLHA and LHAZ sites significantly increases. The diurnal and semi-diurnal amplitudes of the GNSS ZTD at the WUSH and LHAZ sites decreased to varying degrees. The semi-diurnal wave amplitude of the GNSS ZTD at the DLHA did not change significantly. The amplitude of the 9-month significant variation period of GNSS ZTD at the WUSH site decreases as judged by the partial trend lines of the 9-month significant variation period. The same analysis is done for the CP-El Niño-2. The results of the analysis of the effects of the two CP-El Niño events on the GNSS ZTD significant variation periods of all stations in China are summarized in Figures 3.23–3.24.

According to Figures 3.23–3.24, under the influence of the CP-El Niño event, the amplitude of the 9-month significant variation period of the GNSS ZTD decreases at most stations in the mainland China region, while the amplitude of the significant variation cycle within 1.2–3 months increases to

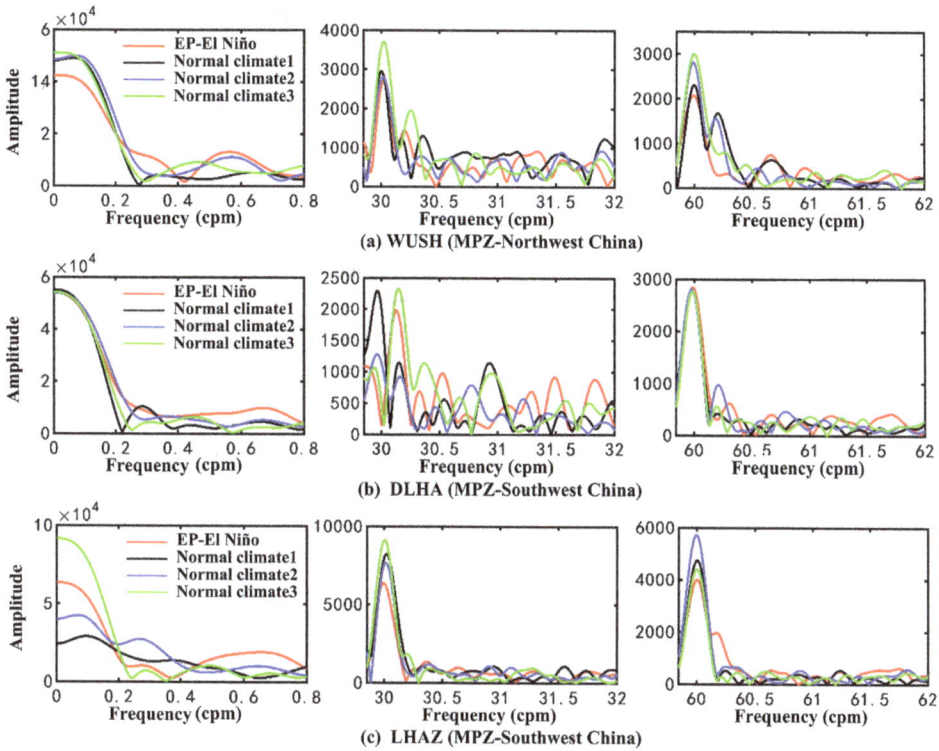

FIGURE 3.22 The effect of EP-El Niño on the GNSS ZTD significant change period in the MPZ.

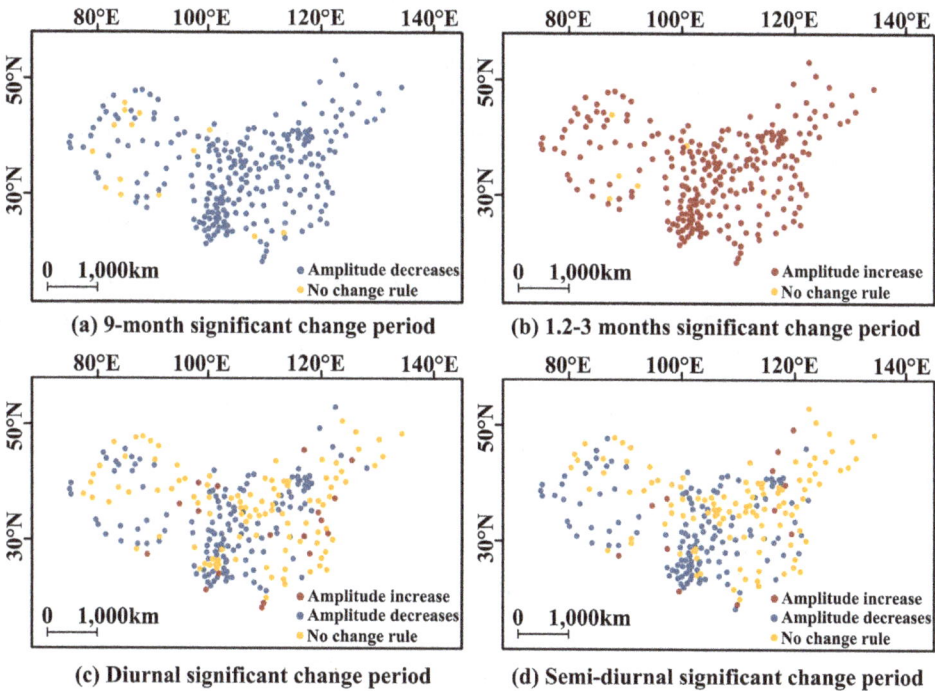

FIGURE 3.23 Response patterns of GNSS ZTD significant change period amplitudes to CP-El Niño-1 events.

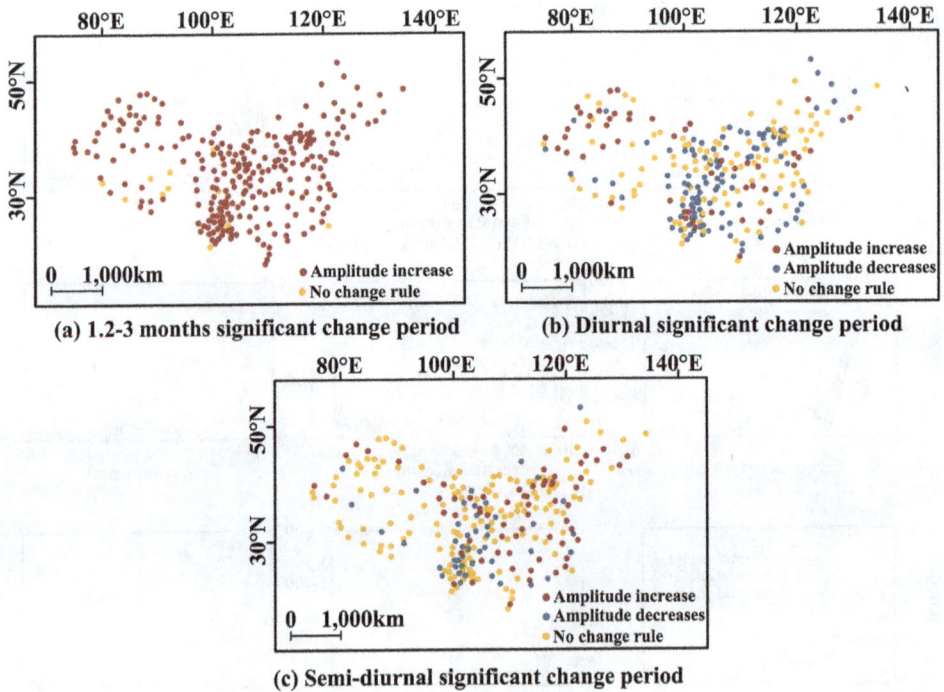

(a) 1.2-3 months significant change period

(b) Diurnal significant change period

(c) Semi-diurnal significant change period

FIGURE 3.24 Response patterns of GNSS ZTD significant change period amplitudes to CP-El Niño-2 events.

different degrees. Only some stations have no variation pattern, mainly concentrated in the MPZ. The amplitude changes of GNSS ZTD diurnal and semi-diurnal periods are varied in response to the two CP-El Niño events. Under the influence of the CP-El Niño-1, the amplitude of GNSS ZTD diurnal and semi-diurnal waves decreases at most sites in the MPZ and Sichuan-Yunnan regions, while there is no obvious change pattern in other areas. Under the influence of the CP-El Niño-2, the GNSS ZTD diurnal period amplitude decreases at some stations in the Sichuan-Yunnan region and southeast coastal area and increases at some stations in northwest China. There is no apparent regional pattern in the semi-diurnal period amplitude change. The CP-El Niño-2 has a shorter duration and weaker magnitude and has no significant effect on most sites' small-scale variation period of GNSS ZTD.

3.5.2.2 The Effect of La Niña Events on the GNSS ZTD Significant Change Period

The effects of two EP-La Niña events and one CP-La Niña event on the significant change periods of GNSS ZTD were analyzed separately according to the analysis methods described in the previous section. The results are shown in Figures 3.25–3.27. The EP-La Niña-1 is 12 months, and the FFT results only indicate the 9-month significant change periods of GNSS ZTD and less. Among them, the significant variation periods that exist in response to ENSO events include a 9-month period (corresponding to the frequency of 0.11 cpm) and 1–3-month significant variation periods (corresponding to the frequency interval of [0.3, 1]) as well as diurnal and semi-diurnal periods (corresponding to frequencies of 1/30 cpm and 1/60 cpm, respectively). The EP-La Niña-2 event is 8 months, and the FFT results show only significant change periods of 3 months and less and a partial linear trend of 9-month significant change periods. The significant change periods that respond to ENSO events include 1–3 months (corresponding to the frequency interval [0.3, 1]) and diurnal/semi-diurnal periods. The CP-La Niña event has a duration of 8 months, and the FFT results only show the significant change periods of 3 months and less and a partial linear trend of 9-month significant change periods. The significant variation periods in response to ENSO events include 1.2–3-month significant variation periods (corresponding to a frequency interval of [0.3, 0.8]) and diurnal and semi-diurnal periods (corresponding to frequencies of 1/30 cpm and 1/60 cpm, respectively).

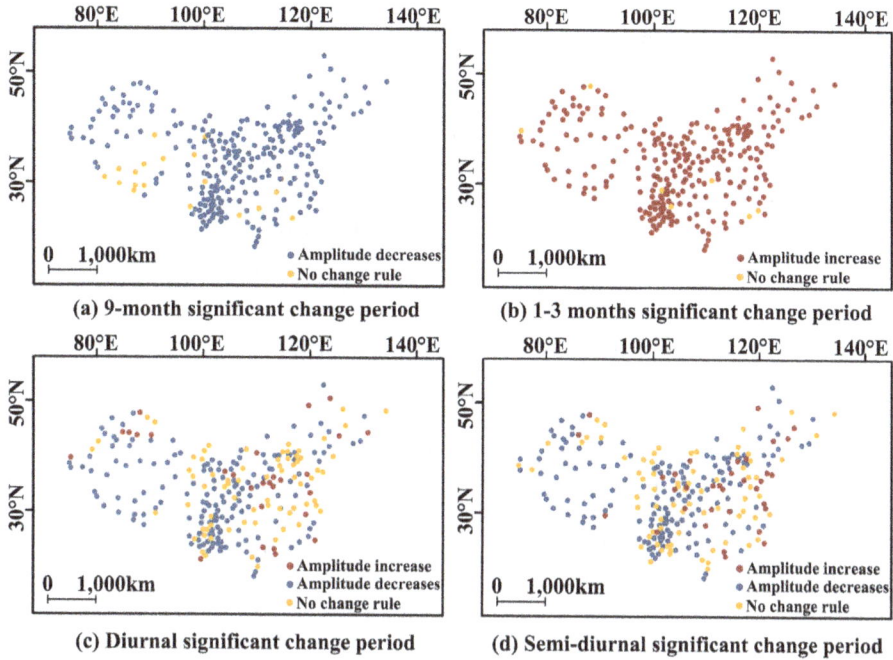

FIGURE 3.25 Response patterns of GNSS ZTD significant change period amplitudes to EP-La Niña-1 events.

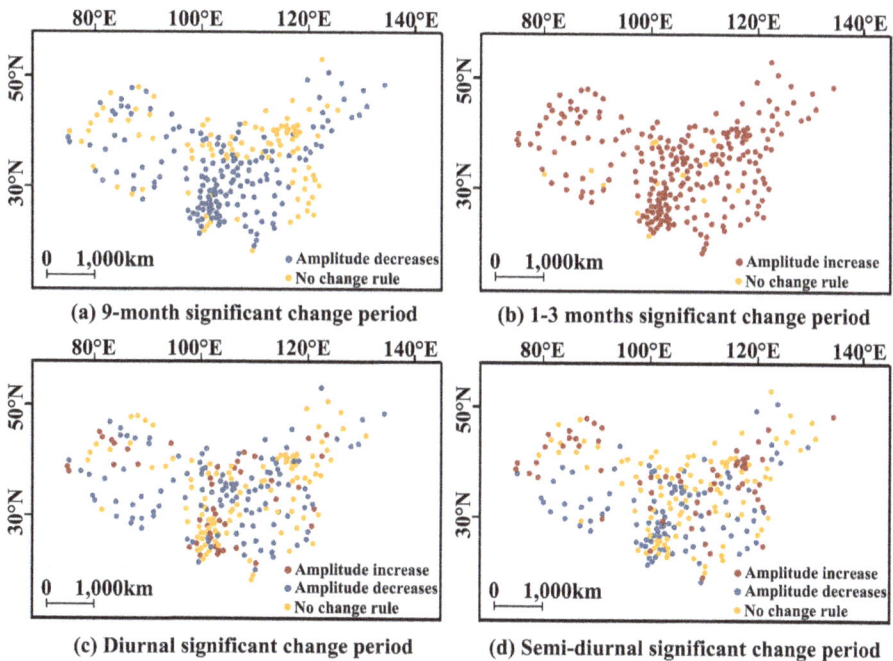

FIGURE 3.26 Response patterns of GNSS ZTD significant change period amplitudes to EP-La Niña-2 events.

According to Figure 3.25, under the EP-La Niña-1, the amplitude of the 9-month significant variation period of GNSS ZTD decreases at most sites in mainland China. In contrast, the amplitude of the significant variation period within 1–3 months increases to different degrees. There is no variation pattern for individual stations in the SMZ and MPZ. The precipitable water vapor base in the SMZ is large and active, which causes the GNSS ZTD significant change period amplitude at individual stations to show no change pattern. The amplitude of GNSS ZTD diurnal and semi-diurnal periods decreases at most stations and shows a certain regional pattern in the MPZ and Sichuan-Yunnan regions.

As shown in Figure 3.26, under the influence of the EP-La Niña-2, the amplitude of the 9-month significant change period of the GNSS ZTD decreases at most stations in main-land China, and the amplitude of the significant change period within 1–3 months increases. Compared with the EP-La Niña-1, the EP-La Niña-2 event is shorter in duration. It has a weaker impact on the 9-month significant period in China, showing a larger number of stations with no variation pattern. The sites with various patterns of GNSS ZTD frequency domain oscillations all show a decrease in amplitude, which is consistent with the analysis results of the EP-La Niña-1. The diurnal and semi-diurnal period amplitudes of the GNSS ZTD decrease at most stations in the southern mountain plateau zone and subtropical monsoon zone, while there is no significant variation pattern in other zones.

Figure 3.27 shows that under the influence of the CP-La Niña event, the 9-month significant change period amplitude of the GNSS ZTD decreases at most stations in mainland China. In contrast, the amplitude of the significant change period within 1.2–3 months increases to differ-ent degrees. The significant period amplitudes of GNSS ZTD in the subtropical monsoon coastal zone, the central temperate continental zone, and some stations in the mountain plateau zone have no variation pattern. Compared with other types of ENSO events, the effect of the studied CP-La Niña events on the 9-month and 1.2–3-month periods of GNSS ZTD is weaker, and there

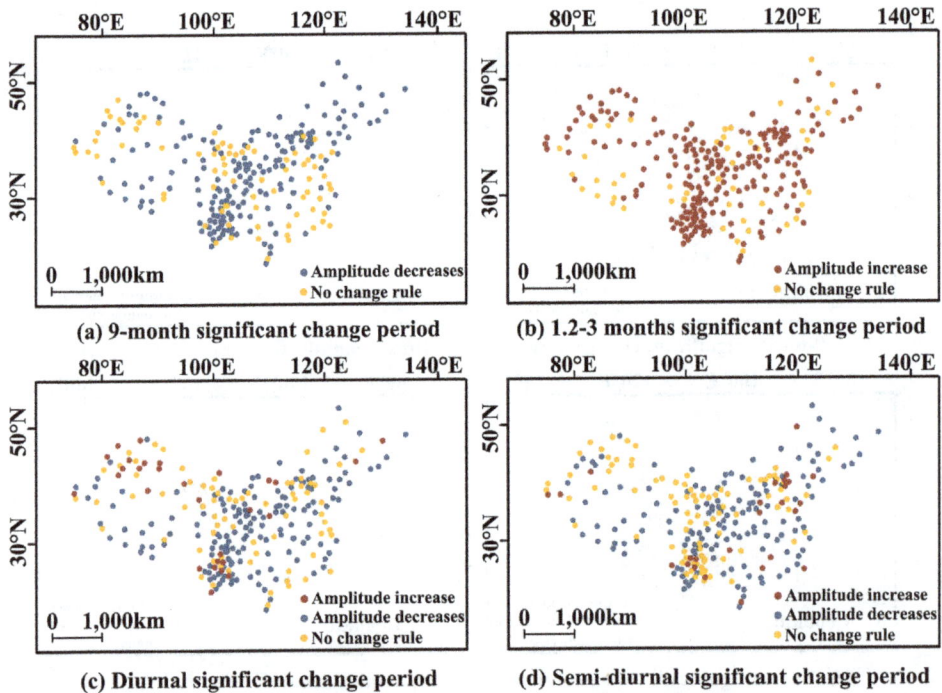

(a) 9-month significant change period

(b) 1.2-3 months significant change period

(c) Diurnal significant change period

(d) Semi-diurnal significant change period

FIGURE 3.27 Response patterns of GNSS ZTD significant change period amplitudes to CP-La Niña events.

are more stations with no variation pattern. The diurnal and semi-diurnal waves of GNSS ZTD decrease in most stations, and the amplitude of individual stations increases, with no apparent regional pattern.

In summary, the occurrence of the ENSO event has an impact on the 9-month significant variation period term, the 0.8–3-month significant variation period term, and the diurnal and semi-diurnal periods of the GNSS ZTD in China. The 9-month significant variation period amplitude of the GNSS ZTD decreases, and the 0.8–3-month significant variation period amplitude of the GNSS ZTD increases. The variability of the diurnal and semi-diurnal periods is not significant.

3.6 CONCLUSION

GNSS ZTD is a good indicator of climate change. This chapter systematically analyzed the spatial-temporal variation characteristics of GNSS ZTD in mainland China and used it to indicate the influence regularity of ENSO events on climate change in different regions of mainland China. In the spatial domain, the GNSS ZTD shows an increasing trend from northwest to southeast. The regional pattern is more consistent with the five major climate-type subdivisions in mainland China. In the time domain, the GNSS ZTD has significant annual periods, 9-month periods, semi-annual periods, seasonal periods and diurnal/semi-diurnal waves, and other considerable change periods. The response patterns of GNSS ZTD time-frequency variation characteristics to ENSO events in mainland China are analyzed and obtained. Different MEI thresholds exist for GNSS ZTD response to ENSO events in different climate type regions of mainland China. The MEI thresholds for the response of GNSS ZTD anomalies to ENSO events are (−1.12, 1.92) in the tropical monsoon zone, (−1.12, 1.61) in the subtropical monsoon zone, (−1.19, 1.62) in the temperate monsoon zone, (−1.26, 1.64) in the temperate continental zone, and (−1.22, 1.72) in the mountain plateau zone. Moreover, the occurrence of ENSO events has an impact on the significant change cycle of GNSS ZTD in China, which will lead to a decrease in the amplitude of the 9-month significant change cycle of GNSS ZTD and an increase in the amplitude of the significant change period in 0.8–3 months. ENSO events also impact the small-scale periods (diurnal and semi-diurnal waves) of GNSS ZTD, but the amplitude changes have no significant regularity. This study can provide a reference for relevant meteorological monitoring departments and disaster prevention and control departments.

3.6.1 ACKNOWLEDGMENT

We would like to thank the First Monitoring and Application Center, China Earthquake Administration for providing CMTEMN zenith tropospheric delays, and the National Oceanic and Atmospheric Administration for providing MEI, SOI, and ONI (https://psl.noaa.gov/enso/).

REFERENCES

Barindelli S, Realini E, Venuti G et al. (2018) Detection of water vapor time variations associated with heavy rain in northern Italy by geodetic and low-cost GNSS receivers. *Earth, Planets and Space* 70(1): 1–18. https://doi.org/10.1186/s40623-018-0795-7.

Bevis M, Businger S, Chiswell S et al. (1994) GPS meteorology: mapping Zenith wet delays onto precipitable water. *Journal of Applied Meteorology and Climatology* 3(33): 379–386. http://www.jstor.org/stable/26186685.

Bevis M, Businger S, Herring TA et al. (1992) GPS meteorology: Remote sensing of atmospheric water vapor using the global positioning system. *Journal of Atmospheric and Oceanic Technology* 97(D14): 15787–15801. https://doi.org/10.1029/92JD01517.

Cai F, Sun FP, Dai HL et al. (2019) Application of wavelet and Fourier transform in time series analysis. *GNSS World of China* 44(4): 40–46. https://doi.org/10.13442/j.gnss.1008-9268.2019.04.006

Cao L, Sun CH, Ren FM et al. (2013) Study of a comprehensive monitoring index for two types of ENSO events. *Journal of Tropical Meteorology* 29(1): 66–74. https://doi.org/10.3969/j.issn.1004-4965.2013.01.008.

Chen M, Li T, Wang X (2019) Asymmetry of atmospheric responses to two-type El Niño and La Niña over northwest pacific. *Journal of Meteorological Research* 33(5): 826–836. https://doi.org/10.1007/s13351-019-9022-0.

Chen W, Ding SY, Feng J et al. (2018) Progress in the study of impacts of different types of ENSO on the East Asian monsoon and their mechanisms. *Chinese Journal of Atmospheric Sciences* 42(3): 640–655. https://doi.org/10.3878/j.issn.1006-9895.1801.17248.

Foster J, Bevis M, Schroeder T et al. (2000) El Niño, water vapor, and the global positioning system. *Geophysical Research Letters* 27(17): 2697–2700. https://doi.org/10.1029/2000GL011429.

Giannaros C, Kotroni V, Lagouvardos K et al. (2020) Assessing the impact of GNSS ZTD data assimilation into the WRF modeling system during high-impact rainfall events over Greece. *Remote Sensing* 12(3): 383. https://doi.org/10.3390/rs12030383.

Guo M, Zhang HW, Xia PF (2021) Analysis of short-time weather forecast based on GNSS zenith troposphere delay. *Science of Surveying and Mapping* 46(4): 28–36. https://doi.org/10.16251/j.cnki.1009-2307.2021.04.005.

Hannachi A, Jolliffffe IT, Stephenson DB (2007) Empirical orthogonal functions and related techniques in atmospheric science: A review. *International journal of climatology* 27(9): 1119–1152. https://doi.org/10.1002/joc.1499.

Hannachi A, Neill A (2001) Atmospheric multiple equilibria and non-Gaussian behavior in model simulations. *Quarterly Journal of the Royal Meteorological Society* 127: 939–958. https://doi.org/10.1002/qj.49712757312.

Hastie TJ, Tibshirani RJ (1986) Generalized additive models. *Statistical Science* 1: 297–310.

He SY, Li GY, Liu HJ et al. (2019) Application and research of optimal wavelet base selection method in seismic data processing. *South China Journal of Seismology* 39(3): 49–56. https://doi.org/10.13512/j.hndz.2019.03.007.

He X, Lin ZS (2017) Interactive effects of the influencing factors on the changes of PM2.5 concentration based on GAM model. *Environmental Science* 38(1): 22–32. https://doi.org/10.13227/j.hjkx.201606061.

Isioye OA, Combrinck L, Botai J (2018) Evaluation of spatial and temporal characteristics of GNSS-derived ztd estimates in nigeria. *Theoretical and Applied Climatology* 132(3–4): 1099–1116. https://doi.org/10.1007/s00704-017-2124-7.

Jiang QC (2000) *The Research on Tropospheric Delay Modeling Method of GNSS Network RTK and Software Development.* Master Thesis, Wuhan University. https://doi.org/10.27379/d.cnki.gwhdu.2020.000380.

Jin SG, Han L, Cho J et al. (2011) Lower atmospheric anomalies following the 2008 Wenchuan Earthquake observed by GPS measurements. *Journal of Atmospheric and Solar-Terrestrial Physics* 73(7–8): 810–814. https://doi.org/10.1016/j.jastp.2011.01.023.

Kao HY, Yu JY (2009) Contrasting Eastern-Pacific and Central-Pacific types of ENSO. *Journal of Climate* 22: 615–632. https://doi.org/10.1175/2008JCLI2309.1.

Lei JH, Huang CJ (2018) Temperature, precipitation and response to ENSO in different regions of China in the past 30 years. *Tropical Geomorphology* 39(2): 12–19. https://doi.org/CNKI:SUN:RDDM.0.2018-02-002.

Li L, Song Y, Zhou JL (2020) Preliminary exploration of GNSS meteorological elements using wavelet transform for rainstorm prediction. *Journal of Geodesy and Geodynamics* 40(3): 225–230. http://doi.org/10.14075/j.jgg.2020.03.002

Liu K, Chen J, Yang R (2019) Connection between two leading modes of autumn rainfall inter-annual variability in southeast china and two types of ENSO-like SSTA. *Advances in Meteorology* (5): 1–14. https://doi.org/10.1155/2019/1762505.

Liu MH, Ren HL, Zhang WJ et al. (2018) Influence of super El Niño events on the frequency of spring and summer extreme precipitation over eastern China. *Acta Meteorologica Sinica* 76(4): 539–553. http://doi.org/10.11676/qxxb2018.021.

McBride JL, Nicholls N (1983) Seasonal relationships between Australian rainfall and the Southern Oscillation. *Monthly Weather Review* 111(10): 1998–2004. https://doi.org/10.1175/1520-0493(1983)111<1998:SRBARA>2.0.CO;2

Niell AE (2001) Preliminary evaluation of atmospheric mapping functions based on numerical weather models. *Physics and Chemistry of the Earth, Part A: Solid Earth and Geodesy* 26(6–8): 475–480. https://doi.org/10.1016/S1464-1895(01)00087-4.

North GR, Bell TL, Cahalan RF et al. (1982) Sampling errors in the estimation of empirical orthogonal functions. *Monthly Weather Review* 110(7): 699. https://doi.org/10.1175/1520-0493(1982)110<0699:SEITEO>2.0.CO;2.

Peng GM, Chen T (2016) Wavelet denoising methods based on Matlab. *Geomatics and Spatial Information Technology* 39(7): 24–26. https://doi.org/10.3969/j.issn.1672-5867.2016.07.007.

Ren FM, Yuan Y, Sun CH et al. (2012) Review of progress of ENSO studies in the past three decades. *Advances in Meteorological Science and Technology* 2(3): 17–24. https://doi.org/10.3969/j.issn.2095-1973.2012.03.002.

Shi C, Zhou LH, Fan L et al. (2022) Analysis of "21·7" extreme rainstorm process in Henan Province using BeiDou/GNSS observation. Chinese. *Chinese Journal of Geophysics* 65(1): 186–196. https://doi.org/10.6038/cjg2022P0706.

Song XM, Zhang CH, Zhang JY et al. (2020) Potential linkages of precipitation extremes in Beijing-Tianjin-Hebei region, China, with large-scale climate patterns using wavelet-based approaches. *Theoretical and Applied Climatology* 141: 1251–1269. https://doi.org/10.1007/s00704-020-03247-8.

Suparta W (2013) Analysis of GPS water vapor variability during the 2011 La Niña event over the western Pacific Ocean. *Annals of Geophysics* 56(3): R0330. https://doi.org/10.4401/ag-6261.

Trenberth KE (1997) The definition of El Niño. *Bulletin of the Chemical Society of Ethiopia* 78(12): 2771–2777. https://doi.org/10.1175/1520-0477(1997)078<2771:TDOENO>2.0.CO;2.

Vey S, Dietrich R, Fritsche M et al. (2009) On the homogeneity and interpretation of precipitable water time series derived from global GPS observations. *Journal of Geophysical Research* 114(D10): 101. https://doi.org/10.1029/2008JD010415.

Vey S, Dietrich R, Fritsche M et al. (2010) Validaton of precipitable water vapor within the NVEP/DOE reanalysis using global observarions from one decade. *Journal of Climate* 23(1): 675–695. https://doi.org/10.1175/2009JCLI2787.1.

Wang SW, Gong DY (1999) ENSO events and their intensity during the past century. *Acta Meteorologica Sinica* 25(1): 9–14. https://doi.org/10.3969/j.issn.1000-0526.1999.01.002.

Wang Y, Liu YP (2012) *Theory and application of ground-based GPS meteorology.* Beijing: Surveying and Mapping Press.

Wang Y, Lou ZS, Liu YP et al. (2019) ZTD long time series characteristics of IGS stations in China and their relationship with annual precipitation. *Journal of Geodesy and Geodynamics* 39(10): 037–1040+1085. https://doi.org/10.14075/j.jgg.2019.10.010.

Wang Y, Yu TL, Liu X et al. (2021) GNSS ZTD time series in Beijing-Tianjin-Hebei regionand their response to El Niño events. *Journal of Nanjing University of Information Science Technology (Natural Science Edition)* 13(2): 170–180. https://doi.org/10.13878/j.cnki.jnuist.2021.02.006.

Wolter K, Timlin MS (1998) Measuring the strength of ENSO events – How does 1997/98 rank? *Weather* 53: 315–324. https://doi.org/10.1002/j.1477-8696.1998.tb06408.x.

Wu ML, Jin SG, Li ZZ et al. (2021) High precision GNSS PWV and its variation characteristics in China based on individual station meteorological data. *Remote Sensing* 13(7): 1296. https://doi.org/10.3390/rs13071296.

Wu P, Ding YH, Liu YJ (2017) A new study of El Niño impacts on summertime water vapor transport and rainfall in China. *Acta Meteorologica Sinica* 75(3): 371–383. https://doi.org/10.11676/qxxb2017.033.

Yao YB, He CY, Zhang B et al. (2013) A new global zenith tropospheric delay model GZTD. *Chinese Journal of Geophysics* 56(7): 2218–2227. https://doi.org/10.6038/cjg20130709.

Yao YB, Luo YY, Zhang JY et al. (2018) Correlation analysis between haze and GNSS tropospheric delay based on Coherent wavelet. *Geomatics and Information Science of Wuhan University* 43(12): 2131–2138. https://doi.org/10.13203/j.whugis20180234.

Yao YB, Zhao QZ, Li ZF et al. (2016) Short-term precipitation forecasting based on the data from GNSS observation. *Advances in Water Science* 27(3): 357–365. https://doi.org/10.14042/j.cnki.32.1309.2016.03.003.

Yu TL, Wang Y, Huang J et al. (2022) Study on the regional prediction model of PM2.5 concentrations based on multi-source observations. *Atmospheric Pollution Research* 13(4): 101363. https://doi.org/10.1016/j.apr.2022.101363.

Yuan Y, Yang H, Li CY (2012) Study of El Niño events of different types and their potential impact on the following summer precipitation in China. *Acta Meteorologica Sinica* 70(3): 467–478. https://doi.org/10.1007/s11783-011-0280-z.

Zhai PM, Yu R, Guo YJ et al. (2016) The strong El Niño in 2015/2016 and its dominant impacts on global and China's climate. *Acta Meteorologica Sinica* (3): 309–321. https://doi.org/10.11676/qxxb2016.049.

Zhang W (2021) *Diversity of East China Summer Rainfall Change in Post-El Niño/La Niña Summers.* Master Thesis, Nanjing University of Information Science and Technology. https://doi.org/10.27248/d.cnki.gnjqc.2021.000273.

Zhang WJ, Lei XB, Geng X et al. (2018) Possible impacts of ENSO on the intra-seasonal variability of precipitation over southern China. *Transactions of Atmospheric Sciences* 41(5): 585–595. https://doi.org/10.13878/j.cnki.dqkxxb.20170101001.

Zhao QZ, Liu Y, Yao WQ et al. (2020) A novel ENSO monitoring method using precipitable water vapor and temperature in Southeast China. *Remote Sensing* 12(4): 649–667. https://doi.org/10.3390/rs12040649.

Zhao QZ, Ma YJ, Li ZF et al. (2021) Retrieval of a high-precision drought monitoring index by using GNSS-derived ZTD and temperature. *IEEE Journal of Selected Topics in Applied Earth Observations and Remote Sensing* 14: 8730–8743. https://doi.org/10.1109/JSTARS.2021.3106703.

Zheng YF (2015) Fast Fourier transform algorithm and Its application. *Science and Technology* 25(29): 144. https://doi.org/10.3969/j.issn.1672-8289.2015.29.132.

Zhou BJ, Li X, Chen YD et al. (2020a) Effect of the GPS ZTD data assimilation on simulation of typhoon "Lekima". *Journal of the Meteorological Sciences* 40(1): 11–21. https://doi.org/10.3969/2019jms.0063.

Zhou J, Zuo ZY, Rong XY (2020b) Comparison of the effects of soil moisture and El Niño on summer precipitation in eastern China. *Science China Earth Sciences* 63: 267–278. https://doi.org/10.1007/s11430-018-9469-6.

Zhou ZM, Yi JJ, Zhou Q (1999) *Measurement Principle and Application of Global Positioning System*. Beijing: Surveying and Mapping Press.

4 Exploration of Cloud and Microphysics Properties from Active and Passive Remote Sensing

Jinming Ge, Chi Zhang, Jiajing Du, Qinghao Li, Yize Li, Zheyu Liang, Yang Xia, and Tingting Chen

4.1 THE IMPORTANCE OF CLOUDS

Clouds play a crucial role in Earth's climate by having an influence on key parameters (Hartmann et al., 1992; Eerme, 2004), such as the radiation budget, the heating of the earth's surface, and the diabatic heating of the atmosphere (Kondragunta and Gruber, 1996). On the one hand, clouds change the radiative balance and atmospheric heating rate by reflecting solar radiation and absorbing surface and atmospheric long wave radiation, thus significantly affecting surface temperature and local atmospheric circulation; on the other hand, cloud formation is accompanied by evaporation, convection, uplift cooling, supersaturation, and condensation of surface moisture, and then clouds return atmospheric moisture to the surface in the form of rain or snow, thus changing moisture transport and distribution (Zhang et al., 2022). This changes the transport and distribution of water, which means that clouds regulate the water cycle by affecting atmospheric moisture transport and precipitation. Changes in water vapor will affect the occurrence, development, and extinction of clouds, and then change the cloud amount, cloud albedo, and cloud microphysical properties, thus changing the long wave and shortwave radiative effects of clouds, and eventually affecting the radiative balance of the earth's atmosphere system.

Clouds can also influence atmospheric radiative heating and latent heat processes, which in turn can alter the thermodynamic state of the atmosphere. Moreover, clouds can influence global and regional atmospheric thermal and dynamical processes through multiple spatial and temporal scale feedbacks (Slingo and Slingo, 1988), and indirectly influence the climate system through interactions with aerosols. However, the role of clouds in climate models is not well characterized in terms of representation and feedback, and is therefore an important factor contributing to the large uncertainty in global climate model (Dolinar et al., 2015; Bony and Dufresne, 2005; Miao et al., 2021; Zhu et al., 2017a, 2017b). Therefore, in order to better understand climate projections, it is important to understand how clouds behave and how they interact with incoming solar radiation and departing longwave radiation (Figure 4.1).

4.1.1 CLOUD RADIATIVE FORCING

In terms of the energy budget, clouds can both cool the climate by reflecting incoming sunlight and warm it by absorbing and reemitting thermal radiation (Ramanathan et al., 1989; Hartmann et al., 1992; Zelinka et al., 2012). The difference between the radiative fluxes at the top of the atmosphere (TOA), in the atmosphere, or at the surface under all sky and clear sky conditions is defined as cloud radiative forcing (also known as cloud radiative effect). According to the waveband, it can be divided into long wave cloud radiative forcing (LWCRF) and shortwave cloud radiative forcing

DOI: 10.1201/9781003363118-4

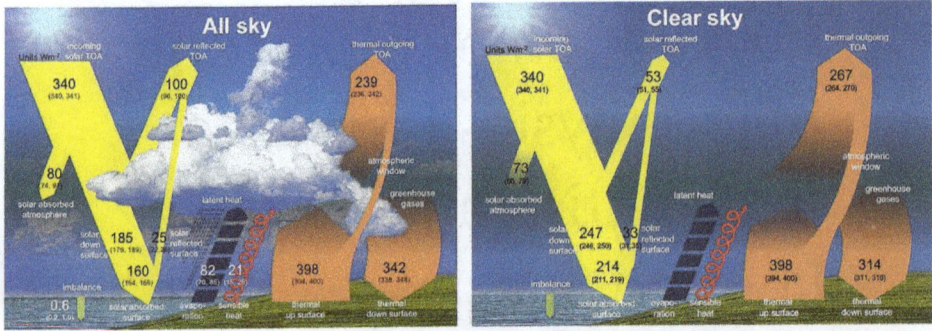

FIGURE 4.1   Impact of clouds on the energy balance of the ground-air system in IPCC AR6 (unit: W/m^{-2}).

(SWCRF). In addition, the definition of net cloud radiative forcing refers to the difference in the net radiative flux at the TOA between clear sky and all the sky.

4.1.2   Cloud Properties

4.1.2.1   Cloud Classification

The formation and distribution of clouds can be divided into two aspects: macroscopic properties and microscopic properties. The macroscopic properties include cloud amount, cloud height, and vertical overlap, while the microscopic properties include cloud water content, cloud and water phase change, optical thickness, cloud particle size, etc. Clouds are complex and variable, any change in cloud parameters may have a significant impact on global climate; the amount of clouds reflects the characteristics of regional weather and climate, and the radiative effect of clouds is closely related to cloud height (Ding et al., 2004). Moreover, the latent heat release, cloud-forming rain, and radiative effects of clouds are all influenced by microphysical processes within clouds, which include the formation, growth, and interactions between warm cloud particles and ice-phase particles.

The classification of clouds depends mainly on the cloud top height and optical thickness. The International Satellite Cloud Climatology Project (ISCCP) classifies clouds into high (CTP ≤ 440 hPa), medium (440≤CTP ≤ 680 hPa), and low (CTP ≤ 680 hPa) clouds according to the cloud-top pressure (CTP), and more detailed classification according to the cloud-top pressure and optical thickness, i.e., 49 types of clouds, including stratocumulus, cirrus, and deep convective clouds, etc.

4.1.2.2   Cloud Cover

Low clouds cover a wide area and have a dominant albedo effect, which contributes the most to the global average net energy balance and has a cooling effect on the earth's atmosphere system. For example, marine low-level clouds efficiently reflect incoming solar radiation back to space while only weakly reducing the emission of terrestrial radiation to space, thereby exerting a strong cooling effect on the planet. As an important part of climate, tropical high clouds can strongly adjust the radiation budget of the earth by interacting with the output long wave radiation and the incoming solar radiation. The optically thin tropical high clouds are relatively transparent to solar radiation, and they typically produce more short wave cloud radiation effects (SWCRE) at the top of the atmosphere to warm the earth (Lee et al., 2009). Conversely, thick tropical high clouds can reflect a lot of solar radiation back into space, thus cooling the earth (Kubar et al., 2007).

4.1.2.3   Optical Thickness

Stratocumulus has influence on both long wave and short wave fluxes. For example, stratocumulus and cirrus clouds of medium optical thickness mainly affect the short wave radiation flux at the top of atmosphere and at the surface, and stratocumulus and upper clouds of medium optical thickness mainly affect the long wave radiation at the surface. In contrast, cirrus clouds, cirrus clouds and deep convective clouds with higher cloud top height make greater contributions to atmospheric top long wave radiation.

4.1.2.4 Cloud Phase

The cloud phase state refers to the liquid or solid state in which the cloud is located, i.e., water or ice clouds. Various inversion models of cloud microphysical parameters are established based on different phase state types. Accurate identification of cloud phase states is especially important to improve the inversion accuracy of cloud optical and microphysical parameters such as optical thickness and effective particle radius (Cho et al., 2009; Riedi et al., 2010). Clouds formed at temperatures above 0°C can be assumed to contain only liquid droplets, while clouds found at temperatures below −40°C are usually composed entirely of ice crystals (Pruppacher et al., 1998). However, at temperatures between −40°C and 0°C, clouds may consist entirely of ice crystals, supercooled liquid water droplets or a mixture of both (called mixed-phase clouds), which complicates the estimation of their radiative effects (Wang et al., 2022). The wide global coverage and complex radiative properties of mixed-phase clouds can influence climate on a global scale. The presence of supercooled liquid in mixed-phase clouds is particularly important because liquid water is opaquer to long wave radiation and increases cloud albedo more than ice crystals, especially at high latitudes, where it is an important driver of radiative flux (Forbes and Ahlgrimm, 2014; Kay et al., 2016; Matus and L'Ecuyer, 2017).

4.1.3 CLOUD FEEDBACK

The global warming effect would be amplified by the doubling of CO_2 concentration. Changes in the radiative effect of clouds directly affect the radiative balance of the earth's atmosphere system and the closely related temperature changes. Additionally, a mere 4% increase in global low cloud cover would be sufficient to offset the 2–3°C global warming caused by a doubling of CO_2 concentration, and vice versa to amplify the corresponding warming effect (Randall et al., 1984). Furthermore, global mean surface temperature cause changes in climate state quantities (e.g., water vapor, clouds, etc.), which then enhance or diminish the initial forcing by influencing radiative processes, a process known as climate feedback.

Cloud feedback is defined as the change in net radiative flux at the top of the atmosphere due to changes in clouds for every 1°C increase in global mean surface temperature. Under warming conditions, a decrease in net radiative flux at the top of the atmosphere due to changes in clouds will partially offset the warming effect caused by the increase in greenhouse gases (negative feedback); conversely, it will enhance the warming effect (positive feedback). As an important component of climate feedbacks, cloud feedbacks are one of the largest sources of uncertainty in modeling current climate and predicting future climate change. According to the assessment of the Intergovernmental Panel on Climate Change (IPCC) Sixth Assessment Report (AR6), the net feedback effect of clouds in the context of global warming is positive, i.e., amplifying the anthropogenic warming effect (IPCC, 2022).

In the context of climate warming, cloudiness is decreasing in most regions, such as most land areas in the mid and low latitudes. It is difficult to determine the reduction of high clouds in the tropics leading to cloud feedbacks, etc., because the model cannot correctly simulate the cloud parameterization process, etc. (Bretherton, 2015; Tobin et al., 2013; Stein et al., 2017). In addition, since the assessment of low cloud feedbacks in the tropical oceans depends on the atmospheric conditions, the response of low clouds will be different if the boundary layer conditions change in the future, and therefore the radiative response of low-latitude ocean boundary layer clouds to global warming is also subject to large uncertainties (Bony and Dufresne, 2005; Chen et al., 2019).

4.1.4 Cloud Interactions

Clouds can also indirectly affect the climate system through interactions with aerosols. The "aerosol-cloud-radiation-precipitation" interaction is one of the most uncertain factors in climate prediction. Aerosols in the atmosphere produce direct radiation effects by absorbing and scattering solar short wave radiation and emitting and capturing long wave radiation; at the same time, aerosols can act as cloud condensation nodules, changing the albedo, lifetime, and microphysical properties of clouds, producing indirect radiation, which can also absorb solar radiation and heat cloud droplets to produce semi-direct radiation effects, affecting radiation income and expenditure and thus climate. Besides, dust aerosols can also affect cloud microphysical properties, radiation income and expenditure, and precipitation through indirect and semi-direct radiation effects (Liu et al., 2021). Latent heat release from clouds is an important source of energy for weather phenomena in the atmosphere at various scales, from single cumulus clouds to mesoscale systems up to the global atmospheric circulation. Moreover, precipitation formed by clouds has an important role and impact on human life and daily activities.

The World Climate Research Program (WCRP) lists understanding clouds, atmospheric circulation and climate sensitivity as one of the key scientific challenges facing the climate community (Bony et al., 2015). Due to the importance of clouds, in the context of current climate warming, studying how clouds behave and their radiative effects and cloud feedback is an important step to improve our confidence in predicting future climate change.

4.2 METHODS OF CLOUD REMOTE SENSING

Cloud remote sensing is a technology for acquiring the radiation information of the cloud without making direct contact with the cloud. It may be split into passive remote sensing and active remote sensing according to the way the detector works (Figure 4.2). Passive remote sensing, which measures the electromagnetic radiation emitted by the cloud within a certain electromagnetic spectrum,

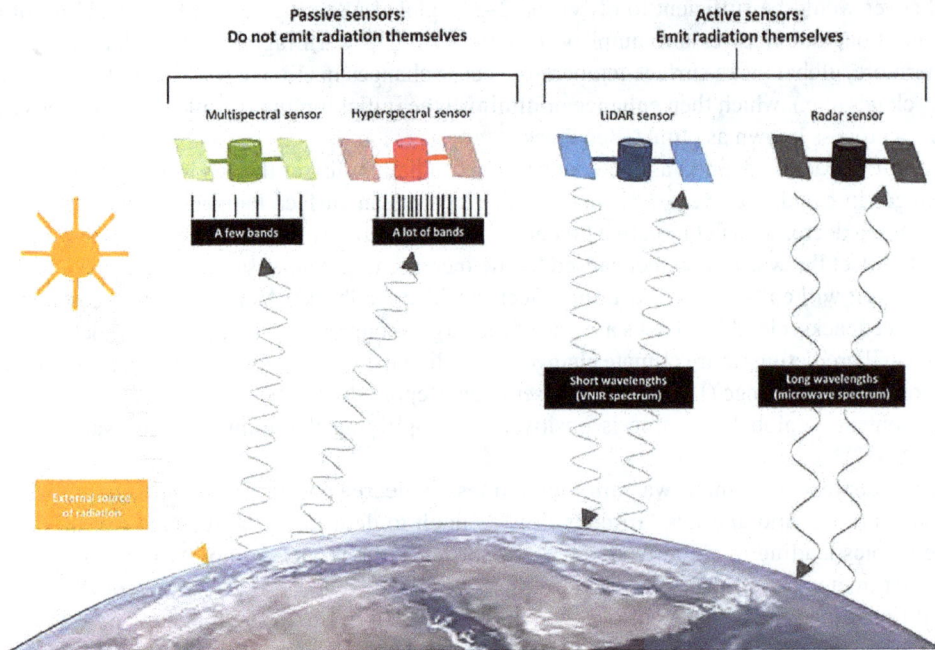

FIGURE 4.2   A schematic of the difference between passive sensors and active sensors.

Source: Adapted from the EO4GEO course which is designed and developed by Carlos Granell.

can provide the properties of the cloud and detect the distribution of the cloud. On the other hand, active remote sensing emits electromagnetic radiation of some certain wavelengths, and then measures the radiation that is reflected or scattered from the cloud. Using the variability of the intensity of this radiation, the properties and distribution of clouds can be inferred.

4.2.1 PASSIVE REMOTE SENSING

Passive remote sensing is a method of atmospheric remote sensing using the long wave radiant energy released by the atmosphere and the surface itself, as well as the reflected and absorbed solar short wave radiant energy. This method directly collects information from nature by using the various radiation energy in nature rather than artificial radiation sources. The most common instrument platform of passive remote sensing is the satellite. It continually monitors the earth's atmosphere from space. Passive sensors are fitted on satellites that record naturally occurring electromagnetic radiation at the top of the atmosphere. Satellite inversion cloud parameters include cloud fraction, cloud height or cloud top temperature, cloud emissivity and reflectance, cloud optical thickness, cloud type, and cloud microphysical properties (i.e., particle effective radius) (Stephens and Kummerow, 2007).

4.2.1.1 Cloud Top Temperature

Radiation $L_{\text{sat}}\left(T_{B\lambda}\right)$ measured by satellite is the sum of radiation $L_\lambda\left(T_C\right)$ emitted by the cloud and radiation $\left(1-\varepsilon_{\lambda c}\right)L_\lambda\left(T_S\right)$ emitted from the surface and transmitting through cloud, which can be expressed by

$$L_{\text{sat}}\left(T_{B\lambda}\right) = L_\lambda\left(T_C\right) + \left(1-\varepsilon_{\lambda c}\right)L_\lambda\left(T_S\right),\tag{4.1}$$

where $\left(1-\varepsilon_{\lambda c}\right)$ is transmittance of cloud. If the cloud is thick and dense, it can be seen as a black body in the infrared spectrum and transmittance is 0. In this case, the radiation measured by the satellite in the infrared window comes from the cloud top. In Eq. (4.1), it is assumed that there is no absorption in the atmosphere, the emissivity of the surface is 1, and all the instantaneous field of view is covered by clouds, regardless of cloud thickness. The cloud top temperature T_C can be calculated, given the emissivity of cloud $\varepsilon_{\lambda c}$ and the surface temperature T_S.

4.2.1.2 Cloud Fraction

The most convenient method to calculate cloud fraction from satellite data is the threshold method (Kazuo and Baba, 1988). If T1 and T2 are the threshold for distinguishing high clouds, low clouds and the surface, NH, NL, and NS are the number of images of high clouds, low clouds and the surface according to the threshold T1 and T2, respectively. The total cloud fraction can be written as

$$S_T = \frac{N_H + N_L}{N_H + N_L + N_S}\tag{4.2}$$

The low cloud fraction can be expressed as

$$S_L = \frac{N_L}{N_H + N_L + N_S}\tag{4.3}$$

The high cloud fraction can be expressed as

$$S_H = \frac{N_H}{N_H + N_L + N_S}\tag{4.4}$$

4.2.1.3   Cloud Emissivity

Mosher (1976) used the luminance data of visible light cloud images to obtain the optical thickness of cloud. The cloud emissivity can be estimated from the cloud luminance of visible light cloud images, since it is related to the optical thickness of cloud. For uniform cloud, the optical thickness is related to the scattering cross section of particles in the cloud σ and the density of particles ρ and the thickness of cloud Δz. The calculation of the scattering cross section of cloud requires the cloud particle spectrum distribution. The optical thickness of cloud τ_c is

$$\tau_c = \sigma\rho\Delta z \tag{4.5}$$

In order to determine the optical thickness of the cloud, the relationship between the brightness and the optical thickness of several typical clouds can be established in advance. According to this relationship, the optical thickness can be obtained from the actual cloud brightness, and then the emissivity of the cloud can be given by

$$\varepsilon = 1 - \exp\left[-\tau_c\right] \tag{4.6}$$

4.2.2   Active Remote Sensing

Active remote sensing needs to emit radiation which is directed toward the cloud from the detector. The radiation reflected from that cloud is detected and measured by the sensor. Active remote sensing can obtain measurements anytime, regardless of the time of day. Active sensors can be used for examining wavelengths that are not sufficiently provided by the sun, such as microwaves. The most important feature of microwave radiation is its ability to penetrate cloud cover, fog, and rainfall, and can penetrate a certain depth of the surface, so microwave radiation can be used to detect the atmospheric conditions in and below clouds (Kollias et al., 2007, 2016). The representative active atmospheric remote sensing detection method is microwave radar detection.

4.2.2.1   Principles of Radar

The transmitter, antenna and receiver are three primary parts of a radar system. Short energy pulses in the radio-frequency region of the electromagnetic spectrum are produced by the transmitter. The antenna narrows these into a beam for usage. They spread out at a speed that is nearly equal to the speed of light. When the pulses intercept an object with different refractive characteristics from air, a current is induced in the object which perturbs the pulse and causes some of the energy to be scattered. In most cases, part of the scattered energy will be returned to the antenna, and if this backscattered component is significant enough, it will be detected by the receiver.

The primary function of radar is to measure the range and direction of backscattering objects. Ranging is accomplished by a timing circuit that counts time between the transmission of a pulse and the reception of a signal. Direction is determined by noting the antenna azimuth and elevation at the instant the signal is received.

4.2.2.2   Radar Equation

The radar equation under Rayleigh scattering condition is

$$P_r = \frac{\pi^3}{1024 ln2} \frac{P_t G^2 h\theta\varphi}{\lambda^2} \frac{1}{R^2} \psi \left|\frac{m^2-1}{m^2+2}\right|^2 Z\cdot10^{-0.2\int_0^R kdR} \tag{4.7}$$

where P_r is received echo power intensity, P_t is transmitted power of radar transmitter, G is the antenna gain, h is the pulse length, θ is horizontal lobe width, φ is vertical lobe width, λ is the radar wavelength, R is the distance from the antenna to the particle, Z is the radar reflectivity factor, m is complex index of refraction, k is the attenuation factor and ψ is filling coefficient, which shows the extent of the radar beam filled with the precipitation particles.

The power intensity of the echo received by the radar is inversely proportional to the radar wavelength and the square of the distance. When the wavelength is shorter and the distance is closer, the echo is stronger. Factors affecting the radar equation mainly include radar parameters, meteorological factors, and the distance factor. (i) Radar parameters are as follows: increasing the pulse length can increase the detection range, but at the same time reduce the range resolution and increase the radar blind area. By decreasing the beam width, the echo power can be increased and the error can be reduced. Increasing the antenna gain can improve the detection capability of radar, but there are some problems with this increase. (ii) Meteorological factors are as follows: mainly scattering and attenuation. The factor $\left| \dfrac{m^2 - 1}{m^2 + 2} \right|^2 Z$ shows the influence of particle size, phase state, shape and temperature on scattering. The factor $10^{-0.2 \int_0^R k dR}$ includes the attenuation of particles such as atmosphere, clouds, rain, snow and hail at different wavelengths and temperatures. (iii) The distance factor is as follows: the size of the echo power is inversely proportional to the square of the distance. Therefore, the precipitation with the same intensity will be considerably weaker at a long distance than at a close distance if there is no range correcting device on the radar, which easily creates an illusion while observing and evaluating the echo strength and movement.

4.2.2.3 Advantage

Microwave is radiation with wavelengths ranging from about one meter to one millimeter. Because of their long wavelengths, compared to the visible and infrared, microwaves have special properties that are important for remote sensing. Longer wavelength microwave radiation can penetrate through cloud cover, haze, dust and all but the heaviest rainfall because the longer wavelengths are not susceptible to atmospheric scattering which affects shorter optical wavelengths. This property allows the detection of microwave energy under almost all weather and environmental conditions so that data can be collected at any time.

The wavelength of the millimeter wave is between the centimeter wave and the light wave, so the millimeter-wave radar has the advantages of microwave navigation and photoelectric navigation. Millimeter-wavelength radars have adequate sensitivity to detect nonprecipitating clouds, such as fair-weather cumuli in the boundary layer and thin cirrus aloft, excellent Doppler velocity resolution to observe the small fall velocities of small ice crystals and cloud droplets, and are unaffected by Bragg scattering and ground clutter. These advantages of millimeter-wavelength radars make them suitable for the detection of weak nonprecipitating clouds that are often not well characterized by radars operating at longer wavelengths.

4.3 CLOUD DETECTION METHODS

To better understand cloud processes to improve parameterization in climate models and reveal their evolution in response to climate change, long-term continuous observations of clouds in terms of both macro- and microphysical properties are essential (Ackerman and Stokes, 2003). Millimeter-wavelength cloud radars (MMCRs) can resolve cloud vertical structure for their occurrence and microphysical properties (Clothiaux et al., 1995). The wavelength of MMCRs is shorter than those of weather radars making them sensitive to cloud droplets and ice crystals and able to penetrate multiple cloud layers (Kollias et al., 2007). Because of their outstanding advantages for cloud research, millimeter-wavelength radars have been deployed on various research platforms including the first space-borne millimeter-wavelength Cloud Profiling Radar (CPR) onboard the CloudSat (Stephens et al., 2002). Ground-based cloud radars are operated at the US Department of Energy's Atmospheric Radiation Program (ARM) observational sites, and a KAZR (Ka-band zenith radar) was deployed in China at the SemiArid Climate and Environment Observatory of Lanzhou University (SACOL) site (Huang et al., 2008), providing an opportunity to observe and reveal the detailed structure and process of the midlatitude clouds over the semi-arid regions of East Asia.

4.3.1 Cloud Detection Research Progress

At present, a lot of research work has been done on cloud detection algorithms, whether it is space-based or ground-based. The main results of the development and testing of the cloud detection method are presented by Kostornaya, AA et al. (2017). The main purpose of the method is the identification and classification of clouds in satellite images with the subsequent retrieval of quantitative characteristics. The method provides digital data sets in the form of maps of cloud classes, cloud top height and cloud top temperature. To improve the dynamic cloud detection threshold value, moving and nesting analysis area methods were used to improve the dynamic cloud detection threshold method by Liu (2010). The analysis results show that the proportion of effective cloud detection threshold is effectively increased and the accuracy of cloud detection is also improved by using the method of moving and nesting analysis area. For image cloud detection, the spectral threshold selection of image cloud detection and the influence of cloud-like ground objects are two vital factors in determining cloud detection results of HSRI. With the purpose of solving these two issues, a novel cloud detection method for HSRI based on the spectrum and texture of superpixels is proposed by Dong et al. (2019). They obtain the adaptive image cloud detection spectral threshold according to the image equalization histogram. Then the initial cloud detection result is obtained based on spectral threshold of the cloud detection and spectral attributes of superpixels, and the initial cloud detection result is refined based on the gray value and angular second moment of the superpixels local binary patterns texture to eliminate the influence of cloud-like ground objects. Finally, the cloud detection result is processed using the region growing algorithm and expansion algorithm to obtain an accurate cloud detection result. Clothiaux et al. (1995, 2000) first developed a classical cloud detection method, which was adopted by ARM as a service algorithm for cloud radar signal recognition and was later adopted to the CPR on board the CloudSat (Marchand et al., 2008).

4.3.2 A Classical Cloud Detection Method for Cloud Radar

This method assumes that the random noise power follows the normal distribution (Figure 4.3) and consists of two main steps. Firstly the mean value (m) and standard deviation (σ) of the noise are calculated by selecting the radar range gate containing only noise. The parameters m and σ are compared with return signals from other radar range gates. If the return signal, that is the power at the receiver antenna output terminals, is greater than $m + \sigma$, it is believed that this range gate may contain a valid return signal and is marked as 1. Otherwise, it is considered radar noise and is set to zero.

Secondly, the cloud signals are highly correlated in both space and time and have more similar values in near pixels, which means there may also be a certain number of cloud signals in other gates near the gate if a range gate is the effective echo signal of the cloud, while the random noise values are not correlated (Ge et al., 2017). A spatial-temporal filter is considered for low-pass filtering, which can effectively remove the misjudgment noise signal. As shown in Figure 4.4, for any range gate, the central pixel and 24 surrounding pixels are selected to form a 5×5 filter. For the 25 pixels covered by the filter, the probability of noise being set to 1 in cloud mask is 0.16, and the probability of noise being set to 0 is 0.84. (The area under a normalized Gaussian curve for the region of values that are greater than one standard deviation above the mean is 0.16.) Therefore, the probability of a certain configuration of zeros and ones occurring in a square of 25 pixels is

$$p = \left(0.84\right)^{n_0} \left(0.16\right)^{n_1} \tag{4.8}$$

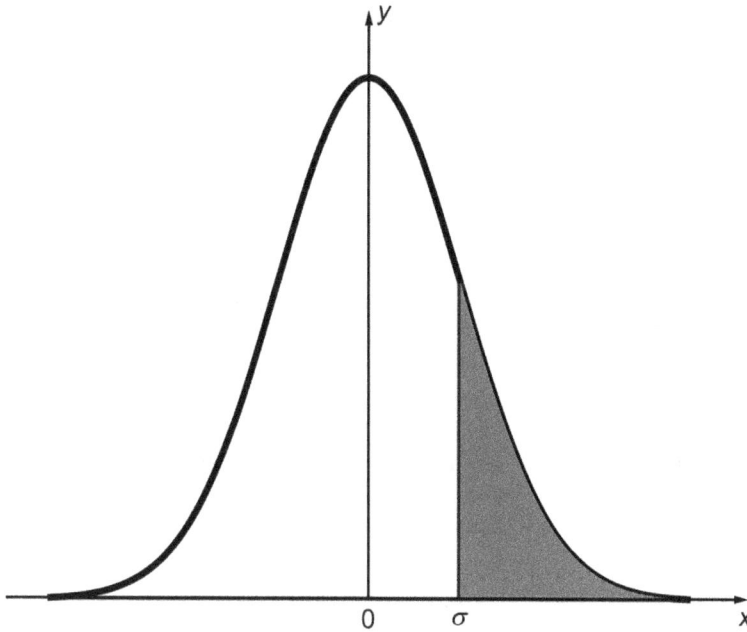

FIGURE 4.3 The normal distribution diagram.

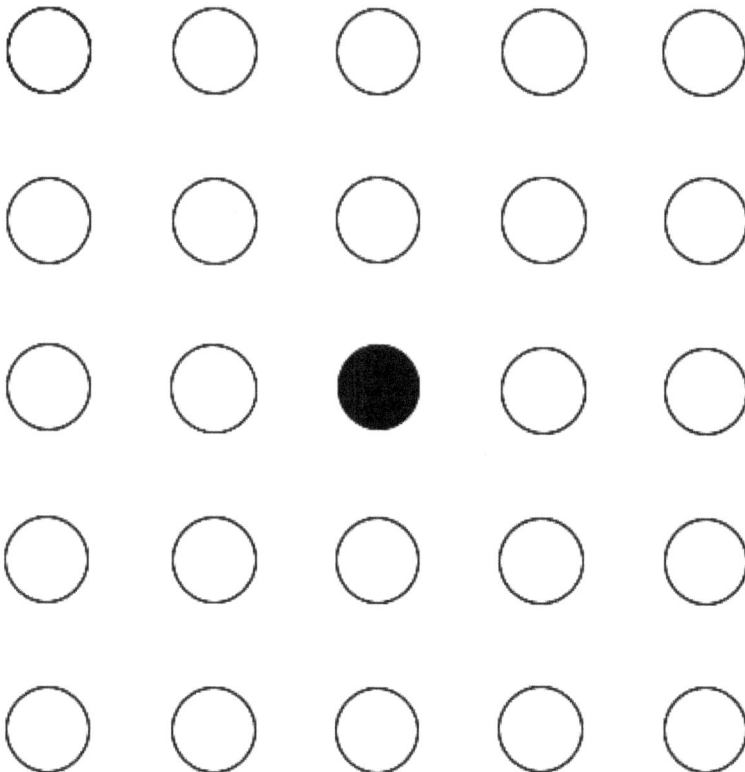

FIGURE 4.4 The 5×5 composed of 0 and 1.

The observed values of 25 pixels are used to adjust and determine whether the central range gate is an effective signal, where n_0 is the number in the box, and n_1 is the number of ones. It can be seen from Eq. (4.8) that the smaller the p value is, the fewer pixels with a value of 0, and the greater the possibility that the center pixel is an effective radar signal. Clothiaux et al. (1995) obtained the threshold value on the basis of many experiments. After several times of filtering, the misjudgment in the first step can be basically filtered out and the final cloud detection value can be obtained (Zhu et al., 2016).

4.3.3 AN IMPROVED CLOUD DETECTION METHOD FOR CLOUD RADAR

The basic assumption of the previous cloud mask method is that the random noise power follows a normal distribution (e.g., Clothiaux et al., 1995; Marchand et al., 2008). However, Ge et al. (2017) found that noise power may not follow Gaussian distribution. Here clear-sky cases in all seasons from KAZR observations were first analyzed for their background noise power distributions, and the probability distribution function (PDF) of the noise power from the KAZR observations of a clear day is displayed in Figure 4.5. The noise power is estimated from the top 30 range gates, which includes both internal and external sources (Fukao and Hamazu, 2014). It has an apparent non-Gaussian distribution with a positive skewness of 1.40. The signal-to-noise ratio (SNR) is defined as

$$SNR = 10\log\left(\frac{P_r}{P_n}\right),\tag{4.9}$$

where P_r is the power received at each range gate in a profile and P_n is the mean noise power that is estimated by averaging the return power in the top 30 range gates. Figure 4.5 shows that the

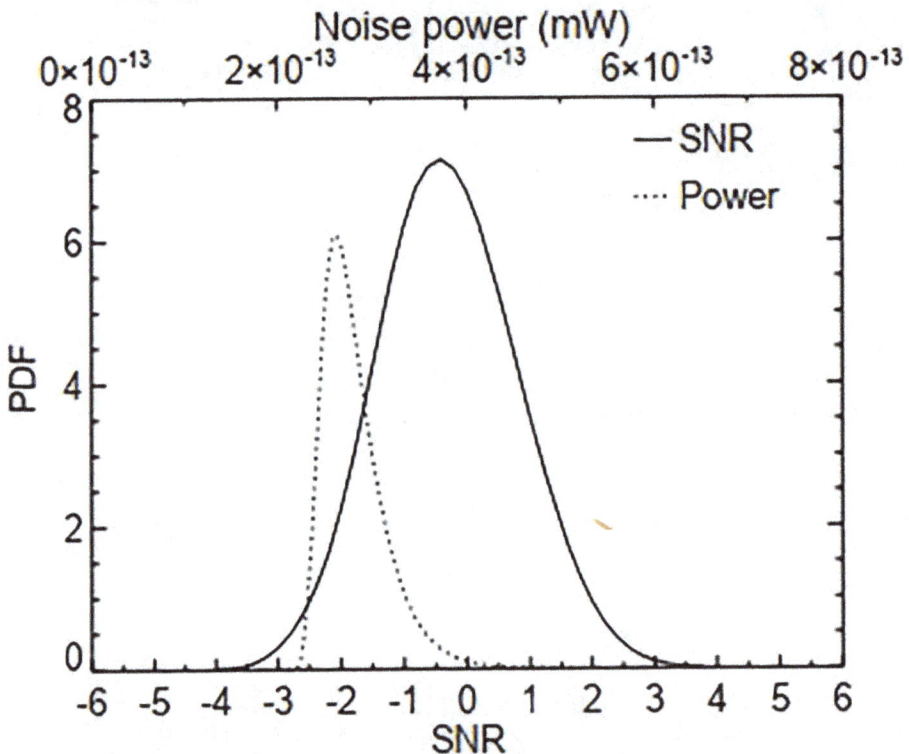

FIGURE 4.5   PDF of the noise power and SNR from the KAZR observations on a clear day.

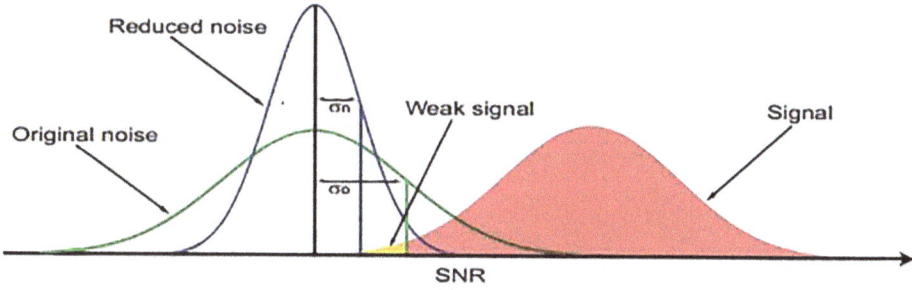

FIGURE 4.6 Comparison of original noise, reduced noise and valid signal distribution.

SNRs for clear skies closely follow a Gaussian distribution. Instead of using radar-received power, the SNR is used as the input in cloud mask algorithm including estimating the background noise level. This is because in the method of this article the chance of a central range gate being noise or a potential feature relies on the probability of a given range of SNR values following the Gaussian distribution.

In image processing described previously, the random noise can be smoothed out by using a low-pass filter, which gives a new value for a pixel of an image by averaging with neighboring pixels. Figure 4.6 shows a schematic comparison of the original noise, reduced noise, and valid signal distributions: the low-pass filter could efficiently reduce the original radar noise represented by the green line to a narrow bandwidth (blue line) while keeping the signal preserved. By reducing the standard deviations of noise, which shrinks the overlap region of signal and noise and enhances their contrast, the weak signals (yellow area) that cannot be detected based on original noise level may become distinguished.

Following this idea, Ge et al. (2017) developed a non-iterative hydrometeor detection algorithm by applying a noise reduction and a central-pixel weighting schemes. For given mean SNR values (S_0) and 1 standard deviation (σ_0) of the original background noise, the input SNR data set is first separated into two groups. The group with values greater than $S_0 +3\sigma_0$ is considered to be the cloud features that can be confidently identified. Another group with values between S_0 and $S_0+3\sigma_0$ may potentially contain moderate ($S_0+\sigma_0 < SNR \leq S_0+3\sigma_0$) to weak ($S_0 < SNR \leq S_0+\sigma_0$) cloud signals, which will further go through a noise reduction process.

The noise reduction process is performed by convolving radar SNR time–height data with a low-pass filter. The Gaussian filter is one of the most common functions of the noise reduction filter. For the high frequency noise with random distribution, the Gaussian filter can make it more concentrated, so as to effectively reduce the standard deviation of noise distribution. A 2-D Gaussian distribution kernel can be expressed as

$$G(i,j) = \frac{1}{2\pi\sigma^2} exp\left(-\frac{i^2 + j^2}{2\sigma^2}\right),$$

(4.10)

where i and j are the indexes in a filter window and are 0 for the central pixel, and σ is the standard deviation of the Gaussian distribution for the window size of the kernel. This formula is used to filter the radar SNR image. However, this method is only aimed at the ideal case of pure noise. In actual observations, noise and signal pixels cannot be distinguished in advance. If noise and signal pixels are averaged together, the large echo energy gradient existing at the boundary between signal and noise will be blurred; that is, the signal is falsely extended. The large gradient of echo power corresponds to the edge of cloud, and the blurring of cloud boundary means the increase of false detection rate of cloud detection. To solve this problem, a bilateral filtering idea proposed

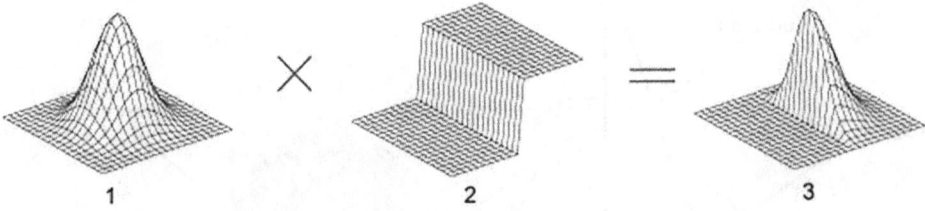

FIGURE 4.7 Illustration of bilateral filtering process: (1) Gaussian kernel distribution in space; (2) δ function; and (3) bilateral kernel by combining Gaussian kernel with δ function.

by Tomasi and Manduchi (1998) is adopted (Figure 4.7). First of all, the noise and signal should be pre-distinguished. A $C_w \times C_h$ matrix is selected considering the space-time correlation of the signal. According to the normal distribution, the radar range bin with signal values significantly higher than noise value (SNR> $S_0 + 3\sigma_0$) is less likely to be noise (the number of these bins is N_s). This part of the signal value is directly retained as the real signal. If the remaining range bins are all noises, the range bin number (N_m) with SNR greater than $S_0 + \sigma_0$ should be about equal to an integral number (N_t) of $0.16 \times (C_w \times C_h - N_s)$ where 0.16 is the probability for a remaining range bin to have a value greater than $S_0 + \sigma_0$ for a Gaussian noise. Thus when N_m is equal to or smaller than N_t, all the $(C_w \times C_h - N_s)$ range bins could only contain pure noise and/or some weak cloud signals. In this case, the δ function is set to 1 for all the $(C_w \times C_h - N_s)$ bins. When N_m is found to be larger than N_t, the $(C_w \times C_h - N_s)$ range bins might contain a combination of moderate signal, noise, and/or some weak clouds. In this case, $S_0 + \sigma_0$ is selected as a threshold to determine whether the pixels are on the same side of the central pixel. If the central pixel has a value greater than $S_0 + \sigma_0$, the δ function is assigned to 1 for the $(C_w \times C_h - N_s)$ pixels with SNR $\geq S_0 + \sigma_0$, but 0 for the bins with SNR $< S_0 + \sigma_0$. If the central pixel is less than $S_0 + \sigma_0$ the δ function is assigned to 1 for the pixels with SNR $< S_0 + \sigma_0$, but 0 for the $(C_w \times C_h - N_s)$ bins with SNR $\geq S_0 + \sigma_0$. The bilateral filter represented by Eq. (4.11) is constructed from the 2-D Gaussian function and the δ function:

$$B(i,j) = G(i,j) \cdot \delta(i,j) \tag{4.11}$$

Marking function $\delta(i,j)$ can be computed as follows:

$$\delta(i,j) = \begin{cases} 1, sign\left[SNR(i,j) - (m+\sigma) \right] = sign\left[SNR(0,0) - (m+\sigma) \right] \\ 0, sign\left[SNR(i,j) - (m+\sigma) \right] \neq sign\left[SNR(0,0) - (m+\sigma) \right] \end{cases} \tag{4.12}$$

where x is a sign function

$$sign(x) = \begin{cases} 1, x \geq 0 \\ -1, x < 0 \end{cases} \tag{4.13}$$

This bilateral filter is applied in the matrix containing signal and noise to compress the noise, which can avoid the false expansion of the boundary caused by the mixture of noise and signal, and compress the noise effectively at the same time.

In conclusion, the group which is considered to be the confidently identified cloud feature directly performs initial cloud detection and spatial filtering to obtain the final mask. The part that may contain signals uses a bilateral filter and the pure noise part uses the Gaussian filter directly to make noise reduction and reduce the standard deviations of noise. Then the noise level is recalculated for initial cloud detection and spatial filtering.

4.4 CLOUD PHASE DETECTION

Clouds are suspended bodies composed of single or mixed forms of water droplets or condensed ice crystals liquefied by water vapor in the atmosphere. According to the microscopic particle structure, clouds are classified as water clouds, ice clouds, and mixed-phase clouds (Figure 4.8). Water clouds are mainly composed of water droplets. Water clouds are divided into two types according to the temperature inside the clouds: warm water clouds when the temperature is greater than 0°C, and supercooled water clouds when the temperature is less than 0°C. Clouds mainly composed of ice crystal particles are called ice clouds. Clouds between ice clouds and water clouds are called mixed-phase clouds (Intrieri et al., 1993). Clouds are an important link in the global water cycle and in the radiative transfer between the earth and the air (Stephens, 2005). Cloud liquid water and cloud ice water affect Earth's atmosphere radiation by influencing the degree of reflection and absorption of the sun by clouds transport. In many atmospheric studies, cloud liquid water and cloud ice water content is an important indicator of cloud properties. Therefore, it is very important to realize the continuous and high precision detection of cloud liquid water and cloud ice water in the field of weather and climate change, disaster weather prediction, and so on.

LIDAR is an active remote sensing technology that has been applied to atmospheric sounding since the 1970s. It has the advantages of high temporal and spatial resolution, which can reach the resolution in meters and seconds. Meanwhile, it also has the characteristics of high sensitivity, which can achieve the detection of very small amount of molecules in a small area. Therefore, it is widely used in the measurement of atmospheric meteorological parameters. At present, the technical means for cloud phase identification at home and abroad are mainly polarization LIDAR, and a few units have carried out research on Raman LIDAR phase water.

4.4.1 POLARIZATION LIDAR

As a unique active remote sensing detection instrument, polarization LIDAR can detect the particle properties in clouds more effectively and is now widely used for cloud and aerosol particle microphysical properties detection (Qi et al., 2021). Polarization detection can expand the detection information from intensity, spectrum, and space according to the polarization occurrence degree, polarization azimuth angle, polarization ellipticity, and rotation direction, so as to improve the multi-directional detection ability of LIDAR. Using the depolarization properties of particles, it is

FIGURE 4.8 Water clouds (a) and ice clouds (b).

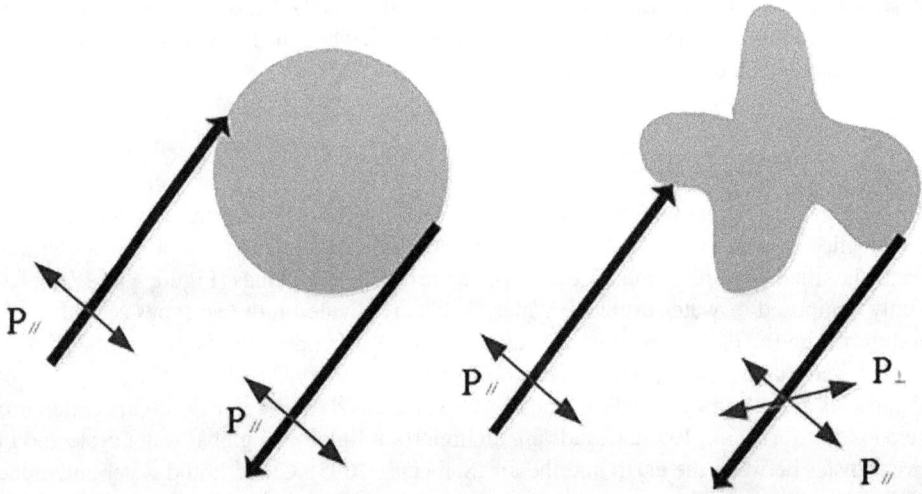

FIGURE 4.9 Schematic diagram of depolarization principle.

possible to distinguish spherical particles in water clouds from non-spherical particles in ice clouds (Figure 4.9), and the depolarization ratio is currently an important basis for studying the composition and interaction mechanisms of clouds and aerosols (Hu, 2007). Therefore, as an important detection method of LIDAR, polarization LIDAR is of great significance in the study of cloud and aerosol detection.

At present, the identification of cloud phase states mainly relies on the depolarization ratio. The depolarization characteristics of particles in clouds are obtained by a polarization LIDAR system, and the laser with linear polarization characteristics interacts with particles of different shapes, and the scattered laser polarization state is changed. The ideal laser is a line of polarized light, and the light has wave-particle duality; the particles in different states of the cloud can not only affect the laser scattering energy intensity, but also change the polarization state. The different states of particles in the cloud can not only affect the intensity of laser scattering energy, but also change the polarization state of the laser. When a beam of linearly polarized light is irradiated to a cloud particle, different polarization degrees are generated depending on the shape of the cloud particle, and the direction of the resulting polarization state is perpendicular to the direction of the polarization state of the incident light. This phenomenon suggests the use of a LIDAR system with polarization channels to measure and calculate the ratio of the vertical and parallel components of the scattered light in the echo signal to measure the cloud particle morphology (Biele et al., 2000). The polarization LIDAR emits line polarized light, which enters the atmosphere and scatters, thus changing the polarization characteristics of the laser. The backscattered echo signal is received through the telescope. After passing through a polarizing beam splitter, the echo signal is divided into two linear directions perpendicular to each other. According to the polarization direction of the echo signal with respect to the laser source, the echo signals of the two channels are divided into vertical echo signal $P_\perp$ and parallel echo signal $P_\parallel$. The parallel echo signal reflects the detection information of the meter scattering LIDAR, and the ratio of the vertical echo signal to the parallel echo signal is the declination ratio δ. Depolarization ratio is an important physical parameter for cloud research because it can reflect the microscopic characteristics of particles in clouds. At present, the size of depolarization ratio is one of the criteria to distinguish clouds from aerosols, water clouds from ice clouds, and aerosol compositions and types. When the polarization LIDAR is used to detect the atmosphere, the parallel and vertical components of the received atmospheric backscattered echo signal power direct components, which can be given by the LIDAR equations, are expressed as:

$$P_\perp(z) = P_0 C_\perp z^{-2} \beta_\perp(z) exp\left[\int_0^z \left(\alpha_\perp(z') + \alpha_\parallel(z')\right) dz'\right] \tag{4.14}$$

$$P_\parallel(z) = P_0 C_\parallel z^{-2} \beta_\parallel(z) exp\left[-2\int_0^z \alpha_\parallel(z') dz'\right] \tag{4.15}$$

where P_0 is the laser transmitting power, $P_\parallel(z)$ and $P_\perp(z)$ represent the parallel and vertical components of the atmospheric return power of the backscattered light received by the LIDAR at height z, respectively. $C_\parallel$ and $C_\perp$ represent the channel constants of the parallel and vertical components, respectively. $\beta_\parallel(z)$ and $\beta_\perp(z)$ represent the parallel and vertical components of the atmospheric backward scattering coefficient at height z. $\alpha_\parallel(z)$ and $\alpha_\perp(z)$ represent the parallel and vertical components of the atmospheric extinction coefficient at height z, respectively.

According to the definition of the depolarization ratio, the profile $\delta(z)$ of the depolarization ratio with height can be obtained using the echo signal of the volume channel, denoted as:

$$\delta(z) = \frac{P_\perp(z)/C_\perp}{P_\parallel(z)/C_\parallel} = \frac{\beta_\perp(z)}{\beta_\parallel(z)} exp[\int_0^z (\alpha_\perp(z') - \alpha_\parallel(z'))dz'] \tag{4.16}$$

In general, $K = C_\perp / C_\parallel$ so Eq. (4.16) can be written as

$$\delta(z) = \frac{P_\perp(z)/C_\perp}{P_\parallel(z)/C_\parallel} = K\frac{P_\perp(z)}{P_\parallel(z)} = \frac{\beta_\perp(z)}{\beta_\parallel(z)} \tag{4.17}$$

From Eq. (4.17), it can be seen that the polarization LIDAR declination ratio is not only related to the intensity of the echo signals of the vertical and parallel channels, but also depends on the gain ratio of the two channels, K which is also called the calibration factor of the polarization LIDAR system. Therefore, the calculation of the declination ratio needs to consider the measurement of K value.

In the existing research of polarization LIDAR technology, most scientific research institutions have carried out the research of cloud phase recognition. Table 4.1 shows the corresponding relationship between the depolarization ratio range and the cloud phase state obtained in the previous experiment (Sassen et al., 1992; Sassen and Zhao, 1992; Sassen, 2005; Sassen et al., 2008; Sassen et al., 2009). As can be seen from Table 4.1, when the depolarization ratio is less than 0.15, the particles in the cloud are mainly spherical particles, and the cloud phase state is judged to be water cloud. When the depolarization ratio is greater than 0.15 and less than 0.38, the cloud phase is

TABLE 4.1

Correspondence between Depolarization Ratio and Cloud Phase

Depolarization Ratio	Cloud Phase
$\delta < 0.15$	water cloud
$0.15 < \delta < 0.38$	mixed-phase cloud
$0.38 < \delta < 0.5$	ice cloud

judged to be mixed. When the depolarization ratio is less than 0.5 and greater than 0.38, the cloud phase is identified as ice cloud.

4.4.2 RAMAN LIDAR

Different material of molecules and atoms interact with light, and scattering of light absorption/ dispersion degree is different. Light through the uneven material will scatter around the material; the scattered light and incident light wavelength is called elastic dispersion. When the wavelength of the scattered light is different from that of the incident light, it is called inelastic scattering. Raman scattering light refers to the Stokes line whose frequency is lower than the incident light and the anti-Stokes line whose frequency is higher than the incident light when light interacts with matter. The Raman shift $\Delta \upsilon$ is the difference between the Stokes line or the anti-Stokes line and the frequency of the incident light. The Raman shift depends on the nature of the molecule itself and is independent of the incident light frequency, so it can be used as a characteristic parameter of the molecule (Cooney, 1970). When the wavelength of the incident laser is λ_0, the corresponding wave digit υ_0, the center wavelength λ of the scattered light after Raman frequency shift generated by the interaction between laser and atmospheric substances, can be expressed as follows:

$$\lambda = \frac{1}{\upsilon_0 - \Delta \upsilon} \tag{4.18}$$

There are many dense transition lines between different phase water molecules, which leads to no relatively independent structure of Raman spectra of phase water, and is also the fundamental reason for the overlap between phase water Raman spectra. The overlapping information between phase water Raman spectra is the key to high precision extraction and accurate inversion of echo signals, and it is also one of the sources of detection error of atmospheric temperature and humidity. Figure 4.10 shows the Raman scattering spectra of liquid water and ice water measured at an excita-

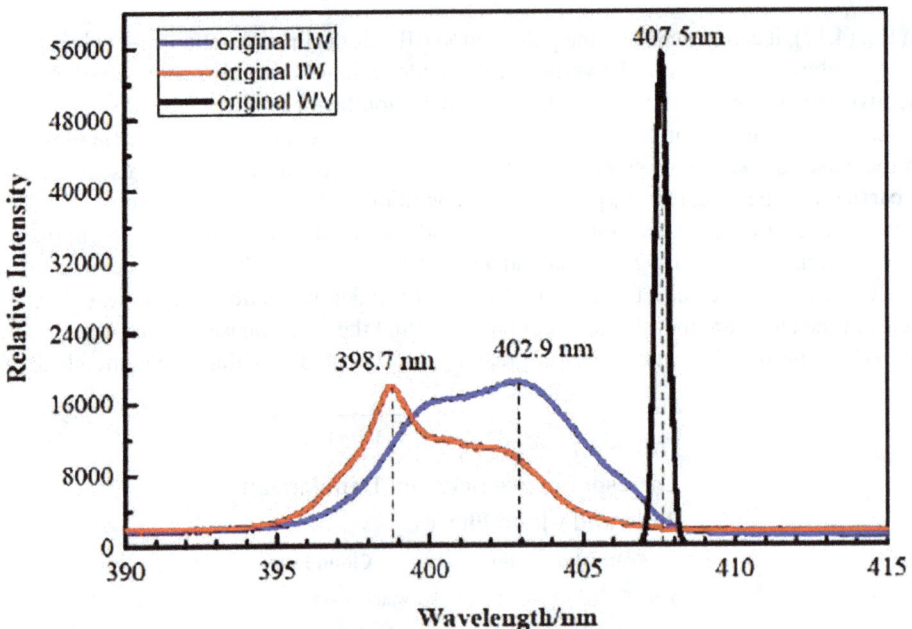

FIGURE 4.10   Raman spectroscopy of three-phase water.

tion wavelength of 355 nm under laboratory conditions (room temperature 25°C, humidity 32%). It can be seen from the figure that the Raman spectra of liquid water and ice water are continuous in the wavelength range of 395–408 nm, and there is an obvious overlap region. The peak value of the Raman spectra of liquid water is near 402.9 nm, and the bandwidth is about 8 nm. The peak value of Raman spectra of liquid water is around 398.7 nm, and the bandwidth is about 6 nm. Meanwhile, the theoretical curve of water vapor vibration Raman spectrum is given in Figure 4.10. The central wavelength is 407.5 nm, and the bandwidth is narrow, about 1 nm.

4.5 CLOUD MICROPHYSICS PROPERTIES

Cloud microphysical properties, determined by the phase, size, shape, and number concentration of cloud droplets, play an important role in cloud radiation feedback and have significant impact on the fields of weather and climate change. The cloud droplet spectral distribution, the cloud water content, the cloud water path, and the cloud droplets effective radius are all basic physical variables that characterize the microphysical properties of clouds. It is necessary to obtain values on the basis of instruments observation and retrieval algorithm for improving cloud characterization in climate models. The basic definitions and retrieval algorithm of cloud microphysical properties are introduced in this section.

4.5.1 BASIC THEORY

The particle size is the basic parameter to describe the clouds microphysical properties, which can be characterized by the diameter or radius for spherical or nearly spherical particles and the equivalent diameter or maximum size for non-spherical particles. It is difficult to measure the size of each particle for clouds with a large number of droplets, so the droplet size distribution should be quantified using a density distribution function. That is, dividing the cloud droplets size into equal or unequal bins (the width of the i_{th} bin in the unit volume is Δr_i) and then counting the number of cloud droplets in each bin (for the i_{th} bin in the unit volume is ΔN_i). The number of droplets in each bin of the unit volume can be obtained and defined as:

$$n_i = \frac{\Delta N_i}{\Delta r_i},$$

(4.19)

where the ΔN_i is the number density (or concentration) in the i_{th} bin. In order to avoid the difficulty in comparing the size distribution of various particles with different bin widths, the bin width can be further reduced to be small enough by applying the concept of differentiation to obtain the particle number density distribution function:

$$n(r) = \frac{dN}{dr}$$

(4.20)

where n(r) is the number density distribution function and dN represents the number density in the bin of $[r + \Delta r]$. The total density N can be further gained by integrating the number density over all bins as follows:

$$N = \int_0^\infty n(r)\,dr$$

(4.21)

4.5.1.1 Cloud Droplet Spectral Distribution

The cloud droplet spectral distribution is the density distribution function of the cloud, and the specific function expression can be obtained by fitting the measured data. However, it is difficult to obtain accurate expression because the distribution varies with the environment (region, temperature, humidity, etc.) and the cloud development stage. There are two typical cloud droplet spectral distribution functions commonly used in cloud physics research, described here.

Deirmendjian (1969) proposed that the cloud droplet spectral distribution constructed based on the power exponent of the radius is close to the distribution fitted by the observational data. The modified Gamma distribution is expressed as:

$$n(r) = ar^{\mu} \exp\left(-br^{\gamma}\right) \tag{4.22}$$

and the total particle density is defined as:

$$N = \int_0^{\infty} n(r)\,dr = \frac{a}{\gamma} b^{-\frac{\mu+1}{\gamma}} \Gamma\left(\frac{\mu+1}{\gamma}\right) \tag{4.23}$$

where r is the particle radius; a, b, μ, and γ are the four non-negative control parameters that can be adjusted with the particle properties. The disadvantage of the modified Gamma distribution is that there are too many parameters to be adjusted in the function, which means the distribution function has high uncertainty and complexity.

In addition to the Gamma distribution, the lognormal distribution is also one of the distribution functions widely used in cloud physics. The distribution function is expressed as

$$n(r) = \frac{N}{\sqrt{2\pi}\, r \ln \sigma_g} \exp\left[-\frac{\left(\ln r - \ln r_g\right)^2}{2\ln^2 \sigma_g}\right] \tag{4.24}$$

where N is the total number of cloud droplets per unit volume; r is the droplet radius; r_g and σ_g are the geometric mean radius and standard deviation of the cloud, respectively.

$$\ln r_g = \overline{\ln r} \tag{4.25a}$$

$$\sigma_g^2 = \overline{\left(\ln r - \ln r_g\right)^2} \tag{4.25b}$$

which are the key parameters to determine the spectral shape, and vary with different algorithms for various instruments.

4.5.1.2 Cloud Droplet Radius

The cloud droplet radius is a key physical variable to quantify particle size, and its mean and variance are often used to characterize the cloud droplet spectral distribution properties. The series of representative statistics variable related to particle radius are defined as follows:

(i) The average radius $\overline{r}$

$$\overline{r} = \int_0^{\infty} rn(r)\,dr \Big/ \int_0^{\infty} n(r)\,dr = \frac{1}{N} \int_0^{\infty} rn(r)\,dr \tag{4.26}$$

(ii) The median radius r_{med}

$$\int_0^{r_{med}} n(r)\,dr = \frac{1}{N} \tag{4.27}$$

(iii) The mode radius r_c

$$\left[\frac{dn(r)}{dr}\right]_{r_c} = 0 \tag{4.28}$$

(iv) The volume-mean radius r_V

$$r_V^3 = \frac{1}{N}\int_0^{\infty} r^3 n(r)\,dr \tag{4.29}$$

(v) The effective radius r_e

$$r_e = \frac{\displaystyle\int_0^{\infty} r^3 n(r)\,dr}{\displaystyle\int_0^{\infty} r^2 n(r)\,dr} \tag{4.30}$$

4.5.1.3 Cloud Water Content

Cloud water content is the mass of liquid or solid water contained per unit volume, including liquid water content (LWC) for warm cloud and ice water content (IWC) for cold cloud. On the basis of the cloud droplet spectral distribution, the LWC and the IWC can be respectively expressed as

$$LWC = \int_0^{\infty} \rho_w N(r)\frac{4}{3}\pi r^3\,dr \tag{4.31}$$

and

$$IWC = \int_0^{\infty} \rho_i \frac{\pi}{6} N(D) D^3\,dD, \tag{4.32}$$

where ρ_w and ρ_i are the density of liquid and ice, respectively. r is the droplet radius, and D is the equivalent diameter of the ice crystal. The liquid water path (LWP) or ice water path (IWP) can be further defined as the columnar cloud water content or the integral of the water content over the path, that is:

$$LWP = \int_{z_{base}}^{z_{top}} LWC(z)\,dz \tag{4.33}$$

and

$$IWP = \int_{z_{base}}^{z_{top}} IWC(z)\,dz. \tag{4.34}$$

4.5.2 RETRIEVAL ALGORITHM FOR CLOUD WATER CONTENT

There are two important ways to obtain cloud microphysical properties. One is direct observation with instruments; for example, the aircraft equipped with the particle measurement system can directly provide data of microphysical properties such as cloud water content, cloud droplet effective particles, and cloud water path. And the other is to use appropriate algorithms for retrieval based on the observation data. The observation equipment includes active and passive remote sensing instruments such as radar, spectrometer, microwave radiometer, etc. Observations are introduced in detail in the previous section on cloud remote sensing, so the retrieval algorithm of cloud microphysical properties are mainly illustrated in this section.

The algorithms for retrieving cloud water content mainly include the empirical relationship method based on single-wavelength radar, the differences with reflectivity at two frequencies based on dual-wavelength radar, and the combination of radar and other instruments.

4.5.2.1 Single-Wavelength Radar

Based on the definitions of the radar reflectivity and cloud water content, they are related to the sixth power and the third power of the particle radius, respectively. Therefore, the relationship between radar reflectivity and cloud water content can be established for warm or cold cloud as

$$LWC = aZ^b \tag{4.35a}$$

$$IWC = cZ^d \tag{4.35b}$$

where a, b and c, d are the empirical coefficient corresponding to warm or cold cloud, respectively. The coefficients are affected by various factors such as the cloud location, cloud type and whether there is precipitation in the cloud. In addition to the empirical coefficient method, there are several other algorithms for obtaining the cloud water content using single wavelength radar. For example, the forward mode applied to CloudSat can retrieve LWC and IWC based on 94 GHz cloud radar (Austin and Stephens, 2001). Additionally, the LWC retrieval algorithm can also be constructed using the intrinsic relationship between the attenuation, LWC and radar reflectivity on the basis that the attenuation caused by cloud absorption is proportional to LWC. The algorithm does not rely on the use of assumed coefficients to calculate the cloud droplet spectral distribution function, and is therefore free from the limitations of empirical coefficients.

For the coefficients of liquid water content, Atlas (1954) proposed that the coefficients for non-precipitation stratus clouds are $a = 4.564$ and $b = 0.5$ by fitting the aircraft observation data and the radar reflectivity; Sauvageot and Omar (1987) obtained $a = 14.54$ and $b = 0.76$ based on the observational experiments and first proposed that the threshold of precipitation and non-precipitation cloud is about -15 dBZ; Fox and Illingworth (1997) proposed $a = 9.27$, $b = 0.64$ using the radar data of stratocumulus cloud in the North Atlantic; Baedi et al. (2000) obtained the empirical coefficient of $a = 0.457$ and $b = 0.19$ for the cloud with drizzle based on the experimental data of CLARE98 (Cloud LIDAR and Radar Experiment in 1998), and Krasnov and Russchenberg (2002) obtained the empirical coefficient $a = 0.258$ and $b = 0.633$ for the cloud with precipitation particles by combining the test results of the convective precipitation clouds with the data of CLARE98. The retrieved results based on these empirical coefficients are shown in Figure 4.11.

Similarly, for the ice cloud, Sassen (1984) proposed empirical coefficients with $c = 0.037$ and $d = 0.629$ for ice clouds appearing at the poles; Liu and Illingworth (2000) obtained the coefficient $c = 0.097$ and $d = 0.59$ using data from several field experiments, and Protat et al. (2007) proposed that the coefficients applicable to global ice clouds, ice clouds at mid-high latitude and tropical ice clouds are $c = 0.09$, $d = 0.58$; $c = 0.082$, $d = 0.54$; $c = 0.103$, $d = 0.6$ respectively.

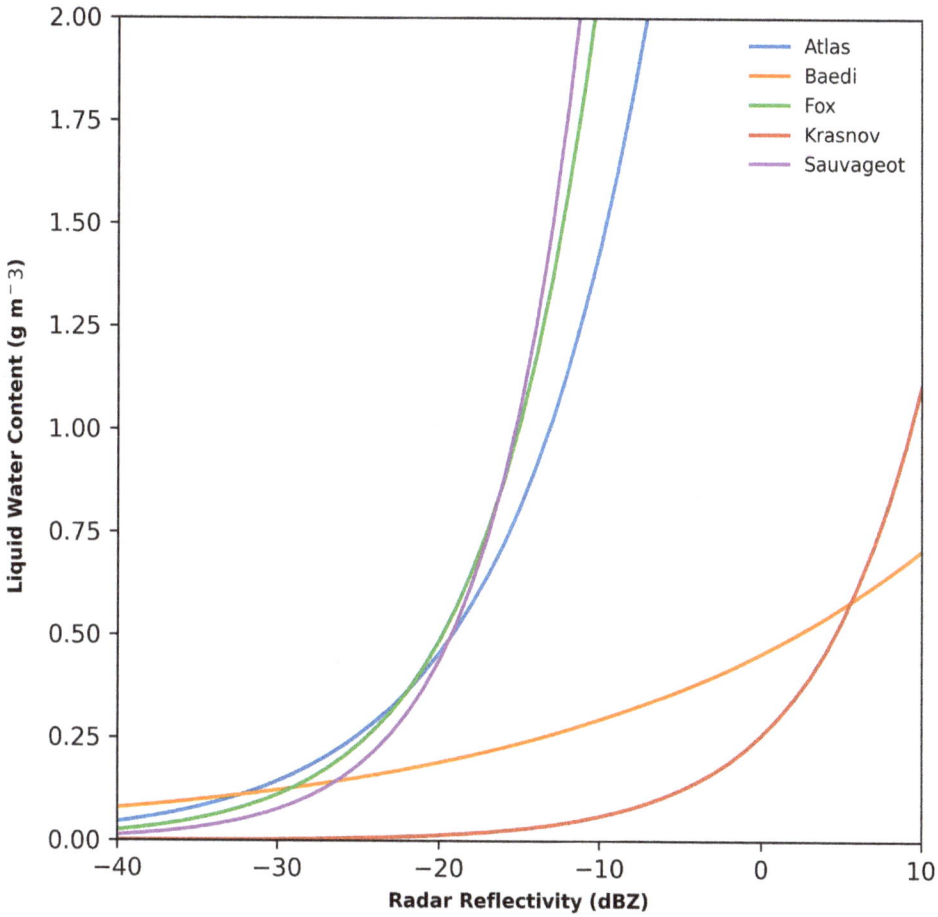

FIGURE 4.11 Diagram of retrieving cloud liquid water content based on Z-LWC empirical relationship.

4.5.2.2 Dual-Wavelength Radar

In the Rayleigh approximation regime, the attenuation caused by the cloud absorption is proportional to the LWC and increases with the frequency. And the attenuation coefficient difference of the cloud radar at two frequencies (f_1 and f_2) is proportional to the difference with height in dual-wavelength ratio (DWR), so it is possible to apply the DWR to retrieving the LWC. The dual-wavelength ratio in logarithmic units is defined as $DWR = Z_{f_1} - Z_{f_2}$ and the LWC at each range gate between the height of h_1 and h_2 can be calculated by

$$\text{LWC} = \frac{1}{\kappa_{f_1} - \kappa_{f_2}} \left[\frac{DWR_2 - DWR_1}{2(h_2 - h_1)} - \alpha_{f_1} + \alpha_{f_2} \right] \tag{4.36}$$

where α_f is the one-way attenuation coefficient resulting from the atmospheric gases and κ_f is the attenuation coefficient of liquid water, which is defined as

$$\kappa_f = 4.343 \times 10^3 \frac{6\pi}{\lambda \rho_l} Im(-K) \tag{4.37}$$

where λ is the radar wavelength, ρ_l is the density of liquid water and K is the complex dielectric constant ε by $K = (\varepsilon - 1)/(\varepsilon + 2)$.

4.5.2.3 Combination of Radar and Microwave Radiometer

Millimeter-wave radar combined with other remote sensing equipment is one of the important manners to retrieve cloud microphysical properties. The microwave radiometer is the commonly used passive remote sensing instrument for providing accurate LWP, which can be regarded as a constraint on the relationship between Z and LWC. The LWC at i_{th} range gate can be calculated by (Frisch et al., 1995):

$$LWC(i) = \frac{LWPZ(i)^{\frac{1}{2}}}{\sum_{i=1}^{m} Z(i)^{\frac{1}{2}} \Delta h}$$ (4.38)

where m is the total number of radar gate within the cloud layer and Δh is the height between different radar gates.

4.5.3 RETRIEVAL ALGORITHM FOR CLOUD DROPLET EFFECTIVE RADIUS

There is also an exponential relationship between the cloud droplets effective radius r_e and the radar reflectivity factor Z for retrieval, which is expressed as

$$r_e = AZ^B$$ (4.39)

where A and B are empirical coefficients. Table 4.2 lists several relationships between reflectivity and cloud droplet effective radius obtained from previous studies.

Furthermore, if the cloud droplet spectral distribution is assumed to satisfy the lognormal distribution, the effective radius can be obtained by (Frisch et al., 1995; Wu et al., 2014)

$$r_e = 50\exp\left(-0.5\sigma^2\right) N^{-1/6} Z^{-1/6}$$ (4.40)

where σ is the logarithmic spread of the droplet spectral distribution.

TABLE 4.2
The List of Algorithms for Cloud Droplet Effective Radius Retrieval in Previous Studies

Cloud Phase	Reference	Relationship
Liquid Cloud	Atlas (1954)	$r_e = 21.99Z^{0.33}$
	Sauvageot and Omar (1987)	$r_e = 52.72Z^{0.31}$
	Fox and Illingworth (1997)	$Z = 56.5\log r_e - 77.3$
Ice Cloud	Zhao et al. (2016)	$r_e = 59.8Z^{0.06}$
	Matrosov et al. (2003)	$r_e = 39.9\exp\left(-0.16\right) Z^{0.16(1-f)}$
	Fu (1996)	$\log IWC = 0.06Z_e - 0.0212T - 1.92$ $r_e = 192IWC^{0.331}$

For the precipitation cloud, the particle effective radius can be calculated on the basis of the radar radial velocity V_r; the relationship between the radius and velocity is (Liu et al., 2012)

$$r_e = aV_T + b \qquad (4.41)$$

where V_T is the falling velocity of the particle. It can be calculated from the radial velocity V_r using Eq. (4.42) as follows:

$$V_r = -\frac{\int_{D_{min}}^{D_{max}} V_T(D) \times N(D) \times D^6 dD}{\int_{D_{min}}^{D_{max}} N(D) \times D^6 dD} \qquad (4.42a)$$

$$N(D) = \frac{N_0}{\sqrt{2\pi} r \sigma_x} \exp\left[\frac{(x - x_0)^2}{2\sigma_x^2}\right] \qquad (4.42b)$$

where N is the density of precipitation particles; N_0, x_0 and σ_x are the total number density, the logarithmic droplet size and spread of the droplet spectral distribution, respectively; D is the particle diameter; $x = \ln D$ and $x_0 = \ln R_0$, where R_0 is the average droplet size. Liu et al. (2012) proposed that the influence of the air vertical velocity on the falling velocity of precipitation particles can be ignored in stable cloud layer, so the falling velocity of precipitation particles can be replaced by the radial velocity detected by radar. The effective radius can be calculated by radar Doppler velocity directly and the coefficients of a and b in Eq. (4.41) are $a = 1.2 \times 10^{-4}$ s and $b = 1.0 \times 10^{-5}$ m, respectively.

4.6 MICROWAVE REMOTE SENSING OF PRECIPITATION AND CLOUD

Earth observation and interstellar exploration using radars, earth satellites and spacecraft have been carried out for 64 years since 1958. Many discoveries have been made in studying the planets of the solar system, the deep space and stars of the universe, and a macroscopic, real, rapid, and dynamic understanding of the land-ocean-atmosphere system on which mankind depends for survival has been achieved, which is a great achievement in human scientific and technological development history. In both interstellar exploration and earth observation, we mainly use electromagnetic waves as a medium, and remote sensing to obtain target information (Figure 4.12). The visible spectrum (0.45 ~ 0.80 μm) is used to obtain the visible image of the target; the infrared band (0.8 ~ 1000 μm) is used to obtain the infrared image of the target; the microwave band (1 mm ~ 1 m) is used to obtain the microwave data and images of the target.

Remote sensors in visible and infrared wavelengths have been the main remote sensors in remote sensing technology because they have high spatial resolution and can obtain images consistent with human visualization. However, the necessity of daylight conditions (except thermal infrared) and the absence of cloud cover are their weaknesses, resulting in low image acquisition rates, which prevent the advantages of remote sensing, such as real-time dynamic monitoring, from being fully exploited. Visible light and near-infrared remote sensors are used to detect the amount of scattering of sunlight by an object, which only reflects the surface condition of the object, so the information obtained about the characteristics of the observed object is not rich enough.

Microwave remote sensing employs microwave radiation using wavelengths that range from about 1 mm to 1 m, in frequency interval from 300 GHz to 300 MHz, which enables observation in all weather conditions and get a 3-dimensional structure of cloud and rain all day.

Microwave remote sensing can be divided into two categories according to its working principle: active remote sensing and passive remote sensing (Figure 4.13). Active remote sensing is used to

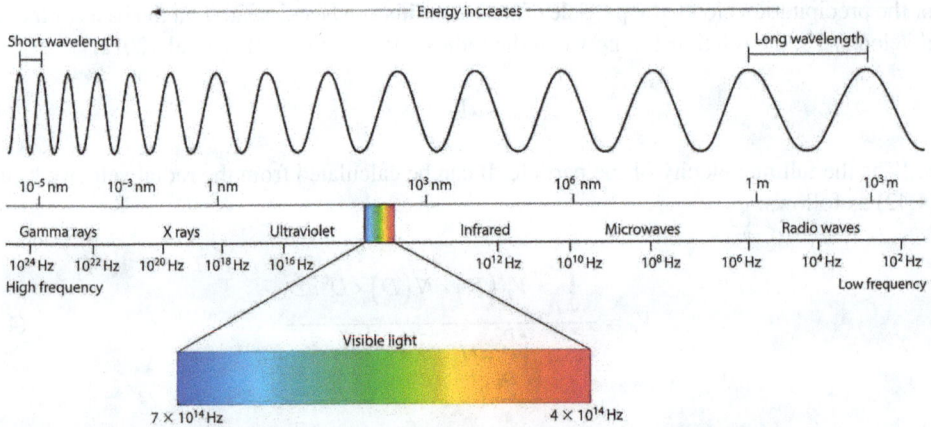

FIGURE 4.12 Electromagnetic spectrum.

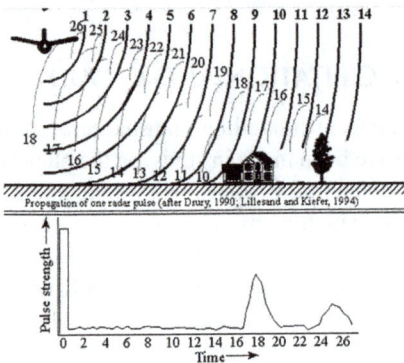

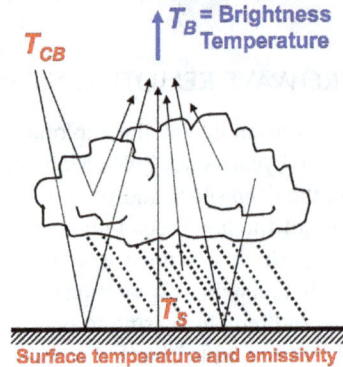

FIGURE 4.13 Difference between active and passive remote sensing.

transmit microwave signals to the detection target through sensors (mainly radar) and receive the backscattered signals from the interaction with the target to form a digital or analog image of remote sensing. Passive remote sensing uses sensors such as microwave radiometers or microwave scatter meters to receive the reflected and emitted microwaves from the ground under natural conditions, which usually cannot form images.

4.6.1 ACTIVE MICROWAVE REMOTE SENSING

Weather radar, cloud radar and earth satellite as active microwave remote sensing system has been used in atmospheric science to observe cloud and precipitation since the 1940s. Especially in the past 20 years, microwave remote sensing technology has made great progress. Weather radar, cloud radar, and meteorological satellites provide continuous observation of cloud, precipitation, and deep convective system, effectively making up for the shortcomings of the traditional ground observation network.

Among them, weather radars in S-, C-, and X-band can detect the occurrence and development of precipitation structures and convective systems with high spatial and temporal resolution without attenuation of raindrops and have promoted the development of quantitative precipitation estimation (QPE) technology. The methods of estimating precipitation rates by weather radars can be broadly classified into two categories. The first one is based on the Z-I relationship to determine precipitation intensity. The Z-I relationship is a direct relationship between the statistical characteristics of radar reflectivity and precipitation rate without calibration of rain gauges. In the 1940s, Ryde (1941) first established the relationship between radar reflectivity and particles such as clouds, rain, fog, and hail for weather forecasting. Then the concept of radar precipitation measurement was proposed and the relationship between convective storm volume and precipitation rate was discussed and established (Bent, 1943; Byers, 1948; Doneaud et al., 1984; Atlas et al., 1990a and 1990b; Kedem et al., 1990). Besides, the probability matching method (PMM) was proposed, which thought that the probability of a certain rain rate and the probability of a certain value Z should be equal in a homogeneous climate area and the relationship can be statistically obtained from the climate average (Calheiros, R. V and I Zawadzki., 1987).

The second one is based on the combination of accurate measurements of rain gauges at points and radar to estimate the precipitation rates. The method of rain gauge calibration assumes that the precipitation measured by the rain gauge at the point is accurate, and the calibration aims at minimizing the deviation between the radar and the rain gauge measurements and extracting the calibration field spatially using objective analysis methods to obtain the precipitation analysis field. Scientists import complicated mathematic methods to calibrate the error such as optimization, the Kriging method, and the Kalman filter (Ahnert et al., 1986; Krajewski, W. F, 1987; Zhang et al., 1992). However, the QPE is still unclarified from region to region because of the great spatial and temporal variation of precipitation. So, with the development of radar technology, since the 1980s, countries such as Europe and the United States began to develop dual-polarization Doppler weather radar and put it into operational use. The introduction of the dual polarization variables significantly helps us calculate the variation of the raindrop size distribution (DSD) and reduces the quantitative precipitation estimation (Seliga and Bringi, 1976; Ryzhkov et al., 2005).

High spatial and temporal resolution cloud observations are essential for better understanding of various cloud dynamics and microphysical processes, improving cloud parameterization schemes, improving the reasonable characterization of clouds in models, evaluating and constraining model simulation results, and reducing the cloud large uncertainties in the model. Cloud radar in millimeter wavelengths such as Ka- and W-band has high sensitivity to cloud droplet particles, has good penetration of thick and precipitation clouds, and can more accurately observe turbulence and other microphysical processes in clouds based on the Doppler effect. The US Department of Energy's Atmospheric Radiation Measurement (ARM) program has deployed several millimeter-wave cloud radars in different climatic areas all over the world since the late 1980s to provide long-term continuous observations of clouds and their properties (Stokes and Schwartz, 1994; Ackerman and Stokes, 2003). Many studies about the macro and microphysical properties of clouds and their relationship with weather and climate have been based on the data. For example, Dong et al. (1997) studied the radiative effects of marine boundary layer clouds in the Azores and continental boundary layer clouds in Oklahoma and assessed their uncertainties with the help of radiative transfer models; Shupe et al. (2001) studied the microphysical and radiative properties of clouds in the Arctic Ocean region and raised concerns about the inversion methods and radiative effects of mixed-phase clouds in the Arctic; Mace and Benson-Troth (2002) analyzed and compared regional differences in cloud vertical overlap properties using millimeter wave cloud radar observations from multiple locations. The cloud overlap characteristics in the tropics, mid-latitudes, and the Arctic are to some extent consistent with the random overlap hypothesis, while at mid-latitudes, the cloud overlap characteristics show more pronounced seasonal variations. In the mid-latitudes, the cloud overlap characteristics show more obvious seasonal variations.

Ground-based millimeter wavelength radar provides local long-term continuous cloud observations, while satellite-based millimeter-wavelength cloud radar offers the vertical structure of global clouds. Thanks to the development of radar detection technology and aerospace technology, more and more satellites with millimeter-wave radar have been launched or are in the planning stage of operation in recent years. The National Aeronautics and Space Administration (NASA) launch the CloudSat satellite which carries the first W-band Cloud Profiling Radar (CPR) to offer vertical structure of tropical clouds and precipitation (Im et al., 2006; Lebsock and L'Ecuyer, 2011). The Japanese Aerospace Exploration Agency (JAXA) and NASA jointly launched the Global Precipitation Measurement (GPM) core satellite which carries the second generation of the dual-frequency precipitation radar (DPR) which uses the Ku-band radar (KuPR) at 13.6 GHz and Ka-band radar (KaPR) at 35.5 GHz to achieve global precipitation observation (Iguchi et al., 2012; Hou et al., 2014).

4.6.2 Passive Microwave Remote Sensing

Global geophysical measurements from passive microwave radiometers provide key variables for scientists and forecasters. The continuous daily measurements of sea surface temperature (SST), wind speed, water vapor, cloud liquid water, rain rate, and, in the future, sea surface salinity (SSS) over the oceans have provided data sets used to significantly improve our understanding of daily variation, macro and microphysical parameters, and development mechanism of cloud and precipitation.

The inversion of sea surface rainfall rates using satellite-based microwave radiometer observations dates back to the 1970s. Allison et al. (1974) earlier demonstrated that the spaceborne microwave radiometer can be used to identify different rain rate distributions under the control of tropical cyclones. Subsequently, Wilheit et al. (1984) simulated the quantitative relationship between bright temperature and rainfall rate for the 19.35 GHz channel based on radiative transfer models. In the simulation of bright temperature, the study not only considered the absorption of oxygen, water vapor, liquid water in clouds, and raindrops, but also calculated the effect of raindrop scattering using the Mie scattering theory. A good dependence of the simulated bright temperature on the height of the ice crystal layer was found, and an important "beam filling effect", i.e., the influence of the non-uniform distribution of rainfall on the inversion results within the surface metric of the microwave radiometer, was also found in this study.

In 1997, the world's first TRMM satellite for measuring rainfall in the tropics was launched with active and passive instruments, opening a new era of microwave remote sensing of precipitation (Kummerow et al., 1998). The researchers combined the active radar and passive microwave instruments on board the TRMM to conduct a lot of joint active-passive rainfall inversion studies (Haddad et al., 1997; Grecu et al., 2002). In 2014, the GPM core satellite was launched with a precipitation radar and microwave radiometer to observe the global precipitation distribution.

4.7 SUMMARY

This paper introduces the importance of clouds, methods of cloud remote sensing, cloud detection methods, cloud phase detection, cloud microphysical property retrieval, and microwave remote sensing applications for clouds and precipitation from six different perspectives, which comprehensively reveal the cloud physical properties and their significant effects on the climate system and radiation balance.

Part I introduces the importance of clouds in the Earth's climate system from different perspectives, focusing mainly on the macroscopic and microscopic physical properties of clouds and their radiative effects on the system, and further introducing the concept of cloud feedback and the role of the greenhouse effect on clouds. Differences in cloud height, optical thickness, and phase state result in large differences in cloud radiation properties, as do differences in geographic location. In addition, the interaction between clouds and aerosols and precipitation likewise has a significant impact on the climate system; the aerosol-cloud-radiation-precipitation interaction is one of the most uncertain factors in climate prediction.

Part II focuses on the means of application of remote sensing for the observation of cloud properties and radiation characteristics. Remote sensing can be divided into active remote sensing and passive remote sensing according to the working principle. The former emits microwave signals to the detection target through sensors and receives scattered signals to form remote sensing images, while the latter uses microwave radiometers or microwave scatter meters to receive reflected and emitted microwaves from the ground, which usually cannot form images. Active remote sensing can be used to detect microwaves, because the wavelength of microwaves is long, so it is not easy to receive the influence of atmospheric scattering, microwave detection is not limited by time, and now the mainstream of atmospheric remote sensing detection method is microwave radar detection. Compared with active remote sensing, passive remote sensing is usually installed on satellites to continuously monitor the entire atmosphere and record the electromagnetic radiation emitted at the top of the atmosphere.

Long-term continuous observations of clouds in terms of macro- and microphysical properties are important in order to better understand cloud processes to improve the parameterization of climate models. Therefore, a detailed description of cloud detection algorithms for ground-based radar is given in Part III, which lists two cloud detection methods: the classical cloud detection method for ground-based cloud radar and a novel cloud detection method improved by Ge et al. (2017). The classical cloud detection method assumes that the random noise power follows a normal distribution, based on which the mean and standard deviation of the noise in a radar ranging gate containing only noise are calculated, then the valid signals in the ranging gate are filtered by these two covariates. Since cloud signals are highly correlated in time and space and have more similar values in neighboring pixels, low-pass filtering using spatial-temporal filters is required. The improved ground-based radar cloud detection method considers the presence of a non-Gaussian distribution of random noise power. The method takes the signal-to-noise ratio (SNR) as the input term, and the input data set is divided into two groups, where the data to be noise-reduced are processed by convolution. After processing the noisy data, it is bilaterally filtered to obtain the final cloud mask.

In many atmospheric studies, cloud liquid water and cloud ice water content are important indicators of cloud properties, and it is important to achieve high precision detection of cloud phase states. Part IV focuses on the detection of liquid and ice phases of clouds using LIDAR in active remote sensing technology. Currently, the main use is polarization LIDAR to observe the phase state of clouds. When the linearly polarized light emitted from the polarization LIDAR is irradiated onto the cloud particles, the resulting polarization state direction is perpendicular to the polarization state direction of the incident light, so the phase state of the cloud particles can be calculated by calculating the ratio of the vertical and parallel components of the scattered light in the echo signal. By using this method, the depolarization rate can be calculated in combination with the radar calibration factor to better distinguish the cloud phases. In addition, there are a few scientific departments using Raman LIDAR for scientific research, which mainly calculates the overlap information between phase water Raman spectra to achieve high precision extraction and accurate inversion of the echo signal.

Cloud microphysical properties have important effects on weather, climate change, and radiative feedback, and obtaining cloud microphysical property values using instrumental observations and retrieval algorithms can effectively improve cloud properties in climate models. Part V focuses on the basic definitions of cloud microphysical properties and retrieval algorithms. The basic parameters of cloud microphysics are mainly characterized by the particle size of cloud droplets, which in turn can be expressed by the spectral distribution of cloud droplets, the radius of cloud droplets, and the water content of clouds. The cloud droplet spectral distribution refers to the density distribution function of the particles in the cloud, mainly using the modified gamma distribution function and the log-normal distribution function. The cloud droplet radius is a key physical quantity to characterize the particle size, and its mean and variance are mainly used to characterize the cloud droplet spectral distribution. The main approach to retrieving the cloud droplet radius is to establish a connection with the radar reflectivity factor. Finally, cloud water content refers to the mass of liquid or solid water contained in a unit volume, including liquid water content and ice water content, for which the empirical relationship method based on single-wavelength radar, the difference method based on dual-frequency reflectivity of dual-wavelength radar and the combination method of radar and other instruments are mainly used to retrieve.

Microwave remote sensing plays an important role in the observation of cloud and precipitation structures; therefore, Part VI focuses on the applications of active and passive remote sensing in cloud and precipitation observation. Weather radar, cloud cover radar and meteorological satellites from active remote sensing systems are able to provide continuous observations of clouds, precipitation and deep convective systems. Weather radar in S-band, C-band and X-band can detect precipitation structures and convective systems with high spatial and temporal resolution, while weather radar can provide good estimates of precipitation rates. Passive microwave radiometer global geophysical measurements provide scientists and forecasters with continuous daily data sets of ocean surface temperature (SST), wind speed, water vapor, cloud liquid water, rainfall rates, and future sea surface salinity (SSS), significantly improving our understanding of daily variability, macro- and microphysical parameters, and cloud and precipitation development mechanisms.

REFERENCES

Ackerman, T. P., Stokes, G. M. The atmospheric radiation measurement program. Physics Today, 2003, 56(1), 38–44. doi: 10.1063/1.1554135.

Ahnert, P. R, Krajewskl, W. F., Johnson, E. R. Kalman filter estimation of radar-rainfall field bias. In: preprint 23rd Conf on radar, meteor, JP33–JP37, 1986.

Allison, L. J., Rodgers, E. B., Wilheit, T. T., . . . Fett, R. W. Tropical cyclone rainfall as measured by the Nimbus Electrically Scanning Microwave Radiometer. Bulletin of the American Meteorological Society, 1974, 55(9), 1074–1089.

Atlas, D. The estimation of cloud parameters by radar. Journal of Meteorology, 1954, 11, 309–317.

Atlas, D., Rosenfeld, D., Short, D. A. The estimation of convective rainfall by area integrals: 1. The theoretical and empirical basis. Journal of Geophysical Research, 1990a, 90(3), 2153–2160.

Atlas D., Rosenfeld D., Wolff, D. B. Climatologically tuned reflectivity-rain rate relations and links to area-time integrals. Journal of Applied Meteorology, 1990b, 29(11), 1120–1135.

Austin, R. T., Stephens, G. L. Retrieval of stratus cloud microphysical parameters using millimeter-wave radar and visible optical depth in preparation for CloudSat: 1. Algorithm formulation. Journal of Geophysical Research: Atmospheres, 2001, 106(D22), 28233–28242.

Baedi, R. J. P., De Wit, J. J. M., Russchenberg, H. W. J., . . . Baptista, J. P. V. Estimating effective radius and liquid water content from radar and lidar based on the CLARE98 data-set. Physics and Chemistry of the Earth, Part B: Hydrology, Oceans and Atmosphere, 2000, 25(10–12), 1057–1062.

Bent, A. E. Radar echoes from atmospheric phenomena. Mit Radiation Laboratory Rep., 1943.

Biele, J., Beyerle, G., Baumgarten, G. Polarization lidar: Corrections of instrumental effects. Optics Express, 2000, 7(12), 427–435.

Bony, S., Dufresne, J. L. Marine boundary layer clouds at the heart of tropical cloud feedback uncertainties in climate models. Geophysical Research Letters, 2005, 32, L20806.

Bony, S., Stevens, B., Frierson, D. M. W., . . . Webb, M. J. Clouds, circulation and climate sensitivity. Nature Geoscience, 2015, 8, 261–268.

Bretherton, C. S. Insights into low-latitude cloud feedbacks from high-resolution models. Philosophical Transactions of the Royal Society A, 2015, 373, 20140415.

Byers, H. R, 1948: The use of radar in determining the amount of rain falling over a small area. Trans. Am. Geophys. Union, 29, 187–196.

Calheiros, R. V., Zawadzki, I. Reflectivity-rain rate relationships for radar hydrology in Brazil. Journal of Applied Meteorology and Climatology, 1987, 26, 118–132.

Chen, X. L., Guo, Z., Zhou, T. J., . . . Su, J. Climate sensitivity and feedbacks of a new coupled model CAMS-CSM to idealized CO2 forcing: A comparison with CMIP5 models. Journal of Meteorological Research, 2019, 33, 31–45.

Cho, H. M., Nasiri, S. L., Yang, P. Application of CALIOP measurements to the evaluation of cloud phase derived from MODIS infrared channels. Journal of Applied Meteorology and Climatology, 2009, 48(10), 2169–2180.

Clothiaux, E. E., Miller, M. A., Albrecht, B. A., . . . Syrett, W. J. An evaluation of a 94-GHz radar for remote sensing of cloud properties. Journal of Atmospheric and Oceanic Technology, 1995, 12(2), 201–229. doi: 2.0.CO;2">10.1175/1520-0426(1995)012<0201:AEOAGR>2.0.CO;2.

Clothiaux, E. E., Ackerman, T. P., Mace, G. G., . . . Martner, B. E. Objective determination of cloud heights and radar reflectivities using a combination of active Remote sensors at the ARM CART sites. Journal of Applied Meteorology, 2000, 39(5), 645–665. doi: 2.0.co;2">10.1175/1520-0450(2000)039<0645:odocha>2.0.co;2.

Cooney, J. Remote measurements of atmospheric water vapor profiles using the Raman component of laser backscatter. Journal of Applied Meteorology (1962–1982), 1970, 9(1), 182–184.

Deirmendjian, D. Electromagnetic scattering on spherical polydispersions. Santa Monica, CA: Rand Corp, 1969.

Ding, S., Shi, G., Zhao, C. Analysis of global cloud variability of different cloud types in the last 20 years and its possible influence on climate using ISCCP D2 data. Science Bulletin, 2004, 49(11), 1105–1111.

Dolinar, E. K., Dong, X., Xi, B., . . . Su, H. Evaluation of CMIP5 simulated clouds and TOA radiation budgets using NASA satellite observations. Climate Dynamics, 2015, 44(7–8), 2229–2247.

Doneaud, A. A., Ionescu-Niscov, S., Priegnitz, D. L., . . . Smith, P. L. The ama-time integral as an indicator for convective rain volumes. Journal of Applied Meteorology and Climatology, 1984, 23, 555–561.

Dong, X., Ackerman, T. P., Clothiaux, E. E., . . . Han, Y. Microphysical and radiative properties of boundary layer stratiform clouds deduced from ground-based measurements. Journal of Geophysical Research: Atmospheres, 1997, 102(D20), 23829–23843.

Dong, Z., Wang, M., Li, D., . . . Zhang, Z. Cloud detection method for high resolution remote sensing imagery based on the spectrum and texture of superpixels. Photogrammetric Engineering & Remote Sensing, 2019, 85(4), 257–268. doi: 10.14358/pers.85.4.257.

Eerme, K. Changes in spring–summer cirrus cloud amount over Estonia, 1958–2003. International Journal of Climatology: A Journal of the Royal Meteorological Society, 2004, 24(12), 1543–1549.

Forbes, R. M., Ahlgrimm, M. On the representation of high-latitude boundary layer mixed-phase cloud in the ECMWF global model. Monthly Weather Review, 2014, 142(9), 3425–3446. doi:10.1175/MWR-D-13-00325.1.

Fox, N. I., Illingworth, A. J. The retrieval of stratocumulus cloud properties by ground-based cloud radar. Journal of Applied Meteorology, 1997, 36(5), 485–492.

Frisch, A. S., Fairall, C. W., Snider, J. B. Measurement of stratus cloud and drizzle parameters in ASTEX with a Ka-band doppler radar and a microwave radiometer. Journal of the Atmospheric Sciences, 1995, 52(16), 2788–2799.

Fu, Q. An accurate parameterization of the solar radiative properties of cirrus clouds for climate models. Journal of Climate, 1996, 9(9), 2058–2082.

Fukao, S., Hamazu, K. Electromagnetic waves. In: Radar for meteorological and atmospheric observations (pp. 7–31), 2014. doi:10.1007/978-4-431-54334-3_2.

Ge, J., Zhu, Z., Zheng, C., . . . Fu, Q. An improved hydrometeor detection method for millimeter-wave-length cloud radar. Atmospheric Chemistry and Physics, 2017, 17(14), 9035–9047. doi: 10.5194/acp-17-9035-2017.

Grecu, M., Anagnostou, E. N. Use of passive microwave observations in a radar rainfall-profiling algorithm. Journal of Applied Meteorology, 2002, 41(7), 702–715.

Haddad, Z. S., Short, D. A., Durden, S. L., . . . Black, R. A. A new parametrization of the rain drop size distribution. IEEE Transactions on Geoscience and Remote Sensing, 1997, 35(3), 532–539.

Hartmann, D. L., Ockert-Bell, M. E., Michelsen, M. L. The effect of cloud type on Earth's energy balance: Global analysis. Journal of Climate, 1992, 5(11), 1281–1304.

Hou, A. Y., Kakar. R. K., Neeck, S., . . . Iguchi, T. The global precipitation measurement mission. Bulletin of the American Meteorological Society, 2014, 95(5), 701–722.

Hu, Y. Depolarization ratio–effective lidar ratio relation: Theoretical basis for space lidar cloud phase discrimination. Geophysical Research Letters, 2007, 34(11).

Huang, J., Zhang, W., Zuo, J., . . . Chou, J. An overview of the semi-arid climate and environment research observatory over the Loess plateau. Advances in Atmospheric Sciences, 2008, 25(6), 906–921. doi: 10.1007/s00376-008-0906-7.

Iguchi, T., Seto, S., Meneghini, R., . . . Liang, L. An overview of the precipitation retrieval algorithm for the Dual-frequency Precipitation Radar (DPR) on the Global Precipitation Measurement (GPM) mission's core satellite. Earth Observing Missions and Sensors: Development, Implementation, and Characterization II, 2012, 85281C, doi:10.1117/12.977352

Im, E., Durden, S. L., Tanelli, S. CloudSat: The Cloud Profiling Radar Mission. 2006, 1–4.

Intrieri, J. M., Stephens, G. L., Eberhard, W. L., . . . Uttal, T. A method for determining cirrus cloud particle sizes using lidar and radar backscatter technique. Journal of Applied Meteorology and Climatology, 1993, 32(6), 1074–1082.

IPCC. Contribution of Working Group II to the sixth assessment report of the intergovernmental panel on climate change. In: H. -O. Pörtner, D. C. Roberts, M. Tignor, E. S. Poloczanska, K. Mintenbeck, A. Alegría, M. Craig, S. Langsdorf, S. Löschke, V. Möller, A. Okem, B. Rama (eds.) Climate change 2022: Impacts, adaptation, and vulnerability. Cambridge University Press, 2022.

Kay, J. E., Bourdages, L., Miller, N. B., . . . Eaton, B. Evaluating and improving cloud phase in the Community Atmosphere Model version 5 using spaceborne lidar observations, Journal of Geophysical Research: Atmospheres, 2016, 121, 4162–4176. doi:10.1002/2015JD024699.

Kazuo, S., Baba, A. A statistical relation between relative humidity and the GMS observed cloud amount, Journal of the Meteorological Society of Japan, 1988, 66, 1, 187–192.

Kedem, B., Chiu, L. S., Kami, Z. An analysis of the threshold method for measuring am-average rainfall. Journal of Applied Meteorology and Climatology, 1990, 29, 3–20.

Kollias, P., Clothiaux, E. E., Ackerman, T. P., . . . Mace, G. G. Development and applications of ARM millimeter-wavelength cloud radars. Meteorological Monographs, 2016, 57, 17.1–17.19. https://doi.org/10.1175/amsmonographs-d-15-0037.1.

Kollias, P., Clothiaux, E. E., Miller, M. A., . . . Ackerman, T. P. Millimeter-wavelength radars: New frontier in atmospheric cloud and precipitation research. Bulletin of the American Meteorological Society, 2007, 88(10), 1608–1624. doi: 10.1175/bams-88-10-1608.

Kondragunta, C. R., Gruber, A. Seasonal and annual variability of the diurnal cycle of clouds. Journal of Geophysical Research: Atmospheres, 1996, 101(D16), 21377–21390. doi: 10.1029/96JD01544.

Kostornaya, A. A., Saprykin, E. I., Zakhvatov, M. G., . . . Tokareva, Y. V. A method of cloud detection from satellite data. Russian Meteorology and Hydrology, 2017, 42(12), 753–758. doi: 10.3103/s1068373917120020.

Krajewski, W. F. Co-knging radar-rainfall and rain gauge data, Journal of Geophysical Research, 1987, 92(d8), 9571–9580.

Krasnov, O. A., Russchenberg, H. W. Retrieval of water cloud microphysical parameters from simultaneous RADAR and LIDAR measurements. International Unin of Radio Science, 2002, 20(3), 101–115.

Kubar, T. L., Hartmann, D. L., Wood, R. Radiative and convective driving of tropical high clouds. Journal of Climate, 2007, 20, 5510–5526. https://doi.org/10.1175/2007jcli1628.1

Kummerow C., Barnes W., Kozu T., . . . Simpson, J. The tropical rainfall measuring mission (TRMM) sensor package. Journal of Atmospheric and Oceanic Technology, 1998, 15(3), 809–817.

Lebsock, M. D., L'Ecuyer, T. S. The retrieval of warm rain from CloudSat, Journal of Geophysical Research, 2011, 116, D20209, doi:10.1029/2011JD016076

Lee, J., Yang, P., Dessler, A. E., . . . Platnick, S. Distribution and radiative forcing of tropical thin cirrus clouds. Journal of the Atmospheric Sciences, 2009, 66, 3721–3731. https://doi.org/10.1175/2009jas3183.1.

Liu, C., Illingworth, A. J. Toward more accurate retrievals of ice water content from radar measurements of clouds. Journal of Applied Meteorology, 2000, 39(7), 1130–1146.

Liu, J. FY-2云检测中动态阈值提取技术改进方法研究," 红外与毫米波学报, 2010, 29(4), 288–292, Aug. Accessed: Sep. 27, 2022. [Online]. Available: http://journal.sitp.ac.cn/hwyhmb/hwyhmbcn/article/abstract/29526?st=search

Liu, L. P., Zong, R., Qi, Y. B., . . . Jian, L. Microphysical parameters retrieval by cloud radar and comparing with aircraft observation in stratiform cloud. Engineering Science, 2012, 14(9), 64–71.

Liu, Y., Jia, R., Hua, S. Effects of aerosol-cloud interactions on energy balance and precipitation on the Tibetan Plateau. China Science and Technology Achievements, 2021, 22(24), 3.

Mace, G. G., Benson-Troth, S. Cloud-layer overlap characteristics derived from long-term cloud radar data. Journal of Climate, 2002, 15(17), 2505–2515.

Marchand, R., Mace, G. G., Ackerman, T., . . . Stephens, G. Hydrometeor detection using cloudsat—An earth-orbiting 94-GHz cloud radar. Journal of Atmospheric and Oceanic Technology, 2008, 25(4), 519–533. doi: 10.1175/2007jtecha1006.1.

Matrosov, S. Y., Shupe, M. D., Heymsfield, . . . Paquita, Z. Ice cloud optical thickness and extinction estimates from radar measurements. Journal of Applied Meteorology, 2003, 42(11), 1584–1597.

Matus, A. V., L'Ecuyer, T. S. The role of cloud phase in Earth's radiation budget, Journal of Geophysical Research: Atmospheres, 2017, 122, 2559–2578, doi:10.1002/2016JD025951.

Miao, H., Wang, X., Liu, Y., . . . Wu, G. A regime-based investigation into the errors of CMIP6 simulated cloud radiative effects using satellite observations. Geophysical Research Letters, 2021, 48(18), e2021GL095399.

Mosher, F. R. Cloud height determination, In: Proceedings of COSPAR symposium on meteorological observations from space and their contribution to FGGE. COSPAR, 201–204, 1976.

Protat, A., Delano, E. J., Bouniol, D., . . . Brown, P. R. A. Evaluation of ice water content retrievals from cloud radar reflectivity and temperature using a large airborne in situ microphysical database. Journal of Applied Meteorology and Climatology, 2007, 46(5), 557–572.

Pruppacher, H. R., Klett, J. D., Wang, P. K. Microphysics of clouds and precipitation, Aerosol Science and Technology, 1998, 28(4), 381–382.

Qi, S., Huang, Z., Ma, X., . . . Shi, J. Classification of atmospheric aerosols and clouds by use of dual-polarization lidar measurements. Optics Express, 2021, 29(15), 23461–23476.

Ramanathan, V., Cess, R. D., Harrison, E. F., . . . Hartmann, D. Cloud-radiative forcing and climate: Results from the earth radiation budget experiment. Science, 1989, 243(4887), 57–63.

Randall, D. A., Coakley, J. A. Jr, Fairall, C. W., . . . Lenschow, D. H. 1984. Outlook for research on subtropical marine stratiform clouds. Bulletin of the American Meteorological Society, 65, 1290–1301.

Riedi, J., Marchant, B., Platnick, S., . . . Dubuisson, P. Cloud thermodynamic phase inferred from merged POLDER and MODIS data. Atmospheric Chemistry and Physics, 2010, 10(23), 11851–11865.

Ryde, J. W. Echo intensities and attenuation due to clouds, rain, hail, sand, and dust storms at centimeter. Rep. No. 7831, General Electric Research Laboratory, Wembley England, 1941.

Ryzhkov, A. V., Giangrande, S. E., Schuur, T. J. Ramfall estimation with a polarimetric prototype of WSR-88D. Journal of Applied Meteorology and Climatology, 2005, 44(4), 502–515.

Sassen, K. Deep orographic cloud structure and composition derived from comprehensive remote sensing measurements. Journal of Applied Meteorology and Climatology, 1984, 23(4), 568–583.

Sassen, K. Polarization in lidar. In Lidar (pp. 19–42). New York, NY: Springer, 2005.

Sassen, K., Wang, Z., Liu, D. Global distribution of cirrus clouds from CloudSat/Cloud-Aerosol lidar and infrared pathfinder satellite observations (CALIPSO) measurements. Journal of Geophysical Research: Atmospheres, 2008, 113(D8).

Sassen, K., Wang, Z., Liu, D. Cirrus clouds and deep convection in the tropics: Insights from CALIPSO and CloudSat. Journal of Geophysical Research: Atmospheres, 2009, 114(D4).

Sassen, K., Zhao, H. Polarization lidar liquid cloud detection algorithm for winter mountain storms. In NASA. Langley Research Center, Sixteenth International Laser Radar Conference, Part 1, 1992, July.

Sassen, K., Zhao, H., Dodd, G. C. Simulated polarization diversity lidar returns from water and precipitating mixed phase clouds. Applied Optics, 1992, 31(15), 2914–2923.

Sauvageot, H., Omar, J. Radar reflectivity of cumulus clouds. Journal of Atmospheric and Oceanic Technology, 1987, 4(2), 264–272.

Seliga, T. A., Bringi, V. N. Potential use of radar differential reflectivity measurements at orthogonal polarizations for measuring precipitation. Journal of Applied Meteorology, 1976, 15(1), 69–76.

Shupe, M. D., Uttal, T., Matrosov, S. Y., . . . Frisch, A. S. Cloud water contents and hydrometeor sizes during the FIRE Arctic Clouds Experiment. Journal of Geophysical Research: Atmospheres, 2001, 106 (D14): 15015–15028.

Slingo, A., Slingo, J. M. The response of a general circulation model to cloud longwave radiative forcing. I: Introduction and initial experiments. Quarterly Journal of the Royal Meteorological Society, 1988, 114: 1027–1062.

Stein, T. H. M., Holloway, C. E., Tobin, I., . . . Bony, S. Observed relationships between cloud vertical structure and convective aggregation over tropical ocean. Journal of Climate, 2017, 30, 2187–2207.

Stephens, G. L. Cloud feedbacks in the climate system: A critical review. Journal of Climate, 2005, 18(2), 237–273.

Stephens, G. L., Vane, D. G., Boain, R. J., . . . The CloudSat Science Team. The cloudsat mission and the A-train, Bulletin of the American Meteorological Society, 2002, 83(12), 1771–1790. doi: 10.1175/bams-83-12-1771.

Stephens, G. L., Kummerow, C. D. The remote sensing of clouds and precipitation from space: A review, Journal of the Atmospheric Sciences, 64, 3742–3765, https://doi.org/10.1175/2006JAS2375.1, 2007.

Stokes, G. M., Schwartz, S. E. The Atmospheric Radiation Measurement (ARM) program: Programmatic background and design of the cloud and radiation test bed. Bulletin of the American Meteorological Society, 1994, 75(7), 1201–1221.

Tobin, I., Bony, S., Holloway, C. E., . . . Roca, R. Does convective aggregation need to be represented in cumulus parameterizations? Journal of Advances in Modeling Earth Systems, 2013, 5, 692–703.

Tomasi, C., Manduchi, R. Bilateral filtering for gray and color images, IEEE Xplore, Jan. 01, 1998. https://ieeexplore.ieee.org/document/710815

Wang, Y., Li, J., Zhao, Y., . . . Wu, X. Distinct diurnal cycle of supercooled water cloud fraction dominated by dust extinction coefficient. Geophysical Research Letters, 2022, 49, e2021GL097006. https://doi.org/10.1029/2021GL097006.

Wilheit, T. T., Greaves, J. R., Gatlin, J. A., . . . Chang, E. S. Retrieval of ocean surface parameters from the Scanning Multifrequency Microwave Radiometer (SMMR) on the Nimbus7 satellite. IEEE Transactions on Geoscience and Remote Sensing, 1984, (2), 133–143.

Wu, J., Wei, M., Hang, X., . . . Li, N. The first observed cloud echoes and microphysical parameter retrievals by China's 94-GHz cloud radar. Journal of Meteorological Research, 2014, 28(3), 430–443.

Zelinka, M. D., Klein, S. A., Hartmann, D. L. Computing and partitioning cloud feedbacks using cloud property histograms. Part I: Cloud radiative Kernels. Journal of Climate, 2012, 25(11), 0778.

Zhang, H., Wang, F., Wang, F. Progress of cloud radiative feedback in global climate change. China Science: Earth Science, 2022, 52(3), 18.

Zhang, P., Tiepei, D., Dengyan, W., . . . Lin, B. Derivation of the Z-I relationship by optimization and the accuracy in the quantitative rainfall measurement. Journal of the Meteorological Sciences, 1992, (3), 333–338.

Zhao, C., Liu, L., Wang, Q., . . . Fan, T. Toward understanding the properties of high ice clouds at the Naqu site on the Tibetan Plateau using ground-based active remote sensing measurements obtained during a short period in July 2014. Journal of Applied Meteorology and Climatology, 2016, 55(11), 2493–2507.

Zhu, Z., Zheng, C., Ge, J., . . . Fu, Q. Cloud macrophysical properties from KAZR at the SACOL. Chinese Science Bulletin, 2017a, 62(8), 824–835. doi: 10.1360/n972016-00857.

Zhu, Z., Zheng, C., Ge, J., . . . Fu, Q. Study on the macroscopic properties of SACOL station clouds using KAZR cloud radar, Science Bulletin, 2017b, (8), 824–835.

5 Remote Sensing of Cloud Properties Using Passive Spectral Observations

Chao Liu, Shiwen Teng, Yuxing Song, and Zhonghui Tan

5.1 INTRODUCTION

Clouds are ubiquitous in the atmosphere with a globally averaged occurrence larger than 60%. It strongly modulates atmospheric circulation, radiative transfer, and energy budget of the earth's atmosphere system through their interactions with radiation from solar and terrestrial sources (Liou, 1986; Shupe et al., 2006; Baker and Peter, 2008). On the one hand, clouds reflect solar radiations to space and thereby have a cooling effect on the earth's atmosphere; at the same time, clouds lead to a warming effect on the atmosphere by absorbing longwave radiations emitted from the surface; meanwhile, clouds emit longwave radiations outward. Besides, clouds are associated with extreme weather phenomena such as rainstorms, hail, tornadoes, and so on. Thus, clouds are of great concern for weather prediction and climate change research. However, clouds show complex spatiotemporal variations and microphysical properties during their formation and evolution, which greatly limit our understanding of clouds and their radiative forcing (Lenaerts et al., 2017). The fifth assessment report of the Intergovernmental Panel on Climate Change (IPCC, 2013) also indicated clouds are one of the largest uncertainty sources in future climate change. Therefore, accurately estimating cloud properties is highly important and conducive to improving our understanding of their roles in the Earth's atmosphere (Wetherald and Manabe, 1988).

To date, a large number of cloud studies, relying on multi-platforms, have been going about to improve our understanding of cloud macro- and micro-properties (Yang et al., 2015). However, their low spatial sampling and temporal resolutions limit corresponding applications based on laboratory or *in situ* platforms. The term "remote sensing", differentiated from *in situ* measurements, refers to the acquisition of information about an object without physical contact and involves measurements of electromagnetic radiation and retrievals based on passive or active sensors. Currently, remote sensing retrievals of cloud properties have played one of the most crucial and meaningful roles in cloud studies (Loeb et al., 2009; Kato et al., 2006). The sensors can take measurements onboard satellite, aircraft, or ground platforms. Ground- or aircraft-based remote sensing can hardly be used to collect observed data over the whole earth at an acceptable temporal resolution. Thus, remote sensing based on satellite measurements becomes the most practical and powerful choice for retrieving cloud properties over large spatiotemporal scales (Yang et al., 2015).

Satellite-based sensors, both active and passive ones, are vital in inferring cloud properties from space and can provide long-term, high-resolution, and stable cloud information for weather and climate studies. Those space-board instruments receive radiation scattered or emitted by clouds from visible to microwave. Active sensors, for example, Cloud-Aerosol LIDAR with Orthogonal Polarization (CALIOP) (Vaughan et al., 2005) and Cloud Profiling Radar (CPR) (Heymsfield et al., 2018), can better detect cloud vertical structures, which are significant for understanding cloud three-dimensional properties. Nevertheless, due to the limitation of current technologies, the sampling efficiency of cloud radar and LIDAR is low, which cannot be applied in associated meteorological applications, and, again, remote sensing of cloud properties using satellite-based passive

DOI: 10.1201/9781003363118-5

spectral observations is the most practicable. Both polar-orbiting and geostationary meteorology satellites have advantages in cloud detection. The former can achieve higher spatial resolutions because of their relatively lower orbital altitude, and the latter is able to monitor a particular area with higher temporal resolutions. Both two types of satellites cooperate with each other to form a comprehensive observation system for global cloud detection.

Key variables that quantitatively describe clouds include cloud top properties (e.g., cloud top pressure [CTP], cloud top temperature [CTT], and cloud top height [CTH]), cloud optical and microphysical properties (e.g., cloud mask, cloud phase, cloud optical thickness [COT], cloud effective radius [CER], and cloud water path [CWP]) during both day and night, which can be derived by satellite passive sensors (Wang et al., 2012; Iwabuchi et al., 2014; Roebeling et al., 2015; Platnick et al., 2017). Clouds vary considerably on horizontal and vertical scales (Rossow et al., 1985; Stowe et al., 1989); thus, a more accurate understanding of cloud properties as well as their spatiotemporal distributions and variations is crucial to global climate change studies (Wetherald and Manabe, 1988; Yang et al., 2015). Satellite remote sensing shows the characteristics of cloud detection with a large spatial range and long-time series, and passive-sensor-based cloud retrieval algorithms typically depend on visible and infrared bands due to their higher spatiotemporal resolutions than hyperspectral and microwave ones. At present, a constellation of more than a dozen satellites, equipped with passive sensors with visible and infrared spectral bands suitable for deriving cloud properties, has facilitated an unprecedented global view from space.

A variety of algorithms have been proposed and applied to infer pixel-level cloud properties utilizing satellite-based sensor measurements during the past few decades. The series of Advanced Very High-Resolution Radiometer (AVHRR) sensors provide one of the most comprehensive satellite records, which have been in continuous operation on NOAA polar-orbiting platforms since 1978. Based on the AVHRR imagers, decadal records of the derived cloud properties are available currently (Karlsson et al., 2013; Stengel et al., 2015). The Moderate Resolution Imaging Spectroradiometer (MODIS) onboard the Terra and Aqua satellites, with relatively high radiometric performance and spatiotemporal resolution, is one of the most widely used and reliable sensors in cloud research and related applications, as well as the ability to offer nearly global spatial coverage by combining the multiple daily satellite overpasses (King et al., 1992; Baum et al., 2012; Platnick et al., 2017). The Visible Infrared Imaging Radiometer Suite (VIIRS) onboard the Suomi National Polar-Orbiting Partnership (Suomi NPP) satellite also can take moderate-resolution atmospheric measurements, which possess the capability of inferring aerosol, cloud, and surface properties (Platnick et al., 2020). Furthermore, a number of geostationary satellite sensors with adequate spectral bands and high calibration accuracy are currently applied by many remote sensing research groups around the world to detect clouds and derive cloud properties, such as the Advanced Himawari Imager (AHI), the Advanced Baseline Imager (ABI), the Geostationary Operational Environmental Satellite-R (GEOS-R), and so on (Menzel and Purdom, 1994; Schmit et al., 2005; Goodman et al., 2013; Bessho et al., 2016; Letu et al., 2020).

This chapter reviews the development of satellite-based passive remote sensing on clouds and introduces some well-established retrieval algorithms to derive cloud properties including cloud mask, cloud phase, CTH, CTP, COT, and CER. Meanwhile, this chapter also notes concerns in terms of significant differences in retrieved results using different methods or sensor measurements and summarizes the main and potential causes. Accurate description and research of these cloud properties are of great significance for cloud observations and simulation, and play an essential role in further studies of the cloud effects on the radiative transfer of Earth's atmosphere system.

5.2 CLOUD DETECTION

Accurate cloud detection is a fairly basic and essential step for associated applications in satellite remote sensing (Nicoll et al., 2012), which is able to lay a solid foundation for subsequent cloud classification, cloud properties retrievals, and further scientific research (Liu et al., 2021). The physical foundation of cloud detection is that it has relatively high reflectance but low brightness temperature (BT) in the visible and near-infrared (NIR) bands, and there are differences between the reflectances and BTs over different underlying surface types. Then the radiances received by passive sensors are categorized to detect whether the area is cloudy or clear. Researchers have proposed various approaches for cloud detection utilizing passive radiometer measurements, and these methods can be broadly classified into three categories, i.e., the threshold-based method, statistical method, and machine learning method.

During the 1980s and early 1990s, the threshold-based approach is the most straightforward approach for cloud detection by applying a set of dynamic or static thresholds of reflectance, BT, and BT difference (BTD) (Ackerman et al., 1998; Key and Barry, 1989). However, when the cloud system is complex and two different classes have no significant BTD, the threshold cannot be confirmed; thus, the threshold-based method is no longer suitable for cloud detection (Liu et al., 2009). The statistical methods include the histogram method, the clustering method, the statistical estimation method, and so on (Berendes et al., 2008; Ebert, 1989; Ruprecht, 1985). Based on more information for available bands, these statistical methods are supposed to be more reliable than threshold-based ones. However, traditional statistical methods always fail in the presence of overlapping clouds (Liu et al., 2009). At present, many studies have acknowledged that machine learning approaches have provided impressive results and good prospects for cloud detection. This section focuses on introducing threshold-based and machine learning approaches.

5.2.1 THRESHOLD-BASED METHOD

Most of the classical and commonly used cloud detection methods belong to threshold-based algorithms. The algorithm refers to the use of multiple channels of meteorological satellite sensors based on empirical or statistical values to determine the threshold in each channel. It can combine cloud images of different channels for analysis according to different spectral characteristics of visible and infrared channels, and the key to threshold-based methods is the selection of appropriate thresholds for reflectance or BT, and the differences among various threshold-based algorithms lie in the different combinations of selected channels and the corresponding threshold values. Literature has reported a variety of threshold-based algorithms for different satellite sensors, geographical locations, and underlying surface types.

Current threshold-based methods are categorized into the spectral combined (Wang et al., 2011) and the frequency combined threshold-based method (Gao et al., 2014), and the former takes advantage of the strong reflection properties of clouds at visible wavelengths. Early threshold-based methods include the ISCCP (the International Satellite Cloud Climatology Project) (Rossow et al., 1989), the CLAVR (the NOAA Cloud AVHRR) (Stowe et al., 1994), and the CASPR (the Cloud and Surface Parameter Retrieval) algorithm (Key, 2002). At present, a cloud mask algorithm proposed and developed by the MODIS team of NASA (National Aeronautics and Space Administration) is relatively popular internationally (Ackerman et al., 1998). The algorithm uses the MODIS measurements of 22 bands and sets several characteristic values for threshold determination. The cloud detection results are mainly represented by four categories, i.e., confident clear, probably clear, probably cloudy, and cloudy. The calculation formulas of cloud mask confidence are as follows:

$$G_{j=1,N} = \min[F_i]_{i=1,m} \tag{5.1}$$

$$Q = N\sqrt{\prod_{i=1}^{N} G_j} \tag{5.2}$$

where F_i is the confidence level of an individual spectral test, m is the number of tests in a given group, j is the group index, and N is the number of groups. According to the cloud mask output based on these formulas, there are four confidence levels contained including confident clear

FIGURE 5.1   True-color composite (Band: 1–4-3) from MODIS data on 18 July 2001 at 15:30 UTC from Platnick et al. (2003).

($Q > 0.99$), probably clear ($Q > 0.95$), uncertain/probably cloudy ($Q > 0.66$), and cloudy ($Q \leq 0.66$), respectively. Figure 5.1 shows a true-color composite of a study case from MODIS data on 18 July 2001 at 15:30 UTC, and the cloud detection results using the threshold-based method for the granule of Figure 5.1 are shown in Figure 5.2 (Platnick et al., 2003).

However, certain thresholds vary with the seasons and the elevation of the sun, which leads to variational selected thresholds for all regions and times (Mahajan and Fataniya, 2020). Meanwhile, the spatiotemporal climates and underlying surfaces change on a world-wide range; thus, the variation of thresholds with various influences, including surface type, temperature, atmospheric humidity, viewing geometry, and so on, will affect cloud mask results of remote sensing (Dybbroe et al., 2005). Therefore, most threshold-based approaches are only appropriate to specific conditions or radiometers with poor universality and easily lead to omissions or misjudgments, and it is often difficult to select proper thresholds during the process of cloud detection.

With the improvement of cloud detection accuracy requirements, the threshold-based algorithm has gradually developed from the early fixed threshold to the dynamic threshold, adaptive threshold-based, and multi-spectral combination threshold. For example, Wei et al. (2016) developed a general dynamic threshold-based method using the land surface reflectance database (LSRD), which is on the basis of 8-day synthetic MODIS surface albedo products. According to viewing geometry, atmospheric, and aerosol models, the relationship between apparent reflectance and surface albedo can be simulated. In detail, the 6S model is used to consider the aerosol scattering properties, and the least square method is used to establish the new cloud detection model.

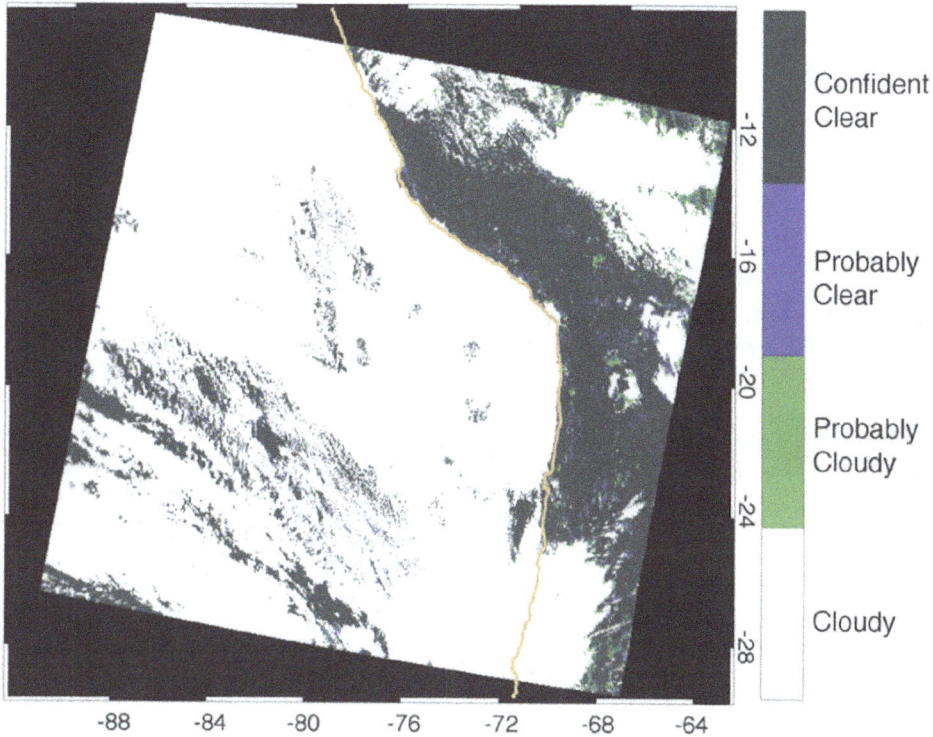

FIGURE 5.2 The spatial distribution of MODIS-derived cloud mask based on the threshold-based method for the study case of Figure 5.1 from Platnick et al. (2003).

Furthermore, MODIS Collection 6 updates to the clear sky restoral (CSR) algorithm, which is primarily focused on optimizing the skill of the "Not Cloudy" (i.e., CSR = 2) category (Platnick et al., 2017). It acknowledged that the spatial variability tests employed in MODIS Collection 5 still have some issues. Obtaining an aerosol-like spatial variability signature from very uniform optically thin marine stratus clouds is possible; thus, the CSR algorithm often created "holes" in cloud regions where retrievals should have been attempted. To remedy this issue, a neural net-based fast aerosol optical depth (AOD) retrieval algorithm was implemented with code from the Goddard Modeling and Assimilation Office used in GEOS-5 aerosol data assimilation. The algorithm was designed to operate in cloud-free conditions, which was used internally by GEOS-5. Based on the CSR algorithm, two distinct pixel populations emerge while it is applied to all MODIS "Not Cloudy" pixels. One population has a reasonable AOD retrieval, while the other gives large values outside the expected range. For present purposes, optical depth values with $\log (AOD + 0.01) > 0.95$ are assumed to be associated with cloudy scenes, and the CSR category is reset to cloudy. Figure 5.3 shows the CSR results of a case study on 9 April 2005 at 10:50 UTC from Platnick et al. (2017). It is evident that CSR results are more consistent with the true distribution of clouds.

5.2.2 MACHINING LEARNING ALGORITHMS

Since the 1990s, machine learning technology has made remarkable achievements in a lot of fields, such as computers and image recognition. Machine learning algorithms are also broadly used in cloud detection research due to their good universality in time and region, which can solve the spatiotemporal limitations of the threshold-based method to a certain extent by training data sets (Liu et al., 2009). Generally speaking, machine learning can be categorized into supervised and

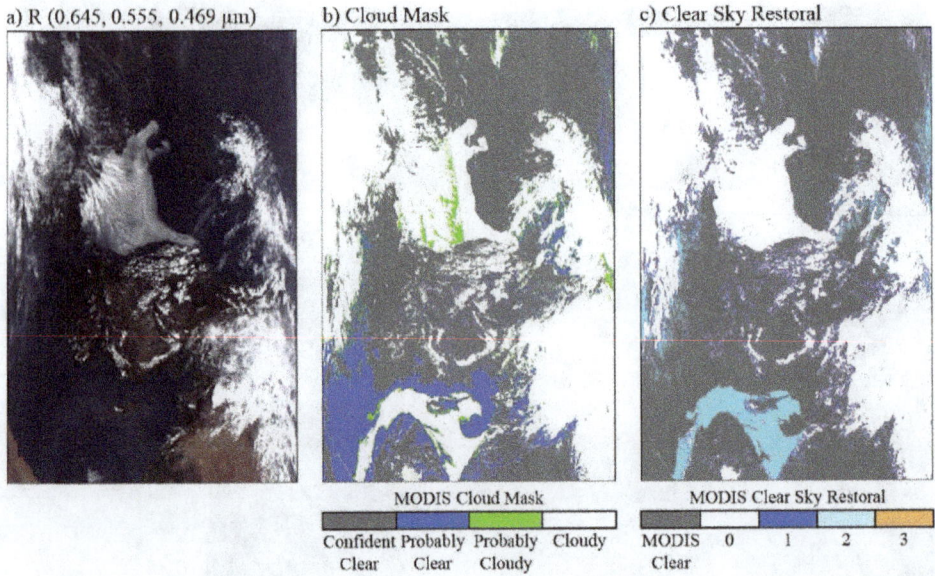

a) R (0.645, 0.555, 0.469 µm)   b) Cloud Mask   c) Clear Sky Restoral

MODIS Cloud Mask

Confident Clear   Probably Clear   Probably Cloudy   Cloudy

MODIS Clear Sky Restoral

MODIS Clear   0   1   2   3

FIGURE 5.3 (a) True-color-image (Band: 1–4–3) from MODIS data on 9 April 2005 at 10:50 UTC, (b) MOD35 cloud mask results, and (c) MOD06 Collection 6 CSR algorithm results (0-overcast, 1-cloud edge, 2-restored to the clear sky, and 3-partly cloudy) from Platnick et al. (2017).

unsupervised learning, the former being more popular among kinds of cloud detection algorithms. Various researchers used an artificial neural net with many variations, such as support vector machines, fusing multi-scale convolution features, deep learning, decision tree, Bayesian classification, random forest-based methods, and object-based convolution neural network models.

Machine learning-based cloud detection algorithms are independent of multi-spectral thresholds and have good application prospects. For different passive remote sensors, different methods are used to carry out a lot of research. Bankert (1994) used the probabilistic neural network method to carry out cloud detection for AVHRR, which started the prologue of the artificial intelligence cloud detection method. Heidinger et al. (2012) adopted the naive Bayes method to develop six Bayesian classifiers for seven underlying surface types, and the accuracy of cloud detection reached over 90% for surface types of the ocean, desert, and snowless land. Thampi et al. (2017) described the new CERES (Clouds and the Earth's Radiant Energy System) algorithm for improving the clear/cloudy scene classification, which is based on the atmospheric top radiation flux in the cloud and earth radiation system using machine learning algorithms. Chen et al. (2018) developed a threshold-free cloud mask algorithm based on a neural network classifier driven by extensive radiative transfer simulations, which made significant progress in cloud detection on the underlying surface of the snow. Ishida et al. (2018) proposed an adjustable cloud detection algorithm, which incorporates a support vector machine to satisfy the requirements so as to realize cloud detection under various scenes. Wang et al. (2020) developed two machine learning models incorporating the random forest algorithm for VIIRS onboard the Suomi NPP.

Due to significant differences in surface albedo and emission characteristics, surface type is a key contributor that should be taken into consideration for the development of cloud detection algorithms (Platnick et al., 2003). Currently, most machine learning-based cloud detection algorithms establish independent classifiers accounting for different surface types, and several studies also treat the surface type as an additional variable in the input parameters, but there is also the problem of how to parameterize the surface type. Therefore, for the purpose of addressing the impact of surface types on the stability and accuracy of cloud detection algorithms, Liu et al. (2021) proposed three models with different treatments of the surface for AHI onboard the Himawari-8 geostationary satellite. Instead of developing independent machine learning-based algorithms, the researchers added

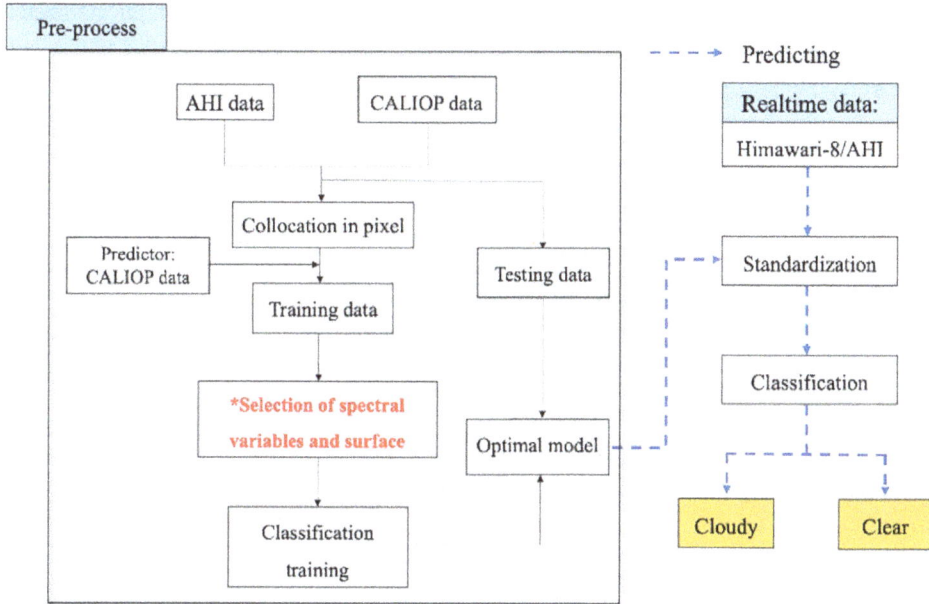

FIGURE 5.4 Flowchart of the machine learning-based cloud detection algorithm development and prediction for AHI from Liu et al. (2021).

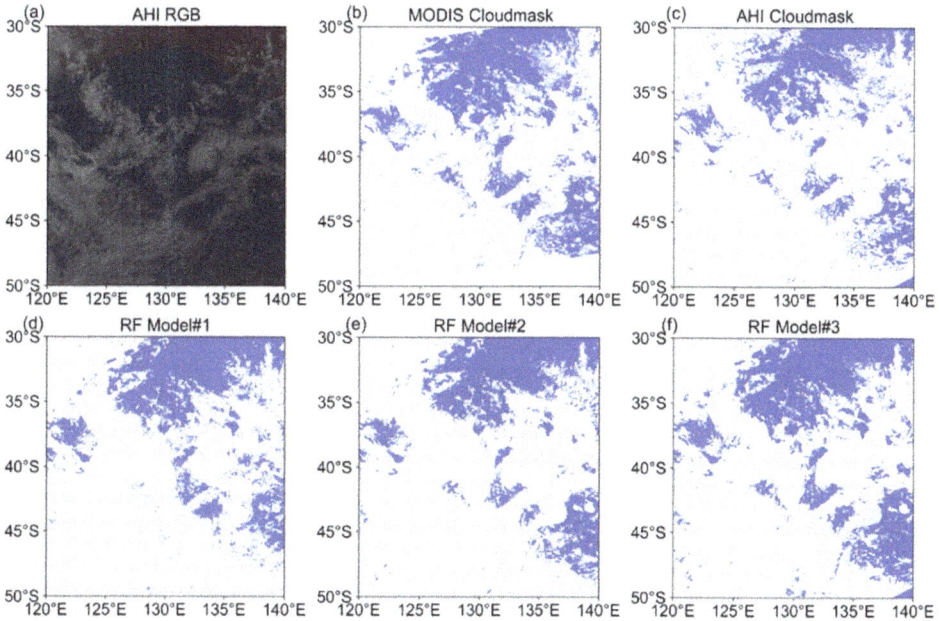

FIGURE 5.5 A study case for the MODIS and AHI operational cloud product and machine learning results on 4 June 2018 at 05:10 UTC from Liu et al. (2021). (a) RGB image, (b) MODIS cloud mask, (c) AHI cloud mask, and (d-f) random forest-based results. Cloudy pixels are marked gray, and clear ones are marked blue.

surface variables in a binary way, which enhanced the accuracy of cloud detection by ~5%. Figure 5.4 illustrates the flowchart of the machine learning-based cloud detection algorithm development and prediction for AHI, and Figure 5.5 shows a study case of cloud mask results from the AHI, MODIS operational products, and the daytime random forest-based algorithms (Liu et al., 2021).

In a word, machine learning-based algorithms are more flexible and less complicated since this kind of method simulates decisions on training data only but are not consistent due to model training relying on the input data set.

5.2.3 OTHER APPROACHES

Other than the aforementioned two series of methods, the literature shows new ones, such as statistical and texture analysis approach. Statistical methods are mainly divided into the statistical equation and cluster analysis methods. The statistical equation methods use the existing data set samples to establish the corresponding simulation formula to calculate the BT or reflectance of clouds and then judge whether clear or cloudy. The cluster analysis method mainly uses unused ground objects to present differences in pixel values to achieve cloud detection. However, when the samples of cloud detection images are large, it is difficult to obtain a consistent clustering result, which requires human intervention and greatly affects the efficiency of cloud detection.

In addition, cloud detection algorithm based on the visible bands of the satellite image is proposed by Tian et al. (2019). The method considers the differences between the cloud and the ground, including various gray levels, and accounts for a reference satellite image, which introduces a reference satellite image by comparing the variance corresponding to the reference. This method detects multiple cloud regions and determines whether or not the cloud exists in an image described. Zhang et al. (2019) proposed a cloud detection method for high resolution satellite images using ground multi-features of objects such as color and shape, then the texture is extracted utilizing a multi-scale decomposition on the basis of the domain transform. The overall accuracy of this method is high while it is easy to be misidentified if the surface albedo of the non-cloudy region is relatively high. Kwan et al. (2020) summarized straightforward and well-performed algorithms for cloud and shadow detection, which are based on an inverted map to convert an image into a greyscale image. The inverted map is the threshold to generate the shadow mask, but this method will lead to false cloud mask when improper thresholds are grasped. A novel cloud detection algorithm using superpixel segmentation (SPS) is employed for all-sky images (Liu et al., 2014). Moreover, methods based on texture features (Cao et al., 2007) and statistical features (Shan et al., 2009) already have been actually applied in the China-Brazil Earth Resources 02B Satellite.

Cloud detection is of vital importance in the process of satellite-based remote sensing, and the stability and accuracy of various algorithms are significantly improved, whereas the reports illustrate that there is much work required to achieve the desired correctness of cloud detection results for further studies.

5.3 CLOUD THERMODYNAMIC PHASE

Cloud thermodynamic phase, as a prerequisite for inferring cloud properties such as COT, CER, and CWP, is often divided into four categories including ice, liquid water, mixed-phased, and uncertain. All kinds of cloud optical and microphysical parameter retrieval methods are developed according to the different phase types, and cloud phase classification is a critical initial step: accurate results improve the accuracy of retrievals.

Passive spectral sensors mainly receive the radiation that is reflected and emitted by the surface, atmosphere, and clouds in a cloudy field of view (FOV). For optically thicker clouds, the radiation observed by the satellite mainly comes from the top of the cloud and atmosphere, which is between the top of the cloud and satellite, and the radiation from the cloud top mainly depends on the size, shape, number density, and phase of the cloud particles (Yang et al., 2003). For the electromagnetic wave of a specific incident frequency, the cloud particles will show a variety of radiation characteristics. For inferring the cloud phase, it is necessary to select a wavelength with different optical properties between the liquid water and ice clouds. According to the different selected wavelengths, this section summarizes the cloud phase retrieval algorithms into three categories, i.e., thermal infrared band method, visible and shortwave infrared (SWIR) band method, and all-spectral-band-based method.

5.3.1 INFRARED-BAND-BASED METHOD

The purpose of the infrared-band-based method for cloud phase classification is to implement an infrared-only-based technique that works independently of solar illumination conditions. Originally, a tri-spectral infrared technique based on spectral bands at 8.5, 11, and 12 μm was developed to infer cloud phase (Ackerman et al., 1990). The BTD between 11- and 12-μm channels is taken as the horizontal coordinate, the BTD between 8- and 11-μm channels is regarded as the vertical coordinate, and the scatter plot of the BTD is made to distinguish the ice cloud from the liquid water cloud. However, misjudgments are often made for optically thin liquid water and ice clouds, which require correction for changes in the water vapor content in the atmosphere.

With MODIS in orbit, the tri-spectral method has been simplified. Baum et al. (2000) found that for ice COT greater than 1, the BTD of 8.5- and 11-μm channels tends to be positive, while for optically thick water clouds, the BTD tends to be negative, usually less than -2 K, and the BTD is sensitive to atmospheric absorption, especially water vapor absorption. Figure 5.6 shows the decision tree of the bi-spectral method from Platnick et al. (2003), which has become one of the algorithms for cloud phase classification by MODIS. Through this algorithm, the infrared-based

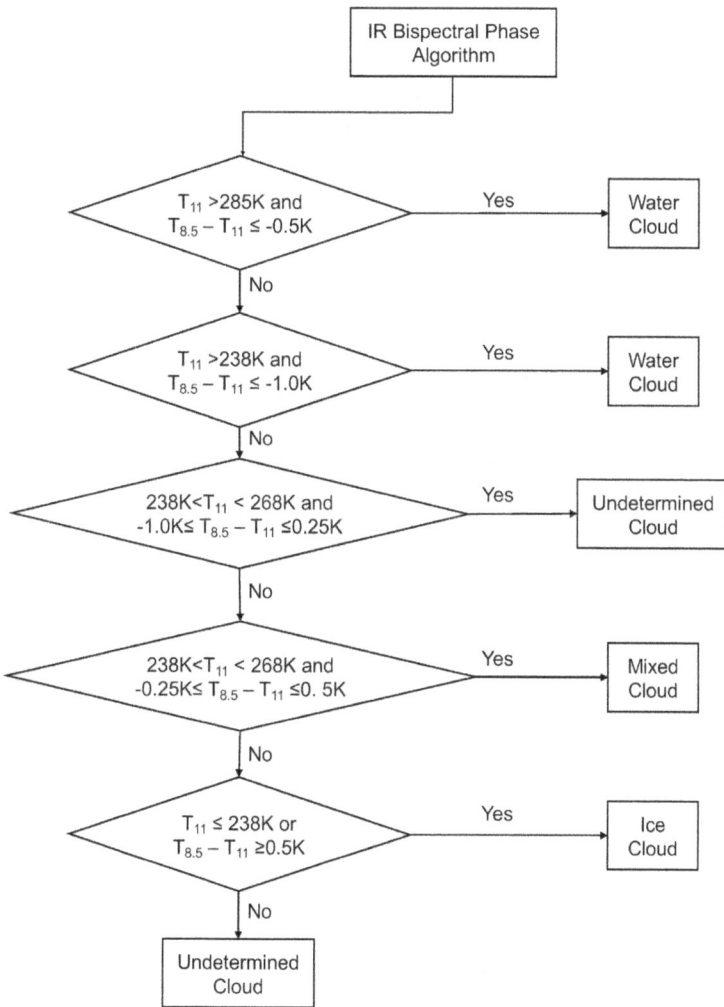

FIGURE 5.6 Decision tree for MODIS Collection 5 infrared bi-spectral cloud phase determination from Platnick et al. (2003).

retrieval algorithm classified cloud phases as ice, water, mixed-phase, and uncertain. There are two primary shortcomings of this method: (1) optically thin cirrus may not be classified as ice phase, and (2) cloud classification for supercooled water clouds (SWCs) or mixed-phase ones may be problematic if only using infrared measurements (Platnick et al., 2003).

The MODIS team further enhanced the bi-spectral cloud phase algorithm with the addition of a third infrared channel and the use of emissivity ratios (Baum et al., 2012). Because infrared absorption bands are sensitive to CTH, the 7.3-μm channel is introduced to effectively discriminate optically thin ice clouds and low-level clouds. Moreover, the use of emissivity ratios (i.e., β) is employed primarily to improve the discrimination of optically thin and high-level clouds as being ice; it is not very useful to improve the discrimination of liquid water clouds. In the MODIS Collection 6 infrared-based cloud phase classification algorithm, three different band pairs are used: 7.3 and 11 μm, 8.5 and 11 μm, and 11 and 12 μm. Specifically, the information contained in the 7.3–11 μm pair helps to separate high clouds from low clouds, the 8.5–11 μm band pair is primarily sensitive to ice clouds, and the 11–12 μm band pair is related to cloud opacity.

A comparison between the MODIS Collection 5 and Collection 6 infrared cloud phase results is presented in Figure 5.7 (Baum et al., 2012). For the results derived from MODIS Collection 5, pixels

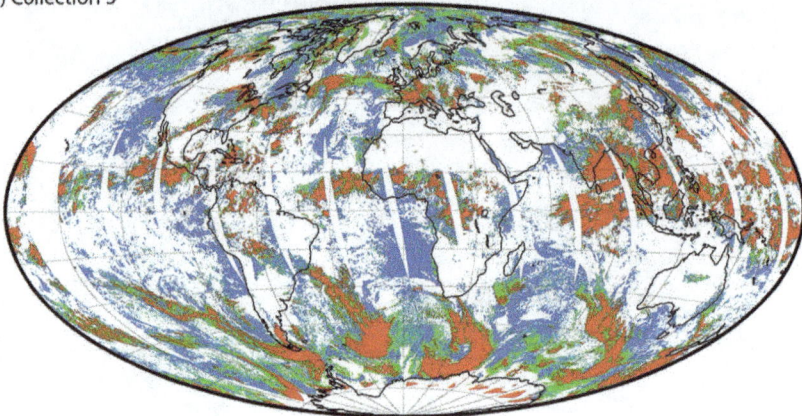

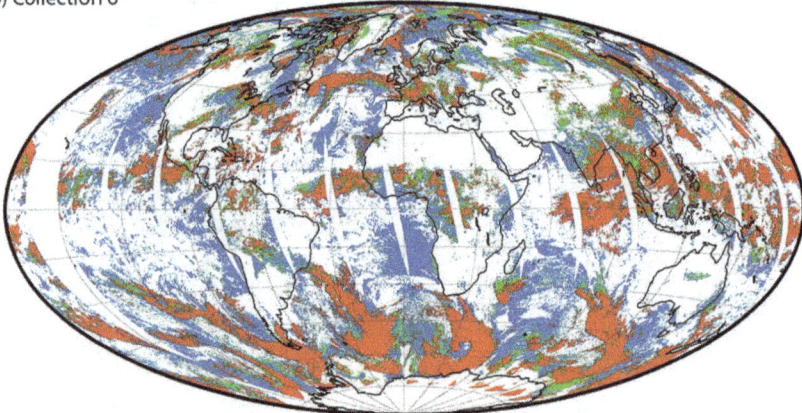

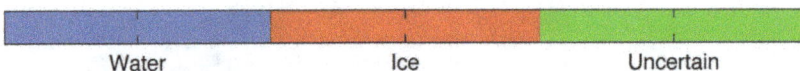

FIGURE 5.7   Snap-to-grid daytime results for infrared-based cloud phase on 28 August 2006 for (a) MODIS Collection 5 and (b) MODIS Collection 6 infrared cloud phase results from Baum et al. (2012).

that are not positively identified as either ice or liquid water are labeled as uncertain. The MODIS Collection 6 algorithm greatly reduces the percentage of pixels that are classified as uncertain, and much of pixel-level results are often labeled as ice.

The cloud phase classification method based on thermal infrared multi-spectral data is not restricted by solar radiance and can be used for continuous retrieval during the day and night. However, the measurements are sensitive to surface emissivity and atmospheric absorption, especially for water vapor absorption. In the condition of optically thin clouds and discrete cloud fragmentation, surface radiation contributes a lot to satellite observations, which is easy to cause differences in cloud phase classification.

5.3.2 VISIBLE AND SWIR BAND-BASED METHOD

Both liquid water and ice clouds have non-negligible absorption in the SWIR channels, and for MODIS 1.6- and 2.1-μm channels, the absorption of the liquid water cloud is obviously less than that of the ice cloud. There is almost no absorption for both liquid water and ice clouds in visible bands. Therefore, assuming all other conditions are equal except phase state, the reflectance ratio of liquid water clouds between the SWIR and the visible band is greater than that of ice clouds. And the ratio of reflectance of MODIS SWIR channels (centered at 1.6 and 2.1 μm) to that of the visible channel (centered at 0.66 μm) is used by King et al. (1992) to complete the classification of cloud phase, and its threshold is related to particle size and solar zenith angle (SZA). Optically thin cirrus is easily misjudged under high reflectance surface conditions. Another algorithm takes advantage of constant optical differences between liquid water and ice clouds in selected SWIR bands, which are centered at 1.6 and 2.1 μm, respectively.

According to the difference in the reflectance of liquid water and ice clouds in the SWIR band (Figure 5.8), some indices can also be defined to identify the phase states of clouds. Knap et al. (2002) completed the classification of cloud phase states by judging the shape parameter threshold based on the fact that ice particles have a larger absorption than water particles at the 1.6-μm

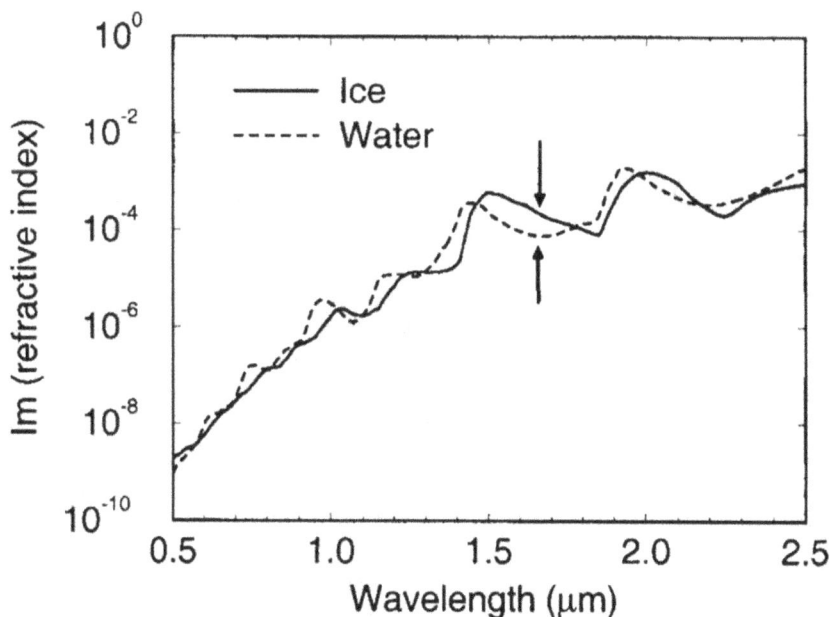

FIGURE 5.8 Imaginary part of the refractive index of ice and water as a function of wavelength between 0.5 and 2.5 μm from Knap et al. (2002). The arrows indicate the spectral slopes for liquid water and ice are highly different on the 1.67-μm channel.

channel, and the change curve of the imaginary part of the complex refraction index of ice cloud particles has a larger slope while the slope of water particles is approximately 0. This method is not effective for the retrieval of optically thin cirrus clouds on the surface snow anymore and is not sensitive to other surface types, so it is not suitable for global cloud phase classification.

Although the retrieval technology based on solar reflectance is not affected by the cloud temperature deviation and the geometric sampling effect of the system, it can only be used in the daytime and is sensitive to the SZA and satellite zenith angle (VZA). When the size of liquid water cloud particles is too large or ice ones is too small, the classification of the cloud phase will become uncertain.

5.3.3 ALL-SPECTRAL-BAND-BASED METHOD

So as to improve the accuracy of cloud phase classification, visible, NIR, and thermal infrared bands can be combined for daytime cloud phase determination. Arking and Childs (1985) first used 0.73-, 3.7-, and 11-μm channels from AVHRR data for retrieving cloud properties and obtained cloud optical and microphysical parameters containing cloud thermodynamic phase. According to the sensitivity of the 3.7-μm channel to the cloud phase, the cloud phase can be classified by using this index. Baum et al. (2000) added reflectance information of SWIR 1.6-, 1.9-, and visible 0.65-μm channels to the thermal infrared tri-spectral method for improving the accuracy of optically thin cirrus cloud phase classification. Furthermore, this method used MODIS 1.63-μm channel reflectance and 11-μm channel BT to identify the multi-layer clouds with liquid water clouds on the lower layer and optically thin cirrus clouds on the upper layer.

The MODIS daytime-only cloud phase algorithm uses a combination of visible, SWIR, and infrared channels, and the results are used in the MODIS cloud optical and microphysical property retrievals (Marchant et al., 2016). This algorithm uses four primary tests based on the 1 km CTT, the 1 km infrared cloud phase, the 1.38 μm cirrus detection test from the MOD35 cloud mask, and three spectral CER tests (derived from 1.6-, 2.1-, and 3.7-μm channels). Although these tests reduce computational efficiency, it is evident that the MODIS daytime-only cloud phase algorithm achieves high accuracy through the comparison with CALIOP. Over 90% of cloud phase tested results derived from MODIS Collection 6 show good agreement with those of CALIOP for single-phase cloudy pixels. At the same time, these developments are observed for several surface types, including ocean, land, desert, and snow/ice, and both optically thin and thick clouds.

The AHI cloud phase algorithm is designed on a combination of 0.6-, 1.6-, 8.6-, and 11.2-μm channels (Mouri et al., 2016a), and the flowchart of cloud phase processing is summarized in Figure 5.9. For an opaque cloud, the phase is classified as ice when BT at 11.2-μm channel is below freezing temperature without an ice nucleus and water when BT at 11.2-μm channel is above freezing temperature. The BTD between 8.6 and 11.2 μm is generally smaller for liquid water clouds than that for ice clouds, and a look-up table (LUT) was developed to provide reasonable thresholds for the highest skill score in cloud phase classification. Moreover, the cloud phase determination is also achieved through a ratio of reflectances between 0.64- and 1.6-channel and radiative transfer calculation.

In particular, clouds with temperatures below the freezing point but whose particles are still in the form of liquid water droplets are referred to as SWCs. Zhou et al. (2022) introduced an efficient algorithm to detect SWCs from passive radiometer observations, which combines information from the reflectance difference between 1.61- and 2.25-μm channels, the BTD between the 8.5- and 11-μm channels, and the CTT. Since the channels used for the detection are available in most current operational polar and geostationary satellite radiometers, this SWC detection algorithm can be easily implemented for operations such as cloud monitoring, aviation safety, and SWC-related weather modification. Figure 5.10 gives three examples of SWCs classification results in different seasons and areas from Zhou et al. (2022).

Since the thermal infrared band-based method and visible and SWIR band-based methods have different strengths and shortcomings, a combination of all-spectral-band can be used to

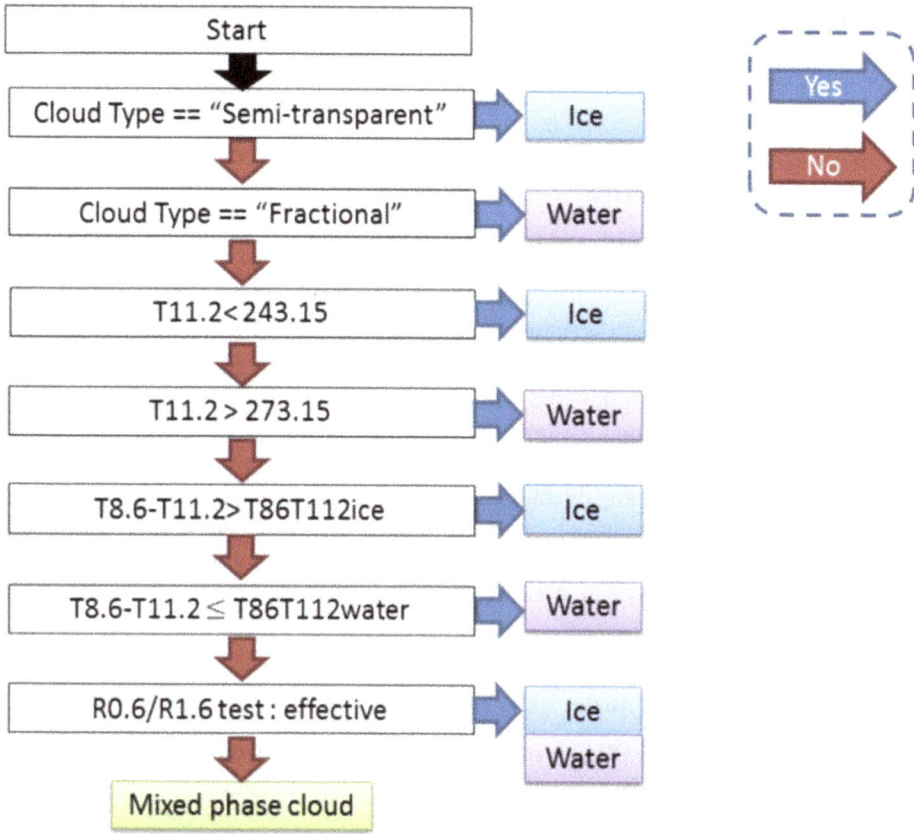

FIGURE 5.9 Flowchart of AHI cloud phase processing from Mouri et al. (2016a).

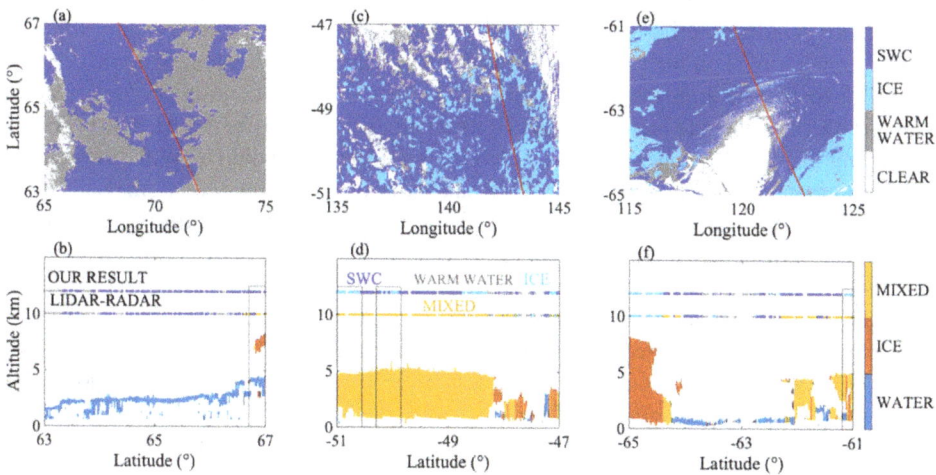

FIGURE 5.10 (Top) Three examples of the supercooled water cloud (SWC) detection results from Zhou et al. (2022). The red lines indicate the collocated CloudSat-CALIPSO tracks. (Bottom) Cloud phase vertical distributions from the LIDAR-radar product, VIIRS SWC results, and SWC results from CALIOP-CPR merged product (i.e., 2B-CLDCLASS-LIDAR) are illustrated on horizontal lines at the upper part of each panel.

provide better cloud phase detection results. However, the all-spectral-band-based method can only be applied to daytime observations, and there are still some challenges that need to be overcome, for example, the identification of multi-layer clouds and dependencies of view and scattering angle.

5.4 CLOUD TOP PROPERTIES

Cloud top properties (i.e., CTP, CTT, and CTH) are of particular importance for determining long-wave radiation at the top of the atmosphere and are especially valuable for aviation safety (Baum et al., 2012). Passive remote sensing has been an important way of cloud top property retrievals because of the advantages of the large FOV and high spatiotemporal and spectral resolution (Baum et al., 2000). There have been numerous methods to retrieve cloud top properties from passive remote sensing observations, including the infrared window algorithm, the CO_2-slicing algorithm, the water vapor window algorithm, the machine learning-based algorithm, and the extrapolation algorithm. This section provides a review of the infrared window method and the CO_2-slicing method, which are adopted by many operational CTH algorithms. In addition, most of the existing approaches are based on the assumption of a single-layer cloud, resulting in inevitable uncertainties when multi-layer clouds are present. To address this problem, considerable efforts have been made. This section also reviews some advances in the retrieval of cloud top properties for multi-layer clouds.

5.4.1 INFRARED WINDOW METHOD

For a given cloud element in a FOV, the radiance observed $R(\lambda)$ in the spectral band centering at wavelength λ can be written as (Baum et al., 2012):

$$R(\lambda) = (1 - NE)R_{clr}(\lambda) + NE\left[R_{bcd}(\lambda, P_c)\right] \tag{5.3}$$

where $R_{clr}(\lambda)$ is the clear-sky radiance, $R_{bcd}(\lambda, P_c)$ is the opaque (black) cloud radiance from pressure level P_c, N is the fraction of the FOV covered with cloud, and E is the cloud emissivity.

The infrared window method assumes a fully covered FOV $N = 1$ and optically thick clouds $E = 1$. The item $(1 - NE)$ on the right side in Eq. 5.3 vanishes, being tantamount to no contribution from the surface and the atmosphere below the cloud. Assuming an atmospheric temperature and humidity profile, the radiance can be calculated using a radiative transfer model. The CTP is found by minimizing the difference between the simulated and observed radiance. With this method and under the aforementioned assumptions, the CTP can be derived by using only a single channel. It is favorable to use a wavelength with a large atmospheric transmissivity to minimize the influence of the atmosphere above the cloud on the retrieval.

For instance, AHI onboard the Himawari-8/9 satellites employs a wavelength of 11.2 μm, which represents a window region for water vapor (Mouri et al., 2016b). Figure 5.11 illustrates a study case of CTH determined using the infrared window method from Mouri et al. (2016b). The vertical profile of radiance at 11.2 μm is calculated using the radiative transfer model. The interpolation ratio of radiance between the two levels sandwiching the observed radiance reflects the interpolated pressure between the two levels. This example shows an observed radiance of 85.2 mW converted to 591.4 hPa.

The infrared window method has been an effective method for retrieving cloud top properties of opaque clouds, especially for low-level water clouds. It is possible to extend this retrieval method by taking cloud cover N and/or the spectral cloud emissivity E explicitly into account (Roebeling, 2006; Menzel et al., 2008). The cloud cover N can be estimated using the high-resolution channel of passive sensors. Furthermore, the spectral emissivity E can be estimated from the approximate 2:1 relationship between the COT at visible and infrared window-channel wavelengths. Rossow and

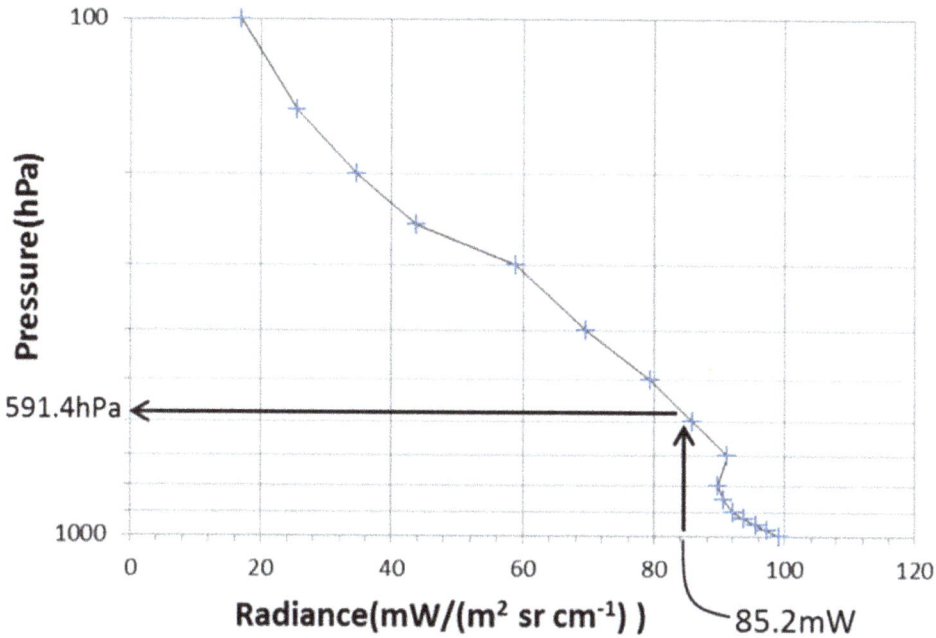

FIGURE 5.11 A study case of CTT determined using the infrared window method from Mouri et al. (2016b).

Schiffer (1999) solve first for the visible optical depth τ_{vis} using the reflected radiance. Then, the emissivity E is computed as:

$$E = 1 - \exp\left(-0.5\tau_{vis} / \mu\right) \tag{5.4}$$

where μ is the cosine of the VZA. Taking the semi-transparency and coverage of the cloud layer into account, the retrieval results of the radiance fitting method can be significantly improved.

5.4.2 CO$_2$-Slicing Method

The CO$_2$-slicing method is based on the atmosphere becoming more opaque resulting from CO$_2$ absorption as the wavelength increases from 13.3 to 15 μm, thereby causing radiances obtained from these spectral bands to be sensitive to a different layer in the atmosphere (Menzel et al., 1983; Wylie and Menzel, 1999). Figure 5.12 shows the weighting functions for the CO$_2$ absorption bands on MODIS from Menzel et al. (1983). Because the peaks in the weighting functions are well into the troposphere, CO$_2$ slicing is most effective for the analysis of mid- to high-level clouds, especially semi-transparent clouds such as cirrus.

The CO$_2$-slicing technique is founded in the calculation of radiative transfer in an atmosphere with a single cloud layer. In the case of semi-transparent clouds, N and E in Eq. 5.3 are unknown. The clear sky radiance can be simulated with a radiative transfer model or estimated by locating clear sky measurements in the vicinity of the observation (Smith and Frey, 1990). The opaque (black) cloud radiance can be calculated from:

$$R_{bcd}\left(\lambda, P_c\right) = R_{clr}\left(\lambda\right) - \int_{P_c}^{P_s} \tau\left(\lambda, p\right) \frac{dB\left[\lambda, T\left(p\right)\right]}{dp} dp \tag{5.5}$$

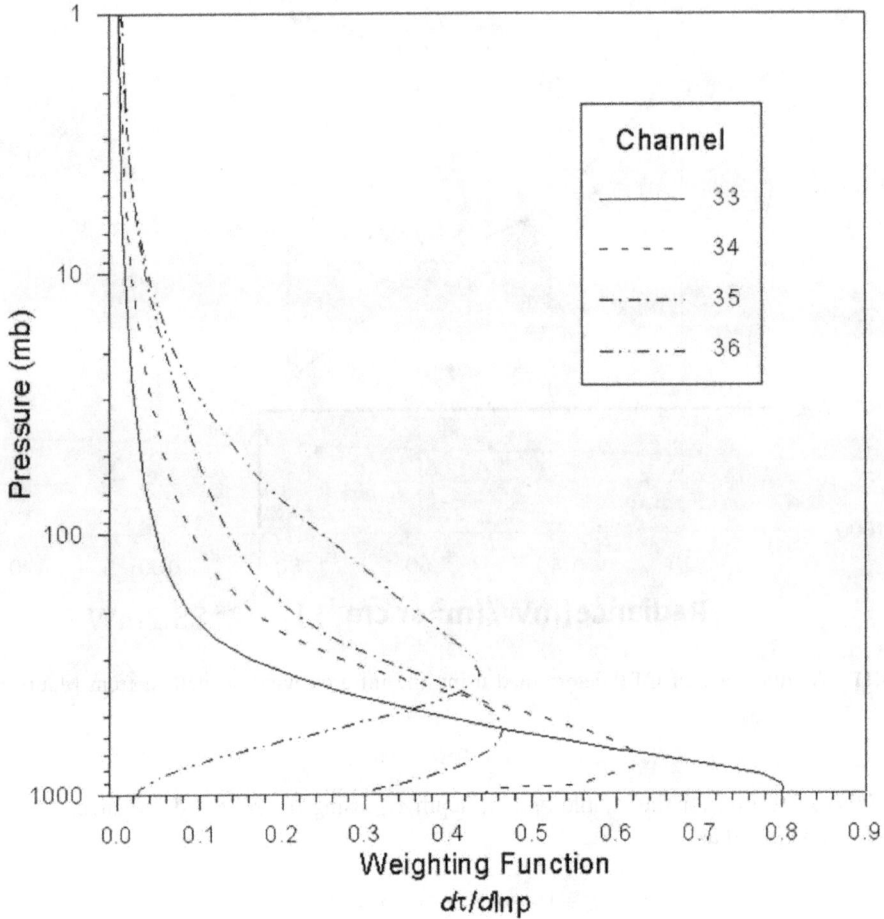

FIGURE 5.12   Weighting functions for the four MODIS bands in the CO_2 absorption band from Menzel et al. (1983).

where P_s is the surface pressure, P_c is the cloud pressure, $\tau(\lambda, p)$ is the fractional transmittance of radiation at frequency λ emitted from the atmospheric pressure level (p) arriving at the top of the atmosphere $(p = 0)$, and $B[\lambda, T(p)]$ is the Planck radiance at frequency λ for temperature $T(p)$. The second term on the right represents the decrease in radiation from clear conditions introduced by the opaque cloud. The inference of cloud-top pressure for a given cloud element is derived from radiance ratios between two spectral bands. The ratio of the deviations in observed radiances $R(\lambda)$ to their corresponding clear-sky radiances $R_{clr}(\lambda)$ for two spectral bands of wavenumber λ_1 and λ_2, viewing the same FOV, is written as:

$$\frac{R(\lambda_1) - R_{clr}(\lambda_1)}{R(\lambda_2) - R_{clr}(\lambda_2)} = \frac{NE_1 \int_{P_c}^{P_s} \tau(\lambda_1, p) \dfrac{dB[\lambda_1, T(p)]}{dp} dp}{NE_2 \int_{P_c}^{P_s} \tau(\lambda_2, p) \dfrac{dB[\lambda_2, T(p)]}{dp} dp} \tag{5.6}$$

For band pairs that are spaced closely in wavelength, the assumption is made that E_1 is approximately equal to E_2. This allows the CTT, CTH, and CTP within the FOV to be inferred when the

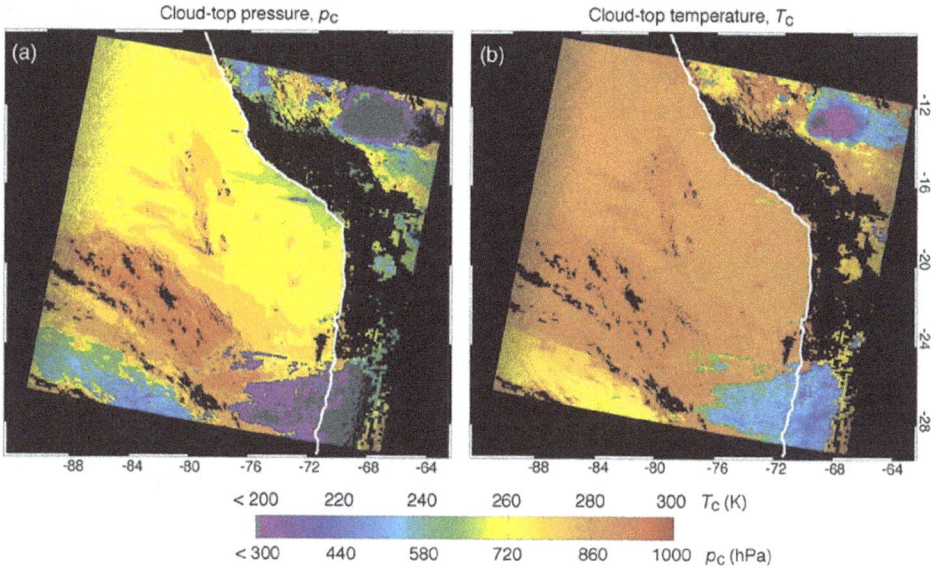

FIGURE 5.13 Example of MODIS (a) CTP and (b) CTT retrievals on 18 July 2001 at 15:30 UTC from Platnick et al. (2003).

atmospheric temperature and transmittance profiles for the two spectral bands are known or estimated. Figure 5.13 shows an example of the MODIS retrievals of CTT and CTP based on the CO_2-slicing method. The image shows widespread boundary layer stratocumulus clouds off the coasts of Peru and Chile on 18 July 2001 at 15:30 UTC, associated with cool upwelling water along the Humboldt current (Platnick et al., 2003).

The CO_2-slicing method performs well for optically thick clouds located at middle to high levels of the atmosphere, whereas it becomes less accurate for optically thin clouds (Zhang and Menzel, 2002; Holz et al., 2006). To date, many operational cloud top properties products have been produced based on the infrared window algorithm combined with the CO_2-slicing algorithm (Baum et al., 2012; Mouri et al., 2016b; Min et al., 2017).

5.4.3 TREATMENT FOR OVERLAPPING CLOUDS

Conventional CTH retrieval algorithms, including the aforementioned infrared window method and CO_2-slicing method, assume clouds to be homogenous and single-layer. However, it is common that multi-layer or overlapping clouds are in the atmosphere, accounting for approximately 25% of world-wide cloud observations (Li et al., 2015). Some validation studies have revealed that the CTH retrievals may be significantly biased when overlapping clouds are present. Figure 5.14 presents the histogram for the differences of MODIS-CALIOP CTH separated by single-layer and multi-layer clouds from Holz et al. (2008). The results indicate that multi-layer clouds dominated the MODIS CTH biases. To improve the accuracy of CTH retrievals, various effective retrieval methods for overlapping clouds have been introduced.

By using ground-based microwave radiometers to constrain lower-layer water clouds, Huang (2005) proposed a multi-layer cloud retrieval algorithm to probe only the upper-layer ice cloud properties with satellite visible and infrared radiances. This algorithm was effective in reducing the large biases of the upper ice CWP and CTH, but the lower-layer water CTH was not quantitatively inferred, and ground-based microwave observations had to be used. Lindstrot et al. (2010) retrieved the multi-layer CTPs by combining oxygen A-band channels for upper

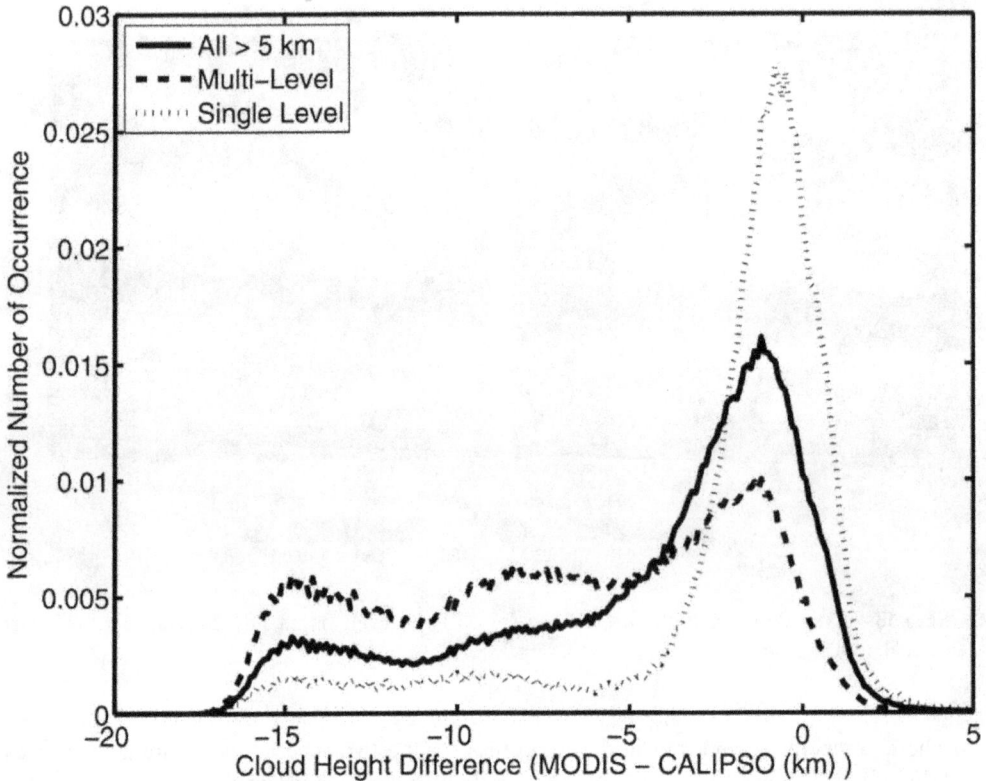

FIGURE 5.14 Histogram of global CTH differences between MODIS and CALIOP is presented filtered by single-layer and multi-layer clouds using CALIOP data from Holz et al. (2008).

layers and an 11-μm window channel for lower layers. Watts et al. (2011) derived the two-layer CTPs of overlapping clouds using only infrared channel measurements by using a simplified assumption that the lower-layer clouds are gray and have a proxy height given by the surface temperature.

Teng et al. (2022) provided an optimal-estimation-based multi-spectral method, which leverages the merits of four shortwave infrared channels (centered at 0.86, 1.6, 2.13, and 2.25 μm) in distinguishing cloud optical and microphysical properties in different phases and the capabilities of longwave infrared channels (centered at 8.6, 11, and 12 μm) for the corresponding CTHs. The method performs effectively for overlapping clouds with an optically thin but detectable ice layer (COT less than ~7) above a liquid water layer. Figure 5.15 presents an example of overlapping CTH retrieved via the multi-spectral algorithm from Teng et al. (2022). The single-layer-based AHI CTHs greatly underestimate the "true" upper-level ice CTHs (ITHs) and overestimate the "true" lower-level water CTHs (WTHs). Compared to the AHI operational CTHs, the CTH retrievals based on the multi-spectral algorithm can better characterize the vertical distribution of overlapping cloud systems.

In addition to radiative transfer-based methods, a statistics-based extrapolation method was proposed to infer upper-layer ITH and lower-layer WTH simultaneously (Tan et al., 2022). Based on the continuity of cloud boundary, this method estimates the CTH of neighboring overlapping clouds by using the well-retrieved CTH of single-layer clouds. Figure 5.16 shows an example of overlapping CTH retrieval based on the extrapolation method from Tan et al. (2022). With the simultaneous retrieval of ITH and WTH, the vertical structures of overlapping clouds become much clearer; thus, the cloud radiative effects could be better evaluated.

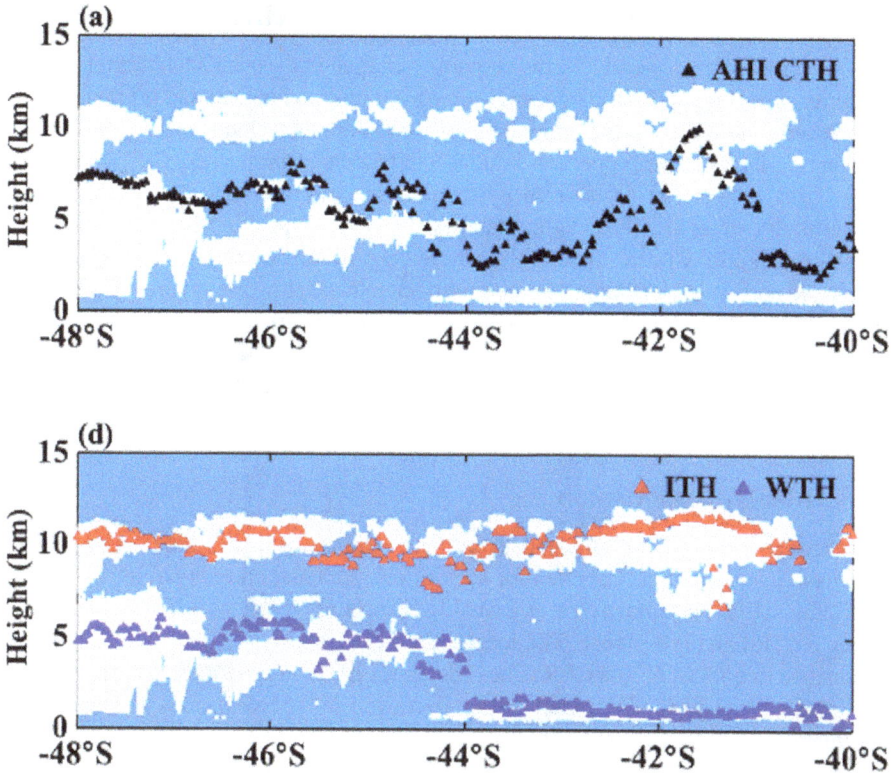

FIGURE 5.15 Example of the CTHs retrieved based on (top) the conventional cloud retrieval method and (bottom) the multi-spectral overlapping cloud retrieval method from Teng et al. (2022). Cloud vertical profiles from the 2B-CLDCLASS-LIDAR are regarded as the truth.

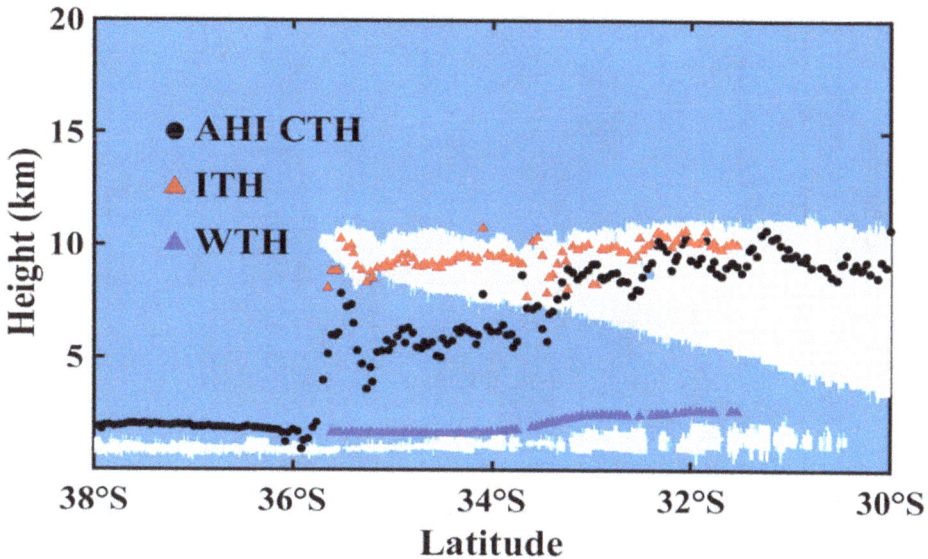

FIGURE 5.16 An example of overlapping CTH retrieval from Tan et al. (2022). Comparison of the CPR-CALIOP cloud profiles (white region) with the AHI operational CTH product (black circles) and the ITH (red triangles) and WTH (blue triangles) derived from the extrapolation algorithm.

5.5   CLOUD OPTICAL AND MICROPHYSICAL PROPERTIES (COT AND CER)

Satellite spectral radiometers measuring the outgoing radiation in the solar and thermal infrared bands is one of the most popular and effective tools for retrieval of cloud optical and microphysical properties. The reflected solar radiation and emitted longwave radiation by cloud particles in the atmosphere largely depend on its radiative properties. Among various cloud properties, COT and CER, as key quantitative variables that can be inferred by satellite spectral radiometers, are primary for cloud radiative effects (Stowe et al., 1989; Yang et al., 2015). So far, two kinds of classical cloud optical and microphysical property retrieval algorithms have been developed and widely applied. One is the solar-band-based retrieval algorithm using measured reflectance at two solar reflective bands and thus can only be used for retrievals during daytime. Another is the infrared-band-based retrieval algorithm using measured BTs at a series of thermal infrared bands which can be widely used during both daytime and nighttime. This section introduces these two retrieval algorithms using solar-band-based remote sensing and infrared-band-based remote sensing of satellite spectral radiometers, respectively.

5.5.1   SOLAR-BAND-BASED METHOD

The solar reflective method, normally a bi-spectral algorithm, was first introduced by Nakajima and King (1990) and has been widely applied for various research and operational retrievals since then to infer COT and CER. The approach uses a weakly absorbing (visible or NIR) band reflectance and an absorbing SWIR band reflectance. The former bands are mainly sensitive to COT, and the latter ones are sensitive to both COT and CER. The reflectance LUTs are pre-calculated for clouds with a wide range of COT and CER. Thus, simulated reflectances at the two bands can be compared with measured reflectances, and the corresponding COT and CER that yields the best agreement with the measurements can be obtained. For efficient retrievals, cloud reflectances are simulated based on 128-stream Discrete-Ordinate-Method Radiative Transfer (Stamnes et al., 1988) forward simulations in many researches (Teng et al., 2022; Liu et al., 2023).

Figure 5.17 shows an example of reflectance LUTs for a pair of bands centered at 0.87 and 2.13 μm for liquid water (red curves) and ice (blue curves) phase clouds over a dark surface for the

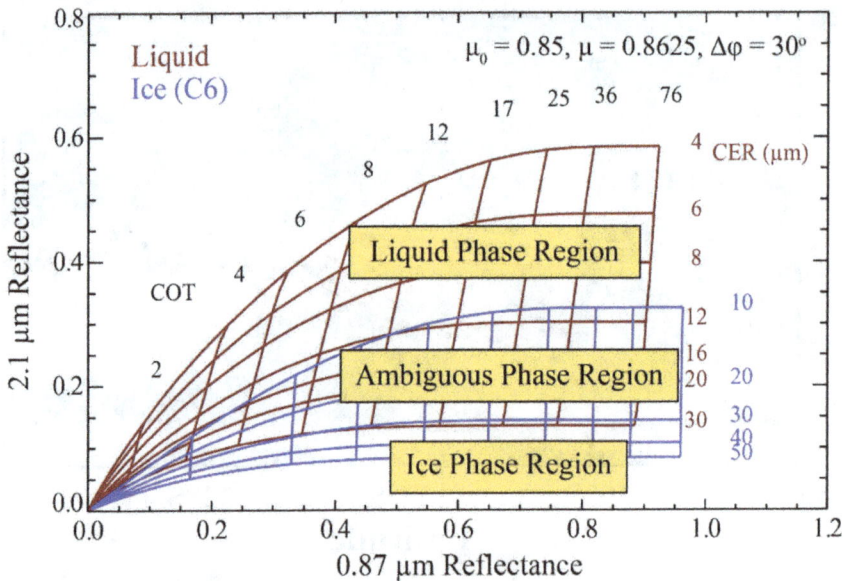

FIGURE 5.17   An example of reflectance LUTs at the 0.87- and 2.13-μm bands with different configurations of COT and CER for liquid water (red lines) and ice (blue lines) clouds from Platnick et al. (2017).

geometry specified in the caption from MODIS (Platnick et al., 2017). It is clear that the reflectance at 0.87 μm (the nonabsorptive band) is a strong function of COT with little dependence on CER, whereas the reflectance at 2.13 μm, in contrast, is sensitive to both. Moreover, some of the solution space is unambiguously liquid water and some is unambiguously ice, but there are overlapping regions in which either phase can yield a viable physical solution. Reflectance measurements can occur in regions of the solution space that are unambiguously liquid water or ice but may also lie in regions that are ambiguous regarding phase. Comparison of liquid water and ice CER retrievals from SWIR bands can reduce ambiguity in the choice of thermodynamic phase. Due to solar bands that satisfy the requirement being included in a lot of satellite sensors, the solar reflective method has been widely employed for cloud optical and microphysical property products. Examples include the MODIS (Platnick et al., 2003; Platnick et al., 2017), the AVHRR (Heidinger et al., 2005), and the Spinning Enhanced Visible and Infrared Imager (SEVIRI) (Roebeling et al., 2006).

The solar reflective method shows powerful performance and wide applications in retrieving COT and CER, but its limitations are also quite obvious. As illustrated in Figure 5.17, the LUT values converge as clouds become optically thin (COT less than ~1), and the reflectance becomes sensitive to the surface condition (i.e., surface albedo characteristics). It indicates that the solar reflective method becomes less accurate and inappropriate for optically thin clouds, especially for ice clouds. Because the single-scattering properties of ice clouds at the solar bands are sensitive to particle microphysical properties, e.g., shape and surface structure, the retrieval becomes extremely sensitive to the assumption of ice cloud habits. For example, only by updating the ice cloud model used for COT and CER retrievals can the MODIS products from different version collections be very different (Platnick et al., 2003; Platnick et al., 2017).

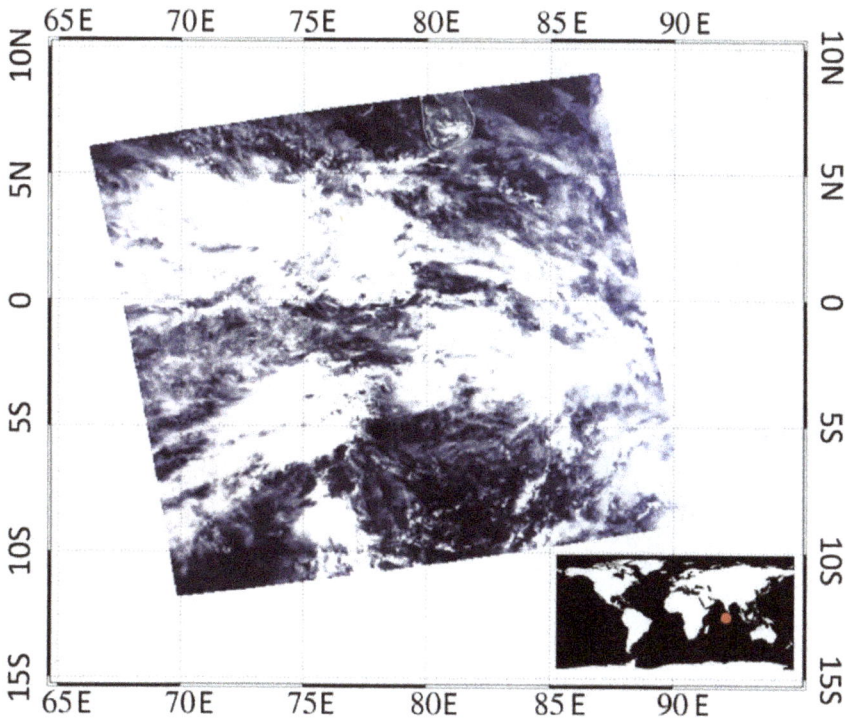

FIGURE 5.18 The true color of Aqua MODIS granule over the Indian Ocean on 10 December 2013 at 08:20 UTC from Yang et al. (2015).

As we have discussed, the solar reflective method is sensitive to the ice habit assumed, and a case study is performed to show its performance and drawbacks. MODIS measurements, i.e., reflectances at bands 2 and 7 (centered at 0.87 and 2.13 μm, respectively), are used to retrieve COT and CER. An Aqua MODIS scene taken on 10 December 2013 at 08:20 UTC is used, and Figure 5.18 shows the true color image of the granule. The scene is taken over the Indian Ocean and mostly covered by high ice clouds. The retrieval is carried out for only pixels over the ocean and identified as ice clouds, and we use MODIS Level 2 Collection 5 cloud product (i.e., MYD06) to give the cloud mask and CTP, and the retrieval algorithm is the same as that used by Bi et al. (2014).

Figure 5.19 illustrates the retrieved COT and cloud effective diameters (D_{eff}) based on the scattering properties assuming sphere and hexagonal column, respectively, and the hexagonal column model and its scattering properties given by Bi et al. (2014) are used. The top panels are the results based on the spherical model, and the bottom ones are from those with the hexagonal column model. Overall, the two models give similar patterns on the retrieved COT, whereas the model based on ice spheres infers larger results than those from the hexagonal columns, and D_{eff} from both retrievals range from 10 to over 100 μm, and the results based on the two models are quite different.

To better compare the retrieved results, Figure 5.20 gives the histograms of occurrence for retrievals based on the two different ice habit models, with red indicating the highest

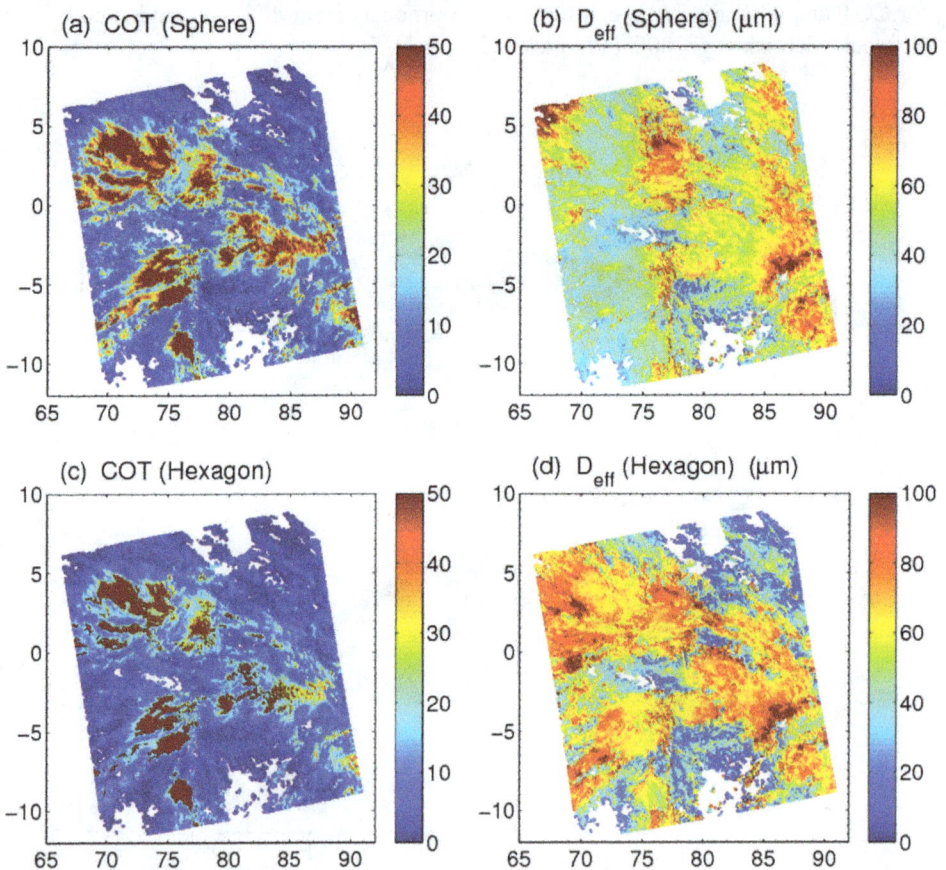

FIGURE 5.19 Retrieved COT and D_{eff} based on the scattering properties from the sphere (top panels) and hexagonal column (bottom panels) models from Yang et al. (2015).

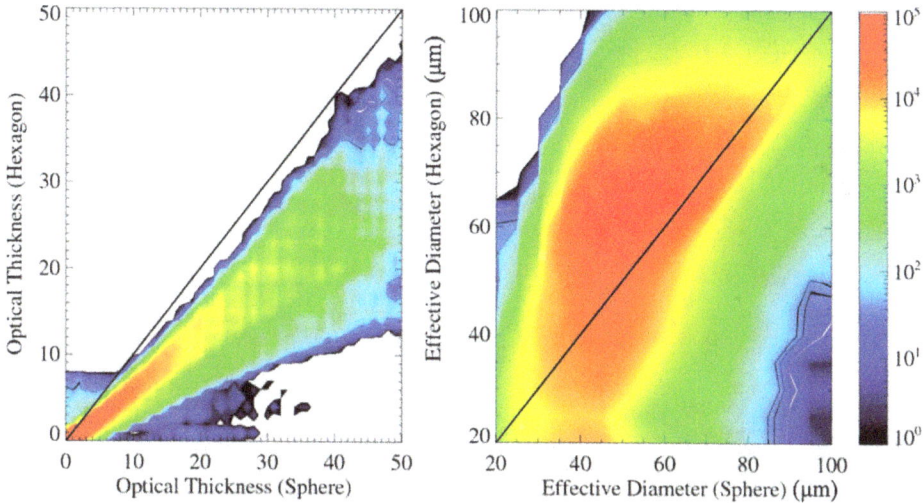

FIGURE 5.20   The histograms of the occurrence frequencies for comparison of retrieved COT and D_{eff} based on the sphere and hexagonal column models from Yang et al. (2015).

occurrence. Black solid lines, i.e., one-to-one ratio lines, are included to ease the interpretation of the results. It clearly shows that the spherical model systematically gives larger values of COT compared with those from the single-column model, and the differences increase and widen as ice clouds become thicker. D_{eff} given by the spherical retrievals can give either larger ice particles or smaller ones, and much wider variations are shown than those of COT. Figure 5.20 clearly shows the importance of assumed ice habits and their scattering properties on inferring cloud optical and microphysical properties using the solar band retrievals and thus indicates the importance of developing more accurate and practical ice cloud models and scattering models.

The retrieved results are not only significantly affected by the ice cloud optical model but also show significant differences among various sensors due to differences in assumed auxiliary parameters, spectral bands, and forward radiative transfer simulations. Figure 5.21 illustrates the pixel-to-pixel comparison of ice COT and CER between the AHI/AGRI (Advanced Geosynchronous Radiation Imager) and the MODIS operational cloud products (Lai et al., 2019). The top and bottom panels are for AHI and AGRI, respectively. Note that only pixels that are classified as ice clouds by all three cloud phase products are considered here for comparison. To quantitatively compare COT and CER, five parameters, i.e., the intraclass correlation coefficient (ICC), the average relative difference (RD), the slope (K), and the intercept (B) from the linear regression, and the standard deviation (Std), are utilized to quantify the relationships between two data sets from different satellites. Figure 5.22 shows the histograms of the occurrence frequencies of liquid water COT and CER (Lai et al., 2019).

Similar to Figures 5.21 and 5.22, Figures 5.23 and 5.24 show pixel-to-pixel comparisons of COTs and CERs, respectively, among the three sensors but using the unified retrieval system (Lai et al., 2019). Here, the "unified retrieval system" refers to using the same cloud optical models, retrieval method, forward radiative transfer model, and auxiliary data. The corresponding statistical analysis shows that the RDs between AHI and MODIS for ice cloud COT and CER decreased by 37% and 50%, respectively, compared with the RDs from the direct comparisons of their operational products. The results of AHI and MODIS liquid water properties are not significantly changed. Moreover, the consistencies of AGRI cloud properties with those of MODIS are significantly improved, and their RDs are optimized by more than 50%.

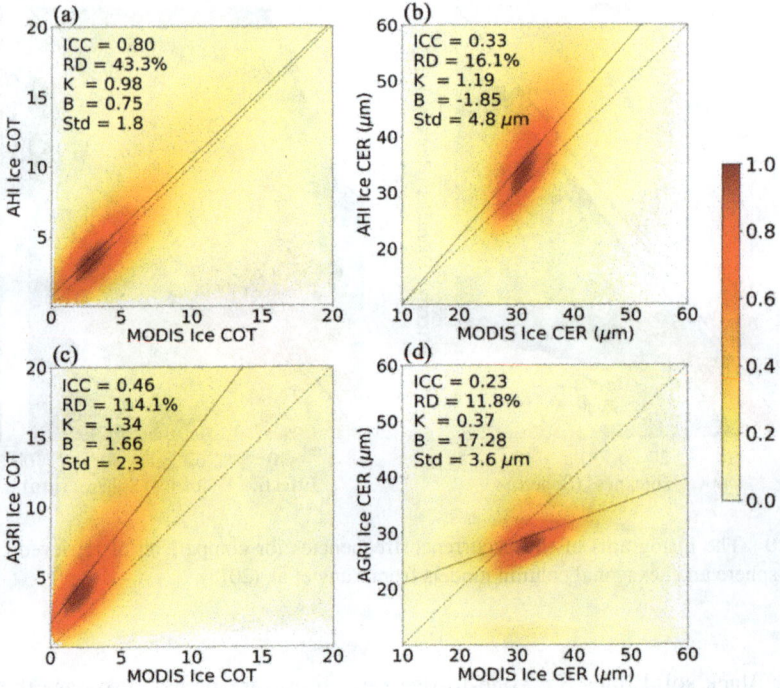

FIGURE 5.21 Two-dimensional histograms of pixel-to-pixel comparisons between MODIS and AHI (top panels)/AGRI (bottom panels) ice cloud optical and microphysical properties from Lai et al. (2019). The dashed lines are the one-to-one ratio lines, and the solid lines are linear regression functions.

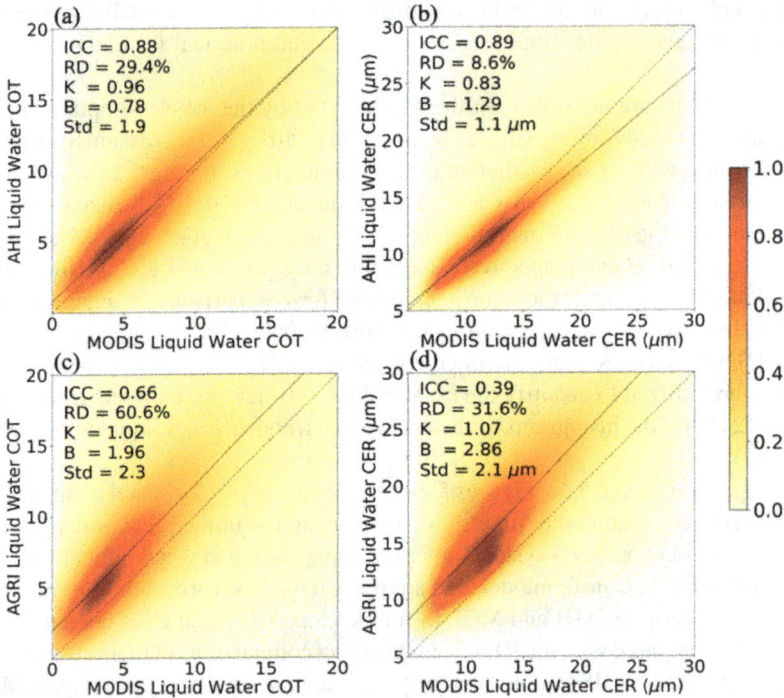

FIGURE 5.22 Same as Figure 5.21 but for liquid water cloud properties from Lai et al. (2019).

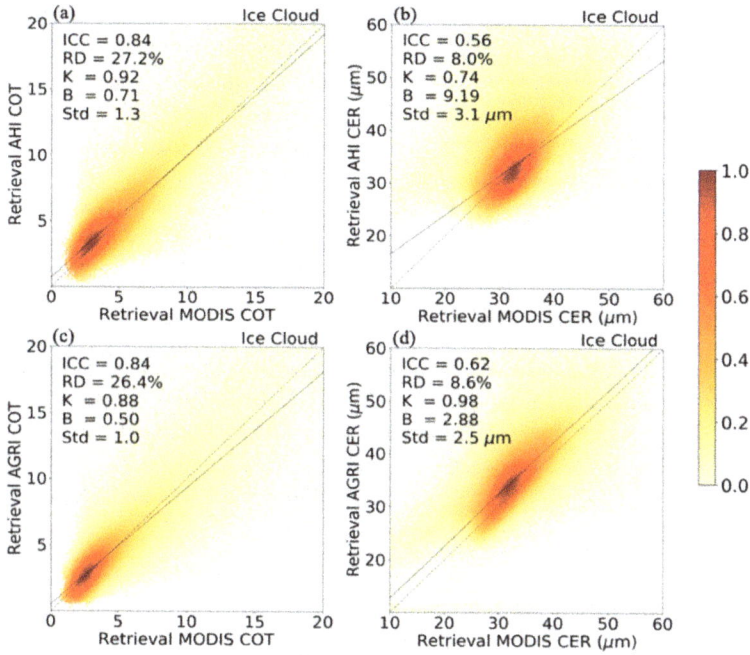

FIGURE 5.23 Two-dimensional histograms of pixel-to-pixel comparisons between MODIS and AHI (top panels)/AGRI (bottom panels) ice cloud properties based on the unified retrieval system from Lai et al. (2019). The dashed lines are the one-to-one ratio lines, and the solid lines are linear regression functions.

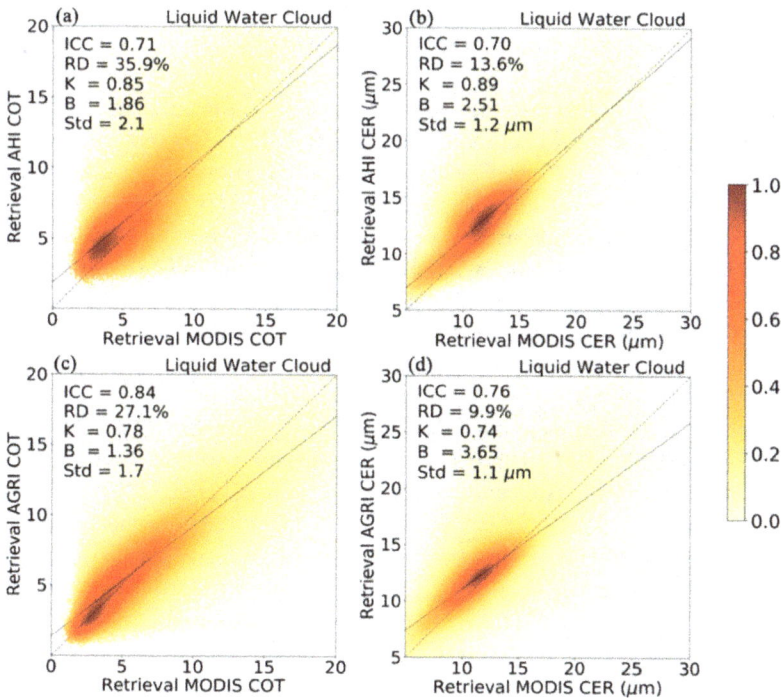

FIGURE 5.24 Same as Figure 5.23 but for liquid water cloud properties from Lai et al. (2019).

Furthermore, using solar radiation as the source, the technique is limited to daytime and is unreliable over poles and places in which the SZA is greater than approximately 80°.

Furthermore, because of strong water vapor absorption at this wavelength, little outgoing radiance from low clouds or surface reaches the satellite, whereas radiance reflected by high ice clouds, above which there is little water vapor, is little attenuated. And this band is long enough that Rayleigh scattering from the atmosphere is negligible. These characteristics of the 1.38-μm band offer a unique and excellent way to derive optically thin ice cloud properties located at the high level of the atmosphere during the daytime (Gao et al., 1998), and, as a result, the 1.38-μm band is included in a lot of sensor band designs, e.g., the MODIS and VIIRS. Cloud properties retrieved based on the 1.38-μm band include CTH, CTP, COT, and CER.

The 1.38-μm band also collaborated with other visible or SWIR channels to have a better quantification of the cloud properties, and it has been used with visible bands for lots of studies. Gao et al. (1998) have proposed an empirical algorithm for cloud reflectance using narrow channels which are near 1.38 and 0.66 μm, which shows good performance. COT of tropical cirrus is retrieved from the MODIS 0.66 and 1.38 μm measurements for each pixel in a granule image (Meyer et al., 2004). Techniques for retrieving COT and CER using the combination of 1.38 and 1.88 μm bands are well established (Gao et al., 2004).

5.5.2 Infrared-Band-Based Method

Another popular approach based on infrared band measurements to infer COT and CER is known as the split-window method, which was introduced by Inoue (1985). The method uses the differences of ice absorption characteristics at the infrared-window bands (e.g., centered at 11 and 12 μm) and has been applied to the infrared interferometer spectrometer (IRIS) (Prabhakara et al., 1988), the AVHRR (Heidinger and Pavolonis, 2009), and MODIS measurements (Wang et al., 2011). The most significant advantage of the split-window method is that it can be applied to all data regardless of solar illumination, i.e., both daytime and nighttime conditions. However, the observed BTs by satellites gradually become saturated when the cloud gets optically thicker (i.e., COT larger than 10), so the split-window band is not sensitive to optically thick ice clouds.

To illustrate the feasibility of the split-window method, Figure 5.25 shows the relationships between the simulated BT and BTD of three VIIRS IR-window bands with central wavelengths of 8.5, 11, and 12 μm (Yang et al., 2015). The example is based on the US standard atmosphere containing ice clouds of different properties, and the surface and CTT are 288 and 223 K, respectively. A VZA of 20° is assumed. Solid curves are isolines of specified particle effective size, and dashed lines are ones with constant COT. The BTs are mainly dependent on COT, whereas the BTDs are sensitive to both COT and CER. Figure 5.25 indicates that the split-window method is more sensitive to optically thin clouds (i.e., COT between 0.5 and 5) with relatively small particle sizes (i.e., CER less than 100 μm). As COT becomes larger than 10, the isolines converge, and the method cannot be applied.

Theoretically, the observations at the IR-window bands can also be used to infer the CTTs because the BTs observed at those bands are not only sensitive to COT and CER, but also CTT and surface temperature. This was recently achieved by Iwabuchi et al. (2014). Their retrieval used the MODIS BTs of 8.5, 11, and 12 μm bands and the optimal estimation method (Rodgers, 2000), which derives an optimal solution from measurements under the constraints of prior information, and COT, CER, CTT, and surface temperature are simultaneously obtained.

5.5.3 All-Spectral-Band Method

The other class of retrieval algorithms for satellite remote sensing of COT uses a combination of a visible band and an infrared-window band (Rossow et al., 1989), and has been applied for the

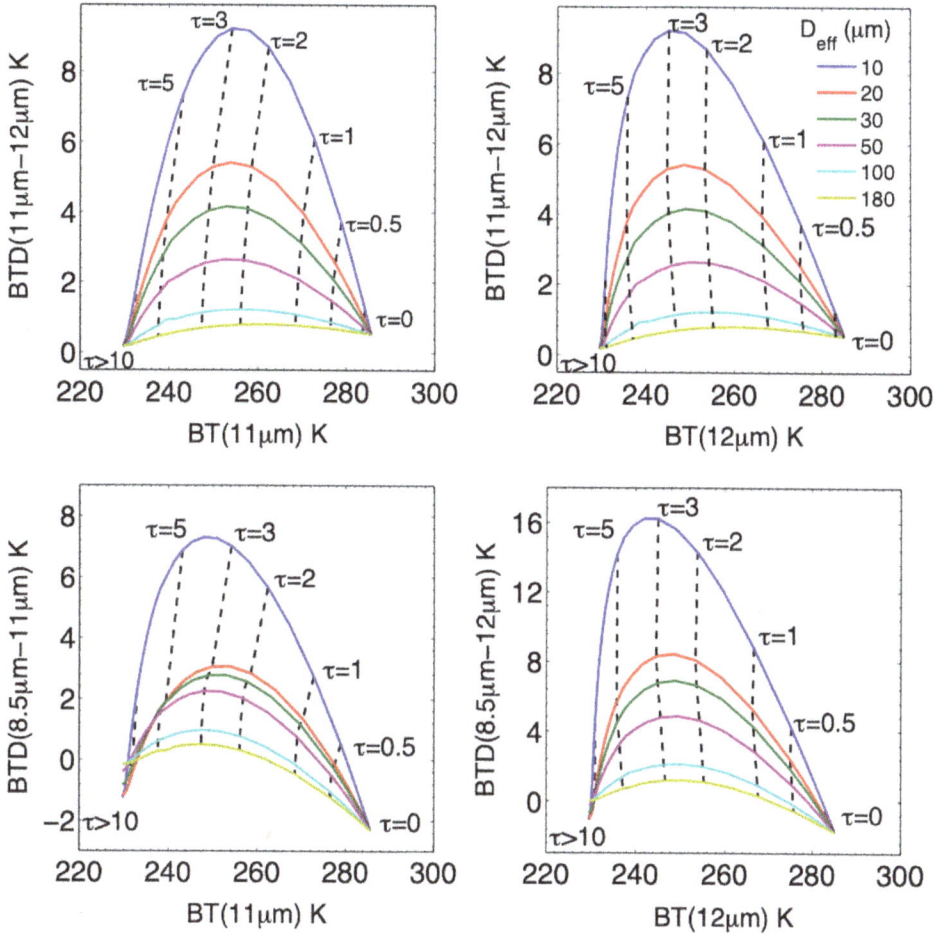

FIGURE 5.25 Relationships between simulated BT and BTD with respect to three VIIRS IR-window bands (centered at 8.5, 11, and 12 μm) from Yang et al. (2015).

AVHRR and the International Satellite Cloud Climatology Project (ISCCP) (Schiffer and Rossow, 1983; Rossow and Schiffer, 1999). The premise of the algorithm is the dependence between infrared emissivity and visible reflectance. The approach utilizes visible reflectance to determine the visible-derived COT by a combination of theoretical or empirical models and also to infer the cloud infrared emissivity based on the visible-derived COT. With the emissivity in the infrared-window known, the CTT also can be inferred using the observed infrared radiance.

Because the visible-light-scattering properties of ice clouds, as well as their relationship with those at the infrared band, are considered, the solar and infrared-window method is sensitive to the ice crystal habits assumed (Minnis et al., 1993). Again, the technique relies on a visible band and is limited to daytime and unreliable over the poles.

5.6 SUMMARY

This chapter has presented an overview of various retrieval algorithms of cloud properties developed for satellite passive spectral sensor measurements based on different wavelengths. We also review some study cases employing the MODIS, VIIRS, AHI, and AGRI radiometers and the factors contributing to the differences in cloud retrievals to illustrate the current development status of remote sensing of clouds. Among various cloud optical and microphysical properties derived from satellite-based remote sensing, cloud mask, cloud phase, CTH, CTP, COT, and CER as key quantitative variables are primary for their radiative effects, and a number of researchers have reported the features of retrieval algorithms.

As the basis of cloud retrievals, cloud detection methods are comparatively mature. The threshold-based algorithm used in cloud detection has gradually developed from the early fixed threshold to the dynamic threshold, adaptive threshold, and multispectral combination threshold. Machine learning algorithms are also often used in cloud detection research, which are independent of multispectral thresholds. Other than these two approaches, the statistical and texture analysis approaches also show good performance in the associated application. Cloud detection is a fairly basic and essential step in satellite-based remote sensing and its accuracy is significantly improved, whereas the literature survey shows that more work is required to achieve the desired accuracy and correctness for cloud detection methods.

Clouds with different thermodynamic phase states have different absorption and scattering properties. All kinds of cloud optical and microphysical property retrieval methods are developed according to the different cloud phases, and accurate classification of the cloud phase would improve cloud properties retrieval accuracy. The tri-spectral retrieval algorithm based on the thermal infrared band is a relatively advanced method for cloud phase classification, and it has been further simplified since the MODIS sensor was in orbit. The thermal infrared-band-based method can also effectively identify cloud phase, which is not restricted by solar radiance and can be used for continuous retrieval during the day and night while the measurements are sensitive to surface emissivity and atmospheric absorption, especially for water vapor absorption. In addition, liquid water and ice clouds have non-negligible absorption in the SWIR channels. To improve the accuracy of the results of cloud phase classification, visible, NIR, and thermal infrared bands can be combined for retrieval, which can effectively separate multi-layer clouds from single-layer clouds. Furthermore, combining information from the reflectance difference between 1.61- and 2.25-μm channels, the BTD between the 8.5- and 11-μm channels, and the CTT is developed to classify SWCs.

Passive remote sensing sensors have been an important way of cloud top properties retrievals. CTP and CTH can be estimated by using the scattering and emission properties of clouds at different heights in different spectral channels. The main algorithms for cloud top properties retrievals include the infrared window method and CO_2-slicing algorithm. The latter relies on the fact that as the wavelength increases from 13.3 to 15 μm, the atmosphere becomes opaquer due to CO_2 absorption, resulting in the radiances obtained from these spectral bands being sensitive to different layers in the atmosphere. With the increase in the number of meteorological satellites for different purposes and cloud detection channels, it has become feasible to improve the retrieval accuracy of cloud top properties by combining other band channels (such as visible, NIR, SWIR, medium-wave IR, millimeter-wave band, etc.) and polar-orbit satellite measurements. Therefore, the retrieval algorithms of cloud top properties based on multiple satellite sensor measurements and multi-spectral channel measurements has become the future development direction.

Satellites measuring the outgoing radiation in the solar bands have been widely used to retrieve COT and CER. The solar reflective method, normally a bi-spectral algorithm, has been widely applied for various research and operational cloud products. However, the solar reflective method becomes less accurate and inappropriate for optically thin clouds, especially for ice clouds, which is due to the single-scattering properties of ice clouds at the solar bands being sensitive to particle microphysical properties. The retrieved results also show significant differences among various sensors due to differences in assumed auxiliary parameters, spectral bands, and forward radiative transfer simulations, and this algorithm is limited to daytime and unreliable over poles and places where the SZA is greater than approximately 80°. The 1.38-μm band offers a unique and excellent way to derive optically thin ice cloud properties during the daytime and also collaborated with other visible or SWIR channels to have a better quantification of the cloud properties. Another popular approach based on infrared bands is known as the split-window method, which has been applied to the IRIS, AVHRR, and MODIS observations. The other class of algorithms uses a combination of a visible and an infrared-window band, which has been applied to the AVHRR and ISCCP.

Although a lot of new algorithms for inferring cloud properties have been developed, there are still some open questions that should be addressed in future research. Firstly, more accurate methods are needed to detect mixed-phase clouds, multi-layer clouds, or SWCs, which may have information close to either ice or liquid water clouds. Secondly, since cloud properties from different sensors over different spectral regions and different platforms are less consistent, algorithms that use observations from wider and more spectral observations should be developed, especially for those of COT and CER. Thirdly, fundamental retrieval algorithms as well as forward radiative transfer models with a more accurate representation of cloud properties are always needed.

REFERENCES

Ackerman, S. A., Smith, W., Revercomb, H., Spinhirne, J., 1990. The 27–28 October 1986 FIRE IFO cirrus case study: Spectral properties of cirrus clouds in the 8–12 μm window. Mon. Weather Rev. 118(11), 2377–2388.

Ackerman, S. A., Strabala, K. I., Menzel, W. P., Frey, R. A., Moeller, C. C., Gumley, L. E., 1998. Discriminating clear sky from clouds with MODIS. J. Geophys. Res-Atmos. 103(D24), 32141–32157.

Arking, A., Childs, J. D., 1985. Retrieval of cloud cover parameters from multi-spectral satellite images. J. Appl. Meteorol. Clim. 24(4), 322–333.

Baker, M. B., Peter, T., 2008. Small-scale cloud processes and climate. Nature. 451(7176), 299–300.

Bankert, R. L., 1994. Cloud classification of AVHRR imagery in maritime regions using a probabilistic neural network. J. Appl. Meteorol. Clim. 33(8), 909–918.

Baum, B. A., Kratz, D. P., Yang, P., Ou, S., Hu, Y., Soulen, P. F., Tsay, S. C., 2000. Remote sensing of cloud properties using MODIS airborne simulator imagery during SUCCESS: 1. Data and models. J. Atmos. Sci. 105(D9), 11767–11780.

Baum, B. A., Menzel, W. P., Frey, R. A., Tobin, D. C., Holz, R. E., Ackerman, S. A., Heidinger, A. K., Yang, P., 2012. MODIS cloud-top property refinements for collection 6. J. Appl. Meteorol. Clim. 51(6), 1145–1163.

Berendes, T. A., Mecikalski, J. R., MacKenzie Jr, W. M., Bedka, K. M., Nair, U. S., 2008. Convective cloud identification and classification in daytime satellite imagery using standard deviation limited adaptive clustering. J. Geophys. Res-Atmos. 113(D20).

Bessho, K., Date, K., Hayashi, M., Ikeda, A., Imai, T., Inoue, H., Kumagai, Y., et al., 2016. An Introduction to Himawari-8/9-Japan's New-Generation Geostationary Meteorological Satellites. J. Meteorol. Soc. Jpn. 94(2), 151–183.

Bi, L., Yang, P., Liu, C., Yi, B., Baum, B. A., Van Diedenhoven, B., Iwabuchi, H., 2014. Assessment of the accuracy of the conventional ray-tracing technique: Implications in remote sensing and radiative transfer involving ice clouds. J. Quant. Spectrosc. Radiat. Transfer. 146, 158–174.

Cao, Q., Zheng, H., Li, X., 2007. A method for detecting cloud in satellite remote sensing image based on texture. Acta Aeronaut. Astronaut. Sin.-Series A and B-. 28(3), 661.

Chen, N., Li, W., Gatebe, C., Tanikawa, T., Hori, M., Shimada, R., Aoki, T., Stamnes, K., 2018. New neural network cloud mask algorithm based on radiative transfer simulations. Remote Sens. Environ. 219, 62–71.

Dybbroe, A., Karlsson, K. G., Thoss, A., 2005. NWCSAF AVHRR cloud detection and analysis using dynamic thresholds and radiative transfer modeling. Part I: Algorithm description. J. Appl. Meteorol. Clim. 44(1), 39–54.

Ebert, E. E., 1989. Analysis of polar clouds from satellite imagery using pattern recognition and a statistical cloud analysis scheme. J. Appl. Meteorol. Clim. 28(5), 382–399.

Gao, B. C., Kaufman, Y. J., Han, W., Wiscombe, W. J., 1998. Corection of thin cirrus path radiances in the 0.4–1.0 μm spectral region using the sensitive 1.375 μm cirrus detecting channel. J. Atmos. Sci. 103(D24), 32169–32176.

Gao, B. C., Meyer, K., Yang, P., 2004. A new concept on remote sensing of cirrus optical depth and effective ice particle size using strong water vapor absorption channels near 1.38 and 1.88/spl mu/m. IEEE Trans. Geosci. Electron. 42(9), 1891–1899.

Gao, X. J., Wan, Y. C., Zheng, S. Y., Yang, Y. W., 2014. Real-time automatic cloud detection during the process of taking aerial photographs. Spectrosc. Spectral Anal. 34(7), 1909–1913.

Goodman, S. J., Blakeslee, R. J., Koshak, W. J., Mach, D., Bailey, J., Buechler, D., Carey, L., Schultz, C., Bateman, M., McCaul, E. Jr, 2013. The GOES-R geostationary lightning mapper (GLM). Atmos. Res. 125, 34–49.

Heidinger, A. K., Goldberg, M. D., Tarpley, D., Jelenak, A., Pavolonis, M. J., 2005. A new AVHRR cloud climatology. Proc. SPIE. 5658, 197–205.

Heidinger, A. K., Evan, A. T., Foster, M. J., Walther, A., 2012. A naive Bayesian cloud-detection scheme derived from CALIPSO and applied within PATMOS-x. J. Appl. Meteorol. Clim. 51(6), 1129–1144.

Heidinger, A. K., Pavolonis, M. J., 2009. Gazing at cirrus clouds for 25 years through a split window. Part I: Methodology. J. Appl. Meteorol. Clim. 48(6), 1100–1116.

Heymsfield, A., Bansemer, A., Wood, N. B., Liu, G. S., Tanelli, S., Sy, O. O., Poellot, M., Liu, C. T., 2018. Toward improving ice water content and snow-rate retrievals from radars. Part II: results from three wavelength radar-collocated in situ measurements and cloudSat-GPM-TRMM radar data. J. Appl. Meteorol. Clim. 57(2), 365–389.

Holz, R. E., Ackerman, S., Antonelli, P., Nagle, F., Knuteson, R. O., McGill, M., Hlavka, D. L., Hart, W. D., 2006. An improvement to the high-spectral-resolution CO_2-slicing cloud-top altitude retrieval. J. Atmos. and Ocen. Tech. 23(5), 653–670.

Holz, R. E., Ackerman, S. A., Nagle, F. W., Frey, R., Dutcher, S., Kuehn, R. E., Vaughan, M., Baum, B. A., 2008. Global Moderate resolution Imaging Spectroradiometer (MODIS) cloud detection and height evaluation using CALIOP. J. Geophys. Res. 113 (D8).

Huang, J., 2005. Advanced retrievals of multilayered cloud properties using multi-spectral measurements. J. Geophys. Res. 110(D15), D15S18.

Inoue, T., 1985. On the temperature and effective emissivity determination of semi-transparent cirrus clouds by bi-spectral measurements in the 10μm window region. J. Meteorol. Soc. Jpn. Ser. II. 63(1), 88–99.

IPCC. 2013. Climate change 2013: The physical science basis. Contribution of working group I to the fifth assessment report of the IPCC. in: Stocker, T. F. et al. (Eds.), Cambridge, UK and New York, NY.

Ishida, H., Oishi, Y., Morita, K., Moriwaki, K., Nakajima, T. Y., 2018. Development of a support vector machine based cloud detection method for MODIS with the adjustability to various conditions. Remote Sens. Environ. 205, 390–407.

Iwabuchi, H., Yamada, S., Katagiri, S., Yang, P., Okamoto, H., 2014. Radiative and microphysical properties of cirrus cloud inferred from infrared measurements made by the Moderate Resolution Imaging Spectroradiometer (MODIS). Part I: Retrieval method. J. Appl. Meteorol. Clim. 53(5), 1297–1316.

Karlsson, K.-G., Riihelä, A., Müller, R., Meirink, J., Sedlar, J., Stengel, M., Lockhoff, M., Trentmann, J., Kaspar, F., Hollmann, R., 2013. CLARA-A1: a cloud, albedo, and radiation dataset from 28 yr of global AVHRR data. Atmos. Chem. Phys. 13(10), 5351–5367.

Kato, S., Hinkelman, L. M., Cheng, A., 2006. Estimate of satellite-derived cloud optical thickness and effective radius errors and their effect on computed domain-averaged irradiances. J. Atmos. Sci. 111(D17).

Key, J., 2002. The Cloud and Surface Parameter Retrieval (CASPR) system for Polar AVHRR User's Guide, Cooperative Institute for Meteorological Satellite Studies, University of Wisconsin, Madison, WI, p. 61.

Key, J., Barry, R., 1989. Cloud cover analysis with Arctic AVHRR data: 1. Cloud detection. J. Geophys. Res.-Atmos. 94(D15), 18521–18535.

King, M. D., Kaufman, Y. J., Menzel, W. P., Tanre, D., 1992. Remote sensing of cloud, aerosol, and water vapor properties from the moderate resolution imaging spectrometer (MODIS). IEEE Trans. Geosci. Electron. 30(1), 2–27.

Knap, W. H., Stammes, P., Koelemeijer, R. B., 2002. Cloud thermodynamic-phase determination from near-infrared spectra of reflected sunlight. J. Atmos. Sci. 59(1), 83–96.

Kwan, C., Hagen, L., Chou, B., Perez, D., Li, J., Shen, Y., Koperski, K., 2020. Simple and effective cloud-and shadow-detection algorithms for Landsat and Worldview images. Signal Image Video Proces. 14(1), 125–133.

Lai, R. Z., Teng, S. W., Yi, B. Q., Letu, H. S., Min, M., Tang, S. H., Liu, C., 2019. Comparison of cloud properties from Himawari-8 and FengYun-4A geostationary satellite radiometers with MODIS cloud retrievals. Remote Sens. 11(14): 1703.

Lenaerts, J. T. M., Van Tricht, K., Lhermitte, S., L'Ecuyer, T. S., 2017. Polar clouds and radiation in satellite observations, reanalyses, and climate models. Geophys. Res. Lett. 44(7), 3355–3364.

Letu, H. S., Yang, K., Nakajima, T. Y., Ishimoto, H., Nagao, T. M., Riedi, J., Baran, A. J., et al., 2020. High-resolution retrieval of cloud microphysical properties and surface solar radiation using Himawari-8/AHI next-generation geostationary satellite. Remote Sens. Environ. 239.

Li, J., Huang, J., Stamnes, K., Wang, T., Lv, Q., Jin, H., 2015. A global survey of cloud overlap based on CALIPSO and CloudSat measurements. Atmos. Chem. Phys. 15, 519–536.

Lindstrot, R., Preusker, R., Fischer, J., 2010. Remote sensing of multi-layer cloud-top pressure using combined measurements of MERIS and AATSR on board Envisat. J. Appl. Meteorol. Climatol. 49, 1191–1204.

Liou, K. N., 1986. Influence of cirrus clouds on weather and climate processes: A global perspective. Mon. Weather Rev. 114(6), 1167–1199.

Liu, C., Song, Y., Zhou, G., Teng, S., Li, B., Xu, N., Feng, L., Zhang, P., 2023. A cloud optical and microphysical property product for the advanced geosynchronous radiation imager onboard China's Fengyun-4 satellites: The first version. Atmos. Oceanic Sci. Lett. 16(3), 100337.

Liu, C., Yang, S., Di, D., Yang, Y., Zhou, C., Hu, X., Sohn, B.-J., 2021. A machine learning-based cloud detection algorithm for the Himawari-8 spectral image. Adv. Atmos. Sci., 39(12), 1994–2007.

Liu, S., Zhang, L., Zhang, Z., Wang, C., Xiao, B., 2014. Automatic cloud detection for all-sky images using superpixel segmentation. IEEE Geosci. Remote Sens. Lett. 12(2), 354–358.

Liu, Y., Xia, J., Shi, C.-X., Hong, Y., 2009. An improved cloud classification algorithm for China's FY-2C multi-channel images using artificial neural network. Sens. 9(7), 5558–5579.

Loeb, N. G., Wielicki, B. A., Doelling, D. R., Smith, G. L., Keyes, D. F., Kato, S., Manalo-Smith, N., Wong, T., 2009. Toward optimal closure of the earth's top-of-atmosphere radiation budget. J. Clim. 22(3), 748–766.

Mahajan, S., Fataniya, B., 2020. Cloud detection methodologies: Variants and development-A review. Complex Intell. Syst. 6(2), 251–261.

Marchant, B., Platnick, S., Meyer, K., Arnold, T., Riedi, J., 2016. MODIS Collection 6 shortwave-derived cloud phase classification algorithm and comparisons with CALIOP. Atmos. Measure. Techn. Discus., 9(4), 1587–1599, 2016.

Menzel, W., Smith, W., Stewart, T., 1983. Improved cloud motion wind vector and altitude assignment using VAS. J. Appl. Meteorol. Clim. 22(3), 377–384.

Menzel, W. P., Frey, R. A., Zhang, H., Wylie, D. P., Moeller, C. C., Holz, R., Maddux, B., Baum, B. A., Strabala, K. I., Gumley, L. E., 2008. MODIS global cloud-top pressure and amount estimation: algorithm description and results. J. Appl. Meteorol. Climatol. 47, 1175–1198.

Menzel, W. P., Purdom, J. F. W., 1994. Introducing GOES-I: The first of a new generation of geostationary operational environmental satellites. Bull. Am. Meteorol. Soc. 75(5), 757–781.

Meyer, K., Yang, P., Gao, B. C., 2004. Optical thickness of tropical cirrus clouds derived from the MODIS 0.66- and 1.375-μm channels. IEEE Trans. Geosci. Electron. 42(4), 833–841.

Min, M., Wu, C., Li, C., Liu, H., Xu, N., Wu, X., Chen, L., Wang, F., Sun, F., Qin, D., 2017. Developing the science product algorithm testbed for Chinese next-generation geostationary meteorological satellites: Fengyun-4 series. J. Meteorol. Res. 31(4), 708–719.

Minnis, P., Liou, K.-N., Takano, Y., 1993. Inference of cirrus cloud properties using satellite-observed visible and infrared radiances. Part I: Parameterization of radiance fields. J. Atmos. Sci. 50(9), 1279–1304.

Mouri, K., Izumi, T., Suzue, H., and Yoshida, R., 2016a. Algorithm theoretical basis document of cloud type/phase product. Meteorol. Satell. Center Tech. Note, 61, 19–31.

Mouri, K., Suzue, H., Yoshida, R., and Izumi, T., 2016b. Algorithm theoretical basis document of cloud top height product. Meteorol. Satell. Center Tech. Note, 61, 33–42.

Nakajima, T., King, M. D., 1990. Determination of the optical thickness and effective particle radius of clouds from reflected solar radiation measurements. Part I: Theory. J. Atmos. Sci. 47(15), 1878–1893.

Nicoll, M. P., Proença, J. T., Efstathiou, S., 2012. The molecular basis of herpes simplex virus latency. FEMS microbiology reviews. 36(3), 684–705.

Platnick, S., King, M. D., Ackerman, S. A., Menzel, W. P., Baum, B. A., Riédi, J. C., Frey, R. A., 2003. The MODIS cloud products: Algorithms and examples from Terra. IEEE Trans. Geosci. Electron. 41(2), 459–473.

Platnick, S., Meyer, K. G., King, M. D., Wind, G., Amarasinghe, N., Marchant, B., Arnold, G. T., et al., 2017. The MODIS cloud optical and microphysical products: Collection 6 updates and examples from terra and aqua. IEEE Trans. Geosci. Electron. 55(1), 502–525.

Platnick, S., Meyer, K. G., Wind, G., Holz, R. E., Amarasinghe, N., Hubanks, P. A., Marchant, B., Dutcher, S., Veglio, P., 2020. The NASA MODIS-VIIRS continuity cloud optical properties products. Remote Sens. 13(1), 2.

Prabhakara, C., Fraser, R., Dalu, G., Wu, M. L. C., Curran, R., Styles, T., 1988. Thin cirrus clouds: Seasonal distribution over oceans deduced from Nimbus-4 IRIS. J. Appl. Meteorol. Clim. 27(4), 379–399.

Rodgers, C. D., 2000. Inverse methods for atmospheric sounding: Theory and practice. World Scientific, Singapore, p. 240.

Roebeling, R., Baum, B., Bennartz, R., Hamann, U., Heidinger, A., Meirink, J. F., Stengel, M., Thoss, A., Walther, A., Watts, P., 2015. Summary of the fourth cloud retrieval evaluation workshop. Bull. Am. Meteorol. Soc. 96(4), ES71-ES74.

Roebeling, R., Feijt, A., Stammes, P., 2006. Cloud property retrievals for climate monitoring: Implications of differences between Spinning Enhanced Visible and Infrared Imager (SEVIRI) on METEOSAT-8 and Advanced Very High Resolution Radiometer (AVHRR) on NOAA-17. J. Atmos. Sci. 111(D20).

Rossow, W. B., Garder, L. C., Lacis, A. A., 1989. Global, seasonal cloud variations from satellite radiance measurements. Part I: Sensitivity of analysis. J. Clim. 2(5), 419–458.

Rossow, W. B., Schiffer, R. A., 1999. Advances in understanding clouds from ISCCP. Bull. Am. Meteorol. Soc. 80(11), 2261–2288.

Rossow, W., Mosher, F., Kinsella, E., Arking, A., Desbois, M., Harrison, E., Minnis, P., Ruprecht, E., Seze, G., Simmer, C., 1985. ISCCP cloud algorithm intercomparison. J. Appl. Meteorol. Clim. 24(9), 877–903.

Ruprecht, E., 1985. Statistical approaches to cloud classification. Adv. Space Res. 5(6), 151–164.

Schiffer, R. A., Rossow, W. B., 1983. The International Satellite Cloud Climatology Project (ISCCP): The first project of the world climate research programme. Bull. Am. Meteorol. Soc. 64(7), 779–784.

Schmit, T. J., Gunshor, M. M., Menzel, W. P., Gurka, J. J., Li, J., Bachmeier, A. S., 2005. Introducing the next-generation Advanced Baseline Imager on GOES-R. Bull. Am. Meteorol. Soc. 86(8), 1079–1096.

Shan, N., Zheng, T., Wang, Z., 2009. High-speed and high-accuracy algorithm for cloud detection and its application. J. Remote Sens. 13(6), 1138–1146.

Shupe, M. D., Matrosov, S. Y., Uttal, T., 2006. Arctic mixed-phase cloud properties derived from surface-based sensors at SHEBA. J. Atmos. Sci. 63(2), 697–711.

Smith, W. L. and Frey R., 1990. On cloud altitude determinations from High Resolution Interferometer Sounder (HIS) observations. J. Appl. Meteorol. 29(7).

Stamnes, K., Tsay, S. C., Wiscombe, W., Jayaweera, K., 1988. Numerically stable algorithm for discrete-ordinate-method radiative transfer in multiple scattering and emitting layered media. Appl. Opt. 27(12), 2502–2509.

Stengel, M., Mieruch, S., Jerg, M., Karlsson, K.-G., Scheirer, R., Maddux, B., Meirink, J., Poulsen, C., Siddans, R., Walther, A., 2015. The Clouds Climate Change Initiative: Assessment of state-of-the-art cloud property retrieval schemes applied to AVHRR heritage measurements. Remote Sens. Environ. 162, 363–379.

Stowe, L., Vemury, S., Rao, A., 1994. AVHRR clear-sky radiation data sets at NOAA/NESDIS. Adv. Space Res. 14(1), 113–116.

Stowe, L. L., Yeh, H. M., Eck, T. F., Wellemeyer, C. G., Kyle, H. L., 1989. Nimbus-7 global cloud climatology. Part II: First year results. J. Clim. 2(7), 671–709.

Tan, Z., Ma, S., Liu, C., Teng S., Xu N., Hu X., Zhang P., Yan, W., 2022. Assessing overlapping cloud top heights: An extrapolation method and its performance. IEEE Trans. Geosci. Remote Sens. 60: 1–11.

Teng, S., Liu, C., Tan, Z., Li, J., Xu, N., Hu, X., Zhang, P., Yan, W., Sohn, B. J., 2022. A multi-spectral method for retrieving overlapping cloud top heights from passive radiometers. Remote Sens. Environ. 286:113425.

Thampi, B. V., Wong, T., Lukashin, C., Loeb, N. G., 2017. Determination of CERES TOA fluxes using machine learning algorithms. Part I: Classification and retrieval of CERES cloudy and clear scenes. J. Atmos. Oceanic Technol. 34(10), 2329–2345.

Tian, P., Qiang, G., Liu, X., 2019. Cloud detection from visual band of satellite image based on variance of fractal dimension. J. Syst. Eng. Electron. 30(3), 485–491.

Vaughan, M. A., Winker, D. M., Powell, K. A., 2005. CALIOP algorithm theoretical basis document, part 2: Feature detection and layer properties algorithms. No. PC-SCI-202 Part 2, Release 1.01. NASA Langley Research Center.

Wang, C., Ding, S., Yang, P., Baum, B., Dessler, A. E., 2012. A new approach to retrieving cirrus cloud height with a combination of MODIS 1.24- and 1.38-μm channels. Geophys. Res. Lett. 39(24).

Wang, C., Platnick, S., Meyer, K., Zhang, Z., Zhou, Y., 2020. A machine-learning-based cloud detection and thermodynamic-phase classification algorithm using passive spectral observations. Atmos. Meas. Tech. 13(5), 2257–2277.

Wang, W., Song, W. G., Liu, S. X., Zhang, Y. M., Zheng, H. Y., Tian, W., 2011. A cloud detection algorithm for MODIS images combining Kmeans clustering and multi-spectral threshold method. Spectrosc. Spectral Anal. 31(4), 1061–1064.

Watts, P. D., Bennartz, R., Fell, F., 2011. Retrieval of two-layer cloud properties from multi-spectral observations using optimal estimation. J. Geophys. Res. 116, D16203.

Wei, J., Sun, L., Jia, C., Yang, Y., Zhou, X., Gan, P., Jia, S., Liu, F., Li, R., 2016. Dynamic threshold cloud detection algorithms for MODIS and Landsat 8 data. IEEE International Geoscience and Remote Sensing Symposium (IGARSS).

Wetherald, R., Manabe, S., 1988. Cloud feedback processes in a general circulation model. J. Atmos. Sci. 45(8), 1397–1416.

Wylie, D. P., Menzel, W. P., 1999. Eight years of high cloud statistics using HIRS. J. Clim. 12(1), 170–184.

Yang, P., Liou, K. N., Bi, L., Liu, C., Yi, B., Baum, B. A., 2015. On the radiative properties of ice clouds: Light scattering, remote sensing, and radiation parameterization. Adv. Atmos. Sci. 32(1), 32–63.

Yang, P., Wei, H.-L., Baum, B. A., Huang, H.-L., Heymsfield, A. J., Hu, Y. X., Gao, B.-C., Turner, D. D., 2003. The spectral signature of mixed-phase clouds composed of non-spherical ice crystals and spherical liquid droplets in the terrestrial window region. J. Quant. Spectrosc. Radiat. Transfer. 79, 1171–1188.

Zhang, H., Menzel, W. P., 2002. Improvement in thin cirrus retrievals using an emissivity-adjusted CO_2 slicing algorithm. J. Atmos. Sci. 107(D17), AAC 2–1-AAC 2–11.

Zhang, J., Zhou, Q., Shen, X., Li, Y., 2019. Cloud detection in high-resolution remote sensing images using multi-features of ground objects. J. Geovis. Spat. Anal. 3(2): 1–9.

Zhou, G., Wang, J., Yin, Y., Hu, X., Letu, H., Sohn, B. J., Yung, Y. L., Liu, C., 2022. Detecting supercooled water clouds using passive radiometer measurements. Geophys. Res. Lett. 49(4): e2021GL096111.

6 Cloud Detection and Aerosol Optical Depth Retrieval from MODIS Satellite Imagery

Jing Wei and Lin Sun

6.1 INTRODUCTION

Clouds pose a difficult challenge in the extraction of atmospheric or surface information using remote sensing satellite data (Greenhough et al., 2005; Nakajima et al., 2011). Clouds affect the radiation energy transmission between land objects and the satellite sensors, seriously decreasing the retrieval accuracy of atmospheric or surface parameters (Li et al., 2011; Kazantzidis et al., 2011, 2013). In addition, the multiple types of clouds and the complexity of land structures hinder the detection of clouds using remote sensing images with high precision (Jedlovec et al., 2008; Hagolle et al., 2010).

Currently, the threshold method, statistical method, artificial neural network method, and object-oriented method are four primary methods that have been widely used for different satellite sensors. The threshold method uses the difference between clear and cloudy pixels to detect clouds, and several cloud detection algorithms have been developed and applied to various projects based on it, such as the ISCCP (International Satellite Cloud Climatology Project) cloud mask algorithm, the APOLLO (AVHRR Processing scheme Over cLouds, Land, and Ocean) cloud mask algorithm, the CLAVR (Clouds from the Advanced Very high Resolution Radiometer) cloud mask algorithm, the CO_2 slicing cloud mask algorithm, and the MODIS (MODerate resolution Imaging Spectroradiometer) cloud mask algorithm.

The ISCCP cloud mask algorithm utilizes the visible narrow band ($0.6\,\mu m$) and infrared window ($11\,\mu m$) channels. A pixel is classified as cloudy only if at least one radiance value is distinct from the inferred clear value by an amount larger than the uncertainty in that clear threshold value. The uncertainty can be caused both by measurement errors and by natural variability. The algorithm is constructed to be cloud conservative, minimizing false cloud detections but missing clouds that resemble clear conditions (Rossow and Schiffer, 1991; Sèze and Rossow, 1991; Rossow and Garder, 1993). The APOLLO cloud mask algorithm uses the first through the fifth AVHRR channels at full spatial resolution and is based on five threshold tests. A pixel is defined as clear if all spectral measures fall on the "clear-sky" sides of the various thresholds or is defined as cloud-contaminated if the pixel fails any single test; thus, this algorithm is clear-sky conservative (Saunders and Kriebel, 1988; Kriebel et al., 2003). The CLAVR cloud mask algorithm uses a series of spectral and spatial variability tests to detect clouds with the cloud and surface parameter retrieval, focusing on the polar areas. The CLAVR algorithm characterizes the variability of scenes, utilizing the fact that uniform scenes are less likely to contain partial or subpixel clouds that other tests fail to detect (Stowe et al., 1991; Liu and Wu, 2004). CO_2 slicing has been used to distinguish transmissive clouds from opaque clouds and clear sky using infrared radiances in the carbon dioxide-sensitive portion of the spectrum (Wylie and Menzel, 1989; Wylie et al., 1994; Hutchinson and Hardy, 1995; Turner et al., 2001; Gao et al., 2003). The MODIS cloud mask algorithm benefits from an extended spectral coverage coupled with high spatial resolution and radiometric accuracy. In this algorithm, an established effective method has been adapted to create a high-quality cloud mask project for the global data obtained from MODIS and mitigate some difficulties experienced by previous algorithms,

DOI: 10.1201/9781003363118-6

such as thin cirrus, fog and low-level stratus at night, and small-scale cumulus, which are difficult to detect because of insufficient contrast with the surface radiance. This algorithm uses 22 of the 36 channels from visible to thermal infrared ranges to detect cloudy pixels (Ackerman et al., 2010).

The statistical methods detect clouds with regression equations established utilizing statistics and analysis of the difference in the apparent reflectance or brightness temperature among the clear and cloudy pixels in the satellite data. This method can effectively detect clouds in specific data. However, this method is not widely used because the sample data used for the regression model are historical; thus, the application is limited to a specific time and area (Molnar and Coakley, 1985; Kärner, 2000). The artificial neural network methods attempt to identify the proper network weights and best thresholds from training samples to achieve cloud detection. This method can achieve automatic cloud detection and great accuracy with self-organizing and self-adapting capabilities. However, because the principle of such a method is unclear, training and validation samples are required to cover most conditions of land surface and cloud type. Because the conditions are not specified during training, this method is less accurate (Karlsson, 1989; Clark and Boyce, 1999; Walder and Maclaren, 2000). The object-oriented method is designed to segment the images into meaningful "objects," which can be described as a set of features, and realizes the "object" classification by the established relations or differences between the object and class structure. This method can achieve multiscale image segmentation and achieve a high level of cloud detection by making full use of related features, including color, shape, texture and level, and multiscale information. However, the object-oriented method is more appropriate for feature extraction of high-resolution images with rich texture features and is slow in feature selection, possibly missing the optimal eigenvalues (Zhu and Woodcock, 2012; Fisher, 2014; Zhang et al., 2014; Zhu et al., 2015).

The threshold method is the most popular method used for cloud detection because of its high accuracy and stable results. These algorithms aim to obtain a series of proper thresholds of apparent reflectance or brightness temperatures via certain channels for different sensors and achieve cloud detection with reliable accuracy. However, for complex land surface composition and cloud types, it is difficult to identify proper thresholds from any wavelength to accurately detect a cloud. In fact, the threshold to separate the clear pixel from the cloudy is closely related to surface features. Thus, a universal dynamic threshold cloud detection algorithm (UDTCDA) supported by a prior surface reflectance database was proposed. A monthly synthesis surface reflectance database was created using the 8-day synthetic MODIS surface reflectance product (MOD09A1) to provide the surface reflectance for cloud detection, and a dynamic thresholds model related to the land surface reflectance, observation geometry, and other parameters was developed based on the simulated relations between the apparent reflectance and the surface reflectance using the 6S (the Second Simulation of the Satellite Signal in the Solar Spectrum) model (Kotchenova et al., 2006). This method effectively improved cloud detection accuracy, particularly for the detection of broken and thin clouds.

6.2 CLOUD DETECTION FOR MODIS IMAGERY

6.2.1 PRINCIPLES

Figure 6.1 shows the spectra of typical features, including vegetation, soil, rock, water, and snow/ice, which were collected from the ASTER spectral library. The ASTER spectral library is a compilation of over 2,400 spectra of natural and manmade materials and was released on 3 December 2008. The library includes data from three other spectral libraries: the Johns Hopkins University Spectral Library, the Jet Propulsion Laboratory Spectral Library, and the United States Geological Survey Spectral Library (Baldridge et al., 2009). Moreover, the spectra of urban areas and clouds were collected from the Airborne Visible Infrared Imaging Spectrometer, a hyperspectral data sensor with 224 spectral bands covering a spectral range of 0.4–2.5 μm with a spectral resolution of 10 nm (Wei et al., 2015).

Figure 6.1 shows that the reflectance of the cloud is much higher than that of most typical objects, including vegetation, water, soil, rock, and urban areas, except snow/ice, particularly in short wavelengths.

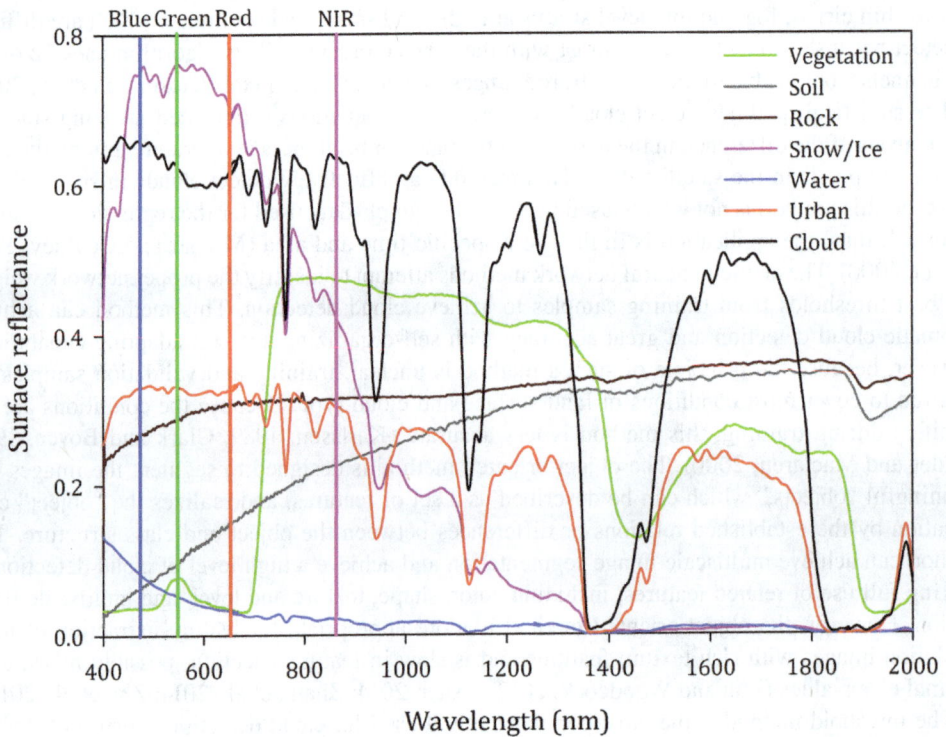

FIGURE 6.1 Spectra of typical objects.

Source: Sun et al. (2016)

The reflectance in the visible wavelength of vegetation, soil, and water is less than 0.2; however, because the cloud reflectance is greater than 0.6, the traditional methods of cloud detection have generally used a fixed threshold in such wavelengths to differentiate the cloudy pixels from the clear sky, such as the ISCCP, APOLLO, and the CLAVR. In addition to the visible bands, the bands of near infrared and short-wave infrared were used to detect clouds because of differences in the reflectance of clouds and other objects. To separate the clouds from other objects, a combination of two or more bands was also used for cloud detection, such as the Sand Dust Index (Hai et al., 2009) to differentiate sand from clouds and the Difference Snow Index or Ratio Snow Index (Lin et al., 2012) to separate snow from clouds.

Figure 6.1 shows that the reflectance is much different between a cloud and typical land objects at certain wavelengths; thus, traditional methods separate the cloudy pixels from the clear pixels. In fact, the reflectance of the satellite data is much more complex than the reflectance of the data. The reflectance only represents the component reflectance in one pixel; however, it is well known that mixed pixels are ubiquitous in remote sensing images. Mixed pixels are a combination of more than one distinct substance. If a pixel is pure pixel, having only one object, the reflectance is nearly identical to the reflectance of the identical object measured on land. A mixed pixel may comprise water and vegetation or soil and snow; a mixed pixel may also comprise water and broken or thin clouds. The reflectance of a mixed pixel is determined by all of the components in the pixel and can be described by a linear equation (Keshava and Mustard, 2002).

The difference between the cloud and the land objects only represents a thick cloud and a pure land object. However, in most cases, the cloudy pixels are covered by thin or broken clouds instead of thick clouds. Thus, the reflectance of the pixel results from the cloud and the land objects together, which can be calculated. The reflectance difference between the cloudy pixels and the land object is not as obvious. In the visible wavelengths, the reflectance of a pixel covered by thin cloud over soil may be lower than the reflectance of the rock pixel, as occurred in the short-wave infrared bands when differentiating pixels of cloud over water from vegetation. The complex land structures rendered it impossible to obtain a proper threshold to separate the cloudy pixels from the clear sky. Traditional methods of cloud detection can accurately detect a thick cloud yet often fail to detect thin or broken clouds, particularly in low reflectance areas. Even for a satellite with many bands, such as MODIS, which uses 22 selected bands from a total of 36 bands for cloud detection, the uncertainty remains.

The traditional method uses a fixed threshold to identify all thin or broken clouds but occasionally misses the cloudy pixels of thin or broken clouds over the low surface reflectance areas or falsely identifies high-reflectance land objects as clouds. The surface reflectance of water, rock, and clouds are 0.05, 0.35, and 0.6 in the red band (approximately at 0.66 μm), respectively. When the area ratio of the thin or broken clouds over a water pixel reaches approximately 40%, the reflectance is approximately 0.27, which is lower than the reflectance of rock. If the threshold is set at greater than 0.35, the pixels with an area ratio of thin or broken clouds of less than 40% over water areas will be missed; however, if the threshold is set lower than 0.35, the clear pixels covered by rock will be mistaken for cloud.

The difficulty in differentiating real land surface from clouds is a primary reason for the failure of thin or broken cloud detection with high precision from satellite data. In such circumstances, it is difficult to determine the appropriate threshold with which to identify clouds. If the reflectance is known prior to cloud detection, the component of underlying surfaces in mixed pixels can be determined and the thresholds can be established according to the real land surface reflectance. Thus, thin or broken clouds over water can be differentiated from clear pixels with relatively high accuracy with a threshold of reflectance greater than 0.05, and the rock pixel will be identified as a clear pixel even if the reflectance reaches 0.35.

Focusing on this problem, a new dynamic threshold cloud detection algorithm with prior surface reflectance support was proposed to improve the accuracy of cloud detection. The surface reflectance is created using the current MODIS surface reflectance products to provide the real surface reflectance for the image to be detected, and then dynamic thresholds related to surface reflectance for cloud detection can be estimated based on the radiative transfer model.

6.2.2 METHODOLOGY

6.2.2.1 Surface Reflectance Database Construction

MOD09A1 data were selected to represent the land surface reflectance supply for the database. The MOD09A1 data set is the 8-day gridded Level 3 product of the MOD09 series of surface reflectance and includes seven bands covering the visible to near-infrared wavelengths at a spatial resolution of 500 m. The MOD09A1 product provides the best possible L2G observations during an 8-day period, and the observations are selected on the basis of high observation coverage, low view angle, absence of clouds, or cloud shadows and aerosol loading, which effectively reduces the effect of surface and cloud contamination. The atmospheric correction accuracy is $\pm(0.005 + 0.05 \times \rho)$ under favorable conditions (Vermote and Vermeulen, 1999; Vermote and Kotchenova, 2008).

In UDTCDA, we assumed that the surface reflectance of most features remains unchanged during a certain period (Levy et al., 2013; Sun et al., 2016). Therefore, a monthly surface reflectance database was created using the minimum synthesis technology for cloud detection for that month. The lowest surface reflectance for each pixel of four images from a given month was chosen to be the pixel for the 1-month series to reduce the effects of cloud and surface contamination. MOD09A1 products for the entire year were collected and processed to construct the prior surface reflectance database. The surface reflectance images included the blue (0.459–0.479 μm), green (0.545–0.565 μm), red (0.620–0.670 μm), and near-infrared (0.841–0.876 μm) bands with a spatial resolution of 500 m, and are resampled to 1 km spatial resolution using the bidirectional linear interpolation method for MODIS. Figure 6.2 shows the false standard color composite image (RGB: band2-band1-band3) of a partial surface reflectance image in Asia (30°–45°N, 100°–125°E) from

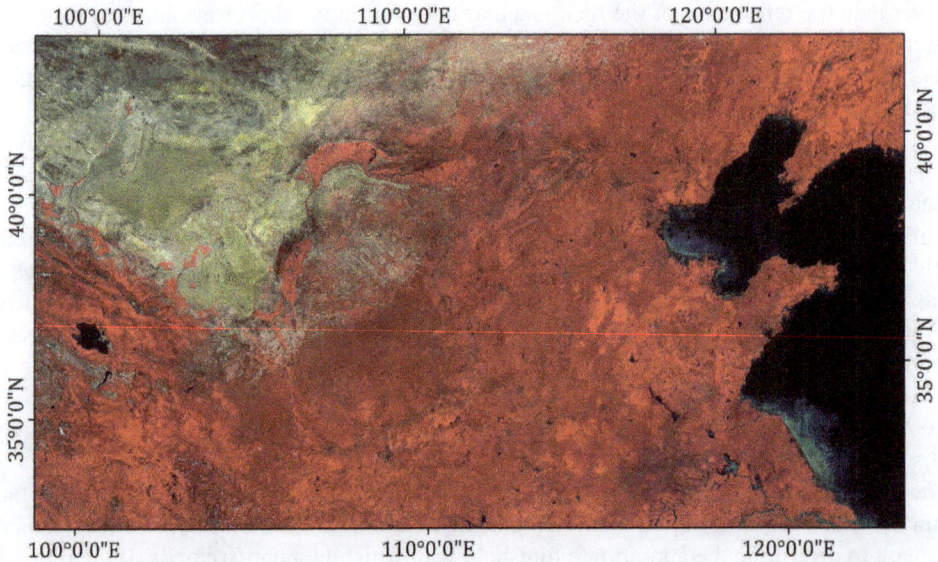

FIGURE 6.2 False standard color composite image (RGB: 214) of a surface reflectance image in July.

Source: Sun et al. (2016)

the synthetic surface reflectance database in July for demonstration. The surface reflectance images can better reflect the land cover type changes and show an overall high quality with less cloud cover, which can provide the actual surface reflectance for underlying surfaces.

6.2.2.2 Estimation of Dynamic Thresholds for Cloud Detection

The estimation of the dynamic thresholds is a key step for cloud detection. Unlike the fixed thresholds in traditional cloud methods, the thresholds used here are related to the real land surface reflectance; thus, the thresholds are called dynamic thresholds. The following work is included in the thresholds estimate: (1) band selection for the cloud detection, (2) analyzing the factors that may affect the relation between cloud and clear pixels, and (3) estimation of dynamic thresholds for cloud detection.

6.2.2.2.1 Band Selection

The greatest difference between the cloud and most land objects is in the wavelengths of visible to near infrared (NIR). Thus, the visible-to-NIR bands—bands 1, 2, 3, and 4 of the MODIS—were chosen for cloud detection. To differentiate snow/ice from cloud, short-wave infrared bands—band 7 of MODIS—were also chosen for the obvious reference difference between cloud and snow/ice.

6.2.2.2.2 Estimation of Dynamic Thresholds

In the land surface-atmosphere system, the cross radiation among different terrain and atmospheric conditions is relatively complex. The apparent reflectance received at the satellite sensor is a combination of atmospheric path reflectance and surface reflectance based on the radiative transfer theory (Levy et al., 2013; Sun et al., 2016). Therefore, to develop the relations between the apparent reflectance change and the surface reflectance, the effects of aerosols, geometric parameters, and atmospheric and aerosol models were first simulated using the 6S model. The 6S code is a basic radiative transfer code used for simulations of satellite observation under clear-sky conditions that carefully considers elevated targets, molecular and aerosol scattering, and gaseous (including H_2O, O_3, O_2, and CO_2) absorption (Vermote et al., 1997a; Kotchenova et al., 2006). Observation parameters such as spectral response function, observation, and geometric parameters; and atmospheric parameters such as observation, geometric, atmospheric, and aerosol models; and AOD were input

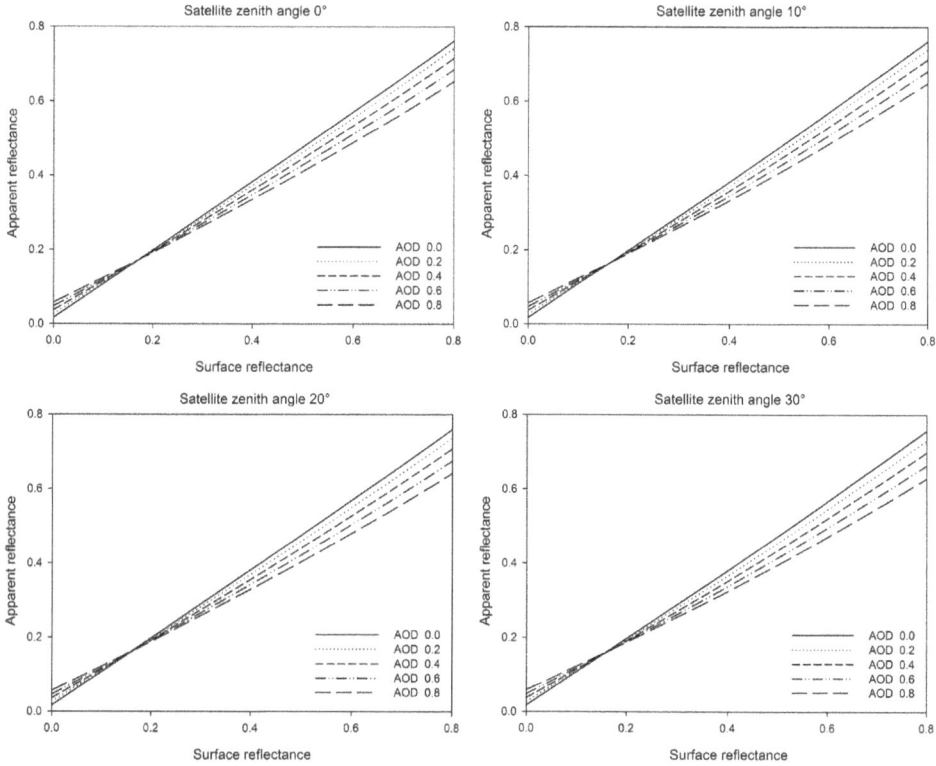

FIGURE 6.3 Simulated relationships between apparent reflectance and surface reflectance.

Source: Sun et al. (2016)

into the simulations using the 6S model. Therefore, the relations between apparent reflectance and surface reflectance under different conditions can be simulated.

The 6S model was used to simulate the influence of these factors on the apparent reflectance. Figure 6.3 shows the simulated relations of the apparent reflectance changing according to the land surface reflectance variation in an AOD of 0.2, 0.4, 0.6, and 0.8 and a satellite zenith of 0°, 10°, 20°, and 30°, respectively. The aerosol model was set to the continental model, and the atmospheric model was middle-latitude summer. The results indicate that the apparent reflectance exhibits certain differences with different AODs, and the changing speeds are much different; when the land surface reflectance is lower than 0.05 or greater than 0.4, the difference is greater.

Using dynamic thresholds based on the prior surface reflectance database is the primary characteristic of the UDTCDA to detect a cloud. In the new algorithm, we assumed that if the apparent reflectance of a pixel exceeded the maximum of the changing simulated apparent reflectance calculated with the changing surface reflectance under different observation and atmosphere conditions, this pixel would be identified as a cloudy pixel. The possible satellite apparent values of the four bands of MODIS, with different land surface reflectance values, were calculated with the 6S model. Solar and satellite zenith angles covered all possible values of the two types of satellite sensors used in this work. The maximum values of apparent reflectance distribution can be simulated and the dynamic thresholds of simulated apparent reflectance can be determined as a function of surface reflectance and observation geometry under possible atmospheric conditions without clouds. The following equations show the corresponding dynamic threshold cloud detection models for MODIS:

$$\rho_B^{*\prime}\left(\theta_s, \theta_v, \varphi\right) = 0.793 \cdot \rho_B + 0.004 \cdot \cos\alpha\cos\beta + 0.158 \tag{6.1}$$

$$\rho_G^{*'}\left(\theta_s,\theta_v,\varphi\right) = 0.807 \cdot \rho_G + 0.025 \cdot cos\alpha cos\beta + 0.125 \qquad (6.2)$$

$$\rho_R^{*'}\left(\theta_s,\theta_v,\varphi\right) = 0.807 \cdot \rho_R + 0.025 \cdot cos\alpha cos\beta + 0.125 \qquad (6.3)$$

$$\rho_{NIR}^{*'}\left(\theta_s,\theta_v,\varphi\right) = 0.928 \cdot \rho_{NIR} + 0.01 \cdot cos\alpha cos\beta + 0.099 \qquad (6.4)$$

$$R_i = \rho_i^* - \rho_i^{*'} > 0, i = B, G, R, NIR \qquad (6.5)$$

$$R = R_B \cup R_G \cup R_R \cup R_{NIR} \qquad (6.6)$$

where $\rho_i^{*'}$ represents the simulated apparent reflectance of different channels, ρ_i represents the surface reflectance, α represents the solar zenith angle, β represents the satellite zenith angle, $\rho_B^{*'}\rho_i^*$ represents the apparent reflectance, R_i represents the corresponding cloud detection results for each channel, and R represents the final cloud detection result.

For similar spectral characteristics, snow/ice is difficult to differentiate from cloud with the four bands from visible to NIR channels used above. The greatest difference between these two types of objects lies in the wavelength of the short-wave infrared band (approximately at 1.6 μm), in which, although clouds still have a higher reflectance, the reflectance of ice and snow is nearly 0. Therefore, the short-wave infrared band was selected to identify clouds and snow/ice. The Normalized Difference Snow Index (NDSI), calculated from a visible and a short-wave infrared band, was selected to distinguish and mask the snow/ice from the cloud. When the NDSI > 0.4, the pixels are deemed snow or ice covered (Hall et al., 1995, 2002; Klein and Barnett, 2003). Figure 6.4 shows the flowchart of the UDTCDA.

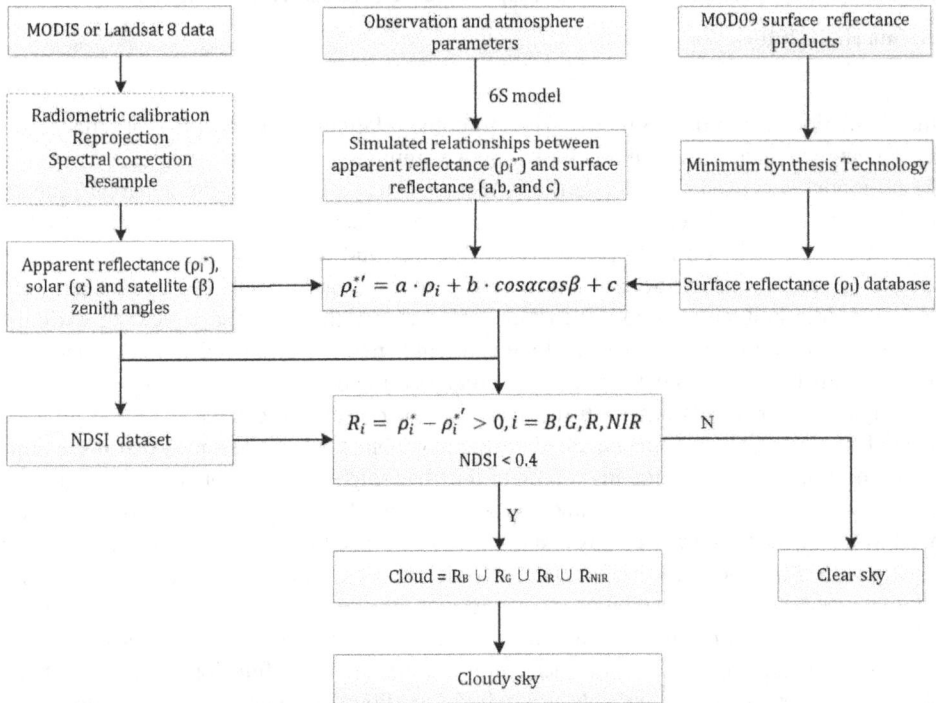

FIGURE 6.4   Flowchart of the UDTCDA

Source: Sun et al. (2016)

6.2.3 Evaluation Methods

In the current study, the cloud amount (CA) was selected to reflect the total cloud content in the remote sensing image, and the cloud amount error (CAE) shows the error difference between the reference and estimated cloud amounts. A CAE greater than 0 indicates an overestimation, and a CAE less than 0 indicates an underestimation. In addition, the correct rate of cloudy pixels (CR), the correct rate of clear-sky pixels (SR), the error rate (ER), and the missing rate (MR), four typical evaluation indices, were selected to evaluate the cloud detection results at the pixel level:

$$CA = \frac{N_{cloud}}{N} \tag{6.7}$$

$$CAE = CA_{product} - CA_{real} \tag{6.8}$$

$$CR = \frac{TP}{N_{real-cloud}} \tag{6.9}$$

$$SR = \frac{TN}{N_{real-clear}} \tag{6.10}$$

$$ER = \frac{FP}{N_{real-clear}} \tag{6.11}$$

$$MR = \frac{FN}{N_{real-cloud}} \tag{6.12}$$

where N_{cloud} represents the total cloudy pixels and N represents the total pixels of the image; $CA_{product}$ and CA_{real} represent the cloud amount of cloud mask product and the reference cloud mask, respectively; $N_{real-cloud}$ represents the total cloudy pixels; $N_{real-clear}$ represents the total clear-sky pixels in the reference cloud mask data; TP (true positive) represents the total number of pixels identified as cloudy pixels in both reference data and cloud detection results; TN (true negative) represents the total pixels identified as clear-sky pixels in both reference data and cloud detection results; FP (false positive) represents the total number of pixels identified as clear-sky pixels in reference data but cloudy pixels in cloud detection results; and FN (false negative) represents the total number of pixels identified as cloudy pixels in reference data but clear-sky pixels in cloud detection results.

6.3 MATERIALS AND EXPERIMENT

In this paper, MODIS Level 1 calibrated radiance data at a 1 km spatial resolution (MOD021KM) were obtained and used to perform cloud detection experiments using the UDTCDA. Moreover, the visual interpretation of clouds, the CALIPSO cloud product, and two MODIS cloud mask products, the MOD35 daily cloud mask product and the MOD04 daily aerosol product, were obtained for validation and comparison purposes.

6.3.1 Visual Interpretation of Clouds

To examine the evaluation of cloud detection results quantitatively, the actual cloud information was extracted using remote sensing visual interpretation method. Based on the standard false-color composite image, the cloud distribution in the images was extracted via artificial quantification using the

ArcGIS software. Then, the reference and estimated cloud mask products were generated through the vector to raster tool, resulting in a binary cloud mask (0=cloud, 1=clear) at a 1,000 m spatial resolution. To reduce the effects of subjective or accidental errors, the cloud information from a total of 36 images from all the Landsat 8 and MODIS data we used in this paper was extracted.

6.3.2 CALIPSO CLOUD PRODUCT

CALIPSO is an environmental satellite constructed by the Cannes Mandelieu Space Center and was launched on 28 April 2006. CALIPSO follows a 705-km, circular polar orbit as part of the Aqua constellation and has a good collocation with the MODIS Aqua Satellite (Poole et al., 2002; Winker et al., 2010). CALIPSO determines the cloud phase based on polarization information derived from ground-based depolarization LIDAR. It provides nearly continuous, highly accurate measurements of the vertical structure and optical properties of clouds and aerosols and has substantially increased the understanding of the climate system and climate change. The LIDAR level 2 cloud products are produced at three horizontal resolutions—1/3 km, 1 km, and 5 km—and have a temporal resolution of 16 days (Powell, 2005; Powell et al., 2009, 2013; Hu, 2007; Hu et al., 2009; Sassen et al., 2008; Chepfer et al., 2010). In this paper, the LIDAR level 2 cloud layer products with a spatial resolution of 1 km are selected for validation purposes.

6.3.3 MODIS CLOUD MASK PRODUCTS

MOD35 is the MODIS Level 2 daily cloud mask product generated at 1 km and 250 m (at nadir) spatial resolutions. The MOD35 data are generated with the MODIS cloud mask algorithm and employs a series of visible and infrared thresholds and consistency tests to specify confidence that an unobstructed view of Earth's surface is being observed (Ackerman et al., 2010; Remer et al., 2012). The data set of "Cloud Mask: MODIS Cloud Mask and Spectral Test Results" at 1 km spatial resolution was selected for this paper. MOD04 is the MODIS Level 2 daily aerosol product and has been updated to the Collection 6 (C6) version. The MOD04 C6 provides an overland cloud mask product that is a combination of tests using absolute magnitude and spatial variability at 0.47 µm (500-m resolution) and 1.38 µm (1-km resolution). The final result is a binary cloud mask at 500-m resolution used to filter pixels for final aerosol retrieval (Levy et al., 2013). The data set of "Aerosol Cloudmask Land Ocean: Aerosol Cloud Mask 500 m resolution 0=cloud 1=clear" at 500-m spatial resolution was selected and resampled into a 1-km resolution using the bidirectional linear interpolation method in this paper.

6.4 MODIS CLOUD DETECTION RESULTS

MODIS data cover different cloud types over different land use types were collected and utilized in cloud detection experiments using the UDTCDA. For evaluation and comparison purposes, cloud detection results covering different cloud types, including broken, thick, and thin clouds over different surface types (e.g., vegetation, water, bare land, and desert), as determined using Google Earth with areas of interest 400×400 pixels in size, were randomly selected from MODIS data.

Figure 6.5 shows typical UDTCDA cloud detection results for broken clouds (a–d) and thick and thin clouds (e–h) over different land use covers (i–l for bright areas) for MODIS data. Table 6.1 shows the corresponding evaluation of cloud detection results for the MODIS data shown in Figure 6.5. Evaluation results indicate that the UDTCDA can identify most broken clouds irregularly distributed in the image (a–d) with higher CR (>91%) and higher SR (>94%) rates and overall lower CAE (<4%), ER (<6%), and MR (<9%) values compared with the reference cloud mask. The UDTCDA has excellent detection accuracy for thick clouds and can also detect most thin clouds as well as the edges of thick clouds over different areas (e–h). The new algorithm shows overall higher CR (>80%) and SR (>96%) values and lower CAE (~0.25–3.26%), ER (<4%), and MR (<20%) values. For bright areas, the UDTCDA can correctly differentiate thick clouds from clear sky and effectively detect broken and thin clouds (i–l) with overall higher CR (~73–93%) and SR (>92%) values and relatively

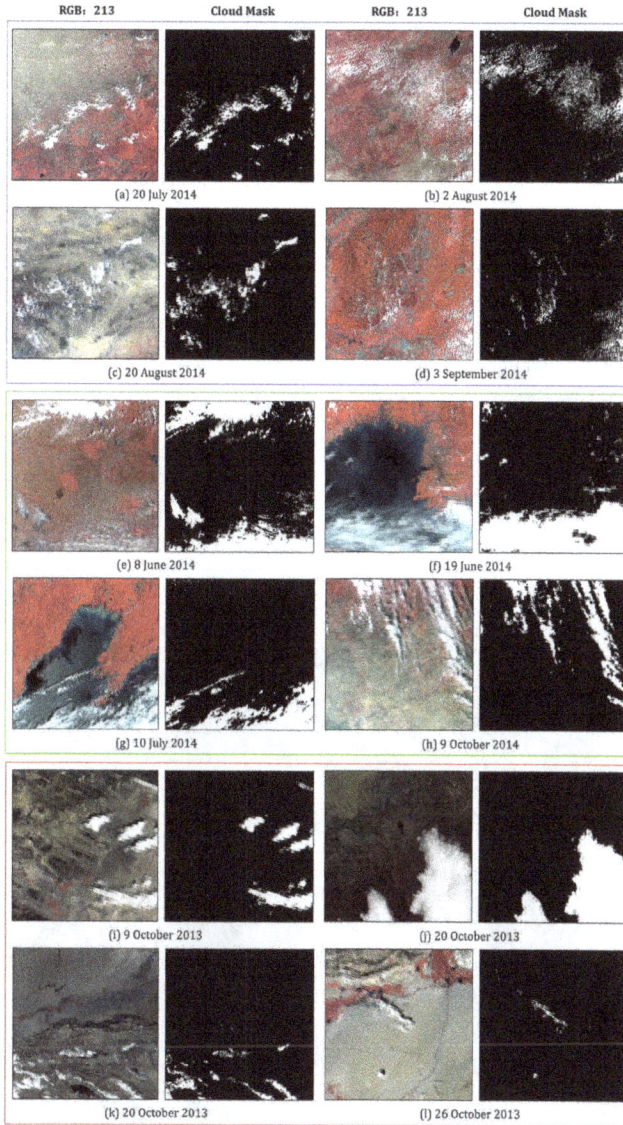

FIGURE 6.5 UDTCDA cloud detection results for different cloud types over different regions based on MODIS data.

Source: Sun et al. (2016)

lower CAE, ER, and MR values. Validation results show that the UDTCDA has an overall higher cloud detection accuracy with higher SR (>92%) and lower CAE (~0.12–3.26%) values for broken, thick, and thin clouds over different land uses, particularly for bright areas.

6.4.1 Compared with MODIS Cloud Mask Products

For comparison purposes, the MOD04 and MOD35 cloud mask products corresponding to the same time and area as the UDTCDA cloud mask retrieved from MODIS images were obtained. The horizontal groups comprise the MODIS standard false-color composite images, the UDTCDA cloud mask, and MOD35 and MOD04 cloud masks. Figure 6.6 compares different cloud mask products for broken clouds. Evaluation results indicate that the MOD35 cloud mask showed overall lower

TABLE 6.1

Evaluation of UDTCDA Cloud Detection Results Based on MODIS Data (Sun et al., 2016)

Cloud Types	No.	CA_{real} (%)	CA_{new} (%)	MAE (%)	CR (%)	ER (%)	MR (%)	SR (%)
Broken cloud	a	4.71	5.13	0.42	91.16	1.37	8.84	98.63
	b	7.63	10.31	2.68	98.08	3.06	1.92	96.94
	c	25.76	28.88	3.12	95.36	5.81	4.64	94.19
	d	43.9	45.24	1.34	95.65	5.79	4.35	94.21
Thick and thin cloud	e	32.91	29.64	−3.26	84.85	2.57	15.15	97.43
	f	31.64	28.95	−2.69	86.83	2.16	13.17	97.84
	g	17.08	17.33	0.25	84.61	3.47	15.39	96.53
	h	10.52	7.5	−3.02	80.21	0.97	19.79	99.04
Bright areas	i	8.27	8.33	0.43	92.48	2.5	7.52	97.5
	g	21.3	21.45	0.15	80.22	7.84	19.78	92.16
	k	6.16	4.31	−1.85	76.85	0.4	23.15	99.6
	l	0.68	0.55	−0.12	73.47	0.13	26.53	99.87

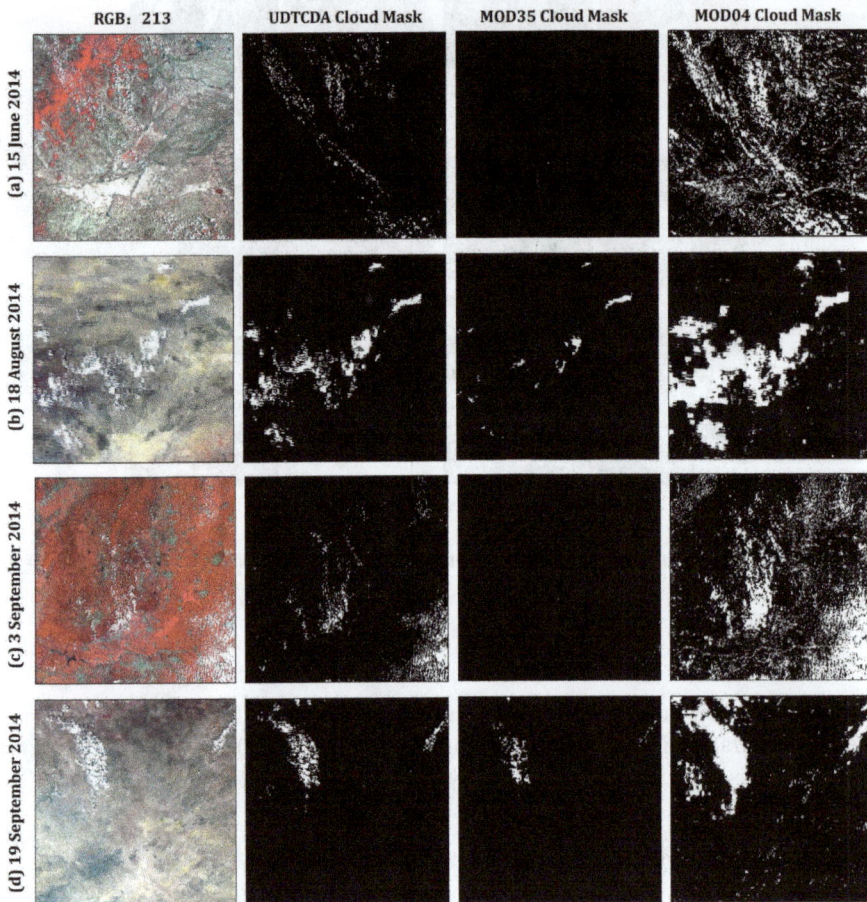

FIGURE 6.6   Comparison of broken cloud detection results between different cloud mask products.

Source: Sun et al. (2016)

cloud detection accuracy with a lower CR less than 30% and seriously underestimated the image cloud content (CAE ~ −0.76 to 4.95%). However, the MOD04 cloud mask showed generally lower detection accuracy with lower SC values less than 53% and seriously overestimated the image cloud content (CAE ~ 1.64 to 19.36%). In addition, the UDTCDA cloud mask showed better cloud detection results for broken clouds with higher CR and SR values greater than 82%; the UDTCDA cloud mask also had a lower uncertainty rate (CAE ~ −0.13 to 0.11%) with smaller ER and MR values of less than 1.2% and 18%, respectively.

Figure 6.7 compares the cloud detection results of different cloud mask products for thick and thin clouds. Clearly, all cloud mask algorithms can better detect the thick clouds correctly in the images but show great differences in detection accuracy for thin clouds. Evaluation results indicate that the MOD35 cloud mask algorithm cannot identify most thin clouds and underestimates the image cloud content (CAE ~ −0.13 to 18.63%), resulting in generally poorer detection accuracy with higher MR (~16.54 to 97.11%) and lower CR values of less than 83%. However, the MOD04 cloud mask algorithm seriously overestimated the cloud content of thin clouds in the image (CAE ~ 4.72 to 35.96%), leading to an overall lower detection accuracy with a higher ER (~11.91 to 55.42%) and lower SC values of less than 53%. However, the UDTCDA showed better cloud detection results for

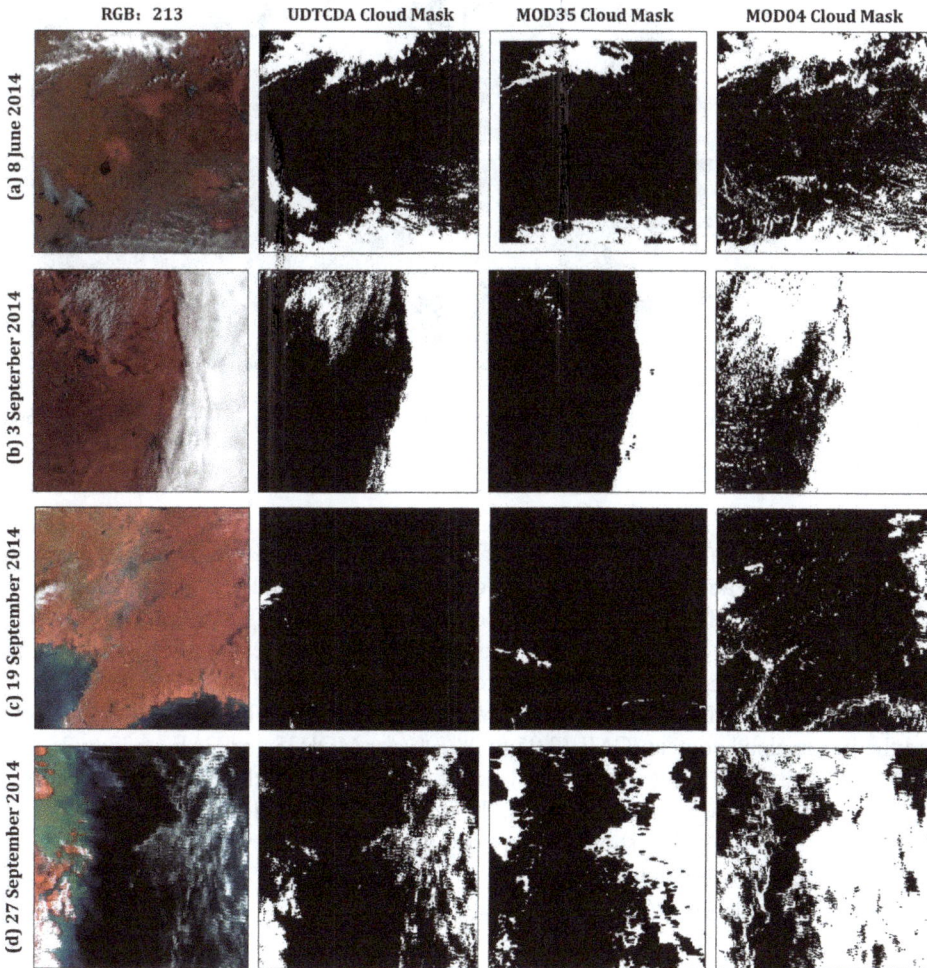

FIGURE 6.7 Comparison of thick and thin clouds detection results between different cloud mask products.

Source: Sun et al. (2016)

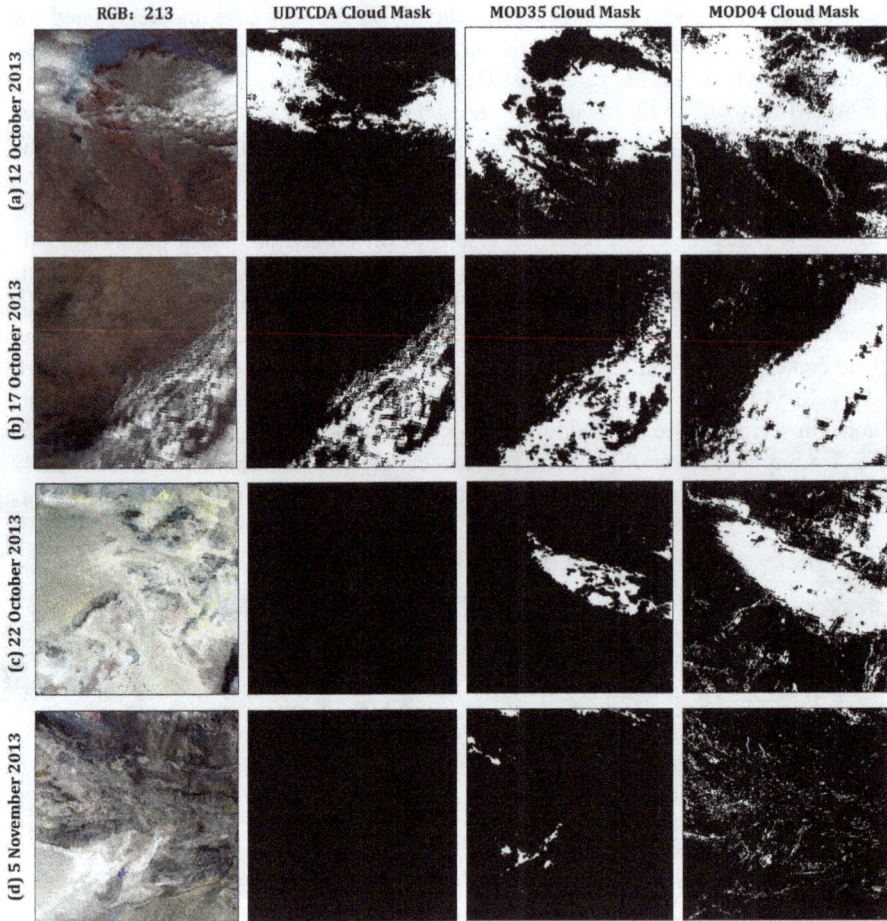

FIGURE 6.8 Comparison of cloud detection results between different cloud mask products over bright areas.

Source: Sun et al. (2016)

thick and thin clouds with higher CR and SR values greater than 82% and lower estimation errors (CAE ~ −3.26 to 0.34%) and ER and MR values of less than 5.79% and 17.6%, respectively.

Figure 6.8 compares different cloud mask products covered with different cloud types over bright areas. Both the UDTCDA and MODIS cloud mask algorithms can detect most thick and thin clouds over bright areas more accurately but show poorer detection results for broken clouds and the edges of clouds. Evaluation results indicate that both MOD35 and MOD04 cloud masks seriously overestimated the image cloud content (CAE ~ 3.05 to 28.79% for MOD35 and CAE ~ 6.32 to 28.32% for MOD04 cloud mask) with higher ER values, leading to overall lower cloud detection accuracy. The estimation uncertainties are serious, particularly in bright areas with higher surface reflectance. Overall, however, the UDTCDA shows better cloud detection accuracy for broken, thick, and thin clouds over bright areas with higher CR and SR values greater than 80% and lower CAE (0 to 1.5%) values, which can effectively reduce the estimation uncertainty in bright areas.

6.4.2 COMPARISON WITH THE CALIPSO CLOUD PRODUCT

The UDTCDA cloud products were generated from the MODIS Aqua data from June to August in 2014 and were compared and validated with the CALIPSO cloud product. The selected data

TABLE 6.2

Evaluation and Comparison of Different MODIS Cloud Products to the CALIPSO Cloud Product (Sun et al., 2016)

Date	UDTCDA Cloud Product (%)				MYD35 Cloud Product (%)				MYD04 Cloud Product (%)			
	CR	ER	MR	SR	CR	ER	MR	SR	CR	ER	MR	SR
12-Jun	90.42	22.24	9(.58	77.76	77.05	0	14	79.97	97.4	50.07	2.6	49.93
17-Jun	82.84	15.71	17.16	84.29	41.52	1.83	43.53	95.81	96.4	85.74	3.6	14.26
23-Jun	83.28	12.5	16.72	87.5	72.32	21.58	26.96	72.98	96.4	85.74	3.6	14.26
5-Jul	81.28	13.08	18.72	86.92	66.43	23.84	33.57	76.16	93.28	61.69	6.72	38.31
10-Jul	83.68	10	16.32	90	60.42	68.31	39.58	31.69	83.91	62.94	16.09	37.06
16-Jul	81.68	19.44	18.32	80.56	60.58	0.53	32.9	79.68	81.93	57.99	18.07	42.01
26-Jul	78.9	28.8	21.1	71.2	35.81	16.08	46.2	80.23	89.08	61.53	10.92	38.47
7-Aug	78.14	18.97	21.86	81.03	64.38	17.51	35.62	82.49	92.53	67.15	7.47	32.85
12-Aug	72.8	22.48	27.2	77.52	59.22	20.68	40.78	79.32	89.11	67.66	10.89	32.34
20-Aug	81.65	28.28	18.35	71.72	65.07	15.62	34.94	84.38	95.78	88.53	4.22	11.47
25-Aug	75.75	27.01	24.25	72.99	68.23	33.76	31.77	66.24	96.22	80.13	3.78	19.87
Average	80.95	19.86	19.05	80.14	61	19.98	34.53	75.36	92	69.92	8	30.08

covered a large area, with a latitudinal range of 5° to 50°N and a longitudinal range of 75° to 135°E, and included different cloud types over different land surfaces. The corresponding MODIS Aqua MYD35 (C6) and MYD04 (C6) cloud mask products were also collected and compared to the CALIPSO cloud product. Four evaluation indexes, i.e., CR, SR, ER, and MR, were calculated, and the results are shown in Table 6.2.

MOD35 cloud products have an overall lower consistency with the CALIPSO cloud products, with an average SR value of 61% and a higher MR value of 34.53%. The higher CR (75.36%) and relatively smaller ER (19.98%) indicate that the MOD35 cloud products seriously underestimated the cloud information in the images. The MOD04 cloud products showed an overall higher SR (92%), a smaller MR (8%), an average CR of less than 40%, and a higher average ER of 69.92%, indicating that it seriously overestimated the cloud information in the images. Compared with the MOD35 and MOD04 cloud products, the UDTCDA cloud products showed a better consistency with CALIPSO cloud products with CR and SR values greater than 80% and ER and MR values less than 20%. The UDTCDA can identify clouds more accurately, thereby effectively reducing the estimation uncertainty compared with the current MODIS cloud products.

6.4.3 Conclusions

To reduce the influence of mixed pixels formed of cloud and ground features on cloud detection and improve the cloud identification ability of land satellites with high spatial resolutions but low spectral resolutions, a new Universal Dynamic Threshold Cloud Detection Algorithm (UDTCDA) with a surface reflectance database support is proposed in this paper. A monthly surface reflectance database was established based on the long-time series of MODIS 8-day synthetic surface reflectance products (MOD09A1). The relation between the apparent reflectance and the surface reflectance was simulated using the 6S model, carefully considering different observation and atmospheric conditions. Then, the dynamic cloud detection models were built and applied to MODIS to perform cloud detection experiments. A visual interpretation of the clouds and the CALIPSO LIDAR cloud estimates were selected to verify the experimental results, and the results were also compared with the current MODIS cloud products.

The evaluation and comparison results indicate that the MOD35 cloud products exhibited an over-all low cloud detection accuracy, with a low correct rate of cloudy pixels (CR) of less than 30%. These products seriously underestimated the cloud content in the images. Similarly, the MOD04 cloud prod-ucts had an overall lower detection accuracy with a lower correct rate of clear-sky pixels (SR) less than 60% and seriously overestimated the cloud content in the images. The UDTCDA cloud products were considerably more consistent with the visual interpretation and CALIPSO-derived cloud estimates and demonstrated overall better cloud detection accuracy, with higher SR and CR values of greater than 80% and lower error rate (ER) and missing rate (MR) values of less than 20%. Therefore, the UDTCDA products exhibit less uncertainty than the MOD04 and MOD35 cloud mask products. The UDTCDA can effectively reduce the effects of mixed pixels and atmospheric factors and can achieve cloud detection from a large-scale area and long-term sequence for different satellite data. These capabilities make the products highly valuable for retrieval of atmospheric and surface parameters.

This study found that the new algorithm shows better and more effective cloud detection results. However, some problems still remain. The UDTCDA with the prior monthly surface reflectance database support is established based on the assumption that the surface reflectance of most features changes little within a certain period. Thus, this algorithm can be limited in some areas where the surface reflectance changes obviously due to snowfall/melt, forest fires, logging, urban expansion, etc. The cloud detection accuracy may decrease due to the lack of terrain correction in the MODIS surface reflectance products over areas of rugged terrain. Due to the lack of ground measurements of clouds, the validation work was performed via comparisons with the remote sensing visual inter-pretations of clouds, which is a more subjective approach. Thus, more comprehensive and effective verification work needs to be performed in future studies.

6.5 AEROSOL RETRIEVAL FROM MODIS IMAGERY

Atmospheric aerosols play an important role in Earth's environment and climate change from local to global scales; in particular, fine particles have a great influence on human health (Li et al., 2011; Solomon et al., 2007; Sun, Wei, Duan, et al., 2016). Therefore, a comprehensive understanding and discussion of the effects of aerosols on the environment and climate are important. Satellite remote sensing has provided an effective way to analyze the spatial distributions and variations of aerosols on long-term and large scales by detecting their main optical properties, such as aerosol optical depth (AOD) and Ångström exponent (α).

AOD is a measure of scattering or extinction of electromagnetic radiation at a given wavelength due to the presence of aerosols in the atmospheric column. For aerosol retrieval in passive remote sensing, the basic principle is to separate the contributions of the atmosphere and the earth's surface from satellite-received signals. The most critical step is to accurately determine the surface reflec-tance. Previous studies showed that 1% estimation errors in surface reflectance could lead to approxi-mately 10% errors in aerosol retrieval when the land surface reflectance (LSR) is less than 0.04; when the LSR increases, estimation errors increase by more than 15% with the same 1% inaccurate estimations for surface reflectance (Kaufman, Tanré, et al., 1997; Wei et al., 2017). For dark-target areas (e.g., vegetation and ocean), surface reflectance can be more accurately estimated due to their homogeneous surfaces and low surface-reflectance characteristics. However, for bright surfaces (e.g., urban areas, arid/semiarid areas, and deserts) other than snow/ice, the sensitivity of aerosol change to top-of-atmosphere (TOA) reflectance decreases with an increase in the LSR. Additionally, diverse underlying surfaces complicate the accurate estimation of LSR and increase the uncertainty of aero-sol retrievals (Hsu et al., 2004; Li et al., 2009; Wei et al., 2017).

Kaufman, Tanré, et al. (1997) and Kaufman, Wald, et al. (1997) found that the LSR over dense vegetation and dark soils was low in blue and red channels and showed nearly fixed ratios with the reflectance at 2.1 μm. Thus, the LSR of blue and red channels could be estimated by the TOA reflec-tance at 2.1 μm, which was minimally affected by atmospheric aerosols, and the dark target (DT) algorithm was developed. Later, the second-generation operational DT algorithm was developed with several main improvements, in which the LSR values of visible channels are estimated via improved dynamic empirical relationships with the TOA reflectance at 2.1 μm related to the normalized difference vegetation index (NDVI) calculated from the shortwave infrared channels (NDVISWIR) and scattering angles (Levy, Remer, & Dubovik, 2007; Levy, Remer, Mattoo, et al., 2007; Levy et al., 2010). The DT algorithm can perform well over dark-target surfaces but not over bright surfaces. However, Hsu et al. (2004, 2006) found that the LSR remained low and stable in deep blue channels over deserts and that the AOD could be retrieved if the LSR could be accurately estimated. The deep blue (DB) algorithm was proposed, in which the LSRs for visible channels are obtained from a precalculated seasonal LSR database using the Sea-Viewing Wide Field-of-View Sensor surface reflectance products. An enhanced DB algorithm was further developed based on several main improvements, including surface reflectance estimation, by adopting three approaches for estimating the LSR over vegetated areas, urban areas, and arid and semiarid regions; aerosol model assumption; and cloud screening schemes (Hsu et al., 2013).

MODerate resolution Imaging Spectroradiometer (MODIS) sensors were successfully launched onboard the Terra and Aqua satellites in December 1999 and May 2002, respectively, and the second-generation operational DT algorithm (for land and ocean; Levy et al., 2010) and the enhanced DB algorithm (only for land; Hsu et al., 2013) have been the main aerosol retrieval algorithms and have produced long-term and global-coverage daily DT and DB AOD products at a spatial resolution of 10 km since collection (C) 6 (MOD04_10K; Levy et al., 2013). MOD04_10K AOD products have been extensively evaluated over land and widely used in studies of atmospheric aerosols from local to global scales (Bilal et al., 2013, 2014; Levy et al., 2013; Li et al., 2007; Wei et al., 2017; Wei & Sun, 2017). However, analyses of the spatial distributions and variations of atmospheric pollutants in small- and medium-scale areas are limited due to their coarse spatial resolutions (Bilal et al., 2013; Li et al., 2005; Wei et al., 2018a). Therefore, recently, a new global-coverage daily aerosol product at a higher spatial resolution of 3 km (MOD04_3K) has been released (Remer et al., 2013).

Moreover, an increasing number of researchers have begun to focus on aerosol retrieval at high spatial resolutions to improve the applications in monitoring the air quality and related aerosol studies at urban or local regions. Li et al. (2005) modified the MODIS algorithm to retrieve AODs at 1-km resolution over Hong Kong, and the results showed that the retrievals exhibited low errors in sun photometer measurements and showed much better correlations with PM10 measurements than did MOD04 AOD products. Wong et al. (2010) proposed a refined aerosol retrieval algorithm and derived AODs from MODIS at a 500-m resolution with good overall accuracies over Hong Kong and the Pearl River Delta. Lyapustin et al. (2011) put forward a new Multi-Angle Implementation of the Atmospheric Correction (MAIAC) algorithm based on a time series of MODIS images to retrieve AODs over both dark and bright surfaces at 1-km resolution. Bilal et al. (2013) developed a Simplified Aerosol Retrieval Algorithm (SARA) to retrieve AODs from MODIS images at 500-m resolution, and the MOD09GA level 2 daily surface reflectance product was used to provide the surface reflectance for the green channel without using a look-up table (LUT). Although these algorithms can produce reliable aerosol data sets, they can be applied only to specific areas with low universality due to excessive dependence on assumptions and measured input parameters (i.e., surface reflectance and aerosol types). Therefore, it is necessary to explore a more suitable aerosol retrieval method at the global scale, which is the main purpose of this study.

6.5.1 Study Area and Data Sources

6.5.1.1 Typical Local Regions

To test and validate the adaptability of the I-HARLS algorithm, four typical regions—central and eastern Europe (42°N–59°N, 0–16°E; Figure 6.9a), central and eastern North America (30°N–50°N, 80°W–100°W; Figure 6.9b), Beijing-Tianjin-Hebei (39°N–41°N, 115°E–118°E; Figure 6.9c), and the Sahara (12°N–40°N, 14°W–16°E; Figure 6.9d)—were selected to perform the aerosol retrieval experiments. Europe is dominated by a temperate marine climate and dense vegetation coverage at low

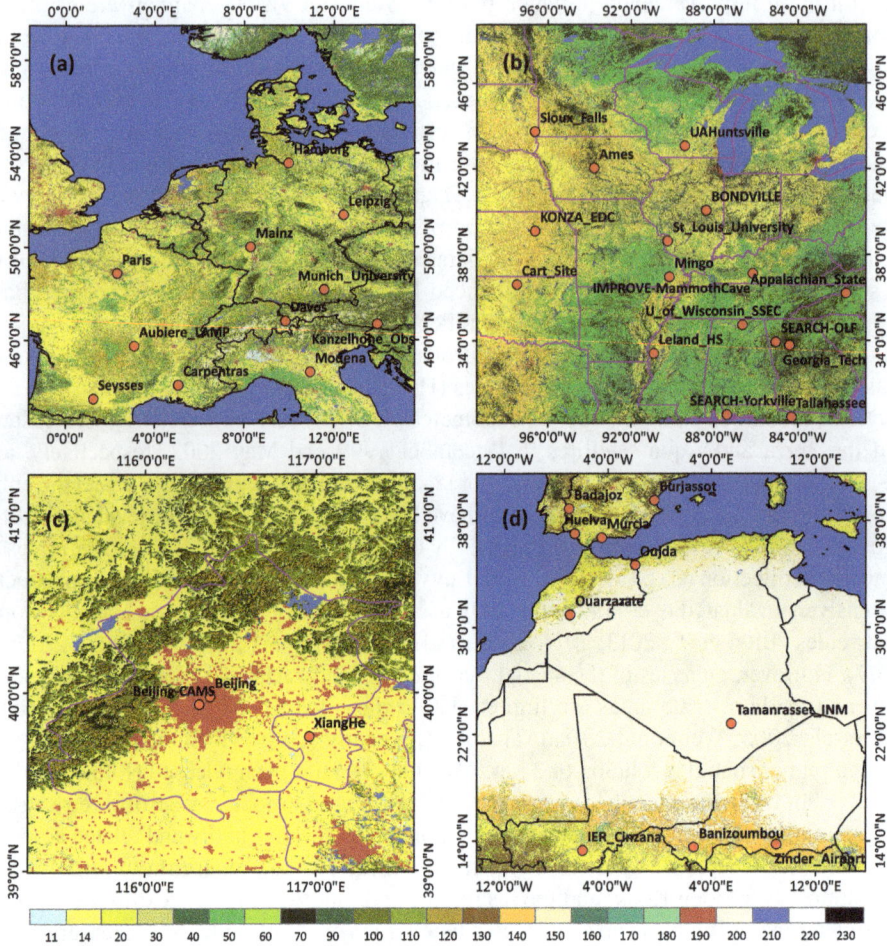

FIGURE 6.9 Locations of selected typical regions: (a) Europe, (b) North America, (c) Beijing-Tianjin-Hebei, and (d) the Sahara. The red spots represent the AERONET sites. Black and purple solid lines represent the national and state borders, respectively. Land use cover is provided by ESA GlobCover product.

Source: Wei et al. (2018b)

elevations. The region has faced environmental pressures in recent years due to air pollution, where the major sources originate from industrial and agricultural production. Aerosols are dominated by fine particles with weak and moderate absorptions (Li et al., 2013). North America has a complex and diverse climate with dense vegetation coverage, and it contains abundant mineral resources with a strong industrial base. Air pollution is inevitable due to early unreasonable industrial development, and the dominant air pollutants are fine particles with weak absorption (Li et al., 2013). Beijing-Tianjin-Hebei is located in eastern China with a large and dense population; the region has experienced increasing air pollution in recent years due to its unreasonable industrial layout and structural pollution and has been a hot spot for urban aerosol retrieval. Aerosols are dominated by fine modes with weak or moderate absorptions (Bilal et al., 2014; Wei et al., 2018a; Wei & Sun, 2017). The Sahara is dominated by an arid subtropical climate with low vegetation cover and sparse human activities; it frequently experiences sandy and dusty weather, and the aerosols are dominated by dust, where the distributions of dust particles extensively vary spatially and temporally because of their short lifetime (Hsu et al., 2004; Li et al., 2013). Figure 6.9 shows the locations of the four typical selected regions.

6.5.1.2 Operational MODIS Products

The MOD09 8-day synthetic surface reflectance products were selected to construct the LSR data-base. In addition, to monitor the atmospheric particle pollution at medium or small scales, the National Aeronautics and Space Administration has released a global daily aerosol product at 3 km (MOD04_3K) based on the second-generation DT aerosol retrieval algorithm. This product is based on the same assumptions for surface reflectance and aerosol types, LUTs, numerical inver-sion, and criteria to determine a good fit as used in the 10-km product (Levy et al., 2013). The main differences between the two products are the way the pixels are organized and the number of pixels required in the retrieval window during the pixel selection. For the MOD04_3K DT algorithm, pixels are organized into 6 × 6 pixels in the retrieval box; the 20% darkest and 50% brightest pixels over land are discarded, and then the measured TOA reflectance of remaining pixels are averaged. The algorithm requires a minimum of 5 pixels over land to make a retrieval (Remer et al., 2013). The MOD04 3-km product is expected to resolve aerosol gradients and pollution sources that are missed with the 10-km product. Because there are more selected pixels in the deselection process at 10 km, dark or bright pixels discarded at 10 km might be retained at 3 km, which makes the 3-km product potentially noisier than the 10-km product. Thus, the expected error (EE) for the MOD04 3-km product over land is [±[0.05% + 20%]), which is slightly less stringent than that of (±[0.05% +15%]) for the 10-km product over land (Levy et al., 2013; Nichol & Bilal, 2016; Remer et al., 2013). A quality assurance (QA) data set is provided to represent the data quality of AOD retrievals, with QA values ranging from 0 to 3 in order from low to high accuracy (Levy et al., 2013). In this paper, MOD04_3K DT AOD retrievals with the highest quality (QA = 3) were selected for comparison.

6.5.1.3 AERONET Ground-Based Measurements

AERONET is a worldwide network of calibrated ground-based aerosol sites where observations were collected using the CE-318 sun photometer measurement. This network has provided a long-term, continuous, and accessible public database of optical properties (i.e., AOD, size distribution, single scattering albedo, and asymmetry parameter) in diverse aerosol regimes. AODs are measured at a wide range of wavelengths from visible to near-infrared channels (0.34–1.02 μm) every 15 min with a low uncertainty of 0.01–0.02. The AOD measurements are computed as three data quality levels (L): L1.0 (unscreened), L1.5 (cloud screened), and L2.0 (cloud screened and quality assured), indicating increasing reliability (Holben et al., 2001; Smirnov et al., 2000).

In this paper, AERONET version 2 level 2.0 AOD measurements were selected to quan-titatively evaluate the reliability of the AOD retrievals. To this end, we collected 11, 16, 3, and 10 AERONET sites over Europe, North America, Beijing-Tianjin-Hebei, and the Sahara, respectively. The spatial locations and site information of each AERONET site are shown in Figure 6.9. However, AERONET does not provide AOD measurements at 550 nm; therefore, they are interpolated with the Ångström exponent algorithm based on the available AOD mea-surements at the two nearest wavelengths among 440, 500, and 675 nm to compare them with the satellite AOD retrievals (Levy, Remer, & Dubovik, 2007; Wei et al., 2018a, 2017; Wei & Sun, 2017).

6.5.2 Methodology

An improved high-spatial-resolution (1 km) aerosol retrieval algorithm with prior land surface parameters database support (I-HARLS) for MODIS images over land is proposed (Wei et al., 2018b). The I-HARLS algorithm requires the TOA reflectance, latitude, longitude, solar zenith/azimuth angles, satellite zenith/azimuth angles, and elevation, which were obtained from MOD02 images at 1-km spatial resolution (MOD021KM). The surface reflectance and aerosol types over land were determined from MODIS surface reflectance (MOD09) products and aerosol (MOD04) products, respectively.

The TOA reflectance received from the satellites contains information from both the atmosphere and surface reflectance and is a function of successive orders of radiation interactions within the coupled surface-atmosphere system, which can be estimated as follows (Tanré et al., 1988; Vermote, El Saleous, et al., 1997a):

$$\rho^*\left(\theta_s,\theta_v,\varphi\right) = \rho_{Aer}\left(\theta_s,\theta_v,\varphi\right) + \rho_{Ray}\left(\theta_s,\theta_v,\varphi\right) + \frac{\rho}{1-\rho*S}T_{\theta_s}T_{\theta_v} \tag{6.13}$$

where $\rho_{Aer(\theta_s,\theta_{v,\varphi})}$ is the aerosol reflectance resulting from multiple scattering in the absence of molecules; $\rho_{Ray(\theta_s,\theta_{v,\varphi})}$ is the multiple Rayleigh reflectance in the absence of aerosols; ρ is the surface reflectance; S is the atmospheric backscattering ratio; $T(\theta_s)$ is the transmission of the atmosphere along the Sun-surface path; $T(\theta_v)$ is the transmission of the atmosphere along the surface-sensor path; and θ_s, θ_v, and φ are the solar zenith angle, view zenith angle, and relative azimuth angle, respectively.

Aerosol reflectance is retrieved at 550 nm by correcting for Rayleigh scattering and the surface function. The aerosol reflectance received from the satellite is a function of the AOD (τ), SSA (ω_0), and aerosol scattering phase function (P) as follows:

$$\rho_{Aer}\left(\theta_s,\theta_v,\varphi\right) = \frac{\omega\tau P\left(\theta_s,\theta_v,\varphi\right)}{4\cos\theta_s\cos\theta_v} \tag{6.14}$$

Rayleigh scattering is a notable factor in the radiation calculation and has a significant impact on the visible channels, especially for blue channels (412–490 nm). The Rayleigh scattering correction for satellite data depends on the determination of the Rayleigh phase function and Rayleigh optical depth (ROD; Mishchenko et al., 1999). At sea level, the ROD caused by Rayleigh scattering is a function of wavelength (Bodhaine et al., 1999; Bucholtz, 1995) as follows:

$$\tau_{Ray}\left(\lambda,z=Z\right) = 0.00877\left[\lambda\left(z=0\right)\exp\left(Z/34\right)\right]^{-4.05}\exp\left(-Z/8.5\right) \tag{6.15}$$

where τ_{Ray} is the ROD, λ is the wavelength (μm), z is the ground elevation above sea level in kilometers (km), and Z is the height (km) of the surface target.

The Rayleigh intrinsic reflectance for actual pressure (P) is determined by adjusting the molecular optical depth at the standard pressure (P_0; 1 atm) level as follows:

$$\tau_{Ray}\left(\lambda,P\right) = {P}\big/{P_0}\,\tau_{Ray}\left(\lambda,P_0\right) \tag{6.16}$$

The surface reflectance is the most important factor and must be estimated accurately in aerosol retrieval from satellite remote sensing images. Moreover, the composition of global aerosol models is constantly changing in different areas, and aerosol model selection is one of the other key issues in AOD retrieval. Therefore, surface reflectance and aerosol model are two important parameters that affect the accuracy of AOD retrieval and need to be carefully considered.

6.5.2.1 Surface Reflectance Estimation over Land

The LSR for vegetated surfaces can vary greatly during growing seasons and remain unchanged for long periods during winter, indicating significant seasonal changes. However, the LSR is relatively high over bright areas (e.g., urban, desert, arid/semiarid, and bare areas), which reduces the sensitivity of aerosol change to TOA reflectance, and no stable relationship between visible and SWIR channels can be estimated. However, the LSRs of bright surfaces do not significantly vary with time, and the effect of the surface's bidirectional reflectance distribution function is weaker than that of vegetated surfaces (Hsu et al., 2013). Therefore, improving LSR estimations for different underlying surfaces requires consideration of dynamic LSR variations. Extensive efforts have focused on these problems, and two typical surface reflectance schemes have been proposed. The pixels over global land are divided into two categories: (1) densely vegetated areas and (2) bright and other areas.

6.5.2.1.1 Densely Vegetated Areas

Previous studies showed that the second-generation operational DT algorithm can retrieve stable and accurate AODs for dark target areas, especially for densely vegetated areas (Levy, Remer, & Dubovik, 2007); thus, the same approach is selected for AOD retrieval over densely vegetated areas in this study. The blue and red channels' LSRs are estimated by the parameters NDVISWIR and scattering angle. Densely vegetated areas are similarly defined as pixels with NDVISWIR greater than 0.75, and the LSRs for visible channels can then be estimated (Levy, Remer, & Dubovik, 2007; Levy et al., 2010) as follows:

$$NDVI_{SWIR} = \frac{\rho^*_{1.24} - \rho^*_{2.12}}{\rho^*_{1.24} + \rho^*_{2.12}} > 0.75 \tag{6.17}$$

$$\rho_{0.47} = g\left(\rho^*_{0.65}\right) = 0.49\rho^*_{0.65} + 0.005 \tag{6.18}$$

$$\rho_{0.65} = f\left(\rho^*_{2.13}\right) = \left(0.21 + 0.002\Theta\right)\rho^*_{2.13} - 0.00025\Theta + 0.033 \tag{6.19}$$

$$\Theta = arccos\left(-cos\theta_s cos\theta_v + sin\theta_s sin\theta_v cos\varphi\right) \tag{6.20}$$

where $\rho^*_{1.24}$ and $\rho^*_{2.13}$ are the TOA reflectance at 1.24 and 2.13 μm, respectively; $\rho_{0.47}$ and $\rho_{0.55}$ are the surface reflectance at 0.47 and 0.65 μm, respectively; and Θ is the scattering angle.

6.5.2.1.2 Bright and Other Areas

For bright and other surfaces, except for snow/ice surfaces, a new approach is proposed to improve LSR estimations for aerosol retrieval. Carefully considering the LSR variations for vegetated areas at the beginning or end of the growing season, we assume that the LSRs of most features remain unchanged for eight days, and a prior eight-day surface reflectance database is constructed based on the MOD09 series of surface reflectance products. For this purpose, MOD09A1 products encompassing the entire years from 2010 to 2014 are collected and mosaiced to construct a global LSR database.

The database provides 44 LSR images in one year at 1-km resolution. Each LSR image covers four spectral bands, including the blue (459–479 nm), green (545–565 nm), red (620–670 nm), and near-infrared (841–876 nm) channels. Figure 6.10 provides the LSR images for the blue (0.47 μm) channel over land on Julian day 161 in 2014. LSRs are apparently bright in the northern parts of the Northern Hemisphere at relatively high latitudes above 60°, certain tropical regions near the equator, and a few high-altitude areas in the mainland. However, the LSRs in most land areas are relatively low and generally less than 0.15 in the blue channel. Previous studies' simulated results illustrated that the TOA reflectance still responds well to aerosol change even when the surface reflectance is much higher than 0.15 in the blue channel (Wei et al., 2017). Meanwhile, the LSR image shows an overall high quality with little cloud contamination and can better reflect the LSR variations at the global scale. Therefore, the eight-day synthetic LSR database is used to provide surface reflectance for cloud detection and aerosol retrieval over land.

6.4.2.2 Assumptions Regarding Aerosol Types over Land

The composition of global aerosol models is constantly changing in different areas, and aerosol model selection is another key issue in AOD retrievals. Thousands of size distribution retrievals

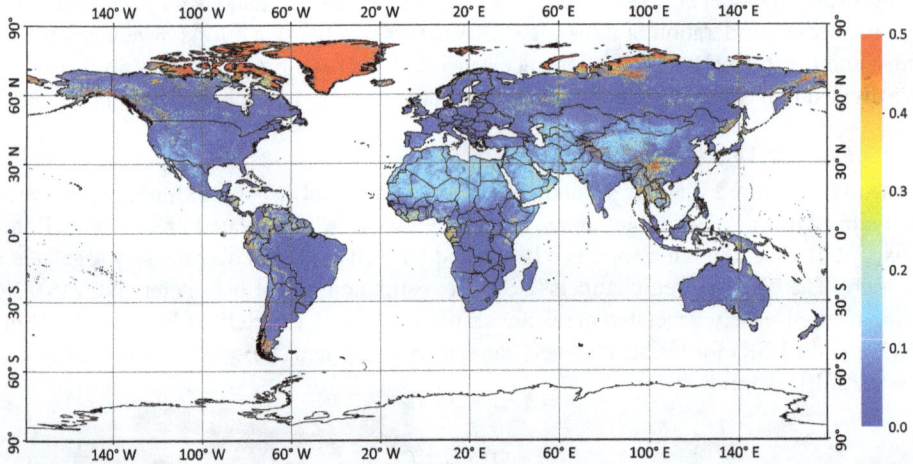

FIGURE 6.10   Surface reflectance image at the blue (0.47 μm) channel on Julian day 161 in 2014 over land.

Source: Wei et al. (2018b)

exist, and additional AERONET sites are available around the world. Operational MODIS aerosol products defined aerosol types based on a cluster analysis of all AERONET almucantar and size distribution retrievals in 2005 (Levy, Remer, Mattoo, et al., 2007). Later, a new cluster analysis was performed using AERONET aerosol optical property measurements in 2010. Most AERONET sites remained unchanged in the overall global pattern (Levy et al., 2013). The land aerosol types included a continental model, three fine models, and a dust model. The fine models were separated into strong, moderate, and weak absorption models employing a global map for four seasons.

The key assumptions include the following: the optical properties of each aerosol type vary little spatially over the region during a short time (Bilal et al., 2013; Levy, Remer, Mattoo, et al., 2007; Levy et al., 2013; Sun, Wei, Wang, et al., 2016), and the dominant aerosol type at each site is a function of the season (Levy, Remer, Mattoo, et al., 2007; Levy et al., 2013; Sun, Wei, Jia, et al., 2016; Wei et al., 2018a). Thus, a prior seasonal aerosol-type database over land is constructed. For this purpose, MOD04_10K daily aerosol-type data sets from 2012 to 2014 are collected and used to construct the aerosol-type database. The mode value of aerosol types for each pixel in all images in one season is chosen to represent the pixel for the one-season series. Then, the synthesized seasonal images are further corrected by aerosol optical properties measured by the AERONET ground-based observations at the same time. The database provides four aerosol-type images for each season at 1-km spatial resolution, containing the continental model, moderate absorption model, strong absorption model, weak absorption model, and dust model.

Figure 6.11 displays the four seasonal land aerosol-type images over land in March-April-May, June-July-August, September-October-November, and December-January-February. The database first provides similar spatial patterns but clearer boundaries of different aerosol types over land compared to that used in the MODIS official aerosol retrieval algorithm (Levy, Remer, Mattoo, et al., 2007; Levy et al., 2013). The aerosol types are as expected in most areas in different seasons. Continental aerosols dominate northern Africa, central Asia, and central Australia. Weakly absorbing aerosols (including urban and industrial aerosols) dominate eastern North America, western Europe, and Southeast Asia, especially in summer and autumn. Strongly absorbing aerosols (including presumably savanna or grassland smoke aerosols) dominate the savannas in South America and Africa. The rest of the world is dominated by moderately absorbing aerosols (including background,

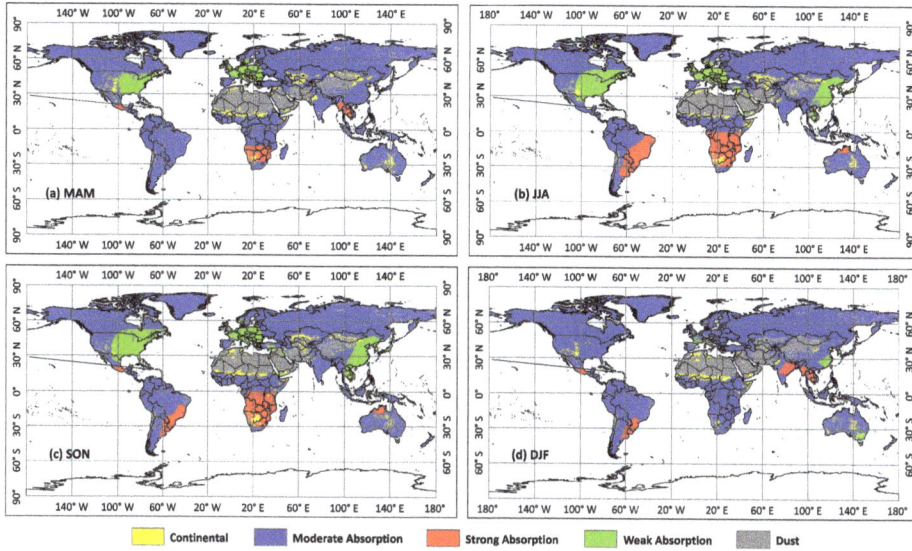

FIGURE 6.11 Spatial distribution of aerosol types over land in (a) MAM, (b) JJA, (c) SON, and (d) DJF.

Source: Wei et al. (2018b)

forest-smoke, and developing-world aerosols). The optical properties (i.e., ω0 and g) of five aerosol models at visible wavelengths are determined from monthly averages of AERONET measurements (Sun, Wei, Jia, et al., 2016).

6.5.2.3 Cloud Screening

The success of an official MODIS aerosol retrieval depends on its ability to discard unsuitable pixels, including clouds, snow, and inland water bodies. The most critical step for aerosol retrieval is accurate cloud detection to mask unsuitable pixels. Failure to remove clouds from images can create cloud contamination, and an excessively strong cloud mask produces insufficient aerosol coverage. The MOD35 cloud mask product is designed to mask pixels that are unsuitable for land surface retrieval (clouds and heavy aerosol loads) and to find suitable pixels for cloud-product retrieval (not aerosols); MOD35 is viewed as overly cloud conservative but not clear-sky conservative enough for aerosol retrieval (Levy et al., 2013; Sun, Wei, Wang, et al., 2016).

The major challenge in cloud detection is identifying thin and broken clouds over land surfaces, particularly in low- or bright-reflectance areas, which create ubiquitous mixed pixels in the remote sensing images. The difficulty in separating real land surfaces from clouds is the primary reason for the failure to detect thin or broken clouds with high accuracy from satellite remote sensing images. If the surface reflectance is known, the underlying surface component in mixed pixels can be determined, and the thresholds for cloud detection can be established. Therefore, the Universal Dynamic Threshold Cloud Detection Algorithm (UDTCDA), which is supported by a prior surface reflectance database, is selected to solve the above problems. The dynamic cloud detection models for the visible to near-infrared channels are built based on the simulated relationships between TOA reflectance and surface reflectance. Validations and comparisons with Cloud-Aerosol LIDAR and Infrared Pathfinder Satellite Observation (CALIPSO) cloud measurements and MODIS cloud mask products indicated that it could more accurately detect different kinds of clouds (Sun, Wei, Wang, et al., 2016). Therefore, the UDTCDA for MODIS images is selected to filter pixels for aerosol retrieval in this paper, which are described in the following section.

6.5.2.4   AOD Retrieval

Similarly, our algorithm uses the LUT approach to provide parameters as MOD04 aerosol retrieval algorithms for AOD retrieval with the Second Simulation of the Satellite Signal in the Solar Spectrum (6S) radiative transfer model (Vermote, Tanré, et al., 1997b). The parameters are calculated for different aerosol loadings from 0.0 to 3.0. The TOA reflectance is calculated for solar and sensor zenith angles from 0 to 60° at intervals of 6° and relative azimuth angles from 0° to 180° in increments of 12°. The LUTs contain five aerosol types in tropical (30°S–30°N), midlatitude (30°N–60°N, 30°S–60°S) summer/winter, and subarctic (>60°N, >60°S) summer/winter atmospheric models, which can be determined via the latitudes. The effective retrieval pixels should meet the following conditions: (1) ensure that all the values (i.e., angles and reflectance) are valid; (2) identify and mask most clouds with the UDTCDA algorithm; and (3) identify and mask snow/ice and inland water via the NDSI and normal difference water index (NDWI; Gao, 1996), respectively. Moreover, due to permanent snow/ice or water cover in the polar regions, our algorithm is not designed to retrieve aerosols over areas with high latitudes (>80°S or 80°N). Here, MOD021KM images covering Europe, North America, and the Sahara during 2012–2014 and Beijing-Tianjin-Hebei during 2010–2014 are downloaded to perform aerosol retrieval experiments.

6.5.2.5   Validation Method

AERONET version 2, level 2.0 ground-based AOD measurements and MOD04_3K AOD products with the same period are collected for validation and comparison. AOD retrievals within a common sampling window of 5×5 pixels around the AERONET site are obtained. To remove the AOD retrievals with large fluctuations and less reliability, the 20% highest and 20% lowest pixels are discarded, and the remaining values are averaged as the retrieved AOD. Then, the average of at least two AERONET AOD measurements at each site within ±30 min of the MODIS satellites' overpass time are calculated as the true value (Bilal et al., 2014; Hsu et al., 2013; Wei et al., 2018a, 2017). To quantify the accuracy, three main evaluation metrics, including the mean absolute error (MAE), root-mean-square error (RMSE), and EE for the MOD04_3K AOD product over land (±[0.05 ± 20%]) (Remer et al., 2013) are selected to evaluate the accuracy and uncertainty.

6.5.3   Results and Discussion

6.5.3.1   Validation with AERONET AOD Measurements

The I-HARLS AOD retrievals at 1-km resolution from MODIS images are first validated against AERONET AODs from 11 sites, 16 sites, 3 sites, and 10 sites over Europe, North America, Beijing-Tianjin-Hebei, and the Sahara, respectively, at both the site and regional scales. Figure 6.12 provides the spatial distributions of I-HARLS AOD retrievals against AERONET AODs as the percentage of the collections falling within the EE (%), MAE, and RMSE for each site. Table 6.3 shows the accuracy statistics for Europe, North America, Beijing-Tianjin-Hebei, and the Sahara.

For European sites, more than 70% of the retrievals meet the requirements of the EE at 10 out of 11 AERONET sites, with average MAE and RMSE values less than 0.06 and 0.09, respectively. However, the Hamburg site shows an overall low accuracy, with 58.46% of the collections falling within the EE and average large MAE and RMSE values of 0.079 and 0.102, respectively. The main reason is that the high-latitude location (53.57°N) and the effects of snow/ice in winter increase the difficulties in estimating the surface reflectance. Moreover, despite good accuracies, AOD retrievals show overall low overestimation uncertainties at most sites (a–c). At the regional scale, a total of 1,264 effective points are collected, and they agree well with AERONET AODs (R = 0.865), with 79.59% of the retrievals falling within the EE, and average small MAE and RMSE values of 0.050 and 0.069, respectively.

The I-HARLS algorithm shows overall good performance over North America at the site scale; more than 60% of the retrievals fall within the EE at 15 out of 16 selected sites, with average MAE

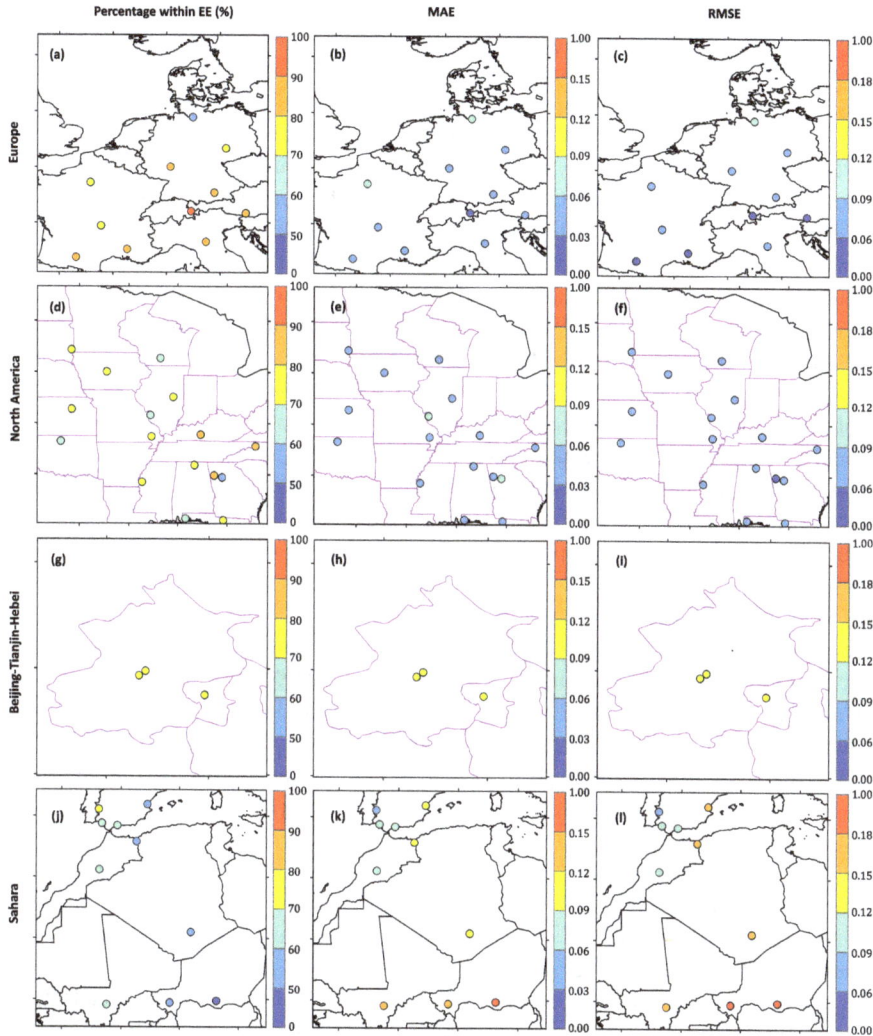

FIGURE 6.12 Spatial distributions of validations of I-HARLS AOD retrievals with AERONET AOD measurements in the percentages of retrievals falling within the EE (%), MAE, and RMSE at each site over (a–c) Europe, (d–f) North America, (g–i) Beijing-Tianjin-Hebei, and (j–l) the Sahara.

Source: Wei et al. (2018b)

TABLE 6.3

Statistical Summary for Validation of I-HARLS AOD Retrievals over Four Typical Regions and Land (Wei et al., 2018b)

Region	N	R	MAE	RMSE	= EE	> EE	< EE
Europe	1264	0.865	0.050	0.069	79.59	18.51	01.90
North America	1439	0.817	0.052	0.072	72.69	23.49	03.82
Beijing-Tianjin-Hebei	1273	0.943	0.090	0.138	74.71	18.46	06.83
The Sahara	1439	0.774	0.104	0.155	61.01	23.84	15.15
Land	5415	0.913	0.074	0.115	71.67	21.24	07.09

and RMSE values less than 0.06 and 0.09, respectively. Despite this, AOD retrievals show certain overestimation uncertainties at most sites (d–f). At the regional scale, we collect a total of 1,439 points; they are highly correlated with the AERONET AODs (R = 0.917), and 72.69% of them meet the requirements of the EE with an average MAE of 0.052 and RMSE of 0.072.

In Beijing-Tianjin-Hebei, 563, 150, and 560 pairs are collected for Beijing, Beijing_CAMS, and XiangHe sites, respectively. For two typical urban sites (Beijing and Beijing_CAMS), the I-HARLS AOD retrievals agree well with AERONET AODs (R = 0.941 and 0.950), and 70.52% and 72.67% of the retrievals fall within the EE, with average MAEs of 0.092 and 0.096 and RMSEs of 0.134 and 0.148, respectively. Moreover, the highest accuracy is found for the XiangHe site, which is located in the suburbs and covered by vegetation, with 79.46% of the collections falling within the EE, and the retrievals agree well with AERONET AOD measurements with an average MAE of 0.086 and RMSE of 0.139, respectively (g–i). Despite the good accuracies, the estimation uncertainty (i.e., MAE and RMSE) increases compared to those in Europe and North America. In general, we collected a total of 1,273 pairs from three sites, and they are highly correlated with AERONET AOD measurements; 74.71% of them meet the acquirements of the EE, with an average MAE of 0.09 and RMSE of 0.138. Furthermore, the overall data qualities of I-HARLS AOD retrievals are obviously improved over those of HARLS AOD retrievals, with 65.41%, 58.95%, and 70.16% of the collections falling within the EE for Beijing, Beijing_CAMS, and XiangHe sites, respectively, as reported in a previous study (Wei & Sun, 2017). This approach improves estimations for surface reflectance and assumptions for aerosol types.

For the Sahara, the I-HARLS algorithm performs poorly at most sites. Only half of the 10 selected sites show considerable accuracies, with more than 60% of collections falling within the EE, showing high estimation uncertainties with large MAE and RMSE values. Especially for the sites located deep in the Sahara Desert, AOD retrievals are poorly correlated with the AERONET AOD measurements, with larger average MAE and RMSE values exceeding 0.09 and 0.15, respectively (j–l). The main reason is the decreasing sensitivity of aerosols changes to TOA reflectance with the increasing surface reflectance over such bright desert surfaces. At the regional scale, a total of 1,439 points are collected, and they show good agreements with AERONET AODs (R = 0.771); 61.1% of them fall within the EE, with an average MAE of 0.104 and RMSE of 0.155. These results illustrate that the aerosol estimation uncertainty for the Sahara is greater than those for Europe, North America, and Beijing-Tianjin-Hebei.

For the whole land, we have collected 5,415 effective data pairs across all selected 40 sites from four typical regions. It is found that our 1-km I-HARLS AOD retrievals are highly consistent (R = 0.913) with AERONET AOD measurements at 550 nm over land. The retrievals are evenly distributed on both sides of the 1:1 line with average MAE and RMSE values of 0.074 and 0.115, respectively. In general, the I-HARLS algorithm performs well, with approximately 71.67% of the retrievals falling within the EE, at the global scale over land.

6.5.3.2 Comparison with MOD04_3K AOD Products

Here, eight images obtained in 2012 on 25 May and 15 June over Europe, 2 and 4 June over North America, 26 May and 4 June over Beijing-Tianjin-Hebei, and 5 and 21 July over the Sahara are selected to display the aerosol spatial distributions of the I-HARLS (1-km) and MOD04 (3-km) AOD products (Figure 6.13). The I-HARLS AODs show consistent spatial distributions similar to those of the MOD04_3K AOD products over Europe and North America, where low aerosol loadings have always been observed. However, I-HARLS AODs could provide wider spatial coverage than MOD04_3K AODs, mainly due to a more accurate cloud mask, given that the MOD04 cloud masks used in the aerosol retrieval overestimated the cloud fraction in the images (Sun, Wei, Wang, et al., 2016). Over bright urban surfaces, the MOD04_3K AOD products show a large number of missing values with poor spatial continuity in central urban areas, while the I-HARLS algorithm could achieve aerosol retrieval in such areas and provide more constant spatial coverage. The MOD04_3K AODs are always higher than I-HARLS AODs in such areas. In addition, in the

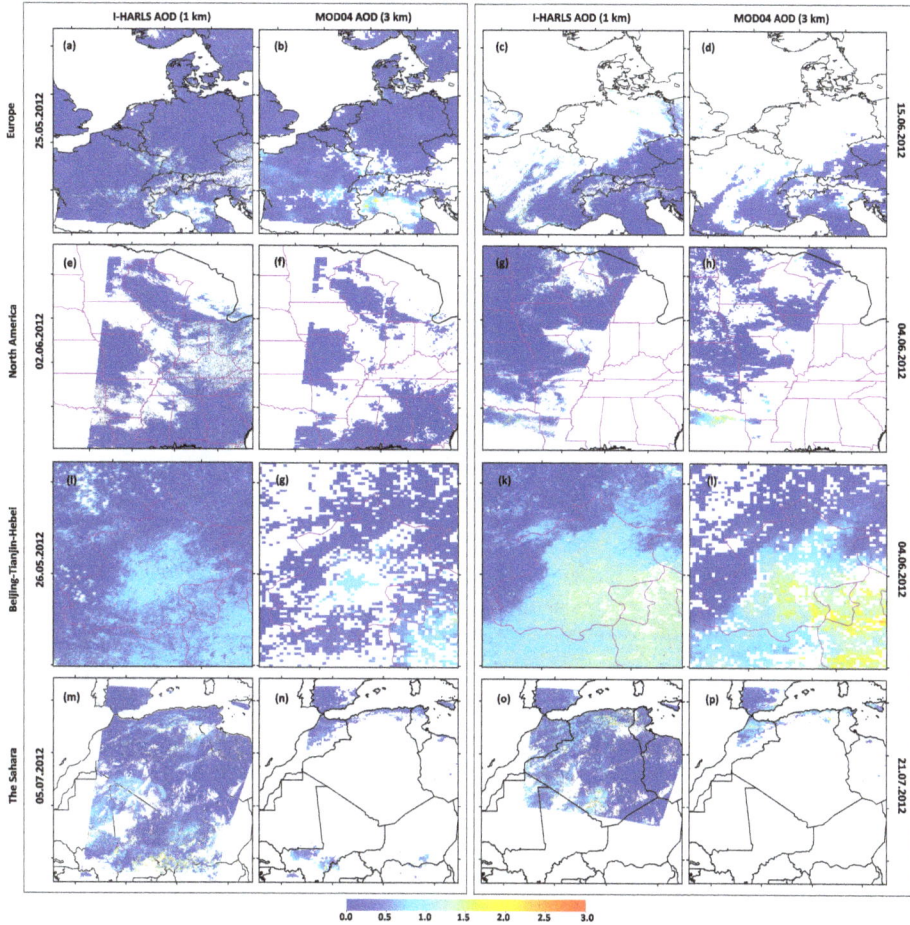

FIGURE 6.13 Comparisons of spatial distributions of I-HARLS AOD (1 km) with MOD04_3K AOD (3 km) over Europe, North America, Beijing-Tianjin-Hebei, and the Sahara on selected days (dd.mm.yyyy). The solid black line represents national borders, and the purple solid line represents state/provincial borders.

Source: Wei et al. (2018b)

surrounding vegetated areas, they show close and continuous spatial coverage. Furthermore, in the deserts, the MOD04_3K AODs show few successful retrievals because the DT algorithm could not achieve aerosol retrievals over such bright surfaces with high surface reflectance; in contrast, the I-HARLS algorithm could achieve much more successful retrievals and provide more constant spatial coverage over the Sahara. More importantly, the I-HARLS AOD product has a higher spatial resolution at 1 km than does the MOD04_3K AOD product, indicating that it can provide more detailed aerosol spatial distributions and variabilities over land.

For comparison purposes, the MOD04_3K AOD retrievals passing the highest-quality assurance (QA = 3) are obtained for four regions. Table 6.4 illustrates the validation and comparison of common retrievals between I-HARLS and MOD04_3K AOD products against AERONET AODs, as well as the validation for unique I-HARLS AOD retrievals during 2010–2014 for four typical regions over land. In Europe, a total of 682 common AOD retrievals for I-HARLS and MOD04_3K AOD products are collected from all sites. The MOD04_3K AODs show close agreements with the AERONET AOD measurements (R = 0.870), yet only 50.73% of them fall within the EE, with an average MAE of 0.100 and RMSE of 0.130. However, serious overestimations with almost half

TABLE 6.4

Comparisons of Common and Unique AOD Collections between I-HARLS (1 km) and MOD04_3K AOD Retrievals (3 km) against AERONET AOD Measurements over Four Regions

Region	Product	N	R	MAE	RMSE	= EE	> EE	< EE
Europe	I-HARLS	682	0.873	0.045	0.062	83.58	14.08	02.34
	MOD04_3K	682	0.870	0.100	0.130	50.73	49.27	00.00
	Unique	582	0.866	0.056	0.075	75.04	23.59	01.37
North America	I-HARLS	644	0.864	0.041	0.056	80.43	14.91	06.66
	MOD04_3K	644	0.837	0.054	0.077	76.86	14.60	08.54
	Unique	795	0.807	0.062	0.083	66.42	30.44	03.14
Beijing-Tianjin-Hebei	I-HARLS	340	0.957	0.078	0.109	77.94	18.53	03.53
	MOD04_3K	340	0.904	0.265	0.331	35.29	64.41	00.29
	Unique	933	0.941	0.093	0.146	72.97	19.13	07.90
The Sahara	I-HARLS	96	0.766	0.112	0.159	61.46	11.46	27.08
	MOD04_3K	96	0.610	0.155	0.185	33.33	31.25	35.42
	Unique	1343	0.780	0.103	0.154	60.98	24.72	14.30

of the collections (49.27%) falling above the EE are observed. Compared to MOD04_3K AODs, the I-HARLS AODs achieve a higher correlation with AERONET AODs (R = 0.873) and an average lower MAE of 0.045 and RMSE of 0.062. Meanwhile, this algorithm can significantly reduce the overestimation uncertainties. In general, 83.55% of the collections fall within the EE, which is approximately 1.65 times greater than that of MOD04_3K AOD retrievals. Furthermore, 582 unique AOD pairs are collected from I-HARLS AOD product, and they are also highly correlated with AERONET AODs, with 75.04% of these collections falling within the EE and average MAE and RMSE values of 0.056 and 0.075, respectively.

In North America, a total of 644 common AOD retrievals between the I-HARLS and MOD04_3K AOD products are collected from all sites. MOD04_3K AOD retrievals show good performance with a high correlation of 0.864 with AERONET AODs, and approximately 76.86% of the collections fall within the EE, with an average MAE of 0.054 and RMSE of 0.077. However, the I-HARLS AOD retrievals achieve a high correlation with AERONET AODs (R = 0.837), and more than 80% of them fall within the EE with an average lower MAE of 0.041 and RMSE of 0.056, showing an overall improvement in the aerosol estimations. Moreover, another 795 unique AOD pairs for I-HARLS retrievals are collected and validated against AERONET AOD measurements. The retrievals agree well with the measurements (R = 0.807), and 66.42% of them are within the EE, with an average MAE of 0.062 and RMSE of 0.083.

Similarly, common retrievals between I-HARLS and MOD04_3K AOD products are extracted, but only 340 effective pairs are obtained in the Beijing-Tianjin-Hebei region. MOD04_3K AODs exhibit poor performance, and only 35.29% of the retrievals fall within the EE, with an average MAE of 0.265 and RMSE of 0.331. Moreover, this algorithm significantly overestimates the aerosol loadings, with more than 64% of the retrievals falling above the EE. However, the I-HARLS algorithm is able to largely reduce the overestimation uncertainties and significantly improve the data quality over this region. The percentage of the retrievals falling within the EE increases to 77.94%, which is approximately 2.2 times more than that of the MOD04_3K retrievals. The retrievals are highly correlated with the AERONET AODs and show a much lower average MAE of 0.078 and RMSE of 0.109. Furthermore, 962 additional AOD pairs are collected from the I-HARLS retrievals, and they agree well with the AERONET

AODs; approximately 73% of them are within the EE, showing average MAE and RMSE values of 0.093 and 0.146, respectively.

The common retrievals between the I-HARLS and MOD04_3K products are extracted from all sites over the Sahara. More serious is that only 96 effective points are obtained, and the MOD04 AOD retrievals show poor performance over the Sahara, with only 33.33% of them falling within the EE and an average MAE of 0.155 and RMSE of 0.185. Similarly, serious estimation uncertainties are observed with 31% and 35% of the retrievals falling above and below the EE, respectively. There are no common retrievals from the desert sites, Tamanrasset_INM and Ouarzazate, which is mainly because the MOD04_3K DT algorithm is unable to retrieve AODs over such bright desert surfaces. However, the I-HARLS algorithm performs much better than the MOD04 DT algorithm, and more than 61.46% of the common retrievals fall within the EE, with decreasing MAE and RMSE values of 0.112 and 0.159, respectively. Moreover, a large number of 1343 unique AOD pairs are collected from the I-HARLS retrievals, and they correlate well with the AERONET AODs, with 60.98% of them falling within the EE and an average MAE of 0.103 and RMSE of 0.154. The comparison results show that the I-HARLS algorithm allows aerosol retrieval from darkest to brightest surfaces, and it not only increases the number of successful retrievals but also improves the aerosol estimations.

6.5.4 Conclusions

Here, an improved high-spatial-resolution aerosol retrieval algorithm with land surface parameter support (I-HARLS) at 1-km resolution for MODIS images is developed. A precalculated global land surface reflectance (LSR) database is constructed using the MODIS 8-day synthetic surface reflectance (MOD09A1) products, and a prior seasonal global land aerosol-type database is created using the MOD04 daily aerosol products. The main aerosol optical properties and types are determined based on the monthly average historical aerosol optical properties from local AERosol RObotic NETwork (AERONET) sites. For cloud screening, the Universal Dynamic Cloud Detection Algorithm (UDTCDA) is selected to mask cloud pixels in remote sensing images. Then, a 1-km-resolution AOD data set is generated based on the I-HARLS algorithm. Successful AOD retrievals are available over dark and bright surfaces. To test and validate the performance of the I-HARLS algorithm, four typical regions (including Europe, North America, Beijing-Tianjin-Hebei, and the Sahara) with different underlying surface and aerosol types are selected for aerosol retrieval experiments. Moreover, AERONET version 2, level 2.0 AOD measurements and MODIS daily AOD products at 3-km resolution (MOD04_3K) are selected for validations and comparisons.

The results show that the I-HARLS algorithm performs well overall at both the site and regional scales, and AOD retrievals are highly correlated with AERONET AOD measurements, with 79.56%, 72.69%, 74.71%, and 61.01% of the collections falling within the EE for the four regions, respectively. However, with an increase in surface reflectance over land, the overall performance of the retrievals decreases with increasing estimation errors, mainly due to the decreasing sensitivity of aerosol change to TOA reflectance. The new AOD products perform better and are less biased than the MOD04_3K AOD product, primarily because of the improvements in LSR estimation and aerosol-type assumption. Furthermore, the generated 1-km-resolution AOD data sets can provide continuous and wide-spatial-coverage AOD distributions over land, which play an important role in quantitative aerosol research and air quality monitoring at the medium and small scales.

This study shows that although the new AOD retrieval algorithm performs well overall over land, certain problems remain. Due to the large amount of data, four representative local regions with three or five years of data are selected for this paper for aerosol retrieval experiments and validations. However, due to the long time series of MODIS data records, longer and wider-scale experiments and validations need to be undertaken. In addition, this paper only performs comparisons with current operational and free-open high-resolution MOD04 aerosol products; therefore, more comprehensive and effective comparison efforts with other high-resolution products (such as MAIAC products) need to be performed in future studies.

REFERENCES

Ackerman, S., R. Frey, K. Strabala, Y. Liu, L. Gumley, B. Baum, and P. Menzel (2010), Discriminating clear-sky from cloud with modis algorithm theoretical basis document (MOD35), 117 pp., Modis Cloud Mask Team Coop. Inst. for Meteorol. Satell. Stud.

Baldridge, A. M., S. J. Hook, C. I. Grove, and G. Rivera (2009), The ASTER spectral library version 2.0. Remote Sensing of Environment, 113, 711–715.

Bilal, M., J. E. Nichol, M. P. Bleiweiss, and D. Dubois (2013). A simplified high resolution MODIS aerosol retrieval algorithm (SARA) for use over mixed surfaces. Remote Sensing of Environment, 136, 135–145. https://doi.org/10.1016/j.rse.2013.04.014

Bilal, M., J. E. Nichol, and P. W. Chan (2014), Validation and accuracy assessment of a simplified aerosol retrieval algorithm (SARA) over Beijing under low and high aerosol loadings and dust storms. Remote Sensing of Environment, 153, 50–60. https://doi.org/10.1016/j.rse.2014.07.015

Bodhaine, B. A., N. B. Wood, E. G. Dutton, and J. R. Slusser (1999), On Rayleigh optical depth calculations. Journal of Atmospheric and Oceanic Technology, 16(11), 1854–1861.

Bucholtz, A. (1995). Rayleigh-scattering calculations for the terrestrial atmosphere. Applied Optics, 34(15), 2765–2773.

Chepfer, H., S. Bony, D. Winker, G. Cesana, J. L. Dufresne, P. Minnis, C. J. Stubenrauch, and S. Zeng (2010), The GCM-oriented CALIPSO cloudproduct (CALIPSO-GOCCP), Journal of Geophysical Research, 115, D00H16.

Clark, C., and J. Boyce (1999), The detection of ship trail clouds by artificial neural network, International Journal of Remote Sensing, 20, 711–726.

Fisher, A. (2014), Cloud and cloud-shadow detection in SPOT5 HRG imagery with automated morphological feature extraction, Remote Sensing, 6, 776–800.

Gao, B. C. (1996). NDWI—A normalized difference water index for remote sensing of vegetation liquid water from space. Remote Sensing of Environment, 58(3), 257–266.

Gao, B. C., P. Yang, and R. R. Li (2003), Detection of high clouds in polar regions during the daytime using the MODIS 1.375-μm channel. IEEE Transactions on Geoscience & Remote Sensing, 41, 474–481.

Greenhough, J., J. J. Remedios, H. Sembhi, and L. J. Kramer (2005), Towards cloud detection and cloud frequency distributions from MIPAS infra-red observations, Adv. Space Res., 36, 800–806.

Hagolle, O., M. Huc, D. V. Pascual, and G. Dedieu (2010), A multi-temporal method for cloud detection, applied to FORMOSAT-2, VEN mu S, LANDSAT and SENTINEL-2 images, Remote Sensing of Environment, 114, 1747–1755.

Hai, Q. S., Y. H. Bao, B. A. O. Alatengtuoya, and L. B. Guo (2009), New method to identify sand and dust storm by using remote sensing technique—With Inner Mongolia autonomous region as example, Journal of Infrared and Millimeter Waves, 2, 129–132.

Hall, D. K., G. A. Riggs, and V. V. Salomonson (1995), Development of methods for mapping global snow cover using moderate resolution imaging spectroradiometer data, Remote Sensing of Environment, 54, 127–140.

Hall, D. K., G. A. Riggs, V. V. Salomonson, N. E. DiGirolamo, and K. J. Bayr (2002), MODIS snow-cover products, Remote Sensing of Environment, 83, 181–194.

Holben, B. N., D. Tanré, A. Smirnov, T. F. Eck, I. Slutsker, N. Abuhassan, W. W. Newcomb, et al. (2001). An emerging ground-based aerosol climatology: Aerosol optical depth from AERONET. Journal of Geophysical Research, 106(D11), 12,067–12,097.

Hsu, N. C., M. -J. Jeong, C. Bettenhausen, A. M. Sayer, R. Hansell, C. S. Seftor, J. Huang, et al. (2013). Enhanced deep blue aerosol retrieval algorithm: The second generation. Journal of Geophysical Research: Atmospheres, 118, 9296–9315.

Hsu, N. C., S. -C. Tsay, M. D. King, and J. R. Herman (2004). Aerosol properties over bright reflecting source regions. IEEE Transactions on Geoscience and Remote Sensing, 42(3), 557–569.

Hsu, N. C., S. -C. Tsay, M. D. King, and J. R. Herman (2006). Deep blue retrievals of Asian aerosol properties during ACE-Asia. IEEE Transactions on Geoscience and Remote Sensing, 44(11), 3180–3195.

Hu, Y. (2007), Depolarization ratio-effective Lidar ratio relation: Theoretical basis for space Lidar cloud phase discrimination, Geophysical Research Letters, 34, L11812.

Hu, Y., et al. (2009), CALIPSO/CALIOP cloud phase discrimination algorithm, Journal of Atmospheric and Oceanic Technology, 26, 2293–2309.

Hutchinson, K. D., and K. R. Hardy (1995), Threshold functions for automated cloud analyses of global meteorological satellite imagery, International Journal of Remote Sensing, 16, 3665–3680.

Jedlovec, G. J., S. L. Haines, and F. J. LaFontaine (2008), Spatial and temporal varying thresholds for cloud detection in GOES imagery, IEEE Transactions on Geoscience & Remote Sensing, 46, 1705–1717.

Karlsson, K. (1989), Development of an operational cloud classification model, International Journal of Remote Sensing, 10, 687–693.

Kärner, O. (2000), A multi-dimensional histogram technique for cloud classification, International Journal of Remote Sensing, 21, 2463–2478.

Kaufman, Y. J., D. Tanré, H. R. Gordon, T. Nakajima, J. Lenoble, R. Frouin, H. Grassl, et al. (1997). Passive remote sensing of tropospheric aerosol and atmospheric correction for the aerosol effect. Journal of Geophysical Research, 102(D14), 16,815–16,830.

Kaufman, Y. J., A. E. Wald, L. A. Remer, B.-C. Gao, R.-R. Li, and L. Flynn (1997). The MODIS 2.1-µm channel-correlation with visible reflectance for use in remote sensing of aerosol. IEEE Transactions on Geoscience and Remote Sensing, 35(5), 1286–1298.

Kazantzidis, A., K. Eleftheratos, and C. S. Zerefos (2011), Effects of cirrus cloudiness on solar irradiance in four spectral bands, Atmospheric Research, 102, 452–459.

Kazantzidis, A., P. Tzoumanikas, A. F. Bais, S. Fotopoulos, and G. Economou (2013), Cloud detection and classification with the use of whole-sky ground-based images. In Advances in Meteorology, Climatology and Atmospheric Physics, pp. 80–88, Springer, Berlin.

Keshava, N., and J. F. Mustard (2002), Spectral unmixing. IEEE Signal Processing Magazine, 19, 44–57.

Klein, A., and A. C. Barnett (2003), Validation of daily MODIS snow cover maps of the Upper Rio Grande River Basin for the 2000–2001 snow year. Remote Sensing of Environment, 86, 162–176.

Kotchenova, S. Y., E. F. Vermote, R. Matarrese, and F. J. Klemm (2006). Validation of a vector version of the 6S radiative transfer code for atmospheric correction of satellite data. Part I: Path radiance. Applied Optics, 45(26), 6762–6774.

Kriebel, K. T., G. Gesell, M. Kästner, and H. Mannstein (2003), The cloud analysis tool APOLLO: Improvements and validations, International Journal of Remote Sensing, 24, 2389–2408.

Levy, R. C., S. Mattoo, L. A. Munchak, L. A. Remer, A. M. Sayer, F. Patadia, and N. C. Hsu (2013). The collection 6 MODIS aerosol products over land and ocean. Atmospheric Measurement Techniques, 6(11), 2989–3034.

Levy, R. C., L. A. Remer, and O. Dubovik (2007). Global aerosol optical properties and application to Moderate Resolution Imaging Spectroradiometer aerosol retrieval over land. Journal of Geophysical Research, 112, D13210.

Levy, R. C., L. A. Remer, R. G. Kleidman, S. Mattoo, C. Ichoku, R. Kahn, and T. F. Eck (2010). Global evaluation of the collection 5 MODIS darktarget aerosol products over land. Atmospheric Chemistry and Physics Discussions, 10(6), 14815–14873.

Levy, R. C., L. A. Remer, S. Mattoo, E. F. Vermote, and Y. J. Kaufman (2007). Second-generation operational algorithm: Retrieval of aerosol properties over land from inversion of Moderate Resolution Imaging Spectroradiometer spectral reflectance. Journal of Geophysical Research, 112, D13211.

Li, C., K. H. Lau, J. Mao, and D. A. Chu (2005). Retrieval, validation, and application of the 1-km aerosol optical depth from MODIS measurements over Hong Kong. IEEE Transactions on Geoscience and Remote Sensing, 43(11), 2650–2658.

Li, Y., Y. Xue, G. D. de Leeuw, C. Li, L. Yang, T. Hou, and F. Marir (2013). Retrieval of aerosol optical depth and surface reflectance over land from NOAA AVHRR data. Remote Sensing of Environment, 133(133), 1–20.

Li, Z., F. Niu, J. Fan, Y. Liu, D. Rosenfeld, and Y. Ding (2011). Long-term impacts of aerosols on the vertical development of clouds and precipitation. Nature Geoscience, 4(12), 888–894.

Li, Z. Q., F. Niu, K. H. Lee, J. Y. Xin, W. M. Hao, B. Nordgren, Y. Wang, et al. (2007). Validation and understanding of moderate resolution imaging spectroradiometer aerosol products (C5) using ground-based measurements from the handheld Sun photometer network in China. Journal of Geophysical Research, 112, D22S07.

Li, Z., X. Zhao, R. Kahn, M. Mishchenko, L. Remer, K.-H. Lee, M. Wang, et al. (2009). Uncertainties in satellite remote sensing of aerosols and impact on monitoring its long-term trend: A review and perspective. Annals De Geophysique, 27(7), 2755–2770.

Lin, J., X. Feng, P. Xiao, H. Li, J. Wang, and Y. Li (2012), Comparison of snow indexes in estimating snow cover fraction in a mountainous area in northwestern China, IEEE Geoscience and Remote Sensing Letters, 9, 725–729.

Liu, C. L., and B. F. Wu (2004), Application of cloud detection algorithm for the AVHRR data, Journal of Remote Sensing, 8, 677–687.

Lyapustin, A., Y. Wang, I. Laszlo, R. Kahn, S. Korkin, L. Remer, R. Levy, et al. (2011). Multi-angle implementation of atmospheric correction (MAIAC): 2. Aerosol algorithm. Journal of Geophysical Research, 116, D03211.

Mishchenko, M. I., Geogdzhayev, I. V., Cairns, B., Rossow, W. B., & Lacis, A. A. (1999). Aerosol retrievals over the ocean by use of channels 1 and 2 AVHRR data: Sensitivity analysis and preliminary results. Applied Optics, 38(36), 7325–7341.

Molnar, G., and J. A. Coakley (1985), Retrieval of cloud cover from satellite imagery data: A statistical approach. Journal of Geophysical Research, 90, 12,960–12,970.

Nakajima, T. Y., T. Tsuchiya, H. Ishida, T. N. Matsui, and H. Shimoda (2011), Cloud detection performance of space borne visible-to-infrared multispectral imagers. Applied Optics, 50, 2601–2616.

Nichol, J., and Bilal, M. (2016). Validation of MODIS 3 km resolution aerosol optical depth retrievals over Asia. Remote Sensing, 8(4), 328.

Poole, L. R., D. M. Winker, J. R. Pelon, and M. P. Mccormick (2002), Calipso: Global aerosol and cloud observations from lidar and passive instruments, Proc. SPIE 4881, Sensors, Systems, and Next-Generation Satellites VI, 419 (April 8, 2003), Crete, Greece, doi:10.1117/12.462519.

Powell, K. A. (2005), The development of the CALIPSO lidar simulator, Optical Radar.

Powell, K. A., et al. (2009), CALIPSO Lidar calibration algorithms. Part I: Nighttime 532-nm parallel channel and 532-nm perpendicular channel. Journal of Atmospheric and Oceanic Technology, 26, 2015–2033.

Powell, K. A., M. Vaughan, D. Winker, K. P. Lee, M. Piyys, and C. Trepte (2013), Cloud-aerosol LIDAR infrared pathfinder satellite observations data management system, data products catalog, Document No: PC-SCI-503, Release 3.6, Sept. 3.

Remer, L. A., Mattoo, S., Levy, R. C., & Munchak, L. A. (2013). MODIS 3 km aerosol product: Algorithm and global perspective. Atmospheric Measurement Techniques, 6(7), 1829–1844.

Remer, L. A., S. Mattoo, R. C. Levy, A. Heidinger, R. B. Pierce, and M. Chin (2012), Retrieving aerosol in a cloudy environment: Aerosol product availability as a function of spatial resolution, Atmospheric Measurement Techniques, 5, 1823–1840.

Rossow, W. B., and L. C. Garder (1993), Cloud detection using satellite measurements of infrared and visible radiances for ISCCP, Journal of Climate, 6, 2341–2369, doi:10.1175/1520-0442(1993)006<2341:CDUSMO>2.0.CO;2.

Rossow, W. B., and R. A. Schiffer (1991), Cloud data products, Bulletin of the American Meteorological Society, 27, 2–20.

Sassen, K., Z. Wang, and D. Liu (2008), Global distribution of cirrus clouds from CloudSat/Cloud-Aerosol Lidar and infrared Pathfinder satellite observations (CALIPSO) measurements, Journal of Geophysical Research, 113, D00A12.

Saunders, R. W., and K. T. Kriebel (1988), An improved method for detecting clear sky and cloudy radiances from AVHRR data, International Journal of Remote Sensing, 9, 123–150.

Sèze, G., and W. B. Rossow (1991), Time-cumulated visible and infrared radiance histograms used as descriptors of surface and cloud variations, International Journal of Remote Sensing, 12, 877–920.

Smirnov, A., B. N. Holben, T. F. Eck, O. Dubovik, and I. Slutsker (2000). Cloud screening and quality control algorithms for the AERONET database. Remote Sensing of Environment, 73(3), 337–349.

Solomon, S., D. Qin, M. Manning, M. Marquis, K. Averyt, M. M. B. Tignor, et al. (Eds) (2007). Climate Change 2007: The Physical Science Basis. Cambridge: Cambridge University Press.

Stowe, L. L., E. P. McClain, R. Carey, P. Pellegrino, G. G. Gutman, P. Davis, C. Long, and S. Hart (1991), Global distribution of cloud cover derived from NOAA/AVHRR operational satellite data, Advances in Space Research, 11, 51–54.

Sun, L., J. Wei, M. Bilal, X. Tian, C. Jia, Y. Guo, and X. T. Mi (2016). Aerosol optical depth retrieval over bright areas using Landsat 8 OLI images. Remote Sensing, 8(1), 23.

Sun, L., J. Wei, D. H. Duan, Y. M. Guo, D. X. Yang, C. Jia, and X. T. Mi (2016). Impact of land-use and land-cover change on urban air quality in representative cities of China. Journal of Atmospheric and Solar-Terrestrial Physics, 142, 43–54.

Sun, L., J. Wei, C. Jia, Y. K. Yang, X. Y. Zhou, et al. (2016). A High-Resolution Global Dataset of Aerosol Optical Depth Over Land from MODIS Data (pp. 5729–5732). Beijing, China: IEEE International Geoscience and Remote Sensing Symposium (IGARSS). https://doi.org/10.1109/IGARSS.2016.7730497

Sun, L., J. Wei, J. Wang, X. Mi, Y. Guo, Y. Lv, et al. (2016). A universal dynamic threshold cloud detection algorithm (UDTCDA) supported by a prior surface reflectance database. Journal of Geophysical Research: Atmospheres, 121, 7172–7196.

Tanré, D., P. Y. Deschamps, C. Devaux, and M. Herman (1988). Estimation of Saharan aerosol optical thickness from blurring effects in thematic mapper data. Journal of Geophysical Research, 93(D12), 15,955–15,964.

Turner, J., G. J. Marshall, and R. S. Ladkin (2001), An operational, real-time cloud detection scheme for use in the Antarctic based on AVHRR data, International Journal of Remote Sensing, 22, 3027–3046.

Vermote, E. F., N. El Saleous, C. O. Justice, Y. J. Kaufman, J. L. Privette, L. Remer, J. C. Roger, et al. (1997a). Atmospheric correction of visible to middle-infrared EOS-MODIS data over land surfaces: Background, operational algorithm and validation. Journal of Geophysical Research, 102(D14), 17,131–17,141.

Vermote, E. F., and S. Y. Kotchenova (2008), MOD09 user's Guide [J/OL]. [Available at http://modis-sr. ltdri.org.]

Vermote, E. F., D. Tanré, J. L. Deuzé, M. Herman, and J. J. Morcette (1997b), Second simulation of the satellite signal in the solar spectrum, 6s: An overview, IEEE Transactions on Geoscience & Remote Sensing, 35, 675–686.

Vermote, E. F., and A. Vermeulen (1999), ATBD: Atmospheric correction algorithm: Spectral reflectances (MOD09) Version 4.0, NASA contract NAS5–96062.

Walder, P., and I. Maclaren (2000), Neural network based methods for cloud classification on AVHRR images, International Journal of Remote Sensing, 21, 1693–1708.

Wei, J., B. Huang, L. Sun, Z. Zhang, L. Wang, and M. Bilal (2017). A simple and universal aerosol retrieval algorithm for Landsat series images over complex surfaces. Journal of Geophysical Research: Atmospheres, 122, 13,338–13,355.

Wei, J., Y. F. Ming, L. S. Han, Z. L. Ren, and Y. M. Guo (2015), [Method of remote sensing identification for mineral types based on multiple spectral characteristic parameters matching], Guang Pu Xue Yu Guang Pu Fen Xi, 35, 2862–2866.

Wei, J., and L. Sun (2017). Comparison and evaluation of different MODIS aerosol optical depth products over the Beijing-Tianjin-Hebei region in China. IEEE Journal of Selected Topics in Applied Earth Observations and Remote Sensing, 10(3), 835–844.

Wei, J., L. Sun, B. Huang, M. Bilal, Z. Zhang, and L. Wang (2018a). Verification, improvement and application of aerosol optical depths in China part 1: Inter-comparison of NPP-VIIRS and aqua-MODIS. Atmospheric Environment, 175, 221–233.

Wei, J., L. Sun, Y. Peng, L. Wang, Z. Zhang, M. Bilal, and Y. Ma (2018b). An improved high-spatial-resolution aerosol retrieval algorithm for MODIS images over land. Journal of Geophysical Research Atmospheres, 123(21), 12291–12307.

Winker, D. M., R. P. Jacques, and P. M. McCormick (2010), The CALIPSO mission: Spaceborne lidar for observation of aerosols and clouds, Bulletin of the American Meteorological Society, 91, 1211–1229.

Wong, M. S., K. H. Lee, J. E. Nichol, and Z. Li (2010). Retrieval of aerosol optical thickness using MODIS, a study in Hong Kong and the Pearl River Delta region. IEEE Transactions on Geoscience and Remote Sensing, 48(8), 3318–3327.

Wylie, D. P., and W. P. Menzel (1989), Two years of cloud cover statistics using VAS. Journal of Applied Meteorology and Climatology, 2, 380–392.

Wylie, D. P., W. P. Menzel, H. M. Woolf, and K. I. Strabala (1994), Four years of global cirrus cloud statistics using HIRS. Jouranl of Climate, 7, 1972–1980.

Zhang, Y., B. Guindon, and X. Li (2014), A robust approach for object-based detection and radiometric characterization of cloud shadow using haze optimized transformation. IEEE Transactions on Geoscience & Remote Sensing, 52, 5540–5547.

Zhu, Z., S. Wang, and C. E. Woodcock (2015), Improvement and expansion of the Fmask algorithm: Cloud, cloud shadow, and snow detection for Landsats 4–7, 8, and Sentinel 2 images. Remote Sensing of Environment, 159, 269–277.

Zhu, Z., and C. E. Woodcock (2012), Object-based cloud and cloud shadow detection in landsat imagery. Remote Sensing of Environment, 118, 83–94.

7 Tropical Belt Widening Observation and Implication from GNSS Radio Occultation Measurements

Mohamed Darrag and Shuanggen Jin

7.1 INTRODUCTION

Numerous studies have indicated a widening of the tropics in observations, model simulations, and reanalysis. This expansion may lead to profound changes in the global climate system, even a minor widening of the tropical belt would have significant implications because the shift of the jet streams and subtropical dry zones poleward has direct impacts on weather and precipitation patterns. The widening of the tropical belt is largely considered to be a response to global warming caused by increased concentrations of greenhouse gases (GHGs) (Davis and Rosenlof, 2012; Davis and Birner, 2013; Staten et al., 2018; Grise et al., 2019; Watt-Meyer et al., 2019; Meng et al., 2021; Pisoft et al., 2021). The majority of previous research determined widening rates ranging from 0.25° to 3.0° latitude per decade, with statistical significance varying greatly depending on the metrics used to estimate the tropical edge latitude (TEL) and the data sets used to determine it. Furthermore, due to their different physics, the metrics used may respond differently to the force driving the widening (Davis and Rosenlof, 2012).

In astronomy and cartography, the edges of the tropical belt are the Tropics of Cancer and Capricorn, at latitudes of ~23.5° north and south, where the sun is directly overhead at solstice. They are determined by the tilt of the earth's axis of rotation relative to the planet's orbital plane, and their location varies slowly, predictably, and very slightly by about 2.5° latitude over 40,000 years (Gnanadesikan and Stouffer, 2006). In climatology, tropics edges vary seasonally, interannually, and in response to climate forcing. They move poleward in the summer and equatorward in the winter (Davis and Birner, 2013). There are several indicators that define the boundaries of the tropical belt. Generally, three main classes of metrics are employed to estimate the tropical belt borders: circulation-based metrics (e.g., based on the Hadley cells and the subtropical jets), temperature-based metrics (e.g., based on tropopause characteristics), and surface climate metrics (e.g., based on precipitation and surface winds) (Waliser et al., 1999). Staten et al. (2018) and Adam et al. (2018) elaborate on the common metrics used for TEL determination. TELs estimated using various metrics do not always yield the same location. Their positions change much more rapidly and in unpredictable ways than the astronomically defined tropics (Lee and Kim, 2003).

In recent years, monitoring the tropopause has received increased attention for climate change studies. Many studies have shown that the tropopause is rising due to tropospheric warming caused by increased GHG emissions in the atmosphere (Davis and Rosenlof, 2012; Davis and Birner, 2013; Staten et al., 2018; Grise et al., 2019; Watt-Meyer et al., 2019 Meng et al., 2021; Pisoft et al., 2021). The tropopause characteristics are critical for understanding the troposphere-stratosphere exchange (Holton et al., 1995). In addition, the chemical, dynamical, and radiative connections between the troposphere and stratosphere are crucial to understanding and predicting climate change worldwide. Exchanges of water, mass, and gases between the troposphere and stratosphere occur through the

DOI: 10.1201/9781003363118-7

tropopause. Several studies have investigated the tropopause over the tropics using different data types and have revealed the problem of the TEL shift (Ao and Hajj, 2013; Tegtmeier et al., 2020; Kedzierski et al., 2020). Global navigation satellite system radio occultation (GNSS-RO) provided high-accuracy remote sensing observations of the thermal structure of the tropopause and was used to investigate the trend and variability of the tropopause (Son et al., 2011). Among the most outstanding advantages of GNSS-RO are its high accuracy of 0.2–0.5 K in estimating temperature in the upper-troposphere lower-stratosphere (UTLS) region and its 200 m vertical resolution. These advantages make GNSS-RO especially appropriate to detect the possible widening of the tropical belt based on the height metrics of the tropopause (Kursinski et al., 1997; Ho et al., 2012). Using tropopause metrics for TEL determination has many advantages because it can be accurately estimated from remotely sensed temperature profiles with sufficient vertical resolution, such as GNSS-RO profiles (Davis and Birner, 2013; Seidel and Randel, 2006).

Some of the earliest, unequivocal signs of climate change have been air and ocean warming, land thawing, and ice melting. In addition, recent studies are showing that the tropics are also changing. Several pieces of evidence show that over the past few decades, the tropical belt has expanded. This expansion may lead to profound changes in the global climate system (Seidel et al., 2007). According to the Intergovernmental Panel on Climate Change (IPCC) Fourth Assessment Report (AR4) (Meehl et al., 2007), increases in GHGs and other human-induced climate forcing would lead to warming of the troposphere, cooling of the stratosphere, a rise of the tropopause, weakening of tropical circulation patterns, poleward migration of midlatitude storm tracks, an increase in tropical precipitation, and other climatic changes. It is not obvious how these changes might relate to variations in the width of the tropical belt, and this question has received much attention. Several recent studies suggest that the tropics have been expanding over the past few decades, and this widening may continue into the future in association with anthropogenic climate change. The widening of the tropics could have far-reaching scientific and societal consequences. The expansion of the tropical belt towards the poles is likely to bring even drier conditions to these densely populated areas, but it may also bring more moisture to other areas, resulting in shifts in precipitation patterns that affect natural ecosystems, agriculture, and water resources.

The study of tropical belt widening is a challenging task due to the complexity and dynamics of the earth's atmospheric system and the data limitations. These limitations are due to the low spatial resolution of radiosonde (RS) data, since it only covers land and its distribution is not symmetrical in both hemispheres. Low vertical resolution plagues both satellite remote sensing technologies and model analyses. Furthermore, reanalysis trends can be biased to reflect changes in both the quality and quantity of the underlying data, and the expansion rates computed from different reanalyses were considerably different (Schmidt et al., 2004; Ao and Hajj, 2013). Nowadays, global navigation satellite systems (GNSS) have provided an exceptional opportunity to retrieve land surface and atmospheric parameters globally (e.g., Jin and Park, 2006; Jin and Zhang, 2016; Wu and Jin, 2014; Jin et al., 2011, 2017), particularly through space-borne GNSS-RO because it has long-term stability and works in all weather conditions, which make it a powerful tool for studying climate variability. The GNSS-RO technique has many advantages, such as uniform global coverage, a higher vertical resolution than any of the existing satellite temperature measurements available for the UTLS, long-term stability, and the ability to work in all weather conditions unaffected by clouds, precipitation, or aerosols. In addition, it is vertically more finely resolved than any of the existing satellite temperature measurements available for the UTLS and now provides a unique data set, so GNSS-RO is well suited for this challenge. Moreover, it is a key component for a broad range of other studies, including equatorial waves, Kelvin waves, gravity waves, Rossby and mixed Rossby–gravity waves, and thermal tides (Bai et al., 2020; Scherllin-Pirscher et al., 2021). A number of studies confirmed the feasibility and excellent eligibility of GNSS-RO measurements for monitoring the atmosphere and for climate change detection (Foelsche et al., 2009; Steiner et al., 2011).

In the previous studies, the problem of the tropical belt expansion was that the rates of expansion were different from one data type to another and from one calculation method to another. The rates range from high to low, raising concerns about the accuracy and reliability of the various data sets and TEL computation methods. Hudson et al. (2006), based on atmospheric ozone concentrations, reported that the tropical region occupying the northern hemisphere (NH) grew at a rate of 1°/decade. Using atmospheric temperature satellite-based microwave observations, Fu et al. (2006) reported tropical belt widening for the period 1979–2005. They estimated a net widening of about 2° latitude. Based on RS and reanalysis data, Seidel and Randel (2007) reported an expansion of 5° to 8° latitude during the period from 1979 to 2005. In addition, Hu and Fu (2007) found a widening of the tropical Hadley circulation system, and estimated its magnitude as 2° to 4.5° latitude during the period from 1979 to 2005. Ao and Hajj (2013) examined the possible expansion of the tropical belt due to climate change using GPS-RO data from 2002 to 2011. Their analysis revealed a statistically significant expansion trend of 1°/decade in the northern hemisphere (NH), but no significant trend in the southern hemisphere (SH). According to the review by Lucas et al. (2014), an assent of the observations suggests that the rate of this expansion since 1979 ranges between 0.5° and 1.0° latitude/decade in both hemispheres. The precise rates of tropical belt expansion and their hemispheric partitions remain significant unknowns. In research from Staten et al. (2018), researchers reviewed the possible causes and rates of observed and projected tropical belt expansion. After accounting for methodological differences, the tropical belt has expanded at a rate of about 0.5°/decade since 1979. However, they reported that it is too early to detect robust anthropogenically induced widening imprints because of large internal variability. Allen and Kovilakam (2017) and Grise et al. (2019) stated that the spatial and temporal patterns of SST play a crucial role in driving the recent tropical belt expansion. Based on observations, numerical experiments, and multi-model simulations, Yang et al. (2020b) find that the tropical belt width closely follows the shift of oceanic midlatitude meridional temperature gradients (MMTG). According to Yang et al. (2020a, 2020b), the entire oceanic and atmospheric circulation is moving poleward. Yang et al. (2022) used an idealized coupled aqua-planet model to explore the mechanism of the circulation shift. They find that ocean surface warming plays a significant role in driving the circulation shift. The expanding tropical warm water causes a poleward shift of the mid-latitude temperature gradient.

The study of tropical belt widening is critical for understanding atmospheric variability, climate dynamics, and change. In this study, we used a group of methods to track atmospheric variability and TEL based on GNSS-RO data since 2001, attempting to provide solutions and accurate results for the studied atmospheric parameters worldwide, with a particular focus on the tropics. Within this study, we worked on filling the gaps and solving problems in the previous studies and also performed a long-term study of the tropics' variability over time as indicators for global climate change.

The main aims of this study are to monitor and investigate the tropics' width and its implication for global climate change. Determination of the tropical belt widening and estimation of its trends and rates include the following parameters:

a. Determination of global tropopause height and temperature Furthermore, determination of their spatial and temporal variations globally and across the study period
b. Establishment of long-term time series for both lapse-rate tropopause (LRT) and cold-point tropopause (CPT)
c. Estimation of the TELs' locations and estimation of the tropical belt widening rates
d. Investigation of the behavior of tropical expansion spatially and temporally
e. Study of the differences in the rate and behavior of the tropical belt widening between the northern and southern hemispheres
f. Study of the trend and spatial-temporal variability of many meteorological parameters, which include carbon dioxide (CO_2) and methane (CH_4), as important drivers of global warming and tropical belt variation. In addition, total column ozone (TCO) can provide information about the tropopause and UTLS status. The changes in surface temperature, precipitation, and drought that may occur as a response to tropical expansion are broadly examined.

7.2 GNSS-RO THEORY AND TROPOPAUSE

7.2.1 GNSS-RO THEORY

A GNSS-RO event happens when a GNSS satellite sets behind or goes up from behind the horizon. Its signals are obscured by the earth from the point of view of the low earth orbiting (LEO) satellite receiver. During an RO event, radio signals transmitted from GNSS satellites and received onboard a LEO satellite are influenced by the refractivity of the atmosphere, resulting in excess propagation and bending of the signals (Figure 7.1). The atmosphere excess phase (AEP) is the main observable which can be calculated with millimeters accuracy (Wickert et al., 2001a). For instance, the AEP estimate is the base to retrieve the bending angle, refractivity, and temperature profiles (Wickert et al., 2004; Xia et al., 2017). The RO technique provides high-quality global observations for the ionosphere, stratosphere, and troposphere. These observations have a high impact on weather forecasting and climate monitoring research.

The GNSS-RO technique was first performed within the US GPS/METeorology experiment for the period from 1995 to 1997 (Kursinski et al., 1997). Also, it has been continuously applied aboard various LEO satellite missions since 2001. These missions are Challenging Mini-satellite Payload (CHAMP) (Wickert et al., 2004; Wickert et al., 2001b); Gravity Recovery and Climate Experiment (GRACE) and Gravity Recovery and Climate Experiment Follow-On (GRACE-FO) (Wickert et al., 2009); Scientific Application Satellite-C/D (SAC-C/D) (Hajj et al., 2004); TerraSAR-X; TanDEM-X; Constellation Observing System for Meteorology, Ionosphere, and Climate (COSMIC/COSMIC-2, also known as FORMOSAT-3/FORMOSAT-7); the Meteorological Operational Satellite Programme-A/B/C (MetOp-A/B/C); FengYun-3C/D (FY-3C/D) (Sun, 2019); Communications/Navigation Outage Forecasting System (C/NOFS); Korea Multi-Purpose Satellite-5 (KOMPSAT-5); the Indian Space Research Organization Spacecraft Ocean Satellite-2 (OceanSat-2); and Spanish PAZ (peace in Spanish). A few missions were retired, such as COSMIC-1, GRACE, CHAMP, and SAC-C/D, and some missions were completed by the end of 2020, such as FY-3C, TanDEM-X/TerraSAR-X, KOMPSAT-5, OceanSat-2, and C/NOFS. More missions like MetOp Second Generation (MetOp-SG), FengYun-3E/F/G/H (FY-3E/F/G/H), TerraSAR-X Next Generation (TSX-NG), Jason Continuity of Service-A/B (JASON-CS-A/B, also known as Sentinel 6A/6B), and Meteor-MP N1/N2 are planned for the future. The future missions will provide around 14,700 RO profiles daily by 2025 (Jin et al., 2013; Oscar, 2020).

The processing of GNSS RO observations can be illustrated by the following chart (Figure 7.2). Where GNSS satellites transmit dual frequency signals at two wavelengths (L1 and L2), these signals are received onboard the LEO satellites. Since the main observable for GNSS-RO calculations is the AEP, it should be calculated accurately by using precise orbit information for

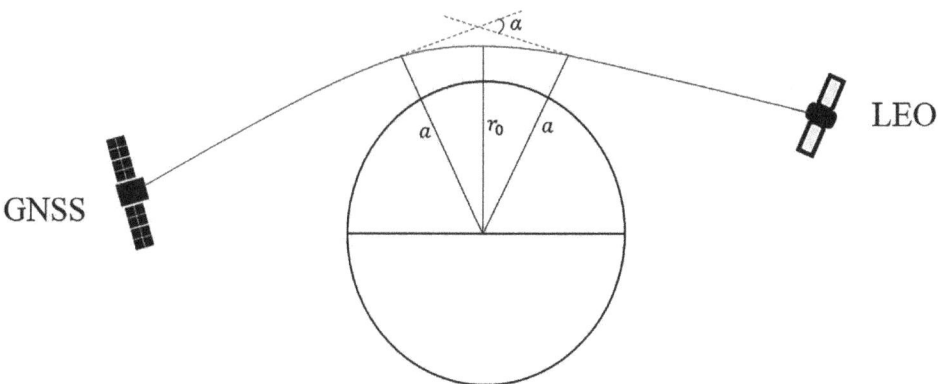

FIGURE 7.1 GNSS-RO principle.

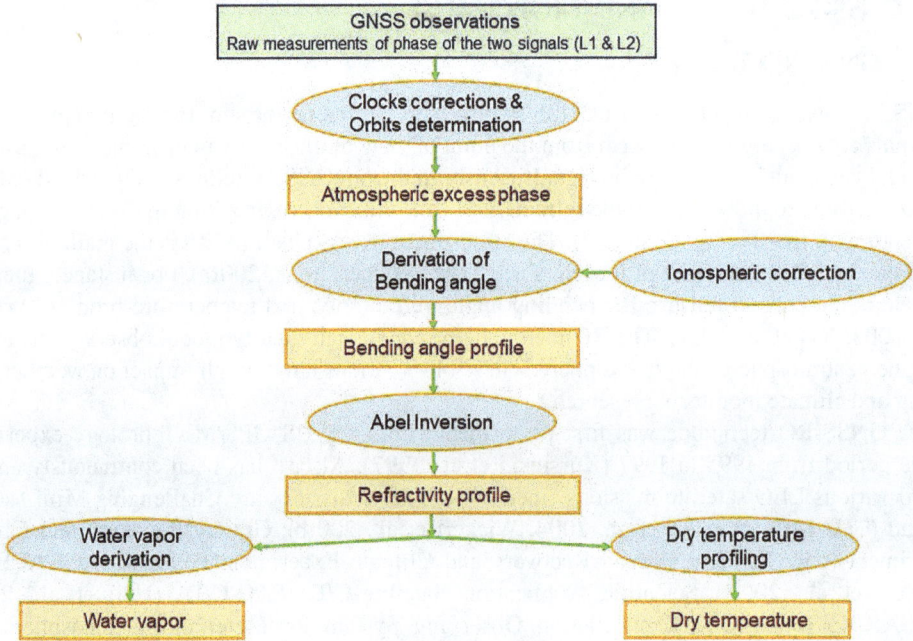

FIGURE 7.2 Flow chart for GNSS-RO data processing steps.

GNSS and LEO satellites. Also, the clock errors of the GNSS and LEO satellites should be removed by the double differencing method using an additional non-occulted GNSS satellite as a reference (Hajj et al., 2002). Then, to get the vertical atmosphere profiles, the first step is the derivation of atmospheric bending angles that are obtained from the AEP time derivation using the Doppler shift equation (Gorbunov et al., 1996). Ionospheric effects are eliminated by a linear combination of bending angles derived from GNSS frequencies, assuming spherical symmetry of the atmosphere (Steiner et al., 1999). The next step is retrieving the atmospheric refraction index (*n*) from bending angle profiles by the inverse Able transform, as shown in equation (1) (Fjeldbo et al., 1971).

$$n(r_0) = \exp\left(\frac{1}{\pi} \int_a^\infty \frac{\alpha(x)}{\sqrt{x^2 - a^2}} \, dx \right) \tag{7.1}$$

Where:

- α (bending angle)
- r_0 (for the given point of the closest approach of the signal path to the earth's surface)
- a (impact parameter)
- x (convenient variable [$x=nr$] [refractive index * radius])

The atmospheric refractivity ($N = (n-1) \cdot 10^6$) is related to pressure, temperature, and water vapor pressure using equation (2) (Smith and Weintraub, 1953)

$$N = 77.6 \frac{p}{T} + 3.73 * 10^5 \frac{P_w}{T^2} \tag{7.2}$$

Where: *P* (air pressure in mbar), *T* (temperature in K), and p_w (water vapor pressure in mbar).

As known, air refractivity (N) is divided into dry refractivity (N_d) caused by dry air and wet refractivity (N_w) caused by wet air. Using dry refractivity, the dry temperature and pressure profiles are given from the hydrostatic equation and the equation of state for an ideal gas (Kursinski et al., 1997). Moreover, the water vapor profiles are derived from the wet refractivity profiles using temperature profiles from meteorological analyses, like those of the ECMWF (European Centre for Medium-Range Weather Forecasts), in an iterative way (Gorbunov et al., 1993). To summarize, the products of RO data processing are atmospheric profiles (temperature and water vapor) with a height resolution of 200 to 1,000 m, a wide height coverage from the earth's surface to 60 km, a nearly uniform geographical distribution, and a high data rate (150–200/day/satellite).

The GNSS-RO technique is critical for climate research because it allows for the generation of climate benchmark data that can be used as a reference data set for other climate observations (Leroy et al., 2006). RO observations are convenient for establishing the stable, long-term record needed for climate monitoring (Scherllin-Pirscher et al., 2012). In addition, GNSS-RO provides accurate input for numerical weather prediction and is a source of data for climate related research (Zus et al., 2014). Many studies comparing the RO analysis results of different processing centers (DMI Copenhagen, EUM Darmstadt, GFZ Potsdam, JPL Pasadena, UCAR Boulder, and WEGC Graz) show excellent agreement in general, independent of processing algorithms for temperature data derivation and included satellite missions (Steiner et al., 2013). The RO technique is also capable of detecting irregularities in electron density in the ionospheric E region (Hocke and Tsuda, 2004). Furthermore, it enables global investigation of sporadic E layer occurrence and intensity (Arras and Wickert, 2018). Moreover, RO temperature profiles are used to derive horizontal and vertical GW parameters in the atmosphere (Schmidt et al., 2016), as small-scale fluctuations of dry temperature profiles can be interpreted as GWs (Marquardt and Healy, 2005).

7.2.2 TROPOPAUSE DEFINITIONS

The tropopause signifies the transition between the troposphere and the stratosphere. which are chemically and dynamically distinct regions (Marshall and Plumb 2008). There are several different definitions of the tropopause, depending on which atmospheric parameters are investigated. This allows for the selection of the definition that is best suited for the area, problem, or situation being analyzed. The most common definitions are the thermal tropopause, dynamical tropopause, and chemical tropopause.

7.2.2.1 The Thermal Tropopause

To define the thermal tropopause, either the lapse rate of the temperature profile or its minimum temperature is used, yielding the LRT or the CPT, respectively. The lapse rate definition is the oldest and most commonly used one. It is defined by the World Meteorological Organization (WMO) (WMO, 1957) as follows:

a) "The first tropopause is defined as the lowest level at which the lapse rate decreases to 2 °C km^{-1} or less, provided also the average lapse rate between this level and all higher levels within 2 km does not exceed 2 °C km^{-1}."
b) "If above the first tropopause the average lapse rate between any level and all higher levels within 1 km exceeds 3 °C km^{-1}, then a second tropopause is defined by the same criterion as under (a). This tropopause may be either within or above the 1 km layer."

$$\Gamma(z_i) = -\frac{\delta T}{\delta z} = \frac{T_{i+1} - T_i}{z_{i+1} - z_i} \tag{7.3}$$

where: Γ is the lapse rate, T and z are the temperatures and heights above mean sea level, respectively.

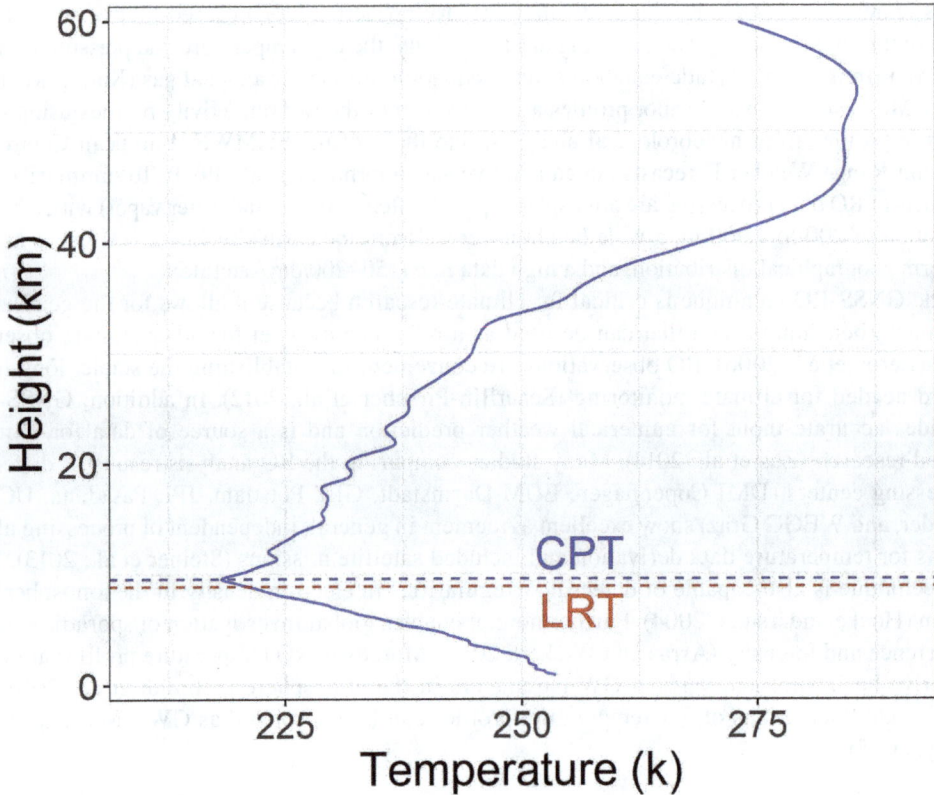

FIGURE 7.3 KOMPSAT5 profile on January 1, 2021 at 00:33; horizontal bars signify LRT (red) and CPT (blue).

The lapse rate definition is very suitable for atmospheric profiles with a high vertical resolution, such as RS or RO data (Figure 7.3). In fact, the vertical resolution of the levels mentioned in the WMO definition is of the order of 1 km (Birner, 2003), so when using this definition, oversampling may slightly influence the result. Because implementation of this definition is rather easy, it is widely used. Anyway, its physical relevance may be limited. For the tropics, Highwood and Hoskins (1998) suggest a definition that reflects the strength of convective processes, which influence troposphere-stratosphere exchange. Hence, the cold point definition is popular in the tropics. It is defined as the first local minimum of the temperature profile above the LRT. Outside the tropics, the CPT can be excessively high, particularly in the winter hemispheric high latitudes.

7.2.2.2 The Dynamical Tropopause

The dynamical tropopause was introduced by Reed (1955). It can separate the air masses based on different compositions or features. So, it effectively divides tropospheric air from stratospheric air. The potential vorticity (PV) dynamical tropopause definition, a common definition for synoptic scale events or climatological studies in the extratropics, uses a subjectively chosen threshold value to determine the tropopause height (Gettelman et al., 2011). The definition requires three-dimensional temperature and wind data, making it effective at determining the tropopause height in global models. The definition uses both static stability and vorticity (Gettelman et al., 2011; Kunz et al., 2011). PV is a conserved quantity for adiabatically frictionless flow (Holton, 2013). Similarly to the Brunt-Väisälä frequency, PV has an abrupt jump in values at the tropopause. This sharp change provides a useful dynamical definition for areas or model runs dominated by environmental vorticity.

7.2.2.3 The Chemical Tropopause

The chemical tropopause uses different concentrations of trace gases for different altitudes to locate the chemical transition layer between the troposphere and the stratosphere, also known as stratospheric tracer tropopause. It is defined using a trace gas in the stratosphere (Pan et al., 2004; Zahn et al., 2004). The ozone mixing ratio is commonly used for the chemical tropopause, but determining the proper threshold value to use can be challenging. Various studies like Zahn et al. (2004) utilize multiple trace gases with sources in both the troposphere and stratosphere for the tropopause determination. In their study, ozone, which has higher concentrations in the stratosphere, and carbon monoxide, which has higher concentrations in the troposphere, were used to determine the tropopause height from in situ measurements. In the vicinity of the chemical tropopause, sharp changes in concentrations of trace gases occur.

7.2.3 TEL Determination Metrics

There are several metrics that are used for the determination of the TEL. Defining the TEL is a matter of choosing a point within the transition zone from the tropics to the extra tropics that is representative for the whole. The methods for TEL determination can be subjective or objective. In addition, there is no definition for the TEL that can be universally applicable across all data sets; each of them sees the transition zone in a different way because the responses of different metrics show different sensitivities. In this section, most of the commonly used TEL metrics are presented (Lucas et al., 2014).

7.2.3.1 Tropopause Based Metrics

In this section, four LRT height metrics used for identifying the TEL are described.

a) The first diagnosing metric based on the LRT height describe the TEL as the latitude at which the tropopause height falls 1.5 km under the average tropopause height between 15°S and 15°N (Davis and Rosenlof, 2012).

b) The second metric signifies the TEL as the latitude at which tropopause height is greater than 15 km for x days per year (where $x = 100$, 200, or 300). The 15-km tropopause height threshold is equivalent to a pressure threshold of ~120 hPa. According to Lu et al. (2009) $x = 200$ days/year. This TEL definition is sensitive to the threshold of LRT height and number of days per year. Although some studies use this method to define annual tropopause, this metric can also be applied for a shorter time average like a season (Seidel and Randel, 2007; Davis and Rosenlof, 2012).

c) The third metric defines the TEL as the latitude of the maximum value of the tropopause meridional gradient. The meridional drop in tropopause height from its tropical to extratropical value undergoes a maximum rate of change in the vicinity of the subtropical jet; this region is often referred to as the tropopause break (Davis and Rosenlof, 2012).

d) The fourth TEL defining metric is the latitude of the maximum dry bulk static stability (Davis and Birner, 2013), which represents the potential temperature difference between the tropopause and the surface.

7.2.3.2 Stream Function-Based Metric

The mean-meridional stream function measures the meridional overturning circulation at a particular latitude and represents the most natural framework through which to study the Hadley cell and the width of the tropical belt. It is calculated as the vertical integral of the mass weighted zonal mean meridional wind between the top of the atmosphere and each pressure level. The edge of the tropics is generally taken as the latitude of the subtropical zero isopleth in the mid-troposphere (Holton, 1994). Based on the mean stream function the TEL has been defined as the latitude at which the mean stream function is 500 hPa (Lu et al., 2007; Frierson et al., 2007), 600 hPa to 400 hPa (Hu and Fu, 2007; Johanson and Fu, 2009), and 700 hPa to 400 hPa (Stachnik and Schumacher, 2011).

$$\psi\left(\phi,p\right)=\frac{2\pi a\cos\left(\phi\right)}{g}\int_{0}^{p}\left[v\right]\partial p \tag{7.4}$$

where: Ψ (φ, p) is the mean-meridional stream function at pressure level p and latitude φ, $g = 9.81$ m/s is the acceleration due to gravity, $a = 6371$ km is the earth's mean radius, and $[v]$ is the monthly-mean, zonal-mean meridional wind.

7.2.3.3 Jet Stream Based Metric

The jet stream position is another metric that is used to examine tropical belt expansion. The subtropical jet is strongly linked to the Hadley circulation in the infrared (Held and Hou, 1980). Based on the subtropical jet, the TEL is defined as the latitude of the most equatorward local maximum in the zonal-mean zonal wind field in UTLS in each hemisphere, capturing the meridional position of the subtropical jet core (Strong and Davis, 2006). According to Fu et al. (2006), the subtropical jet stream can be inferred from the latitudinal tropospheric temperature gradient and thermal wind balance or from the lower stratospheric temperature change with latitude. In addition, following Archer and Caldeira (2008), the subtropical jet stream is calculated from the location of maximum winds aloft. Alternatively, the surface wind can be subtracted from the winds aloft to separate the subtropical jet stream from the eddy-driven jet.

7.2.3.4 Surface Based Metric

There are many surface-based metrics that used to define the TEL such as the following:

a) Precipitation can be used as an independent metric to represent TEL locations. Many studies, which rely upon surface-based variables to investigate tropical widening, used the Global Precipitation Climatology Project (GPCP) monthly data set to examine shifts in the positions and boundaries of the subtropical dry zones (Hu et al., 2010; Zhou et al., 2011; Allen et al., 2012b). Based on the GPCP, the TEL is defined as the latitude of the precipitation minimum, or 2.4 mm/day. Moreover, the TEL can be defined as the latitude of the zero-crossing of precipitation minus evaporation (Zhou et al., 2011).

b) The global position of the subtropical ridge derived from sea level pressure data has also been used to signify the TEL (Hu et al., 2010)

7.2.3.5 Other Metrics

a) The outgoing longwave radiation (OLR) can be used to define the TEL. According to Davis and Rosenlof (2012), the TEL is the latitude at which the OLR is 20 Wm^{-2} below the subtropical maximum. In addition, based on the OLR, the TEL can be defined as the location at which the OLR drops to a threshold of 250 Wm^{-2} on the poleward side of the subtropical maximum in each hemisphere (Hu and Fu, 2007).

b) The total column ozone (TCO) pattern, which is inversely proportional to tropopause height, can give an indication of the tropical belt width. The TCO amount varies with latitude, in part due to the difference in tropopause height between the tropics and midlatitudes. This dependency of TCO on latitude has been used to identify the TEL and to compute their variations over time. The TCO is used to determine the geographic coordinates of the fronts by locating the position of the sharp increase in the total ozone amount due to the decrease in the tropopause height associated with the fronts (Hudson et al., 2003; Hudson et al., 2006; Hudson, 2012; Davis et al., 2018).

7.3 DATA AND METHODOLOGY

7.3.1 DATA SETS

In this study we employed the following data sets.

7.3.1.1 GNSS RO

The main data used in this study is GNSS-RO atmospheric profile data from 12 LEO missions from June 2001–November 2020. The data (CDAAC, 2021) is available at the COSMIC Data Analysis and Archive Center (CDAAC). The GNSS-RO data availability and its time span are shown in Figure 7.4.

7.3.1.2 ERA5 Temperature Data on Pressure Levels

ERA5 is the fifth generation of the ECMWF reanalysis of the global climate and weather. Monthly averaged temperature data on pressure levels from ERA5, which provides global coverage for the period from June 2001–November 2020, are used to calculate the LRT tropopause height and temperature. The horizontal resolution of the ERA5 data is $0.25° \times 0.25°$, while the vertical coverage ranges from 1,000 hPa to 1 hPa, with a vertical resolution of 37 pressure levels (Hersbach et al., 2019a).

7.3.1.3 AIRS LRT Height and Temperature

The Atmospheric Infrared Sounder (AIRS) is the spectrometer onboard the second Earth Observing System (EOS) polar-orbiting platform, Aqua. In combination with the Advanced Microwave Sounding Unit (AMSU), AIRS constitutes an innovative atmospheric sounding instrument with infrared and microwave sensors. LRT height and temperature data provided by AIRS (AIRX3STM v7.0) are provided monthly and have global coverage, with a horizontal resolution of $1° \times 1°$ (Aumann et al., 2003; AIRS, 2019a). In this study, we use data for the period from September 2002 to November 2020. The data is available at (AIRS, 2019a).

7.3.1.4 Total Column Ozone (TCO)

The Modern-Era Retrospective Analysis for Research and Applications Version 2 (MERRA-2) provides TCO at a global scale, monthly, and has a spatial resolution of $0.5° \times 0.625°$. In this work, we use data from June 2001 to November 2020. The data is to be compared with the LRT height from GNSS-RO. In addition, TCO can provide information about the tropics behavior and can help in emphasizing the GNSS-RO outputs (GMAO, 2015).

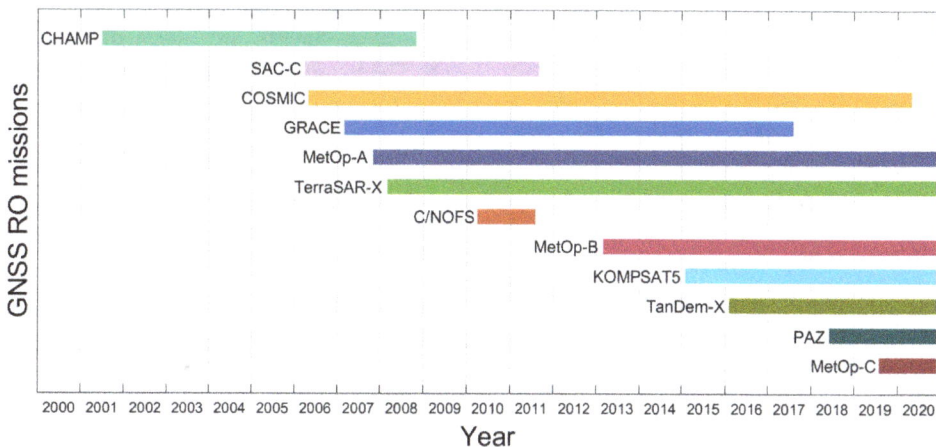

FIGURE 7.4 GNSS-RO data used in this study.

7.3.1.5 Carbon Dioxide (CO_2)

CarbonTracker is a CO_2 measurement and modeling system developed by NOAA Earth System Research Laboratories (ESRL) to keep track of CO_2 sources and sinks throughout the world. Monthly column average CO_2 data with global coverage from June 2001 to March 2019 is used in this study (Jacobson et al., 2020). The data has a spatial resolution of 2° x 3°. Here we use this data to study the behavior and trend of CO_2. This is the most important GHG and the largest forcing component of climate change.

7.3.1.6 Methane (CH_4)

AIRS provides monthly measurements of CH_4 at 24 pressure levels with a spatial resolution of 1° x 1°. (AIRS, 2019b). We use data from September 2002 to November 2020. As one of the primary GHGs driving long-term climate change, CH_4 plays a critical role in global warming.

7.3.1.7 Surface Temperature

Global monthly average surface temperature data from ERA5 reanalysis has a horizontal resolution of 0.25° x 0.25° (Hersbach et al., 2019b). In this study, we utilize data from June 2001 to November 2020. The purpose of using this data is to study the impacts of the variability in the tropics on global climatological parameters.

7.3.1.8 Global Precipitation Climatology Project (GPCP)

Monthly average precipitation data is available from the Global Precipitation Climatology Project (GPCP) at a horizontal resolution of 2.5° x 2.5° (Adler et al., 2016). We use data from June 2001 to November 2020. The purpose of this data is to investigate the relation between the tropical belt width and the corresponding precipitation pattern.

7.3.1.9 Precipitation and Potential Evapotranspiration (PET)

Global monthly average precipitation and PET at a horizontal resolution of 0.5° x 0.5° are available from the Climatic Research Unit (CRU) Time-Series (TS). This data is employed to compute the Standardized Precipitation Evapotranspiration Index (SPEI), meteorological drought index. We use data ranging from June 2001–November 2020. The data is available at Harris et al. (2020). The SPEI drought index was calculated following the indications of Vicente-Serrano et al. (2010) and Beguería et al. (2013).

7.3.2 Methodology

In this study, atmospheric profiles from 12 GNSS-RO missions are used for the first time together. Before the analysis, we compared the different missions' profiles to investigate the consistency between data from different sources at CDAAC web. The profile pairs are spaced 3 hours apart and separated by 230 kilometers. After that, the GNSS-RO temperature profiles with uniform coverage worldwide were used to calculate the tropopause height and the tropopause temperature, applying both tropopause definitions (LRT and CPT). According to the definition of the World Meteorological Organization (WMO), "The thermal LRT is defined as the lowest level at which the lapse rate decreases to 2°C/km or less, provided also that the average lapse rate between this level and all higher levels within 2 km does not exceed 2°C/km" (WMO, 1957). While the CPT is indicated by the minimum temperature in a vertical profile of temperature (Holton et al., 1995), Here, in order to avoid outliers, the tropopause height values of both definitions are limited to 6–20 km. The results of both the LRT and CPT definitions are then gridded into 5° x 5° grids. In addition, the spatial and temporal variability of the studied parameters are widely investigated using empirical orthogonal function (EOF) technique, also known as the principal component analysis (PCA) (Calabia and Jin, 2016; Calabia and Jin, 2020). This technique is commonly applied for climate variables' spatial and temporal analysis. It provides the spatial patterns of variability and expansion coefficient time series for a single geophysical variable in addition to the contribution of each mode of variability to the total variance. Because the variability of the studied parameters is mostly driven by the annual variation, the first PCA component is the only one considered for each variable. In addition, the seasonal variation of tropopause parameters is widely examined.

The locations of TEL are estimated from the monthly zonal average of LRT height derived from GNSS-RO, ERA5, and AIRS data. The ERA5 and AIRS tropopause parameters are resampled at the same resolution as GNSS-RO to avoid uncertainty caused by different resolution data. The zonal average LRT height is spline interpolated as a function of latitude (Ao and Hajj, 2013), and the TEL is determined at each hemisphere independently using two tropopause height metrics. The first method relies on a subjective criterion; according to the first method, TEL is defined as the latitude at which the LRT height falls 1.5 km below the tropical average ($15°S–15°N$) LRT height (Davis and Rosenlof, 2012). The second method is an objective criterion in which the TEL is defined as the latitude of the maximum LRT height meridional poleward gradient (Davis and Rosenlof, 2012). Furthermore, the decadal rate of expansion and/or contraction of the tropical belt is estimated independently for each hemisphere using both calculation methods. In addition, the trend and spatial-temporal variability of CO_2 and CH_4, as important drivers of global warming and TEL variability, are investigated. Furthermore, the trend and spatial-temporal pattern of TCO that give information about the tropical belt width are investigated. TCO has a high negative correlation with LRT height, and it is highly indicative of the positions of the TELs. Finally, we broadly examine the surface temperature, which is a proposed driver for the tropical belt expansion. Moreover, we deeply investigated the trends and spatial-temporal pattern of precipitation and the meteorological SPEI drought index as meteorological parameters that may have changed behavior as a response to tropical expansion.

7.4 RESULTS AND DISCUSSION

7.4.1 Assessment of GNSS-RO Temperature Profiles

In several previous studies, multiple GNSS-RO missions were utilized together for the purpose of obtaining high spatial resolution. In addition, the assessment of using different GNSS-RO missions together showed a high level of consistency (Hajj et al., 2004; Li et al., 2017; Tegtmeier et al., 2020; Xian et al., 2021). In our study, the atmospheric profiles from all used GNSS-RO missions are compared together to signify the high level of consistency and compatibility between RO missions available on the CDAAC web, as well as the ability to merge them together in our study as a single data set. COSMIC mission profiles are used as a fixed member in the intercomparison of all utilized RO missions since it is the most abundant in terms of profile density and its time span overlaps with all other missions. The results of the conducted intercomparison show high agreement and consistency between the profiles of the collocated pairs (Figure 7.5). Table 7.1 demonstrates the results of the collocated GNSS profile pairs. The correlation coefficient between the collocated profile pairs ranges from 0.97 to 0.99, and the temperature mean difference ranges from 0.1 to 0.5 K.

7.4.2 Tropopause Characteristics from GNSS-RO

Figure 7.6 depicts the global GNSS LRT and CPT parameters from June 2001 to November 2020. As shown in Figure 7.6, the CPT height is always greater than the LRT height. The average distance between them is approximately 2.62 km, and there is a correlation of approximately 0.66 between LRT and CPT height. Previous studies have reported that the average CPT height is between 0.5 and 1 km higher than the LRT average (Munchak and Pan, 2014). The LRT temperature is higher than the CPT temperature. The mean difference between them is 4.02 K, and the correlation coefficient between them is 0.61. Our results are consistent with previous studies that displayed a global increase in the tropopause height from radiosonde observations (Seidel and Randel, 2006) and reanalysis (Santer et al., 2004).

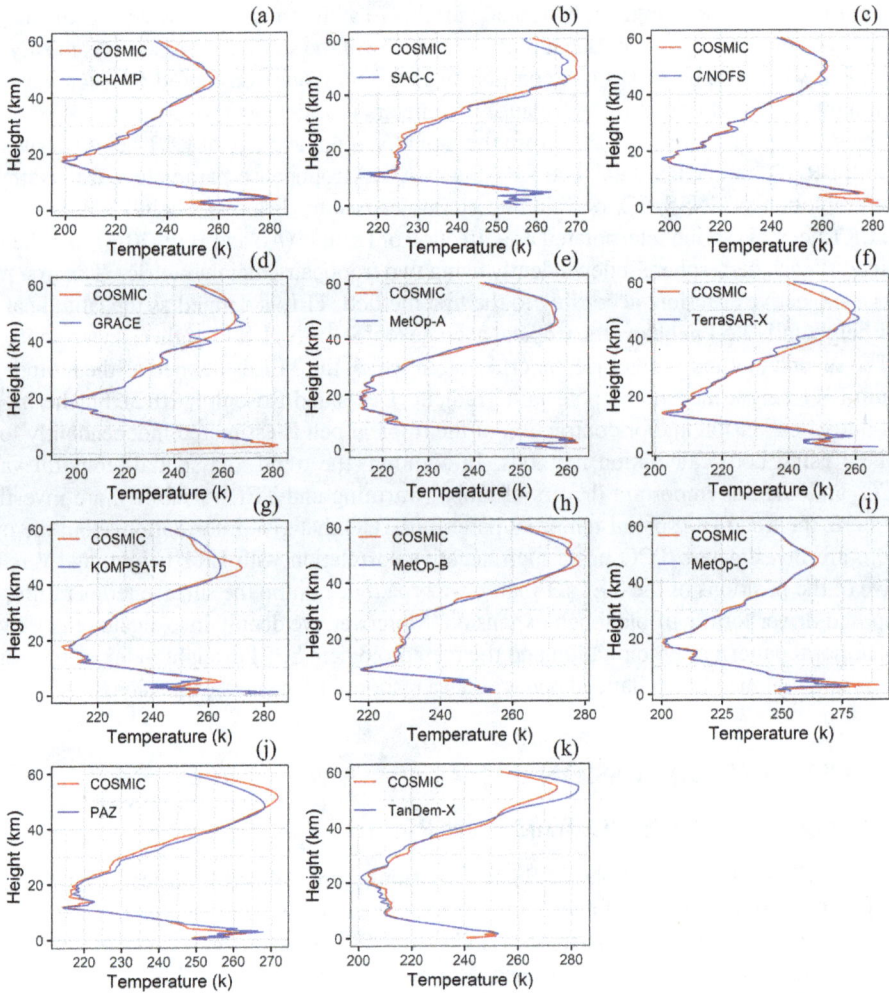

FIGURE 7.5 Intercomparison of collocated GNSS profile pairs.

TABLE 7.1
Intercomparison of Collocated GNSS Profile Pairs

Mission	Correlation Coefficient	Mean Difference (K)
(a) COSMIC—CHAMP	0.99	0.5
(b) COSMIC—SAC-C	0.99	0.2
(c) COSMIC—C/NOFS	0.99	0.32
(d) COSMIC—GRACE	0.99	0.1
(e) COSMIC—MetOp-A	0.99	0.28
(f) COSMIC—TerraSAR-X	0.98	0.22
(g) COSMIC—KOMPSAT5	0.97	0.13
(h) COSMIC—MetOp-B	0.99	0.14
(i) COSMIC—MetOp-C	0.99	0.47
(j) COSMIC—PAZ	0.98	0.33
(k) COSMIC—TanDem-X	0.99	0.47

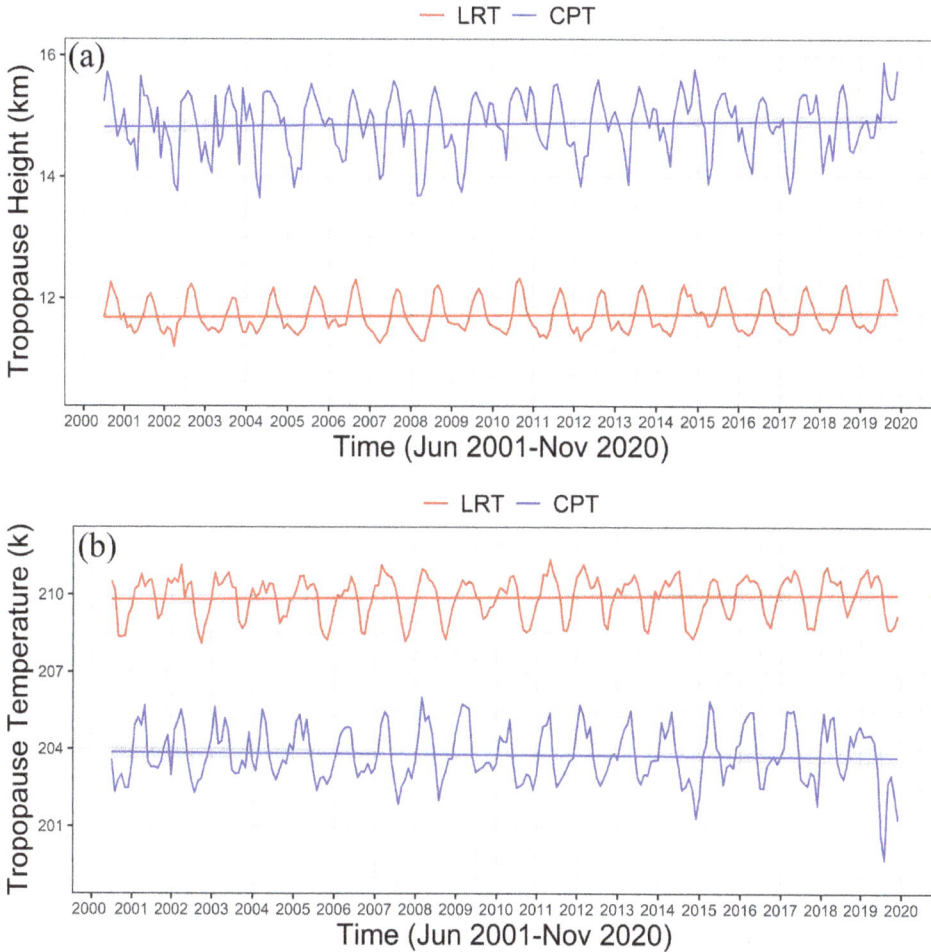

FIGURE 7.6   The LRT and CPT (a) height and (b) temperature is shown from 2001 to 2020.

Our analysis shows a global increasing trend of LRT height of 36 m/decade since 2001, and this is in good agreement with that of Schmidt et al. (2008) which showed an upward trend of global LRT height of 39–66 m/decade. The LRT temperature shows an increase of 0.09 K/decade. For the LRT definition, the correlation coefficient between the LRT height and temperature is -0.78. In the case of the CPT definition, the global trend of CPT height has increased by 60 m/decade since 2001, but that of CPT temperature has decreased by 0.09 K/decade. The correlation coefficient between the CPT height and temperature is -0.82.

7.4.3   COMPARISON BETWEEN GNSS, ERA5, AND AIRS

In this study, TEL in each hemisphere is estimated from the monthly zonal average tropopause height retrieved from the LRT definition. This is done because the LRT represents the location of the point of thermal transition between the troposphere and the stratosphere. Furthermore, it reacts to both tropospheric and stratospheric temperature changes. Many studies (Seidel and Randel, 2006; Santer et al., 2004) have shown that LRT height is a good climate change indicator. Figure 7.7 shows the LRT height and temperature values derived from GNSS, ERA5, and AIRS. In general, AIRS shows the highest values of LRT height, while GNSS shows the lowest values. The trends show that ERA5 data has the highest increasing rate of LRT height, at 48 m/decade since June 2001.

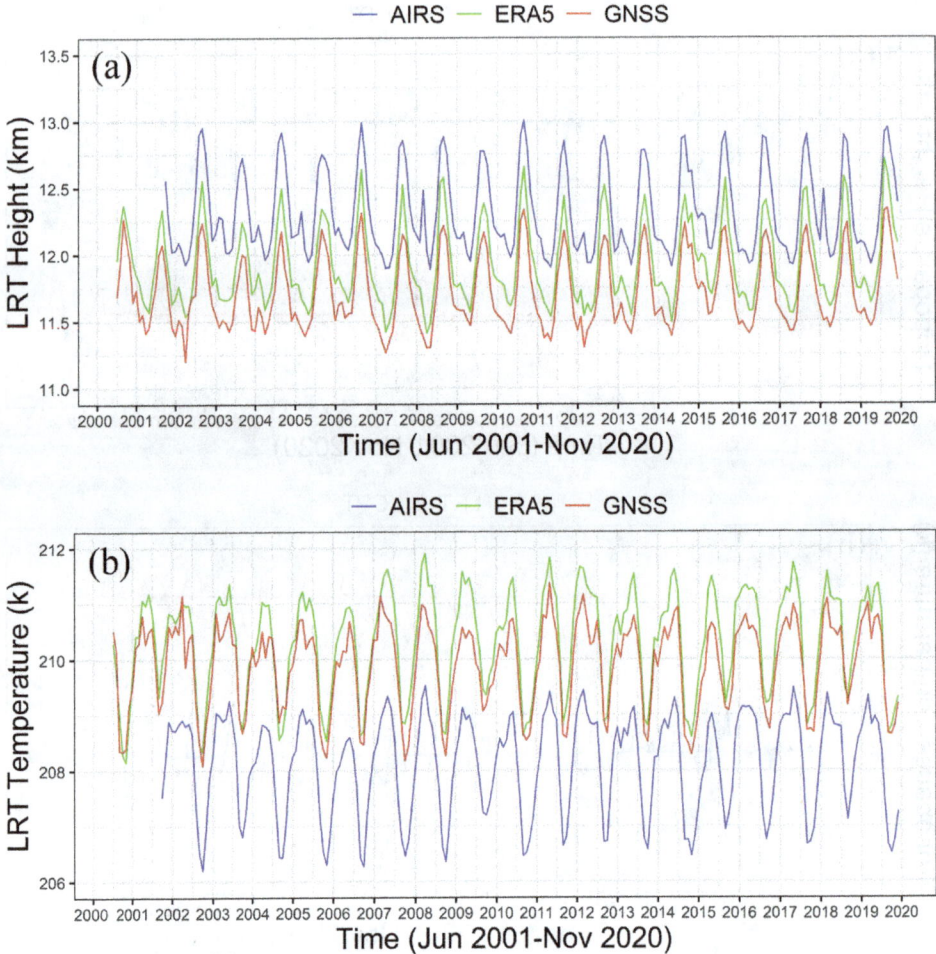

FIGURE 7.7   GNSS, ERA5, and AIRS (a) LRT height and (b) LRT temperature.

In contrast, AIRS has the lowest rates for LRT height, showing an increase of 12 m/decade since September 2002. For the LRT temperature, ERA5 shows the highest values, while AIRS shows the lowest values. ERA5 has the highest increasing rate of the LRT temperature of about 0.18°C/decade. In contrast, AIRS has the lowest upward trend of the LRT temperature of about 0.072°C/decade.

The zonal mean of LRT height for the 3 data sets during January, April, July, and October of 2008 is shown in Figure 7.8. In January 2008, the high LRT covered higher latitudes in the SH than in the NH. The opposite occurs in July. In April 2008, the high LRT covered roughly the same area in both hemispheres. In October, the area covered with high tropopause in NH is larger than that of the SH, but not as wide as the coverage in July. This suggests that the warmer the air, the wider the area covered by the high tropopause. As stated in Section 3, the TEL at NH and SH have been estimated using two tropopause height metrics. The results are discussed in detail in the following.

7.4.3.1   Subjective Criterion for TEL

According to the subjective criterion (Davis and Rosenlof, 2012), the TEL in each hemisphere is the latitude at which the tropopause height is 1.5 km below the tropical average tropopause height (15°S–15°N). As shown in Figure 7.9 and Table 7.2, based on GNSS data, the tropical belt has expanded 0.41°/decade in the NH and 0.08°/decade in the SH since 2001. Using GNSS-RO data, the tropical belt expansion trends in NH and SH agree with the results of Ao and Hajj (2013). According

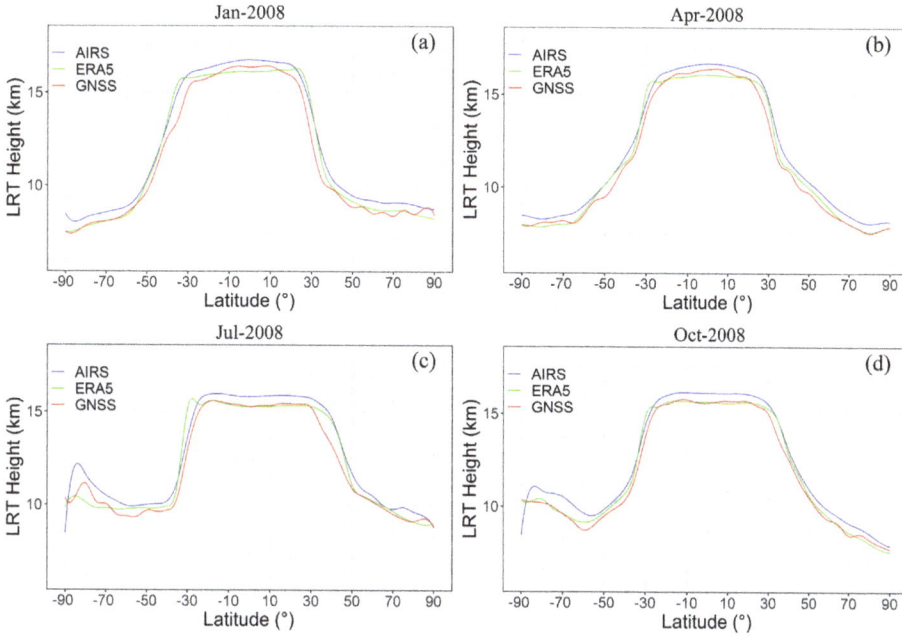

FIGURE 7.8 Monthly zonal average LRT height from GNSS, ERA5, and AIRS.

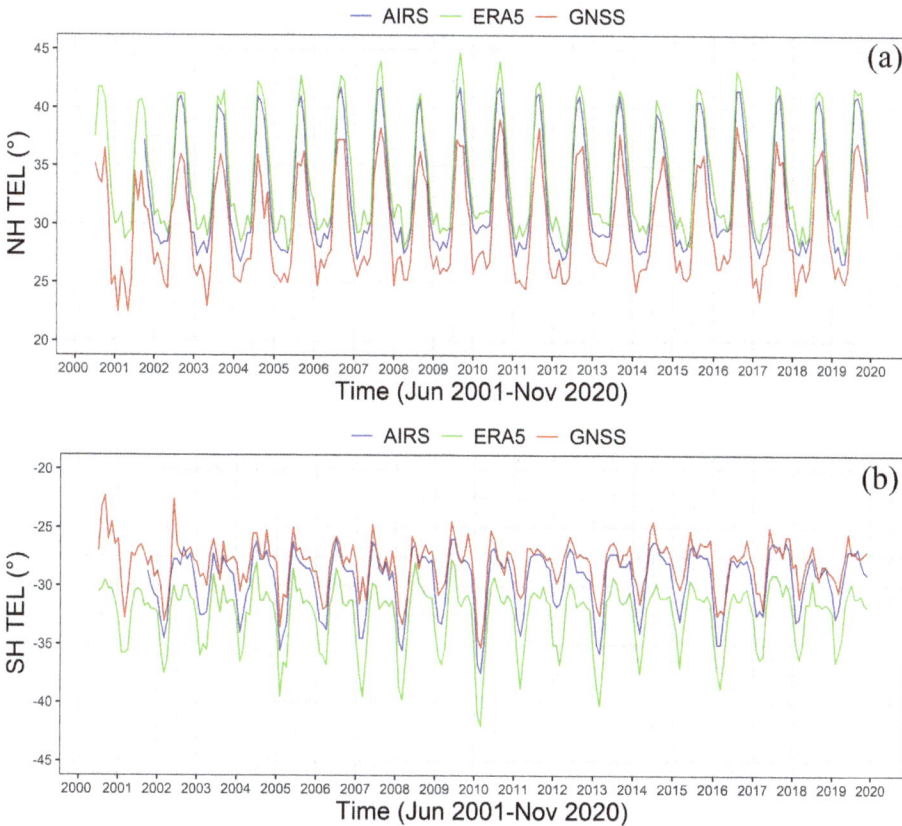

FIGURE 7.9 TEL using the subjective criterion, (a) NH and (b) SH.

TABLE 7.2

Tropical Belt Expansion and Contraction Rates Based on the Subjective Criterion

Source	Duration	NH		SH	
GNSS	June 2001–November 2020	0.41	± 0.09	0.08	± 0.04
ERA5	June 2001–November 2020	−0.01	± 0.1	−0.04	± 0.05
AIRS	September 2002–November 2020	0.34	± 0.11	−0.48	± 0.05

to Meng et al. (2021), the highest trend of LRT height is covering the latitudinal band 30°N to 40°N and this is possibly caused by the tropical widening and subtropical jet poleward shift over the past four decades (Staten et al., 2018), and this corresponds with our study findings. In the case of ERA5, there is no significant expansion or contraction in both hemispheres, while AIRS expands by about 0.34°/decade at the NH and contracts by about -0.48°/decade at the SH.

7.4.3.2   Objective Criterion for TEL

According to the objective criterion (Davis and Rosenlof, 2012), TEL in each hemisphere is the latitude of the maximum poleward gradient of tropopause height. As shown in Figure 7.10 and Table 7.3, the tropical belt based on GNSS has expanded about 0.13°/decade in the NH since 2001, but there has been no significant expansion or contraction in the SH. In the case of ERA5, there is no significant trend in NH, while SH has a minor contraction of approximately -0.08°/decade. AIRS

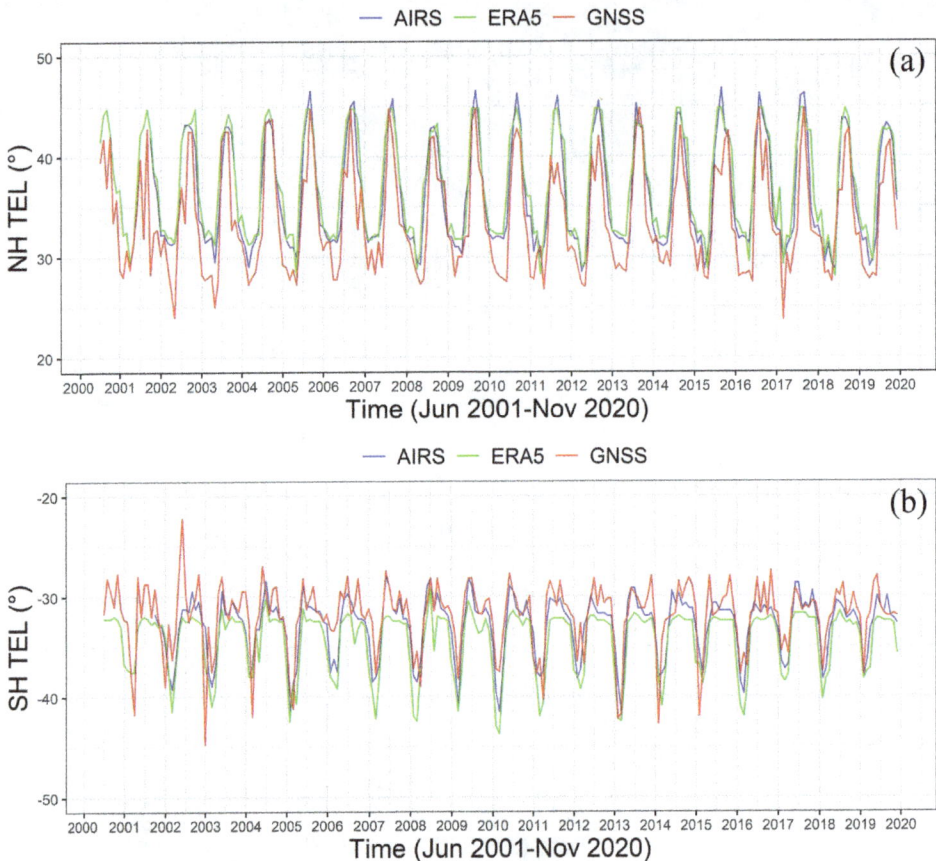

FIGURE 7.10   TEL using the objective criterion, (a) NH and (b) SH.

TABLE 7.3

Tropical Belt Expansion and Contraction Rates Based on the Objective Criterion

Source	Duration	NH		SH	
GNSS	June 2001–November 2020	0.13	± 0.1	−0.03	± 0.06
ERA5	June 2001–November 2020	−0.06	± 0.1	−0.08	± 0.06
AIRS	September 2002–November 2020	0.13	± 0.04	−0.37	± 0.06

has an expansion of 0.13°/decade in NH and a strong contraction in SH of -0.37°/decade. It is clear from these results that the rates of expansion and contraction using the objective criterion are less than those using the subjective criterion. In the case of the objective method, TEL are more poleward than in the case of the subjective method.

7.4.4 Spatial and Temporal Variability of LRT

In this section, the GNSS LRT height and temperature between 50°N and 50°S are investigated (Figure 7.11). In the NH, the LRT height has increased by about 48 m/decade since 2001, and this is consistent with the results of Meng et al. (2021), which show an increase in LRT height of around 44.4 m/decade over 20°N to 80°N for the period from 2001 to 2020. In contrast, LRT height in the SH shows a slight decrease of -2.4 m/decade. Regarding LRT temperature, it has increased about 0.21 K/decade in NH and 0.34 K/decade in SH. Both hemispheres' LRT temperature time series show increasing rates higher than the global one (0.09 K/decade). Figure 7.11 also shows the temporal

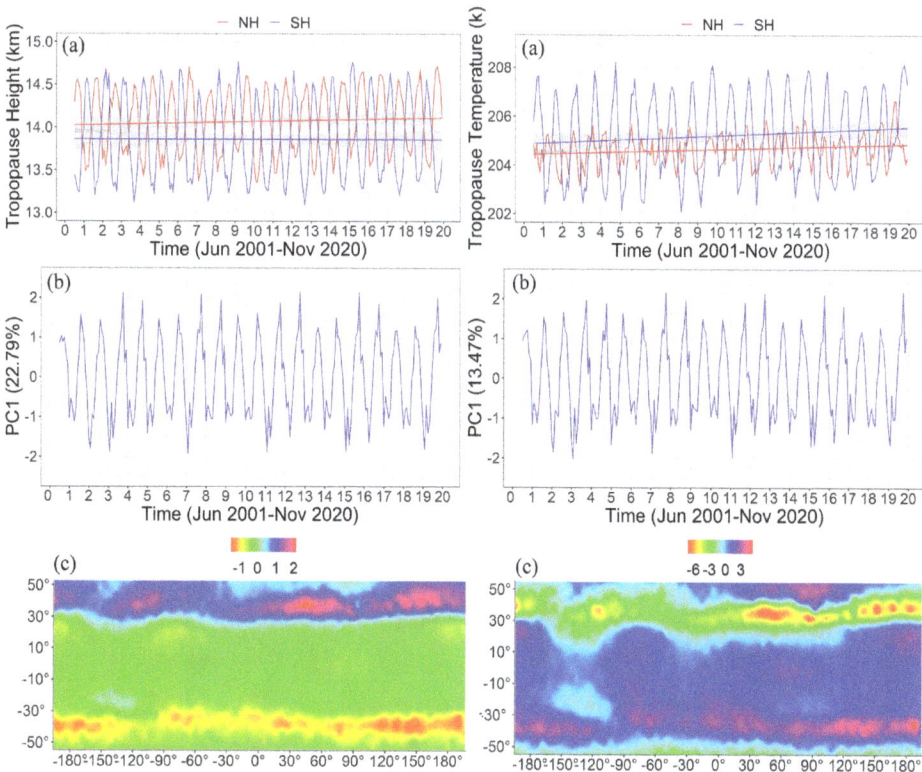

FIGURE 7.11   GNSS-RO based LRT height (left) and temperature (right). In (a) temporal time series (b) temporal variability given by PCA1, and (c) spatial variability map given by PCA1.

and spatial variability given by the first PCA. The temporal variability for LRT height explains 22.79% of the total variance. For the LRT temperature, PCA1 describes 13.47% of the total variability. These values are relatively small, showing that the variability spreads along lower degree PCA modes. We can clearly see the annual force. The spatial variability shows similar patterns for LRT height and temperature. The signal at the NH is stronger and wider than that at the SH.

7.4.5 SEASONAL VARIATION OF LRT HEIGHT AND TEMPERATURE

Tropopause height and temperature, calculated based on multi-mission GNSS data, have a clear seasonal variation. Figure 7.12 depicts the LRT height monthly averaged over the period from June 2001–November 2020 in both hemispheres. As can be seen, the LRT height median in NH gradually increases from January to August, then decreases. In SH, the median LRT height is maximum in January, then falls until June, after that rises until August, then falls again. Figure 7.12 depicts the minimum range of LRT height in both hemispheres in July and August. In the NH, it ranges from 8.48 to 17 km in July and from 8.28 to 16.82 km in August. In the case of the SH, LRT height ranges from 8.58 to 16.08 km in July and from 8.56 to 16.1 km in August. The distribution of the LRT height values shows three main modalities: the first is the high values that represent the tropics; the second is the intermediate values that represent the transitional zone from the tropics, where tropopause is high, to midlatitudes and poles, where tropopause is low; and the third is the low values that mainly represent the high latitudes.

Figure 7.13 depicts the LRT temperature averaged over the period from June 2001–November 2020 in both hemispheres. As shown, in the NH the LRT temperature, like the LRT height, increases gradually from January to August, then decreases. In contrast, in the SH, the median LRT temperature peaks in January and then falls until August, then it rises again. In the NH, the distribution of the LRT temperature values shows three main modalities: the first are the high values that represent the low latitudes; the second are the intermediate values that represent the transitional zone from the tropics, where the tropopause temperature is low, to the midlatitudes and poles, where the tropopause temperature is high; and the third are the low LRT temperature values that mainly represent

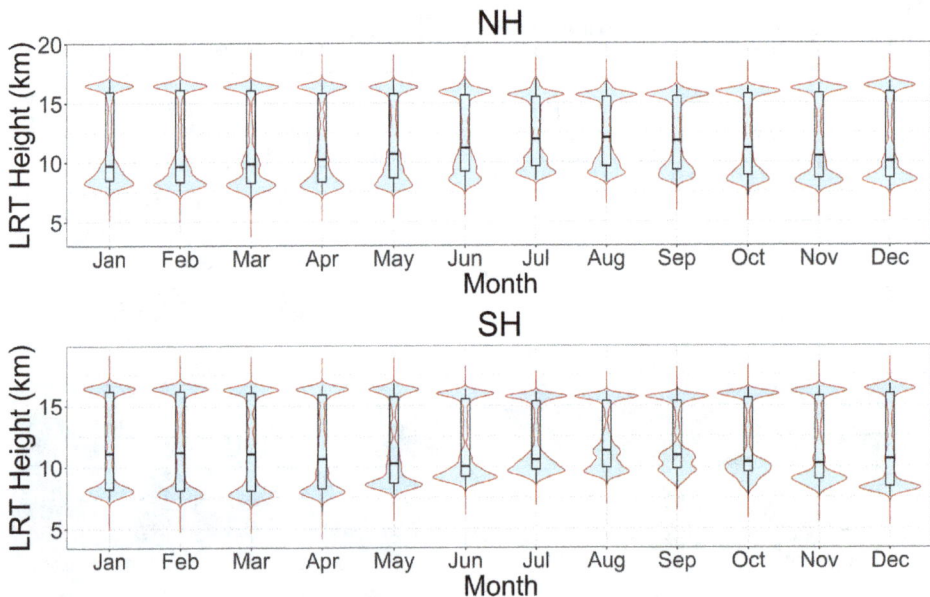

FIGURE 7.12 Monthly GNSS LRT height in NH and SH averaged over the period from 2001 to 2020.

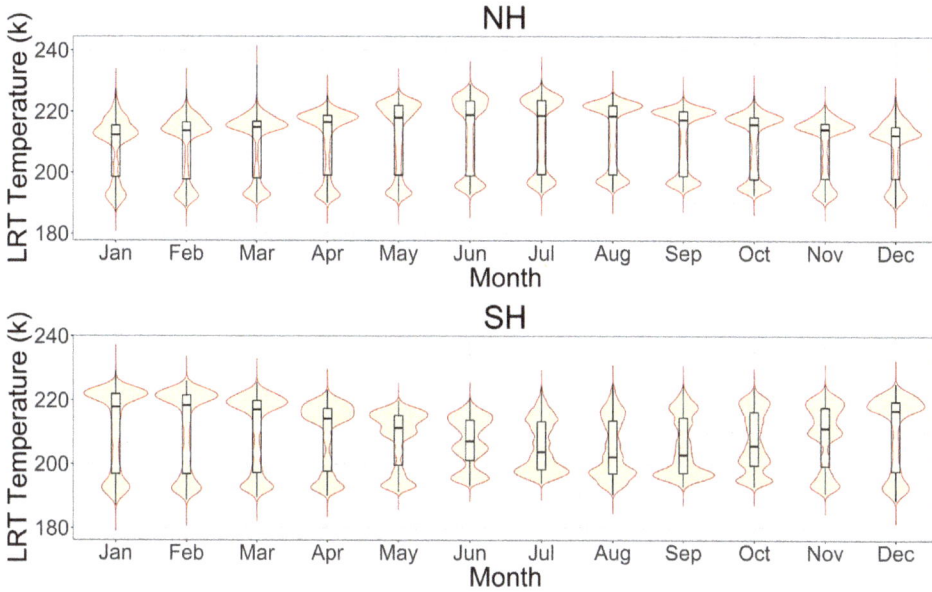

FIGURE 7.13 Monthly GNSS LRT temperature in the NH and the SH averaged over the period from 2001 to 2020.

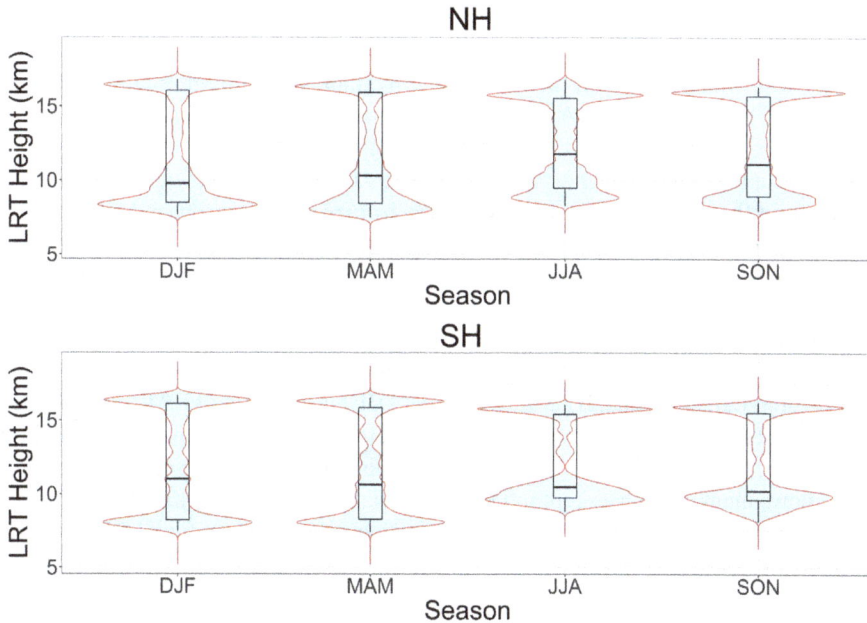

FIGURE 7.14 Seasonal GNSS LRT height in the NH and the SH averaged over the period from 2001 to 2020.

the tropics. In the case of the SH, LRT temperature values show three main modalities for most of the year, but there are more modalities in June, October, and November. In both hemispheres, the LRT temperature value distribution is inversely proportional to the LRT height value distribution.

Figures 7.14 and 7.15 signify the seasonal variation of both LRT height and temperature, respectively. In the NH, the LRT height median is maximum in JJA about 11.77 km and minimum in

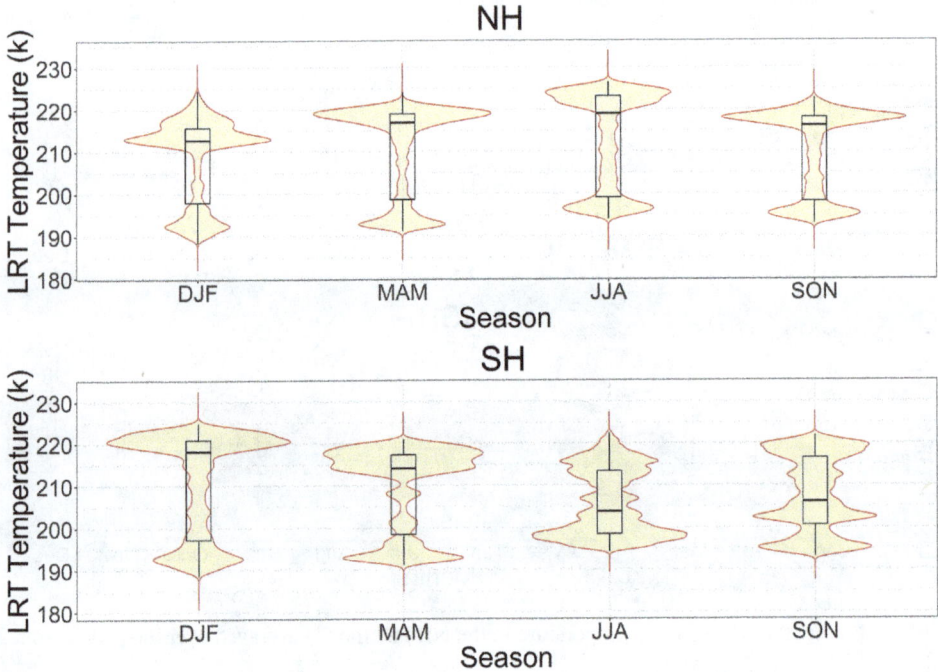

FIGURE 7.15 Seasonally GNSS LRT temperature in the NH and the SH averaged over the period from 2001 to 2020.

DJF about 9.78 km, while in the SH, the LRT height median is maximum in DJF about 11 km and minimum in SON about 10.22 km. As is clear in Figure 7.14, in both hemispheres the density of high LRT height values is less than that of low LRT height, and that is because the high tropopause only covers the tropical zone. The intermediate zone in JJA of the NH is wide, and this indicates a high tropopause in the transitional zone between the tropics and extratropics that is caused by the high surface and air temperatures caused by warming in JJA. While the NH intermediate zone in DJF appears to be so narrow, this is due to a cold atmosphere and a steep transition from the tropics to the extratropics. In contrast, in the SH, the intermediate zone between high and low LRT height modals is wide in DJF and very narrow in JJA.

As shown in Figure 7.15, the median LRT temperature in the NH is highest in JJA at around 219 K and lowest in DJF at around 212.8 K. In the SH, DJF has the highest LRT temperature median at about 218.21 K and JJA has the lowest LRT temperature median at about 203.81 K. In general, the NH LRT temperature values show 3 main modals: high, intermediate, and low; but the SH LRT temperature shows many modals except in DJF, which has 3 main modals like those of NH. In contrast to the LRT height in Figure 7.14, the density of values that represent high LRT temperatures is higher than that of values that represent low LRT temperatures in both hemispheres.

The global distribution of LRT height demonstrates geographic changes with latitude from month to month over the course of the year. Figure 7.16 shows the variation in monthly LRT height that averaged for the period from June 2001–November 2020. The LRT height is greatest in the tropical region, which extends between 30°N and 30°S. In both hemispheres, the LRT height in the transitional zone from the tropics to the extratropics is less than that of the tropical zone. Moreover, the high latitudes of the NH and SH have the lowest LRT height values. Except for January, February, and March, the tropical tropopause cover area in the NH is larger than that in the SH throughout the year.

In contrast to LRT height, the LRT temperature has minimum values in the tropical zone and maximum values in the polar zones (Figure 7.17). The spatial structure of the tropopause pattern

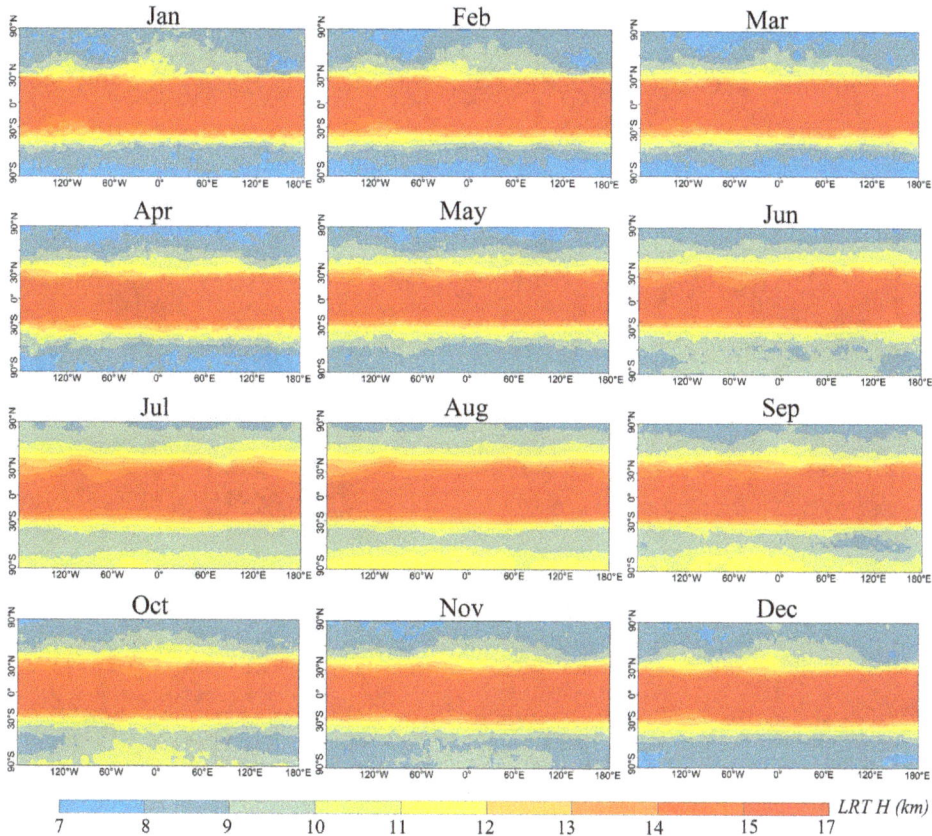

FIGURE 7.16 Global spatial distribution of monthly GNSS LRT height averaged over the period from 2001 to 2020.

revealed that the tropical tropopause covers a larger area in the NH than in the SH. The tropopause temperature in the NH has a single modality from the TEL to the pole. On the other hand, in the SH from the TEL towards the pole, there is bimodality, with a band of high tropopause temperature mainly in the mid-latitude followed by a band of lower tropopause temperature in the high latitudes. The bimodality of the tropopause in the SH is so significant from June to October (Figure 7.17).

Figures 7.18 and 7.19 depict the seasonally averaged LRT height and temperature, respectively, over the period from 2001 to 2020. The tropopause height is maximum over the tropics at 17 km and minimum at the poles at 7 km. In all seasons, the transitional zone from the tropics to the extratropics covers a wider area in the NH than that covered in the SH. In the SH, the tropopause height is bimodally distributed spatially in mid- and high-latitude regions, as seen in JJA and SON (Figure 7.18). The tropopause temperature (Figure 7.19) is minimum at the tropics (188 K) and maximum at mid- and high latitudes (226 K). Furthermore, the tropical tropopause layer covers a larger geographic area in the NH than it does in the SH. In all seasons, there is a clear bimodal pattern in the SH at mid- and high latitudes. The LRT temperature values are low in the tropics, then they become high in the mid-latitudes, and after that, a lower tropopause temperature spatial pattern appears in the high latitudes.

As shown in Figure 7.20, there is a clear difference between the LRT height seasonal variability in the NH and that in the SH. In the NH, all seasons have observed upward trends and all seasons show significant variability. JJA has the highest LRT height increasing trend of about 75 m/dec., and DJF has the lowest increasing rate of about 27 m/dec. The tropopause height values in NH cover a

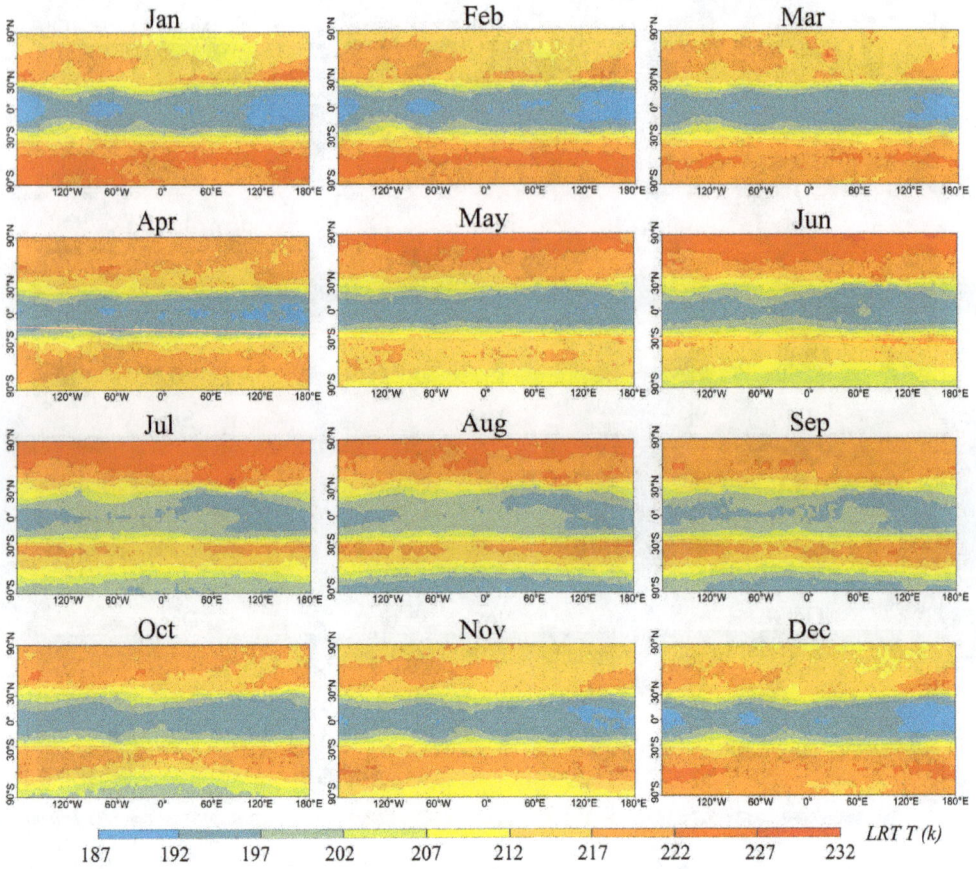

FIGURE 7.17 Global spatial distribution of monthly GNSS LRT temperature averaged over the period from 2001 to 2020.

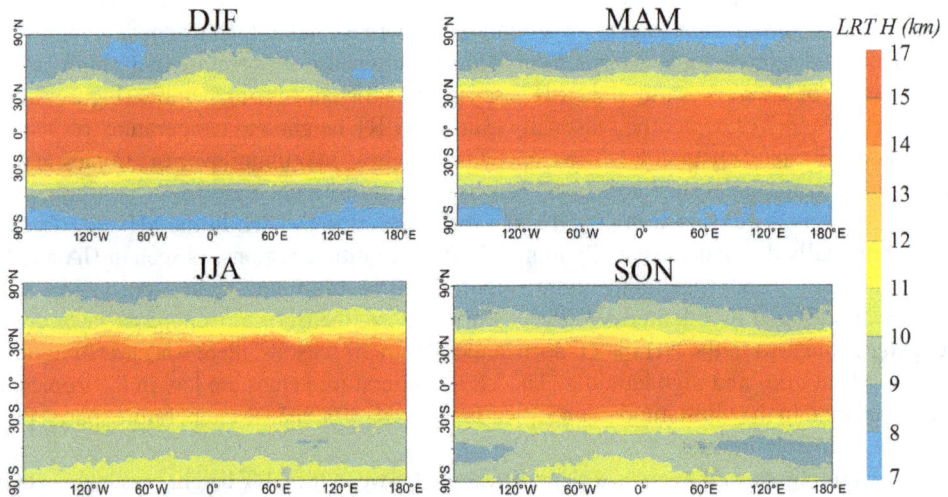

FIGURE 7.18 Global spatial distribution of seasonal GNSS LRT height averaged over the period from 2001 to 2020.

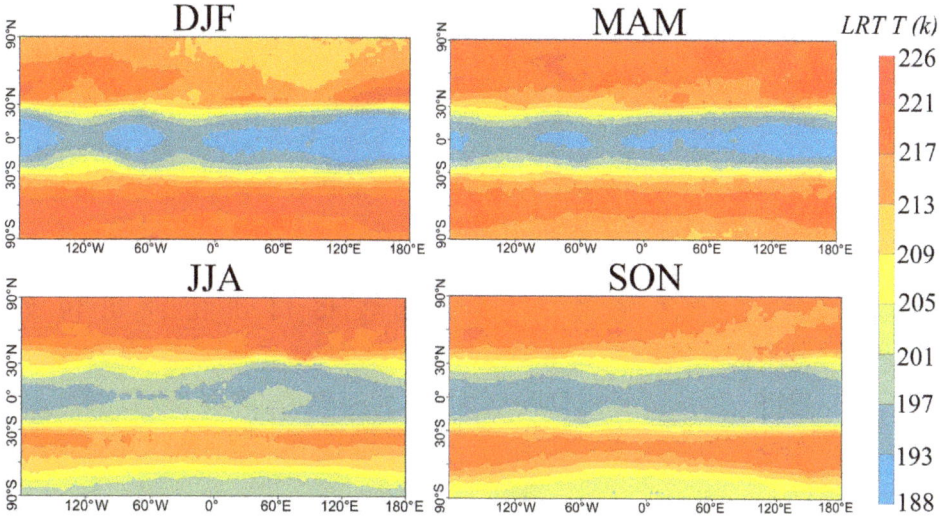

FIGURE 7.19 Global spatial distribution of seasonal GNSS LRT temperature averaged over the period from 2001 to 2020.

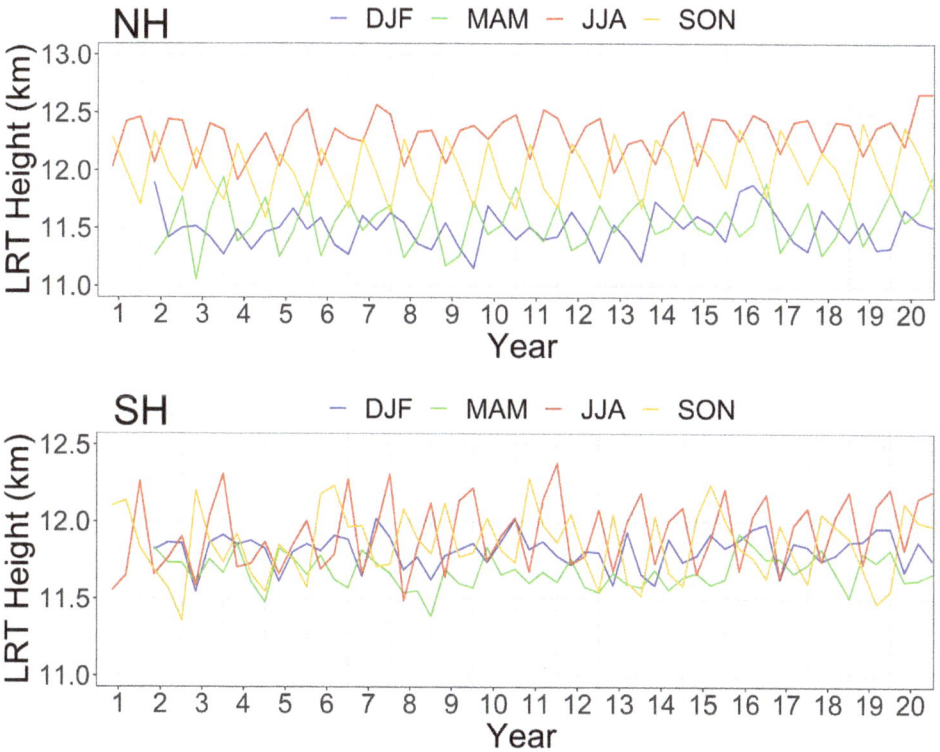

FIGURE 7.20 Time series for LRT height seasons months in the NH and SH for the period from 2001 to 2020.

range from 11.05 km to 12.66 km, which is higher than that of the SH, which is from 11.35 km to 12.38 km. In the SH, LRT height values show a significant increasing trend only in JJA of about 105 m/dec. In contrast, SON has a decreasing trend of about -6 m/dec.

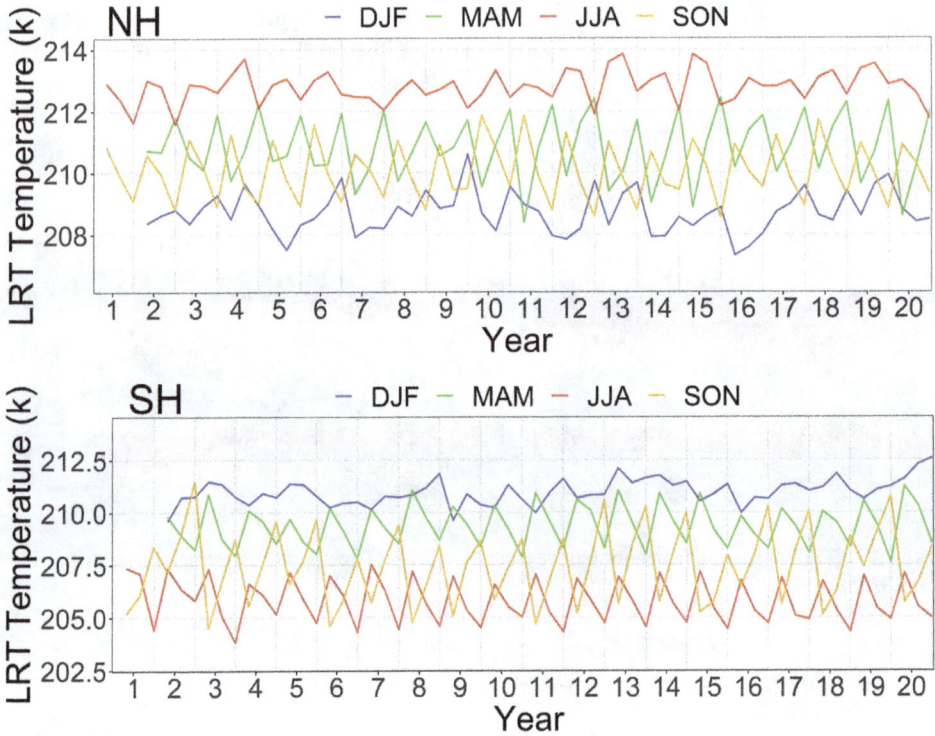

FIGURE 7.21   Time series for LRT temperature seasons months in the NH and SH for the period from 2001 to 2020.

Figure 7.21 illustrates that there is a clear difference between the LRT temperature seasonal pattern in the NH and that in the SH. In the NH, all seasons have increasing trends. JJA has the highest LRT temperature upward trend of about 0.15 K/dec, and SON has the lowest increasing rate of about 0.02 K/dec. The tropopause temperature values in the NH cover a range from 207.32 K to 213.84 K, which is lower than that of the SH, which is from 203.83 K to 212.55 K. In the SH, LRT temperature values show a significant increasing trend in DJF of about 0.44 K/dec. In contrast, JJA has a decreasing trend of about -0.3 K/dec.

7.4.6   TOTAL COLUMN OZONE (TCO), CARBON DIOXIDE (CO₂), AND METHANE (CH₄)

Figure 7.22 shows that since 2001, TCO has had a global increase of 0.7 DU/decade. TCO has a strong negative correlation of -0.64 with the LRT height. This corresponds with the results of previous work, which clarified that the TCO pattern is inversely proportional to tropopause height and can give an indication of the tropical belt width (Hudson et al., 2003; Hudson et al., 2006; Hudson, 2012; Davis et al., 2018). TCO has increased by 0.06 DU/decade and 1.05 DU/decade in the NH and SH, respectively. Shangguan et al. (2019) reported asymmetric ozone trends in the middle stratosphere of both hemispheres, with significant ozone decrease in the NH and ozone increase in the SH. In our results, the PCA1 of TCO represents 66.68% of the total variability. The spatial map of PCA1 shows a stronger signal in the NH than in the SH. The NH signal is located more poleward than that of the SH. Comparisons with the GNSS-RO LRT height spatial and temporal pattern suggest the TCO expansion in the NH and a weak expansion or non-significant contraction in the SH.

Several studies have indicated to an increase in the tropopause height as a result of the troposphere warming caused by the rise of the GHGs concentrations in the atmosphere (Meng et al.,

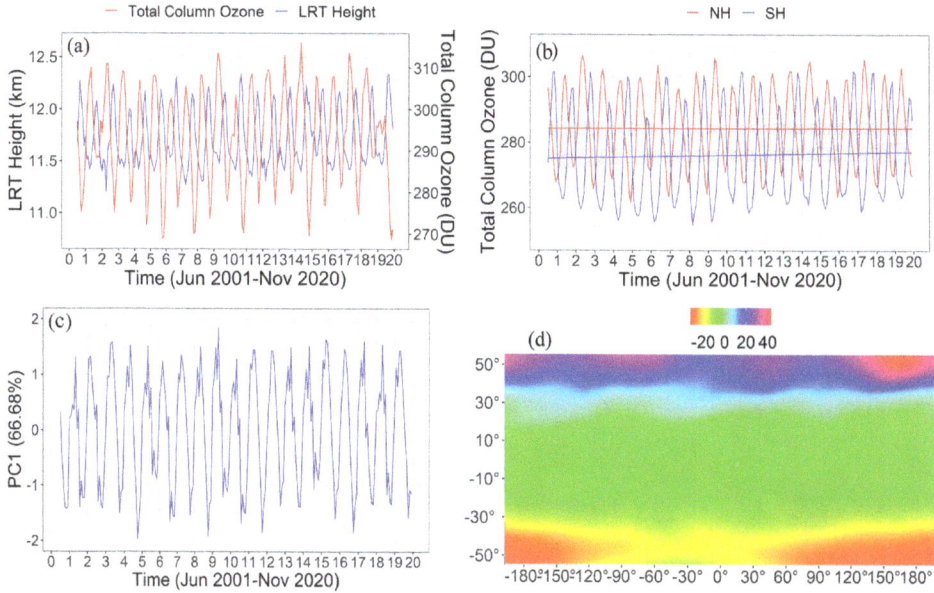

FIGURE 7.22 TCO: (a) global time series against GNSS LRT height, (b) temporal time series in NH and SH, (c) temporal variability given by PCA1, and (d) spatial variability map given by PCA1.

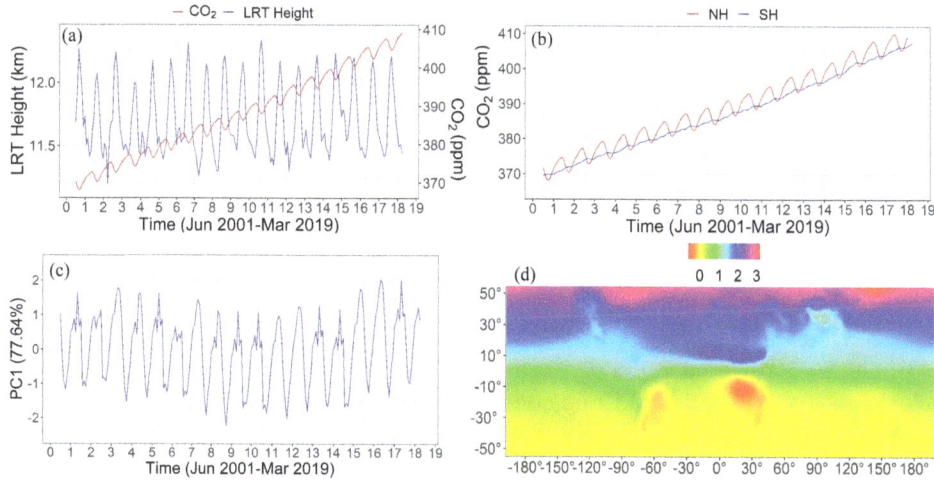

FIGURE 7.23 CO_2: (a) global time series against GNSS LRT height, (b) temporal time series in NH and SH, (c) temporal variability given by PCA1, and (d) spatial variability map given by PCA1.

2021; Pisoft et al., 2021). CO_2 is the most important GHG, and it is considered the main driver of global warming. Figure 7.23 depicts a time series of CO_2 levels. Since 2001, CO_2 levels have risen at a rate of 21.38 ppm per decade. It has a correlation of -0.05 with GNSS LRT height. The CO_2 column average in both the NH and SH has the same increasing rate of 21.6 ppm/decade. This is higher than the global rate. The NH has a higher CO_2 standard deviation (STD) of 11.38 than the SH, which is 10.90. The temporal variability given by the PCA1 represents 77.64% of the total variability. PCA1 shows an increasing trend and large variability with time. The map of PCA1 variability shows a shift towards the north pole. This seems to be related to the coverage of the tropical belt, i.e., the TEL occurrence at the NH is more poleward than that of the SH.

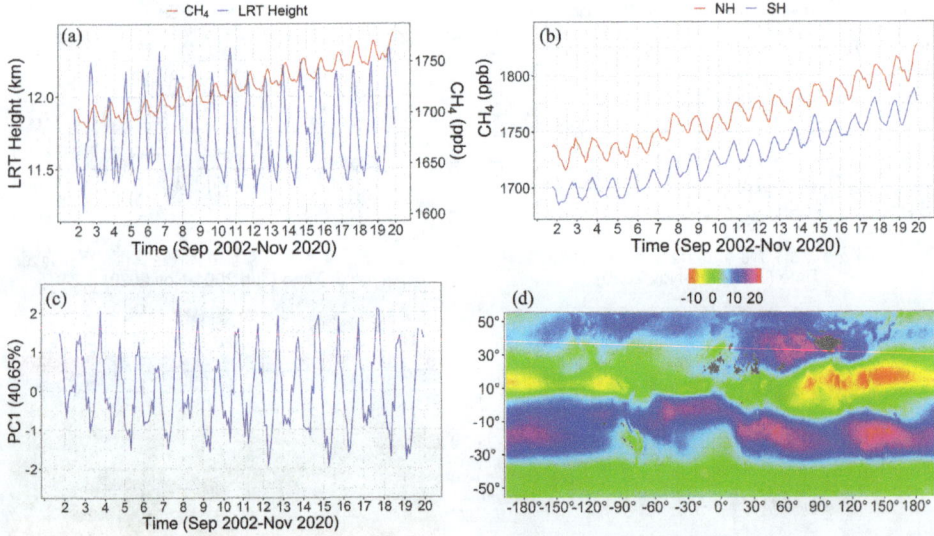

FIGURE 7.24   CH$_4$: (a) global time series against GNSS LRT height, (b) temporal time series in NH and SH, (c) temporal variability given by PCA1, and (d) spatial variability map given by PCA1.

CH$_4$ is one of the main GHGs, which is considered a long-term driver of climate change. The global time series of CH$_4$ column average (Figure 7.24a) shows an increasing trend of 39 ppb/decade since 2001. This variable has a correlation of 0.23 with GNSS-RO LRT height. The CH$_4$ column average in both the NH and the SH shows equal increasing trends of 46.8 ppb/decade. This is higher than the global rate. The CH$_4$ STD in the NH is similar to that in the SH (25.91). The temporal variability of PCA1 explains 40.65% of the total variance. It shows a non-significant trend, but its range increases with time. The map of PCA1 shows a more poleward signal in the NH than its equivalent in the SH. The NH variability pattern reaches 30°N, while that of the SH does not reach the limit of 30°S. This is clearly in line with the GNSS TEL results, showing that the tropical condition in the NH covers a wider area than that in the SH.

7.4.7   Surface Temperature and GPCP Precipitation

Many studies have revealed a link between surface temperature, tropopause height, and tropical belt expansion. Thuburn and Craig (1997, 2000) found the simulated tropopause height to be sensitive to the surface temperature. Figure 7.25 shows that the global surface temperature has increased by 0.3 K/decade since 2001. A clear correlation between the surface temperature and the GNSS-RO LRT height is seen, with a value of 0.81. The surface temperature in both the NH and the SH shows increasing trends of 0.23 K/decade and 0.18 K/decade, respectively. The surface temperature in the NH has a STD of 3.5, while that of the SH has a STD of 1.5. The PCA1 accounts for 84.41% of the total variance. The PCA1 shows an increasing trend and amplitude with time. The PCA1 map has a signal in the SH weaker than that in the NH. The results of surface temperature agree with those of GNSS-RO TEL. For instance, the NH shows more expansive behavior than the SH, which shows a minor expansion using the subjective criterion and a non-significant contraction using the objective criterion. Gao et al. (2015) indicated that the correlation coefficient between global tropopause height anomalies and the El Niño 3.4 sea surface temperature index is 0.53, with a maximum correlation coefficient of 0.8 at a lag of three months. Fomichev et al. (2007) also found that an increase in sea surface temperature resulted in a tropopause height increase in a coupled chemistry climate model simulation. Hu and Fu (2007) suggested that an increase in sea surface temperatures in the tropics could result

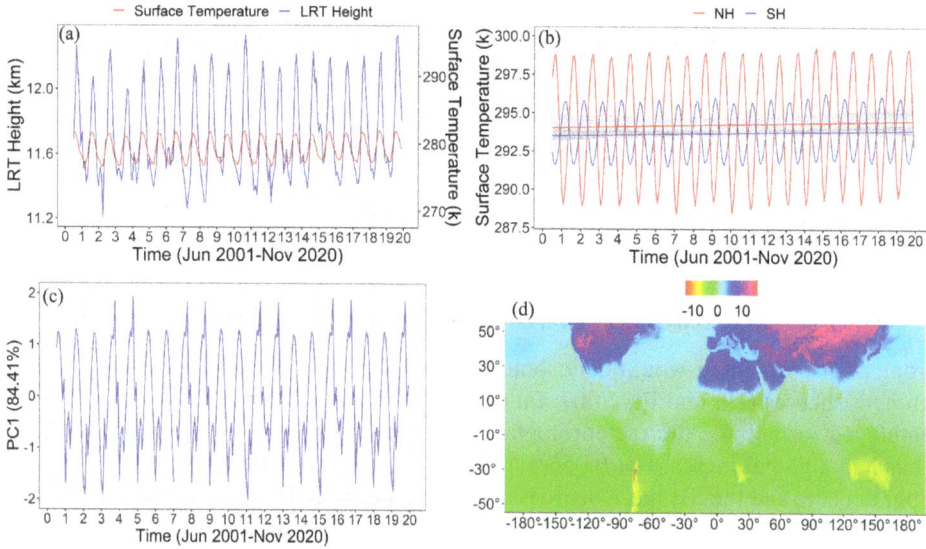

FIGURE 7.25 Surface temperature: (a) global time series against GNSS LRT height, (b) temporal time series in NH and SH, (c) temporal variability given by PCA1, and (d) spatial variability map given by PCA1.

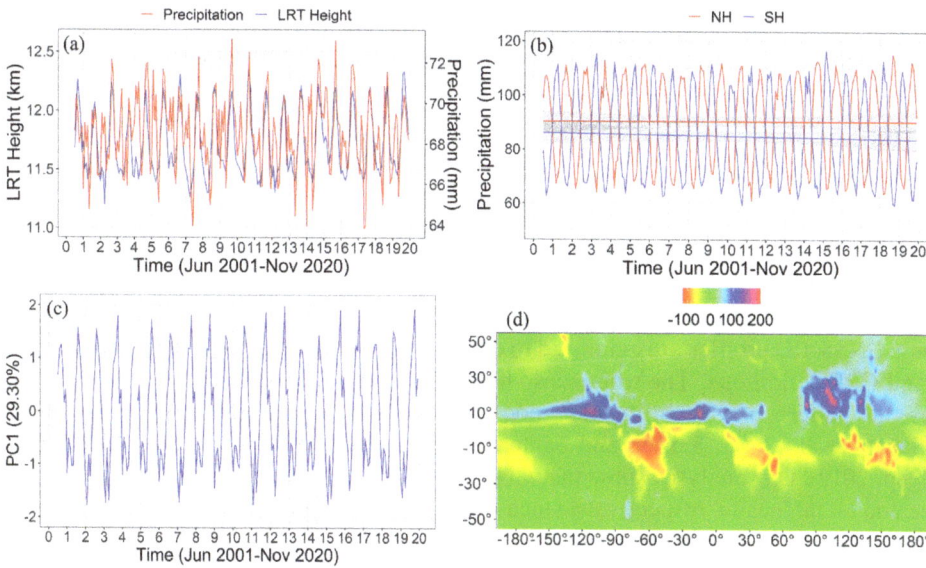

FIGURE 7.26 GPCP precipitation: (a) global time series against GNSS LRT height, (b) temporal time series in NH and SH, (c) temporal variability given by PCA1, and (d) spatial variability map given by PCA1.

in an increase in the tropopause height and a wider Hadley circulation (tropics width). In addition, our results support surface temperature as a proposed driver for tropical expansion (Allen et al., 2012a; Adam et al., 2014).

The spatial and temporal variability of precipitation are investigated to determine the effects of TEL variability on precipitation behavior (Figure 7.26). Since 2001, global GPCP precipitation has decreased by -0.04 mm per decade. The precipitation behavior has a strong correlation of 0.61 with the GNSS LRT height. The GPCP precipitation in the NH shows a minor decreasing trend of

-0.02 mm/decade whereas the SH shows a significant decreasing trend of -1.3 mm/decade. The precipitation in the NH has a STD of 15.84, and the SH has a STD of 16.47. The PCA1 explains 29.30% of the total variability. PCA1 has an upward trend and increases in amplitude with time. The PCA1 map shows a pattern in the NH that is stronger and more poleward than that in the SH. The precipitation can be used as an independent metric in identifying the TELs' locations. Many studies, relying on surface-based variables to investigate tropical widening, used the GPCP monthly data set to examine shifts in the positions and boundaries of the subtropical dry zones (Hu et al., 2010; Zhou et al., 2011; Allen et al., 2012b).

7.4.8   STANDARDIZED PRECIPITATION EVAPOTRANSPIRATION INDEX (SPEI)

The tropical belt widening would contribute to increasing the mid-latitude drought frequency in both hemispheres (Hu and Fu, 2007; Fu et al., 2006; Seidel et al., 2007). The SPEI is usually employed to monitor the meteorological drought status. As is clear in Figure 7.27, the SPEI has had a global increase of 0.056 per decade since 2001. The NH shows an increase of 0.035 per decade, and the SH has a decrease of -0.005 per decade. The SPEI has no correlation with GNSS LRT height (-0.002). Because the study area is wide and extends through many continents, the SPEI in our study only provides information about the dry and wet conditions. Figure 7.27 shows the spatial pattern of SPEI in September 2019 and the areas by category of no drought, moderate drought, severe drought, and extreme drought. Figure 7.28 shows the number of cells covered with drought and its corresponding classification from Figure 7.27. The total number of cells covered by drought in the NH nearly doubles its value in the SH. Both hemispheres have a decreasing trend in the number of cells covered by drought. The decrease rate is 510 cell/decade in the NH and 373 cell/decade in the SH. The drought does not show any spatial pattern associated with the locations of TELs.

7.5   CONCLUSIONS

The GNSS-RO is a well-established technique to derive atmospheric temperature structure in the UTLS region. In this study, GNSS-RO data from 12 RO missions is combined to examine the possible expansion of the tropical belt. When used together in our analysis, the intercomparison of GNSS-RO profiles from the various RO missions demonstrates a high level of consistency. GNSS-RO profiles are employed to derive tropopause height and temperature based on LRT and CPT definitions. The tropopause height becomes crucial in climate change research because its pattern of variability has good accordance with the global warming phenomenon (Santer et al., 2003; Sausen and Santer, 2003; Seidel and Randel, 2006; Mohd Zali and Mandeep, 2019). Our analyses show that GNSS LRT and CPT height have increased by 36 m/decade and 60 m/decade, respectively, since June 2001. There is a high correlation between the tropopause height and temperature, being -0.78 and -0.82 for LRT and CPT, respectively. The LRT height from ERA5 shows an increase of 48 m/decade since June 2001, and that derived from AIRS has a smaller increasing rate of 12 m/decade since September 2002. Both GNSS LRT height and temperature show clear seasonal variations. The NH shows variation throughout all four seasons. The variability is high in JJA and low in DJF. In contrast, the SH shows significant LRT height variation only in the summer, but it has variation in LRT temperature in all seasons. The LRT height range in the NH is wider than in the SH. On the other hand, the SH has a wider range of LRT temperature values than the NH.

The reported tropics widening rates in most previous studies range from 0.25° to 3.0° latitude per decade, and their statistical significance varies greatly depending on the metrics used to estimate the TEL as well as the data sets used for its derivation (Davis and Rosenlof, 2012). In our study, TEL in each hemisphere is estimated using two tropopause height metrics. Applying the first method, subjective criterion, there are higher expansion and contraction rates

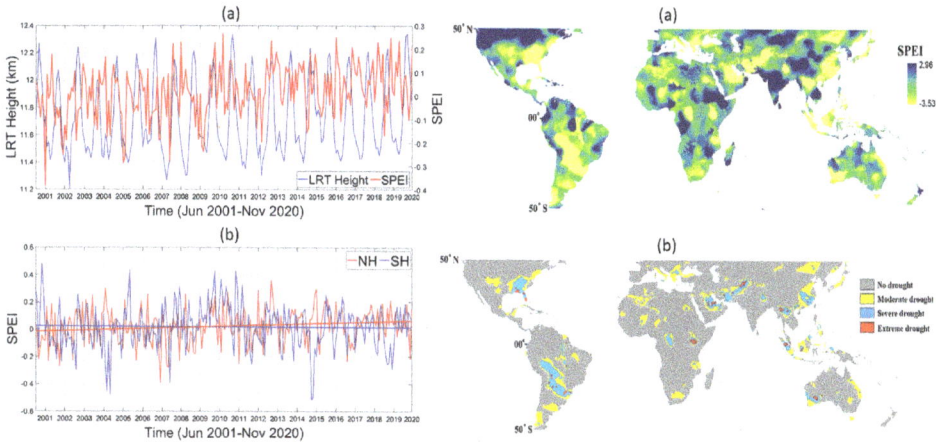

FIGURE 7.27 On the left, SPEI drought index: (a) global SPEI time series in comparison with LRT height and (b) SPEI for two latitudinal bands 0°–50°N and 0°–50°S. On the right: (a) SPEI drought index in September 2019 and (b) SPEI drought categories in September 2019.

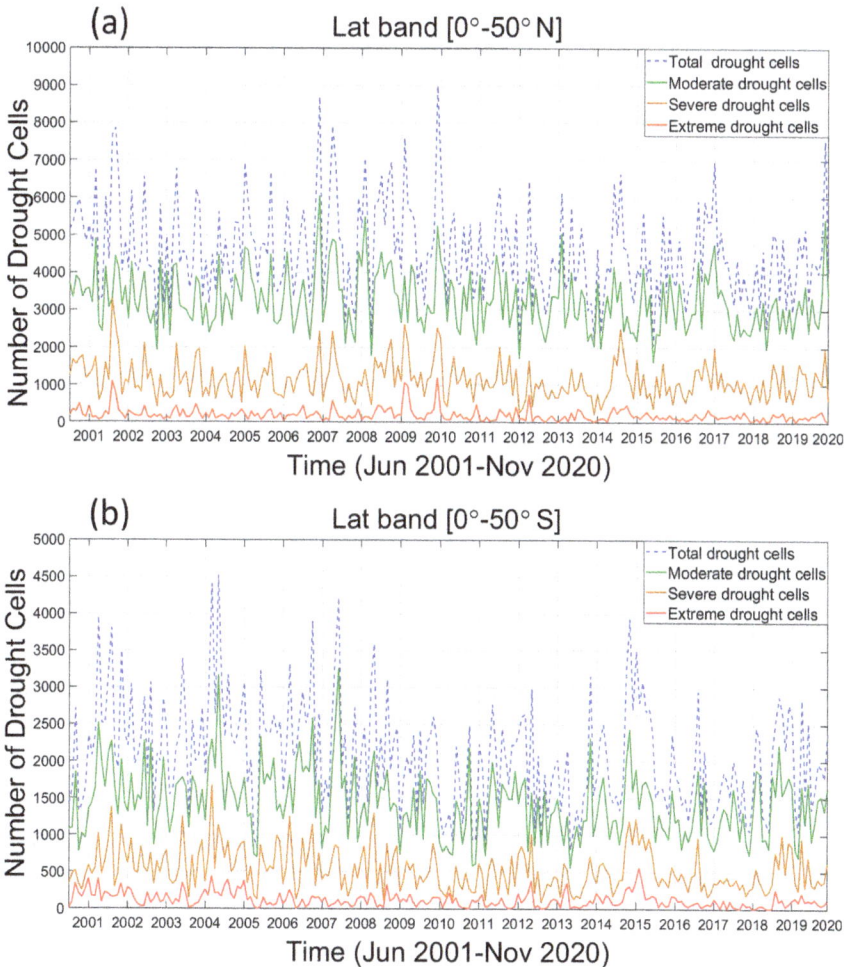

FIGURE 7.28 Number of cells covered with drought at (a) NH and (b) SH.

than those from the second method, objective criterion. While using the objective criterion, the locations of TELs in both hemispheres are more poleward than those from the subjective criterion. Based on the subjective method, tropical width results from GNSS-RO have an expansive behavior in the NH with about 0.41°/decade, and a minor expansion trend in the SH with 0.08°/decade. ERA5 has non-significant contractions in both hemispheres. In the case of the AIRS data, there is a clear expansion behavior in the NH with about 0.34°/decade and a strong contraction in the SH with about -0.48°/decade. Based on the objective method, GNSS-RO has an expansive behavior in the NH with about 0.13°/decade, but there is no significant expansion or contraction in the SH. For ERA5, there is no significant trend for the TEL results in the NH, while there is a minor contraction of about -0.08°/decade in the SH. The AIRS data show an expansion of 0.13°/decade in the NH and a strong contraction of -0.37°/decade in the SH. Results of several studies, based on different data sets and metrics, have shown an expansive behavior of the tropical belt in the NH higher than that of the SH, and this broadly agrees with our GNSS-RO-based results (Hu and Fu, 2007; Archer and Caldeira, 2008; Hu et al., 2010; Zhou et al., 2011; Allen et al., 2012b). From all data sets, the TEL is located more poleward in the NH than in the SH. For both subjective and objective methods, the TELs reach latitudes of 44.75°N and 46.75°N, respectively, at the NH. Meanwhile, at the SH, the TELs reach latitudes of 42°S and 44.75°S for subjective and objective methods, respectively. In both hemispheres, the variability of tropopause parameters (temperature and height) is greatest around the TEL locations.

The TCO shows increasing rates globally. The rate in the SH is higher than that in the NH. The ozone variability agrees well with the spatial and temporal modes of TEL estimated from GNSS-RO LRT height, and this supports GNSS-RO TEL estimates over those of ERA5 and AIRS. In addition, CO_2 and CH_4, as the main GHGs responsible for global warming, cause a rise in tropopause height as concentrations increase (Meng et al., 2021; Pisoft et al., 2021). In our analysis, both CO_2 and CH_4 display an upward global rate. Their trends at the NH and the SH are nearly the same. The patterns of TCO and CO_2 display good agreement with the TEL locations at the NH and SH. They show more poleward occurrence with time, and their variability in the NH is higher than that in the SH. In addition, CH_4 has a signal at the NH that occurs more poleward than that at the SH. The surface temperature and precipitation both increase with time and have a strong correlation with LRT height. Both variables show an increasing rate at the NH that is higher than at the SH. The surface temperature shows a strong spatial variability pattern that broadly agrees with the TEL locations from GNSS-RO. The spatial pattern of precipitation shows a northward orientation. The SPEI meteorological drought index shows an increasing rate globally. It has an upward trend in the NH while having a decreasing trend in the SH. Since SPEI is multivariate, it has no direct response to TEL behavior. In both hemispheres, the number of cells covered by drought has decreased since 2001. It can be concluded that the rates of tropical belt widening differ from one data set to another and from one metric to another. Furthermore, TEL behavior in the NH differs from that in the SH. In addition, the variability of meteorological parameters agrees with GNSS TEL results more than with other data sets. The study's findings highlight the importance of monitoring the tropopause and TEL parameters, which can accurately indicate global climate variability and change.

REFERENCES

Adam, O., Grise, K. M., Staten, P., Simpson, I. R., Davis, S. M., Davis, N. A., Waugh, D. W., Birner, T., and Ming, A.: The TropD software package (v1): standardized methods for calculating tropical-width diagnostics. Geoscientific Model Development, 11(10), 4339–4357. https://doi.org/10.5194/gmd-11-4339-2018, 2018.

Adam, O., Schneider, T., and Harnik, N.: Role of changes in mean temperatures versus temperature gradients in the recent widening of the Hadley circulation. Journal of Climate, 27(19), 7450–7461. https://doi.org/10.1, 2014.

Adler, R., Wang, J., Sapiano, M., Huffman, G., Chiu, L., Xie, P., Ferraro, R., Schneider, U., Becker, A., Bolvin, D., Nelkin, E., Gu, G., and NOAA CDR Program.: Global Precipitation Climatology Project (GPCP) Climate Data Record (CDR), Version 2.3 (Monthly). National Centers for Environmental Information. doi:10.7289/V56971M6, 2016.

AIRS project: Aqua/AIRS L3 Monthly Standard Physical Retrieval (AIRS-only) 1 degree x 1 degree V7.0, Greenbelt, MD, USA, Goddard Earth Sciences Data and Information Services Center (GES DISC), 10.5067/UBENJB9D3T2H, 2019a.

AIRS project: Aqua/AIRS L3 Monthly Standard Physical Retrieval (AIRS+AMSU) 1 degree x 1 degree V7.0, Greenbelt, MD, USA, Goddard Earth Sciences Data and Information Services Center (GES DISC), 10.5067/KUC55JEVO1SR, 2019b.

Allen, R. J., and Kovilakam, M.: The role of natural climate variability in recent tropical expansion. Journal of Climate, 30(16), 6329–6350. https://doi.org/10.1175/jcli-d-16-0735.1, 2017.

Allen, R. J., Sherwood, S. C., Norris, J. R., and Zender, C. S.: The equilibrium response to idealized thermal forcings in a comprehensive GCM: implications for recent tropical expansion. Atmospheric Chemistry and Physics, 12(10), 4795–4816. https://doi.org/10.5194/acp-12-4795-2012, 2012a.

Allen, R. J., Sherwood, S. C., Norris, J. R., and Zender, C. S.: Recent Northern Hemisphere tropical expansion primarily driven by black carbon and tropospheric ozone. Nature, 485(7398), 350–354. https://doi.org/10.1038/nature11097, 2012b.

Ao, C. O., and Hajj, J. A.: Monitoring the width of the tropical belt with GPS radio occultation measurements. Geophysical Research Letters, 40, 6236–6241, doi: 10.1002/2013GL058203, 2013.

Archer, C. L., and Caldeira, K.: Historical trends in the jet streams. Geophysical Research Letters, 35(8). https://doi.org/10.1029/2008gl033614, 2008.

Arras, C., and Wickert, J.: Estimation of ionospheric sporadic E intensities from GPS radio occultation measurements. Journal of Atmospheric and Solar-Terrestrial Physics, 171, 60–63, https://doi.org/10.1016/j.jastp.2017.08.006, 2018.

Aumann, H., Chahine, M., Gautier, C., Goldberg, M., Kalnay, E., McMillin, L., Revercomb, H., Rosenkranz, P., Smith, W., Staelin, D., Strow, L., and Susskind, J.: AIRS/AMSU/HSB on the aqua mission: design, science objectives, data products, and processing systems. IEEE Transactions on Geoscience and Remote Sensing, 41(2), 253–264. https://doi.org/10.1109/tgrs.2002.808356, 2003.

Bai, W., Deng, N., Sun, Y., Du, Q., Xia, J., Wang, X., Meng, X., Zhao, D., Liu, C., Tan, G., Liu, Z., and Liu, X.: Applications of GNSS-RO to numerical weather prediction and tropical cyclone forecast. Atmosphere, 11(11), 1204. https://doi.org/10.3390/atmos11111204, 2020.

Beguería, S., Vicente-Serrano, S. M., Reig, F., and Latorre, B.: Standardized precipitation evapotranspiration index (SPEI) revisited: parameter fitting, evapotranspiration models, tools, datasets and drought monitoring. International Journal of Climatology, 34(10), 3001–3023. https://doi.org/10.1002/joc.3887, 2013.

Birner, T.: The extratropical tropopause region. PhD Thesis. University of Munich, Germany, 2003.

Calabia, A., and Jin, S.: New modes and mechanisms of thermospheric mass density variations from GRACE accelerometers. Journal of Geophysical Research: Space Physics, 121(11), 11,191–11,212. https://doi.org/10.1002/2016ja022594, 2016.

Calabia, A., and Jin, S.: New modes and mechanisms of long-term ionospheric TEC variations from global ionosphere maps. Journal of Geophysical Research: Space Physics, 125(6). https://doi.org/10.1029/2019ja027703, 2020.

CDAAC: https://cdaac-ww.cosmic.ucar.edu/cdaac/products.html, last access: 20 March 2021.

Davis, N. A., and Birner, T.: Seasonal to multidecadal variability of the width of the tropical belt. Journal of Geophysical Research: Atmospheres, 118(14), 7773–7787. https://doi.org/10.1002/jgrd.50610, 2013.

Davis, S. M., Hassler, B., and Rosenlof, K. H.: Revisiting ozone measurements as an indicator of tropical width. Progress in Earth and Planetary Science, 5(1). https://doi.org/10.1186/s40645-018-0214-5, 2018.

Davis, S. M., and Rosenlof, K. H.: A multidiagnostic intercomparison of tropical-width time series using reanalyses and satellite observations. Journal of Climate, 25(4), 1061–1078. https://doi.org/10.1175/jcli-d-11-00127.1, 2012.

Fjeldbo, G., Kliore, A. J., and Eshleman, V. R.: The neutral atmosphere of Venus as studied with the Mariner V radio occultation experiments, The Astronomical Journal, 76, 123–140, 1971.

Foelsche, U., Pirscher, B., Borsche, M., Kirchengast, G., and Wickert, J.: Assessing the climate monitoring utility of radio occultation data: From CHAMP to FORMOSAT-3/COSMIC. Terrestrial, Atmospheric and Oceanic Sciences, 20(1), 155. https://doi.org/10.3319/tao.2008.01.14.01(f3c), 2009.

Fomichev, V. I., Johnson, A. I., de Grandpre, J., Beagley, S. R., McLandress, C., Semeniuk, K., and Shepherd, T. G.: Response of the middle atmosphere to CO_2 doubling: Results from the Canadian middle atmosphere model. Journal of Climate, 20, 1121–1144, 2007.

Frierson, D. M. W., Lu, J., and Chen, G.: Width of the Hadley cell in simple and comprehensive general circulation models. Geophysical Research Letters, 34(18). https://doi.org/10.1029/2007gl031115, 2007.

Fu, Q., Johanson, C. M., Wallace, J. M. and Reichler, T.: Enhanced midlatitude tropospheric warming in satellite measurements. Science 312, 1179, 2006.

Gao, P., Xu, X., and Zhang, X.: Characteristics of the trends in the global tropopause estimated from COSMIC radio occultation data. IEEE Transactions on Geoscience and Remote Sensing, 53(12), 6813–6822. https://doi.org/10.1109/tgrs.2015.2449338, 2015.

Gettelman, A., Hoor, P., Pan, L. L., Randel, W. J., Hegglin, M. I., and Birner, T.: The extratropical upper troposphere and lower stratosphere. Reviews of Geophysics, 49(3). https://doi.org/10.1029/2011rg000355, 2011.

GMAO (Global Modeling and Assimilation Office), MERRA-2 instM_2d_asm_Nx: 2d,Monthly mean,Single-Level,Assimilation,Single-Level Diagnostics V5.12.4, Greenbelt, MD, USA, Goddard Earth Sciences Data and Information Services Center (GES DISC), 10.5067/5ESKGQTZG7FO, 2015.

Gnanadesikan, A., and Stouffer, R. J.: Diagnosing atmosphere-ocean general circulation model errors relevant to the terrestrial biosphere using the Köppen climate classification. Geophysical Research Letters, 33(22). https://doi.org/10.1029/2006gl028098, 2006.

Gorbunov, E. M., and Sokolovskiy, V. S.: Remote sensing of refractivity from space for global observations of atmospheric parameters, Report 119 (Max Planck Institute for Meteorology, Hamburg), 1993.

Gorbunov, E. M., Sokolovskiy, V. S., and Bengtsson, L.: Space refractive tomography of the atmosphere: Modeling of direct and inverse problems, Report 210 (Max Planck Institute for Meteorology, Hamburg), 1996.

Grise, K. M., Davis, S. M., Simpson, I. R., Waugh, D. W., Fu, Q., Allen, R. J., Rosenlof, K. H., Ummenhofer, C. C., Karnauskas, K. B., Maycock, A. C., Quan, X. W., Birner, T., and Staten, P. W.: Recent tropical expansion: Natural variability or forced response? Journal of Climate, 32(5), 1551–1571. https://doi.org/10.1175/jcli-d-18-0444.1, 2019.

Hajj, G. A., Ao, C. O., Iijima, B. A., Kuang, D., Kursinski, E. R., Mannucci, A. J., Meehan, T. K., Romans, L. J., de la Torre Juarez, M., and Yunck, T. P.: CHAMP and SAC-C atmospheric occultation results and intercomparisons. Journal of Geophysical Research: Atmospheres, 109(D6), n/a. https://doi.org/10.1029/2003jd003909, 2004.

Hajj, G. A., Kursinski, R. E., Romans, J. L., Bertiger, I. W., and Leroy S. S.: A technical description of atmospheric sounding by GPS occultation. Journal of Atmospheric and Solar-Terrestrial Physics, 64, 451–469, doi:10.1016/S1364–8826(01)00114–6, 2002.

Harris, I., Osborn, T. J., Jones, P. et al.: Version 4 of the CRU TS monthly high-resolution gridded multivariate climate dataset. Scientific Data, 7, 109: https://doi.org/10.1038/s41597-020-0453-3, 2020.

Held, I. M., and Hou, A. Y.: Nonlinear axially symmetric circulations in a nearly inviscid atmosphere. Journal of the Atmospheric Sciences, 37, 515–533, doi:10.1175/1520-0469(1980)037<0515:NASCIA>2.0.CO;2, 1980.

Hersbach, H., Bell, B., Berrisford, P., Biavati, G., Horányi, A., Muñoz Sabater, J., Nicolas, J., Peubey, C., Radu, R., Rozum, I., Schepers, D., Simmons, A., Soci, C., Dee, D., and Thépaut, J-N.: ERA5 monthly averaged data on pressure levels from 1979 to present. Copernicus Climate Change Service (C3S) Climate Data Store (CDS). 10.24381/cds.6860a573, 2019a.

Hersbach, H., Bell, B., Berrisford, P., Biavati, G., Horányi, A., Muñoz Sabater, J., Nicolas, J., Peubey, C., Radu, R., Rozum, I., Schepers, D., Simmons, A., Soci, C., Dee, D., and Thépaut, J-N.: ERA5 monthly averaged data on single levels from 1979 to present. Copernicus Climate Change Service (C3S) Climate Data Store (CDS). 10.24381/cds.f17050d7, 2019b.

Highwood, E. J., and Hoskins, B. J.: "The tropical tropopause." Quarterly Journal of the Royal Meteorological Society, 124.549, pp. 1579–1604. https://doi.org/10.1002/qj.49712454911, 1998.

Ho, S. P., Hunt, D., Steiner, A. K., Mannucci, A. J., Kirchengast, G., Gleisner, H., Heise, S., von Engeln, A., Marquardt, C., Sokolovskiy, S., Schreiner, W., Scherllin-Pirscher, B., Ao, C., Wickert, J., Syndergaard, S., Lauritsen, K. B., Leroy, S., Kursinski, E. R., Kuo, Y. H., and Gorbunov, M.: Reproducibility of GPS radio occultation data for climate monitoring: Profile-to-profile inter-comparison of CHAMP climate records 2002 to 2008 from six data centers. Journal of Geophysical Research: Atmospheres, 117(D18), n/a. https://doi.org/10.1029/2012jd017665, 2012.

Hocke, K., and Tsuda, T.: Application of GPS radio occultation data for studies of atmospheric waves in the middle atmosphere and ionosphere. Journal of the Meteorological Society of Japan, 82(1B), 419–426, 2004.

Holton, J. R.: An Introduction to Dynamic Meteorology, Academic Press, New York, 1994.

Holton, J. R.: An Introduction to Dynamic Meteorology (fifth edition), Academic Press, 2013.

Holton, J. R., Haynes, P. H., McIntyre, M. E., Douglass, A. R., Rood, R. B., and Pfister, L.: Stratosphere-troposphere exchange. Reviews of Geophysics, 33(4), 403–439, doi:10.1029/95RG02097, 1995.

Hu, Y., and Fu, Q.: Observed poleward expansion of the Hadley circulation since 1979. Atmospheric Chemistry and Physics, 7(19), 5229–5236. https://doi.org/10.5194/acp-7-5229-2007, 2007.

Hu, Y., Zhou, C., and Liu, J.: Observational evidence for poleward expansion of the Hadley circulation. Advances in Atmospheric Sciences, 28(1), 33–44. https://doi.org/10.1007/s00376-010-0032-1, 2010.

Hudson, R. D.: Measurements of the movement of the jet streams at mid-latitudes, in the Northern and Southern Hemispheres, 1979 to 2010. Atmospheric Chemistry and Physics, 12(16), 7797–7808. https://doi.org/10.5194/acp-12-7797-2012, 2012.

Hudson, R. D., Andrade, M. F., Follette, M. B., and Frolov, A. D.: The total ozone field separated into meteorological regimes – Part II: Northern Hemisphere mid-latitude total ozone trends. Atmospheric Chemistry and Physics, 6(12), 5183–5191. https://doi.org/10.5194/acp-6-5183-2006, 2006.

Hudson, R. D., Frolov, A. D., Andrade, M. F., and Follette, M. B.: The total ozone field separated into meteorological regimes. Part I: Defining the Regimes. Journal of the Atmospheric Sciences, 60(14), 1669–1677, 2003.

Jacobson, A. R., Schuldt, K. N., Miller, J. B., Oda, T., Tans, P., Arlyn Andrews, Mund, J., Ott, L., Collatz, G. J., Aalto, T., Afshar, S., Aikin, K., Aoki, S., Apadula, F., Baier, B., Bergamaschi, P., Beyersdorf, A., Biraud, S. C., Bollenbacher, A., . . . Miroslaw Zimnoch.: CarbonTracker CT2019B. NOAA Global Monitoring Laboratory. https://doi.org/10.25925/20201008, 2020.

Jin, S., Cardellach, E., and Xie, F.: GNSS Remote Sensing: Theory, Methods and Applications (Remote Sensing and Digital Image Processing, 19) (2014th ed.). Springer, 2013.

Jin, S. G., Han, L., and Cho, J.: Lower atmospheric anomalies following the 2008 Wenchuan Earthquake observed by GPS measurements. JJournal of Atmospheric and Solar-Terrestrial Physics, 73(7–8), 810–814, doi: 10.1016/j.jastp.2011.01.023, 2011.

Jin, S. G., Jin, R., and Kutoglu, H.: Positive and negative ionospheric responses to the March 2015 geomagnetic storm from BDS observations. Journal of Geodesy, 91(6), 613–626, doi: 10.1007/s00190-016-0988-4, 2017.

Jin, S. G., and Park, P.: Strain accumulation in South Korea inferred from GPS measurements, Earth Planets Space, 58(5), 529–534, doi: 10.1186/BF03351950, 2006.

Jin, S. G., and Zhang, T.: Terrestrial water storage anomalies associated with drought in Southwestern USA derived from GPS observations. Surveys in Geophysics, 37(6), 1139–1156, doi: 10.1007/s10712-016-9385-z, 2016.

Johanson, C. M., and Fu, Q.: Hadley cell widening: Model simulations versus observations. Journal of Climate, 22(10), 2713–2725. https://doi.org/10.1175/2008jcli2620.1, 2009.

Kedzierski, R., Matthes, K., and Bumke, K.: New insights into Rossby wave packet properties in the extratropical UTLS using GNSS radio occultations. 10.5194/acp-2020–124, 2020.

Kunz, A., Konopka, P., Müller, R., and Pan, L. L.: Dynamical tropopause based on isentropic potential vorticity gradients. Journal of Geophysical Research, 116(D1). https://doi.org/10.1029/2010jd014343, 2011.

Kursinski, E. R., Hajj, A. G., Hardy, R. K., Schofield, T. J., and Linfield, R.: Observing the Earth's atmosphere with radio occultation measurements using the Global Positioning System, Journal of Geophysical Research, 102, 23,429–23,465, doi: 10.1029/97JD01569, 1997.

Lee, S., and Kim, H.: The Dynamical Relationship between Subtropical and Eddy-Driven Jets. Journal of the Atmospheric Sciences, 60, 1490–1503, 2003.

Leroy, S. S., Anderson, J. G., and Dykema, J. A.: Testing climate models using GPS radio occultation: A sensitivity analysis, Journal of Geophysical Research, 111, https://doi.org/10.1029/2005jd006145, 2006.

Li, W., Yuan, Y. B., Chai, Y. J., Liou, Y. A., Ou, J. K., and Zhong, S. M.: Characteristics of the global thermal tropopause derived from multiple radio occultation measurements. Atmospheric Research, 185, 142–157. https://doi.org/10.1016/j.atmosres.2016.09.013, 2017.

Lu, J., Deser, C., and Reichler, T.: Cause of the widening of the tropical belt since 1958. Geophysical Research Letters, 36(3), n/a-n/a. https://doi.org/10.1029/2008gl036076, 2009.

Lu, J., Vecchi, G. A., and Reichler, T.: Expansion of the Hadley cell under global warming. Geophysical Research Letters, 34, L06805, doi:10.1029/2006GL028443, 2007.

Lucas, C., Timbal, B., and Nguyen, H.: The expanding tropics: a critical assessment of the observational and modeling studies. WIREs Climate Change, 5(1), 89–112. https://doi.org/10.1002/wcc.251, 2014.

Marquardt, C., and Healy, B. S.: Measurement noise and stratospheric gravity wave characteristics obtained from GPS occultation data. Journal of the Meteorological Society of Japan, 83(3), 417–428, 2005.

Marshall, J., and Plumb R. A.: Atmosphere, Ocean and Climate Dynamics. USA: Elesvier Academic Press, 2008.

Meehl, G. A. et al.: In Climate Change 2007. The Physical Science Basis. Contribution of Working Group I to the Fourth Assessment Report of the Intergovernmental Panel on Climate Change (eds Solomon, S. et al.), Cambridge University Press, Cambridge, UK, 2007.

Meng, L., Liu, J., Tarasick, D. W., Randel, W. J., Steiner, A. K., Wilhelmsen, H., Wang, L., and Haimberger, L.: Continuous rise of the tropopause in the Northern Hemisphere over 1980–2020. Science Advances, 7(45). https://doi.org/10.1126/sciadv.abi8065, 2021.

Mohd Zali, R., and Mandeep, J. S.: The tropopause height analysis in equatorial region through the GPS-RO. E3S Web of Conferences, 76, 04002. https://doi.org/10.1051/e3sconf/20197604002, 2019.

Munchak, L. A., and Pan, L. L.: Separation of the lapse rate and the cold point tropopauses in the tropics and the resulting impact on cloud top-tropopause relationships. Journal of Geophysical Research: Atmospheres, 119(13), 7963–7978. https://doi.org/10.1002/2013jd021189, 2014.

Oscar: https://www.wmo-sat.info/oscar/gapanalyses?mission=9, last accessed on 15 August 2020.

Pan, L. L., Randel, W. J., Gary, B. L., Mahoney, M. J., and Hintsa, E. J.: Definitions and sharpness of the extratropical tropopause: A trace gas perspective. Journal of Geophysical Research: Atmospheres, 109(D23). https://doi.org/10.1029/2004jd004982, 2004.

Pisoft, P., Sacha, P., Polvani, L. M., Añel, J. A., de la Torre, L., Eichinger, R., Foelsche, U., Huszar, P., Jacobi, C., Karlicky, J., Kuchar, A., Miksovsky, J., Zak, M., and Rieder, H. E.: Stratospheric contraction caused by increasing greenhouse gases. Environmental Research Letters, 16(6), 064038. https://doi.org/10.1088/1748-9326/abfe2b, 2021.

Reed, R. J.: A study of a characteristic type of upper-level frontogenesis. Journal of Meteorological Research, 12, 226–237. DOI: 10.1175/1520–0469(1955)012<0226:ASOACT>2.0.CO;2, 1955.

Santer, B. D., Sausen, R., Wigley, T. M. L., Boyle, J. S., AchutaRao, K., Doutriaux, C., Hansen, J. E., Meehl, G. A., Roeckner, E., Ruedy, R., Schmidt, G., and Taylor, K. E.: Behavior of tropopause height and atmospheric temperature in models, reanalyses, and observations: Decadal changes, Journal of Geophysical Research, 108, D14002, doi:10.1029/2002JD002258, 2003.

Santer, B. D., Wigley, T. M. L., Simmons, A. J., Kållberg, P. W., Kelly, G. A., Uppala, S. M., Ammann, C., Boyle, J. S., Brüggemann, W., Doutriaux, C., Fiorino, M., Mears, C., Meehl, G. A., Sausen, R., Taylor, K. E., Washington, W. M., Wehner, M. F., and Wentz, F. J.: Identification of anthropogenic climate change using a second-generation reanalysis. Journal of Geophysical Research: Atmospheres, 109(D21), n/a. https://doi.org/10.1029/2004jd005075, 2004.

Sausen, R., and Santer, B. D.: Use of changes in tropopause height to detect human influences on climate. Meteorologische Zeitschrift, 12, 131–136, doi:10.1127/0941-2948/2003/0012-0131, 2003.

Scherllin-Pirscher, B., Deser, C., Ho, -P. S., Chou, C., and Randel, W., Kuo, -H. Y.: The vertical and spatial structure of ENSO in the upper troposphere and lower stratosphere from GPS radio occultation measurements, Geophysical Research Letters, 39(L20801), 1–6, 2012.

Scherllin-Pirscher, B., Steiner, A. K., Anthes, R. A., Alexander, M. J., Alexander, S. P., Biondi, R., Birner, T., Kim, J., Randel, W. J., Son, S. W., Tsuda, T., and Zeng, Z.: Tropical temperature variability in the UTLS: New insights from GPS radio occultation observations. Journal of Climate, 34(8), 2813–2838. https://doi.org/10.1175/jcli-d-20-0385.1, 2021.

Schmidt, T., Alexander, P., and de la Torre.: A Stratospheric gravity wave momentum flux from radio occultations. Journal of Geophysical Research, 121, 9, 4443–4467. DOI: http://doi.org/10.1002/2015JD024135., 2016.

Schmidt, T., Wickert, J., Beyerle, G., and Heise, S.: Global tropopause height trends estimated from GPS radio occultation data. Geophysical Research Letters, 35(11). https://doi.org/10.1029/2008gl034012, 2008.

Schmidt, T., Wickert, J., Beyerle, G., and Reigber, C.: Tropical tropopause parameters derived from GPS radio occultation measurements with CHAMP. Journal of Geophysical Research: Atmospheres, 109(D13), n/a. https://doi.org/10.1029/2004jd004566, 2004.

Seidel, D. J., Fu, Q., Randel, W. J., and Reichler, T. J.: Widening of the tropical belt in a changing climate. Nature Geoscience, 1(1), 21–24. https://doi.org/10.1038/ngeo.2007.38, 2007.

Seidel, D. J., and Randel, W. J.: Variability and trends in the global tropopause estimated from radiosonde data. Journal of Geophysical Research, 111(D21). https://doi.org/10.1029/2006jd007363, 2006.

Seidel, D. J. and Randel, W. J.: Recent widening of the tropical belt: Evidence from tropopause observations, Journal of Geophysical Research, 112, D20113, https://doi.org/10.1029/2007jd008861, 2007

Shangguan, M., Wang, W., and Jin, S.: Variability of temperature and ozone in the upper troposphere and lower stratosphere from multi-satellite observations and reanalysis data. Atmospheric Chemistry and Physics, 19(10), 6659–6679. https://doi.org/10.5194/acp-19-6659-2019, 2019.

Smith, E., and Weintraub, S.: The constants in the equation for atmospheric refractive index at 20 radio frequencies, Proceedings of the IRE, 41, 1035–1037, 1953.

Son, S. W., Tandon, N. F., and Polvani, L. M.: The fine-scale structure of the global tropopause derived from COSMIC GPS radio occultation measurements. Journal of Geophysical Research, 116(D20). https://doi.org/10.1029/2011jd016030, 2011.

Stachnik, J. P., and Schumacher, C.: A comparison of the Hadley circulation in modern reanalyses. Journal of Geophysical Research: Atmospheres, 116(D22), n/a-n/a. https://doi.org/10.1029/2011jd016677, 2011.

Staten, P. W., Lu, J., Grise, K. M., Davis, S. M., and Birner, T.: Re-examining tropical expansion. Nature Climate Change, 8(9), 768–775. https://doi.org/10.1038/s41558-018-0246-2, 2018.

Steiner, A. K., Kirchengast, G., and Ladreiter, P. H.: Inversion, error analysis, and validation of GPS/MET occultation data. Annales Geophysicae, 17, 122–138, 1999.

Steiner, A. K., Lackner, B. C., Ladstädter, F., Scherllin-Pirscher, B., Foelsche, U., and Kirchengast, G.: GPS radio occultation for climate monitoring and change detection. Radio Science, 46(6). https://doi.org/10.1029/2010rs004614, 2011.

Steiner, K. A., Hunt, D., Ho, -P. S., Kirchengast, G., Mannucci Mannucci, J. A., Scherllin-Pirscher, B., Gleisner, H., von Engeln, A., Schmidt, T., Ao, C., Leroy, S. S., Kursinski, R. E., Foelsche, U., Gorbunov, M., Heise, S., Kuo, -H. Y., Lauritsen, B. K., Marquardt, C., Rocken, C., Schreiner, W., Sokolovskiy, S., Syndergaard, S., and Wickert, J.: Quantification of structural uncertainty in climate data records from GPS radio occultation, Atmospheric Chemistry and Physics, 13, 1469–1484, doi: 10.5194/acp-13-1469-2013, 2013.

Strong, C., and Davis, R. E.: Temperature-related trends in the vertical position of the summer upper tropospheric surface of maximum wind over the Northern Hemisphere. International Journal of Climatology, 26(14), 1977–1997. https://doi.org/10.1002/joc.1344, 2006.

Sun, Y., Liu, C., Tian, Y., Liu, C., Li, W., Zhao, D., Li, F., Qiao, H., Wang, X., Du, Q., Bai, W., Xia, J., Cai, Y., Wang, D., Wu, C., and Meng, X.: The status and progress of Fengyun-3e GNOS II mission for GNSS. Remote Sensing, 5181–5184.10.1109/IGARSS.2019.8899319, 2019.

Tegtmeier, S., Anstey, J., Davis, S., Dragani, R., Harada, Y., Ivanciu, L., Kedzierski, R., Krüger, K., Legras, B., Long, C., Wang, J., Wargan, K., and Wright, J.: Temperature and tropopause characteristics from reanalyses data in the tropical tropopause layer. Atmospheric Chemistry and Physics, 20, 753–770.10.5194/acp-20–753–2020, 2020.

Thuburn, J., and Craig, G. C.: GCM tests of theories for the height of the tropopause, J. Atmos. Sci., 54, 869–882, 1997.

Thuburn, J., and Craig, G. C.: Stratospheric influence on tropopause height: The radiative constraint, Journal of the Atmospheric Sciences, 57, 17–28, 2000.

Vicente-Serrano, S. M., Beguería, S., and López-Moreno, J. I.: A multiscalar drought index sensitive to global warming: The standardized precipitation evapotranspiration index. Journal of Climate, 23(7), 1696–1718. https://doi.org/10.1175/2009jcli2909.1, 2010.

Waliser, D. E., Shi, Z., Lanzante, J. R., and Oort, A. H.: The Hadley circulation: assessing NCEP/NCAR reanalysis and sparse in-situ estimates. Climate Dynamics, 15(10), 719–735. https://doi.org/10.1007/s003820050312, 1999.

Watt-Meyer, O., Frierson, D. M. W., and Fu, Q.: Hemispheric asymmetry of tropical expansion under CO_2 forcing. Geophysical Research Letters, 46(15), 9231–9240. https://doi.org/10.1029/2019gl083695, 2019.

Wickert, J., Galas, R., Beyerle, G., König, R., and Reigber, C.: GPS ground station data for CHAMP radio occultation measurements. Physics and Chemistry of the Earth, Part A: Solid Earth and Geodesy, 26(6–8), 503–511. https://doi.org/10.1016/s1464-1895(01)00092-8, 2001a.

Wickert, J., Michalak, G., Schmidt, T., Beyerle, G., Cheng, C. Z., Healy, S. B., Heise, S., Huang, C. Y., Jakowski, N., Köhler, W., Mayer, C., Offiler, D., Ozawa, E., Pavelyev, A. G., Rothacher, M., Tapley, B., and Köhler, C.: GPS radio occultation: Results from CHAMP, GRACE and FORMOSAT-3/COSMIC. Terrestrial, Atmospheric and Oceanic Sciences, 20(1), 35. https://doi.org/10.3319/tao.2007.12.26.01(f3c), 2009.

Wickert, J., Reigber, C., Beyerle, G., König, R., Marquardt, C., Schmidt, T., Grunwaldt, L., Galas, R., Meehan, T. K., Melbourne, W. G., and Hocke, K.: Atmosphere sounding by GPS radio occultation: First results from CHAMP. Geophysical Research Letters, 28(17), 3263–3266. https://doi.org/10.1029/2001gl013117, 2001b.

Wickert, J., Schmidt, T., Beyerle, G., König, R., Reigber, C., and Jakowski, N.: The radio occultation experiment aboard CHAMP: Operational data analysis and validation of vertical atmospheric profiles. Journal of the Meteorological Society of Japan. Ser. II, 82(1B), 381–395. https://doi.org/10.2151/jmsj.2004.381, 2004.

WMO: Meteorology—A three dimensional science: Second session of the commission for aerology. Geneva: World Meteorological Organization (WMO). (WMO Bull. No. 4), 1957.

Wu, X. R., and Jin, S. G.: GNSS-Reflectometry: Forest canopies polarization scattering properties and modeling. Advances in Space Research, 54(5), 863–870, doi: 10.1016/j.asr.2014.02.007, 2014.

Xia, P., Ye, S., Jiang, K., and Chen, D.: Estimation and evaluation of COSMIC radio occultation excess phase using undifferenced measurements. Atmospheric Measurement Techniques, 10(5), 1813–1821. https://doi.org/10.5194/amt-10-1813-2017, 2017.

Xian, T., Lu, G., Zhang, H., Wang, Y., Xiong, S., Yi, Q., Yang, J., and Lyu, F.: Implications of GNSS-Inferred Tropopause Altitude Associated with Terrestrial Gamma-ray Flashes. Remote Sensing, 13(10), 1939. https://doi.org/10.3390/rs13101939, 2021.

Yang, H., Lohmann, G., Krebs-Kanzow, U., Ionita, M., Shi, X., Sidorenko, D., Gong, X., Chen, X., and Gowan, E. J.: Poleward shift of the major ocean gyres detected in a warming climate. Geophysical Research Letters, 47(5). https://doi.org/10.1029/2019gl085868, 2020a.

Yang, H., Lohmann, G., Lu, J., Gowan, E. J., Shi, X., Liu, J., & Wang, Q.: Tropical expansion driven by poleward advancing midlatitude meridional temperature gradients. Journal of Geophysical Research: Atmospheres, 125, e2020JD033158. https://doi.org/10.1029/2020JD033158, 2020b.

Yang, H., Lu, J., Wang, Q., Shi, X., & Lohmann, G.: Decoding the dynamics of poleward shifting climate zones using aqua-planet model simulations. Climate Dynamics, 58(11–12), 3513–3526. https://doi.org/10.1007/s00382-021-06112-0, 2022.

Zahn, A., Brenninkmeijer, C. A. M., and van Velthoven, P. F. J.: Passenger aircraft project CARIBIC 1997–2002, Part I: the extratropical chemical tropopause. Atmospheric Chemistry and Physics Discussions, v.4, 1091–1117 (2004). 4.10.5194/acpd-4-1091–2004, 2004.

Zhou, Y. P., Xu, K. M., Sud, Y. C., and Betts, A. K.: Recent trends of the tropical hydrological cycle inferred from Global Precipitation Climatology Project and International Satellite Cloud Climatology Project data. Journal of Geophysical Research, 116(D9). https://doi.org/10.1029/2010jd015197, 2011.

Zus, F., Beyerle, G., Heise, S., Schmidt, T., and Wickert, J.: GPS radio occultation with TerraSAR-X and TanDEM-X: sensitivity of lower troposphere sounding to the Open-Loop Doppler model, Atmos. Meas. Tech. Discuss, doi: 10.5194/amtd-7-12719-2014, 2014.

8 Upper Atmospheric Mass Density Variations and Space Weather Responses from GNSS Precise Orbits

Andres Calabia and Shuanggen Jin

8.1 INTRODUCTION

Monitoring and understanding geophysical processes in the earth's thermosphere is primordial for applications such as low earth orbit (LEO) satellite tracking and upper-atmosphere research. Unfortunately, the existing models are incapable of predicting the variability as required for practical operations, largely due to the limited quality and quantity of observations, and the lack of comprehensive approaches for calibrating the models. Nowadays, accelerometers onboard LEO artificial satellites can measure non-gravitational accelerations and derive thermospheric mass density and wind velocity estimates with unprecedented detail, but the high costs and technical issues of the dedicated space missions have only provided a very limited amount of data. Fortunately, with the increasing number of LEO satellites equipped with high-precision Global Navigation Satellite System (GNSS) receivers, precise orbit products could be used to obtain comparable data sets. These new density estimates from LEO GNSS are a promising data source that can help to better characterize the upper atmosphere and to improve the existing models.

The global nature of satellite measurements can increase the understanding of the whole upper atmosphere system, and the combination of physical models can help to identify unknown mechanisms. Unfortunately, the existing quality and quantity of measurement data is limited, and it is difficult to calibrate the models. This results in satellite tracking errors far too great to meet the operational requirements (e.g., Anderson et al. [2009]; Calabia et al. [2020]). Therefore, the present upper atmosphere models are unable to predict the variability as accurately and efficiently as required, and the resulting processes from geomagnetic storms, solar flares, and solar wind are still not well understood. During the last decade, mass density changes with variable geophysical conditions have been investigated with satellite measurements to a great extent (e.g., Müller et al. [2009]; Sutton et al. [2009]; Liu et al. [2011, 2010]; Chen et al. [2014]; Cnossen and Förster [2016]; Calabia and Jin [2016a]; Calabia and Jin [2019]). These studies have exposed the limitations of the existing models to predict neutral density variability. For instance, upper atmosphere empirical models have been improving with the use of new techniques, algorithms, and proxies (e.g., Calabia et al. [2020]; Bowman et al. [2008]; Bruinsma [2015]; Picone et al. [2002]). These empirical models are simple and suitable for routine applications such as Precise Orbit Determination (POD). In many applications, for example remote sensing or satellite altimetry and gravity, a dynamic POD scheme estimates the orbital position and velocity of a satellite with an accuracy of a few millimeters. This is achieved by propagating the orbital trajectory through a double integration and linearization of the Newton-Euler's equation of motion (Montenbruck and Gill [2013]). In addition, by combining the dynamic POD with empirical observations, for example, laser-ranging, Doppler, accelerometer, or GNSS measurements, the position and velocity can be stochastically estimated with high accuracy (Tapley et al., [2004]). The main force-models involved

DOI: 10.1201/9781003363118-8

in POD include the variable gravity field, the atmospheric drag, and the solar and Earth-albedo radiation pressures. Among these, the variable atmospheric drag strongly harms the lifetime of LEO missions. Atmospheric drag mainly depends on atmospheric expansion/contraction driven by the solar and geomagnetic activity. Increases in atmospheric drag reduce the orbital velocity of a satellite and its nominal altitude, and shortens its lifetime. For instance, the contribution of atmospheric drag pressure to the position of a satellite orbiting at an altitude of around 450 km may drag around 3 m per revolution in the along-track axis, limiting the satellite's lifetime to about 5–10 years.

Atmospheric drag effects in LEO satellites are highly variable and not well modeled, resulting in large difficulties for orbital tracking, collision analysis, and re-entry calculations. Accelerometers onboard satellites can provide accurate and globally distributed atmospheric drag data, but these are exposed to high instrumental costs and other technical issues [Bruinsma et al., 2004; Doornbos et al., 2010; Calabia et al., 2015; Siemes et al., 2016; Calabia and Jin, 2017; Jin et al., 2018]. These limitations have forced these payloads to be carried by only few satellites, viz. Gravity Recovery and Climate Experiment (GRACE), Challenging Mini-Satellite Payload (CHAMP), Gravity Field and Steady-State Ocean Circulation Explorer (GOCE), and Swarm. Other techniques also have other specific drawbacks, including problems with accuracy, resolution, coverage, calibration, complexity, etc. Besides high-precision accelerometers onboard satellites (Marcos and Forbes, 1985; Bruinsma et al., 2004), other measurement techniques include the semi-major axis variation (Picone et al., 2005), the stochastic mass density estimation within the POD scheme (Ijssel And Visser, 2007; McLaughlin et al., 2013; Visser et al., 2013; Kuang et al., 2014), mass spectrometers (Tang et al., 2020), incoherent scatter radars (Nicolls et al., 2014), Broglio drag balance instruments (Santoni et al., 2010), miniaturized pressure gauge instruments (Clemmons et al., 2008), ultraviolet remote sensing (Meier and Picone, 1994), and the techniques of atmospheric occultation (Determan et al., 2007; Aikin et al., 1993).

8.2   UPPER ATMOSPHERE AND OBSERVATIONS

Earth's upper atmosphere is highly variable in space and time, and its physical processes and coupled mechanisms are very complex and still not well understood. The upper atmosphere is mainly composed of two parts, the thermosphere and the ionosphere. In the thermosphere, photoabsorption, photoionization, and photodissociation of molecules through extreme ultraviolet radiation (EUV) create the ions of the ionosphere, and thermal energy transfer from ions to neutrals drives the regular dynamics during solar-flux, diurnal, and annual cycles. In addition, waves from the lower atmosphere including atmospheric tides and planetary waves can feed into ionospheric electrodynamics, and consequently to the magnetosphere-thermosphere-ionosphere system. Figure 8.1 shows a schematic representation of the known processes in the coupled magnetosphere-thermosphere-ionosphere system.

Solar flares and geomagnetic storms can produce thermospheric Joule heating and particle precipitation along the earth's magnetic field lines, generating short-term and abrupt changes in the coupled system, compared to that given by the regular dynamics. Solar flares are ejections of clouds of electrons, ions, and atoms through the corona of the sun into space. In the thermosphere, solar flares increase the X-ray and EUV irradiance, causing immediate energy absorption, ionization, and dissociation of molecules. Some of the effects result in rapid thermospheric Joule heating and particle precipitation along the earth's magnetic field lines.

Unfortunately, the existing quality and quantity of mass density observations is still very limited, and the exact understanding of solar-terrestrial coupling processes and solar-wind/magnetosphere effects that lead to Joule heating enhancement and disturbances to different parameters, including, e.g. composition, temperature, ionospheric plasma, mass density, and wind velocity, is a big challenge (e.g., Heelis and Maute [2020], Maute et al. [2021]). The magnetosphere-ionosphere coupling is crucial in modeling the thermosphere-ionosphere response to geomagnetic activity, and advances

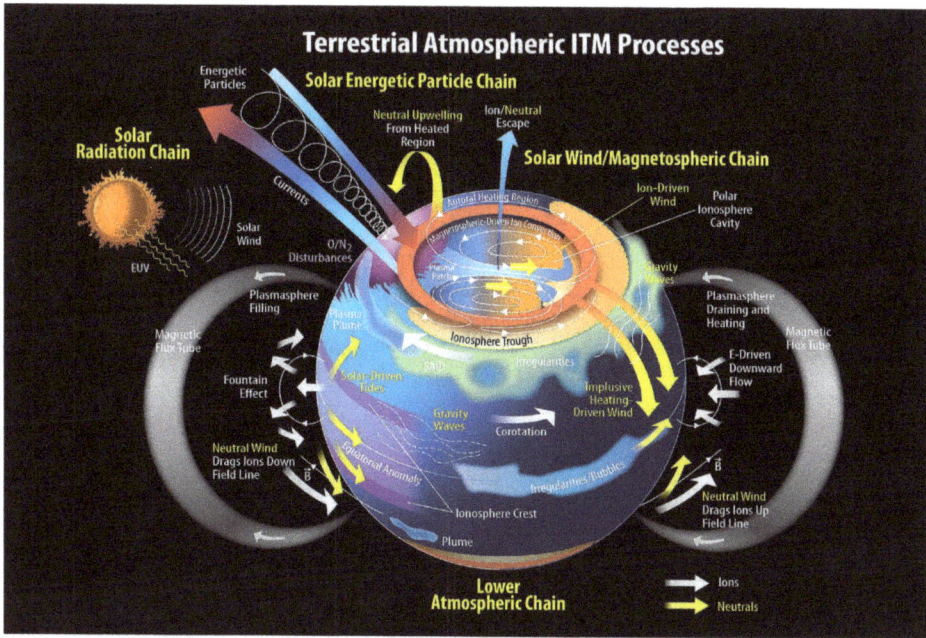

FIGURE 8.1 Known processes in the coupled magnetosphere-thermosphere-ionosphere system.

Source: NASA's Scientific Visualization Studio, https://svs.gsfc.nasa.gov/4641.

in observing thermospheric mass density and wind-velocity along with plasma density and electric currents is crucial to understanding upper-atmosphere processes. Figure 8.2 shows a simplified description of the complex connection between ionospheric plasma and thermospheric mass density and winds.

The exact connection between mass density, wind velocity, ionospheric plasma, and solar and magnetospheric forcing is still unclear, but physics-based models such as Thermosphere-Ionosphere-Electrodynamics General Circulation Model (TIEGCM) (Richmond et al., 1992), which generally use empirical parameterizations and boundary conditions to solve 3-dimensional fluid equations, can provide an approximated solution with clear prognostic variables. Compared to empirical models, which are simple and suitable for routine applications such as dynamic POD and GNSS positioning, the physics-based models are more complex, but can help to better understand the physical mechanisms responsible for the observed variability, and have a greater potential for predictions and projections of future states: for instance, processes of energy absorption, ionization, and dissociation of molecules due to variable X-ray and EUV solar radiance, as well as high-latitude thermospheric Joule heating and precipitation of energetic particles due to solar wind and magnetospheric forcing.

Figure 8.3 shows the satellite missions that have shown capabilities to measure aerodynamic accelerations and subsequently estimate mass density at a high resolution. Currently, an emerging technique based on GNSS precise orbits of LEO satellites has proven the ability to measure high-cadence non-gravitational accelerations and to estimate thermospheric mass density at a high resolution (Calabia and Jin, 2017, 2021a, 2021b; Li and Lei, 2020; Yuan et al., 2019; Calabia et al., 2015). With the increasing number of LEO satellites being equipped with high-precision GNSS receivers and more enhanced data processing and orbit determination strategies, GNSS-based mass density may become an essential data source to effectively monitor global thermosphere fluctuations at a high resolution. Here, in this paper, a detailed review of the current state of the art on thermosphere monitoring with GNSS precise orbits is presented.

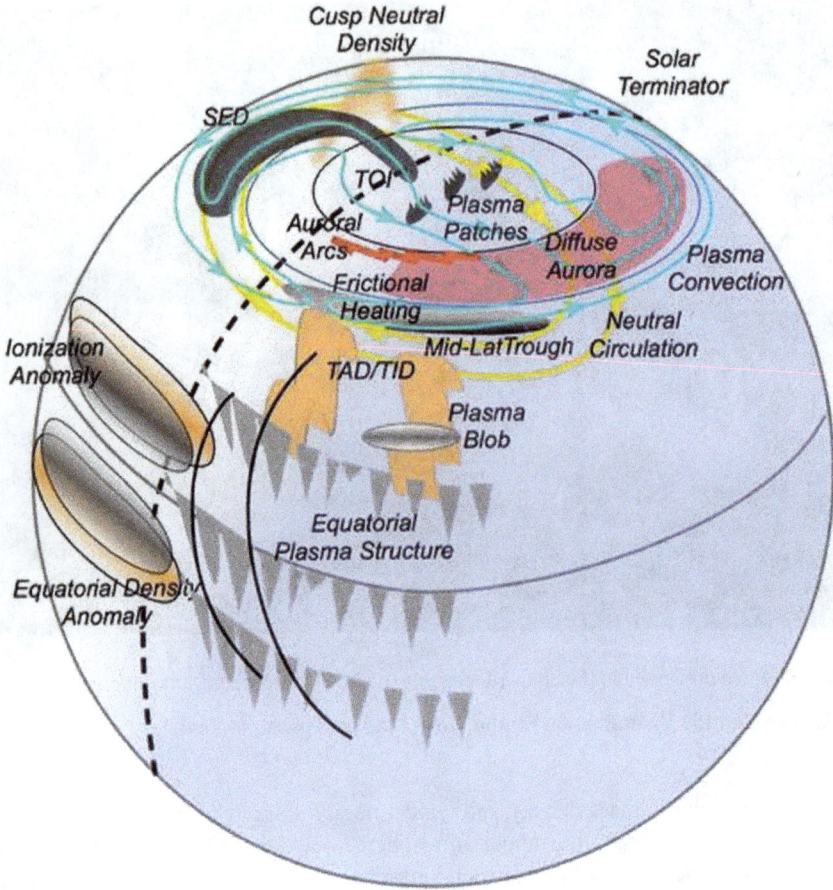

FIGURE 8.2 A simplified description of the connection between ionosphere and thermosphere.

Source: Heelis and Maute [2020].

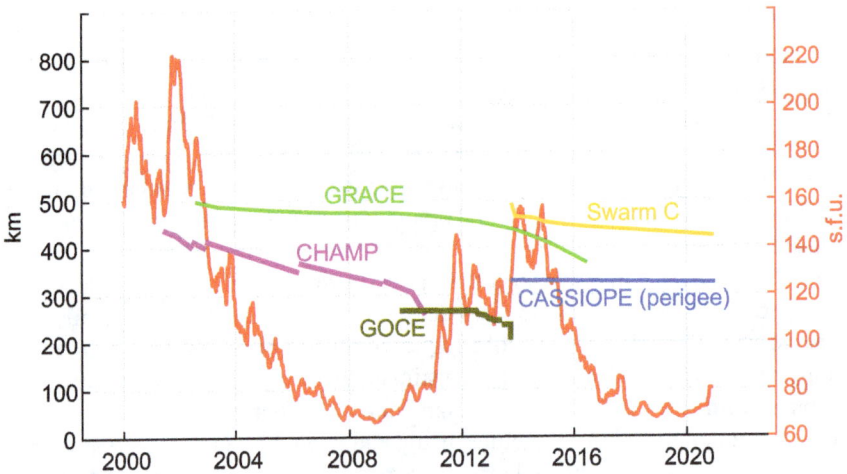

FIGURE 8.3 Orbital altitudes (left axis) of the satellites capable of estimating high-resolution thermospheric mass densities. CASSIOPE does not contain an accelerometer. The 81-day mean solar flux F10.7 index is shown in red (right axis).

8.3 THEORY AND METHODOLOGY

Geodetic grade GNSS position and velocity products from LEO spacecraft can be used to obtain high-precision mass density estimates. Firstly, GNSS-based total accelerations are retrieved through numerical interpolation of the precise orbit velocity. Subsequently, mass density and wind velocity estimates are retrieved from atmospheric drag forces, which are obtained by removing gravitational and radiation-pressure force-models from total accelerations.

8.3.1 Atmospheric Drag Accelerations

Instantaneous total accelerations (a_T) are calculated through numerical differentiation of the GNSS-based precise orbit velocities. The 8-data point piece-wise Lagrange interpolation and a time interval of 0.05 s is recommended for the numerical differentiation (Calabia et al., 2015). At LEO altitudes, these settings allow the obtainment of an unbiased accuracy in the arc-to-chord threshold approach of approximately 10^{-9} m/s². Then, the instantaneous GNSS-based atmospheric drag accelerations (a_D) are retrieved by removing the gravitational (g) and radiation pressure (a_R) accelerations.

$$a_D = a_T - g - a_R \qquad (8.1)$$

The earth's gravitational acceleration is obtained from the combination of the EGM2008 gravity field model with the underlying background for the secular variations (Petit and Luzum, 2010). The geopotential field V in geocentric coordinates (r, φ, λ) is expanded in spherical harmonics with up to degree N as:

$$V(r,\varphi,\lambda) = \frac{GM_{Earth}}{r} \sum_{n=0}^{N} \left(\frac{a_e}{r}\right)^n \sum_{m=0}^{n} \left[\bar{C}_{nm} \cos(m\lambda) + \bar{S}_{nm} \sin(m\lambda) \right] \bar{P}_{nm}(\sin\varphi) \qquad (8.2)$$

where GM_{Earth} and a_e EGM2008 values (398600.4415 km³/s² and 6378136.3 m respectively) should be used as scaling parameters with its gravitational potential coefficients. In this equation, $\bar{P}_{lm}$ are the normalized associated Legendre functions of degree l and order m, and $\bar{C}_{lm}$ and $\bar{S}_{lm}$ are the normalized Stokes' coefficients of degree l and order m for cosinus and sinus, respectively. In order to use the conventional static gravitational field properly and projected it in time, the secular low degree variations of its $\bar{C}_{20}$, $\bar{C}_{21}$, $\bar{S}_{21}$, $\bar{C}_{30}$, and $\bar{C}_{40}$ coefficients need to be accounted for (Petit and Luzum, 2010).

The gravitational acceleration of a third body (Montenbruck and Gill, 2013) can be described as a difference between the accelerations of the satellite and the earth caused by a third body B:

$$\Delta\ddot{r}_{sat} = GM_B \left(\frac{r_B - r_{sat}}{|r_B - r_{sat}|^3} - \frac{r_B}{|r_B|^3} \right) \qquad (8.3)$$

where r_{sat} and r_B are the geocentric coordinates of the satellite and of a third body of mass M_B. Since accelerations on near-Earth satellites from other planets' actions are relatively small (< 0.1 nm/s²), only lunisolar accelerations can be calculated. Moon and sun coordinates can be interpolated from the solar and planetary ephemerides (DE-421) provided by the Jet Propulsion Laboratory (JPL) in the form of Chebyshev approximations. The evaluation of these polynomials yields Cartesian coordinates in the ICRS for the earth-moon barycenter $b_{Earth,Moon}$ and the sun b_{Sun} with respect to the barycenter of the solar system, while moon positions r_{Moon} are given with respect to the center of the earth. The geocentric position of the sun can be computed as:

$$r_{Sun} = b_{Sun} - b_{Earth,Moon} + \frac{r_{Moon}}{1 + \mu^*} \qquad (8.4)$$

where μ^* denotes the ratio of the earth's and the moon's masses.

Since the changes induced by the earth's solid tides due to its rotation under effects of ellipticity and Coriolis force, can be described in terms of the Love numbers, variations in the low-degree Stokes' coefficients can be easily computed (Petit and Luzum, 2010). Dependent and independent frequency corrections are calculated using lunar and solar ephemerides, the Doodson's fundamental arguments, the nominal values of the earth's solid tide external potential Love numbers, and the in-phase and out-of-phase amplitudes of the corrections for frequency-dependent Love values. To account for the dynamical effects of ocean tides, the periodic variations in the normalized Stokes' coefficients are calculated based on the most recent ocean tide model EOT11a (Mayer-Gürr et al., 2012). Mayer-Gürr et al. (2012) also provided the influences of additional minor tide constituents that are not included in the tide model EOT11a and should not be neglected in LEO. Changes in the geopotential value due to the centrifugal effect of pole motion, known as the earth's solid pole tides, can be readily computed in function of the wobble variables and calculated under sub-daily polar motion variations (Petit and Luzum, 2010). The ocean pole tide, generated by the centrifugal effect of pole motion on the oceans, is calculated as a function of sub-daily wobble variables from the coefficients of the self-consistent equilibrium model given by Desai (2002). For the relativistic corrections, only the main effects are calculated (described by the Schwarzschild field of the earth itself, of approximately 16.5 nm/s²), since the effects of the Lense-Thirring precession (frame-dragging) and the geodesic (de Sitter) precession are two orders of magnitude smaller at a near-Earth satellite orbit (Petit and Luzum, 2010):

$$\Delta \ddot{r}_{sat} = \frac{GM_{Earth}}{c^2 r_{sat}^3} \left\{ \left[4 \frac{GM_{Earth}}{r_{sat}} - \left(\dot{r}_{sat} \cdot \dot{r}_{sat} \right) \right] \cdot r_{sat} + 4 \left(r_{sat} \cdot \dot{r}_{sat} \right) \cdot \dot{r}_{sat} \right\} \tag{8.5}$$

For a LEO satellite, time-varying Stokes' coefficients up to a degree and order 120 should be computed at least, and using an increment of time small enough to desensitize from discontinuities (~400 s). Then, the acceleration due to the variable gravity field can be calculated by using the first derivative of the gravitational potential in Cartesian coordinates. With the substitution of $\bar{P}_{nm} = \bar{P}_{nm}(\sin\varphi)$ and $\bar{P}'_{nm} = \partial \bar{P}_{nm}(\sin\varphi) / \partial\varphi$, the first derivative of the gravitational potential of the Earth in spherical coordinates is calculated as:

$$\frac{dU}{dr} = \frac{-GM_{Earth}}{r^2} \sum_{n=0}^{n_{max}} (n+1) \left(\frac{a_e}{r} \right)^n \sum_{m=0}^{n} \bar{P}_{nm} \left(\bar{C}_{nm} \cos m\lambda + \bar{S}_{nm} \sin m\lambda \right)$$

$$\frac{dU}{d\varphi} = \frac{GM_{Earth}}{r} \sum_{n=0}^{n_{max}} \left(\frac{a_e}{r} \right)^n \sum_{m=0}^{n} \bar{P}'_{nm} \left(\bar{C}_{nm} \cos m\lambda + \bar{S}_{nm} \sin m\lambda \right)$$

$$\frac{dU}{d\lambda} = \frac{GM_{Earth}}{r} \sum_{n=0}^{n_{max}} \left(\frac{a_e}{r} \right)^n \sum_{m=0}^{n} m\bar{P}_{nm} \left(\bar{S}_{nm} \cos m\lambda - \bar{C}_{nm} \sin m\lambda \right) \tag{8.6}$$

Then, the acceleration due to the variable gravity field is as follows:

$$\ddot{x} = \frac{\partial U}{\partial x} = \frac{\partial U}{\partial r} \frac{\partial r}{\partial x} + \frac{\partial U}{\partial \varphi} \frac{\partial \varphi}{\partial x} + \frac{\partial U}{\partial \lambda} \frac{\partial \lambda}{\partial x}$$

$$\ddot{y} = \frac{\partial U}{\partial y} = \frac{\partial U}{\partial r} \frac{\partial r}{\partial y} + \frac{\partial U}{\partial \varphi} \frac{\partial \varphi}{\partial y} + \frac{\partial U}{\partial \lambda} \frac{\partial \lambda}{\partial y}$$

$$\ddot{z} = \frac{\partial U}{\partial z} = \frac{\partial U}{\partial r}\frac{\partial r}{\partial z} + \frac{\partial U}{\partial \varphi}\frac{\partial \varphi}{\partial z} \tag{8.7}$$

As for the radiation pressure accelerations acting on the satellite's surfaces, both direct solar radiation and the earth's albedo need to be accounted for. The direct solar radiation (a_{sr}) was formulated by Luthcke et al. [1997]:

$$a_{sr} = \sum_{i=1}^{n_p} -\frac{E_{sr}A_i\,\hat{n}_i\cdot\hat{s}_{sun}^{sat}}{mc}\left[2\left(\frac{c_{rd,i}}{3} + c_{rs,i}\hat{n}_i\cdot\hat{s}_{sun}^{sat}\right)\hat{n}_i + \left(1 - c_{rs,i}\right)\hat{s}_{sun}^{sat}\right] \tag{8.8}$$

In this equation, n_p is the number of plates, A_i is the plate area, c is the speed of light, $c_{rd,i}$ is the coefficient of diffusive reflectivity, $c_{rs,i}$ is the coefficient of specular reflectivity, m is the satellite mass, $\hat{n}_i$ is the unit plate normal, $\hat{s}_{sun}^{sat}$ is the unit sun-satellite vector, and $E_{sr} = sh\cdot 1366\,(1\mathrm{AU}/s_{sun}^{sat})^2$ is the flux on the earth's atmosphere (1366 W/m²), corrected from the yearly period of Earth's orbit eccentricity and from the planetary eclipse ratio (sh) (Montenbruck and Gill, 2013).

Similarly, the earth albedo acceleration (a_{ea}) can be computed as follows:

$$a_{ea} = \sum_{i=1}^{n_p}\sum_{j=1}^{grid} -\frac{E_{ea,j}A_i\,\hat{n}_i\cdot\hat{s}_j^{sat}}{mc}\left[2\left(\frac{c_{rd,i}}{3} + c_{rs,i}\hat{n}_i\cdot\hat{s}_j^{sat}\right)\hat{n}_i + \left(1 - c_{rs,i}\right)\hat{s}_j^{sat}\right] \tag{8.9}$$

In this equation, the parameter $E_{ea,j} = E_{ea}^R + E_{ea}^{IR}$ is the combination of the radiation reflected at the earth's surface E_{ea}^R and the earth's infrared radiation E_{ea}^{IR}. The earth's reflected solar radiation E_{ea}^R at each satellite position is estimated using the monthly averages of NASA's Total Ozone Mapping Spectrometer (TOMS) reflectivity index (σ) [Bhanderi, 2005]:

$$E_{ea,j}^R = f_j v_j E_{sr}\frac{A_j\left(\hat{n}_j\cdot\hat{s}_j^{sun}\right)\left(\hat{n}_j\cdot\hat{s}_j^{sat}\right)\sigma_j}{\pi\left|\hat{s}_j^{sat}\right|^2} \tag{8.10}$$

In this equation, f_j is field of view of the satellite, v_j is the sunlight function, and A_j is the area of each cell j of TOMS. The reflection angle on each cell is defined by the directions of the satellite $\hat{s}_j^{sat}$, the sun $\hat{s}_j^{sun}$, and the cell normal-vector $\hat{n}_j$. Earth's infrared radiation E_{ea}^{IR} can be modeled as a black body with a surface temperature of 288 K, whose spectrum is mainly IR with an exitance of about 239 W/m² (Taylor, 2005). In a similar way, the IR irradiance $E_{ea,j}^{IR}$ from each visible cell j of the earth's surface has been computed as follows:

$$E_{ea,j}^{IR} = f_j 239\left(\frac{1\mathrm{AU}}{s_{sun}^j}\right)^2 e_{IR,j}\frac{A_j\hat{n}_j\cdot\hat{s}_j^{sat}}{\pi\left|\hat{s}_j^{sat}\right|^2} \tag{8.11}$$

In this equation, the Earth IR radiation $e_{IR,j}$ for each cell j was parameterized in terms of latitude and season by Knocke and Ries (1987):

$$e_{IR} = e_0 + e_1 P_1\sin\varphi + e_2 P_2\sin\varphi$$

$$e_l = k_0 + k_1\cos\left[\omega\left(JD - t_0\right)\right] + k_2\sin\left[\omega\left(JD - t_0\right)\right] \tag{8.12}$$

Here, t_0 is the epoch of 22 December 1981, ω is Earth's orbit rotation rate around the sun ($2\pi/365.5$), φ is the equatorial geocentric latitude, JD is the Julian date, P_1 and P_2 are the Legendre polynomials of degrees 1 and 2, respectively, and $e_0 = 0.68$; $e_2 = -0.18$; $k_0 = 0$; $k_1 = -0.07$, and $k_2 = 0$.

8.3.2 THERMOSPHERIC MASS DENSITY ESTIMATION

GNSS-based density estimates are computed using the drag-force (F_D) formula (Newton [1726]):

$$F_D = a_D m = \frac{1}{2} C_D A \rho v_r^2 \tag{8.13}$$

In this equation, a_D is the acceleration due to atmospheric drag (Equation 1), m is the mass of the satellite, C_D is the drag coefficient, ρ is the mass density, and A is the cross-sectional area perpendicular to the relative velocity of the atmosphere with respect to the spacecraft v_r, which includes the co-rotating atmosphere and the horizontal wind velocity (v_{wind}). Horizontal wind velocity is calculated from the horizontal wind model HWM14 (Drob et al., 2015), and the velocity of the co-rotating atmosphere is computed as the vector product between Earth's angular rotation (ω_{Earth}) and the satellite's position vector (r_{sat}):

$$v_r = v_{sat} + \left(\omega_{Earth} \wedge r_{sat}\right) + v_{wind} \tag{8.14}$$

The most important sources of error in GNSS-based mass density retrieval are caused by the uncertainty in the drag coefficient and the errors in the horizontal wind velocity model (e.g., HWM14 [Drob et al., 2015]). Errors due to zonal and meridional wind velocity are estimated at 1%, and 4% per 100 m/s, respectively (Bruinsma et al., 2006), while drag coefficients are expected to differ by ~15%, during solar minimum conditions, and by ~2–3%, during solar maximum (March et al., 2019; Mehta et al., 2014). The uncertainty of the new GNSS-based thermospheric mass density estimates can be assessed through statistical comparisons to the existing data and models:

- The Space Environment Technologies (SET) Air Force High Accuracy Satellite Drag Model (HASDM) (Tobiska et al., 2021).
- Accelerometer data from GRACE, CHAMP, GOCE, or Swarm missions (see Figure 8.4). Note that space accelerometers need bias calibration (Calabia et al., 2015).
- The JB2008 empirical model (Bowman et al., 2008).
- Drag Temperature Model (DTM) (Bruinsma, 2015).
- The Naval Research Laboratory Mass Spectrometer and Incoherent Scatter Radar (NRLMSISE-00) empirical model (Picone et al., 2002).

8.3.3 SOLAR CYCLE AND SECULAR VARIATIONS

An empirical modeling based in the principal component analysis (PCA) technique can overcome the limitation of the sparse nature of the observations. The aim of a PCA technique (Pearson, 1901) is to determine a new set of bases that capture the largest variance in the data, based on eigen value decomposition of the covariance matrix. The starting data sets are provided at different temporal and a spatial resolution; thus, to extract the maximum detail of the initial data sets, the covariance matrix $R = F^t F$ for the eigen value problem is represented as a matrix (F) that represents each location (x) by columns, and each epoch (t) by rows. In this step, the spatial patterns of each variable χ, their changes on time, and the measure of their importance, are presented as a low-dimensional space spanned by a set of modes, each of one represented by a pair of time (Γ_i,) and space (Ω_i,) expansion components, from n locations (x) at m epochs (t).

$$\chi(x_n, t_m) = \sum_{i=1}^{i=n} \left[\Gamma_i(t_m) \cdot \Omega_i(x_n) \right] \tag{8.15}$$

Then, it is possible to solve and estimate the best fit in terms of the most representative proxies and secular variations, including solar and magnetospheric indices, and local solar time (LST) and seasonal cycles. In this scheme, a time-domain spectrum analysis can unveil the characteristic periods of secular variations, and a correlation-delay study can reveal the best possible fit. The power spectral density (PSD) estimate (Fourier transform of the biased estimate of the autocorrelation sequence) can be employed to determine the main periodicities in the time series, and the Pearson's linear correlation coefficient will be estimated sequentially to obtain the time delay between two sequences at maximum correlation. In order to better characterize singular events, the least squares fitting of polynomial and Fourier functions to each time-expansion PCA mode (Γ_i) is preferred. This method minimizes the absolute difference of the residuals (least absolute residuals) in a multi-variable parameterization (solar cycle, LST, seasonal, and geomagnetic index). Firstly, the following 2-variable polynomial expression can be employed to fit the first time-expansion PCA component (Γ_1) as follows:

$$\frac{\Gamma_1(t)}{\Upsilon_1(t)} \approx \Psi_1(t) = p_{00} + p_{10} \cdot FLUX(t + \tau_{FLUX}) + p_{01} \cdot MAG(t + \tau_{MAG}) +$$
$$+ p_{20} \cdot \left[FLUX(t + \tau_{FLUX}) \right]^2 + p_{11} \cdot FLUX(t + \tau_{FLUX}) \cdot MAG(t + \tau_{MAG}) \tag{8.16}$$

In this equation, p_{jk}, with j={0,1,2} and k={0,1}, are the fitting coefficients, t is a given epoch to estimate the model, and $FLUX(t + \tau_{FLUX})$ and $MAG(t + \tau_{MAG})$ are the solar and magnetospheric indices evaluated at a given time t, which need to account for a time-delay for each index, τ_{FLUX} and τ_{MAG} respectively (provided in the correlation-delay analysis). In the first iteration, Υ_1 is approximated to 1, and then it is updated with a 2-variable Fourier expression that fits the annual and LST cycles as follows:

$$\frac{\Gamma_i(t)}{\Psi_i(t)} \approx \Upsilon_i(t) = a_0 + a_{11} \cdot \cos(Sa(t)) + b_{11} \cdot \sin(Sa(t)) + a_{12} \cdot \cos(Sl(t)) + b_{12} \cdot \sin(Sl(t)) +$$
$$+ \ldots + a_{jk} \cdot \cos(j \cdot Sa(t)) + b_{jk} \cdot \sin(j \cdot Sa(t)) + a_{jk} \cdot \cos(j \cdot Sl(t)) + b_{jk} \cdot \sin(j \cdot Sl(t)) \tag{8.17}$$

In this equation, a_{jk} and b_{jk} are the fitting coefficients, and Sa and Sl are the tidal constituents derived from Doodson's fundamental arguments and corresponding multipliers for the solar annual and the diurnal cycles (Petit and Luzum, 2010). The final expression for each time-expansion mode i is given as $\Gamma_i = \Psi_i \cdot \Upsilon_i + \varepsilon_i$, where ε_i is the residual disturbance from each PCA_i mode.

8.4 RESULTS AND ANALYSIS

The foregoing methodology to estimate non-gravitational accelerations serves as a reliable reference with unbiased solution for accelerometer calibration and thermospheric mass density retrieval. This section provides a summary of the most important results on thermosphere monitoring and modeling with GNSS precise orbits.

8.4.1 Thermospheric Density Estimation with GNSS Precise Orbits

Satellite accelerometers can be calibrated using GNSS-based non-gravitational accelerations as a true reference value. The GRACE satellites were launched into a nearly circular orbit on

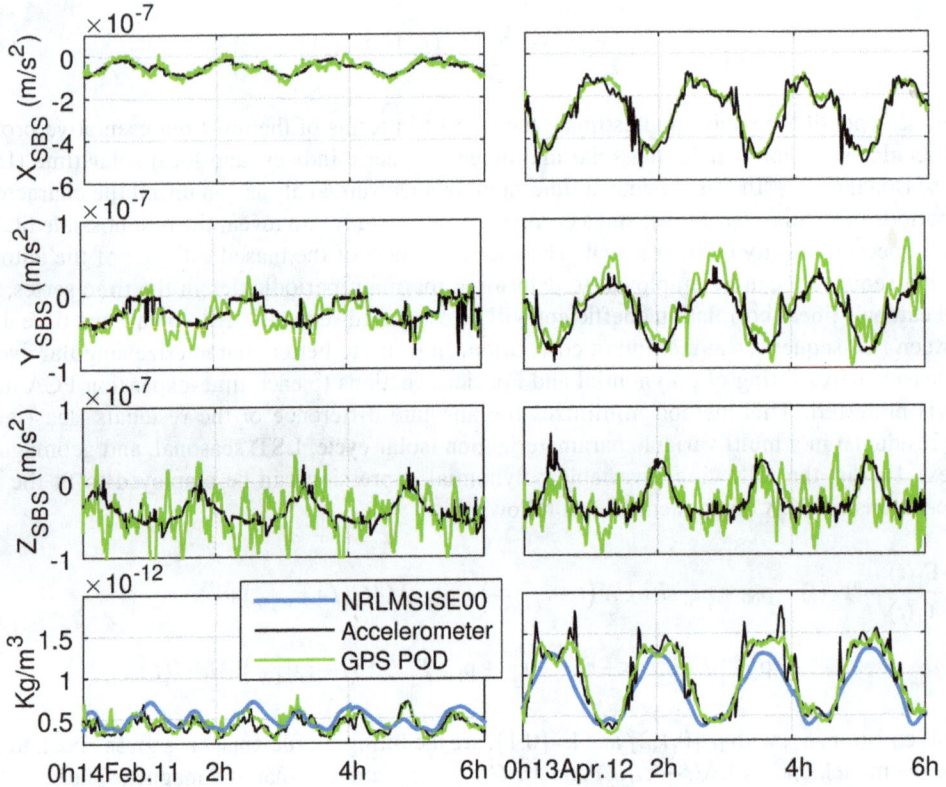

FIGURE 8.4 Non-gravitational accelerations and thermospheric mass densities normalized to 475 km along GRACE orbital path on 14 February 2011 (left) and 13 April 2012 (right). Calibrated accelerometer measurements are in black, and the estimations from GNSS precise orbits are in green. In the bottom panel, the estimations from NRLMSISE-00 are shown in blue.

Source: Calabia and Jin (2017).

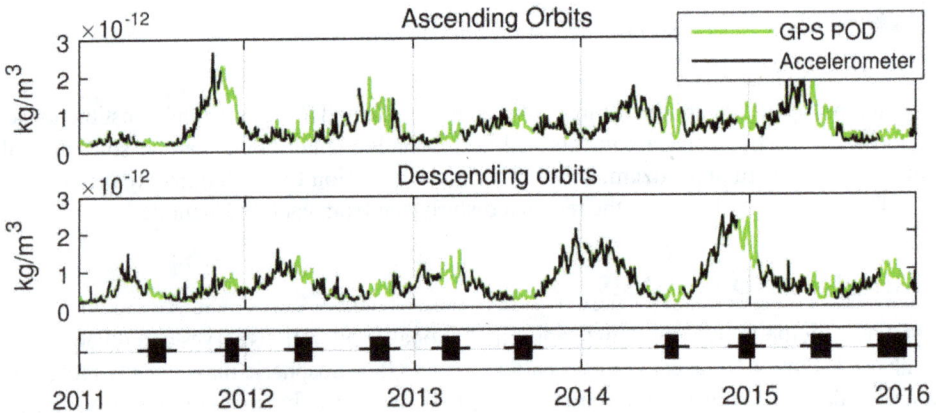

FIGURE 8.5 GRACE-derived thermospheric mass densities from 2011–2016. Global daily averaged densities from GNSS are in green, and the calibrated accelerometer densities are in black, separated in ascending orbits and descending orbits. Bottom timeline shows the days when accelerometer measurements were turned off due to power requirements.

Source: Calabia and Jin (2017).

17 March 2002 with a mean altitude of 475 km. Calabia et al. (2015) estimated the GRACE non-gravitational accelerations from GNSS precise orbits and calibrated the onboard accelerometers. Figure 8.4 shows the density estimates from GRACE GNSS precise orbits and the calibrated accelerometer measurements on 14 February 2011 (low solar activity; right panels) and 13 April 2012 (low solar activity; left panels).

In Figure 8.4, the GNSS-based accelerations in the X-satellite-body-system (X_{SBS}) axis are the main component for density retrieval, showing a little noise when the accelerations are below $\sim 10^{-7}$ m/s^2 (left panel). In the right panels, the accelerations in the X_{SBS} are approximately 5 times larger, and the noise effects have lower impact. In this figure, we also include the density estimates from NRLMSISE-00. In comparison to the NRLMSISE-00 densities, we can observe that GNSS-based densities have an excellent agreement with accelerometer-based densities, mostly for density values above $\sim 10^{-12}$ kg/m^3. In Figure 8.5, the GRACE GNSS-based densities cover the data gaps when the GRACE accelerometers were turned off on several occasions due to power requirements. Section 3.3 provides a clear example of thermospheric density variations during a geomagnetic storm that occurred during a period without GRACE accelerometer data.

8.4.2 Thermospheric Sensing with Commercial-off-the-Shelf GNSS

The CAscade SmallSat and IOnospheric Polar Explorer (CASSIOPE) satellite was launched on September 29, 2013 into an orbit slightly eccentric polar (81° inclination), with a perigee of approximately 325 km altitude and an apogee near 1500 km altitude. The CASSIOPE satellite uses 5 commercial-off-the-shelf, geodetic-grade, dual-frequency GPS receivers L1 C/A and L2 P(Y) tracking up to 12 satellites, to be used for high-precision navigation, attitude determination, time synchronization, and radio occultation measurements. The CASSIOPE satellite does not contain an accelerometer, but the accurate GNSS precise orbits (Montenbruck et al., 2019) can be used to estimate thermospheric mass density. In Figure 8.6, the GNSS-based thermospheric mass density estimates from the CASSIOPE GNSS precise orbits are compared to the existing data and models (Calabia and Jin, 2021b). In this work, Calabia and Jin (2021b) validated 6 years of CASSIOPE density estimates by comparing the High Accuracy Satellite Drag Model (HASDM) density database at altitudes from 325 to 425 km at intervals of 25 km. For density values above $\sim 10^{-12}$ kg/m^3, the CASSIOPE estimates provided standard deviations below 10% of the HASDM background density. Figure 8.7 shows the CASSIOPE densities and those from the NRLMSISE-00 and JB2008 models, relative to the HASDM densities, for the samples at 350 km altitude. During low solar-flux conditions (2017–2020), NRLMSISE-00 largely overestimates both the CASSIOPE and the JB2008 densities by approximately 150%.

8.4.3 Storm-Time Thermospheric Variations

Solar flares and geomagnetic storms can produce thermospheric Joule heating and particle precipitation along the earth's magnetic field lines, generating short-term and abrupt thermospheric mass density variations. Calabia and Jin (2017, 2016b, 2019) showed these events produce a global thermospheric mass density increase during from several hours to several days, and that the magnitude of the disturbances depends on solar cycle and annual season. In Figure 8.8, GNSS-based thermospheric mass densities from GRACE can characterize the anomalous behavior during the geomagnetic storm of 28–29 March 2013. Accelerometer measurements during this storm were unavailable for both GRACE A and B satellites due to instrument power-off.

Calabia and Jin (2021a, 2021b) showed the capability of the CASSIOPE GNSS receivers to retrieve aerodynamic accelerations at high resolution and presented the mass density responses to the February 2014 geomagnetic storm. In Figure 8.9, the CASSIOPE mass density estimates from samples below 400 km altitude are shown for a case study of February 2014. Clear

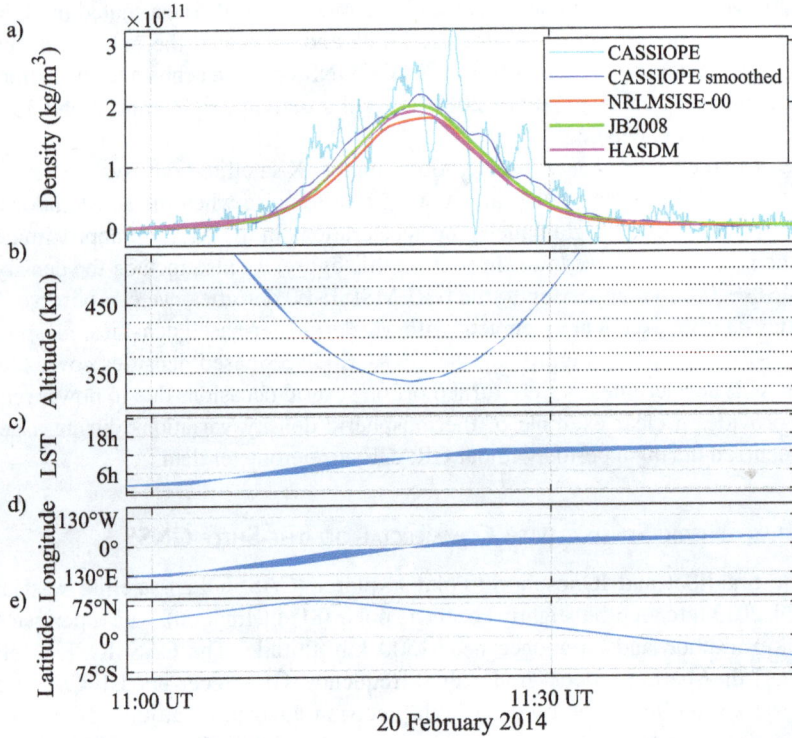

FIGURE 8.6 Thermospheric information with (a) density estimates from CASSIOPE GNSS and models, (b) altitudes, (c) LST, (d) longitude, and (e) latitude along the CASSIOPE orbit on 20 February 2014.

Source: Calabia and Jin (2021b).

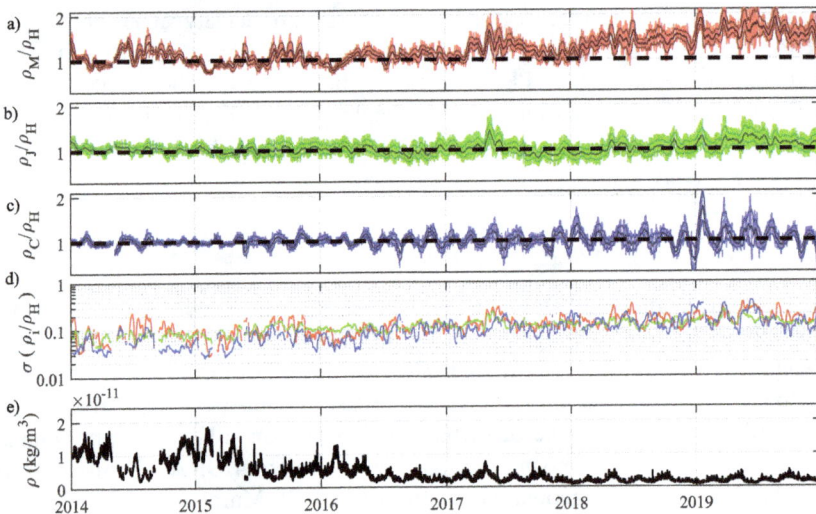

FIGURE 8.7 The (a) NRLMSISE-00, (b) JB2008, and (c) CASSIOPE densities relative to the HASDM densities at 350 km altitude. Panels (a-c) include the 30-day running window median averages (μ) and standard deviations (μ ± σ) of the ratios. Panel (d) shows the standard deviations (i = M, J, C) for comparison. The background density is shown in (e).

Source: Calabia and Jin (2021b).

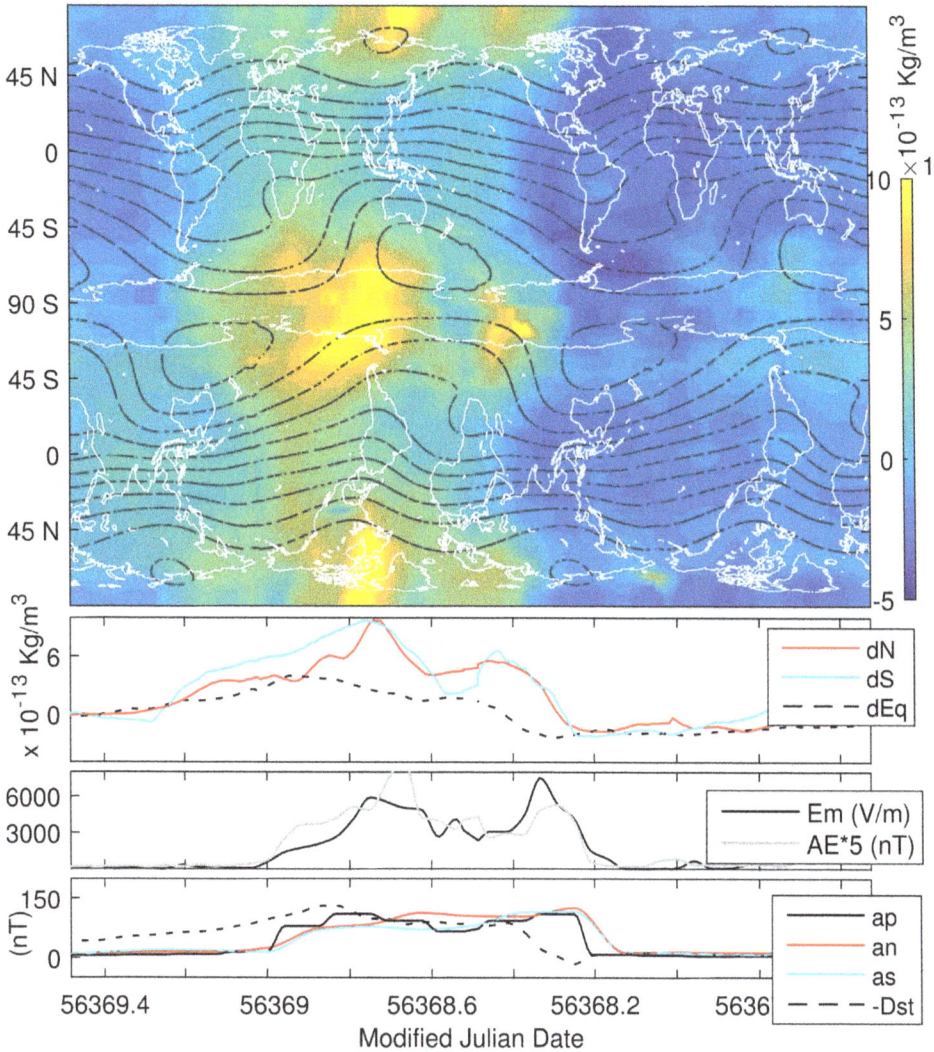

FIGURE 8.8 GRACE GNSS-based thermospheric mass density disturbances during the geomagnetic storm of 28–29 March 2013. Accelerometer-based densities are not available due to instrument power-off during this month. Bottom panels show its profiles at equator (dEq) and poles (dN, dS), plotted along with Em, AE, Ap, An, and As indices.

Source: Calabia and Jin (2017).

differences between the GNSS-based mass density estimates and that from the NRLMSISE-00 can be seen. These differences are mostly caused by inaccuracies in NRLMSISE-00 during the geomagnetic storms. The bottom panels show the agreement between densities at the perigee and the Em and Dst indices. A deeper analysis for this particular storm can be found in Calabia and Jin [2021a].

8.4.4 SOLAR CYCLE AND SECULAR THERMOSPHERIC VARIATIONS

Long-term variations in thermospheric mass density mainly include solar cycle, LST, and annual fluctuations. Calabia and Jin (2016a) investigated and created an empirical model with a 13-year

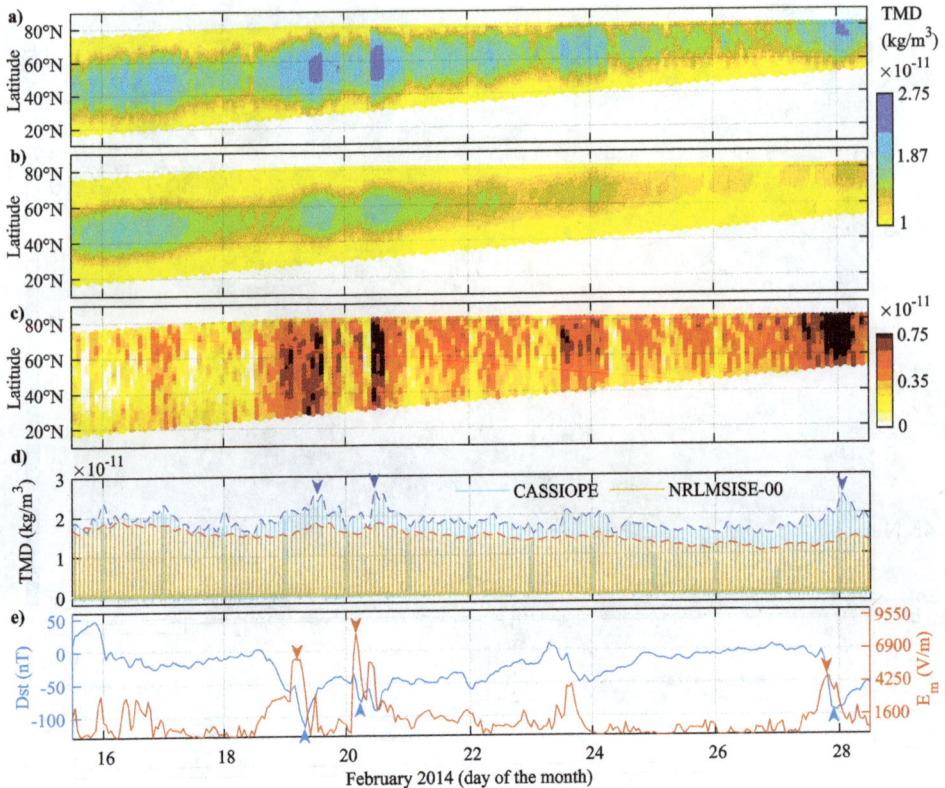

FIGURE 8.9   GNSS-based thermospheric mass density estimates below 400 km altitude from (a) CASSIOPE's precise orbits and from (b) NRLMSISE-00 during the second half of February 2014. The differences are shown in (c). The density estimates at the perigee are plotted with dashed lines in (d). In (e) we show the merging electric field Em and the disturbance storm time Dst index.

Source: Calabia and Jin (2021a).

time series of thermospheric density estimates from GRACE GNSS precise orbits. Figure 8.10 shows the first PCA mode of thermospheric mass densities at 475 km altitude. The first PCA mode represented 90% of the total variability, and the second and third modes provided 4% and 3% of the total variability. New periodic contributions were found at the frequencies of the radiation tides (Munk and Cartwright, 1966), revealing the strong tidal coupling driven by solar radiation. The resulting PCA modes were parameterized in terms of solar cycle, LST, and annual fluctuations to provide an empirical model capable of providing thermospheric mass densities in space and time. The additional fluctuations at the frequencies of the radiation tides were also included in the model.

In Calabia et al. (2020), the authors tested the new thermospheric mass density model (TMDM) of Calabia and Jin (2016a) with 2-year data of APOD and Swarm-C estimates and studied the dynamic orbit propagation of the missions under different mass density input schemes and different magnetospheric activity conditions. The results with TMDM showed similar differences in the dynamically propagated orbits from NRLMSISE-00 and *in situ* observations. Figure 8.11 shows the densities estimated by Swarm and GRACE accelerometers and that estimated by the new TMDM. In this figure, the NRLMSISE-00 model overestimates the actual densities, while the new TMDM provides an unbiased solution.

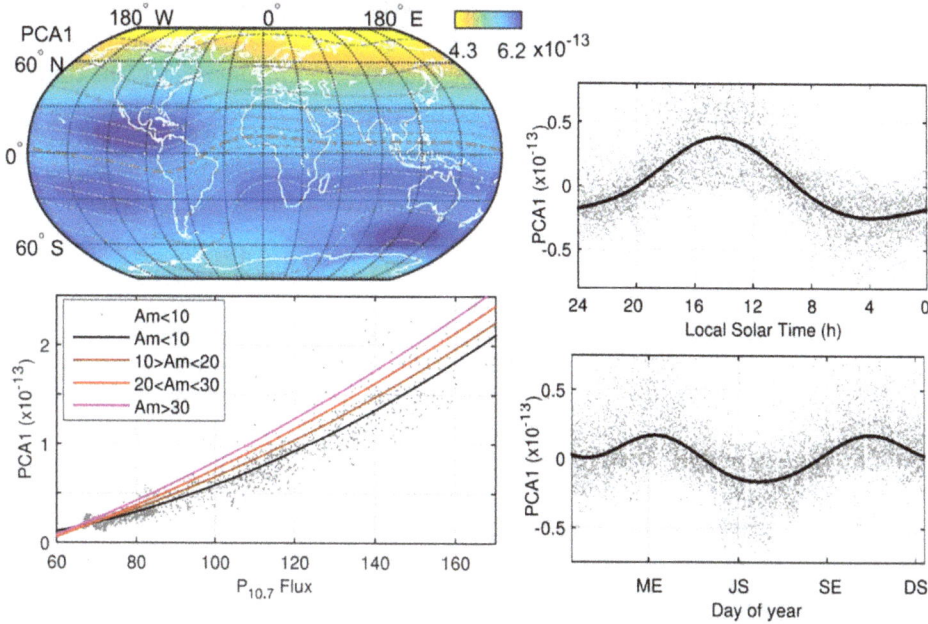

FIGURE 8.10 Solar cycle and secular variations from the first PCA mode of thermospheric mass density from GRACE GNSS precise orbits. This mode corresponds to 90% of the total variability.

Source: Calabia and Jin (2016a).

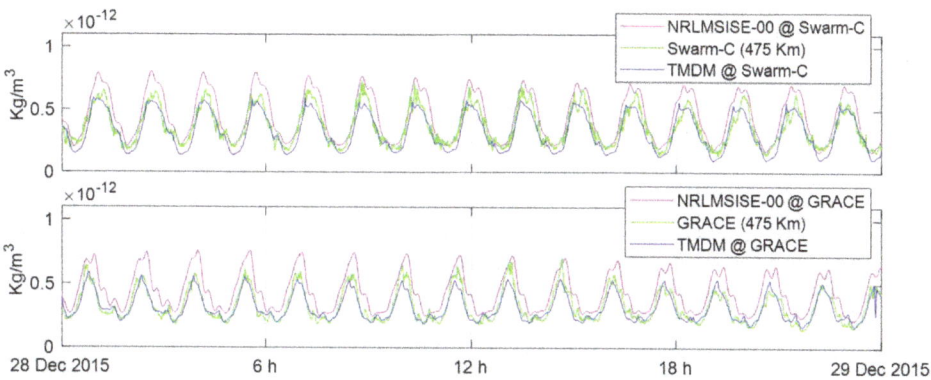

FIGURE 8.11 Thermospheric mass density estimates from Swarm-C and GRACE satellites (normalized to 475 km altitude). NRLMSISE-00 and TMDM are estimated at the same locations and times along the satellite orbits. During this period (28 December 2015) low magnetospheric activity was recorded (Ap = 4).

Source: Calabia et al. (2020).

8.5 CONCLUSIONS

The GRACE and CASSIOPE missions have shown the capability to provide high resolution non-gravitational accelerations from GNSS precise orbits. These accelerations are very valuable for thermospheric mass density estimation, which is a key parameter for accurate satellite tracking and upper atmosphere research. The new GNSS-based density estimates are suitable to study long-term trends and high-frequency disturbances caused by, e.g., geomagnetic storms.

For the GRACE mission, the GNSS-based non-gravitational accelerations have been used to calibrate the accelerometers onboard GRACE and to cover accelerometers data gaps (GRACE accelerometers were turned off on several occasions due to power requirements). The combination of GNSS with accelerometer data is currently the most accurate scheme for thermospheric mass density estimation. GRACE GNSS precise orbits were used to calculate 13 years of mass density estimates, and used as input to generate an empirical model that provides better performance than the existing models. The residual disturbances were employed to investigate short-term variations in detail, globally, and at high temporal and spatial resolution.

For the CASSIOPE mission, the onboard commercial-off-the-shelf, geodetic-grade, dual-frequency GNSS receivers have demonstrated their full capability for thermospheric mass density estimation at affordable cost in a low-budget space mission. The new thermospheric mass density estimates from CASSIOPE GNSS receivers can describe in detail the short-term variations caused by geomagnetic storms. Moreover, 6 years of CASSIOPE density estimates were validated by comparison to the HASDM density database. The validation tests were performed at altitudes from 325 to 425 km at intervals of 25 km. For densities above 10^{-12} kg/m^3, the similarity of the CASSIOPE densities to the HASDM data set were highly significant, with unbiased trends, and smaller deviations than that provided by the present models (e.g., NRLMSISE-00, JB2008). These results represent the validation of the first high-resolution thermospheric mass density estimates inferred from commercial-off-the-shelf GNSS receivers.

Short-term density variations caused by geomagnetic storms have shown irregular and complex patterns, which vary from storm to storm. The irregular patterns seem to depend on space weather and several other factors, and the present models are unable to accurately represent the actual variability. However, we have shown that high-precision GNSS receivers can be used to retrieve thermospheric mass density estimates. These new data and methods are highly important for upper atmosphere research and applications, and for the improvement of the existing models. Upper atmosphere models are very important for numerous applications, including satellite tracking and space weather research. In the near future, GNSS-based densities may overcome the current limitation of lack of data, and the existing upper atmosphere models will be improved with unprecedented details.

REFERENCES

Aikin, AC, AE Hedin, DJ Kendig, and S Drake (1993). Thermospheric molecular oxygen measurements using the ultraviolet spectrometer on the solar maximum mission spacecraft. *J. Geophys. Res.*, 98(A10):17607–17613, doi: 10.1029/93JA01468.

Anderson, RL, HB George, and JM Forbes (2009). Sensitivity of orbit predictions to density variability. *J. Spacecraft Rocket.*, 46(6):1214–1230, doi: 10.2514/1.42138.

Bhanderi, D (2005). Spacecraft Attitude Determination with Earth Albedo Corrected Sun Sensor Measurements. Dissertation, Aalborg University, Denmark.

Bowman, BR, WK Tobiska, FA Marcos, CY Huang, CS Lin, and WJ Burke (2008). A new empirical thermospheric density model JB2008 using new solar and geomagnetic indices, *AIAA/AAS Astrodynamics Specialist Conference*, AIAA 2008–6438, 19 pp.

Bruinsma, S, JM Forbes, RS Nerem, X Zhang (2006). Thermosphere density response to the 20–21 November 2003 solar and geomagnetic storm from CHAMP and GRACE accelerometer data. *J. Geophys. Res.*, 111:A06303.

Bruinsma, S, D Tamagnan, R Biancale (2004). Atmospheric densities from CHAMP/STAR accelerometer observations. *Planet Space Sci.*, 52:297–312, doi: 10.1016/j.pss.2003.11.004.

Bruinsma, SL (2015). The DTM-2013 thermosphere model. *J. Space Weather Space Clim.*, 5:A1.

Calabia, A, and S Jin (2016a). New modes and mechanisms of thermospheric mass density variations from GRACE accelerometers. *J. Geophys. Res. Space Physics.*, 121(11):11191–11212.

Calabia, A, and S Jin (2016b). Thermospheric mass density variations during the March 2015 geomagnetic storm from GRACE accelerometers. *Proceeding of Progress In Electromagnetics Research Symposium (PIERS2016)*, August 8–11, 2016, Shanghai, China, pp. 4976–4980.

Calabia, A, and S Jin (2017). Thermospheric density estimation and responses to the March 2013 geomagnetic storm from GRACE GPS-determined precise orbits. *J. Atmos. Solar Terrest. Phys.*, 154:167–179.

Calabia, A, and S Jin (2019). Solar-flux and asymmetric dependencies of GRACE-derived thermospheric neutral density disturbances due to geomagnetic and solar wind forcing, *Ann. Geophys.*, 37(5):989–1003.

Calabia, A, and S Jin (2021b). Thermospheric mass density disturbances due to magnetospheric forcing from 2014–2020 CASSIOPE precise orbits. *J. Geophys. Res. Space Phys.*, 126. doi: 10.1029/2021JA029540.

Calabia, A, and SG Jin (2021a). Upper-atmosphere mass density variations from CASSIOPE precise orbits. *Space Weather*, 19:e2020SW002645, doi: 10.1029/2020SW002645.

Calabia, A, S Jin, R Tenzer (2015). A new GPS-based calibration of GRACE accelerometers using the arc-to-chord threshold uncovered sinusoidal disturbing signal. *Aerosp. Sci. Technol.*, 45:265–271.

Calabia, A, G Tang, and S Jin (2020). Assessment of new thermospheric mass density model using NRLMSISE-00 model, GRACE, Swarm-C, and APOD observations. *J. Atmos. Solar Terrest. Phys.*, 199:105207.

Chen, GM, et al. (2014). A comparison of the effects of CIR- and CME-induced geomagnetic activity on thermospheric densities and spacecraft orbits: Statistic al studies. *J. Geophys. Res.*, 119:7928–7939.

Clemmons, JH, JH Hecht, DR Salem, and DJ Strickland (2008). Thermospheric density in the Earth's magnetic cusp as observed by the streak mission. *Geophys. Res. Lett.*, 35:L24103, doi: 10.1029/2008GL035972.

Cnossen, I, and M Förster (2016). North-south asymmetries in the polar thermosphere-ionosphere system: Solar cycle and seasonal influences. *J. Geophys. Res. Space Phys.*, 121:612–627.

Desai, SD (2002). Observing the pole tide with satellite altimetry. *J. Geophys. Res.*, 107(C11):3186, doi: 10.1029/2001JC001224.

Determan, JR, SA Budzien, MP Kowalski, MN Lovellette, PS Ray, MT Wolff, KS Wood, L Titarchuk, and R Bandyopadhyay (2007). Measuring atmospheric density with X-ray occultation sounding, *J. Geophys. Res.*, 112:A06323, doi: 10.1029/2006JA012014.

Doornbos, E, J van den IJssel, H Lühr, M Förster, and G Koppen-wallner (2010). Neutral density and cross-wind determination from arbitrarily oriented multiaxis accelerometers on satellites. *J. Spacecraft Rocket.*, 47(4):580–589, doi: 10.2514/1.48114.

Drob, DP, JT Emmert, JW Meriewether, JJ Makela, E Doornbos, M Conde, G Hernandez, J Noto, KA Zawdie, SE McDonald, et al. (2015). An update to the Horizontal Wind Model (HWM): The quiet time thermosphere. *Earth Space Sci.*, 2:301–319, doi: 10.1002/2014EA000089.

Heelis, RA, and A Maute (2020). Challenges to understanding the Earth's ionosphere and thermosphere. *J. Geophys. Res. Space Phys.*, 125:e2019JA027497, doi: 10.1029/2019JA027497.

Ijssel, J van den, and P Visser (2007). Performance of GPS-based accelerometry: CHAMP and GRACE. *Adv. Space Res.*, 39:1597–1603, doi: 10.1016/j.asr.2006.12.027.

Jin, S, A Calabia, and LL Yuan (2018). Thermospheric sensing from GNSS and accelerometer on small satellites. *Proc. IEEE.*, PP(99):1–12.

Knocke P, and J Ries (1987). Earth Radiation Pressure Effects on Satellites, Center for Space Research, Technical Memorandum, CSR-TM-87-01, The University of Texas at Austin.

Kuang, D, S Desai, A Sibthorpe, and X Pi (2014). Measuring atmospheric density using GPS–LEO tracking data. *Adv. Space Res.*, 53(2):243–256, doi: 10.1016/j.asr.2013.11.022.

Li, R, and J Lei (2020). Responses of thermospheric mass densities to the October 2016 and September 2017 geomagnetic storms revealed from multiple satellites observations. *J Geophys Res Space Phys.*, 125:e2020JA028534, doi: 10.1029/2020JA028534.

Liu, R, H Lühr, E Doornbos, and S-Y Ma (2010). Thermospheric mass density variations during geomagnetic storms and a prediction model based on the merging electric field. *Ann. Geophys.*, 28:1633–1645.

Liu R, S-Y Ma, H Lühr (2011). Predicting storm-time thermospheric mass density variations at CHAMP and GRACE altitudes. *Ann. Geophys.*, 29:443–453.

Luthcke SB, JA Marshall, SC Rowton, KE Rachlin, CM Cox, RG Williamson (1997). Enhanced radiative force modeling of the tracking and data relay satellites. *J. Astronaut. Sci.* 45(3):349–370.

March, G, T Visser, PNAM Visser, and EN Doornbos (2019). CHAMP and GOCE thermospheric wind characterization with improved gas-surface interactions modelling. *Adv. Space Res.*, 64(6):1225–1242.

Marcos, FA, and JM Forbes (1985). Thermospheric winds from the satellite electrostatic triaxial accelerometer system. *J. Geophys. Res.*, 90:6543–6552.

Maute, A, AD Richmond, G Lu, DJ Knipp, Y Shi, and B Anderson (2021). Magnetosphere-ionosphere coupling via prescribed field-aligned current simulated by the TIEGCM. *J. Geophys. Res. Space Phys.*, 126:e2020JA028665, doi: 10.1029/2020JA028665.

Mayer-Gürr, T, R Savcenko, W Bosch, I Daras, F Flechtner, and Ch Dahle (2012). Ocean tides from satellite altimetry and GRACE. *J. Geodynam.*, 59–60:28–38, doi: 10.1016/j.jog.2011.10.00

McLaughlin, CA, T Lechtenberg, E Fattig, and D Mysore Krishna (2013). Estimating density using precision satellite orbits from multiple satellites. *J. Astronaut. Sci.*, 59(1–2):84–100, doi: 10.1007/s40295-013-0007-4.

Mehta PM, A Walker, CA McLaughlin, and J Koller (2014). Comparing physical drag coefficients computed using different gas–surface interaction models. *J. Spacecraft Rocket.*, 51(3):873–883, doi: 10.2514/1.A32566.

Meier, RR, and JM Picone (1994). Retrieval of absolute thermospheric concentrations from the far UV dayglow: an application of discrete inverse theory. *J. Geophys. Res.*, 99:6307–6320, doi: 10.1029/93JA02775.

Montenbruck, O, and E Gill (2013). Satellite orbits: Models, methods and applications. Berlin: Springer.

Montenbruck, O, A Hauschild, RB Langley, and C Siemens (2019). CASSIOPE orbit and attitude determination using commercial off-the-shelf GPS receivers. *GPS Solut.*, 23(114), doi: 10.1007/s10291-019-0907-2

Müller, S, et al. (2009). Solar and geomagnetic forcing of the low latitude thermospheric mass density as observed by CHAMP. *Ann. Geophys.*, 27:2087–2099.

Munk, W. H., and D. E. Cartwright (1966). Tidal spectroscopy and prediction. *Phil. Trans. R. Soc. Lond.*, A259:533–581.

Newton, I (1726), Philosophiæ naturalis principia mathematica. Cambridge, MA, Harvard UP.

Nicolls, MJ, H Bahcivan, I Häggström, and M Rietveld (2014). Direct measurement of lower thermospheric neutral density using multifrequency incoherent scattering. *Geophys. Res. Lett.*, 41, doi: 10.1002/2014GL062204.

Pearson, K (1901), On lines and planes of closest fit to systems of points in space. *Philos. Mag.*, 2(11):559–572.

Petit, G, and B Luzum (2010). IERS conventions (2010). IERS technical note 36, International Earth Rotation and Reference Systems Service (IERS), Frankfurt am Main: Verlag des Bundesamts für Kartographie und Geodäsie.

Picone, JM, et al. (2002). NRLMSISE-00 empirical model of the atmosphere: Statistical comparisons and scientific issues, *J. Geophys. Res.*, 107(A12), 1468.

Picone, JM, JT Emmert, and JL Lean (2005). Thermospheric densities derived from spacecraft orbits: accurate processing of two-line element sets, *J. Geophys. Res.*, 110:A03301. http://dx.doi.org/10.1029/2004JA010585.

Richmond, AD, et al. (1992). A thermosphere/ionosphere general circulation model with coupled electrodynamics. *Geophys. Res. Lett.*, 19(6):601–604.

Santoni, F, F Piergentili, F Graziani (2010). Broglio Drag Balance for neutral thermosphere density measurement on UNICubeSAT. *Adv. Space Res.*, 45(5):651–660, doi: 10.1016/j.asr.2009.10.001.

Siemes, C, J de Teixeira da Encarnação, E Doornbos, J van den Ijssel, J Kraus, R Pereštý, L Grunwaldt, G Apelbaum, J Flury, and PE Holmdahl Olsen (2016). Swarm accelerometer data processing from raw accelerations to thermospheric neutral densities. *Earth Planet Space.*, 68(1):92, doi: 10.1186/s40623-016-0474-5.

Sutton EK, et al. (2009). Rapid response of the thermosphere to variations in Joule heating. *J. Geophys. Res.*, 114:A04319.

Tang, G, X Li, J Cao, S Liu, G Chen, H Man, X Zhang, S Shi, J Sun, Y Li, and A Calabia (2020). APOD mission status and preliminary results. *Sci. China Earth Sci.*, 63:257–266, doi: 10.1007/s11430-018-9362-6.

Tapley, BD, BE Schutz, and GH Born (2004). Statistical orbit determination. San Diego: Elsevier, Academic Press.

Taylor, FW (2005), Elementary climate physics. Oxford: U Press.

Tobiska, W. K., B. R. Bowman, D. Bouwer, A. Cruz, K. Wahl, M. Pilinski, P. M. Mehta, and R. J. Licata (2021). The SET HASDM database. *Space Weather.*, 19:e2020SW002682. doi: 10.1029/2020SW002682.

Visser, P, E Doornbos, J van den IJssel, and J de Teixeira da Encarnação (2013). Thermospheric density and wind retrieval from Swarm observations. *Earth Planets Space.*, 65:1319–1331, doi: 10.5047/eps.2013.08.003.

Yuan, LL, S Jin, and A Calabia (2019). Distinct thermospheric mass density variations following the September 2017 geomagnetic storm from GRACE and SWARM precise orbits. *J. Atmos. Solar Terrest. Phys.*, 184:30–36.

9 Estimation and Variations of Surface Energy Balance from Ground, Satellite, and Reanalysis Data

Usman Mazhar and Shuanggen Jin

9.1 INTRODUCTION

The climate cycle is governed by the energy that the earth receives from the sun and, in response, sends it back in other forms of energy. This energy cycle triggered the general atmospheric circulations, e.g., upper air cold fronts and jet air. Top-of-atmosphere (TOA) energy balance accounts only for the shortwave and longwave radiations. On the other hand, the earth's surface energy balance (SEB) is the balance between the total energy coming into the earth's surface and the energy going out of the surface. SEB is the principle deriving force of the earth's climate. Global warming or cooling trends are observed as the response to SEB variations [1]. The impact of SEB is not only retained in the temperature because it is observed that global heat and hydrological cycles vary with the varying SEB patterns [2, 3].

SEB comprises radiative and non-radiative energy fluxes. The radiation balance is the prime factor in the determination of SEB. Other factors, such as biophysical and biogeochemical processes, play an important role in characterizing SEB. The change in biophysical parameters, e.g., clearing forests or changing one land cover into another, causes energy fluxes to exhibit varying patterns [4]. These patterns also determine the local heat and water cycle. The impact of forest and cropland changes has been widely studied in recent years [5–8]. The geographical distribution plays a significant role in SEB variations as varying responses of energy fluxes are observed over different latitudinal levels [9].

SEB is not only affected or controlled by the surface parameters, but the atmospheric composition and environmental constituents also influence various surface energy fluxes. These factors primarily affect radiation balance and cause radiative forcing, a term used to describe changes in radiation balance due to an external agent such as climatic changes or anthropogenic factors. Air pollution is an important atmospheric parameter and an increased risk for human health as well as the global climate. Aerosols, fine suspended solid particles or liquid droplets, play a key role in air pollution and serve as radiative forcing agents. Changes in atmospheric pollutants resulted in the perturbations of radiation balance due to changing proportion of absorbing and scattering particles [10, 11]. Anthropogenic activities such as biomass burning, vehicle smoke, industrial combustion, and atmospheric pollution caused by power plants largely influence the atmospheric composition and consequently affect radiation balance. Greenhouse gasses, especially carbon dioxide (CO_2), methane (CH_4), and nitrous oxide (N_2O), are the other atmospheric constituents that significantly perturb atmospheric radiation balance [12–15]. Cloud cover is one of the most important parameters that define the amount of shortwave and especially longwave radiation directed to or away from the earth's surface. Longwave radiations on a cloudy day are largely different from those on a clear sky day. The cloud radiative effect (CRF) is a greenhouse effect that blocks the outgoing longwave radiations from escaping into the upper atmosphere.

DOI: 10.1201/9781003363118-9

The role of meteorological parameters in defining regional SEB is significant and well established. Water vapor pressure and relative humidity (a measure of atmospheric moisture) largely influenced shortwave and longwave radiations. These parameters are an integral part of estimating various surface energy fluxes [16]. Air temperature is often considered a measure to analyze the impacts of SEB variations; however, this important meteorological parameter is a fundamental ingredient in defining and estimating SEB [17, 18]. Air temperature also indirectly affects varying air density, which eventually perturbs regional surface energy fluxes. Another meteorological parameter, wind speed (also called wind velocity in literature), affects the surface energy fluxes by varying heat energy transportation between the surface and the nearby atmosphere. Higher wind speed resulted in rapid energy transformation [19]. Precipitation has two-fold indirect effects on SEB over any specific region. Firstly, it increases the relative humidity, and secondly, it increases the soil moisture that varies the ground response to absorb or radiate thermal energy.

A few studies highlight the interesting relationship between various geological, geophysical, and socioeconomic factors and surface energy fluxes. A study established a connection between water salinity and the evaporation rate and found its effect on surface energy flux [20]. One study analyzed the 1991 eruption at Mount Pinatubo (classified as the second-largest volcanic eruption of the twentieth century) and found that it produced a large cloud of sulfur dioxide (SO_2) which ultimately produced sulfate aerosols. Eventually, it reduced the radiation balance of the northern hemisphere by 4 Wm^{-2} [21]. A study showed that the reduced social and economic activity due to the global pandemic Coronavirus Disease 2019 (COVID-19) significantly reduced surface radiative forcing [10]. Before analyzing SEB's variations, mechanisms, and impacts on climate, it is necessary to observe or compute various surface energy fluxes with acceptable accuracy.

Efforts have been made for a long time to compute SEB on regional and global scales with acceptable accuracy. Multiple technologies and methods have been developed for this purpose. Observing SEB from ground-based instruments is a widely used and reliable technique. However, observing different surface energy parameters on a uniformly distributed scale is impossible. As an alternative, satellite remote sensing and reanalysis data sources are used to estimate global and regional SEB. Various mathematical, climatological, and physical methods have also been developed to estimate SEB. In the following sections, each component of SEB is discussed in detail first and then various observation and estimation methods are described. In the final section, the results of a recent study are included as a case study of SEB estimation and analysis of variance.

9.2 FUNDAMENTALS OF SURFACE ENERGY BALANCE

SEB involves radiative and non-radiative fluxes, at Earth's surface. Energy is transported between the surface and aloft air in radiative energy as well as other forms such as heat and evapotranspiration. The principal SEB equation is written as follows:

$$R_N \text{-} G = SH + LE \tag{9.1}$$

where R_N is net radiation, G is ground or soil heat flux, SH is sensible heat flux, and LE is latent heat flux. The right-hand side of Eq. 9.1 describes the available energy at the surface, while the left-hand side corresponds to the turbulent fluxes that involve the escaping of energy from the land surface to the aloft air. Eq. 9.1 follows the principle of the law of conservation of energy which is an ideal case. However, in real time, this principle does not hold in its absolute state. Thus, the more realistic form of Eq. 9.1 is:

$$Q = R_N \text{-} G \text{-} SH \text{-} LE \tag{9.2}$$

where Q is the heat anomaly or canopy heat storage, which is not computed in any of the energy mentioned above flux. It is the amount of energy stored in biomass, especially in the vegetation canopy.

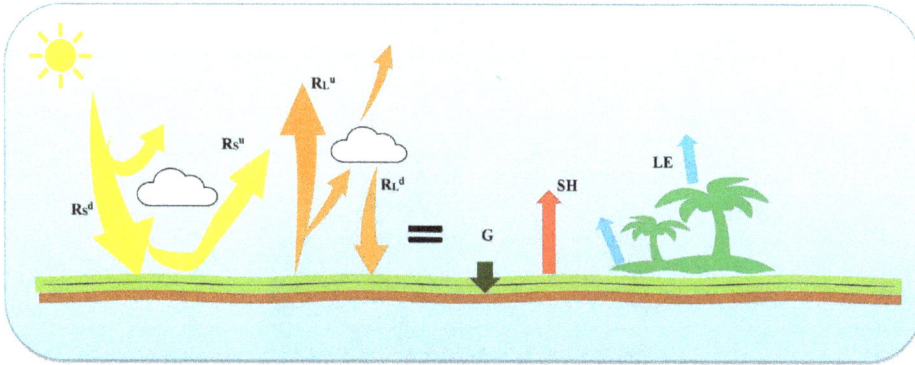

FIGURE 9.1 Schematic diagram of surface energy balance.

Uncertainties in the measurements of other energy fluxes also contribute to this heat anomaly and enlarge its real-time value. Figure 9.1 shows the schematic diagram of surface energy balance.

Net radiation (hereafter R_N), also known as radiation balance, is the balance between the downward (incoming) and upward (outgoing) radiation in both shortwave and longwave spectra. It is a basic quantitative measure to analyze any perturbation in climate [22]. At TOA, it is often referred to as radiation imbalance [23]. The radiation balance is obtained between downward solar radiation, reflected solar radiation, downward atmospherically emitted thermal radiation, and emitted thermal radiation from the surface. The sign convention of shortwave and longwave radiations is such that downward radiations are considered positive, while upward radiations are considered negative. Eq. 9.3 describes the R_N at the surface.

$$R_N = R_S^d - R_S^u + R_L^d - R_L^u \qquad (9.3)$$

where R_S^d and R_S^u are the downward shortwave and longwave radiations, respectively, and R_L^d and R_L^u are the upward shortwave and longwave radiations respectively. Downward solar radiation is the primary energy source for the earth and its climate. It serves as the prime governing factor for Eq. 9.3. The earth's surface absorbs part of the downward solar radiation while reflecting almost one third of these radiations. These reflected shortwave radiations are termed upward shortwave radiation. From here, the contribution of albedo starts, which is the ratio of the reflected shortwave to incident shortwave radiation. Albedo is a dimensionless quantity and is often denoted by α. Thus, the upward shortwave is the function of albedo and downward shortwave radiation. Albedo has a range between 0 and 1. It largely depends upon the underlying land cover, e.g., for clear water, the albedo is less than 0.01. For fresh snow, it might exceed 0.9 [24]. On average planetary albedo is considered to be around 0.3 [3, 25].

The longwave part of Eq. 9.3 comprises thermal radiation emitted from different features of Earth's climate. downward longwave are the radiations absorbed by clouds, atmospheric particles, and greenhouse gases. These radiations are emitted from the absorbing body following Max Planck's black body radiation law. The clear sky longwave radiation mainly depends on air temperature and atmospheric emissivity. For cloudy skies, cloud cover/fraction of clouds are the major contributors to the downward longwave radiation. The upward longwave radiations are the emitted radiations that the earth absorbed earlier in the daytime and then radiated back. It depends upon the surface emissivity and surface temperature.

Figure 9.2 shows the contribution of each radiation flux in the composition of R_N. Although the individual percentages are greater in the case of longwave radiations, the almost equal magnitude of downward and upward longwave radiations canceled each other's effects. Consequently, downward shortwave radiations seminally dominate R_N. While upward radiations are primarily controlled by

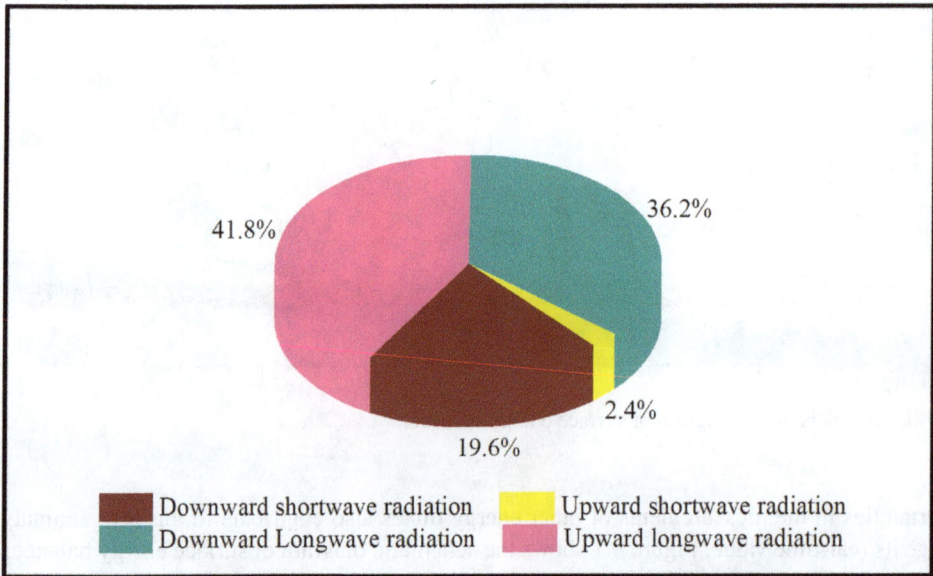

FIGURE 9.2 Individual radiation flux contribution in the composition of R_N. Percentages are based on 20 years of global mean data from the Cloud and Earth's Radiant Energy System (CERES). The negative sign indicates the upward direction of the corresponding flux.

land cover and surface properties, the downward radiations are affected by atmospheric compositions e.g., aerosol optical depth, and concentrations of trace and greenhouse gases [10].

Ground heat flux (hereafter G), also referred to as soil heat flux is an ignored yet important parameter of the SEB equation. It is the amount of thermal energy transported within a certain soil area [26]. When gained, incident solar energy on the surface is stored in the underlying layers of soil, which is eventually released during the night or in the absence of solar energy [27]. After the R_N, G is the available energy source for plants' metabolism and other biogeochemical processes. It affects the micrometeorology of a particular region since G is responsible for the energy transfer process between soil and aloft air.

Heat transforms between the layers of the surfaces through thermal conductance. The thermal conductivity of soils largely depends upon soil type, soil temperature, wetness or dryness of the soil, land cover of that particular area, and downward solar radiation. Although the surface absorbs solar radiation during the daytime and radiates through the night, enabling G to be near zero on a diurnal cycle [28], the equilibrium is disturbed seasonally. Depending upon the certain surface and solar conditions, G can be comparable to or even exceeds the sensible heat flux over dry soils in the day hours of the summer [29]. Soils stored the least thermal energy over wet lands, moist soils, and snow cover, and under the canopy of the dense plants. G represents 1 to 10% of R_N in the growing season. This percentage can be exceeded by up to 50% over barren land with almost zero canopy cover [30–32].

The fluxes on the right-hand side of Eq. 9.1 are often termed turbulent fluxes. The concept of sensible heat flux (hereafter SH) can be understood from the isobaric process of a thermo-dynamical system. There is a change in the system's temperature without changing any other parameter, e.g., volume, pressure, and state of matter (e.g., liquid to water). On the surface, when available energy from R_N or G is converted into thermal or heat energy, energy transportation is termed SH. This transportation of energy raises the near-surface temperature. It often serves as a climate warming measure. The temperature difference between surface and aloft air is the transporting force that makes the heat travel between these layers [33]. Other factors that significantly influence SH are wind speed and canopy height. These factors influence aerodynamic resistance and hence alter heat transportation between surface and nearby air. The significance of SH is not only restricted to thermal studies, but the global precipitation cycle and the hydrological cycle are also largely affected by

SH [2]. The magnitude of SH varies seasonally as well as over various land covers. For dry areas, it reaches the maximum and is nearly equal to the left-hand side of Eq. 9.1. For wet areas and water bodies, SH approaches zero. For all other land covers, SH remains between the two extreme cases [17]. Varying land covers resulted in the different responses of SH. Change of land cover, e.g., converting the forest into crops, introduces the changes in SH. Multiple studies have been carried out to analyze these responses. The changes in land cover induced changes in land surface temperature (LST), which eventually resulted in SH variation [4, 8, 9, 34].

The turbulent flux other than SH is the latent heat of evaporation or latent heat flux (hereafter LE). LE corresponds to the transfer of energy when there is a change in the state of matter [33]. Evapotranspiration consists of two processes: evaporation and transpiration. Evaporation is the process whereby liquid water changes into vapors in the presence of some external energy. Transpiration is restricted to plants when water in the leaves gets removed in the form of vapor through stomata. Evaporation and transpiration combined form evapotranspiration, occurring simultaneously [35]. LE is an important parameter in crop management studies and governs the micrometeorology of crops throughout the growing season. LE is also important for agricultural studies, such as analyzing the crop cycle, crop yield, and crop water requirement. LE depends on many factors, including wetness and dryness of the soil, canopy cover, leaf area index, wind speed, water vapor pressure, and relative humidity. The magnitude of LE is higher over wetlands and fully grown canopies, and in contrast, approaches zero over dry, barren regions. Like SH, LE is equally responsible for the precipitation and humidity of the atmosphere. Higher evaporation through crop leaves increases the evaporative cooling and vice versa. Figure 9.3 shows the mean annual spatial distribution of surface energy fluxes for the year 2020.

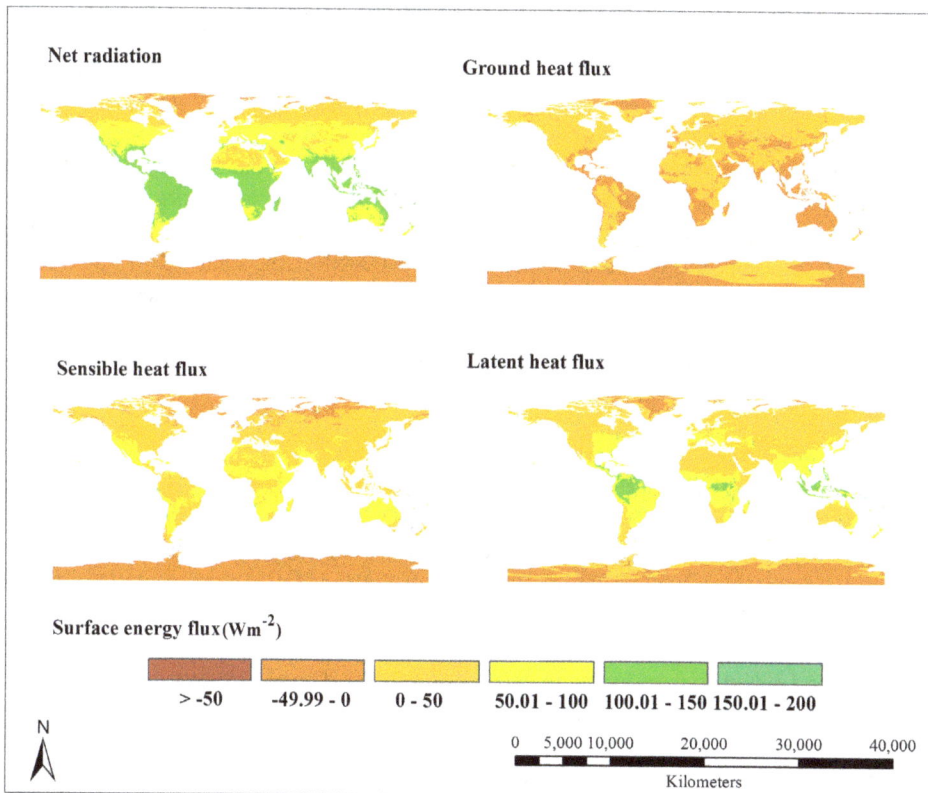

FIGURE 9.3   Spatial distribution of surface energy fluxes. This map shows the annual mean fluxes of 2020 obtained from European Centre for Medium-Range Weather Forecasts (ECMWF) Reanalysis 5th Generation (ERA-5) data set.

Climate scientists have agreed on the seminal importance of SEB for a long time and put a lot of effort into accurately estimating/observing various parameters of SEB. This chapter discussed some of the significant progress made in the estimation and observation of SEB including ground observing SEB, as well estimations from satellite remote sensing and reanalysis data. For reference, long-term global means that values of various surface energy fluxes are also included here from different data sources.

9.3 OBSERVATION AND ESTIMATIONS OF SEB

9.3.1 GROUND OBSERVATIONS OF SEB

The computation of SEB is the prime interest to climate scientists and has seminal importance for climate and meteorological studies. Efforts have been made for a long time in this regard. There are many *in situ* ground observational sites dedicated to observing SEB globally. The method used in the majority of these ground observing towers is eddy covariance [36]. It is a standardized method worldwide to estimate atmospheric and surface parameters. The method observes through vertical turbulent fluxes and target mixing ratios over each site/node. Each node represents a micrometeorological tower equipped with sensors and located over various land covers worldwide [37].

One of the largest ground-observing tower networks is the FLUXNET tower network, started in 1991 and still functioning worldwide. FLUXNET observes many parameters such as air temperature, wind speed, solar and thermal radiation, LE, SH, G, atmospheric pressure, and carbon dioxide. FLUXNET is a network of various regional networks including, Integrated Carbon Observation System (ICOS), OxFLUX, AmeriFLUX, and AsiaFLUX. Each regional network is dedicated to a certain geographical region. ICOS primarily covers Europe, OxFLUX covers Australia and New Zealand, AmeriFLUX covers North and South America, and AsiaFLUX covers Asia. Each site or node has a specific observational life depending upon many factors [38]. FLUXNET has two modes of data availability: the first is free available data under the CC-BY-4.0 license; the other is the FLUXNET-CH4 community product, which is restricted to license availability. By now, the FLUXNET2015 database is available under the aforementioned data policy. The open-access data set FLUXNET2015 is a useful source of information that provides *in situ* SEB observations at hourly, daily, weekly, monthly, and annual temporal resolutions, covering 206 sites globally. Figure 9.4 shows the locations of available sites of the FLUXNET2015 data set.

Another global network of ground observing stations for surface energy fluxes is the European Fluxes Database Cluster. The network primarily uses the eddy covariance method for flux observations. Like FLUXNET, the European Fluxes Database Cluster also provides data under restricted as well as public data distribution policies.

Another ground-based wide-range network is the Solar Radiation Network (SolRad-Net), a companion network of the Aerosol Robotic Network (AERONET). The network is a collaborative program of the National Aeronautics and Space Administration (NASA) and PHOtométrie pour le Traitement Opérationnel de Normalisation Satellitaire; Univ. of Lille 1, CNES, and CNRS-INSU (PHOTONS). AERONET is intended to observe aerosol (suspended fine solid particles or liquid droplets) in the atmosphere but its companion network SolRad-Net is used to observe downward solar radiation through a pyranometer; an instrument that observes downward shortwave radiations. SolRad-Net is a useful ground observing network which provides ground-based quasi-real-time solar radiation and is widely used by the scientific community. The solar radiation data is limited to specific AERONET sites and is free to use. Table 9.1 summarizes those ground-observing networks that are still operational.

Along with a wide range of continuous data-providing ground-based networks, many project-specific programs have been launched to understand the global and regional SEB. The observations of such programs are available for a limited time span. The Coordinated and Enhanced Observation Period (CEOP) was intended to understand the global hydrological and energy cycle.

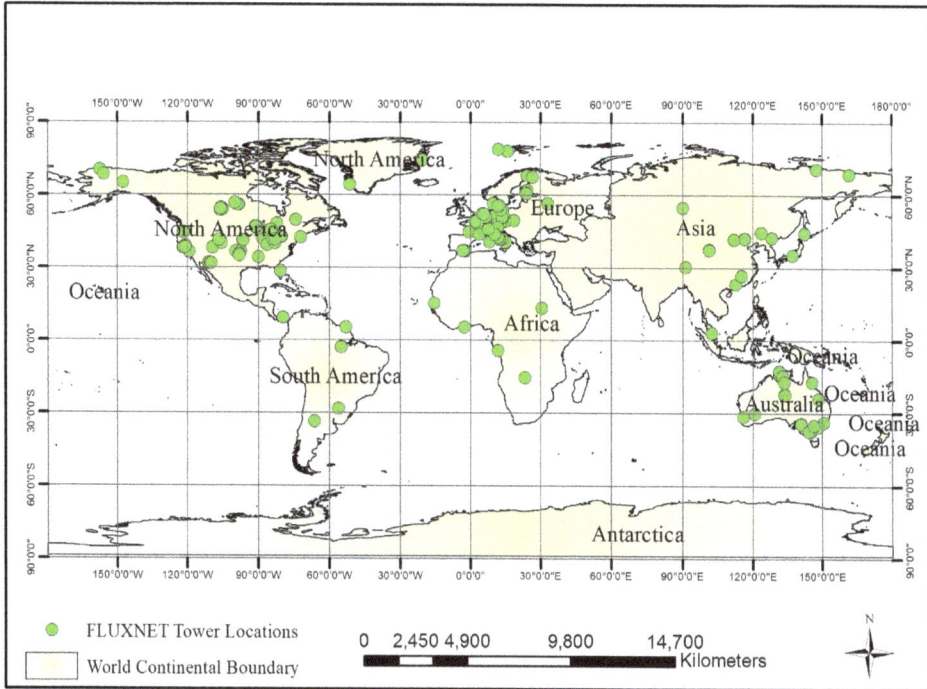

FIGURE 9.4 Location map of FLUXNET towers included in FLUXNET2015 data.

TABLE 9.1
Summary of Ground Observing Networks

Flux Observing Network	Spatial Coverage	Temporal Coverage	Official Weblink
FLUXNET	Global	1991-present	https://fluxnet.org/
AmeriFLUX	North and South America	1996-present	https://ameriflux.lbl.gov/
AsiaFLUC	Asia	1999-present	https://www.asiaflux.net/
OxFLUX	Australia and New Zealand	2001-present	https://www.ozflux.org.au/
ICOS	Europe	2008-present	https://www.icos-cp.eu/
European Fluxes Database Cluster	Global		http://www.europe-fluxdata.eu/
SolRad-Net	Global	1992-present	https://solrad-net.gsfc.nasa.gov/

The project was initiated under the World Climate Research Program (WCRP) in cooperation with the World Meteorological Organization (WMO) and Committee on Earth Observation Satellites (CEOS) under the framework of the Integrated Global Observing Strategy Partnership (IGOS-P). CEOP's objectives include understanding monsoons, global and regional water and energy cycles, establishing an extensive ground observation system, and validating energy fluxes observed through satellite data.

CEOP Asia-Australia Monsoon Project (CAMP) is a part of the vast CEOP project which was initiated to i) monitor and analyze the hydro-meteorological cycle triggered by the Asian monsoon, ii) monitor and analyze the observed regional energy cycle, and iii) establish an observational mechanism for water and energy cycles through satellites [39]. CAMP provides valuable ground observations for the regional energy fluxes. Another such program is Global Energy and Water Exchanges (GEWEX), a component of the WCRP. GEWEX was mainly initiated to model the global hydrological cycle and analyze the energy that the earth receives. Initially, observations from Special Sensor Microwave/Imager (SSM/I) on board the Defense Meteorological Satellite Program (DMSP) were used as the base input [37–39].

GEWEX Asian Monsoon Experiment (GAME) is a regional program that aims to understand the role and significance of the Asian monsoon in the global hydrological and energy cycle [40–43]. The program consists of two phases. The first phase was launched in 1996 and ended in 2001. This phase covers the Siberian, Huaihe Basin Experiment (Hubex), Tibet, and tropics regions. Phase 2 expanded globally and studied many aspects of climate, such as land surface process, precipitation, and monsoon system modeling.

9.3.2 ESTIMATIONS FROM SATELLITE REMOTE SENSING

Ground observations have time and space limitations despite the accurately observed surface energy fluxes. The ground observing stations are sparse and not linearly distributed, and cover small footprint areas. Also, many regions of the world have no ground observing stations. The best and most effective alternative to this problem is the use of satellite remote sensing data [44]. The scientific community has tried to observe the earth's energy balance through satellite remote sensing data since its evolution in the 1960s [3]. These efforts are ever increasing with the advent of modern technology, since with the advancement of technology, many ground observing satellites are providing valuable data sets related to energy fluxes.

Multiple satellite-based sensors have been launched to monitor several parameters of the global energy balance. NASA's Surface Radiation Budget (SRB) contributes to the GEWEX experiment. The sensor provides 3-hourly, daily, and monthly shortwave and longwave fluxes at TOA and surface levels. Its operation time spans from 1983 to 2007. SRB uses International Satellite Cloud Climatology Project (ISCCP) data for input. A subproject of the GEWEX Baseline Surface Radiation Network (BSRN) is used to validate SRB data.

In 1984, NASA started an Earth Radiation Budget Experiment (ERBE) project and launched three satellites, Earth Radiation Budget Satellite (ERBS), National Oceanic and Atmospheric Administration NOAA-9, and NOAA-10. This multi-satellite experiment was aimed to estimate the earth's radiation balance [45]. ERBS provides monthly solar irradiance data at 2.5° resolution. The satellite provided continuous data from 1984 to almost 1990 [46].

After the discontinuity of ERBS, NASA initiated another project and launched Clouds and the Earth's Radiant Energy System (CERES), which started observing data in March 2000. CERES's sensor is mounted on Terra, Aqua, Suomi National Polar-Orbiting Partnership (S-NPP), and NOAA-20 satellites [47]. Each CERES product measures filtered radiances in shortwave from 0.3 to 5.0 μm, a window from 8 to 12 μm, and a total from 0.3 to 200 μm wavelengths [48]. Radiance-to-flux conversion uses empirical distribution models [49]. CERES provides accurate and continuous instantaneous, daily, and monthly data at TOA and surface level at 1° resolution. CERES global coverage and accurate observations of shortwave and longwave radiation enable this satellite to be widely used in the scientific community [47, 48].

CERES Energy Balance and Filled (EBAF) is a level 3b product that provides monthly TOA and surface radiation fluxes. EBAF provides continuous data, since March 2000 to date. Apart from downward/upward shortwave and longwave fluxes, this product also provides cloud radiative effects and other cloud parameters such as cloud area fraction, cloud visible optical depth, cloud effective pressure, and cloud effective temperature. The radiation fluxes from EBAF are available for all sky, clear sky for cloud-free areas, and clear sky for total regions. downward shortwave and longwave radiation fluxes are accurate up to 4 Wm^{-2} and 6 Wm^{-2} respectively [47, 50]. Figure 9.5 shows the two decadal (2001–2020) global mean values of downward/upward shortwave and longwave radiation fluxes computed from CERES EBAF data.

CERES synoptic TOA and surface fluxes and clouds (SYN1deg) level 3 product provides data at hourly, 3-hourly, daily, and monthly temporal resolution. SYN1deg product provides TOA and

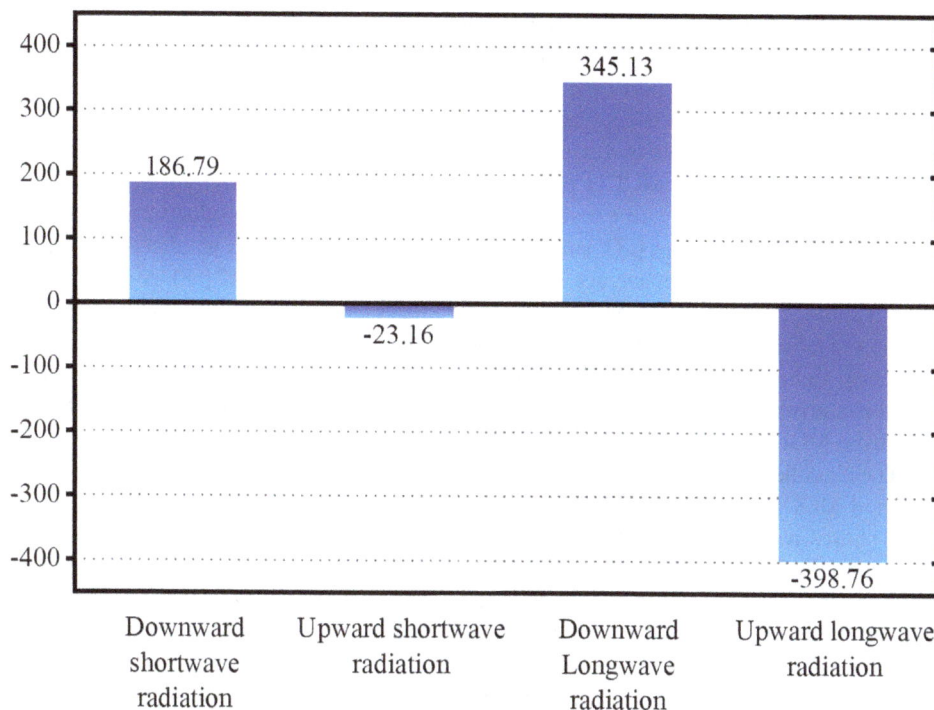

FIGURE 9.5 Mean global surface downward/upward shortwave and longwave flux. These values are computed from 20 years (2001 to 2020) of CERES EBAF data. Values are shown in Wm^{-2}. Negative values correspond to the upward radiation fluxes.

surface initial and computed fluxes along with various surface parameters including solar zenith angle, elevation above sea level, ocean fraction coverage, and snow/ice percent coverage. Both EBAF and SYN1deg are widely used reliable sources for shortwave and longwave radiation fluxes. Another level 3 product, cloud type histogram (CldTypHist) is a monthly CERES product that provides various cloud and meteorological parameters at 1° spatial resolution. CldTypHist replaces CERES's International Satellite Cloud Climatology Project (ISCCP) product. Other CERES products are the single scanner footprint (SSF) level 2, SSF1deg level 3, flux by cloud type (FluxByCldTyp) level 3, fast longwave and shortwave flux (FLASHFlux), etc.

Next on the list is the Moderate Resolution Imaging Spectroradiometer (MODIS). MODIS is a multispectral sensor aboard Terra (EOS AM-1) and Aqua (EOS PM-1). Southward Terra orbit is designed to pass zenith at 10:30 AM and Northward Aqua orbit passes zenith at 01:30 PM local time. MODIS acquires data in 36 spectral bands with spatial resolution ranging from 250 m to 1 Km. MODIS is not a SEB-centric sensor, but it observes through multiple wavelengths ranging from shortwave to longwave, including window regions. Various scientific teams have developed multiple products of MODIS such as albedo, LST, surface emissivity, and evapotranspiration, which are widely used to estimate surface fluxes [51–54]. MODIS products are a reliable source of information and are widely used in recent literature [51].

Surface albedo is the ratio between incidents and the reflected SW radiation. MODIS's albedo product series MCD43 provides reliable albedo data. These product series compute albedo from MODIS bands 1 to 7, including three broad bands, visible (0.3 to 0.7 µm), near-infrared (0.7 to 5.0 µm), and SW (0.3 to 5.0 µm). It contains directional hemispherical reflectance (direct or black sky albedo) and bi-hemispherical reflectance (diffuse or white sky albedo). Blue sky albedo is the overall reflectance effect of SW radiation and can be computed as a function of white sky and black sky albedo [55, 56]. Blue sky albedo corresponds to the ratio of black sky and white sky albedo, as the actual albedo range lies between these two extreme cases. The use of both black and white sky albedo for the computation of shortwave radiation flux is evident in recent literature [9]. The high-quality MODIS albedo product is accurate within 5% at local solar noon [57].

LST and surface emissivity are two essential parameters for computing upward longwave radiation. In this study, MODIS provides MCD11 product series for LST and emissivity data on the instantaneous, daily, 8-day, and monthly time scale. MODIS LST and emissivity products used a generalized split-window algorithm to obtain LST values. Collection 6 LST product is validated against various land cover sites and reported an average standard deviation error of 0.5 K [58]. A study conducted over Tibet validated MODIS LST and emissivity products using ground-based measurements at a semi-desert site in western Tibet [59]. Emissivity is a dimensionless quantity representing the ratio of thermal energy radiating from a surface to that radiating from a black body. MODIS LST and emissivity products provide emissivity from various narrow bands in the middle and thermal infrared spectrum from MODIS bands 29 (8.4 to 8.7 μm), 31 (10.78 to 11.28 μm), and 32 (11.77 to 12.22 μm) [52, 58]. The classification-based emissivity method is used to obtain emissivity [60, 61].

Combined MODIS product MCD18 is a set of data products that provide 3-hourly and daily downward shortwave radiation and photosynthetically active radiation (PAR) at 5 km and 0.05° spatial resolution [62]. Basic input parameters are TOA reflectance, surface reflectance, geo-location data, surface albedo, total column water vapor, and surface elevation [63]. Reflectance parameters are obtained from various MODIS products, Modern-Era Retrospective Analysis for Research and Applications (MERRA), and GTOPO30 DEM.

For the LE estimation, a widely used satellite product is the MODIS 8-day evapotranspiration and LE product (MOD16A2) at 500 m resolution. The product is based on a modified Penman-Monteith method; the modified version of the method involves several parameters, including saturated water vapor pressure, available energy in the form of R_N, air density, the specific heat capacity of air, aerodynamic resistance, and surface resistance. Saturated water vapor pressure can be obtained from air temperature [64]. Mu et al. [65, 66] used the MODIS remote sensing products of Fraction of Photosynthetically Active Radiation (FPAR), Leaf Area Index (LAI), land cover, albedo, and NDVI, with tower observations for meteorological data inputs. Obtained results were tested against eddy covariance flux towers from the AmeriFLUX network. The mean absolute error of daily MODIS ET was observed as 0.33 mm day^{-1}. This product has a limitation that it does not provide LE values over barren lands.

Global Land Surface Satellite (GLASS) is a product suite that provides various parameters including broadband albedo, broadband emissivity, downward solar radiation, R_N, and evapotranspiration based on multiple satellite data sets and look-up table methods. GLASS provides long-term (1982–2018) data on spatial resolutions of 250 m, 500 m, 1 km, 0.1°, 0.25°, and 0.05° [23, 67, 68]. Another algorithm suite, the Global Land Evaporation Amsterdam Model (GLEAM), is dedicated to providing global evapotranspiration and soil moisture data [69, 70]. Table 9.2 summarizes the remote sensing data sources that are mentioned in this section.

TABLE 9.2

Summary of Satellites That Provide SEB Parameters

Sensor	Parameters	Spatial Resolution	Temporal Resolution	Temporal Coverage
SRB	Radiation	1°	3 hourly, daily	1983–2007
ERBS	Radiation	2.5°	monthly	1984–1990
CERES	Radiation	1°	Hourly, daily, monthly	2003–present
MODIS	Radiation, LE, albedo, LST, emissivity, LE	250 m, 500 m, 1 km, 5 km, 0.05°	Instantaneous, daily, 8-day, monthly, annual	2000–present

9.3.3 Estimations from Reanalysis Data

Apart from ground observations and satellite remote sensing data, another option for the SEB observations/estimations is reanalysis data. Reanalysis data is the combination of observations and numerical models that simulate any particular parameter. Reanalysis data sets provide long-term, continuous climatic parameters on a global scale. Various reanalysis data sources provide reliable and widely used products of surface energy fluxes. Japanese 55-Year Reanalysis (JRA-55) is a project conducted by the Japanese Meteorological Agency (JMA) that provides a wide range of atmospheric parameters. The 55-year project was completed from 1958 to 2013, and since then it provides data on a continuing real-time basis to date.

The European Centre for Medium-Range Weather Forecasts (ECMWF) under the Copernicus Climate Change Service (C3S) provides a long-term reanalysis data set ECMWF Reanalysis 5th Generation (ERA5). ERA5 provides highly accurate hourly data at a spatial resolution of 31 km compared to the other reanalysis sources [69, 71]. Along with many other meteorological, atmospheric, and surface parameters, ERA5 provides hourly and monthly surface radiation flux, SH, and LE data at 0.1° spatial resolution from 1950 to date which is available online at https://cds.climate.copernicus.eu/#!/home. ERA5 replaces another ECMWF product ERA-Interim whose temporal coverage is from January 1979 to August 2019. ERA-Interim provides monthly data four times a day at a spatial resolution of almost 80 km.

A reanalysis data set was developed by the National Centers for Environmental Prediction (NCEP) and the National Center for Atmospheric Research (NCAR), known as NCEP/NCAR reanalysis, one of the longest reanalysis records from 1948 to the present. The data is available online through the National Oceanic and Atmospheric Administration (NOAA) Physical Sciences Laboratory (PSL) (https://psl.noaa.gov/data/gridded/data.ncep.reanalysis.html). The data is available on multiple pressure levels, at 4 times daily, daily, and monthly temporal resolutions. The products are also available after various statistical operations e.g., anomalies, monthly mean, long-term mean, and standard deviation. Figure 9.6

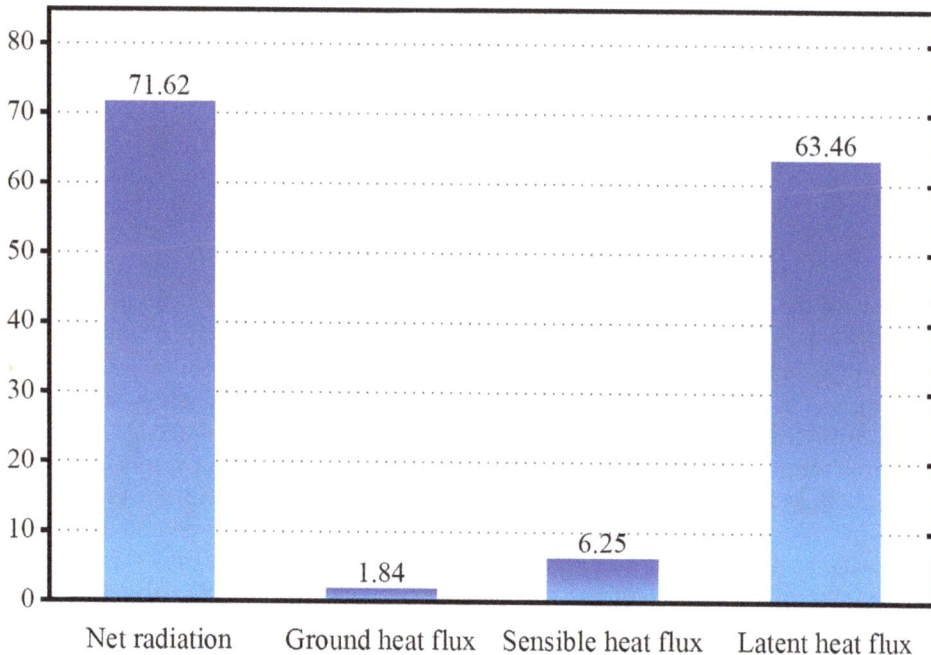

FIGURE 9.6 Thirty years (1991–2020) global mean energy fluxes. All values are in Wm^{-2} and computed directly from the NCEP/NCAR long-term mean product of the corresponding energy flux. Energy fluxes provided in this product cover the whole earth including ocean and land surface.

TABLE 9.3

Summary of Available Reanalysis Data for Surface Energy Fluxes

Product	Provided By	Spatial Resolution	Temporal Resolution	Temporal Range
JRA-55	JPA	1.25°×1.25°	Monthly	1958–present
ERA-5	ECMWF	0.1°°×0.1°°	Hourly, monthly	1950–present
ERA-Interim	ECMWF	80 km	4 times a day, monthly	1979–2019
MERRA-2	GMAO	0.625°×0.5°		1980–present
NCEP/NCAR	NCEP, NCAP	2.5°×2.5°	4 times a day, daily, monthly	1948

shows the long-term (1991–2020) global mean of R_N, G, SH, and LE obtained from NCEP/NCAR data. Table 9.3 summarizes the reanalysis data sources that are mentioned in this section.

The Modern-Era Retrospective Analysis for Research and Applications, Version 2 (MERRA-2) is a project by the Global Modeling and Assimilation Office (GMAO), NASA. MERRA-2 provides various climatic parameters on 0.625°×0.5° (longitude by latitude) spatial resolution. The temporal range of MERRA-2 is from 1980 to the present on a global scale. MERRA-2 assimilation scheme includes the Goddard Earth Observing System (GEOS) model and analysis scheme to provide long-term, wide-range, and accurate data [72]. The MERRA-2 data are available online through the Goddard Earth Sciences (GES) Data and Information Services Center (DISC) (http://disc.sci.gsfc.nasa.gov/mdisc/).

9.3.4 ESTIMATIONS FROM SEB ALGORITHMS

Apart from remote sensing and reanalysis products of surface energy fluxes, many estimation methods and algorithms have been developed to estimate these fluxes. A few significant and widely used studies are discussed in this section. Bastiaanssen et al. [17] developed the widely used Surface Energy Balance Algorithm for Land (SEBAL). SEBAL uses remote sensing data from visible and thermal infrared radiations, along with a few meteorological parameters such as wind speed, humidity, and air temperature. The algorithm was developed for clear sky conditions only and deals with a variety of land covers from barren land to wetlands even with no vegetation cover. In the SEBAL, LE is estimated as the residual of the SEB as:;

$$LE = f(\alpha, R_S^d, \varepsilon_s, \varepsilon_a, T_S, G, z_{0m}, KB^{-1}, \upsilon, L, \delta T) \tag{9.4}$$

where α is albedo, R_S^d is the downward shortwave radiation, ε_s is surface emissivity, ε_a is atmospheric emissivity, Ts is LST, G is ground heat flux, z_{0m} is surface roughness length for momentum transfer, KB^{-1} is the relation between surface length for momentum transfer with that of heat transfer, υ is friction velocity, L is the Monin-Obukhov length, and δT is the near-surface vertical air temperature difference. SEBAL also proposed a novel empirical parameterization for estimating G that is widely used in the later literature. SEBAL was tested and validated using the Landsat thematic mapper over various locations and ground observations such as the large-scale field experiments EFEDA (Spain), HAPEX-Sahel (Niger), and HEIFE (China) [73]. The evaporative fraction produced by SEBAL was compared with the ground-observed evaporative fraction. The results show an RMSE of 0.1 to 0.2 only. SEBAL was also used for vast evapotranspiration mapping using MODIS data [74].

Roerink et al. [75] proposed an algorithm called Simplified Surface Energy Balance Index (S-SEBI). The model was developed using LANDSAT-TM data and verified against the ground observations of a field experiment over Piano de Rosia in Tuscany (Italy) in August 1997. They estimate surface downward longwave radiation using an empirical relation based on downward longwave radiations at TOA. The main feature of the algorithm is that it estimates surface turbulent fluxes using evaporative fraction, which was computed using the extremes temperature. The two extremes were classified as $T\lambda_E$ for wet pixels (H = 0) and T_H for dry pixels (LE = 0). The word "simplified" in the name of the algorithm S-SEBI specifies that $T\lambda_E$ and T_H can be measured from the remote sensing image itself, and this would be possible when the whole scene has uniform atmospheric conditions and wet and dry pixels are available in the sufficient quantity. The validation process showed the accuracy of estimated evaporative fractions within 8% [75].

Su [18] developed the Surface Energy Balance System (SEBS) algorithm. SEBS proposed a new parameterization for estimating evaporative fractions. This study also discriminates between the atmospheric boundary layer (ABL) and the atmospheric surface layer (ASL). The latter was defined as the bottom 10% of the former. The algorithm was tested using the data collected from the cotton fields, shrubs, and grasslands and one remote sensing data from the EFEDA experiment that uses the thematic mapper simulator (TMS-NS001) [76].

Allen et al. [77] proposed an algorithm for evaporation mapping based on the energy balance method. The algorithm was named Mapping Evapotranspiration at High Resolution with Internalized Calibration (METRIC). The method used satellite remote sensing images as the input data. This method inherited the basic concept of the near-surface temperature gradient as the function of radiometric surface temperature from SEBAL developed by Bastiaanssen et al. [17]. The internal calibration enhanced the applicability of this method by eliminating the need for an estimated surface temperature; air temperature also reduces the impacts of surface roughness. A significant feature of this study is that they consider the special case of the hilly/mountainous regions and developed elevation-specific methods for some of the parameters.

The method was verified using MODIS and LANDSAT images, a digital elevation model, and field observations of meteorological parameters. The method was primarily developed and validated to generate evapotranspiration maps on various temporal scales. A few disadvantages were reported by the authors: the requirement of high-quality and fine temporal data (hourly data), and of a trained operator that can handle the spatial images and correctly select the appropriate wet and dry pixels. A dedicated toolbox named METRIC-GIS for the ArcGIS software was developed and deployed by Ramírez-Cuesta et al. [78]. The tool was tested against the semi-arid environmental conditions of a region in southern Spain. The tool gives 100% accurate results (R^2 = 1) when compared to METRIC results.

The Penman-Monteith evapotranspiration equation is a method evolved from two studies by Penman [79] and Monteith [80]. Eq. 9.5 presents the mathematical form of the Penman-Monteith method.

$$ET = \frac{\Delta\left(R_N - G\right) + \rho C_P (e_s - e_a)/r_a}{(\Delta + \gamma(1 + \frac{r_s}{r_a}))} \tag{9.5}$$

where Δ is the slope of the saturation vapor pressure vs. temperature curve; ρ is air density; C_P is the specific heat of dry air; e_s is the saturation vapor pressure of the air; e_a is the actual vapor pressure of the air; r_a is aerodynamic resistance; r_s is a bulk surface resistance; and Γ is the psychrometric constant. The equation is used and modified by United Nations (UN) Food and Agriculture Organization (FAO) [81]. The modified Penman-Monteith equation for reference evapotranspiration (ET_o) is presented in Eq. 9.6 in which all other parameters are the same as defined for Eq. 9.5 and u_2 is the wind speed at a 2-meter height [35].

$$ET_o = \frac{0.408\Delta\left(R_N - G\right) + \gamma \dfrac{900}{T + 273} u_2 (e_s - e_a)}{\Delta + \gamma(1 + 0.34u_2)} \tag{9.6}$$

Priestley-Taylor evapotranspiration is a modified version of the Penman-Monteith equation. In the Priestley-Taylor equation, they eliminated the field requirement (e.g., surface roughness) and introduced a new constant α [82]. The equation is a simpler version of the Penman-Monteith equation. Eq. 9.7 presents the actual Priestley-Taylor equation.

$$PE = \alpha \frac{s}{s+\gamma} (R-G) \tag{9.7}$$

where PE is potential evapotranspiration; s is the slope of the saturation vapor density curve, and α is the model coefficient. In the original study, α was assigned the value of 1.26 [83]; however, the original study mentioned that the value of the coefficient α may vary according to the moisture conditions.

9.4 SURFACE ENERGY FLUXES AND VARIATIONS

In the previous sections, we discussed various data sources for the observations/estimations of surface energy fluxes. Then, a few SEB estimating algorithms were described that were developed and applied in the recent literature. In this section, we present the summary of a recently published study that estimated SEB and analyzed its long-term variations over the Tibetan Plateau or simply Tibet (hereafter TP). This study was recently published [51] by the authors of this chapter. This study analyzed variations of SEB from 2001 to 2019, while input parameters were obtained from various data sources including CERES, MODIS, and ERA5. CERES's downward shortwave and longwave products were used along with the albedo, emissivity, LST, and LE products obtained from MODIS to estimate surface energy fluxes. For reanalysis data, the ERA5 product generated by ECMWF was used. Selective and most related results of the published article are added here.

TP is one of the largest ice reservoirs and is referred to as the third pole of the world [1]. Due to its vast area and exceptional height (>4000 m above mean sea level), it plays a major role in shaping the regional climate especially for the east and south Asian monsoon [2, 3]. TP is a climate-sensitive region with different climate change behavior from the surrounding regions [4, 5]. According to some studies, the warming rate of TP is 0.3°C per decade, which is more than any other part of the

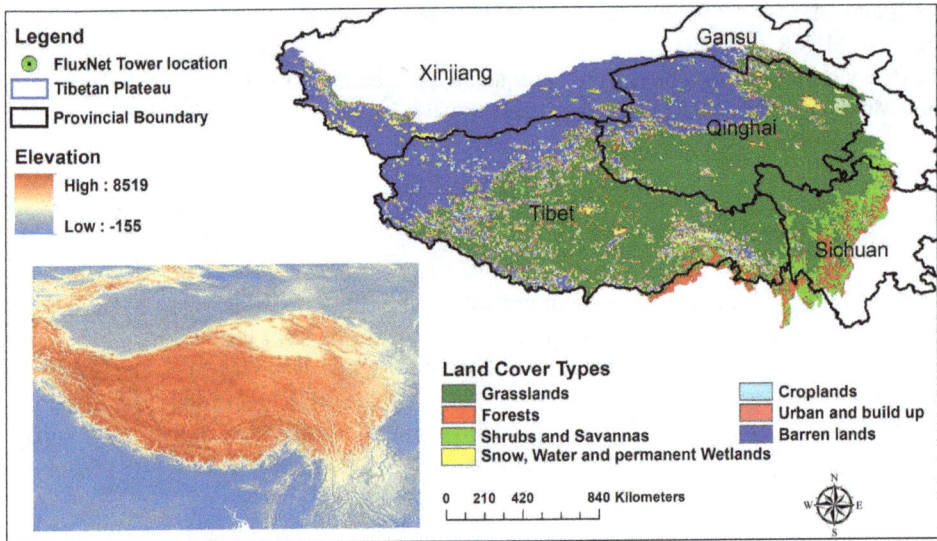

FIGURE 9.7 Administrative and land cover map of the Tibetan Plateau (TP). The land cover map is prepared from MODIS MCD12C1 data; 17 IGBP land cover classes are reclassified and merged into 7 major land covers in TP; insight shows the elevation map of TP using SRTM30 data set (download from https://earthexplorer.usgs.gov/) and locations of FluxNet tower sites.

Source: Figure courtesy of Mazhar et al. [51].

world. This affects permafrost and snow melting in TP. Eventually, freshwater reservoirs deplete at an alarming rate. Moreover, any change in the uplifted land affects the Asian monsoon upon which millions of people and thousands of hectares of crops depend [1, 2, 4, 6–8]. The major land cover of TP is grasslands that cover approximately two thirds of the total TP followed by deserted barren lands which cover approximately 14% of TP [84]. As an enormous uplifted land parcel, TP hugely impacts Asian monsoon and local climatic conditions. The area is highly affected by climate change and high solar radiation activity [85, 86]. Thus, the SEB analysis is important to understand its regional climatic phenomena. Figure 9.7 shows the elevation, land cover, and administrative map of TP.

9.4.1 Estimation of Surface Energy Fluxes

The first step is the estimation of various surface energy fluxes. The downward shortwave and longwave radiations were obtained from the CERES EBAF monthly product. For upward shortwave radiation Eq. 9.8 was used in which α is the albedo obtained from the MODIS MCD43C3 product, while for upward longwave radiation, Eq. 9.9 is used in which emissivity (ε) and LST (T) were obtained from the MODIS MOD11C3 product, and σ is the Stephan Boltzmann's constant (5.67×10^{-8} Wm^{-2}. K^{-4}). Also, by replacing upward shortwave and longwave radiations values in Eq. 9.3, R_N was computed as through Eq. 9.10.

$$R_S^u = \alpha R_S^d \tag{9.8}$$

$$R_L^u = \sigma \varepsilon T^4 \tag{9.9}$$

$$R_N = (R_S^d - \alpha R_S^d) + (R_L^d - \sigma \varepsilon T^4) \tag{9.10}$$

LE was obtained from the MODIS MOD16A2 product. G can be ignored over large temporal durations since it has a very small amplitude compared to other energy fluxes. Moreover, it is balanced to zero over the annual cycle [28]. Thus, the remaining energy flux SH was computed as the residual of energy balance. By replacing all the corresponding values and rearranging Eq. 9.1, SH was computed as:

$$SH = \{(R_S^d - \alpha R_S^d) + (R_L^d - \sigma \varepsilon T^4)\} - LE \tag{9.11}$$

Along with the estimated energy fluxes using remote sensing data, the study also used the same fluxes, i.e., R_N, SH, and LE obtained from ECMWF ERA-5 data, and then compared the validation results and variation outputs from the data sources.

9.4.2 Validation of Surface Energy Fluxes

Before the analysis of SEB, first the accuracy of data as well as method shall be verified. For this purpose, FluxNet tower observations were used. Figure 9.8 describes the relationship between satellite, ERA5, and *in situ* ground observations. Linear regression slope (LRS), Pearson's correlation coefficient (r-value), mean bias error (MBE), and mean absolute error (MAE) were used to check the accuracy of satellite and ERA5 data. Table 9.4 provides a summary of statistical analyses for the validation of SEB parameters. It is noteworthy that the significance of the correlation values (r-value and LRS) was obtained at a 95% confidence level. Any value which did not satisfy this condition is mentioned as insignificant.

R_N observations from satellite and ERA5 data sets are 99% significant with r-values of 0.87 and 0.88, respectively. All four statistical indicators validate the accuracy of the R_N from both data sets. LE values are 99% significant and show very less MBE −0.03 and 5.55 Wm^{-2} and MAE 18.98 and

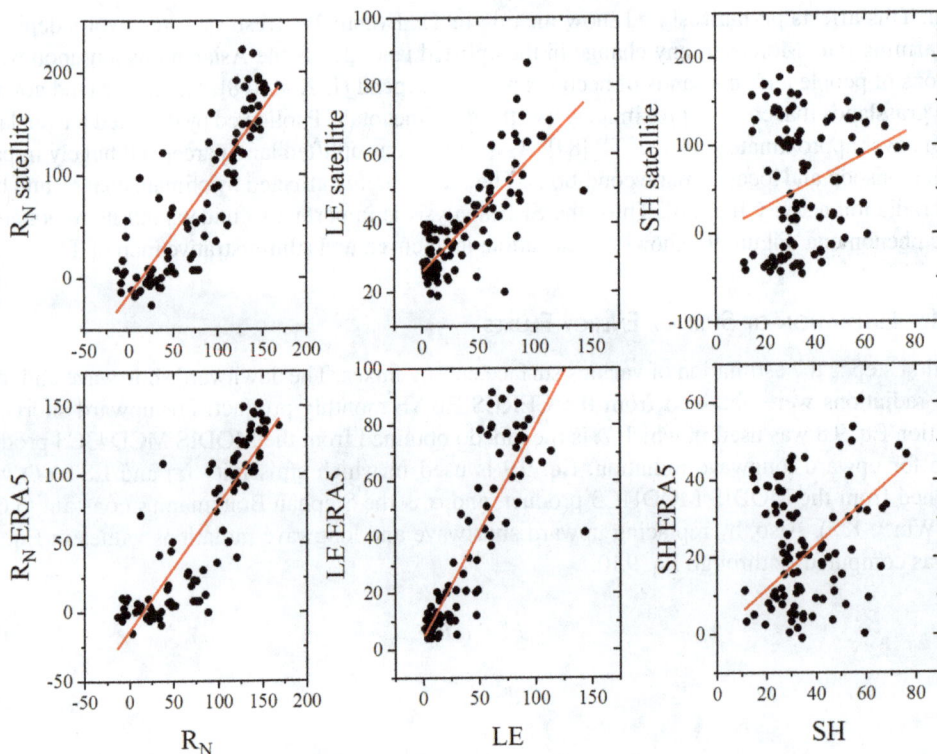

FIGURE 9.8 Correlation of satellite and ERA5 data with FluxNet observations. Red line describes the linear regression (LR) slope. Each black dot corresponds to a mean monthly value of surface energy budget (SEB) parameters measured in Wm^{-2}.

Source: Figure courtesy of Mazhar et al. [51].

TABLE 9.4
Validation Statistics of Satellite and ERA5 Data with Respect to FluxNet Observations

Statistical Analysis	R_N (Wm^{-2})		LE (Wm^{-2})		SH (Wm^{-2})	
	ERA5	satellite	ERA5	satellite	ERA5	satellite
LR Slope	0.91**	1.19**	0.75**	0.35**	0.75*	1.49
Pearson's r	0.88**	0.87**	0.86**	0.79**	0.81*	0.63
MBE (Wm^{-2})	20.53	0.33	5.55	−0.37	13.19	−21.8
MAE (Wm^{-2})	26.39	30.03	11.59	18.98	18.93	62.85

Note: For LR slope and Pearson's r, ** describes 99% significance while * describes 95% significance. Without * values are insignificant.
Source: Table courtesy of Mazhar et al. [51].

11.59 Wm^{-2} for satellite and ERA5 data, respectively. The LR slope value of satellite LE is the only indicator showing a weak relationship with FluxNet observed LE. However, based on the other three statistical parameters, the accuracy of satellite LE is established. For SH, ERA5 observations are validated from 95% significance with 0.81 r-value and 13.19 and 18.92 Wm^{-2} MBE and MAE, respectively. Satellite observed SH is insignificant (less than 95% significance level). Although the slope value is very high (1.49), yet the low r-value 0.63, and very high MAE 62.85 Wm^{-2}, make this parameter less accurate.

9.5   VARIATIONS OF SEB

9.5.1   Spatio-Temporal Analysis

For spatio-temporal trends of SEB over TP, the Mann-Kendall (MK) trend test was performed on mean annual raster images of R_N, LE, and SH obtained from satellite and ERA5 data. The results of MK tau (τ) values are shown in Figure 9.9. Positive values (in green and blue) represent increasing trends while negative values (in red and yellow) represent a decreasing trend. R_N shows a positive trend in east and southeast TP in satellite and ERA5 data. Major land covers in these regions are forest and shrub lands as shown in Figure 9.7. A few pixels in southeast TP show more increasing trends (trend values between 0.5 and 1) in ERA5 data while such an increasing trend is not observable in satellite data. The central regions of TP exhibit a decreasing trend in both data sets. For spatio-temporal trends in the east, southeast, and central TP, satellite and ERA5 data are approximately in agreement, but the upper region, the northern part of TP, shows an opposite trend from both data sets. In ERA5 data, these regions show an increasing R_N trend, with positive MK τ values, but satellite R_N exhibits a decreasing trend (<0 trend values). The major land cover in northern TP is barren land. An overall net positive spatio-temporal trend is dominant in the entire TP.

LE shows an increasing trend in both data sets, especially satellite LE exhibits a highly increasing trend (MK τ values approaching 1) in northeast TP. From satellite LE the only decreasing trend is observed in the central region of TP. One limitation of the MODIS LE product is that the product does not produce any values for barren lands; thus, the northern part of TP cannot be observed from satellite LE data, while ERA5 data covers the whole region. In ERA5 LE, a few regions from the northeast and southwest exhibit decreasing trends. The main contradiction is observed in the northeast where satellite LE shows a prominent increasing trend (0.5 to 1 trend value). In contrast, ERA5 LE shows a decreasing trend (−0.5 to 0 trend values). Overall, both data sets observe a significant increasing trend for LE. The limitation of LE data from the MODIS product prolongs SH since Eq. 9.11 was used to calculate SH. SH shows the only visibly negative trend amongst all SEB parameters. From satellite SH a minor increasing trend in east and central regions of TP is observed, while from ERA5, east, northwest, and a few central regions of TP show increasing trends. In satellite SH

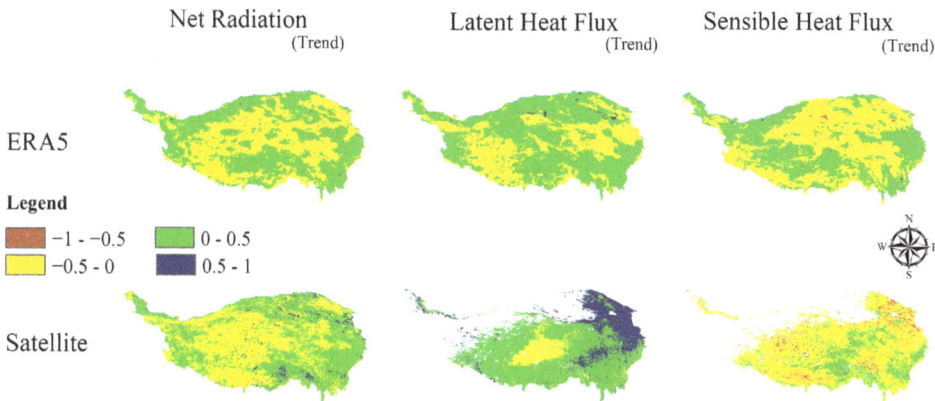

FIGURE 9.9   Spatio-temporal trends of R_N, LE, and SH from ERA5 and satellite data using MK τ value. Negative values (red and yellow) correspond to the decreasing trend while positive values (green and blue) correspond to the increasing trend; white regions in satellite LE and SH are the missing values.

Source: Figure courtesy of Mazhar et al. [51].

some regions from the northeast and southwest show a significant decreasing trend (−1 to −0.5 trend values). In ERA5 SH, such low trend values are not observed.

9.5.2 TEMPORAL ANALYSIS

For temporal analysis, mean annual aerial averages of R_N, LE, and SH were computed over TP. On these aerial averages, LRS and Sen's slope (SS) were computed on a decadal scale and for the whole study duration (2001–2019). These slope values were calculated at a 95% confidence level. Figure 9.10a, c and e show the ERA5 R_N, LE, and SH trends, while Figure 9.10b, d, and f show the satellite R_N, LE, and SH temporal trends. LRS lines for decadal trends (2001–2010 and 2011–2019) are shown in blue while temporal trends from 2001–2019 are shown in the red line. Table 9.5 summarizes the temporal trend values of the LRS and SS. Decadal trends of ERA5 R_N show negative trends in both LRS and SS for both decades. However, when the temporal trend is computed for complete study duration (2001–2019) LRS and SS show increasing trends with the value of 0.01 and 0.02, respectively. For satellite R_N, temporal trends for each decade as well as for total study duration show an increasing trend. The overall temporal trend for satellite R_N from LRS and SS is 0.01 and 0.03, respectively. Temporal trends of R_N from satellite and ERA5 data are in agreement for total study duration and show a nominal increasing trend.

For LE, all trends, including the decadal scale and for the total study duration from both data sets, show increasing trends in both statistical slopes. However, ERA5 LE shows an overall minor increasing trend with the LRS and SS value of 0.03, but for the same duration, satellite LE exhibits 0.25 LRS and SS values which exhibit the prominent increasing trends. For the decadal scale, satellite LE shows the more increasing temporal trend for the second decade (2011–2019). This effect is more pronounced in ERA5 LE in which the second decade shows a prominently

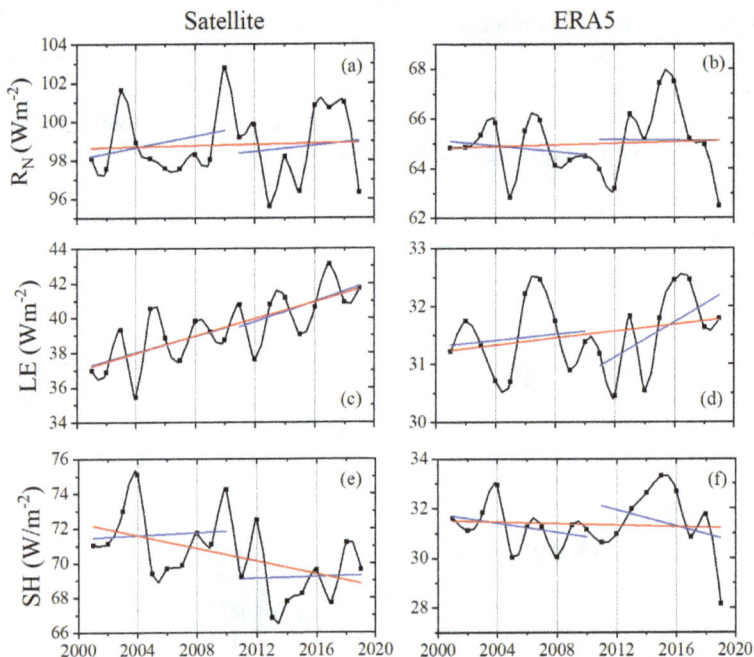

FIGURE 9.10 Temporal variation of RN, LE, and SH based on aerial average over TP. (a), (c), and (e) represent trends observed from satellite data while (b), (d), and (f) represent trends observed from ERA5 data. Decadal trends are shown in blue and the overall temporal trend from 2001–2019 is shown in red.

Source: Figure courtesy of Mazhar et al. [51].

TABLE 9.5

Decadal and Total Temporal Trends of RN, LE, and SH from ERA5 and Satellite Data Sets

		ERA5		Satellite	
		LRS	SS	LRS	SS
R_N	2001–10	−0.06 ±0.1	−0.04	0.15 ±0.2	0.007
	2011–19	−0.005 ±0.2	−0.03	0.08 ±0.2	0.18
	2001–19	0.01 ±0.05	0.02	0.01 ±0.08	0.03
LE	2001–10	0.02 ±0.7	0.02	0.24 ±0.1	0.19
	2011–19	0.15 ±0.08	0.15	0.29 ±0.1	0.25**
	2001–19	0.03 ±0.02	0.03	0.25 ±0.05 **	0.25**
SH	2001–10	−0.09 ±0.9	−0.05	0.04 ±0.2	0.11
	2011–19	−0.16 ±0.2	0.006	0.02 ±0.2	0.15
	2001–19	−0.02 ±0.05	0.005	−0.18 ±0.08 *	−0.18*

Note: LRS and SS are used for temporal trend analysis. ** describes p-value <0.01 while * describes p-value <0.05 at a 95% confidence level. The values after ± are standard errors in LRS.

Source: Table courtesy of Mazhar et al. [51].

increasing temporal trend (LRS and SS value 0.15) with respect to the first decade (LRS and SS value 0.02).

SH shows the mixed decadal and overall temporal trends from both data sets. In the LRS, ERA5 SH shows a continuously decreasing trend for all the temporal durations. Overall, the ERA5 LRS value is −0.016, which exhibits a weak decreasing trend. In the SS, the first decade of ERA5 SH shows a decreasing trend. In contrast, the second decade shows a very weak increasing trend. For the total duration, the SS value for ERA5 SH is 0.005, exhibiting a weak increasing trend. For satellite SH, LRS values for the first and second decade are 0.04 and 0.02, respectively, and for SS, these values are 0.1 and 0.14, respectively. All these four values exhibit a weak but increasing trend. For the total study duration, both LRS and SS values are −0.18, which exhibits a weak negative trend. Increasing decadal temporal trends but decreasing trends over a longer duration signify the importance of SEB analysis over a longer duration.

9.6 SUMMARY AND PERSPECTIVE

The earth's energy balance controls global atmospheric circulation and the climatological cycle. While in its simplistic form at TOA where only shortwave and longwave radiations are involved, energy balance is a complex phenomenon at the earth's surface. SEB comprises radiative and non-radiative energy fluxes. The radiative part of SEB includes downward and upward shortwave and longwave radiation, while the non-radiative part includes other energy fluxes viz. G, SH, and LE. At the surface, several factors are involved in determining SEB, including atmospheric, meteorological, biophysical, and biogeochemical factors. Also, anthropogenic activities effects largely on SEB. Land cover change, deforestation, thawing of permafrost, melting of glaciers, greenhouse gas emissions, and rapid urbanization are the significant factors responsible for SEB variations. It was also observed that the response to these changes on SEB is different on various latitudinal levels.

The accurate estimation and observation of SEB is an integral requirement to understand global and regional climate. This chapter summarizes various progress made in the observations/estimations of SEB. Various global ground observing networks, e.g., FLUXNET provide accurate and valuable SEB data using the eddy covariance method. FLUXNET is also referred to as the network of networks because it includes many regional networks such as AmeriFLUX, AsiaFLUX, ICOS, and OxFLUX. Since the ground observing stations have the limitation that they cannot evenly cover the whole earth, an effective alternative is satellite remote sensing data. Since the evolution of space technology, climate scientists have initiated many projects/missions to observe SEB from space. This chapter discussed SRB, ERBS, CERES, and MODIS as significant data-providing sources for SEB. Whereas CERES provides large-scale radiation data at TOA and surface level, MODIS provides many parameters that help estimate SEB with acceptable accuracy. A few satellite-based data suites are also available to provide global long-term radiation and LE data, e.g., GLASS and GLEAM. While the ground observing stations are limited to certain point locations, satellite data sources are limited to the satellite technology that has evolved a few decades ago. Reanalysis data is another suitable option to estimate/compute long-range SEB data. Various well-established and widely used reanalysis data products such as ERA-5, MERRA-2, JRA-55, and NCEP/NCAR are valuable SEB data. Each reanalysis data product uses a different algorithm for the SEB estimation and hence differs in the outputs.

With the advent of complex flux observing instruments, modern space technology, and more sophisticated computing algorithms, today we have fairly clear knowledge about Earth's energy balance. However, the complex and multi-dimensional climatic processes are yet to be explored in deeper detail. As discussed, each observing/estimating method/technology has its own limitation. Also, it is observed that no two data sources provide the same flux values due to the different sources of input parameters and used algorithms. Consequently, definite energy flux values are not available. Also, the effects of abrupt anthropogenic activities on energy fluxes are complex in nature and need detailed research. Thus, with the evolution of human knowledge, new aspects of energy fluxes and their influencing factors are analyzed every day. While scientists have covered a long distance, there is still a long way to go to understand the earth's climate for which cutting-edge technology and mathematical sophistication are required.

REFERENCES

[1] M. Wild, A. Ohmura, and K. Makowski, "Impact of global dimming and brightening on global warming," Geophys. Res. Lett., vol. 34, no. 4, pp. 1–4, 2007.

[2] G. Myhre, B. H. Samset, O. Hodnebrog, et al., "Sensible heat has significantly affected the global hydrological cycle over the historical period," Nat. Commun., vol. 9, no. 1, pp. 1–9, 2018.

[3] G. L. Stephens, J. Li, M. Wild, et al., "An update on Earth's energy balance in light of the latest global observations," Nat. Geosci., vol. 5, no. 10, pp. 691–696, 2012.

[4] R. G. Anderson, J. G. Canadell, J. T. Randerson, et al., "Biophysical considerations in forestry for climate protection," Front. Ecol. Environ., vol. 9, no. 3, pp. 174–182, 2011.

[5] G. B. Bonan, "Forests and climate change: Forcings, feedbacks, and the climate benefits of forests," Science, vol. 320, no. 5882, pp. 1444–1449, 2008.

[6] G. Duveiller, J. Hooker, and A. Cescatti, "The mark of vegetation change on Earth's surface energy balance," Nat. Commun., vol. 9, no. 1, pp. 679, 2018.

[7] J. Houspanossian, M. Nosetto, and E. G. Jobbágy, "Radiation budget changes with dry forest clearing in temperate Argentina," Glob. Chang. Biol., vol. 19, no. 4, pp. 1211–1222, 2013.

[8] X. Lee, M. L. Goulden, D. Y. Hollinger, et al., "Observed increase in local cooling effect of deforestation at higher latitudes," Nature, vol. 479, no. 7373, pp. 384–387, 2011.

[9] Y. Li, M. Zhao, S. Motesharrei, Q. Mu, E. Kalnay, and S. Li, "Local cooling and warming effects of forests based on satellite observations," Nat. Commun., vol. 6, no. 1, pp. 1–8, 2015.

[10] U. Mazhar, S. Jin, M. Bilal, M. Arfan Ali, and R. Khan, "Reduction of surface radiative forcing observed from remote sensing data during global COVID-19 lockdown," Atmos. Res., vol. 261, p. 105729, 2021.

[11] M. A. Ali, M. M. Islam, M. N. Islam, and M. Almazroui, "Investigations of MODIS AOD and cloud properties with CERES sensor based net cloud radiative effect and a NOAA HYSPLIT Model over Bangladesh for the period 2001–2016," Atmos. Res., vol. 215, pp. 268–283, 2019.

[12] M. Etminan, G. Myhre, E. J. Highwood, and K. P. Shine, "Radiative forcing of carbon dioxide, methane, and nitrous oxide: A significant revision of the methane radiative forcing," Geophys. Res. Lett., vol. 43, no. 24, pp. 12,614–12,623, 2016.

[13] V. Masson-Delmotte, P. Zhai, H. O. Pörtner, et al., Summary for Policymakers, Global Warming of 1.5 °C—IPCC Special Report on the Impacts of Global Warming of 1.5 °C above Pre-Industrial Levels and Related Global Greenhouse Gas Emission Pathways, Cambridge University Press, 2022.

[14] E. Lewis, O. Chamel, M. Mohsenin, E. Ots, and E. T. White, "Intergovernmental Panel on Climate Change," in Sustainaspeak, 2018, pp. 153–154.

[15] K. P. Shine, "Radiative forcing of climate change," Space Sci. Rev., vol. 94, no. 1–2, pp. 363–373, 2000.

[16] D. Brunt, "Notes on radiation in the atmosphere. I," Q. J. R. Meteorol. Soc., vol. 58, no. 247, pp. 389–420, 1932.

[17] W. G. M. Bastiaanssen, M. Menenti, R. A. Feddes, and A. A. M. Holtslag, "A remote sensing surface energy balance algorithm for land (SEBAL): 1. Formulation," J. Hydrol., vol. 212–213, no. 1–4, pp. 198–212, 1998.

[18] Z. Su, "The Surface Energy Balance System (SEBS) for estimation of turbulent heat fluxes," Hydrol. Earth Syst. Sci., vol. 6, no. 1, pp. 85–99, 2002.

[19] W. P. Kustas, B. J. Choudhury, M. S. Moron, et al., "Determination of sensible heat flux over sparse canopy using thermal infrared data," Agric. For. Meteorol., vol. 44, no. 3–4, pp. 197–216, 1989.

[20] A. Abdelrady, J. Timmermans, Z. Vekerdy, and M. S. Salama, "Surface energy balance of fresh and saline waters: AquaSEBS," Remote Sens., vol. 8, no. 7, pp. 583, 2016.

[21] S. Self, J. X. Zhao, R. E. Holasek, R. C. Torres, and A. J. King, "The Atmospheric Impact of the 1991 Mount Pinatubo Eruption," No. NASA/CR-93-207274.1993.

[22] U. S. Nair, D. K. Ray, J. Wang, et al., "Observational estimates of radiative forcing due to land use change in southwest Australia," J. Geophys. Res. Atmos., vol. 112, no. 9, pp. 1–8, 2007.

[23] S. Liang, D. Wang, T. He, and Y. Yu, "Remote sensing of earth's energy budget: synthesis and review," Int. J. Digit. Earth, vol. 12, no. 7, pp. 737–780, 2019.

[24] K. Keegan, "Snow Metamorphism and Firn Compaction," in Treatise on Geomorphology, 2nd ed., vol. 4, Academic Press, 2022, pp. 39–51.

[25] G. L. Stephens, D. O'Brien, P. J. Webster, P. Pilewski, S. Kato, and J. L. Li, "The albedo of earth," Rev. Geophys., vol. 53, no. 1. pp. 141–163, 2015.

[26] T. J. Sauer and R. Horton, "Soil Heat Flux," Micrometeorol Agric. Syst., vol. 47, pp. 131–154, 2005.

[27] A. J. Purdy, J. B. Fisher, M. L. Goulden, and J. S. Famiglietti, "Ground heat flux: An analytical review of 6 models evaluated at 88 sites and globally," J. Geophys. Res. Biogeosciences, vol. 121, no. 12, pp. 3045–3059, 2016.

[28] W. G. M. Bastiaanssen, M. J. M. Cheema, W. W. Immerzeel, I. J. Miltenburg, and H. Pelgrum, "Surface energy balance and actual evapotranspiration of the transboundary Indus Basin estimated from satellite measurements and the ETLook model," Water Resour. Res., vol. 48, no. 11, pp. 1–16, 2012.

[29] M. Fuchs and A. Hadas, "The heat flux density in a non-homogeneous bare loessial soil," Boundary-Layer Meteorol., vol. 3, no. 2, pp. 191–200 1972.

[30] B. E. Clothier, K. L. Clawson, P. J. Pinter, M. S. Moran, R. J. Reginato, and R. D. Jackson, "Estimation of soil heat flux from net radiation during the growth of alfalfa," Agric. For. Meteorol., vol. 37, no. 4, pp. 319–329, 1986.

[31] S. B. Idso, J. K. Aase, and R. D. Jackson, "Net radiation – soil heat flux relations as influenced by soil water content variations," Boundary-Layer Meteorol., vol. 9, no. 1, pp. 113–122, 1975.

[32] B. J. Choudhury, S. B. Idso, and R. J. Reginato, "Analysis of an empirical model for soil heat flux under a growing wheat crop for estimating evaporation by an infrared-temperature based energy balance equation," Agric. For. Meteorol., vol. 39, no. 4, pp. 283–297, 1987.

[33] P. K. Taylor, "Air–sea interaction I Momentum, heat, and vapor fluxes," Encycl. Atmos. Sci., vol. 1, pp. 93–100, 2003.

[34] W. Ma, G. Jia, and A. Zhang, "Multiple satellite-based analysis reveals complex climate effects of temperate forests and related energy budget," J. Geophys. Res., vol. 122, no. 7, pp. 3806–3820, 2017.

[35] R. G. Allen, L. S. Pereira, D. Raes, and M. Smith, "FAO Irrigation and Drainage Paper No. 56 – Crop Evapotranspiration," Food and Agriculture Organization of United Nations, Rome, vol. 300, no. 9, pp. D05109, 1998.

[36] M. Aubinet, T. Vesala, and D. Papale, Eddy Covariance: A Practical Guide to Measurement and Data Analysis. Springer Science & Business Media, 2012.

[37] G. Pastorello. C. Trotta, E. Canfora, et al., "The FLUXNET2015 dataset and the ONEFlux processing pipeline for eddy covariance data," Sci. Data, vol. 7, no. 1, p. 225, 2020.

[38] D. Baldocchi, E. Falge, L. Gu, et al., "FLUXNET: A new tool to study the temporal and spatial variability of ecosystem-scale carbon dioxide, water vapor, and energy flux densities," Bull. Am. Meteorol. Soc., vol. 82, no. 11, pp. 2415–2434, 2001.

[39] Y. Ma, Z. Su, T. Koike, et al., "On measuring and remote sensing surface energy partitioning over the Tibetan Plateau-from GAME/Tibet to CAMP/Tibet," Phys. Chem. Earth, vol. 28, no. 1–3, pp. 63–74, 2003.

[40] M. T. Chahine, "GEWEX: The Global Energy and Water Cycle Experiment," Eos, Trans. Am. Geophys. Union, vol. 73, no. 2, pp. 9–14, 1992.

[41] Y. Ding, W. Hu, Y. Huang, and F. Chen, "Major scientific achievements of the first China-Japan cooperative GAME/HUBEX experiment: A historical review," Journal of Meteorological Research, vol. 34, no. 4. 2020.

[42] K. E. Trenberth, "25-year SSM/I and SSMIS global data record: A valuable resource for global water cycle studies," Publ. by Int. GEWEX Proj. Off., vol. 23, no. 3, pp. 3–5, 2012.

[43] K. Nakamura, "GAME Letter," Publ. by GAME Int. Proj. Off., 2002.

[44] U. Mazhar, S. Jin, T. Hu, M. Bilal, M. A. Ali, and L. Atique, "Long-time Variation and Mechanism of Surface Energy Budget over Diverse Geographical Regions in Pakistan," IEEE J. Sel. Top. Appl. Earth Obs. Remote Sens., vol. 15, no. 1, pp. 1–13, 2022.

[45] B. R. Barkstrom, "The Earth Radiation Budget Experiment (ERBE).," Bull. – Am. Meteorol. Soc., vol. 65, no. 11, pp. 1170–1185, 1984.

[46] T. He, D. Wang, and Y. Qu, "Land surface albedo," Compr. Remote Sens., vol. 1–9, pp. 140–162, 2017.

[47] N. G. Loeb et al., "Clouds and the Earth'S Radiant Energy System (CERES) Energy Balanced and Filled (EBAF) top-of-atmosphere (TOA) edition-4.0 data product," J. Clim., vol. 31, no. 2, pp. 895–918, 2018.

[48] N. G. Loeb, D. R. Doelling, H. Wang, et al., "Determination of unfiltered radiances from the clouds and the earth's radiant energy system instrument," J. Appl. Meteorol., vol. 40, no. 4, pp. 822–835, 2001.

[49] W. Su, J. Corbett, Z. Eitzen, and L. Liang, "Next-generation angular distribution models for top-of-atmosphere radiative flux calculation from CERES instruments: Methodology," Atmos. Meas. Tech., vol. 8, no. 2, pp. 611–632, 2015.

[50] S. Kato, F. G. Rose, D. A. Rutan, et al., "Surface irradiances of edition 4.0 Clouds and the Earth's Radiant Energy System (CERES) Energy Balanced and Filled (EBAF) data product," J. Clim., vol. 31, no. 11, pp. 4501–4527, 2018.

[51] U. Mazhar, S. Jin, W. Duan, M. Bilal, M. A. Ali, and H. Farooq, "Spatio-temporal trends of surface energy budget in tibet from satellite remote sensing observations and reanalysis data," Remote Sens., vol. 13, no. 2, pp. 1–20, 2021.

[52] K. Wang, Z. Wan, P. Wang, et al., "Estimation of surface long wave radiation and broadband emissivity using moderate resolution imaging spectroradiometer (MODIS) land surface temperature/emissivity products," J. Geophys. Res. D Atmos., vol. 110, no. 11, pp. 1–12, 2005.

[53] Z. Wan, "New refinements and validation of the MODIS Land-Surface Temperature/Emissivity products," Remote Sens. Environ., vol. 112, no. 1, pp. 59–74, 2008.

[54] C. B. Schaaf et al., "First operational BRDF, albedo nadir reflectance products from MODIS," Remote Sens. Environ., 2002.

[55] C. B. Schaaf, F. Gao, A. H. Strahler, et al., "First operational BRDF, albedo nadir reflectance products from MODIS," vol. 83, no. 1, pp. 135–148, 2002.

[56] M. O. Román C. B. Schaaf, P. Lewis, et al., "Assessing the coupling between surface albedo derived from MODIS and the fraction of diffuse skylight over spatially-characterized landscapes," Remote Sens. Environ., vol. 114, no. 4, pp. 738–760, 2010.

[57] C. B. Schaaf, J. Liu, F. Gao, and A. H. Strahler, "Aqua and terra MODIS albedo and reflectance anisotropy products," Remote Sens. Digital Image Process., vol. 11, pp. 549–561, 2011.

[58] Z. Wan, "New refinements and validation of the collection-6 MODIS land-surface temperature/emissivity product," Remote Sens. Environ., vol. 140, pp. 36–45, 2014.

[59] K. Wang, Z. Wan, P. Wang, M. Sparrow, J. Liu, and S. Haginoya, "Evaluation and improvement of the MODIS land surface temperature/emissivity products using ground-based measurements at a semi-desert site on the western Tibetan Plateau," Int. J. Remote Sens., vol. 28, no. 11, pp. 2549–2565, 2007.

[60] Z. Wan, "A generalized split-window algorithm for retrieving land-surface temperature from space," IEEE Trans. Geosci. Remote Sens., vol. 34, no. 4, pp. 892–905, 1996.

[61] W. C. Snyder and Z. Wan, "BRDF models to predict spectral reflectance and emissivity in the thermal infrared," IEEE Trans. Geosci. Remote Sens., vol. 36, no. 1, pp. 214–225, 1998.

[62] S. Liang, T. Zheng, R. Liu, H. Fang, S. C. Tsay, and S. Running, "Estimation of incident photosynthetically active radiation from Moderate Resolution Imaging Spectrometer data," J. Geophys. Res. Atmos., vol. 111, no. 15, pp. 1–13, 2006.

[63] D. Wang, S. Liang, Y. Zhang, X. Gao, M. G. L. Brown, and A. Jia, "A new set of modis land products (Mcd18): Downward shortwave radiation and photosynthetically active radiation," Remote Sens., vol. 12, no. 1, 2020.

[64] J. Huang, "A simple accurate formula for calculating saturation vapor pressure of water and ice," J. Appl. Meteorol. Climatol., vol. 57, no. 6, pp. 1265–1272, 2018.

[65] Q. Mu, M. Zhao, and S. W. Running, "Improvements to a MODIS global terrestrial evapotranspiration algorithm," Remote Sens. Environ., vol. 115, no. 8, pp. 1781–1800, 2011.

[66] Q. Mu, F. A. Heinsch, M. Zhao, and S. W. Running, "Development of a global evapotranspiration algorithm based on MODIS and global meteorology data," Remote Sens. Environ., vol. 111, no. 4, pp. 519–536, 2007.

[67] S. Liang, X. Zhao, S. Liu, et al., "A long-term Global LAnd Surface Satellite (GLASS) data-set for environmental studies," Int. J. Digit. Earth, vol. 6, no. SUPPL1, pp. 5–33, 2013.

[68] S. Liang, X. Zhang, Z. Xiao, J. Cheng, Q. Liu, and X. Zhao, Global LAnd Surface Satellite (GLASS) Products: Algorithms, Validation and Analysis. Springer Science & Business Media, 2013.

[69] B. Martens, D. L. Schumacher, H. Wouters, J. Muñoz-Sabater, N. E. C. Verhoest, and D. G. Miralles, "Evaluating the land-surface energy partitioning in ERA5," Geosci. Model Dev., vol. 13, no. 9, pp. 4159–4181, 2020.

[70] D. G. Miralles, T. R. H. Holmes, R. A. M. De Jeu, J. H. Gash, A. G. C. A. Meesters, and A. J. Dolman, "Global land-surface evaporation estimated from satellite-based observations," Hydrol. Earth Syst. Sci., vol. 15, no. 2, pp. 453–469, 2011.

[71] H. Hersbach, B. Bell, P. Berrisford, et al., "The ERA5 global reanalysis," Q. J. R. Meteorol. Soc., vol. 146, no. 730, pp. 1999–2049, 2020.

[72] R. Gelaro, W. McCarty, M. J. Suárez, et al., "The modern-era retrospective analysis for research and applications, version 2 (MERRA-2)," J. Clim., vol. 30, no. 14, pp. 5419–5454, 2017.

[73] W. G. M. Bastiaanssen, H. Pelgrum, J. Wang, et al., "A remote sensing surface energy balance algorithm for land (SEBAL): 2. Validation," J. Hydrol., vol. 212–213, no. 1–4, 1998.

[74] X. C. Zhang, J. W. Wu, H. Y. Wu, and Y. Li, "Simplified SEBAL method for estimating vast areal evapotranspiration with MODIS data," Water Sci. Eng., vol. 4, no. 1, pp. 24–35, 2011.

[75] G. J. Roerink, Z. Su, and M. Menenti, "S-SEBI: A simple remote sensing algorithm to estimate the surface energy balance," Phys. Chem. Earth, Part B Hydrol. Ocean. Atmos., vol. 25, no. 2, pp. 147–157, 2000.

[76] H. Bolle, J. C. Andre, J. L. Arrue, et al., "EFEDA: European Field Experiment in a Desertification-threatened Area," Ann. Geophys., vol. 11, no. 2, 1988.

[77] R. G. Allen, M. Tasumi, and R. Trezza, et al., "Satellite-Based Energy Balance for Mapping Evapotranspiration with Internalized Calibration (METRIC)—Applications," J. Irrig. Drain. Eng., vol. 133, no. 4, pp. 395–406, 2007.

[78] J. M. Ramírez-Cuesta, R. G. Allen, D. S. Intrigliolo, et al., "METRIC-GIS: An advanced energy balance model for computing crop evapotranspiration in a GIS environment," Environ. Model. Softw., vol. 131, p. 104770, 2020.

[79] H. L. Penman, "Natural Evaporation from Open Water, Bare Soil and Grass," Proc. R. Soc. Lond. A. Math. Phys. Sci., vol. 193, no. 1032, pp. 120–145, 1948.

[80] J. L. Monteith, "Evaporation and environment.," Symposia of the Society for Experimental Biology, vol. 19, 1965.

[81] R. G. Allen, M. Smith, A. Perrier and L. S. Pereira, "An update for the defination of reference evapotranspiration," ICID Bulletin, vol. 43. no. 2, pp. 1–34, 1994.

[82] A. L. Flint and S. W. Childs, "Use of the Priestley-Taylor evaporation equation for soil water limited conditions in a small forest clearcut," Agric. For. Meteorol., vol. 56, no. 3–4, pp. 247–260, 1991.

[83] C. H. B. Priestley and R. J. Taylor, "On the Assessment of Surface Heat Flux and Evaporation Using Large-Scale Parameters," Mon. Weather Rev., vol. 100, no. 2, pp. 81–92, 1972.

[84] X. Cui and H. F. Graf, "Recent land cover changes on the Tibetan Plateau: A review," Clim. Change, vol. 94, no. 1–2, pp. 47–61, 2009.

[85] J. Duan, J. Esper, U. Buntgen, et al., "Weakening of annual temperature cycle over the Tibetan Plateau since the 1870s," Nat. Commun., vol. 8, pp. 1–7, 2017.

[86] X. Wang, D. Zheng, and Y. Shen, "Land use change and its driving forces on the Tibetan Plateau during 1990–2000," Catena, vol. 72, no. 1, pp. 56–66, 2008.

10 Aboveground Carbon Dynamics from SMOS L-Band Vegetation Optical Depth

Lei Fan and Jean-Pierre Wigneron

Aboveground biomass (AGB) is an important proxy for productivity, carbon sequestration and carbon balance capacity in terrestrial ecosystems. Accurate estimation of AGB in terrestrial ecosystems is fundamental for quantifying carbon emissions and removals due to land use and climate change. Remote sensing is poised to advance the mapping of vegetation structure and quantify the stocks and changes of aboveground carbon (AGC) in vegetation. Vegetation optical depth (VOD), retrieved from passive microwave satellite observations and related to the water content of vegetation mass, offers opportunities for monitoring the AGC dynamics due to its key features. In this chapter, the new VOD product, hereafter L-VOD, has been produced using low-frequency (L-band, 1.4 GHz) microwave observations from the Soil Moisture and Ocean Salinity (SMOS) satellite. Spatial changes in AGC are derived from the L-VOD product over 2010–2017 across the pantropics. The L-VOD data set allowed us to gain new insights into the dynamics of tropical AGC and the co-variation with climate, anthropogenic forest cover disturbances and changes in the global atmospheric CO^2 concentration.

10.1 INTRODUCTION

Aboveground biomass (AGB), defined as the total amount of aboveground living organic matter in vegetation, is an important proxy for productivity, carbon sequestration and carbon balance capacity in terrestrial ecosystems [1, 2]. Accurate estimation of AGB in terrestrial ecosystems is fundamental for quantifying carbon emissions and removals due to land use and climate change [3–5]. It is therefore critical to monitor the AGB stocks and its dynamics to mitigate climate change.

Tropical terrestrial biomes contribute to the interannual variability of the global terrestrial carbon balance which in turn is essential to changes in the global atmospheric CO_2 concentration [6]. Thus, accurate monitoring of temporal and spatial changes in carbon stocks over the tropics is key for better predicting the evolution of the atmospheric CO_2 over the coming century. However, at present no method exists for spatially explicit quantification of the tropical land sink/source. Current observational tools are impeded by signal saturation in dense forests [7] and sparse spatial or temporal sampling [8] so that the spatial distribution and the trends of carbon sources and sinks across the tropics remains poorly resolved [6].

Results from top-down atmospheric inversions that are consistent with vertical CO_2 profiles [9] indicate that the long term tropical net CO_2 flux is close to zero, but *in situ* surface CO_2 stations are too scarce to separate carbon sinks from tropical forest regrowth and carbon sources from deforestation. Bottom-up approaches using ground forest inventory and satellite data suggest that tropical deforestation represents large emissions 0.57–1.3 Pg C yr^{-1} [10–12]. A more diffuse carbon sink is observed in undisturbed and re-growing forests [13], but also a decline of the forest carbon sink in the Amazon [14] and a strong reduction of this sink during extreme El Niño events have been reported [15]. However, forest inventory data are also scarce in the tropics [6] and semi-arid woody biomes are critically under-sampled, although they cover 40% of the tropical land area [16].

DOI: 10.1201/9781003363118-10

The interannual variability of carbon fluxes from tropical land to the atmosphere is also coupled with climatic conditions, and the increased frequency of drought events is a threat to tropical forest biomes [17]. Major droughts in 2005, 2010 and 2015–2016 represent a testing ground for understanding how the frequency of extreme climatic events may affect the carbon balance in the future. Recent studies suggest that the tropics switched to a net source during the 2015–2016 El Niño [18–20], which is supported by model simulations [21]. However, observations of the spatial distribution of this major flux anomaly are still unavailable, limiting the attribution of the El Niño anomaly to specific tropical continents and biomes [6].

Remote sensing is poised to advance the mapping of vegetation structure and quantify the stocks and changes of aboveground carbon (AGC) in vegetation [8, 10, 22]. Although static maps of AGC have been produced from remote sensing [8, 10, 23, 24], these maps generally differ both in terms of magnitude and spatial patterns and are available only for a single epoch, and therefore cannot be used to assess interannual variations in carbon stocks [22].

Vegetation optical depth (VOD), retrieved from passive microwave satellite observations and related to the water content of vegetation mass [25–27], offers opportunities for monitoring the AGC dynamics [20, 28, 29] due to its key features: frequent observations providing daily tropical coverage and independence of the effects of atmospheric and cloud contamination [28]. The new VOD product used in this study, hereafter L-VOD, has recently been produced using low-frequency (L-band, 1.4 GHz) microwave observations from the Soil Moisture and Ocean Salinity (SMOS) satellite [30, 31]. The radiometer onboard the SMOS satellite has superior sensitivity to carbon density than previous higher-frequency passive microwave VOD products, and is able to retrieve the overall aboveground carbon stocks even in dense tropical ecosystems [20, 32, 33]. In contrast, high-frequency VOD products [34] saturate in vegetation with carbon stocks higher than 100 Mg C ha^{-1} [28].

Here, we used the L-VOD product to derive spatially explicit representations of changes in AGC (methods) over 2010–2017 across the pantropics (consisting of tropical America, Africa and Asia between 23.45°N and 23.45°S, excluding Australia), known to play a pivotal role in the global terrestrial carbon sink [6]. The L-VOD data set allowed us to gain new insights into the dynamics of tropical AGC and the co-variation with climate, anthropogenic forest cover disturbances and changes in the global atmospheric CO_2 concentration.

10.2 DATA SETS AND METHODS

The L-VOD index used in this study is sensitive to the total vegetation water content (VWC, Mg ha^{-1}) [33]. The relationship between L-VOD and VWC is nearly linear [30, 35]. L-VOD for woody vegetation is mainly sensitive to the water content of stems and branches, so the effects of leaves can be neglected to a first order [33]. Moreover, a specificity of SMOS is its multi-angular capability which allows a robust decoupling of the effects of soil moisture and vegetation opacity (parameterized by L-VOD) [33]. This capability arises from the design of the synthetic aperture imaging antenna of the SMOS L-band microwave radiometer and is exploited in the SMOS-IC algorithm, which is based on the original SMOS algorithm [36] as defined for the ESA Earth Explorer mission call. The principle of the algorithm is to retrieve simultaneously both SM and L-VOD for "rich" SMOS observational configurations (e.g., when a large range of multi-angular observations are available) and to benefit from the slow time variations of L-VOD for "poor" SMOS observational configurations (e.g., when a narrow range of multi-angular observations are available). The high accuracy of both the SMOS-IC SM and L-VOD products have been evaluated in several recent studies [20, 21, 32, 33, 37].

We assumed that the yearly average of the moisture content (%) of stems/branches for woody vegetation at the spatial scale of the SMOS grid (25 km × 25 km) was relatively constant between years, so that the yearly average of vegetation water content and dry biomass would be strongly correlated over time. This assumption is supported by several studies reporting the strong relationship

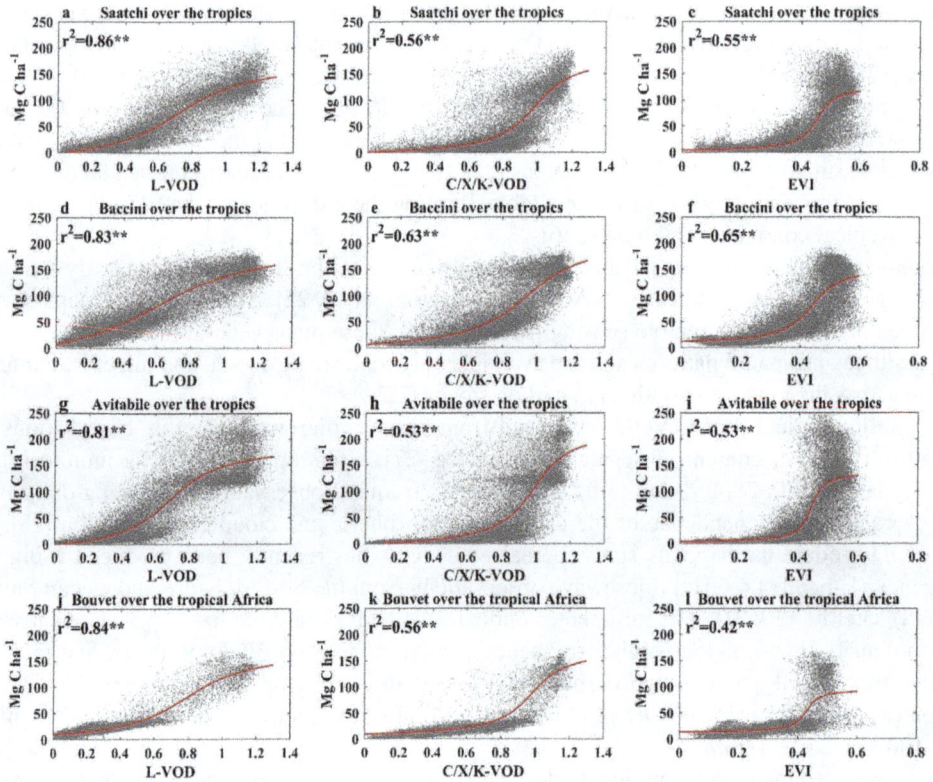

FIGURE 10.1 Scatterplots between benchmark AGC density (Mg C ha^{-1}) maps and yearly mean L-VOD values (a, d, g and j), C/X/K-VOD (b, e, h and k) (from Liu et al., 2015) and EVI (c, f, i and l) in 2011 for the tropics based on the Saatchi, Baccini, Avitabile and Bouvet-Mermoz reference data sets. Fitted relationships of the scatterplots (using Eq. 10.1 in main text) are indicated in red.

between L-VOD and biomass for woody vegetation being almost linear and independent of the year of calculation [20, 32]. The yearly average of L-VOD, by its strong link to vegetation water content, can thus be considered as a robust proxy of biomass. Other remotely sensed estimates/proxies of biomass have been used to estimate the annual changes in AGC at continental scales, such as LIDAR estimates of canopy height [8, 10], high-frequency VOD [28] or radar backscattering [8]. Radar backscattering was strongly sensitive to forest structure, but its relationship to biomass is highly nonlinear at L-band [24]. The computation of L-VOD in the SMOS-IC version is independent of the use of these indexes, making it a new and complementary tool for monitoring AGC.

L-VOD is more closely related to AGC density ($r^2 = 0.81$–0.86), as compared to C/X/K-VOD ($r^2 = 0.53$–0.63) and EVI ($r^2 = 0.42$–0.65) over the tropics (Figure 10.1), which is in line with previous findings over Africa [20, 32]. The relationship between AGC and C/X/K-VOD (Figure 10.1b, e, h and k) has a similar shape to that of AGC versus L-VOD Figure 10.1a, d, g and j) but C/X/K-VOD shows a stronger saturation at high AGC values relatively to L-VOD. EVI shows some sensitivity to AGC for low AGC values (with a low slope) but clear saturation effects are found for medium or high AGC values (Figure 10.1c, f, i and l).

AGC was firstly retrieved from the L-VOD product based on an empirical calibration of the spatial relationships linking L-VOD to reference AGC gridded data sets, as in Brandt et al. [20]. The reference AGC data sets were obtained from static benchmark maps (corresponding to average values over a few years). Assuming that a good calibration can be achieved, the SMOS L-VOD product adds a temporal dimension to static maps provided that a "space for time" substitution holds

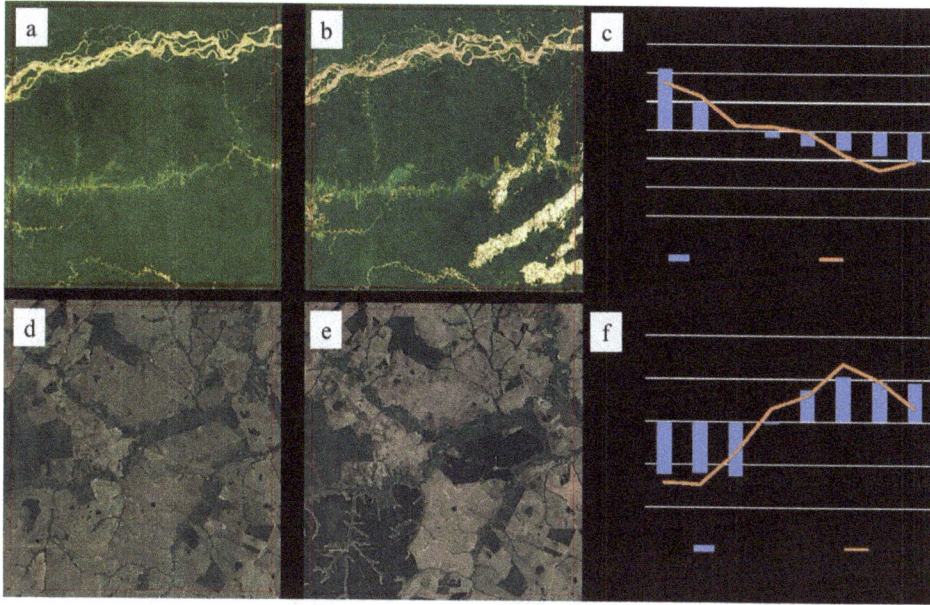

FIGURE 10.2 Comparison of time series of MOD100 annual forest area (%) and AGC density over (a-c) a forest-dominated SMOS-gridcell (25 km²) in Peru (12.875°S, 70.125°W), corresponding to area and AGC losses and (d-f) a forest and other vegetation mixed SMOS-gridcell (25 km²) in southern Brazil (19.625°S, 52.625°W), corresponding to area and AGC gain, during 2010–2017. (a-b) Landsat images within the Peruvian SMOS-gridcell acquired in December 2009 and December 2016 showing forest area loss caused by mining. (c) Z-score analysis between annual forest area and AGC over the Peruvian site with forest area loss (r = 0.94 between annual AGC density and forest area). (d-e) Landsat images within the Brazilian SMOS-gridcell acquired in December 2009 and December 2016 showing reforestation/afforestation. (f) Z-score analysis of annual forest area and AGC over the Brazilian site with forest area gain (r = 0.94 between annual AGC density and forest area). MOD100 annual forest maps are generated based on time series of MOD09A1 images [38]. (a, b, d, e) Landsat images were provided from the Map data: Google, Image Landsat/Copernicus.

true [20]. Annual changes in AGC are quantified as explained in the following and compared with several vegetation and climatic variables to analyze the response of AGC to deforestation and recent climatic events.

As an illustration of the ability of L-VOD to capture deforestation/degradation/forest regrowth events, comparisons with forest dynamics information resolved with higher spatial resolution (Landsat- and MODIS-based information) have been conducted. Large forest area losses caused by mining can be observed between December 2009 and December 2016 by Landsat imagery (Figure 10.2a-b) as well as from the MOD100 forest area data set (Figure 10.2c). The estimates of the AGC changes retrieved from L-VOD (Figure 10.2c) are strongly correlated with MODIS derived forest area (r = 0.94, P < 0.01, n = 8). Similarly, the high sensitivity of AGC to changes in forest area was also found in a region with afforestation and forest regrowth (Figure 10.2d-f, r = 0.94, P < 0.01, n = 8).

10.2.1 BENCHMARK MAPS OF AGC DENSITY

Brandt et al. [20] have used the maps produced by Baccini et al. [10] for calibrating the L-VOD/AGC relationship for Africa. We used here four static AGC benchmark maps (Figure 10.3a-d) to calibrate L-VOD and retrieve AGC in order to decrease the dependence of our results on the accuracy of a single biomass map. These maps include three pantropical maps published by Saatchi et al. [8],

Avitabile et al. [23] and Baccini et al. [10], hereafter referred to as the "Saatchi", "Avitabile" and "Baccini" maps, respectively. The Saatchi map used in the present study is an updated version which represents AGC circa 2015 [8, 39]. A fourth map covering only Africa was produced by extending the data set by Bouvet et al. [24] to higher AGC values using the data set by Mermoz et al. [40], described by Rodriguez-Fernandez et al. [32], hereinafter referred to as the "Bouvet-Mermoz" map. The original units of aboveground biomass density (Mg ha^{-1}) were converted to AGC density (Mg C ha^{-1}) by multiplying the original values by a factor of 0.5 [10]. All AGC maps were aggregated to 25 km spatial resolution to match the spatial resolution of the SMOS data by averaging AGC pixels within the SMOS grid cells.

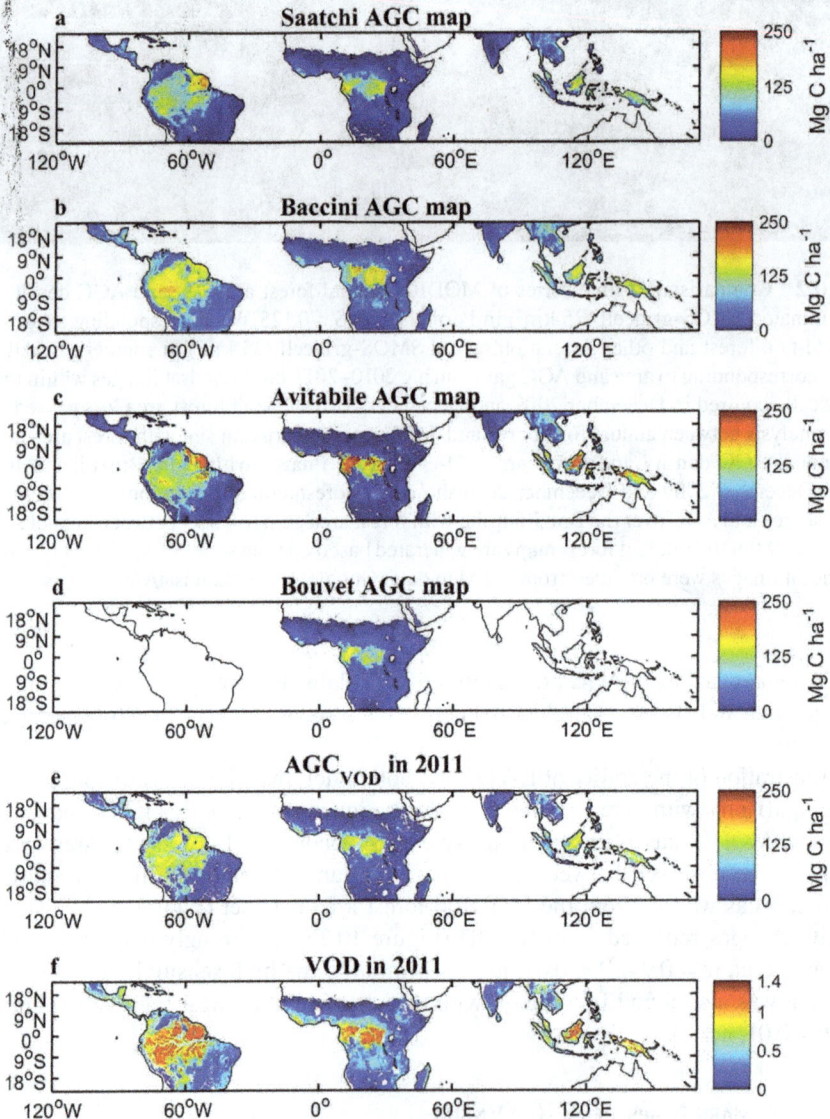

FIGURE 10.3   Maps of the reference AGC data sets, the AGC estimates and L-VOD in the tropics. Four reference AGC density maps are illustrated at a spatial resolution of 25 km: (a) Saatchi (circa 2015), (b) Baccini (circa 2007–2008), (c) Avitabile (circa 2000), and (d) Bouvet-Mermoz (circa 2010). (e) AGC in 2011 estimated by L-VOD in 2011 (f) using ten sets of calibrated parameters (method).

10.2.2 SMOS-IC Soil Moisture, L-VOD and the Retrieved AGC Products

Changes in AGC were estimated from the L-VOD product using SMOS data sets in the SMOS-IC version. The SMOS-IC product provides data for global daily L-VOD and soil moisture (SM) data from the descending and ascending orbits covering the period from January 12, 2010 to December 31, 2017 at a spatial resolution of 25 km (Table 10.1 and Figure 10.3f) [31]. The SMOS-IC L-VOD and SM data were retrieved simultaneously from a two-parameter inversion of the L-band Microwave Emission of the Biosphere (L-MEB) model from the multi-angular and dual-polarized SMOS observations [41]. In the newly developed SMOS-IC algorithm, L-VOD and SM are retrieved without external vegetation or hydrologic products as inputs in the L-MEB inversion model. L-VOD retrievals thus depend only on temperature fields from the European Centre for Medium-Range Weather Forecasts (ECMWF) for calculating the effective surface temperature, and are independent of any vegetation index, contrary to previous VOD products from SMOS [42].

The root mean square error (RMSE) between the measured and simulated brightness temperature (referred to as RMSE-TB) associated with the SMOS-IC product was used to filter out observations affected by radio frequency interference (RFI), which perturbs the natural microwave emission from the earth's surface measured by passive microwave systems [45, 46]. We excluded daily observations, influenced by RFI effects, for which RMSE-TB was larger than 8 K [20]. Robust estimates of annual L-VOD and SM were then obtained as the medians of all high-quality ascending and descending retrievals with more than 30 valid observations per year. This filtering left a large fraction of the original SMOS pixels available for the analysis in tropical America (85.2%), Africa (86.6%) and Asia (68.5%). Relatively to tropical America and Africa, many regions in tropical Asia were much more affected by RFI effects, especially for ascending orbits. After filtering RFI effects using a RMSE-TB threshold of 8 K, many SMOS observations for ascending orbits were filtered out so that a lower percentage of pixels was available over Asia in this study (pixels with N < 30 for at least one year were filtered out).

The yearly L-VOD data were ranked from low to high VOD values and were pooled into bins of 250 grid cells. The mean of the corresponding AGC distribution in the reference map was calculated

TABLE 10.1
Main Features of the AGC Maps and Auxiliary Data Sets Used in This Study

Data Sets	Spatial Cover	Available Period	Original Spatial Resolution
AGC maps			
Saatchi [8, 39]	Tropics	Circa 2015	1 km
Baccini [10]	Tropics	Circa 2007–2008	500 m
Avitabile [43]	Tropics	Circa 2000	1 km
Bouvet-Mermoz [24, 32]	Africa	Circa 2010	25 m for savanna and woodlands and 500 m for forests
Forest loss map			
the "lossyear" map (v1.5) [44]	Tropics	2000–2017	30 m (yearly)
Vegetation and climatic variables			
VOD and SM (SMOS-IC, v105)	Tropics	2010–2017	25 km (daily)
MEI	Tropics	2010–2017	Monthly
EVI (MODIS, MOD13C2)	Tropics	2010–2017	0.05° (monthly)
Skin temperature (ECWMF)	Tropics	1979–2017	0.25° (monthly)
TRMM Precipitation (3B43, v7)	Tropics	1998–2017	0.25° (monthly)
Terrestrial water storage (GRACE)	Tropics	2003–2017	1° (monthly)

for each L-VOD bin, obtaining an AGC curve as a function of L-VOD [20]. The curve was fitted using the four-parameter function [28]:

$$AGC = a \times \frac{\arctan\left(b \times (VOD - c)\right) - \arctan\left(-b \times c\right)}{\arctan\left(b \times (Inf - c)\right) - \arctan\left(-b \times c\right)} + d \tag{10.1}$$

where a, b, c and d are four best-fit parameters and VOD is the yearly L-VOD data. The yearly L-VOD data calculated for 2011 (Figure 10.3f) was used in Eq. 10.1, as described by Rodríguez-Fernández et al. [32], because 2011 was the first complete year after the SMOS commissioning phase.

We converted the yearly L-VOD map into maps of yearly AGC density (Mg C ha^{-1}) for 2010–2017 using Eq. 10.1. Regional AGC stocks were obtained by multiplying the AGC density by the area of the corresponding L-VOD pixels.

AGC benchmark maps contain uncertainties and bias, and none can be considered reliable, as outlined previously. We used all the different maps to fit Eq. 10.1 for tropical America, tropical Africa and the entire tropical region separately. Benchmark maps in tropical Asia were not used in this calibration process due to the limited number of SMOS observations in the region. Ten calibrations of Eq. 10.1 were thereby obtained (Table 10.2). We used all ten calibrations to create ten maps of AGC stocks. We used the median of these ten maps to calculate yearly tropical AGC maps during 2010–2017. The minima and maxima were also reported, because they provide estimates in the uncertainty of retrieved AGC estimates used in this study that relates to systematic errors in the reference biomass maps. A description of the computation of the uncertainties associated with the AGC estimates is given in the following section.

TABLE 10.2
Fitted Parameters in Eq. 10.1 from Four Benchmark Maps

Abbreviation	AGC	Region	A	B	C	D	r²
$P_{African\ Saatchi}$	Saatchi	Tropical Africa	158.858	4.228	0.673	0.912	0.999**
$P_{American\ Saatchi}$	Saatchi	Tropical America	215.548	2.090	0.738	−7.390	0.997**
$P_{tropical\ Saatchi}$	Saatchi	Tropics	183.635	2.822	0.718	−1.117	0.997**
$P_{African\ Baccini}$	Baccini	Tropical Africa	251.969	1.661	0.760	4.838	0.998**
$P_{American\ Baccini}$	Baccini	Tropical America	162.904	2.812	0.614	20.800	0.997**
$P_{tropical\ Baccini}$	Baccini	Tropics	203.168	1.964	0.586	6.458	0.997**
$P_{African\ Avitabile}$	Avitabile	Tropical Africa	204.163	7.173	0.720	0.962	0.998**
$P_{American\ Avitabile}$	Avitabile	Tropical America	200.663	2.709	0.666	−4.644	0.999**
$P_{tropical\ Avitabile}$	Avitabile	Tropics	195.103	3.965	0.685	−0.990	0.999**
$P_{African\ Bouvet}$	Bouvet-Mermoz	Tropical Africa	185.119	2.707	0.823	5.600	0.990**

Notes: */** indicate significant correlations at $P < 0.05/0.01$. Fitted parameters (a, b, c, d) in Eq. 10.1 in main text for the relationship between L-VOD in 2011 and AGC from the four benchmark maps for the various tropical areas (Africa, America and the entire tropical region).

10.2.3 Uncertainties Associated with the AGC Product

It is difficult to use independent data sets for validating the L-VOD derived AGC estimates because most reference biomass data sets are based on the same LIDAR (ICESat GLAS) data set for areas of relatively high vegetation biomass.

We used a bootstrap and cross validation approach to evaluate the "internal" uncertainties (corresponding to sampling and calibration errors) associated with the L-VOD derived AGC estimates. To account for "external" uncertainties (uncertainties associated with the "reference" biomass maps) we used a very conservative approach in which the AGC estimates were derived as the median values of ten L-VOD derived AGC estimates. The ten estimates were computed from four reference biomass data sets (Baccini, Saatchi, Avitabile and Bouvet) calibrated against L-VOD over three different areas (the whole tropics, tropical Africa and tropical America). We used this subset of the Saatchi, Baccini and Avitabile data sets calibrated over three different areas and applied over the whole tropics, in an attempt to account in a realistic way for the uncertainties associated with the parameters in Eq. 10.1. Then the range (or spread) in the ten L-VOD derived AGC estimates is used as an indicator of the "external" uncertainties associated with the AGC estimates. In a final step, we combined both external and internal uncertainties, to get a more realistic estimate of the uncertainties associated with our calculation of AGC and AGC changes. A summary of the main conclusions of the analysis is given next.

Based on a bootstrap cross-validation method, we found that internal errors (due to errors associated with sampling strategies and calibration errors) are almost negligible in comparison to external errors (due to uncertainties associated with the reference maps, and estimated here using a set of ten calibration functions). There is an order of magnitude between uncertainties coming from internal and external errors.

Considering combined internal and external errors, the relative uncertainties associated with the AGC stocks and changes in the AGC stocks over the tropics are on the order of 20–30%. Similar orders of magnitude were found at continental scales. We consider that this relative value is realistic, since it is based on a cross-validation approach considering sampling errors and a large set of ten different calibration functions.

Because internal errors are almost an order of magnitude lower than external errors, and to simplify the computations of uncertainties, only external errors are considered in this study to compute uncertainties associated with the AGC stocks and AGC changes herein.

10.2.4 Additional Uncertainties in the AGC Product

The coarse spatial resolution of the AGC product failed to separate pixel-scale carbon gains and losses due to deforestation, regeneration, livestock pressure, conservation, fires and other events [20]. Moreover, the period of analysis covering two extreme climatic events (the 2011 La Niña and the 2010 and 2015/16 El Niños) corresponding to strong carbon sinks and losses increased the uncertainty in the trend analysis of the carbon changes in Figure 10.4b. The main results of this study, however, do not rely on trend analysis but on spatial and temporal changes in carbon stocks. Open water bodies can affect the retrievals of L-VOD and SM data [34, 47], although SMOS-IC pixels where the sum of the water fractions are > 10% have been filtered out using quality control flags provided by the SMOS-IC products [32].

10.2.5 Vegetation and Climatic Products

The types of vegetation cover in the present study included forest, shrubland, savanna, grassland, cropland and a mosaic of cropland/natural vegetation, which were identified using a 25-km International Geosphere–Biosphere Programme (IGBP) land-cover classification map [48] (Figure 10.5). The 25-km IGBP map was produced by aggregating the 500-m Moderate-Resolution

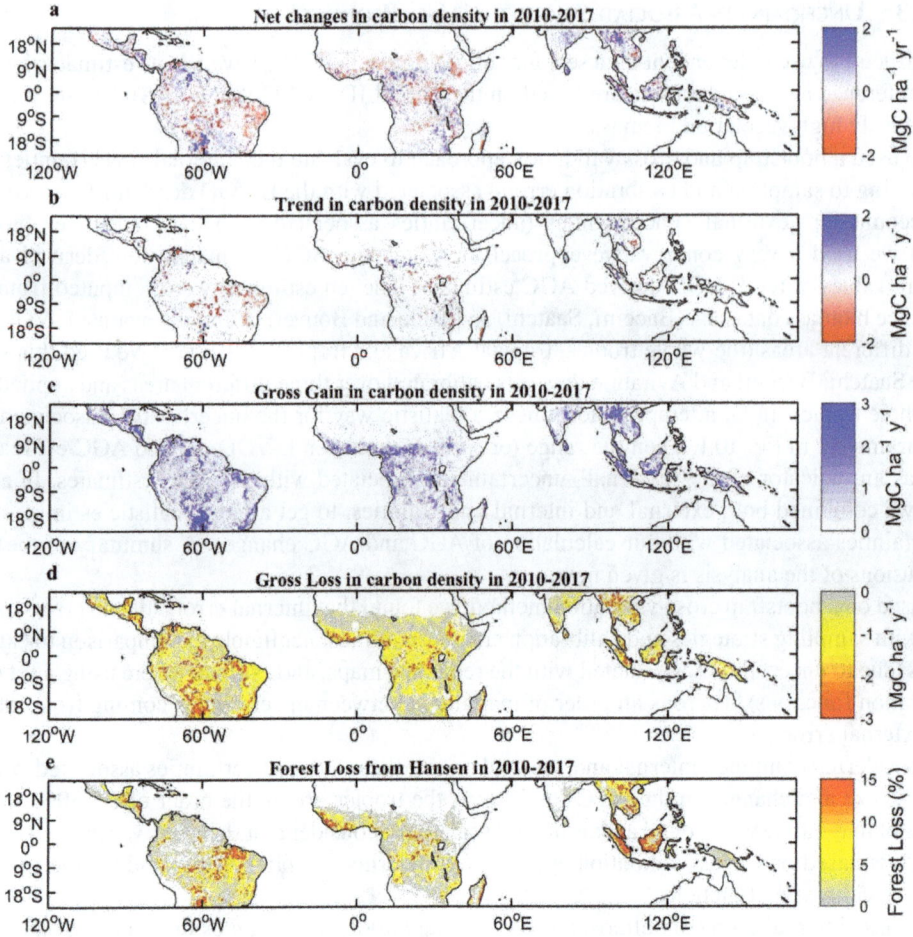

FIGURE 10.4 Yearly net changes (a), trends (b), gross gains (c) and gross losses (d) in AGC and yearly net changes in forest-loss rates [44] (e) for 2010–2017. Yearly net changes, trends and gross gains/losses in AGC were estimated based on the medians of the changes in AGC estimated by ten sets of the fitted relationships between L-VOD and AGC (n = 51,395, 11,992, 51,361 and 47,199 for [a–d0, respectively). Yearly trends in AGC are represented by significantly positive and negative trends (linear trend; P < 0.05). Gross losses in AGC are calculated by cumulating negative changes in AGC for consecutive years from 2010 to 2017.

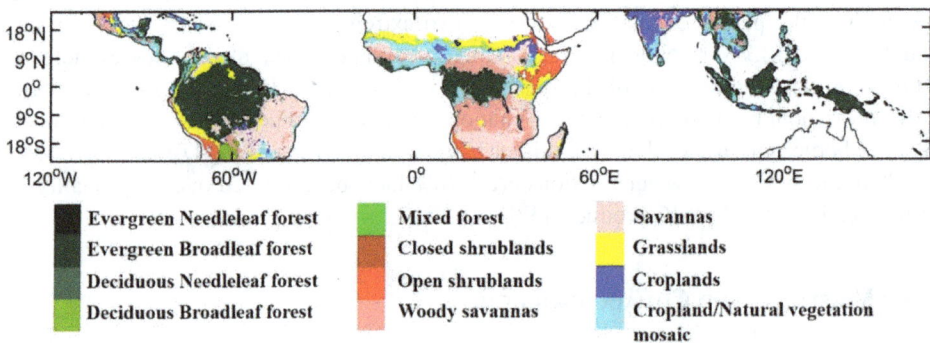

FIGURE 10.5 Biome classes for 2001–2010 based on the MODIS IGBP products over tropics, aggregated to 25 km by dominant class.

Imaging Spectroradiometer (MODIS) IGBP product into a 25-km resolution by dominant class within each SMOS L-VOD grid resolution cell. Tropical semi-arid biomes include shrubland, woodland and savanna regions based on the 25-km IGBP map [49].

We used the "yearloss" forest area loss map produced by Hansen et al. (2013) to calculate forest loss rates. Forest loss was defined as a stand-replacement disturbance, or a change from a forest to a non-forest state [44, 50]. Each 30-m pixel in the "yearloss" Landsat data was labeled with a loss year representing the loss of forest (defined as tree higher than 5 m) cover detected primarily during 2000–2017. Here, forest percentage loss rates during the study period of 2010–2017 were calculated at the resolution of SMOS as the proportion of the summed areas of forest loss (detected by the "yearloss" map) within each SMOS grid cell (~25 km) during 2010–2017.

The data used to compute trends in the annual average MODIS LAI (2010 to 2017) at a spatial resolution of 0.05° are provided by Chen et al., 2019, who used the Mann–Kendall test to calculate the LAI trends based on the MODIS LAI product (MOD15A2H and MYD15A2H). Greening and browning are defined as statistically significant increases and decreases, respectively, in the annual average green leaf area for a given pixel over 2010–2017 [51].

The MOD100 annual forest area product used in this study (spatial resolution of 500 m) was produced from information on canopy phenology from the analyses of EVI and a land surface water index derived from the MOD09A1 product [52]. The MOD100 product is a recent product using all the observations in a year (dense time series) from MOD09A1, and has shown excellent performance when compared against the official Brazilian deforestation data set (PRODES) and Global Forest Watch (GFW) [38].

We used data for the annual mean global CO_2 growth rate data for 2010–2017, based on globally averaged marine surface data, compiled and published by the National Oceanic and Atmospheric Administration (NOAA) Earth System Research Laboratory (ESRL) in Colorado.

Several vegetation and climate variables (Table 10.1) were used for further investigating the response of AGC to climate events. These variables include (1) the multivariate El Niño/Southern Oscillation (ENSO) Index (MEI) [53]; (2) enhanced vegetation index (EVI) from the MODIS Vegetation Index product (MOD13C2 Climate Modeling Grid) [54]; (3) land surface temperature from skin temperature data produced by ECMWF atmospheric reanalysis ERA-Interim [55]; (4) precipitation from data sets of the Tropical Rainfall Measuring Mission (TRMM 3B43 v7) [56]; and (5) terrestrial water storage (TWS) measured by the twin satellites of the Gravity Recovery and Climate Experiment (GRACE) providing the total relative water storage, including groundwater, SM, surface water, snow and water stored in the biosphere [57, 58]. Monthly TWS was calculated as a simple arithmetic mean of three data sets, the monthly 1-degree GRACE TWS products released by JPL, CSR and GFZ, and was then aggregated to yearly TWS [59].

EVI, precipitation and land surface temperature were aggregated to an annual composite at 25-km spatial resolution by averaging or bilinear interpolation from their original resolution to match the L-VOD grid.

10.2.6 STATISTICAL METRICS

We calculated two goodness-of-fit metrics between pairs of reference benchmark map and AGC map: the coefficient of determination (r^2) and the root-mean-squared error (RMSE, Mg C ha^{-1}). Trend estimates were calculated using linear regression slope. Linear correlation coefficients (Pearson's r) were calculated to quantify the concurrent association between time series. The levels of statistical significance (P values) were estimated throughout this analysis, and the correlation coefficients r were considered to be statistically significant if the P values were less than 0.05.

10.3   RESULTS AND ANALYSIS

10.3.1   NEUTRAL CARBON BALANCE AND HIGHLY DYNAMIC AGC STOCKS IN TROPICAL BIOMES

During 2010–2017, tropical AGC change represents a small net increase of +0.11 [+0.08, +0.13] Pg C yr^{-1} (the range represents the minimum and maximum of AGC changes estimated by ten calibrations; a positive value indicates net accumulation (sink) of carbon in aboveground biomes; Figure 10.6a). This net carbon budget is composed of gross losses of –2.86 [–2.31, –3.05] PgC yr^{-1}

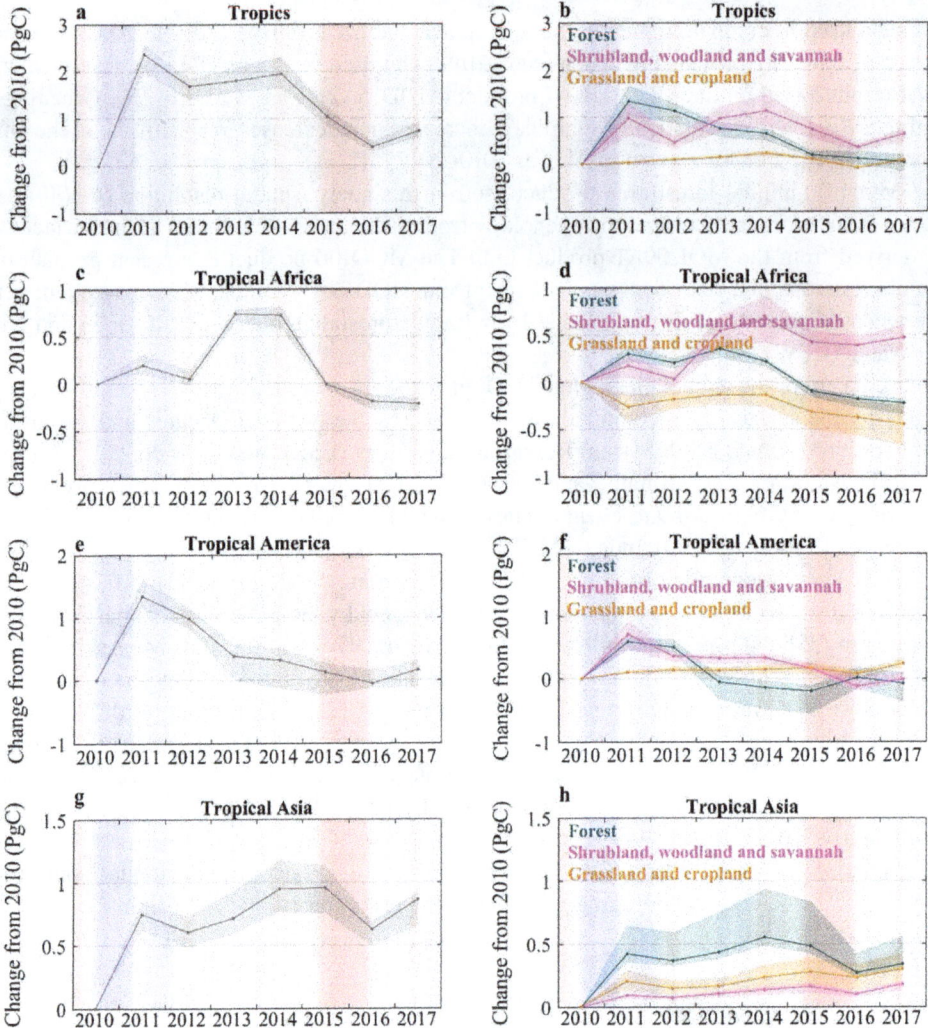

FIGURE 10.6   Temporal variations in annual AGC in the tropics (continents and biomes), expressed as the difference from 2010 values. Annual variations in AGC in (a) the tropics (n = 51,395), and in the tropical regions of (c) Africa (n = 25,058), (e) America (n = 19,777) and (g) Asia (n = 6,560), respectively. Corresponding changes in AGC (b, d, f, h) are shown for three biomes (forest; shrubland, woodland and savanna; grassland and cropland). The ranges represent the minimum and maximum of AGC changes estimated by ten calibrations (Table 10.2). The background shading shows the intensity of La Niña (blue) and El Niño (red) events defined by Multivariate ENSO Index (MEI).

offset by gross gains of +2.97 [+2.41, +3.15] Pg C yr^{-1} estimated at the spatial resolution of the SMOS-grid (25 × 25 km). Tropical Asia was a net mean sink of +0.12 [+0.09, +0.13] Pg C yr^{-1} (Figure 10.6g), and tropical Africa and South America were almost neutral with a flux of -0.03 [–0.04, –0.02] Pg C yr^{-1} (Figure 10.6c) and +0.02 [–0.02, +0.05] Pg C yr^{-1} (Figure 10.6e) respectively. Carbon stocks increased slightly in woodland, shrubland and savanna regions, particularly in tropical Africa, whereas changes in forest, grassland and cropland were close to zero (Figure 10.6b).

Over the study period, AGC peaked in 2011 in response to the strong La Niña event and decreased subsequently over the tropics (Figure 10.6a). Strong La Niña conditions prevailed from late 2010 to early 2012 [60] (Figure 10.6a) resulting in a transient increase of tropical AGC of +2.36 [+1.97, +2.57] Pg C, mainly from tropical America: +1.34 [+1.13, +1.61] Pg C (Figure 10.6e) and Asia: +0.75 [+0.61, +0.84] Pg C Figure 10.6g). In tropical America, the peak of AGC in 2011 is mainly observed in forests and shrublands/savannas and suggests recovery of vegetation following the 2010 drought (Figure 10.6a), mainly driven by a wet climatic anomaly (Figure 10.7a).

A strong El Niño event developed in mid-2015 and persisted until mid-2016 (Figure 10.6a) [18]. This event caused a drop of tropical AGC of –0.95 [–1.00, –0.76] Pg C in 2015, including -0.74 [–0.86, –0.62] Pg C in Africa (Figure 10.6c) and –0.20 [–0.26, –0.1] Pg C in America (Figure 10.6e), mainly attributed to extremely dry and warm climatic conditions (Figure 10.7a). The 2015 loss in Africa occurred in all biomes, with the largest losses in woodland, shrubland and savanna regions. By contrast, carbon losses and gains were evenly balanced in tropical Asia in 2015. Interestingly, AGC losses continued in 2016, with a biomass loss of –0.65 [–0.82, –0.38] Pg C, mostly in Asia (–0.35 [–0.50, –0.26] Pg C) followed by Africa and America (–0.19 [–0.22, –0.15] Pg C and –0.12 [–0.3, +0.11] Pg C, respectively), in response to more severe anomalies in both surface soil moisture and land surface temperature in 2016 as compared to 2015 (Figure 10.7a). Lumping the two years 2015–2016 together, the average AGC carbon losses (–0.80 [–0.59, –0.96] Pg C yr^{-1}) are in the range of the net land-atmosphere abnormal CO_2 source simulated by land surface models (–1.1 [–2.5, +0.1] Pg C yr^{-1}) [21].

10.3.2 CARBON UPTAKE IN NON-DEFORESTED REGIONS OFFSETS DEFORESTATION CARBON LOSSES

Pixels with more than 5% forest losses (covering 16% of the tropics) as identified by Hansen et al. [44] (methods), displayed a net carbon loss of –0.09 [–0.14, –0.07] Pg C yr-1 in the aboveground vegetation compartment for the period 2010–2017 (Table 10.3). Net carbon losses due to deforestation were offset by a net carbon uptake of +0.20 [+0.14, +0.24] Pg C yr^{-1} across pixels with less than 5% deforestation. This sink was found mainly in tropical Asia (+0.10 [+0.06, +0.13] Pg C yr^{-1}) and America (+0.09 [+0.06, +0.12] Pg C yr^{-1}). Trends for the period 2010–2017 displayed carbon losses in the Arc of Deforestation of southern Amazonia, in the Democratic Republic of Congo and in Indonesia (Figure 10.4a and 4b). The carbon uptake was found in the Central African Republic and in the northernmost regions of tropical Asia and Central America (Figure 10.4a and 4b).

We defined gross carbon losses as cumulated yearly losses, excluding regrowth years. Overall, gross carbon loss from areas of deforestation (forest losses >5%) was –0.78 [–0.61, –1.04] Pg C yr^{-1} (Table 10.3 and Figure 10.8b). Areas with high gross carbon loss (Figure 10.4d) matched well areas where tropical forest cover decreased (Figure 10.4e) in the data set of Hansen et al. [44] (methods) (as an illustration, results obtained over a deforestation and an afforestation site are shown in Figure 10.2). Carbon gains in Central America, southern and northern regions of tropical America, Central African Republic and in the northernmost regions of tropical Asia and India reflect high recovery rates Figure 10.4c) offsetting carbon losses (Figure 10.4d) leading to an overall net carbon

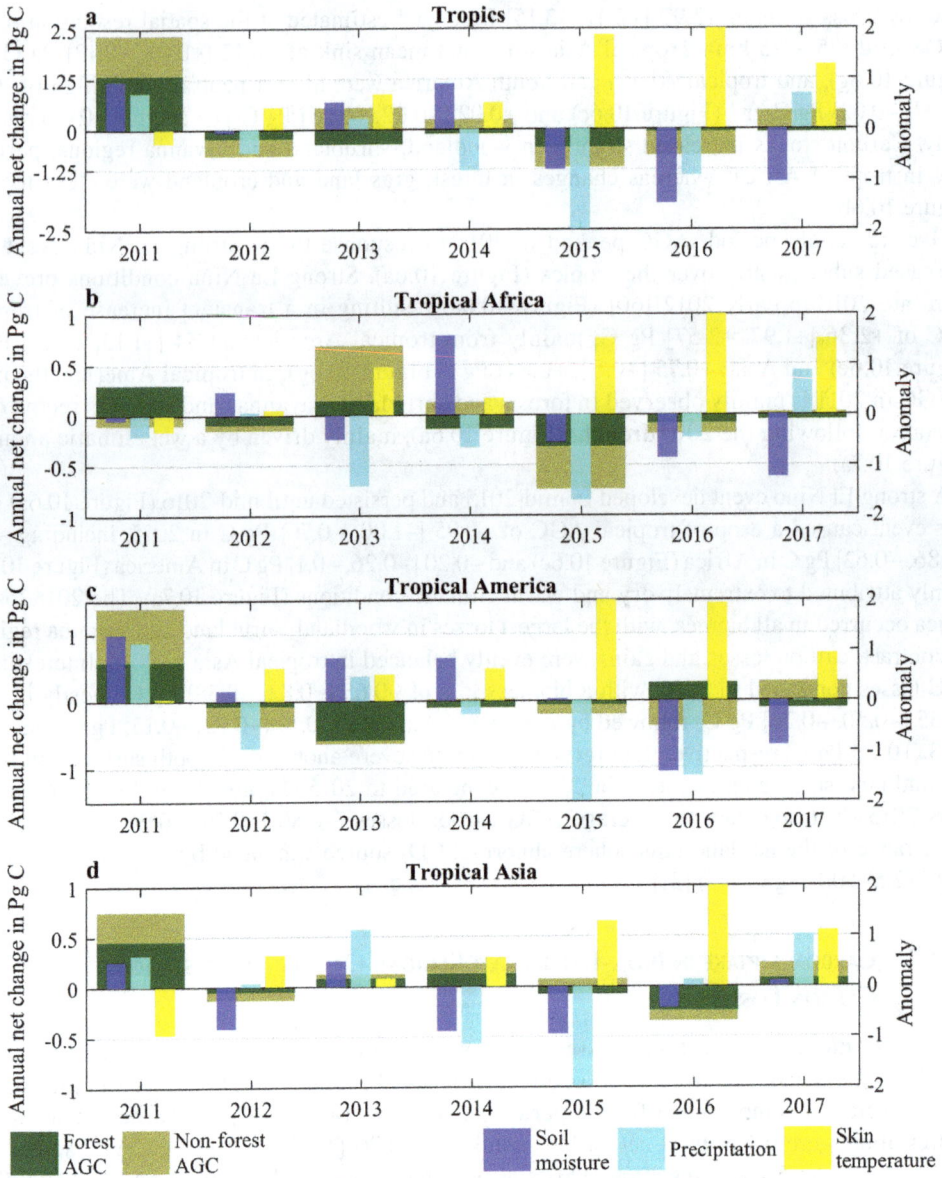

FIGURE 10.7   Interannual variations in tropical aboveground carbon. Non-forested areas include shrubland, savanna, grassland, cropland and cropland/natural vegetation mosaic based on the MODIS IGBP land-cover map. Annual net changes for individual years (compared to the previous year) in forest and non-forest AGC are displayed together with, anomalies (z-score) in surface soil moisture, precipitation and land surface temperature for the tropics (a), and tropical regions of Africa (b), America (c) and Asia (d). Yearly anomalies were calculated using the z score: (value – mean)/standard deviation.

storage in these regions (Figure 10.4a). The spatial patterns of the areas showing carbon sinks agree well with greening regions as evaluated by Chen et al., 2019 [51]. In parallel, a spatial agreement between regions showing browning trends and carbon losses was found in eastern tropical Africa and Madagascar's tropical rainforests.

TABLE 10.3
Changes in Tropical AGC over Deforestation and Non-Deforested Regions

		Net Changes (Pg C yr⁻¹)	Gross Loss (Pg C yr⁻¹)	Gross Gain (Pg C yr⁻¹)
Tropics	Total	+0.11	−2.86	+2.97
	Deforestation	−0.09	−0.78	+0.69
	Non-deforestation	+0.20	−2.08	+2.28
Tropical Africa	Total	−0.03	−1.09	+1.06
	Deforestation	−0.05	−0.22	+0.18
	Non-deforestation	+0.01	−0.87	+0.88
Tropical America	Total	+0.02	−1.29	+1.31
	Deforestation	−0.07	−0.35	+0.29
	Non-deforestation	+0.09	−0.94	+1.03
Tropical Asia	Total	+0.12	−0.47	+0.60
	Deforestation	+0.03	−0.20	+0.23
	Non-deforestation	+0.10	−0.27	+0.37

Note: Net and gross (cumulative gain or loss of the consecutive years) changes in the three continents of tropics and deforestation zones (forest loss rates >5%).

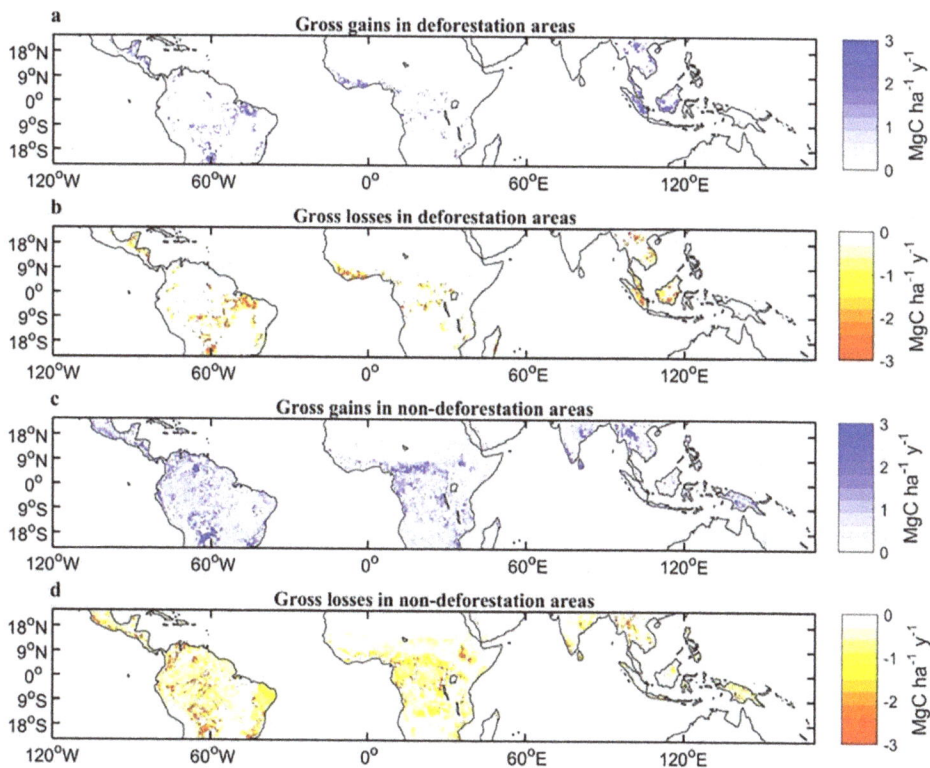

FIGURE 10.8 Yearly gross gains and losses in AGC in deforestation (forest losses >5%, estimated by Hansen et al. [44]) and non-deforestation (forest losses ≤5%) regions for 2010–2017.

10.3.3　Pivotal Role of Tropical AGC Stocks in the Global Carbon Budget

There is an ongoing debate about the role of humid versus semi-arid tropical biomes in controlling the global atmospheric CO_2 growth rate (CGR) [61, 62]. A strong association was found between yearly de-trended global atmospheric CGR measured from the National Oceanic and Atmospheric Administration (NOAA/ESRL) [63] and annual tropical AGC fluxes as inferred earlier (r = 0.86, P = 0.03, n = 7, Figure 10.3a), supporting previous findings [64–66] that tropical biomes dominate the interannual variability in atmospheric CGR. Carbon losses of biomass (–1.6 [–1.82, –1.14] Pg C) during the severe 2015–2016 El Niño accounted for 90% of the anomaly in atmospheric CGR (1.7 Pg C).

We evaluated the contribution of different biomes to the inter-annual variability of aboveground carbon by separating tropical forests [49], semi-arid biomes (shrubland, woodland and savanna), cropland and grasslands. The contribution of semi-arid biomes accounts for the largest fraction of the interannual variability of tropical AGC fluxes (55.5%), with a smaller contribution of forests

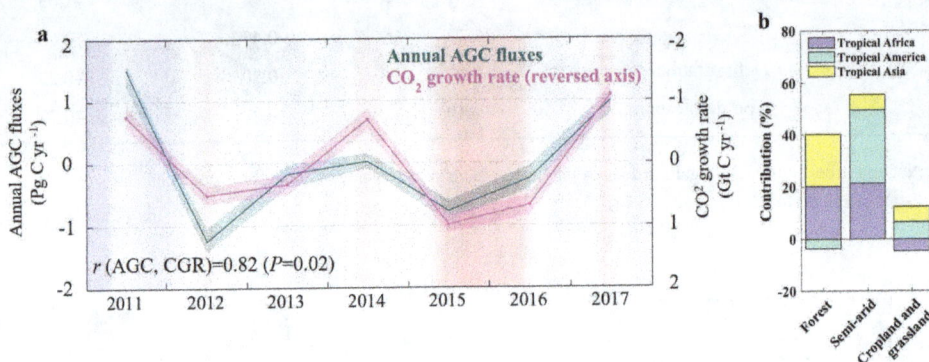

FIGURE 10.9 Interannual variability of global atmospheric CO_2 growth rate and tropical AGC fluxes. (a) The interannual variability in the atmospheric CGR and AGC fluxes was calculated by removing trends from annual atmospheric CGR and AGC fluxes over 2011–2017, respectively. The tropical annual AGC fluxes were calculated using net AGC changes for individual years (compared to the previous year; n = 7). The vertical axis is inverted for the de-trended CGR so that positive (downwards) anomalies indicate a weaker land carbon sink. (b) Contribution (%) of land cover classes and continent to the interannual variability in tropical AGC fluxes. Tropical semi-arid biomes consist of shrubland, woodland and savanna [49]. The contributions (%) of the different biomes and regions to the tropical AGC fluxes were estimated using the method described in Ahlström et al. [49].

TABLE 10.4
Variation and Covariation of AGC Fluxes in Tropical Forest and Semi-Arid Biomes

	Tropics (Pg C yr^{-1})^2	Tropical Africa (Pg C yr^{-1})^2	Tropical America (Pg C yr^{-1})^2	Tropical Asia (Pg C yr^{-1})^2
Variation in AGC fluxes of both forest and semi-arid biomes	0.94	0.32	0.15	0.07
Variation in AGC fluxes of forests	0.35	0.12	0.04	0.05
Variation in AGC fluxes of semi-arid biomes	0.22	0.09	0.05	0.00
Covariation in AGC of forest and semi-arid biomes	0.19	0.05	0.03	0.01

(36.6%), and croplands and grasslands (7.9%) (Figure 10.9b), suggesting that semi-arid ecosystems are one of the most important components of the interannual variability in the tropical AGC [60]. Interannual variability in the tropical AGC fluxes is controlled predominantly by semi-arid biomes from tropical America and Africa and by forests from tropical Africa and Asia. Likewise, the positive covariation (+0.19 [Pg C yr^{-1}]; Table 10.4) of AGC fluxes from tropical forests and semi-arid biomes suggests that both biomes act in phase to control interannual variability in AGC.

10.4 DISCUSSION

The L-VOD satellite data set provided insights into recent spatial changes of the carbon cycle in the tropics in relation to deforestation and tropical extreme climatic events. The data set was used to quantify both AGC losses in the tropics during the 2010 and 2015/16 El Niño events, and the subsequent recoveries in 2011 and 2017. Altogether, the results show a neutral contribution of the tropics to the global carbon budget between 2010 and 2017. Yet, L-VOD revealed that the recovery in 2017 was weaker than in 2011, which could be partly attributed to the warm climatic conditions in 2017 (Figure 10.7), which negatively impacted the terrestrial carbon uptake [66, 67]. Using the year of 2011 as a reference for comparison [18], our estimations of AGC losses caused by the 2015/16 El Niño were generally lower than estimates from OCO-2 [18] (which include soil carbon, aboveground biomass and river CO_2 fluxes) over tropical America (−1.41 versus −1.60 Pg C for OCO-2), Africa (−0.40 versus −0.70 Pg C for OCO-2) and Asia (−0.13 versus −1.00 Pg C for OCO-2). This difference could be partly attributed to the fact that our estimations of AGC do not account for ecosystem respiration rate [21] and peat fires [68], especially in tropical Asia, associated to large carbon losses from soils [69, 70].

Furthermore, we were able to quantify AGC losses from areas of deforestation, which were fully compensated by carbon uptakes over undisturbed forests over the whole tropics. The L-VOD based estimation of emissions from deforestation (0.78 Pg C y^{-1}) match very well with previous one (e.g., 0.81 Pg C y^{-1} obtained by Harris et al. [11] between 2000 and 2005), suggesting that the flux from gross tropical deforestation is within 0.6–0.8 Pg C yr^{-1} since the early 2000s [71, 72]. Moreover, we estimated AGC losses from other processes than deforestation to be 2.08 Pg C y^{-1}, caused by natural disturbances, climate-induced mortality and forest degradation, including selective removals from within forested stands (not currently included in deforestation estimates based on optical satellite data [71]). This suggests that other processes than deforestation is responsible for about twice the amount of carbon release as compared to deforestation, however with large regional variations [22, 73]. In addition, parts of the losses in carbon may be caused by the reduction of AGC following the extreme La Niña (return to normal conditions) [74] and subsequent El Niño.

We further showed that non-deforested regions act as a carbon sink, which is supported by measurements from forest inventory plots [75–77]. The increasing AGC trend over intact, non-disturbed forests may be attributed to CO_2 fertilization effect on tree growth, consistent with no strong signal from widespread disturbance recovery [78] in forest plots and with model-based attribution of the recent greening trend over the tropics [79]. The carbon sink of the Sahel and South Africa are primarily driven by increasing precipitation [79, 80], whereas human land-use management may be a dominant driver of carbon sink in India and northern tropical Asia [51]. Here, L-VOD data resolves the spatial distribution of this uptake over the whole tropics showing that the net sink density in non-deforested regions was rather low between 2010 and 2017 (+0.05 Mg C ha-1 yr^{-1}). This low carbon accumulation rate could be partly explained by a long-term increase in mortality rates [15] and the recent El Niño events [81]. This result is in contrast to the high carbon accumulation ([+1.33, +3.05] Mg C ha^{-1} yr^{-1}) that was estimated from individual field plots across Amazonian secondary and managed forests [75, 82]. The disagreement could also stem from the fact that the coarse spatial resolution of L-VOD (25 km × 25 km) merges all aboveground biomes including disturbed forests and non-forest ecosystems, which have lower rates of gain than secondary and managed forests [22]. While carbon changes in both deforested and non-deforested areas are expected, our estimates are admittedly conservative as a result of the coarse spatial resolution of the L-VOD data, which averages gross C sources and sinks at scales smaller than 25 km. Gross gain and loss could thus be larger at higher spatial resolution.

AGC fluxes estimated from L-VOD, which are independent from process-based models, are consistent with the phase and amplitude of global CO_2 growth rates anomalies. This suggests that litter and soil carbon fluxes have a smaller variability than AGC fluxes, and highlights that changes in the tropical AGC balance dominate changes in the global carbon balance. The observed spatial patterns of the contribution of terrestrial ecosystems to the total tropical inter-annual variability in AGC fluxes agreed with model results [49]. This supports the model-based findings that semi-arid biomes can have profound impacts on the inter-annual variability of the global carbon cycle [60]. From observational data we revealed spatial patterns over recent years showing (i) the main positive contributions are found in the eastern and northern regions of the Amazon basin, south-eastern regions of Africa, and Asia; (ii) the main regions with negative contributions are found in forested regions in tropical America (for instance in the arc of deforestation in the Amazon basin), and non-forested regions (e.g. semi-arid biomes and croplands and grasslands) in tropical Africa. These negative contributions could be mainly attributed to both human activities (e.g., deforestation and high population growth) [83] and the different sensitivity of biomes to climate variations among regions [49].

The L-VOD data provide direct and spatially explicit remote sensing information that scales up to annual tropical AGC anomalies. This product overcomes several of the limitations of current tools used to estimate the tropical land sink. The coarse resolution (25 km × 25 km) of L-VOD limits its applicability for detailed regional analysis, but is however not a limitation to address the critical role of the terrestrial land sink on the changing atmospheric characteristics. Based on L-VOD, a first direct observational estimate of the pantropical carbon sink could be clearly related, in terms of correlation and magnitude, to the observed CO_2 growth rate in the recent years. The results show the applicability of L-VOD to monitor, in near-real time, spatio-temporal changes in AGC to reveal hotspot areas of changes due to human activity (deforestation) and climate variability (such as El Niño–Southern Oscillation) at large scale. The data and results shown here hold promises for data-informed process-based Earth system models to better predict the future of land carbon sinks, and to further reconcile divergent estimates of carbon sources/sinks derived from modeling approaches (bottom-up [10–12] as well as top-down [81]) and observational systems [22].

10.5 CONCLUSIONS

Changes in terrestrial tropical carbon stocks play an important role in the global carbon budget. However, current observational tools do not allow accurate and large-scale monitoring of the spatial distribution and dynamics of carbon stocks. Here, we used low frequency L-band passive microwave observations to compute a direct and spatially explicit quantification of annual aboveground carbon (AGC) fluxes and show that the tropical net AGC budget was approximately in balance during 2010 to 2017, the net budget being composed of gross losses offset by gross gains between continents. Large interannual and spatial fluctuations of tropical AGC were quantified during the wet 2011 La Niña year and throughout the extreme dry and warm 2015–2016 El Niño episode. These interannual fluctuations, controlled predominantly by semi-arid biomes, were shown to be closely related to independent global atmospheric CO_2 growth-rate anomalies (Pearson's r = 0.86), highlighting the pivotal role of tropical AGC in the global carbon budget.

REFERENCES

[1] R. A. Houghton, "Aboveground forest biomass and the global carbon balance," Global Change Biology. vol. 11, no. 6, pp. 945–958, Jun, 2005.
[2] A. Tyukavina, A. Baccini, M. C. Hansen, P. V. Potapov, S. V. Stehman, R. A. Houghton, A. M. Krylov, S. Turubanova, and S. J. Goetz, "Aboveground carbon loss in natural and managed tropical forests from 2000 to 2012," Environmental Research Letters. vol. 10, no. 7, Jul, 2015.

[3] W. Li, P. Ciais, S. S. Peng, C. Yue, Y. L. Wang, M. Thurner, S. S. Saatchi, A. Arneth, V. Avitabile, N. Carvalhais, A. B. Harper, E. Kato, C. Koven, Y. Y. Liu, J. Nabel, Y. D. Pan, J. Pongratz, B. Poulter, T. A. M. Pugh, M. Santoro, S. Sitch, B. D. Stocker, N. Viovy, A. Wiltshire, R. Yousefpour, and S. Zaehle, "Land-use and land-cover change carbon emissions between 1901 and 2012 constrained by biomass observations," Biogeosciences. vol. 14, no. 22, pp. 5053–5067, Nov, 2017.

[4] S. Besnard, S. Koirala, M. Santoro, M. Bao, O. Cartus, F. Gans, M. Jung, T. Trautmann, and N. Carvalhais, "Constraining carbon allocation in a terrestrial ecosystem model using long-term forest biomass time series." p. 10523.

[5] A. Baccini, W. Walker, L. Carvalho, M. Farina, D. Sulla-Menashe, and R. A. Houghton, "Tropical forests are a net carbon source based on aboveground measurements of gain and loss," Science. vol. 358, no. 6360, pp. 230–234, Oct 13, 2017.

[6] E. T. A. Mitchard, "The tropical forest carbon cycle and climate change," Nature. vol. 559, no. 7715, pp. 527–534, 2018/07/01, 2018.

[7] M. C. Hansen, P. Potapov, and A. Tyukavina, "Comment on "Tropical forests are a net carbon source based on aboveground measurements of gain and loss"," Science. vol. 363, no. 6423, pp. eaar3629, 2019.

[8] S. S. Saatchi, N. L. Harris, S. Brown, M. Lefsky, E. T. Mitchard, W. Salas, B. R. Zutta, W. Buermann, S. L. Lewis, and S. Hagen, "Benchmark map of forest carbon stocks in tropical regions across three continents," Proceedings of the National Academy of Sciences. vol. 108, no. 24, pp. 9899–9904, 2011.

[9] B. Gaubert, B. B. Stephens, S. Basu, F. Chevallier, F. Deng, E. A. Kort, P. K. Patra, W. Peters, C. Rödenbeck, T. Saeki, D. Schimel, I. Van der Laan-Luijkx, S. Wofsy, and Y. Yin, "Global atmospheric CO2 inverse models converging on neutral tropical land exchange, but disagreeing on fossil fuel and atmospheric growth rate," Biogeosciences. vol. 16, no. 1, pp. 117–134, 2019.

[10] A. Baccini, S. Goetz, W. Walker, N. Laporte, M. Sun, D. Sulla-Menashe, J. Hackler, P. Beck, R. Dubayah, and M. Friedl, "Estimated carbon dioxide emissions from tropical deforestation improved by carbon-density maps," Nature Climate Change. vol. 2, no. 3, pp. 182, 2012.

[11] N. L. Harris, S. Brown, S. C. Hagen, S. S. Saatchi, S. Petrova, W. Salas, M. C. Hansen, P. V. Potapov, and A. Lotsch, "Baseline map of carbon emissions from deforestation in tropical regions," Science. vol. 336, no. 6088, pp. 1573–1576, 2012.

[12] F. Achard, R. Beuchle, P. Mayaux, H. J. Stibig, C. Bodart, A. Brink, S. Carboni, B. Desclée, F. Donnay, and H. D. Eva, "Determination of tropical deforestation rates and related carbon losses from 1990 to 2010," Global Change Biology. vol. 20, no. 8, pp. 2540–2554, 2014.

[13] R. L. Chazdon, E. N. Broadbent, D. M. Rozendaal, F. Bongers, A. M. A. Zambrano, T. M. Aide, P. Balvanera, J. M. Becknell, V. Boukili, and P. H. Brancalion, "Carbon sequestration potential of second-growth forest regeneration in the Latin American tropics," Science Advances. vol. 2, no. 5, pp. e1501639, 2016.

[14] Y. Yang, S. S. Saatchi, L. Xu, Y. Yu, S. Choi, N. Phillips, R. Kennedy, M. Keller, Y. Knyazikhin, and R. B. Myneni, "Post-drought decline of the Amazon carbon sink," Nature Communications. vol. 9, no. 1, pp. 3172, 2018.

[15] R. J. Brienen, O. L. Phillips, T. R. Feldpausch, E. Gloor, T. R. Baker, J. Lloyd, G. Lopez-Gonzalez, A. Monteagudo-Mendoza, Y. Malhi, and S. L. Lewis, "Long-term decline of the Amazon carbon sink," Nature. vol. 519, no. 7543, pp. 344, 2015.

[16] Y. Pan, R. A. Birdsey, J. Fang, R. Houghton, P. E. Kauppi, W. A. Kurz, O. L. Phillips, A. Shvidenko, S. L. Lewis, and J. G. Canadell, "A large and persistent carbon sink in the world's forests," Science, pp. 1201609, 2011.

[17] Y. Malhi, "The productivity, metabolism and carbon cycle of tropical forest vegetation," Journal of Ecology. vol. 100, no. 1, pp. 65–75, 2012.

[18] J. Liu, K. W. Bowman, D. S. Schimel, N. C. Parazoo, Z. Jiang, M. Lee, A. A. Bloom, D. Wunch, C. Frankenberg, and Y. Sun, "Contrasting carbon cycle responses of the tropical continents to the 2015–2016 El Niño," Science. vol. 358, no. 6360, pp. eaam5690, 2017.

[19] C. Yue, P. Ciais, A. Bastos, F. Chevallier, Y. Yin, C. Rödenbeck, and T. Park, "Vegetation greenness and land carbon-flux anomalies associated with climate variations: a focus on the year 2015," Atmospheric Chemistry and Physics. vol. 17, no. 22, pp. 13903–13919, 2017.

[20] M. Brandt, J.-P. Wigneron, J. Chave, T. Tagesson, J. Penuelas, P. Ciais, K. Rasmussen, F. Tian, C. Mbow, and A. Al-Yaari, "Satellite passive microwaves reveal recent climate-induced carbon losses in African drylands," Nature Ecology & Evolution. vol. 2, no. 5, pp. 827, 2018.

[21] A. Bastos, P. Friedlingstein, S. Sitch, C. Chen, A. Mialon, J.-P. Wigneron, V. K. Arora, P. R. Briggs, J. G. Canadell, P. Ciais, F. Chevallier, L. Cheng, C. Delire, V. Haverd, A. K. Jain, F. Joos, E. Kato, S. Lienert, D. Lombardozzi, J. R. Melton, R. Myneni, J. E. M. S. Nabel, J. Pongratz, B. Poulter, C. Rödenbeck, R. Séférian, H. Tian, C. van Eck, N. Viovy, N. Vuichard, A. P. Walker, A. Wiltshire, J. Yang, S. Zaehle, N. Zeng, and D. Zhu, "Impact of the 2015/2016 El Niño on the terrestrial carbon cycle constrained by bottom-up and top-down approaches," Philosophical Transactions of the Royal Society B: Biological Sciences. vol. 373, no. 1760, 2018.

[22] A. Baccini, W. Walker, L. Carvalho, M. Farina, D. Sulla-Menashe, and R. Houghton, "Tropical forests are a net carbon source based on aboveground measurements of gain and loss," Science. vol. 358, no. 6360, pp. 230–234, 2017.

[23] V. Avitabile, M. Herold, G. B. Heuvelink, S. L. Lewis, O. L. Phillips, G. P. Asner, J. Armston, P. S. Ashton, L. Banin, and N. Bayol, "An integrated pan-tropical biomass map using multiple reference datasets," Global Change Biology. vol. 22, no. 4, pp. 1406–1420, 2016.

[24] A. Bouvet, S. Mermoz, T. Le Toan, L. Villard, R. Mathieu, L. Naidoo, and G. P. Asner, "An above-ground biomass map of African savannahs and woodlands at 25m resolution derived from ALOS PALSAR," Remote Sensing of Environment. vol. 206, pp. 156–173, 2018.

[25] A. G. Konings, and P. Gentine, "Global variations in ecosystem-scale isohydricity," Global Change Biology. vol. 23, no. 2, pp. 891–905, 2017.

[26] A. Konings, A. Williams, and P. Gentine, "Sensitivity of grassland productivity to aridity controlled by stomatal and xylem regulation," Nature Geoscience. vol. 10, no. 4, pp. 284, 2017.

[27] J.-P. Wigneron, Y. Kerr, A. Chanzy, and Y.-Q. Jin, "Inversion of surface parameters from passive microwave measurements over a soybean field," Remote Sensing of Environment. vol. 46, no. 1, pp. 61–72, 1993.

[28] Y. Y. Liu, A. I. Van Dijk, R. A. De Jeu, J. G. Canadell, M. F. McCabe, J. P. Evans, and G. Wang, "Recent reversal in loss of global terrestrial biomass," Nature Climate Change. vol. 5, no. 5, pp. 470, 2015.

[29] Y. Y. Liu, A. I. van Dijk, M. F. McCabe, J. P. Evans, and R. A. de Jeu, "Global vegetation biomass change (1988–2008) and attribution to environmental and human drivers," Global Ecology and Biogeography. vol. 22, no. 6, pp. 692–705, 2013.

[30] J.-P. Wigneron, T. Jackson, P. O'neill, G. De Lannoy, P. De Rosnay, J. Walker, P. Ferrazzoli, V. Mironov, S. Bircher, and J. Grant, "Modelling the passive microwave signature from land surfaces: A review of recent results and application to the L-band SMOS & SMAP soil moisture retrieval algorithms," Remote Sensing of Environment. vol. 192, pp. 238–262, 2017.

[31] R. Fernandez-Moran, A. Al-Yaari, A. Mialon, A. Mahmoodi, A. Al Bitar, G. De Lannoy, N. Rodriguez-Fernandez, E. Lopez-Baeza, Y. Kerr, and J.-P. Wigneron, "SMOS-IC: An alternative SMOS soil moisture and vegetation optical depth product," Remote Sensing. vol. 9, no. 5, pp. 457, 2017.

[32] N. J. Rodríguez-Fernández, A. Mialon, S. Mermoz, A. Bouvet, P. Richaume, A. Al Bitar, A. Al-Yaari, M. Brandt, T. Kaminski, and T. Le Toan, "An evaluation of SMOS L-band vegetation optical depth (L-VOD) data sets: high sensitivity of L-VOD to above-ground biomass in Africa," Biogeosciences. vol. 15, no. 14, 2018.

[33] F. Tian, J.-P. Wigneron, P. Ciais, J. Chave, J. Ogée, J. Peñuelas, A. Ræbild, J.-C. Domec, X. Tong, and M. Brandt, "Coupling of ecosystem-scale plant water storage and leaf phenology observed by satellite," Nature Ecology & Evolution, 2018.

[34] Y. Y. Liu, R. A. de Jeu, M. F. McCabe, J. P. Evans, and A. I. van Dijk, "Global long-term passive microwave satellite-based retrievals of vegetation optical depth," Geophysical Research Letters. vol. 38, no. 18, 2011.

[35] T. Jackson, and T. Schmugge, "Vegetation effects on the microwave emission of soils," Remote Sensing of Environment. vol. 36, no. 3, pp. 203–212, 1991.

[36] J.-P. Wigneron, P. Waldteufel, A. Chanzy, J.-C. Calvet, and Y. Kerr, "Two-dimensional microwave interferometer retrieval capabilities over land surfaces (SMOS mission)," Remote Sensing of Environment. vol. 73, no. 3, pp. 270–282, 2000.

[37] A. Al-Yaari, J.-P. Wigneron, W. Dorigo, A. Colliander, T. Pellarin, S. Hahn, A. Mialon, P. Richaume, R. Fernandez-Moran, and L. Fan, "Assessment and inter-comparison of recently developed/reprocessed microwave satellite soil moisture products using ISMN ground-based measurements," Remote Sensing of Environment. vol. 224, pp. 289–303, 2019.

[38] Y. Qin, X. Xiao, J. Dong, Y. Zhang, X. Wu, Y. Shimabukuro, E. Arai, C. Biradar, J. Wang, Z. Zou, F. Liu, Z. Shi, R. Doughty, and B. M. III, "Improved estimates of forest cover and loss in the Brazilian Amazon in 2000–2017," Nature Sustainability. vol. 2, no. 8, pp. 764–772, 2019.

[39] J. M. Carreiras, S. Quegan, T. Le Toan, D. H. T. Minh, S. S. Saatchi, N. Carvalhais, M. Reichstein, and K. Scipal, "Coverage of high biomass forests by the ESA BIOMASS mission under defense restrictions," Remote Sensing of Environment. vol. 196, pp. 154–162, 2017.

[40] S. Mermoz, T. Le Toan, L. Villard, M. Réjou-Méchain, and J. Seifert-Granzin, "Biomass assessment in the Cameroon savanna using ALOS PALSAR data," Remote Sensing of Environment. vol. 155, pp. 109–119, 2014.

[41] J.-P. Wigneron, Y. Kerr, P. Waldteufel, K. Saleh, M.-J. Escorihuela, P. Richaume, P. Ferrazzoli, P. De Rosnay, R. Gurney, and J.-C. Calvet, "L-band microwave emission of the biosphere (L-MEB) model: Description and calibration against experimental data sets over crop fields," Remote Sensing of Environment. vol. 107, no. 4, pp. 639–655, 2007.

[42] Y. H. Kerr, P. Waldteufel, P. Richaume, J. P. Wigneron, P. Ferrazzoli, A. Mahmoodi, A. Al Bitar, F. Cabot, C. Gruhier, and S. E. Juglea, "The SMOS soil moisture retrieval algorithm," IEEE Transactions on Geoscience and Remote Sensing. vol. 50, no. 5, pp. 1384–1403, 2012.

[43] V. Avitabile, M. Herold, G. Heuvelink, S. L. Lewis, O. L. Phillips, G. P. Asner, J. Armston, P. S. Ashton, L. Banin, and N. Bayol, "An integrated pan-tropical biomass map using multiple reference datasets," Global Change Biology. vol. 22, no. 4, pp. 1406–1420, 2016.

[44] M. C. Hansen, P. V. Potapov, R. Moore, M. Hancher, S. Turubanova, A. Tyukavina, D. Thau, S. Stehman, S. Goetz, and T. Loveland, "High-resolution global maps of 21st-century forest cover change," Science. vol. 342, no. 6160, pp. 850–853, 2013.

[45] R. Oliva, E. Daganzo, Y. H. Kerr, S. Mecklenburg, S. Nieto, P. Richaume, and C. Gruhier, "SMOS radio frequency interference scenario: Status and actions taken to improve the RFI environment in the 1400–1427-MHz passive band," IEEE Transactions on Geoscience and Remote Sensing. vol. 50, no. 5, pp. 1427–1439, 2012.

[46] Y. H. Kerr, A. Al-Yaari, N. Rodriguez-Fernandez, M. Parrens, B. Molero, D. Leroux, S. Bircher, A. Mahmoodi, A. Mialon, and P. Richaume, "Overview of SMOS performance in terms of global soil moisture monitoring after six years in operation," Remote Sensing of Environment. vol. 180, pp. 40–63, 2016.

[47] L. Fan, J.-P. Wigneron, Q. Xiao, A. Al-Yaari, J. Wen, N. Martin-StPaul, J.-L. Dupuy, F. Pimont, A. Al Bitar, and R. Fernandez-Moran, "Evaluation of microwave remote sensing for monitoring live fuel moisture content in the Mediterranean region," Remote Sensing of Environment. vol. 205, pp. 210–223, 2018.

[48] P. D. Broxton, X. Zeng, D. Sulla-Menashe, and P. A. Troch, "A global land cover climatology using MODIS data," Journal of Applied Meteorology and Climatology. vol. 53, no. 6, pp. 1593–1605, 2014.

[49] A. Ahlström, M. R. Raupach, G. Schurgers, B. Smith, A. Arneth, M. Jung, M. Reichstein, J. G. Canadell, P. Friedlingstein, A. K. Jain, E. Kato, B. Poulter, S. Sitch, B. D. Stocker, N. Viovy, Y. P. Wang, A. Wiltshire, S. Zaehle, and N. Zeng, "The dominant role of semi-arid ecosystems in the trend and variability of the land CO2 sink," Science. vol. 348, no. 6237, pp. 895–899, 2015.

[50] M. C. Hansen, S. V. Stehman, and P. V. Potapov, "Quantification of global gross forest cover loss," Proceedings of the National Academy of Sciences. vol. 107, no. 19, pp. 8650–8655, 2010.

[51] C. Chen, T. Park, X. Wang, S. Piao, B. Xu, R. K. Chaturvedi, R. Fuchs, V. Brovkin, P. Ciais, and R. Fensholt, "China and India lead in greening of the world through land-use management," Nature Sustainability. vol. 2, no. 2, pp. 122, 2019.

[52] Y. Qin, X. Xiao, J. Dong, Y. Zhou, J. Wang, R. B. Doughty, Y. Chen, Z. Zou, and B. Moore III, "Annual dynamics of forest areas in South America during 2007–2010 at 50-m spatial resolution," Remote Sensing of Environment. vol. 201, pp. 73–87, 2017.

[53] K. Wolter, and M. S. Timlin, "El Niño/Southern Oscillation behaviour since 1871 as diagnosed in an extended multivariate ENSO index (MEI. ext)," International Journal of Climatology. vol. 31, no. 7, pp. 1074–1087, 2011.

[54] A. Huete, C. Justice, and W. Van Leeuwen, "MODIS vegetation index (MOD13)," Algorithm Theoretical Basis Document. vol. 3, pp. 213, 1999.

[55] D. P. Dee, S. M. Uppala, A. Simmons, P. Berrisford, P. Poli, S. Kobayashi, U. Andrae, M. Balmaseda, G. Balsamo, and d. P. Bauer, "The ERA-Interim reanalysis: Configuration and performance of the data assimilation system," Quarterly Journal of the Royal Meteorological Society. vol. 137, no. 656, pp. 553–597, 2011.

[56] G. J. Huffman, D. T. Bolvin, E. J. Nelkin, D. B. Wolff, R. F. Adler, G. Gu, Y. Hong, K. P. Bowman, and E. F. Stocker, "The TRMM multisatellite precipitation analysis (TMPA): Quasi-global, multiyear, combined-sensor precipitation estimates at fine scales," Journal of Hydrometeorology. vol. 8, no. 1, pp. 38–55, 2007.

[57] J. Wahr, M. Molenaar, and F. Bryan, "Time variability of the Earth's gravity field: Hydrological and oceanic effects and their possible detection using GRACE," Journal of Geophysical Research: Solid Earth. vol. 103, no. B12, pp. 30205–30229, 1998.

[58]   S. Swenson, D. Chambers, and J. Wahr, "Estimating geocenter variations from a combination of GRACE and ocean model output," Journal of Geophysical Research: Solid Earth. vol. 113, no. B8, 2008.

[59]   Y. Y. Liu, A. I. van Dijk, D. G. Miralles, M. F. McCabe, J. P. Evans, R. A. de Jeu, P. Gentine, A. Huete, R. M. Parinussa, and L. Wang, "Enhanced canopy growth precedes senescence in 2005 and 2010 Amazonian droughts," Remote Sensing of Environment. vol. 211, pp. 26–37, 2018.

[60]   B. Poulter, D. Frank, P. Ciais, R. B. Myneni, N. Andela, J. Bi, G. Broquet, J. G. Canadell, F. Chevallier, and Y. Y. Liu, "Contribution of semi-arid ecosystems to interannual variability of the global carbon cycle," Nature. vol. 509, no. 7502, pp. 600, 2014.

[61]   V. Humphrey, J. Zscheischler, P. Ciais, L. Gudmundsson, S. Sitch, and S. I. Seneviratne, "Sensitivity of atmospheric CO_2 growth rate to observed changes in terrestrial water storage," Nature. vol. 560, no. 7720, pp. 628, 2018.

[62]   M. Jung, M. Reichstein, C. R. Schwalm, C. Huntingford, S. Sitch, A. Ahlström, A. Arneth, G. Camps-Valls, P. Ciais, and P. Friedlingstein, "Compensatory water effects link yearly global land CO_2 sink changes to temperature," Nature. vol. 541, no. 7638, pp. 516, 2017.

[63]   K. A. Masarie, and P. P. Tans, "Extension and integration of atmospheric carbon dioxide data into a globally consistent measurement record," Journal of Geophysical Research: Atmospheres. vol. 100, no. D6, pp. 11593–11610, 1995.

[64]   J. Wang, N. Zeng, and M. Wang, "Interannual variability of the atmospheric CO_2 growth rate: roles of precipitation and temperature," Biogeosciences. vol. 13, no. 8, pp. 2339–2352, 2016.

[65]   N. Zeng, A. Mariotti, and P. Wetzel, "Terrestrial mechanisms of interannual $CO2$ variability," Global Biogeochemical Cycles. vol. 19, no. 1, 2005.

[66]   W. R. Anderegg, A. P. Ballantyne, W. K. Smith, J. Majkut, S. Rabin, C. Beaulieu, R. Birdsey, J. P. Dunne, R. A. Houghton, and R. B. Myneni, "Tropical nighttime warming as a dominant driver of variability in the terrestrial carbon sink," Proceedings of the National Academy of Sciences. vol. 112, no. 51, pp. 15591–15596, 2015.

[67]   M. Fernández-Martínez, J. Sardans, F. Chevallier, P. Ciais, M. Obersteiner, S. Vicca, J. Canadell, A. Bastos, P. Friedlingstein, and S. Sitch, "Global trends in carbon sinks and their relationships with CO_2 and temperature," Nature Climate Change, pp. 1, 2018.

[68]   S. Lohberger, M. Stängel, E. C. Atwood, and F. Siegert, "Spatial evaluation of Indonesia's 2015 fire-affected area and estimated carbon emissions using Sentinel-1," Global Change Biology. vol. 24, no. 2, pp. 644–654, 2018.

[69]   V. Huijnen, M. J. Wooster, J. W. Kaiser, D. L. Gaveau, J. Flemming, M. Parrington, A. Inness, D. Murdiyarso, B. Main, and M. van Weele, "Fire carbon emissions over maritime southeast Asia in 2015 largest since 1997," Scientific Reports. vol. 6, pp. 26886, 2016.

[70]   Y. Yin, P. Ciais, F. Chevallier, G. R. Werf, T. Fanin, G. Broquet, H. Boesch, A. Cozic, D. Hauglustaine, and S. Szopa, "Variability of fire carbon emissions in equatorial Asia and its nonlinear sensitivity to El Niño," Geophysical Research Letters. vol. 43, no. 19, 2016.

[71]   A. Tyukavina, A. Baccini, M. Hansen, P. Potapov, S. Stehman, R. Houghton, A. Krylov, S. Turubanova, and S. Goetz, "Aboveground carbon loss in natural and managed tropical forests from 2000 to 2012," Environmental Research Letters. vol. 10, no. 7, pp. 074002, 2015.

[72]   D. J. Zarin, N. L. Harris, A. Baccini, D. Aksenov, M. C. Hansen, C. Azevedo-Ramos, T. Azevedo, B. A. Margono, A. C. Alencar, and C. Gabris, "Can carbon emissions from tropical deforestation drop by 50% in 5 years?," Global Change Biology. vol. 22, no. 4, pp. 1336–1347, 2016.

[73]   C. M. Ryan, N. J. Berry, and N. Joshi, "Quantifying the causes of deforestation and degradation and creating transparent REDD+ baselines: A method and case study from central Mozambique," Applied Geography. vol. 53, pp. 45–54, 2014.

[74]   G. E. Ponce-Campos, M. S. Moran, A. Huete, Y. Zhang, C. Bresloff, T. E. Huxman, D. Eamus, D. D. Bosch, A. R. Buda, and S. A. Gunter, "Ecosystem resilience despite large-scale altered hydroclimatic conditions," Nature. vol. 494, no. 7437, pp. 349, 2013.

[75]   L. Poorter, F. Bongers, T. M. Aide, A. M. A. Zambrano, P. Balvanera, J. M. Becknell, V. Boukili, P. H. Brancalion, E. N. Broadbent, and R. L. Chazdon, "Biomass resilience of Neotropical secondary forests," Nature. vol. 530, no. 7589, pp. 211, 2016.

[76]   S. L. Lewis, G. Lopez-Gonzalez, B. Sonké, K. Affum-Baffoe, T. R. Baker, L. O. Ojo, O. L. Phillips, J. M. Reitsma, L. White, and J. A. Comiskey, "Increasing carbon storage in intact African tropical forests," Nature. vol. 457, no. 7232, pp. 1003, 2009.

[77]   O. L. Phillips, Y. Malhi, N. Higuchi, W. F. Laurance, P. V. Núñez, R. M. Vásquez, S. G. Laurance, L. V. Ferreira, M. Stern, and S. Brown, "Changes in the carbon balance of tropical forests: evidence from long-term plots," Science. vol. 282, no. 5388, pp. 439–442, 1998.

[78] M. Gloor, O. L. Phillips, J. Lloyd, S. L. Lewis, Y. Malhi, T. R. Baker, G. LÓPEZ-GONZALEZ, J. Peacock, S. Almeida, and A. A. de Oliveira, "Does the disturbance hypothesis explain the biomass increase in basin-wide Amazon forest plot data?," Global Change Biology. vol. 15, no. 10, pp. 2418–2430, 2009.

[79] Z. Zhu, S. Piao, R. B. Myneni, M. Huang, Z. Zeng, J. G. Canadell, P. Ciais, S. Sitch, P. Friedlingstein, and A. Arneth, "Greening of the Earth and its drivers," Nature Climate Change. vol. 6, no. 8, pp. 791, 2016.

[80] M. Brandt, P. Hiernaux, K. Rasmussen, C. J. Tucker, J.-P. Wigneron, A. A. Diouf, S. M. Herrmann, W. Zhang, L. Kergoat, and C. Mbow, "Changes in rainfall distribution promote woody foliage production in the Sahel," Communications Biology. vol. 2, no. 1, pp. 133, 2019.

[81] D. Schimel, B. B. Stephens, and J. B. Fisher, "Effect of increasing CO2 on the terrestrial carbon cycle," Proceedings of the National Academy of Sciences. vol. 112, no. 2, pp. 436–441, 2015.

[82] E. Rutishauser, B. Hérault, C. Baraloto, L. Blanc, L. Descroix, E. D. Sotta, J. Ferreira, M. Kanashiro, L. Mazzei, and M. V. d'Oliveira, "Rapid tree carbon stock recovery in managed Amazonian forests," Current Biology. vol. 25, no. 18, pp. R787-R788, 2015.

[83] M. Brandt, K. Rasmussen, J. Peñuelas, F. Tian, G. Schurgers, A. Verger, O. Mertz, J. R. Palmer, and R. Fensholt, "Human population growth offsets climate-driven increase in woody vegetation in sub-Saharan Africa," Nature Ecology & Evolution. vol. 1, no. 4, pp. 0081, 2017.

11 Sea Ice Thickness Estimation from Spaceborne GNSS-R

Qingyun Yan, Shuanggen Jin, Yunjian Xie, and Weimin Huang

11.1 INTRODUCTION

Obtaining accurate information about the Arctic sea ice thickness (SIT) and its change not only helps to conduct research on climate change, environmental change, and ecological security at regional and global scales, but also has important practical significance for marine resources development, maritime transportation, shipping, and polar expeditions [1–3]. However, the *in situ* SIT measuring in the field is very cumbersome and limited by spatial coverage. Instead, remote sensing gives a more efficient and economical option.

So far, large-scale SIT data that people use extensively are generally from remote sensing satellites. With the help of passive microwave sensors [4, 5], scatterometers [6, 7], radar altimeters [8] and synthetic aperture radar (SAR) [9], one can estimate SIT. However, on one hand, the spatial of resolution passive microwave sensors and scatterometers is lower (usually 25–50 km). On other hand, SAR and radar altimeters can provide better resolution, but they are more costly [10] and the revisit time is long. Besides, interpreting SAR images is often time-consuming and subjective, and the empirical retracking used for altimeters lacks actual physical models. Since the concept of Global Navigation Satellite System Reflectometry was proposed, it has been successfully applied to various remote sensing missions, such as sea surface wind [11, 12] and roughness monitoring [13], sea surface height observation [14], snow depth estimation [15, 16], soil moisture [17, 18] and vegetation sensing [19], etc. GNSS-R uses the bistatic scattering mode, in which the transmitter and receiver are in different positions. Theoretically, a transmitter can be any GNSS satellite; for example, GPS and China's BeiDou Navigation Satellite System, etc. The transmitted signal is reflected by the earth's surface (e.g., ocean, land or ice) and thus carries information about the reflected surface (e.g., roughness, etc.). The reflected signal is then captured by one or more GNSS-R receivers. In addition, most GNSS-R receivers can collect multiple reflected signals at the same time. It is also worth mentioning that as a passive instrument, the GNSS-R receivers usually have the characteristics of low cost, low weight and low power consumption, and can be deployed flexibly. Depending on the platform on which the receiver is mounted, GNSS-R can be divided into three categories: specifically spaceborne, airborne and ground-based. The first one is mainly used for large-scale or global monitoring, and the latter two are usually used for regional observations. Therefore, by deploying multiple GNSS-R receivers on various platforms, one can realize time-intensive coverage on both global and regional scales. It should be noted that GNSS-R uses the L-band with low attenuation by the atmosphere and it is not influenced by day or night, cloud cover and weather conditions, thus is particularly suitable for surface remote sensing [20]. In the part of spatio-temporal resolution, taking the CYGNSS system as an example, its global average revisit time is 4 hours. The spatial resolution in the case of incoherent scattering is about 10 km, while the resolution in the case of coherent scattering is about 500 m [21]. Spatial resolution of spaceborne GNSS-R is comparable to or better than radar altimeter (Envisat nominal circular footprint of the altimeter is 2–10 km in diameter, while the CryoSat-2 altimeter has a footprint of 1.65 km × 0.30 km along the track [8]).

Compared to other remote sensing methods, the advantages presented by GNSS-R provide a more promising application for sea ice remote sensing. This chapter will develop and establish

DOI: 10.1201/9781003363118-11

a semi-empirical model to realize the inversion of sea ice thickness based on GNSS-R signals. This basis further improves the accuracy of the results. The development of this study will provide a theoretical basis for the analysis of GNSS-R signals on sea ice cover, and thus give accurate inversion results of sea ice thickness, develop new applications of GNSS-R signals in sea ice remote sensing, and ultimately may provide the theoretical and experimental basis for the promotion and development of sea ice related industries and research.

11.2 CURRENT STATUS AND DEVELOPMENT

As mentioned already, GNSS-R technologies can be classified as spaceborne, airborne and ground-based according to the platform. Here we discuss the current research status of GNSS-R-based sea ice remote sensing according to this classification. It is worth mentioning that the progress of research based on different platforms reflects significant differences due to much more data from spaceborne experiments than the latter two. In addition, we also cover other applications of sea ice remote sensing here because the progress of sea ice thickness inversion is still relatively limited.

11.2.1 STATUS OF AIRBORNE MEASUREMENTS

The first airborne measurements were made in the Beaufort Sea in 1998 by Komjathy et al. [22]. Their results show that the received reflection signal is very sensitive to sea ice, and the received signal consistently shows a sharp and narrow waveform, but its amplitude varies significantly. This illustrates that the roughness of the sea ice in this region varied very little during their experiments, but its reflectivity varied a lot. Subsequently, Rivas et al. proposed a method to invert the dielectric constant and roughness of sea ice from the received time-delayed waveforms [23], whereby the dielectric constant is estimated based on the maximum value of the waveforms while the inversion of the roughness depends on the shape of the delay waveforms. However, available airborne experiments, as well as data, are still limited, so applications based on airborne measurements have not been widely developed.

11.2.2 PROGRESS OF GROUND-BASED EXPERIMENTS

Remote sensing of sea ice based on ground-based experiments has been carried out more actively than airborne measurements. Fabra et al. validated the altimetric application based on ground-based tests conducted at Greenland Island in 2008 and performed a correlation analysis between the polarization rate (ratio of reflected signals from right- and left-handed circular polarization) and sea ice density [24]. In addition, Shao et al. explored the feasibility of GNSS-R signal detection of sea ice through this experimental project [25]. In 2012, the signal-to-noise ratio of the received signal was analyzed at the Onsala Space Laboratory in Switzerland and based on this, a damping factor was derived and the magnitude of this variable was found to be directly related to the presence or absence of sea ice [26].

Based on experiments conducted in Bohai Bay, China, in 2013, Zhang et al. found that the ratio of the direct signal to the received signal is sensitive to the sea ice concentration [10]. Later, the application of the ratio was further broadened by Gao et al. who achieved thickness inversion for sea ice with thicknesses between 10 and 20 cm based on tests in Liaodong Bay in 2016 [27]. In addition, Wang et al. verified the signal simulation capability of the software using the measured reflection signals [28]. Yang et al. found that the emission signals with different polarizations are effective in observing sea ice at different angles [29]. However, it is worth pointing out that ground-based experiments are usually restricted to specific areas. The coverage area is small and the observations are generally not globally generalizable.

11.2.3 DEVELOPMENT OF SPACEBORNE APPLICATIONS

Compared with the aforementioned two platforms, the spaceborne GNSS-R is particularly important because it can provide global coverage and enable sea ice monitoring on a large spatial scale. In 2003, the UK-DMC satellite acquired GNSS-R signals from space for the first time and demonstrated its potential for sea ice remote sensing applications through the literature [30]. The analysis shows that compared to sea ice, the rough seawater decreases the overall coherent specular reflection and leads to a larger glistening zone, resulting in a larger distribution in its corresponding delay-Doppler map (DDM) along the delay-Doppler axes. Despite the limited GNSS-R data available at that time (only two sets totaling 16 seconds long), the results demonstrate the feasibility of using spaceborne GNSS-R to observe sea ice covered areas. In addition, the variation in signal power due to different sea ice concentrations, as well as the variation in time delay and Doppler extension, illustrate the potential of GNSS-R for estimating sea ice concentration. It is also mentioned in [30] that in the case of coherent reflections, one can recover the carrier phase information for accurate surface height measurements. Nevertheless, the practical application of satellite-based GNSS-R sea ice remote sensing was not effectively developed due to the limited data at that time. Subsequently, in 2014, the launch of the TDS-1 satellite raised thousands of DDM data for the public, which offered an opportunity for researchers to further advance the preliminary results in [30]. Since then, a large number of sea ice remote sensing research topics based on TDS-1 data have emerged.

Yan and Huang realized the first application of TDS-1 data in sea ice detection [31, 32]. The results found that the power distribution of sea ice DDM is more concentrated compared to sea water DDM, and thus a sea ice detection method based on DDM data spread evaluation is proposed. Adopting the similar idea, many scholars have also achieved sea ice detection by choosing different observation variables to describe the DDM data spread [33–36]. The method of sea ice detection using observed quantities is simpler and more intuitive to implement, but the accuracy of the method depends on the reasonableness of the selected threshold. In addition, Schiavulli et al. proposed a method based on a 2D truncated singular value decomposition for the inversion of scattering coefficients to achieve sea ice detection [37]. However, the method cannot effectively solve the ambiguity problem in the inversion process. Consequently, Yan and Huang proposed a scheme based on the spatial integration method combined with the dual antenna method to solve the problem of its ambiguity [38]. These methods may provide additional information on sea ice distribution within the glistening zone, but the accuracy and practicality of these applications have not been thoroughly investigated due to limited reference data. The first application of machine learning in the field of GNSS-R, using neural network techniques to achieve sea ice detection and sea ice concentration estimation from DDM data, was introduced in [39]. Subsequently, the use of convolutional neural networks to fully utilize 2D DDM data was proposed [40], and the employment of support vector machines combined with feature extraction further improved the inversion accuracy and reduced the computational complexity [41, 42]. In addition to these two applications, Rodriguez-Alvarez et al. proposed a decision tree-based sea ice classification method [43], but the accuracy needs to be improved. Hu et al. implemented sea ice altimetry based on the study of delayed waveforms [44] and Li et al. performed sea ice altimetry based on phase measurements [45]. Notably, it was found in [45] that the difference between the ice surface altimetry results obtained in [45] and the mean sea surface height correlated well with the local sea ice thickness, reflecting the potential of the GNSS-R technique for measuring sea ice thickness. However, the sea ice altimetry method in [45] requires the use of very scarce raw satellite data and requires special error calibration of the results, such that the technique cannot be a general solution for sea ice thickness inversion for the time being. To further improve the shortcomings of the GNSS-R sea ice thickness inversion method, Yan and Huang proposed a simplified two-layer model [46] and successfully inverse performed this parameter from TDS-1 satellite data. By comparing the reference data, the correlation coefficient and root mean square error obtained were 0.84 and 9.39 cm, respectively.

11.2.4 SUMMARY OF CURRENT STATES

In summary, the research on GNSS-R-based sea ice remote sensing has entered a boom period, and various spaceborne, airborne and ground-based experiments have been carried out continuously. However, the current state of research shows that applications based on spaceborne data are more widespread, in large part because of the large amount of data available from the TDS-1 satellite. In addition, TDS-1 data provide global coverage, including the high polar latitudes, which can ensure remote sensing of sea ice on a large scale. This advantage is difficult to achieve with airborne and ground-based experiments. As such, the development of spaceborne applications is particularly important. From the perspective of current research topic, the research based on spaceborne data mainly focuses on four applications: sea ice detection, density estimation, classification and altimetry, while the work on sea ice thickness inversion is still relatively limited. Although a method based on a simplified two-layer model for sea ice thickness inversion was proposed in [46], it ignores the reflection at the upper partition interface, which can lead to inversion errors that are more significant when the sea ice is thicker or the signal propagation loss is larger, thus leading to its limited applicability.

11.3 THEORY AND METHODS

In reality, most of the sea ice in the Antarctic and Arctic has a large thickness, so the two-layer model cannot realize the large-scale sea ice thickness inversion. Therefore, it is necessary to propose a new GNSS-R sea ice thickness inversion model which can be applied to different sea ice types in a wide range.

11.3.1 THREE-LAYER MODEL

Therefore, to obtain a more generalized inversion model, the reflection at the upper and lower partition interfaces as well as the signal attenuation of the sea ice layer must be considered, and the overall reflection coefficient of the three-layer medium can be obtained by using the following expression:

$$\Re = \frac{R_1 + R_2 e^{-2ikd}}{1 + R_1 R_2 e^{-2ikd}} \tag{11.1}$$

where d is the sea ice thickness (Figure 11.1) and k is the vertical component of the signal propagation vector determined by the signal wavelength λ (known), the satellite incidence angle θ, and the sea ice dielectric constant ε_i

$$k = \frac{2\pi}{\lambda} \cos\theta \left(\operatorname{Re}\sqrt{\varepsilon_i} - j \,|\operatorname{Im}\sqrt{\varepsilon_i}| \right) \tag{11.2}$$

The sea ice dielectric constant ε_i can be obtained from the Vant model [47], i.e.

$$\varepsilon_i = 3.1 + 0.0084V_b + j(a_1 + a_2V_b) \tag{11.3}$$

where V_b is the volume of brine and the coefficients a_1 and a_2 are 0.037 and 0.00445 in the case of first-year ice and 0.003 and 0.00435 in the case of multi-year ice, respectively. V_b can then be obtained from the Ulaby model [48]:

$$V_b = 10^{-3} S(-\frac{49.185}{T} + 0.532) \tag{11.4}$$

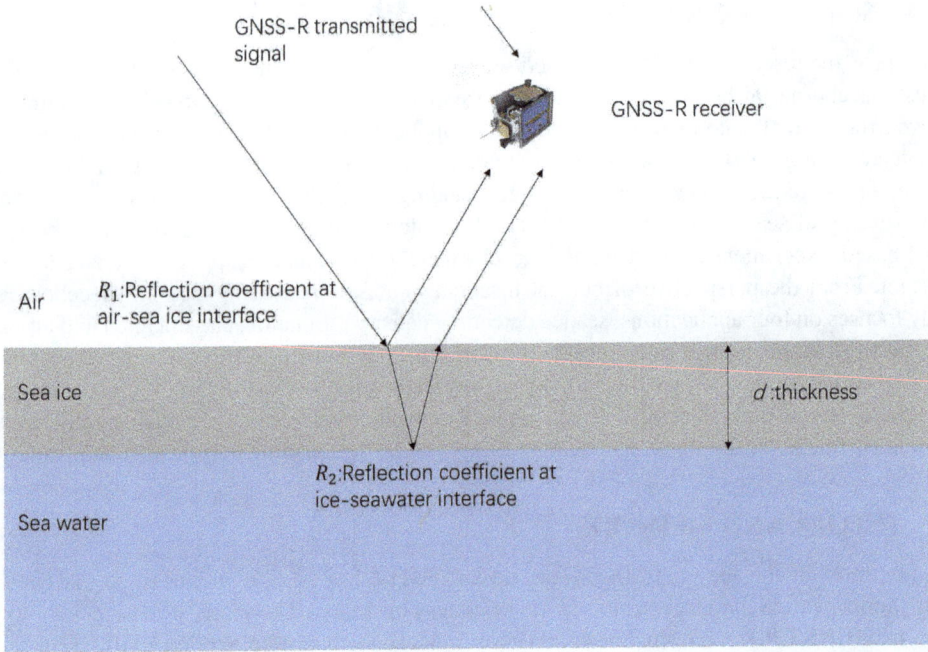

FIGURE 11.1   Schematic of GNSS-R signal reflected from a three-layer model of air, sea ice and seawater.

where S and T are the salinity and temperature of sea ice. In addition, R_1 and R_2 are determined by the dielectric constant of each layer, which can be found by the following general formula:

$$R_{1,2} = \frac{1}{2}(R_{vv} - R_{hh})$$

(11.5)

where R_{vv} and R_{hh} are the vertical polarization and horizontal polarization components, respectively, expressed by

$$R_{vv} = \frac{\varepsilon_{1,2}\cos\theta - \sqrt{\varepsilon_{1,2} - \sin^2\theta}}{\varepsilon_{1,2}\cos\theta + \sqrt{\varepsilon_{1,2} - \sin^2\theta}}$$

(11.6)

$$R_{hh} = \frac{\cos\theta - \sqrt{\varepsilon_{1,2} - \sin^2\theta}}{\cos\theta + \sqrt{\varepsilon_{1,2} - \sin^2\theta}}$$

(11.7)

$$\varepsilon_1 = \frac{1}{\varepsilon_i}$$

(11.8)

$$\varepsilon_2 = \frac{\varepsilon_i}{\varepsilon_w}$$

(11.9)

ε_w is the dielectric constant of seawater, which can be obtained from the Klein-Swift model [49].

In addition, the surface reflectance can be given by the following equation:

$$\Gamma = \mid \Re \mid \cdot \exp[-(\frac{4\pi}{\lambda}\theta_{rms}\sin\theta)^2] \qquad (11.10)$$

where θ_{rms} is the mean squared difference height and the exponential term in the expression represents the effect of the roughness effect. Since the *rms* of the sea ice surface is usually at the centimeter level, the sea ice surface can be considered a smooth surface in the GNSS-R domain, so the value of the roughness term in the expression is approximately equal to 1. Therefore, the reflectivity can be further approximated as

$$\Gamma = \mid \Re \mid^2 \qquad (11.11)$$

Combining the foregoing expressions for the overall reflection coefficient of the three-layer of the medium, we can obtain

$$\Gamma = \mid \frac{R_1 + R_2 e^{-2ikd}}{1 + R_1 R_2 e^{-2ikd}} \mid^2 \qquad (11.12)$$

Therefore, by combining the reflectance with the overall reflectance coefficient, the expression for the relationship between reflectance and sea ice thickness can be obtained, and the expression can be further derived and rewritten, so that the inversion of sea ice thickness from TDS-1 data can be achieved.

11.3.2 CONNECTION BETWEEN PRECISE THREE-LAYER MODEL AND SIMPLIFIED TWO-LAYER MODEL

Firstly, the expression for the simplified two-layer model [46] is listed here:

$$\Re = R_2 e^{-2ikd} \qquad (11.13)$$

The model ignored reflections at the upper air-sea ice interface. This simplification makes the model simple and easy to calculate, but does not apply to thicker or more signal-attenuating sea ice. Simplified two-layer model may create some of these problems:

a) The two-layer model ignores the reflection at the upper partition interface, which can lead to inversion errors that are more significant when the sea ice is thicker or the signal propagation loss is larger.
b) Some of the parameters used in the inversion of sea ice thickness come from other technical means, which may introduce some errors into the results.
c) Simplifying the model will lose some generalizability and will have an impact on the actual results. This is not allowed in large-scale applications.

So the three-layer model provides a more accurate and effective model of the medium, while other effects are improved by alternative auxiliary variables and machine learning, respectively. Thus, an accurate three-layer model is built.

11.3.3 ALTERNATIVES TO AUXILIARY VARIABLES

Based on empirical models of sea ice dielectric constants, we need other sea ice parameters, i.e., sea ice temperature as well as sea ice salinity, to determine their values. Currently, no work has been done to obtain these two sea ice parameters by GNSS-R technology. To eliminate the dependence on it, there are two options: one is to use some fixed sea ice dielectric constant to avoid the use of empirical models and the corresponding sea ice parameters (option 1); the second is to find empirical relationships between these sea ice parameters and sea ice thickness, such as the following Cox-Week model (option 2):

$$S = \begin{cases} 14.24 - 19.39d, & d \leq 0.4 \\ 7.88 - 1.59d, & d > 0.4 \end{cases} \tag{11.14}$$

Application of the model enables the replacement of sea ice salinity. By adopting similar models and thus using sea ice thickness as a proxy for other sea ice parameters, the dependence on other sea ice parameters is eliminated. Correspondingly, the results of sea ice thickness are obtained during the inversion process using, for example, iterative or least squares fitting methods. Although option 1 is simpler and more intuitive to operate, the use of a fixed sea ice dielectric coefficient is not sufficiently universal and thus leads to a large error because its value varies greatly under different sea ice types, temperatures and salinity. Combining the experiments and analyses we've discussed, we will choose option 2, which combines appropriate mathematical methods with sea ice parameter models and finally derives and establishes a set of GNSS-R based sea ice thickness inversion models, and can ensure its universality and high accuracy, and realize independent measurements.

11.3.4 SOLUTIONS TO IMPROVE ACCURACY USING MACHINE LEARNING METHODS

By using a more rational three-layer model, the errors caused by the previous simplified model can be greatly reduced and its generalizability can be improved. However, the errors introduced by the use of empirical models (e.g., sea ice dielectric constant models) can still have a large impact on the actual results. For this reason, here we can introduce a machine learning approach to improve it. Machine learning can directly establish the relationship between inputs and outputs (e.g., GNSS-R data and sea ice thickness) without relying on any empirical formulas or models. Firstly, the modeling results are to be analyzed to validate the contribution of the input variables to the inversion results, so that relevant variables can be selected as inputs for machine learning. A specific model is then determined by training, based on the intrinsic relationship before the data. Validation and performance analysis with test results are provided. Two applications of machine learning in sea ice thickness that have been used in sea ice concentration estimation are presented here, specifically, support vector regression (SVR) and convolutional neural network (CNN). The input here contains four parameters, i.e., reflectivity (Γ), incidence angle (θ), sea ice salinity (S) and sea ice temperature (T), and the output is SIT.

a) SVR-based approach

Given a data set $(x_1, y_1), \ldots, (x_n, y_n)$, it is randomly divided into two roughly equal parts, one as the training set and the other as the test set. The data set x_j is the four-parameter input vector, and for $j = 1, \ldots, n$, y_j is the corresponding reference SIT value. the optimization objective of the SVR is

$$\min_{w,b,\zeta} (\frac{1}{2} \| w \|^2 + C \sum_j (\zeta_j + \zeta_j^*)) \tag{11.15}$$

where w should be subject to

$$\begin{cases} y_j - w\phi(x_i) + b \leq \varepsilon + \zeta_j \\ [w\phi(x_j) + b] - y_j \leq \varepsilon + \zeta_j^*, \\ \zeta_j, \zeta_j^* \geq 0 \end{cases} \tag{11.16}$$

where φ is a mapping function, w is the weight of $\varphi(x_j)$, and C is the regularization parameter. The regularization parameters, ξ_j and ξ_j^*, represent the SIT estimation error that exceeds the error tolerance and b is the bias term. This problem can be solved by introducing Lagrange multipliers $\alpha = \alpha_1, \ldots, \alpha_n$ and $\alpha^* = \alpha_1^*, \ldots, \alpha_n^*$ to solve it. The detailed procedure to obtain the solution can be found in [42].

With solved α, α^* and b, the estimation of SIT can be proceeded through

$$f(x) = \sum_j (\alpha_j - \alpha_j^*) K(x_j, x) + b. \tag{11.17}$$

Through experiments, the radial basis function (RBF) demonstrated excellent accuracy, and was thus adopted here. The RBF is given by

$$K(x_j, x_k) = \exp(-\gamma \| x_j - x_k \|^2), \tag{11.18}$$

where γ is the kernel width.

b) CNN-based Approach

For CNN, the input of four elements is reshaped into a 2×2 image. Correspondingly, the designed CNN framework contains one input layer followed by one convolutional layer, one fully connected layer and one output layer. These layers are of size 2×2, $2 \times 2 \times 7$, 1×1, and 1×1, respectively.

The convolved images resulted from the k th ($k = 1, \ldots, 7$) filter, W^k, can be described by

$$h_{i,j}^k = \phi((W^k * X)_{ij} + b), \ i = 1, 2, j = 1, 2, \tag{11.19}$$

where X and b are the input image and the bias. The convolution operation is denoted by and the activation function by φ. The widely adopted ReLU is chosen for φ, i.e.

$$\phi(z) = \max(0, z). \tag{11.20}$$

The elements in adjacent layers are connected by activation functions with weights. The activation functions in the fully connected and the output layers were given separately by the widely used sigmoid function $\varphi^{(1)}(x) = 1/(1 + e^x)$ and the linear function $\varphi^{(2)}(x) = x$. Weights in the fully connected layer are denoted by $w_{j1}^1, j \in [1, 28]$, and w_{11}^2 for that in the output layers. The weights w_{jk}^i are determined through back-propagation learning [50]. For more details on the implementation of CNN, one can refer to [40].

11.4 RESULTS AND DISCUSSION

Further tests were done on the validation set with the trained CNN and SVR. The test set was analyzed with R of 0.90 and 0.94, RMSD of 7.97 cm and R values of 6.01 cm obtained by the CNN and SVR based methods. The RMSD of 7.97 cm and 6.01 cm, respectively. The decrease in test accuracy is minimal relative to the training set demonstrating the generality of these proposed methods. In addition, the overall estimated SIT is generally consistent with the reference SIT. Comparison with the model-based results. The improvement using the machine learning approach is evident compared to the model-based results (R of 0.90 and RMSD of 8.68 cm). The improvement using the machine learning approach is evident (see Table 11.1). Comparing the overall results (including training and test data), R decreased from 0.95 to 0.93 and RMSD increased from 5.49 to 6.82 cm when TDS-1 data was removed from the inputs. This demonstrates the effect of TDS-1 measurements on SIT estimates. Table 11.1 summarizes the accuracy of the proposed CNN- and SVR-based methods.

The training results obtained by CNN- and SVR-based methods are presented in Figure 11.2 and Figure 11.3, respectively. With the trained CNN and SVR, further tests were done with the validation set (see Figures 11.4 and 11.5). These figures are from [46].

TABLE 11.1

Accuracy of SIT Retrieval

Index	R			RMSD (cm)		
Method	CNN	SVR	Model-Based	CNN	SVR	Model-Based
Training	0.89	0.96		7.79	4.96	
Test	0.90	0.94		7.79	6.01	
Overall	0.90	0.95	0.90	7.79	5.49	8.68

Note: * Three-layer model not included.

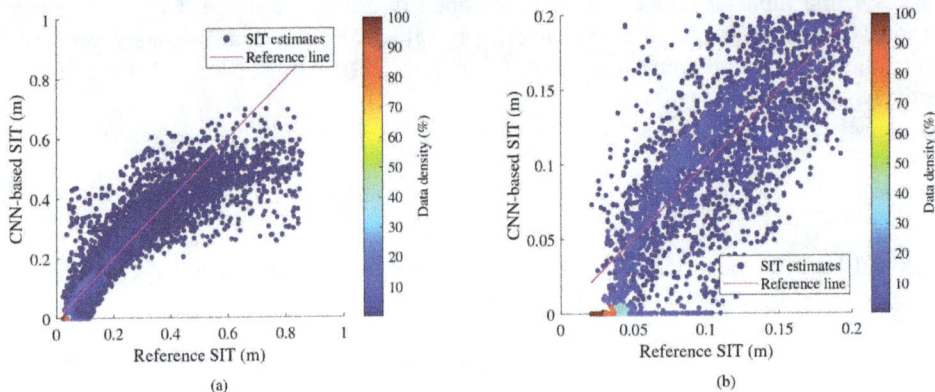

FIGURE 11.2 Density plot comparing SIT from CNN-based training results and SMOS data with the 1:1 reference line (magenta): (a) All results and (b) SMOS SIT below 0.2 m.

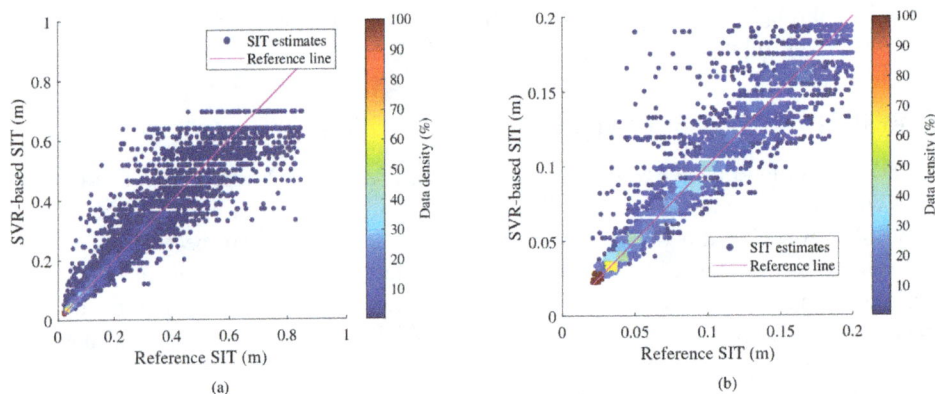

FIGURE 11.3 Density plot comparing SIT from SVR-based training results and SMOS data with the 1:1 reference line (magenta): (a) All results and (b) SMOS SIT below 0.2 m.

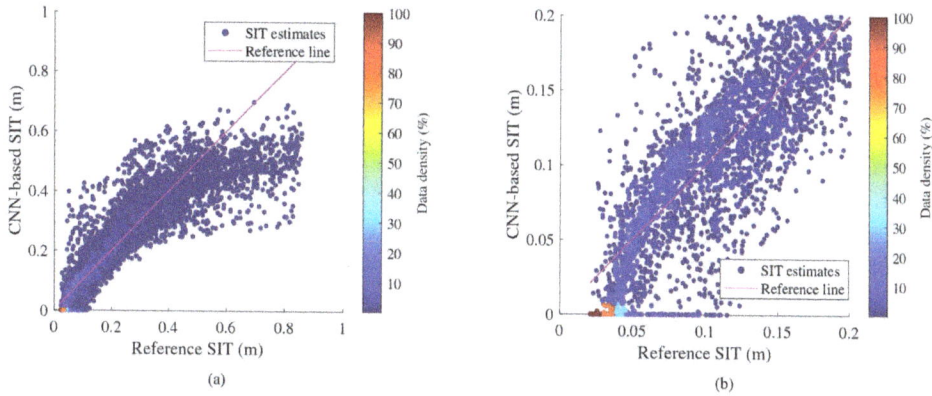

FIGURE 11.4 Density plot comparing SIT from CNN-based test results and SMOS data with the 1:1 reference line (magenta): (a) All results and (b) SMOS SIT below 0.2 m.

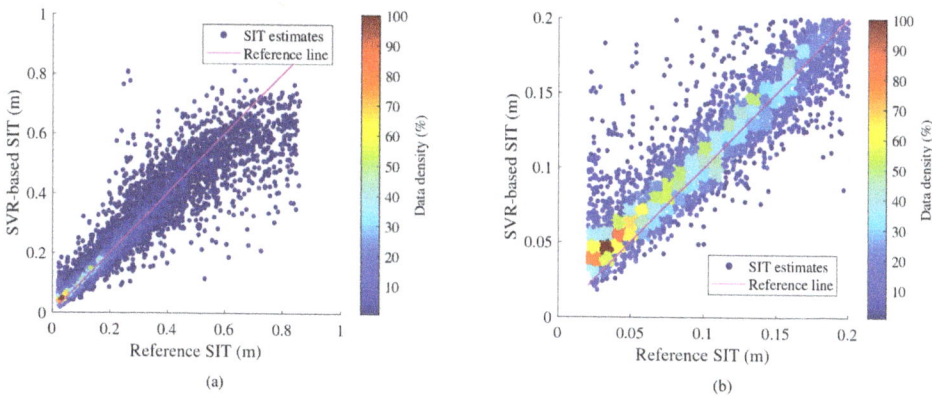

FIGURE 11.5 Density plot comparing SIT from SVR-based test results and SMOS data with the 1:1 reference line (magenta): (a) All results and (b) SMOS SIT below 0.2 m.

It can be expected that the source of sea ice temperature and salinity data as input and its accuracy will have some influence on the SIT estimation, and the use of the auxiliary variable substitution method mentioned earlier in subsequent experiments will improve the quality of SIT estimation. A large number of test experiments are needed to further improve the accuracy of the inversion results, and to achieve a high degree of agreement between the inversion results of sea ice thickness and the reference data on a global scale and over an annual span; the correlation coefficient reaches 0.9 or more, and the RMSE does not exceed 9 cm, improving the accuracy of the existing results in this field.

It is important to note that sea ice characteristics vary globally. In addition, empirical models of sea ice dielectric constants are more complex and are determined by different sea ice types, temperatures and other factors, and therefore the models involve more multiple sea ice parameter variables, adding difficulty to the inversion process. Therefore, the empirical model of sea ice dielectric constant and the effect of surface roughness need to be analyzed and examined in more depth.

When calculating the dielectric constant of sea ice, it is often necessary to use a number of empirical equations, which depend mainly on sea ice temperature and salt content. This information is therefore also necessary in the inversion process. In other words, it may be a new idea to further optimize the inversion model and to achieve the elimination of dependencies on other auxiliary data.

11.5 SUMMARY

In this chapter, based on an existing simplified two-layer model, we optimize modeling errors at the ice and water interface, focus on the derivation and establishment of a precise three-layer model of air-sea ice-seawater which may be more general and universal compared with the previous work. This offers a promising alternative to satellite altimetry for monitoring sea ice thickness.

In addition, a possible scheme to eliminate the dependence on other auxiliary parameters during the retrieving process is suggested, making GNSS-R an independent technique for sea ice thickness measurement. Also to improve the inversion accuracy, a machine learning-based parameter estimation method was introduced. Furthermore, the use of algorithms in this work are shown where applicable.

Sea ice thickness inversion is a difficult and hot issue internationally; current laser and radar thickness measurement techniques and radiometer inversion techniques all have excessive errors. GNSS-R as an emerging alternative remote sensing technology will show its potential in sea ice thickness estimation.

11.6 ACKNOWLEDGMENT

The research was supported by the National Natural Science Foundation of China under Grant 42001362.

REFERENCES

[1] D. A. Rothrock, Y. Yu, and G. A. Maykut, "Thinning of the Arctic sea-ice cover," *Geophys. Res. Lett.*, vol. 26, pp. 3469–3472, Dec. 1999.

[2] D. Hartman, A. Klein Tank, M. Rusicucci, L. Alexander, B. Broenniman, Y. Charabi, F. Dentener, E. Dlugokencky, E. Easterling, A. Kaplan, B. Soden, P. Thorne, M. Wild, P. Zhai, and E. Kent, *Observations: Atmosphere and Surface.* Cambridge University Press, 2013.

[3] Q. Yan and W. Huang, "Sea ice remote sensing using GNSS-R: A review," *Remote Sens.*, vol. 11, p. 2565, Nov. 2019.

[4] Z. I. Petrou and Y. Tian, "Prediction of sea ice motion with convolutional long short-term memory networks," *IEEE Trans. Geosci. Remote Sens.*, vol. 57, pp. 6865–6876, Sept. 2019.

[5] Z. Zhang, Y. Yu, X. Li, F. Hui, X. Cheng, and Z. Chen, "Arctic sea ice classification using microwave scatterometer and radiometer data during 2002–2017," *IEEE Trans. Geosci. Remote Sens.*, vol. 57, pp. 5319–5328, Aug. 2019.

[6] I. Otosaka, M. Belmonte Rivas, and A. Stoffelen, "Bayesian sea ice detection with the ERS scatterometer and sea ice backscatter model at C-Band," *IEEE Trans. Geosci. Remote Sens.*, vol. 56, pp. 2248–2254, Apr. 2018.

[7] M. Belmonte Rivas, I. Otosaka, A. Stoffelen, and A. Verhoef, "A scatterometer record of sea ice extents and backscatter: 1992–2016," *Cryosph.*, vol. 12, pp. 2941–2953, Sept. 2018.

[8] S. K. Rose, O. B. Andersen, M. Passaro, C. A. Ludwigsen, and C. Schwatke, "Arctic ocean sea level record from the complete radar altimetry era: 1991–2018," *Remote Sens.*, vol. 11, p. 1672, July 2019.

[9] F. Gao, X. Wang, Y. Gao, J. Dong, and S. Wang, "Sea ice change detection in SAR images based on convolutional-wavelet neural networks," *IEEE Geosci. Remote Sens. Lett.*, vol. 16, pp. 1240–1244, Aug. 2019.

[10] Y. Zhang, W. Meng, Q. Gu, Y. Han, Z. Hong, Y. Cao, Q. Xia, and W. Wang, "Detection of Bohai bay sea ice using GPS-reflected signals," *IEEE J. Sel. Top. Appl. Earth Obs. Remote Sens.*, vol. 8, pp. 39–46, Jan. 2015.

[11] C. Li and W. Huang, "An algorithm for sea-surface wind field retrieval from GNSS-R delay-doppler map," *IEEE Geosci. Remote Sens. Lett.*, vol. 11, pp. 2110–2114, Dec. 2014.

[12] G. Foti, C. Gommenginger, P. Jales, M. Unwin, A. Shaw, C. Robertson, and J. Rosell´o, "Spaceborne GNSS reflectometry for ocean winds: First results from the UK TechDemoSat-1 mission," *Geophys. Res. Lett.*, vol. 42, pp. 5435–5441, July 2015.

[13] Q. Yan, W. Huang, and G. Foti, "Quantification of the relationship between sea surface roughness and the size of the glistening zone for GNSS-R," *IEEE Geosci. Remote Sens. Lett.*, vol. 15, pp. 237–241, Feb. 2018.

[14] M. P. Clarizia, C. Ruf, P. Cipollini, and C. Zuffada, "First spaceborne observation of sea surface height using GPS-Reflectometry," *Geophys. Res. Lett.*, vol. 43, pp. 767–774, Jan. 2016.

[15] S. Jin, X. Qian, and H. Kutoglu, "Snow depth variations estimated from GPS-reflectometry: A case study in Alaska from L2P SNR data," *Remote Sens.*, vol. 8, p. 63, Jan. 2016.

[16] L. Zhou, S. Xu, J. Liu, and B. Wang, "On the retrieval of sea ice thickness and snow depth using concurrent laser altimetry and L-band remote sensing data," *Cryosph.*, vol. 12, pp. 993–1012, Mar. 2018.

[17] C. Ruf, "Cyclone Global Navigation Satellite System (CYGNSS) and soil moisture product prospects," in *SMAP CalVal Work.*, (Fairfax), 2018.

[18] M. M. Al-Khaldi, J. T. Johnson, A. J. O'Brien, A. Balenzano, and F. Mattia, "Time-series retrieval of soil moisture using CYGNSS," *IEEE Trans. Geosci. Remote Sens.*, vol. 57, pp. 4322–4331, July 2019.

[19] A. Camps, H. Park, M. Pablos, G. Foti, C. P. Gommenginger, P.-W. Liu, and J. Judge, "Sensitivity of GNSS-R spaceborne observations to soil moisture and vegetation," *IEEE J. Sel. Top. Appl. Earth Obs. Remote Sens.*, vol. 9, pp. 4730–4742, Oct. 2016.

[20] S. Jin and A. Komjathy, "GNSS reflectometry and remote sensing: New objectives and results," *Adv. Sp. Res.*, vol. 46, pp. 111–117, July 2010.

[21] C. Ruf, M. Unwin, J. Dickinson, R. Rose, D. Rose, M. Vincent, and A. Lyons, "CYGNSS: Enabling the future of hurricane prediction [Remote Sensing Satellites]," *IEEE Geosci. Remote Sens. Mag.*, vol. 1, pp. 52–67, June 2013.

[22] A. Komjathy, J. Maslanik, V. Zavorotny, P. Axelrad, and S. Katzberg, "Sea ice remote sensing using surface reflected GPS signals," in *IGARSS 2000. IEEE 2000 Int. Geosci. Remote Sens. Symp.*, vol. 7, pp. 2855–2857, IEEE, 2000.

[23] M. Rivas, J. Maslanik, and P. Axelrad, "Bistatic scattering of GPS signals off Arctic sea ice," *IEEE Trans. Geosci. Remote Sens.*, vol. 48, pp. 1548–1553, Mar. 2010.

[24] F. Fabra, E. Cardellach, A. Rius, S. Ribo, S. Oliveras, O. Nogues-Correig, M. Belmonte Rivas, M. Semmling, and S. D'Addio, "Phase altimetry with dual polarization GNSS-R over sea ice," *IEEE Trans. Geosci. Remote Sens.*, vol. 50, pp. 2112–2121, June 2012.

[25] L. Shao, "Method for detecting sea ice using gnss-r signals and its preliminary experimental result," *Remote Sens. Inform.*, no. 2, pp. 12–15, 2013.

[26] J. Strandberg, T. Hobiger, and R. Haas, "Coastal sea ice detection using ground-based GNSS-R," *IEEE Geosci. Remote Sens. Lett.*, vol. 14, pp. 1552–1556, Sept. 2017.

[27] H. Gao, D. Yang, B. Zhang, Q. Wang, and F. Wang, "Remote sensing of sea ice thickness with GNSS Reflected signal," *J. Electron. Inf. Technol.*, vol. 39, pp. 1096–1100, May 2017.

[28] W. Wei, G. Qiang, and B. Wu, "The simulation and analysis of the sea ice characteristics with reflect signal of navigation satellite," *Trans. Oceanol. Limnol.*, no. 4, pp. 1–7, 2015.

[29] M. Yang and Y. Cao, "The following experiment of sea ice obervation using gnss-r signals," *GNSS World of China*, vol. 39, no. 4, pp. 51–54, 2014.

[30] S. Gleason, M. Adjrad, and M. Unwin, "Sensing ocean, ice and land reflected signals from space: Results from the UK-DMC GPS Reflectometry experiment," in *Proc. 18th Int. Tech. Meet. Satell. Div. Inst. Navig. (ION GNSS 2005)*, (Long Beach, CA), pp. 1679–1685, Sept. 2005.

[31] Q. Yan and W. Huang, "Spaceborne GNSS-R sea ice detection using delay-Doppler maps: First results from the U.K. TechDemoSat-1 mission," *IEEE J. Sel. Top. Appl. Earth Obs. Remote Sens.*, vol. 9, pp. 4795–4801, Oct. 2016.

[32] Q. Yan and W. Huang, "Sea ice detection from GNSS-R delay-Doppler map," in *2016 17th Int. Symp. Antenna Technol. Appl. Electromagn.*, pp. 1–2, IEEE, July 2016.

[33] A. Alonso-Arroyo, V. U. Zavorotny, and A. Camps, "Sea ice detection using U.K. TDS-1 GNSS-R data," *IEEE Trans. Geosci. Remote Sens.*, vol. 55, pp. 4989–5001, Sept. 2017.

[34] G. Zhang, J. Guo, D. Yang, F. Wang, and H. Gao, "Sea ice edge detection using spaceborne GNSS-R signal," *Geomatics Inf. Sci. Wuhan Univ.*, vol. 44, no. 5, pp. 668–674, 2019.

[35] J. Cartwright, C. J. Banks, and M. Srokosz, "Sea Ice Detection Using GNSS-R Data From TechDemoSat-1," *J. Geophys. Res. Ocean.*, p. 2019JC015327, Aug. 2019.

[36] Y. Zhu, K. Yu, J. Zou, and J. Wickert, "Sea ice detection based on differential delay-Doppler maps from UK TechDemoSat-1," *Sensors*, vol. 17, p. 1614, July 2017.

[37] D. Schiavulli, F. Frappart, G. Ramillien, J. Darrozes, F. Nunziata, and M. Migliaccio, "Observing sea/ice transition using radar images generated from TechDemoSat-1 delay Doppler maps," *IEEE Geosci. Remote Sens. Lett.*, vol. 14, pp. 734–738, May 2017.

[38] Q. Yan and W. Huang, "Sea ice detection based on unambiguous retrieval of scattering coefficient from GNSS-R delay-Doppler maps," in *2018 Ocean. – MTS/IEEE Kobe Techno-Oceans*, pp. 1–5, IEEE, May 2018.

[39] Q. Yan, W. Huang, and C. Moloney, "Neural networks based sea ice detection and concentration retrieval from GNSS-R delay-Doppler maps," *IEEE J. Sel. Top. Appl. Earth Obs. Remote Sens.*, vol. 10, pp. 3789–3798, Aug. 2017.

[40] Q. Yan and W. Huang, "Sea ice sensing from GNSS-R data using convolutional neural networks," *IEEE Geosci. Remote Sens. Lett.*, vol. 15, pp. 1510–1514, Oct. 2018.

[41] Q. Yan and W. Huang, "Detecting sea ice from TechDemoSat-1 data using support vector machines with feature selection," *IEEE J. Sel. Top. Appl. Earth Obs. Remote Sens.*, vol. 12, pp. 1409–1416, May 2019.

[42] Q. Yan and W. Huang, "Sea ice concentration estimation from TechDemoSat-1 data using support vector regression," in *2019 IEEE Radar Conf.*, (Boston, Massachusetts USA), pp. 1–6, IEEE, 2019.

[43] N. Rodriguez-Alvarez, B. Holt, S. Jaruwatanadilok, E. Podest, and K. C. Cavanaugh, "An Arctic sea ice multi-step classification based on GNSS-R data from the TDS-1 mission," *Remote Sens. Environ.*, vol. 230, p. 111202, Sept. 2019.

[44] C. Hu, C. Benson, C. Rizos, and L. Qiao, "Single-pass sub-meter space-based GNSS-R ice altimetry: Results from TDS-1," *IEEE J. Sel. Top. Appl. Earth Obs. Remote Sens.*, vol. 10, pp. 3782–3788, Aug. 2017.

[45] W. Li, E. Cardellach, F. Fabra, A. Rius, S. Ribó, and M. Martín-Neira, "First spaceborne phase altimetry over sea ice using TechDemoSat-1 GNSS-R signals," *Geophys. Res. Lett.*, vol. 44, pp. 8369–8376, Aug. 2017.

[46] Q. Yan and W. Huang, "Sea ice thickness measurement using spaceborne GNSS-R: First results with TechDemoSat-1 data," *IEEE J. Sel. Top. Appl. Earth Obs. Remote Sens.*, pp. 1–1, 2020.

[47] M. R. Vant, R. O. Ramseier, and V. Makios, "The complex-dielectric constant of sea ice at frequencies in the range 0.1–40 GHz," *J. Appl. Phys.*, vol. 49, pp. 1264–1280, Mar. 1978.

[48] F. T. Ulaby, R. K. Moore, and A. K. Fung, *Microwave Remote Sensing: Active and Passive*, vol. 2. Addison-Wesley Pub. Co., Advanced Book Program/World Science Division, 1986.

[49] L. Kaleschke, N. Maaß, C. Haas, S. Hendricks, G. Heygster, and R. T. Tonboe, "A sea-ice thickness retrieval model for 1.4 GHz radiometry and application to airborne measurements over low salinity sea-ice," *Cryosph.*, vol. 4, pp. 583–592, Dec. 2010.

[50] P. Werbos, *Beyond regression: New tools for prediction and analysis in the behavioral sciences.* PhD thesis, Harvard University, 1974.

12 Global Soil Moisture Retrieval from Spaceborne GNSS-R Based on Machine Learning Method

Yan Jia and Zhiyu Xiao

12.1 BACKGROUND

Soil moisture (SM) plays a very important role in the climate system, which has a great influence on the atmospheric conditions, hydrological environment, and vegetation state of the earth [1–5], and plays an important role in various hydrological and geophysical changes worldwide. In many scientific fields, soil moisture content (SMC) has been regarded as an important environmental factor for land surface dynamic monitoring, energy regulation and water exchange between land and atmosphere, and other hydrological processes [3, 5]. SM is also a major determinant of surface soil permittivity, which affects the scattered signals from the surface.

Global Navigation Satellite System Reflection (GNSS-R) is a microwave remote sensing technology, which uses a variety of satellite constellation systems to receive GNSS signals reflected from the earth's surface based on bistatic geometry for remote sensing monitoring [3–6]. GNSS-R signals are usually L-band signals, which provide high spatial resolution and a long revisit time, showing great potential in remote sensing monitoring and application. In the past two decades, GNSS-R technology has been applied in various geoscience fields, and has attracted great attention in the extraction and measurement of geophysical parameters such as the ocean, ice, and land, and achieved good research results [3].

With the deepening of the research on GNSS-R soil moisture retrieval, the new constellation observation mission with long-time series observation data has become a new approach for GNSS-R SM retrieval. NASA launched the Cyclone GNSS (CyGNSS) satellite in December 2016 to observe tropical cyclones by estimating sea winds between 38°N and 38°S [2, 3–6]. CyGNSS contains eight microsatellites that can receive both direct signals from GPS satellites and reflected signals from the ground. The reflected signal is transmitted by a GPS satellite and then scattered forward along the earth's surface in a specular direction. The reflected signal contains information related to the characteristics of the scattered surface. The GNSS signal reflected from the scattered surface is used to determine the geophysical information of the reflected point surface by cross-correlating the reflected signal with the received GNSS direct signal or the copy of the signal. CyGNSS satellites are being used to complete important scientific research work such as altimetry, sea ice monitoring, biomass estimation, wetland classification, and SM estimation [3]. Through years of research accumulation, SM retrieval using CyGNSS data has become an important method for SM retrieval research.

Assuming that the time of vegetation and roughness changes is longer than the time of SM changes, Al-Khaldi et al. proposed an incoherent CyGNSS measurement method to retrieve soil samples of N time at a given location, and using the ratio of n-1, the RMSE of the retrieval results reached 0.038 cm³/cm³ [1]. There is a strong positive linear relationship between the change of CyGNSS reflectivity and the change of SMAP SM observation value. The sensitivity of CyGNSS

DOI: 10.1201/9781003363118-12

reflectivity to SM changes with space. Chew et al. converted the reflectivity of CyGNSS into the estimated value of SM. The average daily unbiased root mean square error (ubRMSE) between SM derived from CyGNSS and SMAP SM was 0.045 cm³/cm³ [2, 3]. Later, a data set was constructed to calibrate the retrieved values of SMAP SM using the reflectivity observation value of CyGNSS; the ubRMSE of the retrieval result was 0.049 cm³/cm³, and the correlation coefficient between them was 0.4 [3]. In addition, surface reflectivity can be predicted by the soil permittivity model, and there is a clear linear relationship between the soil permittivity and the Fresnel coefficient. Calabia et al. predicted the estimated value of SM based on the Fresnel coefficient measured by GNSS-R, and the predicted result RMSE is 0.05 cm³/cm³ [4]. Clarizia et al. retrieved the estimated daily SM value of the surface within a 36-km grid with an RMSE of 0.07 cm³/cm³ by combining the land signal reflectivity observation value of CyGNSS and the auxiliary information such as surface vegetation and roughness provided by SMAP [5]. Yan et al. used the three-layer air-vegetation-soil model, combined with the surface signal reflectivity provided by CyGNSS data and the vegetation opacity in SMAP data to retrieve SM by linear regression, and the RMSE was 0.07 cm³/cm³ [6]. In addition, Yang et al. proposed a physics-based algorithm coupling surface reflectivity and SMAP luminance temperature estimates to achieve accurate SM estimation, and the RMSE of the estimation results were 0.051 cm³/cm³ [7]. Kim et al. introduced the relative signal-to-noise ratio (rSNR) of the delay-Doppler diagram (DDM) provided by CyGNSS to improve the temporal resolution of SMAP SM and used rSNR to estimate SM. The correlation coefficient R of the estimated results was 0.77, indicating that the collaborative use of CyGNSS observations could improve the SM estimate of SMAP [8].

Traditional methods mainly estimate SM through the principle of linear regression, which believes that there is a linear relationship between the ground features of reflection points and SM [9]. However, there are many ground features that affect SM, and some of them have a nonlinear relationship with SM. For the combination of linear and nonlinear relations, supervised learning ML methods [10], such as a fully connected artificial neural network (ANN), support vector machine (SVM), random forest (RF), and many other ML methods can be used to complete nonlinear regression and SM prediction according to certain learning methods and strategies by taking ground features and other variables as input parameters.

Tang et al. used SVM to retrieve SM of the Wuhan Bao Association based on CyGNSS data and other auxiliary data, and evaluated the influence of DDM and the quality of different ground topography on the estimation results. The results showed that vegetation was an important factor affecting SM estimation [11]. Yang et al. used a back propagation artificial neural network (BP-ANN) to construct a model to retrieve the monthly SM estimate of the target location and evaluated the estimation performance of CyGNSS and British satellite TDS for SM. The results showed that the estimation results of the two were consistent with the actual measured SM and had a good correlation [12, 13]. Eroglu et al. used the observation data of the ISMN site combined with the auxiliary data of SMAP, DEM, and GPS signals to extract 8 ground features as input variables, and used the fully connected ANN algorithm to invert SM, and the correlation of the results was as high as 0.9238 [14], indicating the great potential of ML methods in SM estimation. In the subsequent work, Senyurk et al. extended the research to larger and more diverse data sets and used various ML methods such as ANN, SVM, and RF to conduct the research. From the experimental results, the RF algorithm was the most satisfactory for SM estimation, and the average ubRMSE reached 0.047 cm³/cm³. Other methods are greatly influenced by data quality [9]. Meanwhile, Senyurk et al. proposed an SM estimation model based on CyGNSS machine learning. RF algorithm was used to process CyGNSS data, auxiliary data, and SMAP radiometer brightness temperature data to achieve SM estimation at high time resolution. In the test area, the ubRMSE of SM estimation results reached 0.041 cm³/cm³ [10]. Lei et al., based on the reflectivity and other auxiliary data sets provided by CyGNSS data, combined with SMAP SM, used a machine learning method to retrieve the global daily SM at the spatial resolution of 9 km, and realized the global CyGNSS SM mapping at the spatial resolution of 9 km. The estimation result ubRMSE reaches 0.0543 cm^3/cm^3 [15]. The results show that it is feasible to retrieve SM by machine learning based on a classification regression tree. Yan et al. proposed the bagging random forest method to retrieve SM, taking the CyGNSS data product type, geographical location data, and related climate type information as input parameters, and the RMSE of the estimation result was 0.05 cm^3/cm^3 [16]. SM can be efficiently estimated in near-real time from CyGNSS data with different site locations and climate types. Jia et al. combined an ML algorithm and simple land type (LT) digitalization strategy to retrieve global SM, reduced the number of model input parameters by the extreme gradient boosting (XGBoost) method, and improved the accuracy of SM estimation. The average ubRMSE of the result was 0.041 cm^3/cm^3 [17]. In addition, the ML method combined with a pre-classification strategy was used to retrieve SM, and 10-fold cross-validation technology was used to compare the overall performance difference of SM estimation with/without pre-classification. The results show that different ML algorithms have significantly improved the accuracy of SM estimation retrieved under the pre-classification strategy [18]. It shows that the pre-classification strategy has a positive effect on SM estimation.

12.2   DATA PROCESS AND QUALITY CONTROL

12.2.1   CYGNSS DATA

The CyGNSS mission consists of eight microsatellites, each carrying a four-channel GNSS-R bistatic radar receiver to record reflected GPS signals from the surface. Although its constellation orbits primarily around the tropics and is limited in latitude to ±38°, it has acquired a large number of land observations that provide data support for SM estimation.

To retrieve SM on land, CyGNSS Level-1 (L1) version 2.1 product was used. The key observation data in the CyGNSS L1 data is the delay-Doppler map (DDM), which represents the received surface power of each observed specular reflection point over a range of signals' time delay and Doppler frequencies (bin-by-bin). By inverting the forward scattering model of CyGNSS in L1, DDM takes the non-surface correlation term into account to obtain the bistatic radar cross section (BRCS) area and the effective scattering area of the surface. The bin-by-bin BRCS is provided in the form of 11×17 DDM arrays in L1 data [10]. In addition, the geometric structure and instrumental variables include factors such as the signal incidence angle and the distance between the GPS transmitter and CyGNSS receiver to the specular point, providing detailed acquisition information for each specular reflection point.

Using observations provided by the L1 data, surface reflectivity can be estimated by several methods under coherent and incoherent assumptions. Assuming that the observed GNSS-R signal is mainly reflected coherently, the reflectivity $\Gamma_{RL}(\theta)$ is calculated by using the variable BRCS (σ_{RL}) and the distance term in CyGNSS L1. The calculation can be written as follows:

$$\Gamma_{RL}(\theta) = \left(\frac{4\pi}{\lambda}\right)^2 \frac{P_{RL}^{coh}(R_r + R_t)^2}{P_t G_t G_r} \#$$

(12.1)

where R_t and R_r are the distances from specular reflection points to the GNSS transmitter (*tx_to_sp_range* in L1 data) and GNSS-R receiver (*tx_to_rc_range* in L1 data), respectively. The peak DDM value of BRCS should be used together with the coherence assumption. In addition, the reflectivity delay waveform can be obtained according to the integration of BRCS in the Doppler domain, so as to calculate other CyGNSS observations, such as trailing edge slope (TES) and leading-edge slope (LES). Next, TES and LES are calculated from the reflectivity delay waveform values at delay bin M (peak delay bin) to m + 3 and m to m - 3, respectively. TES and LES are indicators related to coherent or incoherent scattering conditions and provide other supplementary information in addition to CyGNSS reflectivity [9].

12.2.2 SMAP Data

SMAP collects bipolar bright temperature observations from the L-band microwave radiometer and then converts them to SM estimates. The original resolution of SMAP is about 40 km, and the SM estimation results are published to a 36 km × 36 km grid [7–12]. When the proportion of surface water is high and vegetation is dense (VWC > 5 kg/m^2) and urban and mountainous areas, SM estimation results are greatly influenced by ground feature elements, so it is not reliable to compare SM estimation results observed by CyGNSS with SMAP SM in these areas [7, 9–18]. SMAP SM can also be resampled to a grid of 9 km × 9 km. The daily data provided by SMAP contains SM estimates, data quality markers, surface roughness coefficients, vegetation opacity, and other ancillary information that can be gridded on EASE-Grid. To facilitate further comparison and validation, CyGNSS data were also resampled at 36 km × 36 km grid reflection points on EASE-Grid based on the longitude and latitude of SMAP data and the CyGNSS observables at specular reflection points. Therefore, influenced by the resolution of the training data, the spatial resolution of the SM data product is 36 km × 36 km [9].

SMAP Enhanced Radiometer Level 3 SM data (global daily 9 km EASE-Grid SM, version 4) was used as a reference data set for comparison with SM estimated by CyGNSS. In addition, the 16-bit binary string composed of 1 and 0 in the data is called the SMAP Retrieval Quality Flag (RQF), which is an important quality control index in SMAP data [9]. The first position of the string is "recommended quality", which indicates whether the quality of the SM estimate is reliable. The first position of reliable data is 0. The extracted data is filtered according to the first position of the "recommended quality" string to ensure the quality of the data used in modeling calculation and reduce the impact of errors [9–11, 15, 17]. Although this process reduces the overall data volume, the 9 km high-resolution data set can still provide an effective data set for the learning model. Therefore, the data point with the first position of 1 in RQF is eliminated and not used in subsequent calculations. At the same time, to facilitate verification and comparison of estimation results, CyGNSS data were projected into EASE-Grid used for SMAP data, with a distance of 9 km × 9 km [9–12, 15–18]. Therefore, the spatial resolution estimated by SM is considered to be 9 km. In addition, for land areas within a global latitude range of ±45°, SMAP provides SM estimates at an average interval of 3 days, which is two days longer than CyGNSS.

12.2.3 Data Processing and Quality Control

12.2.3.1 Key Data Process Methods

In the process of SM estimation, the main data used include CyGNSS data and ground auxiliary data sets. CyGNSS data mainly include surface reflectivity, incident angle, leading edge slope (LES) and trailing edge slope (TES), etc. The reflectivity obtained after a series of corrections can represent surface information such as soil water content, vegetation roughness, and vegetation opacity of ground reflection points [9, 15]. The calculation of surface reflectivity is the top priority of SM estimation. The observed values provided in CyGNSS L1 data can be estimated by several methods with coherence or incoherence assumptions, or the surface reflectivity can be solved by a bistatic radar equation [9, 10, 14, 15]. The surface reflectivity signal delay of the GNSS-R signal is calculated according to the integration of BRCS in DDM of the Doppler delay map, and the index of coherent or incoherent scattering is derived, which provides supplementary information for SM estimation. Considering the influence of ground features, various time-varying or static ground parameters are also used in SM estimation. The spatial resolution of the CyGNSS observations is related to the nature of the scattering surface and, in the case of coherent scattering, it is determined by the first Fresnel region, a few kilometers. When the area is relatively flat and smooth ground, or there is no vegetation cover, it can be assumed that the reflection occurring is coherent [14]. CyGNSS data usually mesh into regular grid cells with fixed resolution under the assumption of coherence. Grid cells of about 4 km × 4 km centered on specular reflection points generate average topographic features [9, 10, 12–18].

The auxiliary data sets mainly supplement the ground information and also play an important role in the SM estimation process. The 2018 MODIS annual land cover type (MCD12Q1) product generated a 500-m resolution map of major land cover types [11, 12], which provides major land cover types within each grid cell and includes six classification schemes. The International Geosphere-Biosphere Program (IGBP) land cover scheme was selected for further analysis. IGBP contains 17 land cover categories, including water, forest, bush, grassland, arable land, wetland, artificial surface, permanent snow and ice, and bare land [17]. For each 3-km grid, its land type was identified as the land cover type with the largest area. Soil grids are used to represent soil textures, which represent water retention and hydraulic properties such as capillary interactions within the soil profile. In soil grids, the soil profile is vertically discretized into 7 layers with a maximum depth of 2 m. For each layer, the soil is divided into 12 standard soil texture grades based on the ratio of sand, clay, and silt [14, 17, 18]. To be consistent with the L-band signal, the signal penetration depth using the data is 5 cm. The product is available at 250 m, and each specular reflection point is located in a spatial grid size of 3 km × 3 km. The proportions of sand, clay, and silt were spatially averaged, and the major soil texture classes were determined by the percentage of the 12 soil texture classes.

12.2.3.2 Data Quality Control

In order to improve the reliability of SM estimation results, when using CyGNSS data, data should be filtered to achieve data quality control. For stations with an altitude higher than 2,000 m, the data credibility of specular reflection points within the range is low, and the CyGNSS data provided will not be used. At the same time, due to the limitation of the observation height of the L1 band of CyGNSS, the CyGNSS observation results with an altitude higher than 600 m before 2018 have been screened. In order to reduce the influence of noise, data with CyGNSS reflectivity between -5 dB and -30 dB and incident angle of the specular reflection point less than 65° are not used [1–6, 9–18]. The data with elevation angles less than 30° are excluded, which can effectively remove the very weak signal interference from the circularly polarized sidelobe. Because of the very strong coherence of the water surface, the power of the forward scattered signal from the water surface is typically several orders of magnitude higher than the signal scattered from the soil. If the surface water area within the CyGNSS grid is large enough, experimental SM estimation near the water body is not feasible. Therefore, if more than 2% of the 3-km grid centered on the specular reflection point is covered by permanent or seasonal water bodies, the CyGNSS observation data there will not be used [9–18]. At the same time, there is a large amount of artificial ground in urban areas, and the CyGNSS reflected signals in these areas carry little ground feature information, and there is interference in the estimation of SM, so the data in urban areas are not used [9, 10, 16]. When the ground is heavily covered with vegetation water content (VWC>5 kg/m²), the CyGNSS data are too much affected by vegetation, so the data of these stations are not used [9–11, 15, 16].

12.3 SOIL MOISTURE RETRIEVAL APPROACH

12.3.1 Bistatic Radar Equation

An ideal GNSS-R-based SM retrieval method would rely on the calculation of the bistatic radar equation to obtain surface reflectivity. Fresnel reflection coefficients were obtained by correcting surface reflectivity for vegetation cover and surface roughness effects. Then the Fresnel reflection coefficient can be related to SM by means of the Fresnel reflection equation.

The calculation of the coherence component of the received signal of bistatic radar can be written as follows:

$$P_{RL}^{coh} = \left(\frac{\lambda}{4\pi}\right)^2 \frac{P_t G_t G_r}{(R_r + R_t)^2} \Gamma_{RL}(\theta) \#$$

(12.2)

λ is the wavelength, P_t is the peak power of the transmitted GNSS signal, G_t is the gain of the transmitting antenna, and G_r is the gain of the receiving antenna. R_r is the distance between the specular reflection point and the GNSS-R receiver, R_t is the distance between the specular reflection point and the GNSS transmitter, and $\Gamma_{RL}(\theta)$ is the specular reflectivity of SP.

The calculation of incoherent components can be written as follows:

$$P_{RL}^{inc} = \frac{\lambda^2 P_t G_t G_r}{(4\pi)^3 (R_r R_t)^2} \sigma_{RL} \tag{12.3}$$

σ_{RL} is the bistatic radar cross section in m^2 and R_{PL} is the Fresnel coefficient.

The local surface is relatively flat and smooth, and it is considered that the signal is mainly a coherent component, namely $P_{RL}^{coh} = P_{RL}^{inc}$:

$$\Gamma_{RL}(\theta) = \frac{\sigma_{RL}(R_r + R_t)^2}{4\pi R_t^2 R_r^2} \# \tag{12.4}$$

The Fresnel coefficient $R_{RL}(\theta)$ can be derived from the specular reflectivity $\Gamma_{RL}(\theta)$ of the reflection point, and can be written as follows:

$$\Gamma_{RL}(\theta) = R_{RL}(\theta)^2 \gamma^2 \exp{-hcos^2(\theta)} \# \tag{12.5}$$

Parameter h can be directly obtained from SMAP data. Γ is the opacity coefficient of vegetation, $\gamma = exp(-\tau sec(\theta))$, and the optical thickness of vegetation τ is calculated by the ratio of vegetation water content (VWC) and land cover factor [14]. VWC is obtained from the experience of normalized vegetation index (NDVI) by SMAP task, which can be directly obtained from SMAP data.

After obtaining the Fresnel coefficient $R_{RL}(\theta)$, the relationship between reflectivity and the Fresnel coefficient can be written as follows:

$$R_{RL}(\theta) = \frac{1}{2}(R_{VV}(\theta) - R_{HH}(\theta)) \# \tag{12.6}$$

where $R_{VV}(\theta)$ is the vertically polarized component and $R_{HH}(\theta)$ is the horizontally polarized component, both of which are functions related to incident angle θ and soil dielectric constant [4].

The method of bistatic radar equation retrieving SM mainly uses the model method of reflectivity, dielectric constant, and soil moisture, which is a typical method of retrieving SM from linear regression.

12.3.2 MACHINE LEARNING METHOD

12.3.2.1 Artificial Neural Network Model

Fully connected ANNs are also known as multi-layer perceptron (MLP), and the constructed network contains two hidden layers, which can be used to learn complex nonlinear relationships. By running a predetermined number of iterations, the ANN generates a loss function on the training set as the squared error of the model. The ANN parameters are learned by a stochastic gradient descent solver algorithm, whereby the ANN updates the model parameters at each iteration by calculating the partial derivative of the loss function with respect to the ANN parameters (back-propagation). The specific structure of ANN is shown in Figure 12.1.

In a fully connected ANN, neurons in each layer are completely interconnected with other neurons in neighboring layers. Each layer has an array of weights that can be trained by forward and

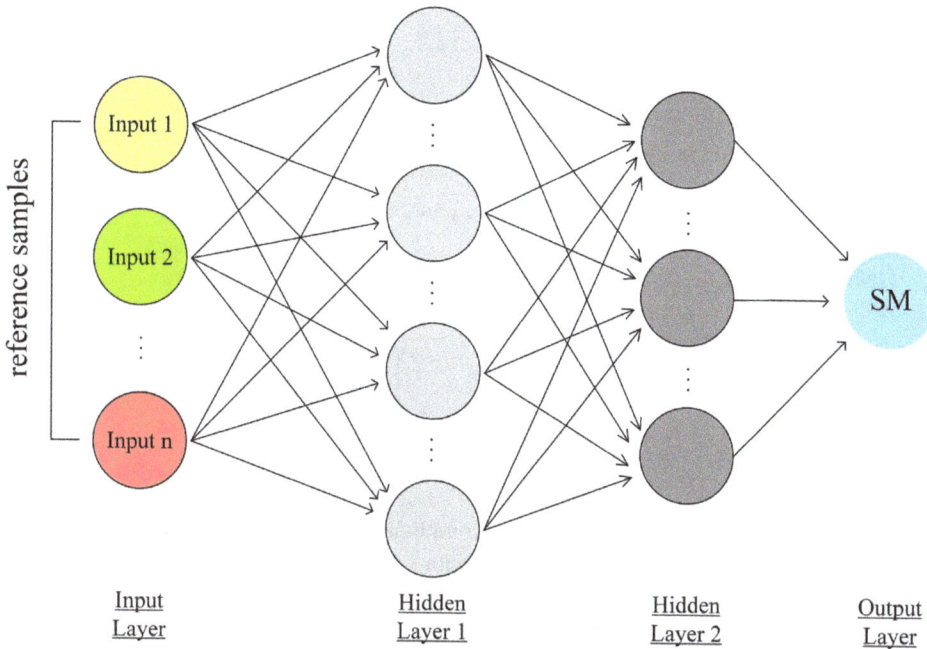

FIGURE 12.1 ANN structure [14].

backward propagation mechanisms. This array controls the linear strength of the connections to the next layer. Assuming that the number of neurons in layer I is N_i, then the weight array size in layer I is $(N_i+1) \times N_i$. The result of matrix multiplication between the weight array and the input array for a particular layer is given as the input for the next layer. To account for bias in linear relationships, trainable bias values are added to the sum of each neuron.

To make the model powerful for solving nonlinear regression problems, ANN has an activation function in each neuron, such as a linear rectification function, logic function, or hyperbolic tangent activation function. These functions are responsible for taking the corresponding offset sum as input and converting it to the new value with the help of the corresponding nonlinearity. This process is repeated on each neuron until the output layer completes the iteration and outputs an estimate of SM. The process of computing from input to output is called forward propagation. The network uses the training data and backpropagates the error information by updating the weights and biases in each layer to minimize the defined loss function with the help of stochastic gradient descent algorithms. The process of the ANN model to predict SM includes one iteration of forward and back propagation until the loss function reaches the minimum threshold or the maximum number of iterations.

In the process of using ANN to retrieve SM, reflectivity, incident angle, and trailing edge slope from CyGNSS data, as well as NDVI, VWC, elevation, slope, and surface roughness from MODIS are input variables for training SM prediction model. SMAP SM is used as the output and the referenced SM data to evaluate the retrieve accuracy [14]. The 10-fold cross-validation method was used to evaluate the accuracy of the SM estimate of the ANN model. Since the quality of the raw data provided by the observation station is not guaranteed, it is necessary to filter the input raw data to achieve quality control so as to improve the quality of SM estimation results [14].

12.3.2.2 Random Forest Model

Random forest is a widely used machine learning model. The prediction result of an artificial neural network is accurate, but it needs a lot of computation in model training and result prediction, which consumes a lot of time. The classification tree algorithm achieves classification or regression by

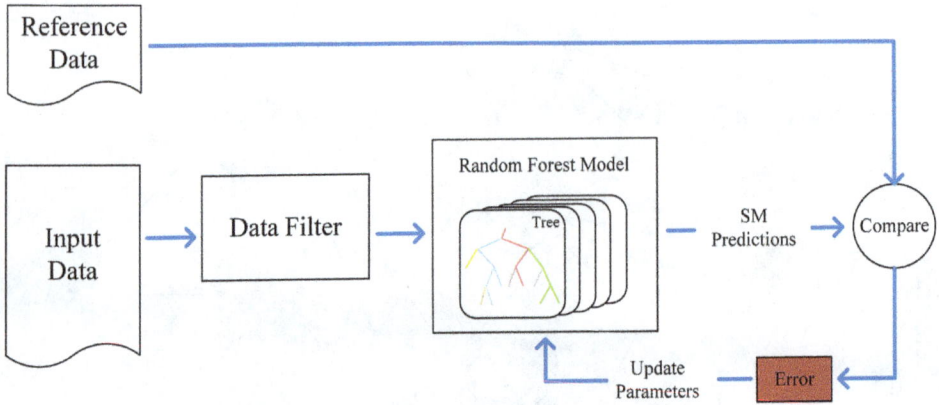

FIGURE 12.2 Random forest model [9].

iteratively dichotomizing the data, making the computational cost of the model greatly reduced. The random forest algorithm generates different data sets by repeatedly resampling the data sets, and trains a classification tree on each data set. The prediction results of all classification trees are integrated through certain combination strategies, which are used as the prediction results of the random forest algorithm. Random forest improves prediction accuracy without a significant increase in computation, and the results are more reliable for missing and unbalanced data because it is insensitive to multicollinearity. The process of RF predicting SM is exhibited in Figure 12.2.

In the process of retrieving SM through RF, the following should be taken as parameters of the RF model: ground reflectivity, incident angle, and slope of trailing edge provided by CyGNSS; NDVI and VWC provided by MODIS; soil clay ratio and sand ratio in soil grids data; elevation and slope in DEM data [9]. The data with poor quality were eliminated by data filtering, and the remaining data were used to construct the training data set for SM retrieval. SM observations at the site were used as reference data to evaluate the estimation results. At the same time, the SM estimates and the SM observations from the site should be averaged to the same spatial resolution for comparison [9].

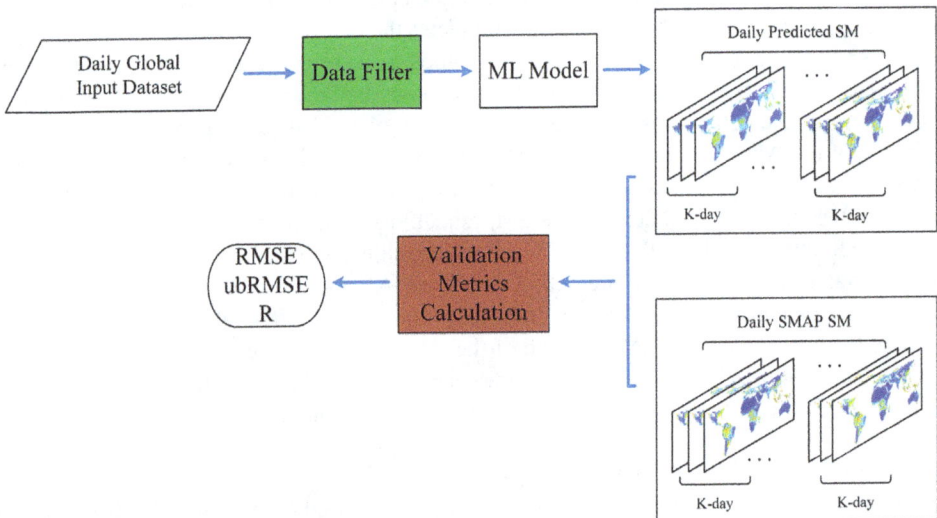

FIGURE 12.3 SM retrieval of K-day average [9].

Senyurek et al. take four types of data from 170 sites as input features, implement data quality control after some data filtering and masking, conduct RF model training, and compare the SM predicted value obtained with the SM observed at sites to obtain errors. The constructed random forest contains 60 classification trees, each of which has no more than 100 nodes. Through the integration strategy of least squares promotion, the learning rate is 0.1 [9].

The SM outputs of the SMAP and random forest algorithms were time-averaged over K consecutive days (i.e., K = 1, 3, 7, and 30 days), and the SM product of the K-day averages were created for both, as shown in Figure 12.3. The daily CyGNSS data set refers to the average data for the same calendar day concurrent with SMAP. In addition, to achieve the same spatial resolution, the 3-km resolution of SM estimation results from the random forest was spatially averaged to the level of SMAP spatial resolution [15].

12.3.2.3 XGBoost Integrated Learning

Extreme gradient boosting (XGBoost) is an optimized distributed gradient boosting library, which is an improvement of the gradient boosting algorithm. It implements a machine learning algorithm in the framework of gradient boosting. XGBoost provides serial tree boosting (also known as GBDT, GBM) that solves many data science problems quickly and accurately. The principle of gradient boosting decision tree (GBDT) is shown in Figure 12.4. XGBoost uses Newton's method to solve the extreme value of the loss function, which expands the loss function Taylor to the second order. At the same time, the loss function also adds a regularization term. Because the error function will make the model fit the data as much as possible to obtain the minimum residual, the phenomenon of overfitting will occur when the error function is directly used to monitor the learning of the model. Regularization will limit the model from becoming too complex, and the result of simple model fitting with limited data will be less random, so as to limit the phenomenon of overfitting.

The essence of XGBoost is an ensemble learning method of classification regression trees, and there are dependencies among classification trees, which is a reverse of boosting. In the learning process of the model, the residuals of the previous classification regression tree will be transferred to the training of the later tree to improve the prediction results of the later tree.

When using XGBoost, you need to configure some hyperparameters to moderate and limit the complexity of model learning, keeping the amount of computation within limits. The depth of the tree and the number of leaf nodes should be set to avoid overfitting caused by excessive learning of local samples. At the same time, setting the minimum sample weight *min_child_weight* when the tree leaves too few nodes will also lead to an underfitting phenomenon [17, 18]. The number of trees

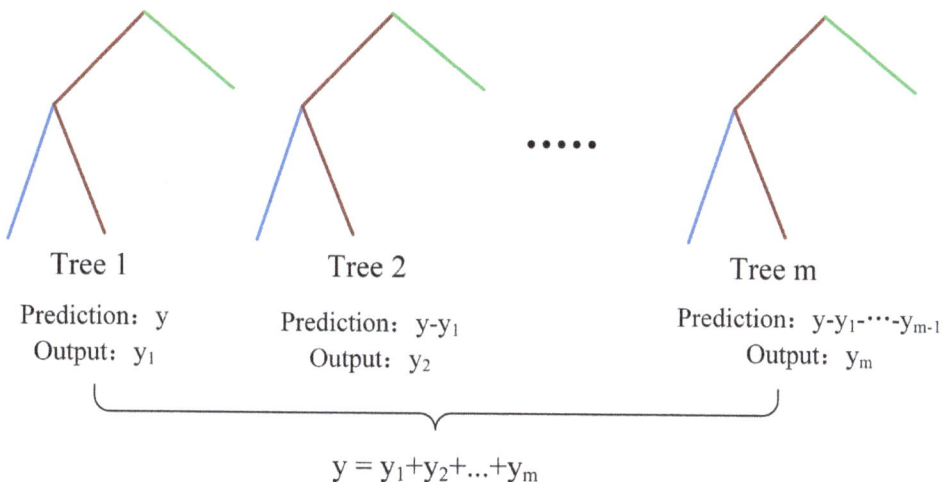

Tree 1   Tree 2   Tree m

Prediction: y Prediction: $y-y_1$ Prediction: $y-y_1-\cdots-y_{m-1}$
Output: y_1 Output: y_2 Output: y_m

$$y = y_1+y_2+\ldots+y_m$$

FIGURE 12.4 GBDT method.

to be built should also be limited to between 200 and 500 to ensure that the model will not be too large and affect the operation speed while meeting the requirements. In addition, the learning rate of the model is set to 0.1 as in most ML methods.

12.4 GLOBAL SOIL MOISTURE RETRIEVAL

In their study, Jia et al. used CyGNSS data and a global land cover map combined with SMAP data to digitize the soil type LT and divide it into 15 types. When the data belonged to one type, the change of SM was always consistent [17]. By inputting digital LT and CyGNSS data into the XGBoost model for learning, combined with the idea of classification, the model makes it easier to identify rules from data and is expected to show better estimation accuracy [17]. When XGBoost was used to predict SM estimation, the trailing edge slope TES, reflectivity, soil permittivity, incidence angle, and land type label were taken as input variables and a 10-fold cross-test was used to verify the estimation accuracy [18]. Too many input variables tend to weaken the essence and mechanism of SM estimation, so the input variables are set to different combinations. The accuracy of SM estimation among combinations was compared, and the variable with the best performance was used as the optimal variable for CyGNSS SM estimation.

The performance of the proposed method is compared with that of other ML methods. The input variables to the model are various combinations of reflectivity (B), trailing edge slope (T), land types (L), incident angle (A), and dielectric constant (D). Depending on the performance of different combinations of variables in the model, the combination of (B+T+L+A) is considered the default combination of input variables. The performance comparison results of different ML methods are shown in Table 12.1.

It is worth noting that the traditional ML methods (XGBoost and RF) give better results than the DL method (ANN). XGBoost performs better than RF in most situation, while the RF method performs comparably to or even slightly better than XGBoost on some specific land categories, such as evergreen needleleaf forests and evergreen broadleaf forests. The small sample size of these two types of LTs indicates that the RF model performs better than the XGBoost algorithm when dealing with small amounts of product data. The expected accuracy of the ANN reached 0.040 cm³/cm³ in LT areas with sparse or barren vegetation and the sample size is more than 9 million. This shows that the ANN method is good at handling large data, while the proposed XGBoost method takes the best accuracy of 0.037 cm³/cm³.

As a whole, both XGBoost and RF performed very well. For the whole data set, the emerging deep learning networks did not guarantee a decisive advantage over the traditional ML methods. This situation may be due to the small number of samples used in the training of the deep neural networks leading to a serious overfitting phenomenon. In general, the LT labeling strategy based on

---

TABLE 12.1

Performance Comparison of Different Methods with the LT-Digitization Strategy for SM Estimation Using 10-Fold

LT	Methods (B+T+L)	RMSE (cm³/cm³)	ubRMSE (cm³/cm³)	Bias (cm³/cm³)
All types	XGBoost - LT	0.0630	0.0630	0.0001
	RF - LT	0.0651	0.0651	0.0004
	ANN - LT	0.0690	0.0690	0.0008
Evergreen needleleaf forest	XGBoost - LT	0.0472	0.0431	0.0192
	RF - LT	0.0447	0.0447	0.0015
	ANN - LT	0.0469	0.0469	0.0006

(Continued)

LT	Methods	RMSE	ubRMSE	Bias
	(B+T+L)	(cm³/cm³)	(cm³/cm³)	(cm³/cm³)
Evergreen broadleaf forest	XGBoost - LT	0.0865	0.0788	0.0358
	RF - LT	0.0772	0.0772	0.0002
	ANN - LT	0.0857	0.0857	0.0007
Deciduous broadleaf forest	XGBoost - LT	0.0699	0.0661	0.0225
	RF - LT	0.0642	0.0641	0.0029
	ANN - LT	0.0721	0.0721	0.0026
Mixed forest	XGBoost - LT	0.0688	0.066	0.0192
	RF - LT	0.0637	0.0637	0.0007
	ANN - LT	0.0700	0.0700	0.0004
Closed shrublands	XGBoost - LT	0.0578	0.0514	0.0265
	RF - LT	0.0470	0.0470	0.0008
	ANN - LT	0.0507	0.0507	0.0001
Open shrublands	XGBoost - LT	0.0452	0.0452	0.0001
	RF - LT	0.0452	0.0452	0.0004
	ANN - LT	0.0495	0.0495	0.0009
Woody savannas	XGBoost - LT	0.0872	0.0872	0.0001
	RF - LT	0.0873	0.0873	0.0001
	ANN - LT	0.0950	0.0950	0.0005
Savannas	XGBoost - LT	0.0753	0.0753	0.0001
	RF - LT	0.0754	0.0754	0.0010
	ANN - LT	0.0824	0.0824	0.0011
Grasslands	XGBoost - LT	0.0855	0.0855	0.0003
	RF - LT	0.0854	0.0854	0.0007
	ANN - LT	0.0946	0.0946	0.0006
Permanent wetlands	XGBoost - LT	0.0906	0.0906	0.0001
	RF - LT	0.0932	0.0930	0.0058
	ANN - LT	0.1005	0.1004	0.0034
Croplands	XGBoost - LT	0.0797	0.0797	0.0001
	RF - LT	0.0798	0.0798	0.0021
	ANN - LT	0.0873	0.0873	0.0010
Urban and built-up	XGBoost - LT	0.0588	0.0587	0.0033
	RF - LT	0.0639	0.0632	0.0093
	ANN - LT	0.0768	0.0744	0.0193
Cropland/Natural vegetation mosaic	XGBoost - LT	0.0971	0.0971	0.0003
	RF - LT	0.097	0.097	0.0005
	ANN - LT	0.1057	0.1057	0.0013
Barren or sparsely vegetated	XGBoost - LT	0.0373	0.0373	0.0001
	RF - LT	0.0373	0.0373	0.0002
	ANN - LT	0.0405	0.0405	0.0007
Water bodies	XGBoost - LT	0.0954	0.0944	0.0139
	RF - LT	0.0941	0.0941	0.0012
	ANN - LT	0.1020	0.1020	0.0010

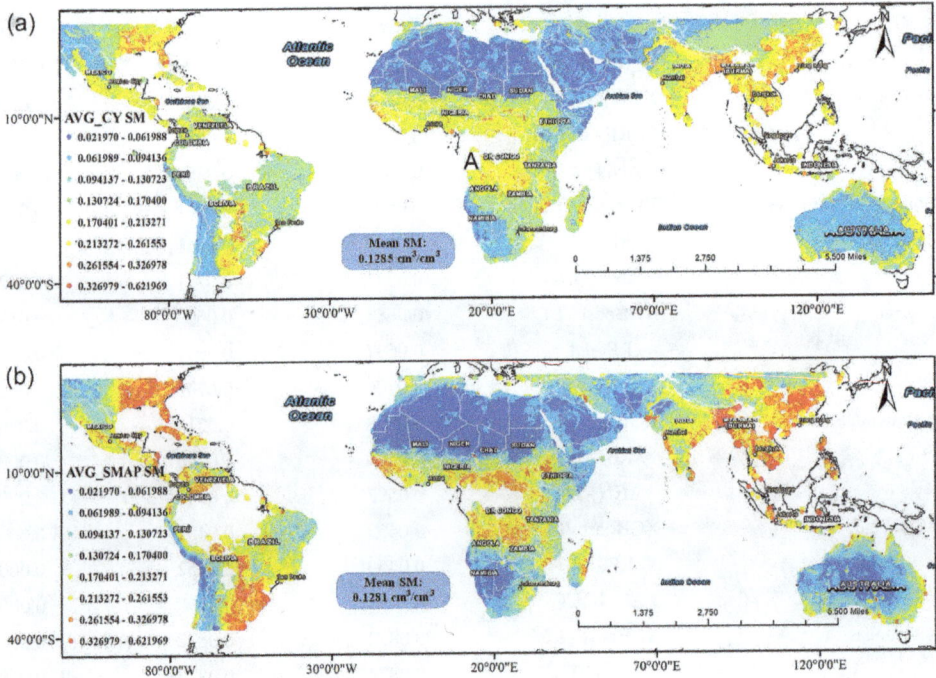

FIGURE 12.5 The predicted daily (a) average SM using XGBoost with the LT strategy, and (b) average SMAP SM for SM estimation using XGBoost with the LT strategy.

XGBoost has a clear advantage over other methods in all statistical metrics. The accuracy of RMSE and ubRMSE of XGBoost over the entire data set improved by 0.002 cm³/cm³ over the RF model. In addition, ANN and RF outperformed some LT types for other LT categories, which means that different ML methods are suitable for different data sets. This finding has been noted previously [10]. Therefore, different ML models are built for training in different LT regions to obtain specialized SM prediction values.

Figures 12.5(a) and (b) show the daily average SM distribution of CyGNSS and SMAP units, respectively. The mean value of SM based on CyGNSS retrieve is 0.1285 cm³/cm³, which is consistent with the reference value of SMAP SM of 0.1281 cm³/cm³. This indicates that the SM retrieve results obtained by CyGNSS using XGBoost LT strategy are consistent with the SMAP reference data. In addition, the legends are classified using a natural break model, in which the classification categories are based on natural groupings inherent in the data. Select the class that best groups similar values and maximizes the difference between classes to identify the breakpoint. These features are divided into classes with boundary sets in which jumps in values are relatively larger. Therefore, Figures 12.5(a) and (b) indicate that CyGNSS tends to have lower estimations when the SM levels are higher [5].

Figure 12.6 shows the density plots of the XGBoost, RF and ANN models, comparing the CyGNSS-based SM estimates with the SMAP SM reference on a logarithmic scale. The figure shows the performance of different ML models for the SM test data set (3 million samples) under the LT digitization strategy. It can be seen from the density plots that the CyGNSS-based SM estimates and the SMAP SM reference agrees quite well in general, especially in the sample-rich case.

It is worth noting that the XGBoost model (Figure 12.6(a)) has the highest R of 0.71. In addition, the data cloud in the density plots looks stacked up by many 1:1 bold horizontal lines. This is caused by using the digitized LTs/labeling strategy. The classification effect of this strategy is remarkable, because it reassigns the samples into groups, which are represented as lines. The

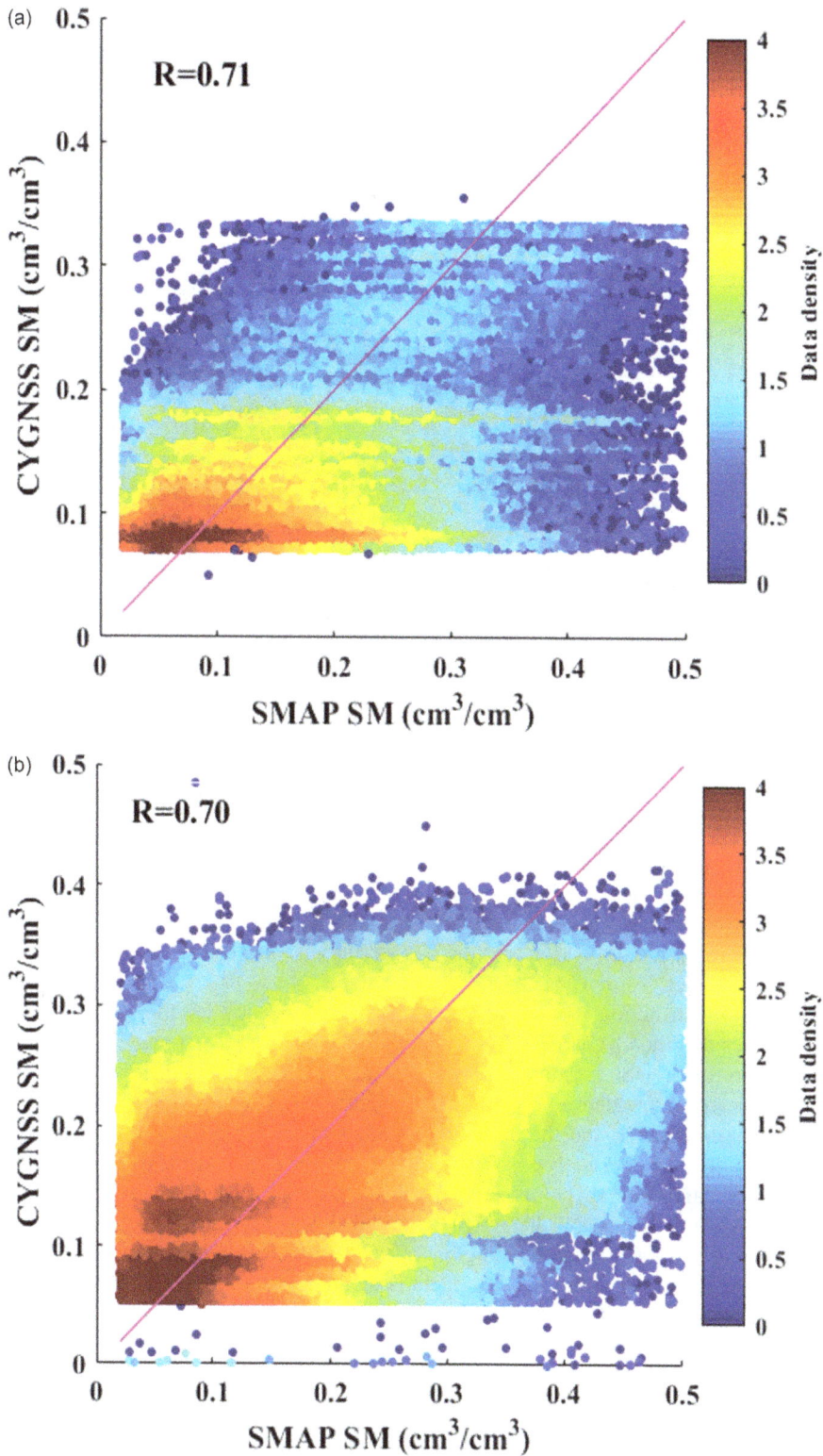

FIGURE 12.6 Density plots of 10-fold CV using LT strategy: (a) SM estimation for XGBoost, (b) SM estimation for RF, and (c) SM estimation for ANN. (*Continued*)

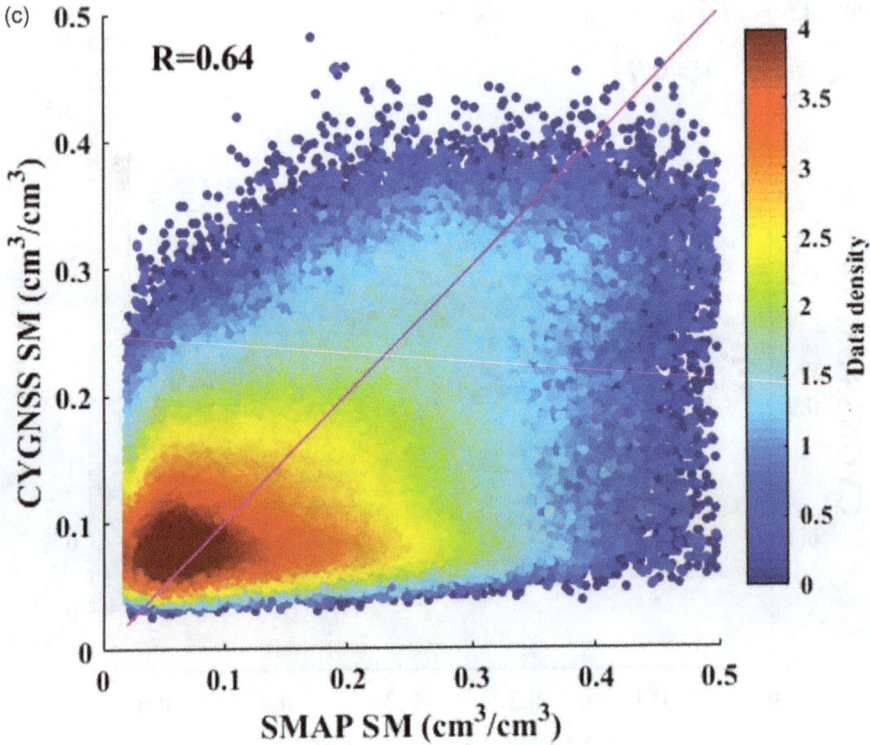

FIGURE 12.6 (*Continued*)

learning model in this way can better establish the relationship between the input and output variables, providing more reliable prediction results. A similar phenomenon is seen in Figure 12.6(b). This phenomenon in Figure 12.6(c) is more ambiguous. The R in Figure 12.6(c) is also the lowest among (a), (b), and (c). It can be seen that the LTs strategy is very effective in improving the correlation of the predicted results.

When the data distribution is more discrete, there is a tendency for the straight line to deviate. This indicates that the CyGNSS SM estimation products are underestimated to some extent, as already shown in the previous results (Figure 12.5) and previously mentioned in [5]. The surfaces with high SM usually grow dense vegetation with high water content, which leads to the increase in the incoherent component and the drops in the coherent component. Under such situation, the change in reflectivity is not perfectly consistent with the change in SM, while a positive correlation is shown between SM and coherent components. Therefore, the SM predicted by the model tends to be underestimated because the ground characteristics of the high SM surface are not extracted correctly.

A new pre-classification strategy to aid global CyGNSS SM estimation was proposed [18]. The performance of different ML regression models is also compared in terms of whether to adopt this strategy or not. The performance of the proposed ML regression model based on the pre-classification strategy is evaluated for all different LTs using as few auxiliary data as possible, showing the simplicity and effectiveness of the model.

Table 12.2 shows the SM estimates retrieved by traditional ML methods (RF, SVM, XGBoost) and emerging DL algorithms (ANN) with/without pre-classification strategies. The most appropriate combination of input variables ($\Gamma_{brcs} + \delta + \tau$) was used to derive the best modeling approach for ML. A detailed demonstration of the proposed ML regression method using pre-classification strategy aided by constructing samples using the annual data to build the model is presented. The results

TABLE 12.2

The Evaluation Performance (RMSE) for SM Estimation Using 10-Fold CV and Different ML with/without Pre-Classification Strategy

Land Type	RMSE (cm³/cm³)							
	XGBoost		RF		ANN		SVM	
	No Pre-cla.	With Pre-cla.	No Pre-cla.	With Pre-cla.	No Pre-cla.	With Pre-cla.	No Pre-cla.	With Pre-cla.
Evergreen Broadleaf Forest	0.0884	0.0822	0.0918	0.0869	0.0901	0.0889	0.0925	0.0903
Deciduous Broadleaf Forest	0.0868	0.0718	0.0881	0.0759	0.0922	0.0798	0.0939	0.0879
Mixed Forest	0.0730	0.0613	0.0662	0.0653	0.0748	0.0723	0.0788	0.0782
Open Shrublands	0.0468	0.0423	0.0454	0.0441	0.0493	0.0488	0.0694	0.065
Woody Savannas	0.0725	0.0657	0.0721	0.0657	0.0745	0.0725	0.0776	0.0757
Savannas	0.0706	0.0612	0.0712	0.0686	0.0723	0.0673	0.0768	0.0728
Grasslands	0.0720	0.0603	0.0668	0.0651	0.0778	0.0682	0.0805	0.0746
Croplands	0.0716	0.0635	0.0721	0.0689	0.0729	0.0717	0.0734	0.0707
Cropland/Natural Vegetation Mosaic	0.0761	0.0641	0.0731	0.0669	0.0779	0.0742	0.0835	0.0768
Barren or Sparsely Vegetated	0.0435	0.0370	0.0425	0.0415	0.0460	0.0455	0.0660	0.0700
Final results	0.0640	0.0520	0.0630	0.0600	0.0700	0.0630	0.0790	0.0680

of each sub-model are weighted averaged to obtain the final results of the strategy, and the performance of the global SM estimation performs well overall. In addition, there is a significant decrease in the RMSE of the prediction results when using this strategy.

According to the IGBP land classification data included in the SMAP release, there are 17 different LTs in the CyGNSS data sample. After processing the CyGNSS and SMAP data according to quality control requirements, for some land categories, the number of data samples included was not sufficient for ML model learning and model building. Therefore, the seven land categories with less than 20,000 annual data were not modeled. The land categories that were not modeled were water bodies, permanent wetlands, snow, ice, and areas with very dense vegetation. It is difficult to construct models to invert the soil moisture in these areas for the heavy influence of vegetation and water bodies.

Among all the traditional ML and DL algorithms, the proposed pre-classification strategy approach yielded good results with small RMSEs in all LTs. For the different ML algorithms, RF performs better than ANN and SVM, which is consistent with the conclusion reached by [10]. In addition, XGBoost has the smallest RMSE of 0.052 cm³/cm³ predicted and the best regression performance, which has not been mentioned in SM estimation on a global scale.

12.5 SUMMARY

The spaceborne CyGNSS provides a very favorable condition for long-time series monitoring of soil moisture on a large scale. It has many advantages, such as real time, wide coverage, low cost, and so on. At the same time, the retrieval of soil moisture is a non-linear complex interaction process. The subtle change of soil moisture could be affected by many factors, so it is difficult to estimate soil moisture under different environments truly and effectively. Artificial intelligence (AI) methods can better mine, express, and use the implicit rules between data, which is a new way to obtain accurate and effective soil moisture values.

This chapter introduces the data set used by the spaceborne CyGNSS soil moisture, several machine learning (ML) methods used in the estimation model, neural network algorithms, and data feature extraction processes. The basic parameters of the design and operation of the CyGNSS satellite system and SMAP satellite as the reference value of soil moisture, as well as the time resolution and spatial resolution, are detailed. Data acquisition, processing, and quality control of CyGNSS data products are reported. The principles, advantages, and disadvantages of artificial intelligence methods such as XGBoost, ANN, and RF are introduced respectively. Finally, the significance of characteristic parameters and the framework of AI algorithm are depicted, and the performance of soil moisture estimation method based on machine learning is evaluated and demonstrated with examples. Soil moisture retrieval based on a machine learning algorithm has achieved many satisfactory results, and further research can be carried out in terms of precision improvement and multiple auxiliary data fusion.

REFERENCES

1. M. M. Al-Khaldi, J. T. Johnson, et al. "Time-series retrieval of soil moisture using CYGNSS." *IEEE Transactions on Geoscience and Remote Sensing*, vol.57, no.7, pp.4322–4331, 2019.
2. C. C. Chew, E. E. Small, "Soil moisture sensing using spaceborne GNSS reflections: Comparison of CYGNSS reflectivity to SMAP soil moisture." *Geophysical Research Letters*, vol.45, no.9, pp.4049–4057, 2018.
3. C. C. Chew, E. E. Small, "Description of the UCAR/CU soil moisture product." *Remote Sensing*, vol.12, pp.1558, 2020.
4. A. Calabia, I. Molina, and S. Jin, "Soil moisture content from gnss reflectometry using dielectric permittivity from fresnel reflection coefficients." *Remote Sensing*, vol.12, no.1, pp.122, 2020.
5. M. P. Clarizia, N. Pierdicca, F. Costantini, and N. Floury, "Analysis of CYGNSS data for soil moisture retrieval." *IEEE Journal of Selected Topics in Applied Earth Observations and Remote Sensing*, vol.12, no.7, pp.2227–2235, 2019.
6. Q. Yan, W. Huang, S. Jin, et al., "Pan-tropical soil moisture mapping based on a three-layer model from CYGNSS GNSS-R data." *Remote Sensing of Environment*, vol.247, pp.111944, 2020.
7. T. Yang, W. Wan, J. Wang, et al., "A Physics-Based Algorithm to Couple CYGNSS Surface Reflectivity and SMAP Brightness Temperature Estimates for Accurate Soil Moisture Retrieval." *IEEE Tranctions on Geoscience and Remote Sensing,* vol.60, pp.409715, 2022.
8. H. Kim, and L. Venkat, "Use of Cyclone Global Navigation Satellite System (CyGNSS) observations for estimation of soil moisture." *Geophysical Research Letters,* vol.45, no.16, pp.8272–8282, 2018.
9. V. Senyurek, F. Lei, D. Boyd, A. C. Gurbuz, R. Moorhead, "Evaluations of a machine learning-based CYGNSS soil moisture estimates against SMAP observations." *Remote Sensing*, vol.12, no.21, pp.3503, 2020.
10. V. Senyurek, F. Lei, D. Boyd, M. Kurum, A. C. Gurbuz, R. Moorhead, "Machine learning-based CYGNSS soil moisture estimates over ISMN sites in CONUS." *Remote Sensing.*, vol.12, pp.1168, 2020.
11. F. Tang, S. Yan. "CYGNSS Soil Moisture Estimations Based on Quality Control." *IEEE Geoscience and Remote Sening Letters,* vol. 19, pp. 8022105, 2022.
12. T. Yang, W. Wan, Z. Sun, et al., "Comprehensive Evaluation of Using TechDemoSat-1 and CYGNSS Data to Estimate Soil Moisture over Mainland China." *Remote Sensing,* vol.12, pp.1699, 2020.
13. M. M. Nabi, V. Senyurek, A. C. Gurbuz and M. Kurum, "Deep learning-based soil moisture retrieval in CONUS using CYGNSS delay–doppler maps," *IEEE Journal of Selected Topics in Applied Earth Observations and Remote Sensing*, vol.15, pp.6867–6881, 2022, doi: 10.1109/JSTARS.2022.3196658.
14. O. Eroglu, M. Kurum, et al., "High spatio-temporal resolution CYGNSS soil moisture estimates using artificial neural networks." *Remote Sensing,* vol.11, no.19, pp.2272, 2019.
15. F. Lei, V. Senyurek, M. Kurum, et al., "Quasi-global machine learning-based soil moisture estimates at highspatio-temporal scales using CYGNSS and SMAP observations." *Remote Sensing of Environment*, vol.276, pp.113041, 2022.
16. Q. Yan, S. Gong, S. Jin, et al., "Near real-time soil moisture in China retrieved from CyGNSS reflectivity." *IEEE Geoscience and Remote Sensing Letters*, vol. 19, pp. 1–5, 2022.
17. Y. Jia, S. Jin, Q. Yan, et al., "An effective land type labeling approach for independently exploiting high-resolution soil moisture products based on CYGNSS data." *IEEE Journal of Selected Topics in Applied Earth Observations and Remote Sensing,* vol. 15, pp. 4234–4247, 2022.
18. Y. Jia, S. Jin, H. Chen, et al., "Temporal-spatial soil moisture estimation from CYGNSS using machine learning regression with a pre-classification approach." *IEEE Journal of Selected Topics in Applied Earth Observations and Remote Sensing,* vol. 14, pp. 4879–4893, 2021.

13 Drought Monitoring and Evaluation in Major African Basins from Satellite Gravimetry

Ayman M. Elameen, Shuanggen Jin, and Mohamed Abdallah Ahmed Alriah

13.1 INTRODUCTION

Droughts have increased in frequency and severity due to climate change throughout the world's river basins in recent decades (Forootan et al., 2019). According to the sixth assessment report of the International Panel for Climate Change (IPCC), global temperatures have risen by ~1°C since industrialization, which may further amplify by 1.5°C between 2030 and 2050 as a result of human activities (IPCC, 2018). As the population grows and water demand increases, droughts are triggered and aggravated by anthropogenic activities such as deforestation and the construction of dams (Schlosser et al., 2014; AghaKouchak et al., 2015; Omer et al., 2020; Sarfo et al., 2022). To prioritize adaptation actions in global hot spots, it is essential to characterize droughts.

Although the continent has abundant water resources with meeting its ecological and agricultural needs, climate extremes are becoming increasingly dangerous, endangering the continent's crucial water supply and millions of lives (Masih et al., 2014; IPCC, 2022). Two of the biggest drought tragedies ever documented in history occurred in the Sahel region in 2007 and the Nile basin in 1984. These droughts caused the death of approximately 750,000 people (Vicente-Serrano et al., 2012). Future projections indicate that the probability of drought occurrence will increase across the entire African continent, leading to significant regional implications (Ahmadalipour and Moradkhani, 2018; IPCC, 2022). Additionally, excessive water demand may lead to the overuse of freshwater resources, which might result in disputes among water users during dry spells. This may increase the risk of hydro-political tension in Africa, because the transboundary rivers represent 64% of the entire region's landmass (United Nations Environment Programme, 2010). Monitoring the drought situation in Africa is crucial for prioritizing adaptations to avert water scarcity and disputes.

Drought monitoring necessitates prolonged and uninterrupted *in situ* hydro-climatic measurements. Yet Africa's land-based observation network has been deteriorating with time, having only one eighth of the minimum density required by the World Meteorological Organization and with only 22% of stations fully meeting the Global Climate Observing System requirements (Dobardzic et al., 2019). Due to the insufficiency of *in situ* data records in Africa, monitoring hydrological drought in the continent's basins has been limited (Ferreira et al., 2018). Additionally, a substantial financial and political commitment is required to record and share *in situ* observations, both of which are frequently missing.

Remote sensing observations represent an alternative source to counter data deficiencies in many data-poor regions worldwide. Moreover, satellite-borne sensors have featured as an effective tool for tracking droughts, considering their capacity to offer regional-to-global coverage (Jiao et al., 2021). Various remote sensing-based systems have been utilized to assess and detect drought conditions around the world. Among these are Moderate Resolution Imaging Spectroradiometer (MODIS)-based evapotranspiration, soil moisture from Sentinel-1 and the Soil Moisture Active Passive radiometer, and the Normalized Difference Vegetation Index from Landsat (West et al., 2019; Modanesi et al., 2020). Although these measurements could deliver valuable information about agricultural and meteorological droughts, the task of assessing hydrological drought remains daunting (Papa et al., 2022) since they can capture only surface and shallow subsurface conditions. Also, it is problematic to evaluate droughts based only on surface measurements (e.g., precipitation and soil moisture), since the reduction of water from the deepest aquifers may continue even after the surface storage has dried up (Leblanc et al., 2009). After launching the Gravity Recovery and Climate Experiment (GRACE) satellite mission in 2002, the potential time-variable gravity measurement offered an integrated perspective for drought monitoring since it can capture vertically integrated terrestrial water storage (TWS) changes (i.e., from the top surface water to the deepest groundwater) (Ndehedehe et al., 2018).

The unique potential of GRACE measurements offered hydrologists a new dimension to develop new GRACE-based drought indices (Hassan and Jin, 2016; Jin and Zhang, 2016). Therefore, numerous studies have applied GRACE-based indicators for drought analysis and monitoring. For example, Kumar et al. (2021) evaluated the drought severity over the Godavari basin using the GRACE Combined Climatologic Deviation Index. Liu et al. (2020b) proposed a GRACE-based Drought Severity Index and assessed the drought variations for China's large basins. Khorrami and Gunduz (2021) proposed an Enhanced Water Storage Deficit Index to observe drought conditions in Turkey. Wu et al. (2021) characterized the drought over southwest China using the GRACE-derived Total Storage Deficit Index. Cui et al. (2021) developed a multiscale Standardized Terrestrial Index of water storage to assess the global hydrological droughts.

Many studies have investigated drought characteristics throughout Africa utilizing GRACE data. For examples, Nigatu et al. (2021) evaluated the drought situation over the Nile basin using GRACE Combined Climatologic Deviation Index (CCDI) and GRACE Water Storage Deficit Index (WSDI). Ferreira et al. (2018) assessed the drought condition over West Africa's river basins utilizing de-seasoned GRACE-TWS. Hulsman et al. (2021) employed GRACE Total Storage Deficit Index (TSDI) to detect drought events over the Zambezi basin. These previous studies, on one side, calculated the drought indices based on accounting the total TWS components (including: surface, soil, ground, snow, and canopy water), and neglected the influencing role of the individual water storage components in drought index. Each water storage component is an essential hydrological variable to comprehend drought occurrences, according to Lopez et al. (2020). Since the TWS-based drought indicator considers all components together, the primary problem is abstract. As a result, this may overestimate or underestimate drought characteristics.

On the other side, most of these studies only verified the capabilities of drought using GRACE data but not the associations between GRACE-based droughts and teleconnection factors. It is clear from earlier studies that telecommunication factors have a major effect on drought (Dai, 2011; Wang et al., 2015). Many worldwide attempts have been made over past years to establish the relationship between climate variability and GRACE-TWS changes. Huang et al. (2016) evaluated the associations between the hydrological drought over the Columbia River and ENSO and Arctic Oscillation (AO). Vissa et al. (2019) assessed the relationship between ENSO-induced groundwater changes derived from GRACE and GLDAS over India. Liu et al. (2020a) explored the role of teleconnections over TWS variations within the Asian and eastern European regions. To the best of our knowledge, most previous studies have focused on the relationship of several atmospheric variables such as precipitation and temperature with teleconnections in Africa (e.g., Alriah et al., 2021; Nkunzimana et al., 2021; Diatta et al., 2020; Bahaga et al., 2019). Limited studies analyzed the relation between teleconnections and GRACE-based drought over the continent.

This present work aims to evaluate the drought condition in major basins in Africa (Figure 13.1) during 2003–2016 using the Weighted Water Storage Deficit Index (WWSDI) (Wang et al., 2020), which was developed from the GRACE-WSDI but also considered the influence of the individual

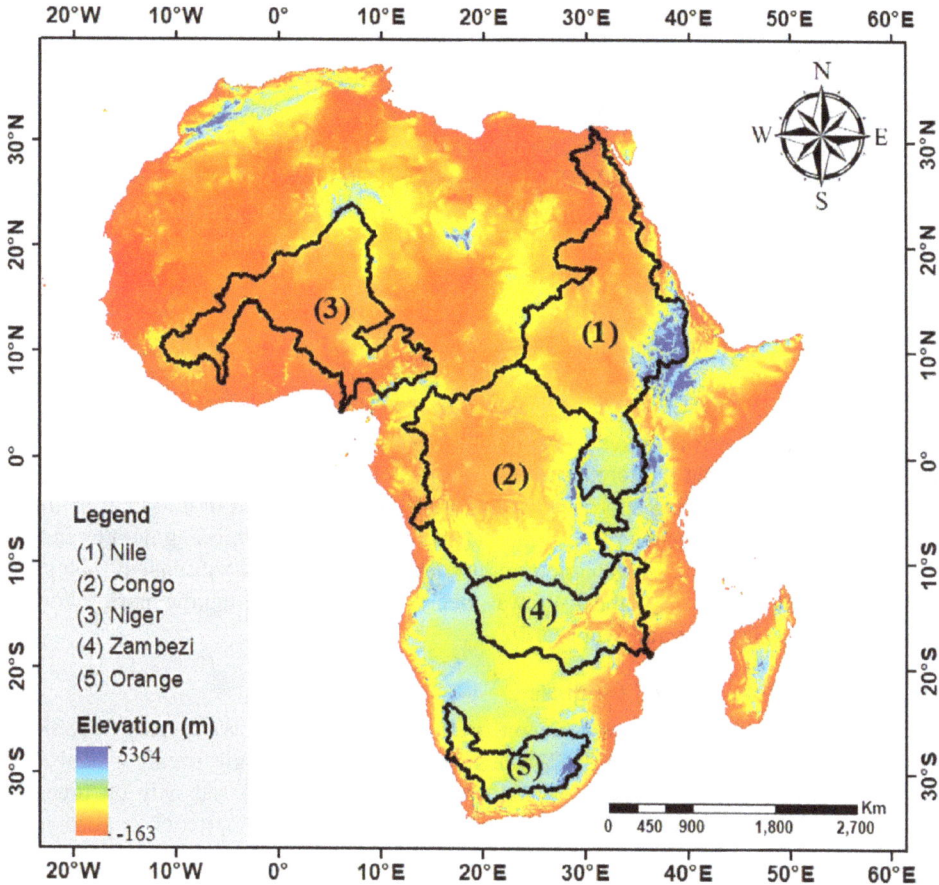

FIGURE 13.1 Map showing the location of river basins selected in this article and the elevations along the basins.

TWS components to provide further reliable droughts assessment. This work also reveals the relationship between the drought identified by the WWSDI and climate teleconnections. The primary objectives of this work are (1) to assess the drought events in major African basins using the WWSDI; (2) to examine the time required for drought recovery in different basins; and (3) to analyze the links between the WWSDI and climate oscillations using the wavelet coherence method.

13.2 MATERIALS AND METHODS

13.2.1 DATA SETS

This study uses monthly precipitation data in gridded form ($0.25° \times 0.25°$) from 2003 to 2016, acquired from the seventh version of the Tropical Rainfall Measurement Mission (TRMM 3B43) (Huffman et al., 2007). The present study also utilizes monthly potential evapotranspiration (PET) retrieved from the MOD16A2 sensor, publicly available worldwide at 8-day temporal resolutions and 500-m spatial resolution (Running et al., 2017). The 8-day PET data were averagely weighted to obtain the monthly PET values for this study. We also utilize monthly self-calibrated Palmer Drought Severity Index (scPDSI) (Wells et al., 2004) time-series (v4.04) data sets for the period 2003–2016, with a spatial resolution of 0.5°. The data sets were collected from the Climate Research Unit (CRU) at the University of East Anglia, United Kingdom. We further employ climate indices time series, namely, El Niño Southern Oscillation Index (ENSO) and Indian Ocean Dipole (IOD). The data was obtained for the period 2003–2016.

This study uses the sixth release of GRACE spherical harmonics coefficient solutions processed by the Center for Space Research (CSR) at the University of Texas at Austin (Chen et al., 2022) to derive gridded terrestrial water storage anomaly (TWSA) from 2003 to 2016 at a spatial resolution of 1°. We also utilize the WaterGAP Global Hydrology Model (WGHM) to separate the components of GRACE-TWS data. The recent model version (WaterGAP 2.2d) at a resolution of 0.5° is used in this study (Müller Schmied et al., 2021). The data are available from January 2000 to December 2016.

13.2.2 Methodology

13.2.2.1 Processing GRACE-Derived Water-Storage Anomalies

The monthly GRACE-derived gravity coefficients were processed by being truncated at degree and order 60. They were then filtered and destriped using a 400-km-radius Gaussian filter. The leakage reduction and averaging approach (Khaki et al., 2018) were used in this study to minimize the leakage error contributions over the understudied river basins. The missing months in the time series were filled using linear interpolation via averaging the prior and subsequent months (Yang et al., 2017). A regional average of the TWS was then computed by defining the mask following the method described in Swenson and Wahr (2002).

13.2.2.1 Standardized Drought Indices

Standardized indices are widely used to quantify droughts worldwide. We employ SPI, SPEI, and scPDSI to assess the effectiveness of WWSDI in characterizing drought events over the chosen basins for this study. SPI is a meteorological drought index that is based only on precipitation (Satish Kumar et al., 2021). To compute SPI, the monthly TRMM precipitation is normalized by utilizing an equal probability function. SPEI is an expansion of SPI, since it includes the influence of evapotranspiration on drought under changing environments. SPEI is computed by subtracting precipitation from potential evapotranspiration using climatic water balance. Hence, TRMM precipitation and MOD16 PET products were employed to calculate SPEI. Both indicators can be obtained at different timescales (1, 3, 6, 9, 12, and 24 months). However, each timescale reflects a distinct condition. For example, 1 month could indicate meteorological types of droughts, 3 months could reflect the soil moisture conditions, 6 months may indicate anomalies in land water storage, and 9 months could depict the agricultural droughts well. Hence, to provide a solid validation for WWSDI performance, the 6-month timescale was employed since it can effectively demonstrate the TWS deficit that was monitored by the WWSDI (Sun et al., 2018; Wang et al., 2020). Another widely used meteorological drought index is the scPDSI, which is developed based on the Palmer Drought Severity Index (PDSI) using a physical water balance model. The scPDSI timescale is fixed unlike the two indices previously described.

13.2.2.2 Water Storage Components Estimation

TWSA is composed of the following:

$$\text{TWSA} = \text{GWSA} + \text{SMSA} + \text{SWEA} + \text{SWSA} + \text{CWSA} \tag{13.1}$$

In this study, TWSA is estimated from GRACE, whereas soil moisture storage anomalies (SMSA), snow water equivalent anomalies (SWEA), surface water storage anomalies (SWSA), and canopy water storage anomalies (CWSA) are the anomalies of SMS, SWE, SWS, and CWS, deduced from the WGHM. Groundwater storage anomalies (GWSA) are estimated via subtracting TWSA from the WGHM-derived components in Equation 13.1. Note that the SWEA and CWSA have minimal contribution to TWSA over African basins. Thus, they are assumed to be negligible and not considered in groundwater storage anomalies computation, as indicated in Equation 13.1. SMSA and SWSA are expanded into the spherical harmonic coefficients, truncated to 60°, ordered, and filtered by Gaussian filter.

13.2.2.3 Component Contribution Ratio

We utilized the component contribution ratio (CCR) to determine the mean percentage contribution of a single water storage component to the temporal variability of the total TWS (Huang et al., 2019). CCR is calculated as the ratio of the mean absolute deviation (MAD) of a storage component to the total TWS variability (TV), as expressed by (Zhang et al., 2019):

$$CCR_S = \frac{MAD_S}{TV} \tag{13.2}$$

where $MAD_S = \frac{1}{N}\sum_{t}^{N}|S_t - |\bar{S}||$, $\sum_{S}^{Storages} MAD_S$, S denotes the single storage components, and TV is the total variability, calculated as summation of all components MAD_s ($\sum_{S}^{Storages} MAD_S$).

13.2.2.4 Evaluation of Hydrological Drought and Its Recovery Time

In this study, in order to depict drought in the five large African basins, we adopted the WWSDI developed by Wang et al. (2020). WWSD is based on WSD, which represents the difference between TWSA time series and the monthly means of TWSA values (Thomas et al., 2014) and is computed as:

$$WSD_{u,v} = TWSA_{u,v} - \overline{TWSA_v} \tag{13.3}$$

where $TWSA_{u,v}$ is the value of TWSA time series for the vth month of the uth year and $\overline{TWSA_v}$ is the mean value of the vth month of TWSA during the study period. A negative deviation represents storage deficits. Furthermore, three continuous negative months or longer is considered a drought event (Thomas et al., 2014). In order to make comparisons against SPI, SPEI, and scPDSI in this study, the WSD is normalized to WSDI by the zero-mean normalization method, based on the expression

$$WSDI = \frac{WSD - \mu}{\sigma} \tag{13.4}$$

where σ and μ indicate standard deviations and the mean of the WSD time series, respectively. In order to construct WWSD, we incorporated different TWS components (i.e., GWS, SWS, and SMS) to the drought index by weighting these components through their CCR using Equation 13.2. We subsequently calculated the water deficit for each component (i.e., groundwater storage deficit [GWSD], surface water storage deficit [SWSD], and soil moisture storage deficit [SMSD]) like the WSD. Thereafter, WWSD was generated by combining these water components' deficits after multiplying them by their respective weights:

$$WWSD = \omega_1 \cdot GWSD + \omega_2 \cdot SWSD + \omega_3 \cdot SMSD \tag{13.5}$$

where ω_i (= 1, 2, 3) represent the derived weight from Equation 13.2. Finally, the WWSDI is achieved by normalizing WWSD, as shown in Equation 13.4.

The hydrological recovery time is calculated based on a statistical method proposed by (Thomas et al., 2014). The rate of change of WWSD during a certain period can be assessed by

$$\frac{\Delta WWSD}{\Delta t} = WWSD_u - WWSDI_{u-1} \tag{13.6}$$

where $\dfrac{\Delta WWSD}{\Delta t}$ represents the rate of change of the water storage deficit, and Δt represents the time interval (one month in this study). If $\dfrac{\Delta WWSD}{\Delta t}$ follows a standard normal distribution according to a one-sample Kolmogorov-Smirnov test, the 68th percentile value C (1 standard deviation) is used as the recovery rate of WWSD for any deficit month. Then, the drought recovery time (R) for each month can be obtained by (Thomas et al., 2014)

$$R = \frac{WWSD}{C} \tag{13.7}$$

where C is the drought recovery rate during the study period, and R (month) is the drought recovery time of the water storage deficit. According to this method, a probabilistic recovery time can be assessed for each hydrological drought.

13.2.2.5 Wavelet Coherence

Within the time-frequency space, wavelet coherence can be used to determine the relationship between the two time series data by estimating the correlation between them that varies between 0 and 1. The coefficient of wavelet coherence between the two sets of time series data can be denoted as follows (Grinsted et al., 2004):

$$R^2(s,\tau) = \frac{\left| S\left(s^{1-} W_{xy}(s,\tau)\right)\right|^2}{S\left(s^{1-}\left|W_x(s,\tau)\right|^2\right) \cdot S\left(s^{1-}\left|W_x(s,\tau)\right|^2\right)} \tag{13.8}$$

where $R^2(s,\tau)$ = coherence coefficient minimum and maximum coherence at 0 and 1, and $W_{xy}(s,\tau)$ = cross wavelet transforms between two series. Equation 13.8 resembles the coefficient of determination equation, and thus the wavelet coherence varies between 0 and 1.

S = smoothing operator represented as given here:

$$S(W) = S_{scale}\left(S_{time}\left(W(s,\tau)\right)\right) \tag{13.9}$$

The smoothing along the wavelet axis is represented as S_{scale} and S_{time}. In the present study, the wavelet coherence was examined at 5% significance level or at the confidence interval > 95%.

13.3 RESULTS AND DISCUSSION

13.3.1 WWSD Generation and Significance

In order to construct the WWSD (Equation 13.5), we first calculated the component contribution ratio (CCR) and the water component storage deficit (WCSD) time series for all basins (Figure 13.2). The obtained result of CCR showed that the highest contribution to total water storage variability over the five river basins was induced mainly by the groundwater storage anomalies (GWSA) accounting (56%, 61%, 47%, 64%, and 78%), followed by soil moisture storage anomalies (SMSA) (34%, 32%, 25%, 26%, and 18%) and surface water storage anomalies (SWSA) (10%, 21%, 14%, 10%, and 4%) for the Nile, Congo, Niger, Zambezi, and Orange basins, respectively. Concerning WCSD (Figure 13.2), groundwater storage deficit (GWSD) found to be highly consistent with terrestrial water storage deficit (WSD), while soil moisture storage deficit (SMSD) and surface water storage deficit (SWSD) followed varied patterns compared to GWSD during different periods in the time series. For example, in the Congo basin (Figure 13.2b), from January 2009 to January 2013, GWSD recorded a declining trend of −1.03 mm, whereas SMSD and SWSD exhibited rising trends of 0.3 mm and 0.15 mm, respectively.

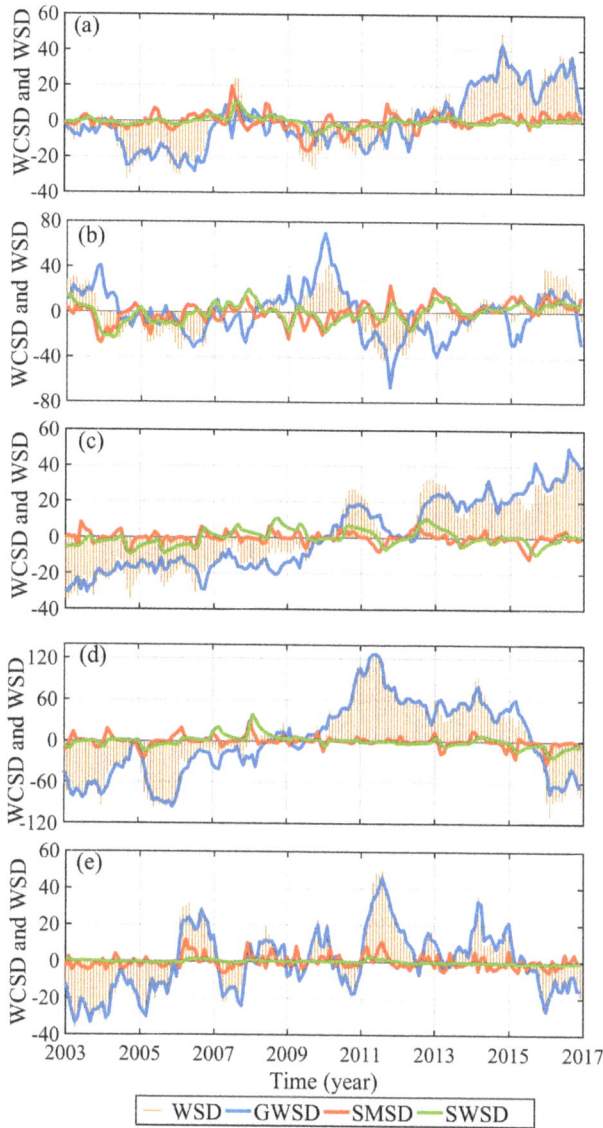

FIGURE 13.2 Water storage deficit (WSD) and water components storage deficit (WCSD) in the (a) Nile, (b) Congo, (c) Niger, (d) Zambezi, and (e) Orange river basins.

However, since these components showed distinguish contribution and variation among them relative to TWS changes, it is natural to wonder whether these water components react differently to the incidence of drought over those basins. To clarify that, water component storage deficit (WCSD) was also utilized as an indicator to identify drought events (Figure 13.3) based on 3 months or more of continuous negative deficits. The results in Figure 13.3 clearly show that different WCSD indicators detected varied onset, duration, and drought occurrences during the study period. For example, in the Nile basin (Figure 13.3a), GWSD exhibited six drought events, whereas SMSD and SWSD exhibited 12 and 7 events, respectively, between January 2003 and December 2016.

Additionally, a remarkable prolonged drought state in terms of groundwater storage (GWSD) was noticed from January 2003 to February 2007, January 2003 to December 2009, January 2003 to July 2008, and January 2003 to January 2006 over the Nile (Figure 13.3a), Niger (Figure 13.33c),

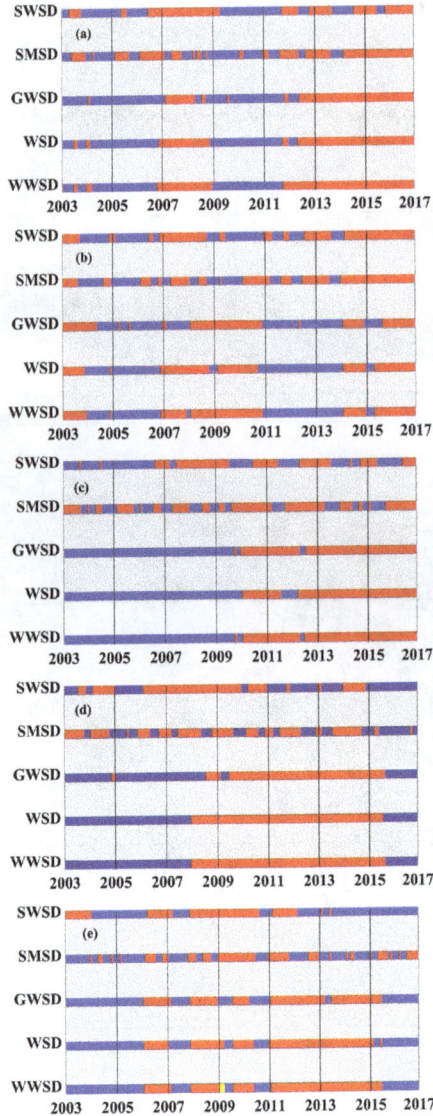

FIGURE 13.3 Identified drought events based on water storage deficit (WSD), water components storage deficit (WCSD), and weighted water storage deficit (WWSD) in the (a) Nile, (b) Congo, (c) Niger, (d) Zambezi, and (e) Orange river basins. The red values denote wet month, while the dark blue values represent drought month.

Zambezi (Figure 13.3d), and Orange (Figure 13.3e) basins, respectively, separated by nearly one wetting month. The late response of GWS to recharge from SWS and/or the increased groundwater withdrawal can support this finding. Furthermore, the Niger basin (Figure 13.3c) had the most prolonged GWS drought state among all the basins recording 7 years. Previous work by Ferreira et al. (2018) on a water storage (TWS) drought signal over West Africa (including the Niger basin) found a long drier period between 2003 and 2008. These findings are consistent with the results presented in this work. According to the analysis of the 2003–2008 period presented in this study, the water storage (TWS)-based drought trend is related to groundwater storage, where most of the TWS (i.e., 61%) in the Niger basin is induced mainly by GWS. Ferreira et al. (2018) reported that the rainfall increasing trend between 2003 and 2008 over West Africa is associated with a drought period.

They attributed this to the unsustainable influencing of rainfall recovery to the water-storage increase across West Africa, in the early 2000s. Consequently, the occurrence of the long GWS drought state in the Niger basin may be attributed to the minimal or late influences of surface water on the groundwater storage in the early 2000s.

The results also demonstrate that SWSD over the Orange basin (Figure 13.3e) exhibited a long drier period from March 2013 to December 2016 except for February and June 2013. This finding is in line with an early study conducted over the South African drying signal (Munday and Washington, 2019). The latter linked the decline in precipitation with local surface temperature change since increased subsidence is linked to clearer skies and higher net solar radiation. Also, the reduction in precipitation magnitude is correlated to the changing patterns of tropical sea surface temperatures. Furthermore, the exceeding demand for surface water may cause the surface water shortage where the water of the Orange basin is heavily utilized, and most of the riparian states rely on the Orange basin's water resources for commercial crop irrigation; in addition, 29 dams are operated over the river (Mgquba and Majozi, 2020), which may cause large water abstractions.

However, since WCSD plays different roles in response to drought events, considering these differences in calculating drought index can provide more realistic and reliable drought evaluation over the continent's river basins. To demonstrate that, we further assessed the performance of weighted water storage deficit (WWSD) and water storage deficit (WSD) in terms of drought events identification as shown in Figure 13.3. From the graph, despite both indicators appearing to behave similarly, the data show some discrepancies in the observed onset and drought duration between WWSD and WSD. For example, in the Nile basin (Figure 13.3a), WWSD recorded one drought between April 2004 and October 2006, whereas WSD monitored the drought from March 2004 to November 2006. In the Congo basin (Figure 13.3b), WSD detected a drought event from November 2008 to January 2009; however, WWSD failed to identify this event. From the foregoing, WWSD has varied sensitivity to drought events compared to WSD. These discrepancies, however, are explained by the weight given to a single TWS component in the WWSD. This result further suggests that accounting for the influencing roles of water storage variables in drought index promises to provide more accurate drought estimation over major basins in Africa.

13.3.2 WWSDI VALIDATION AND DROUGHT DETECTION

Before using WWSDI for characterize drought condition, we assessed its reliability compared to four drought indices including WSDI, SPI, SPEI, and scPDSI as illustrated in Figure 13.4. Graphically, WWSDI exhibited a good agreement with the four drought indices in monitoring drought over most of the basins. We also ran a numerical comparison analysis using Pearson's correlation coefficient between the four drought indices and WWSDI (Table 13.1). From Table 13.1, high positive correlation ranges from 0.95–0.98 were observed between WWSDI and WSDI over the five basins. This strong relation between WWSDI and WSDI is due to their high sensitivity to GRACE-derived TWS and the inclusion of TWS in their calculation procedures. However, the differences in correlation are attributed to the consideration of the weight of a single TWS component in WWSDI. WSDI is based on a single variable (GRACE-TWS); on the other hand, WWSDI is based on combining the TWS estimation from GRACE and WGHM using the CCR of individual TWS compartments as the weight. However, despite the fact that WWSDI and WSDI operate quite similarly, there is a distinction, as discussed previously. The comparison results between WWSDI and other three drought indices showed that WWSDI is significantly correlated with SPI in all basins. The highest positive correlation ($r = 0.69$) between the two indices was observed in the Orange basin, while the lowest was detected in the Congo basin.

Moreover, WWSDI was found to exhibit a significant correlation with the SPEI and scPDSI in the Nile, Congo, Zambezi, and Orange basins, but a weak correlation in the Niger basin.

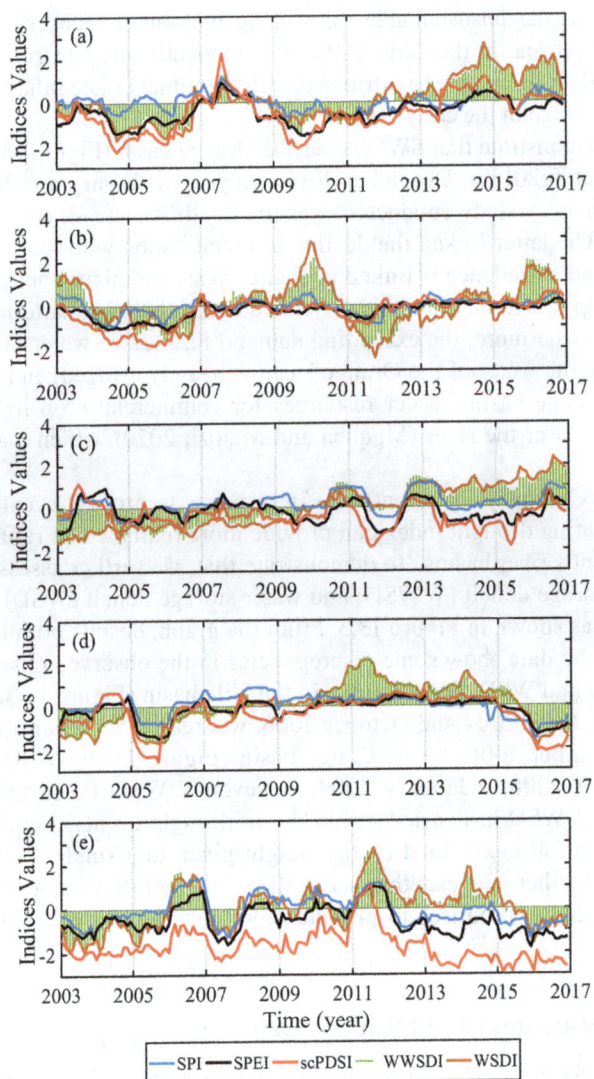

FIGURE 13.4 WWSDI, WSDI, SPI, SPEI, and the scPDSI time series for the (a) Nile, (b) Congo, (c) Niger, (d) Zambezi, and (e) Orange river basins.

TABLE 13.1

Person's Correlation Coefficients between WWSDI and WSDI, SPI, SPEI, and scPDSI over the Major African Basins

Basin	Drought Indices			
	WSDI	**SPI**	**SPEI**	**scPDSI**
Nile	0.98	0.52	0.72	0.78
Congo	0.95	0.5	0.51	0.53
Niger	0.98	0.55	0.1*	0.08*
Zambezi	0.99	0.53	0.73	0.72
Orange	0.98	0.69	0.73	0.72

Note: An asterisk indicates that the correlation is not significant.

To carry out a much more in-depth investigation, we further evaluated the temporal trends of these time series in Figure 13.4 in light of the fact that the association between the WWSDI and SPI, SPEI, and scPDSI was stronger in some situations while being weaker in others. According to Figure 13.4, the performance of WWSDI and its response to climate change correspond to the peaks and troughs of SPI, SPEI, and scPDSI over most basins. For example, all indicators showed that the biggest troughs occurred in the Orange basin in 2003 and across the Nile and Zambezi basins in 2006. However, in several cases, WWSDI was not fitting well with SPI, SPEI, and scPDSI; for example, the drought identified by WWSDI in 2004 over the Niger basin was not detected by SPI, SPEI, and scPDSI. The variations in relationships among SPI, SPEI, scPDSI, and WWSDI are most likely due to the differences in hydrological components and algorithms. For instance, the high correlation between the scPDSI against WWSDI in the Nile basin reflects the significant influence of soil moisture on the TWS. Some recent studies also reported the significant correlation between soil moisture and TWS over the Nile basin (e.g., Abd-Elbaky and Jin, 2019). In contrast, the weak correlation of SPEI and scPDSI with WWSDI in the Niger basin reveals that TWS was not much affected by evapotranspiration and soil moisture. In this context, the Niger basin was previously characterized as having a long-term high reduction in water storage between 2002 and 2008 (Ferreira et al., 2018), which corroborates our findings (Figure 13.3c). Thus, the availability of the stored water was less in the Niger basin, which affects the weak correlation of WWSDI against SPEI and scPDSI. Overall, WWSDI showed a good consistency with SPI, SPEI, and scPDSI in drought monitoring over most of the basins, which may indicate solid evidence on the applicability and capability of WWSDI over the river basins of Africa.

The WWSDI-obtained droughts events for the considering basins from 2003 to 2016 are displayed in Figure 13.5. Table 13.2 represents the magnitude, intensity, and duration characteristics of WWSDI for all the basins. The magnitude is calculated as accumulated WWSDI, and the intensity is calculated as the ratio of magnitude to duration. According to Figure 13.5 and Table 13.2, four drought events were detected in the Nile basin during 2003–2016, whereas the most severe droughts (intensity of −1.15) occurred during 2004–2006 period. This result agreed with the findings of previous studies (Hasan et al., 2021; Nigatu et al., 2021). In the Congo basin, six drought events were observed; more so, the highest frequency of droughts (with an intensity of −1.23) was observed during 2010–2012. Within the Niger basin, two prolonged drought episodes were detected; in addition, the severest drought event (intensity of −1.02) occurred during 2003–2007. Regarding the Zambezi basin, three drought events were observed. The severest drought event (intensity of −1.02) occurred during the 2003–2004 period. Concerning the Orange basin, five drought events were identified; further, the severest drought event (intensity of −1.08) occurred during 2003–2006. Long-term drought occurrence was observed from 2003 to 2006 over the Nile basin, from 2003 to 2009 over the Niger basin, and from 2003 to 2008 over the Zambezi basin, with the inclusion of few wetting months. A dominant wetting tendency was observed for the Nile, Niger, Zambezi, and Orange basins. Also, a mild trend over the Congo basin was detected.

13.3.3 ANALYSIS OF DROUGHT RECOVERY TIMES

The drought recovery time, defined as the time required for drought to recover to normal conditions. Herein, the drought recovery time across the five major basins in Africa is evaluated by a probabilistic approach using the rate of change in WWSD (i.e., ΔWWSD/Δt). The Kolmogorov Smirnov test (KS-test) was used to examine the normality of change rate of the deficits (ΔWWSD/Δt). Table 13.3 shows that the statistic Dmax for the five basins is much smaller than the statistical threshold Da, which indicates that ΔWWSD/Δt follows a standard normal distribution at the significance level of 95%. Therefore, it is appropriate to estimate the recovery time of drought in the considered basins based on the employed approach in this work.

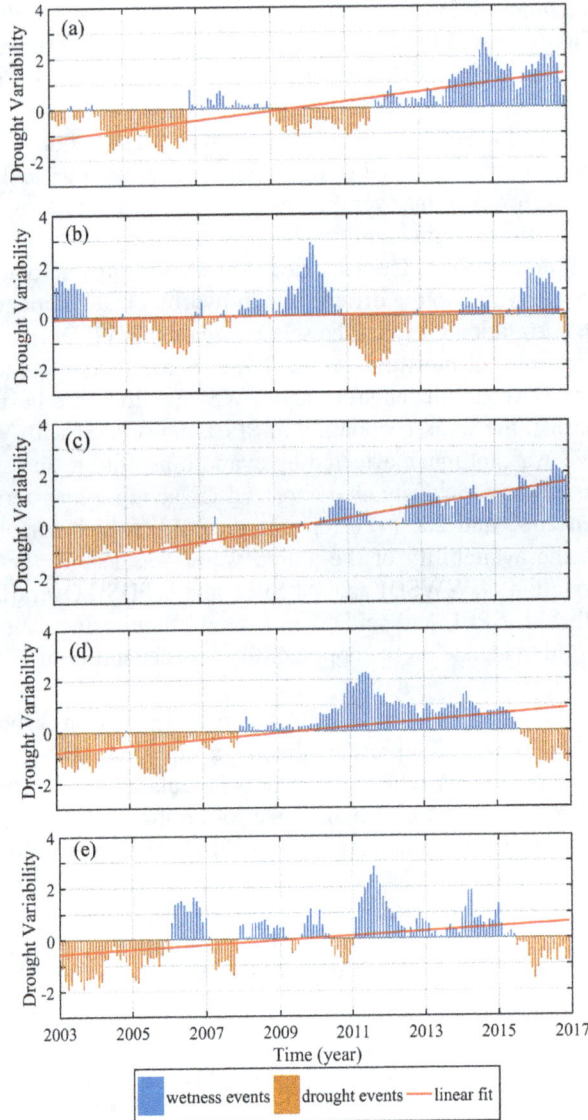

FIGURE 13.5 Statistics of the WWSDI and drought event for the (a) Nile, (b) Congo, (c) Niger, (d) Zambezi, and (e) Orange river basins from 2003–2006.

Table 13.3 lists the drought recovery rates (C) and the maximum and average recovery time (R) obtained for the five basins. The R time series are displayed in Figure 13.6. The drought recovery rate (C) varied in the five basins. The Zambezi basin exhibits the largest drought recovery rate, with the value of 4.1 mm/month, whereas the Nile basin reveals the lowest drought recovery rate, with the value of 2.06 mm/month. A relatively high value of the drought recovery rate (C) indicates a tendency for depleted storage levels to increase rapidly.

For the five basins, the average recovery time (R) for drought ranges from 5.1 to 12.3 months, and the maximum R ranges between 12.2 and 24.5 months. The Zambezi basin takes slightly longer to recover from drought on average, whereas the Niger basin exhibits the longest recovery times. In general, long drought duration leads to longer drought recovery time. Therefore, since the Niger basin exhibits the longest drought duration (54 months), drought in this basin takes the longest recovery time compared to other basins.

TABLE 13.2

Summary Table of Identified Drought Event by WWSDI in Major Basins of Africa

River Basin	No.	Period	Magnitude	Intensity	Duration
Nile	1	February 2003–June 2003	−2.39	−0.48	5
	2	September 2003–December 2003	−1.45	−0.36	4
	3	April 2004–October 2006	−35.78	−1.15	31
	4	January 2009–September 2011	−22.36	−0.68	33
Congo	1	January 2004–November 2004	−4.25	−0.39	11
	2	January 2005–November 2006	−24.85	−1.08	23
	3	March 2007–July 2007	−1.55	−0.31	5
	4	December 2010–October 2012	−28.40	−1.23	23
	5	January 2013–January 2014	−8.01	−0.62	13
	6	January 2015–April 2015	−2.38	−0.60	4
Niger	1	January 2003–May 2007	−54.01	−1.02	54
	2	July 2007–December 2009	−19.64	−0.65	30
Zambezi	1	January 2003–December 2004	−24.34	−1.02	24
	2	January 2005–December 2007	−28.62	−0.79	36
	3	September 2015–December 2016	−15.7	−0.98	16
Orange	1	January 2003–January 2006	−40.07	−1.08	37
	2	March 2007–November 2007	−8.50	−0.95	9
	3	April 2009–July 2009	−1.30	−0.32	4
	4	June 2010–January 2011	−5.05	−0.63	8
	5	July 2015–December 2016	−12.89	−0.72	18

TABLE 13.3

Statistics of Drought Recovery Times for the Five Major Basins in Africa

Basin	KS-Test		Drought Recovery Time		
	Dmax	Da	C (mm/month)	Max R (month)	Ave R (month)
Nile	0.09	0.34	2.06	15.8	7.6
Congo	0.09	0.35	3	16.4	5.5
Niger	0.09	0.27	2.2	24.5	10.7
Zambezi	0.08	0.43	4.1	24.2	12.3
Orange	0.09	0.38	2.2	12.2	5.1

13.3.4 THE CORRELATION BETWEEN WWSDI AND CLIMATE FACTORS

Previous studies have shown that droughts were closely related to climate variables (Liu et al., 2020; Mishra and Singh, 2010; Ferreira et al., 2018). In the present study, ENSO and IOD were chosen to describe the influences of teleconnections over droughts. Moreover, wavelet coherence was employed to evaluate the link between WWSDI and climate factors over major basins in Africa during 2003–2016. The wavelet coherence between monthly WWSDI and climate factors was presented in Figure 13.7.

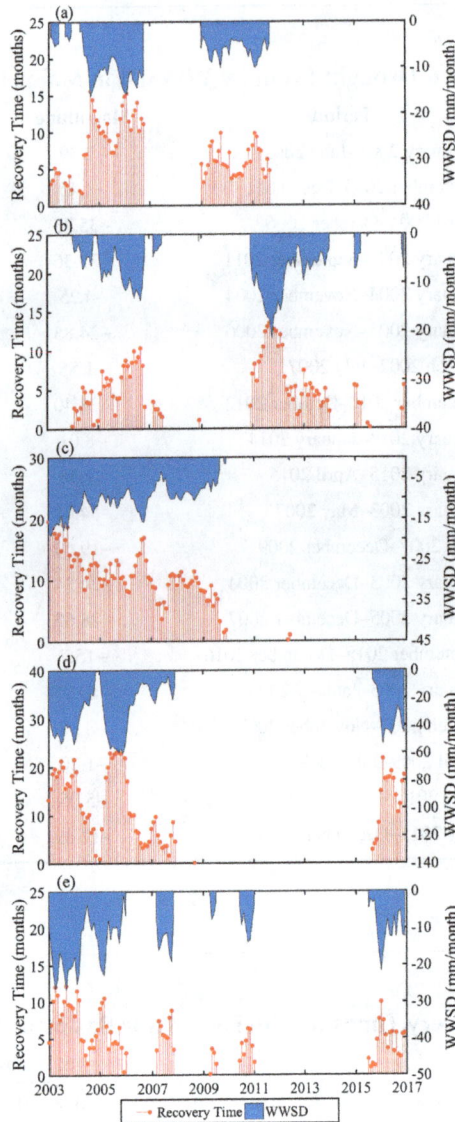

FIGURE 13.6 Time series of drought recovery time (R) in the (a) Nile, (b) Congo, (c) Niger, (d) Zambezi, and (e) Orange river basins.

In the Nile basin (Figure 13.7a, b), the WTC between ENSO and WWSDI shows statistically significant in-phase relationships at 0.25–1-year band over 2004–2006, and at 1–2-year band over 2003–2011 and 2013–2016, respectively. On the other hand, the WTC between IOD and WWDI shows anti-phase relationships with significant coherence at 1–2-year band over 2003–2007, and at 4–5-year band over 2003–2014. In comparison to Figure 13.5a, the strong in-phase relationship between WWSDI and ENSO during 2007–2011 and 2013 to end of the time series was corresponded with the induced of wet condition, which is derived by La Niña events. Furthermore, the shown anti-phase relationship between WWSDI and IOD during 2003–2016 was found to be consistent with the observed recovery tendency, as seen in Figure 13.5a.

For the Congo basin (Figure 13.7c, d), significant in-phase relationships were observed between ENSO and WWSDI over 2004–2007 and 2014–2016 periods at 0.25–1-year band, over 2014 to

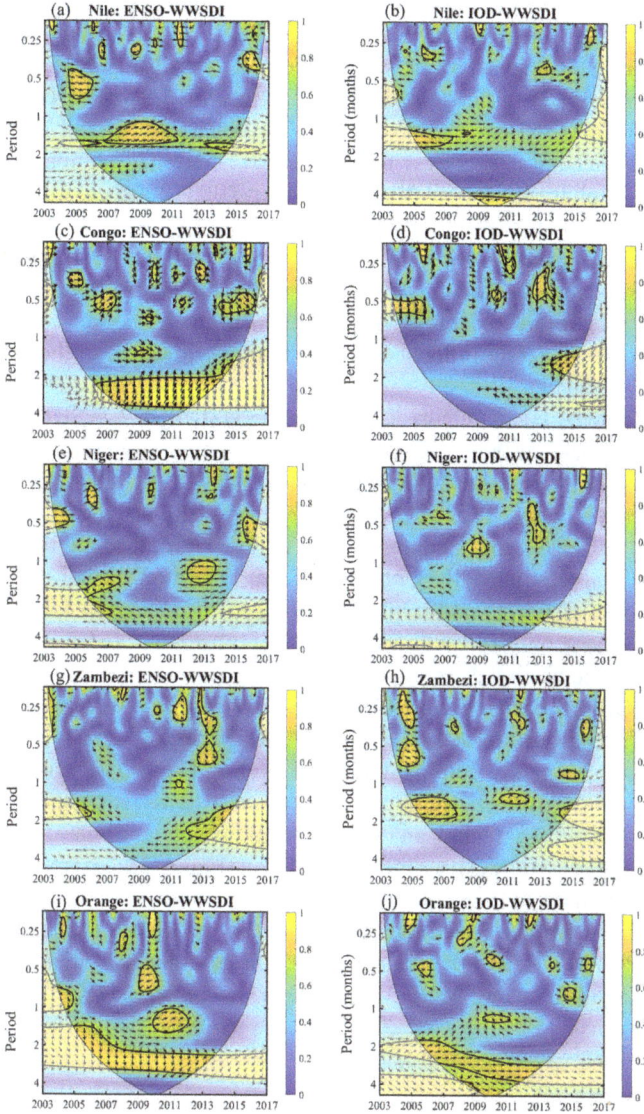

FIGURE 13.7 The left panel represents the wavelet coherence between monthly WWSDI and El Niño Southern Oscillation Index (ENSO). The right panel represents the wavelet coherence between monthly WWSDI and Indian Ocean Dipole (IOD). The 95% confidence level is presented as thick contour and the relative phase relationship is represented by arrows with anti-phase pointing left and in-phase pointing right.

2016 at 1–2-year band, and over 2006 to end of time series at 2–4-year band. WTC plot between IOD and weighted water storage deficit index (WWSDI) shows that the IOD and WWSDI were in anti-relationship at 1–2.5- and 3–4-year band over 2013 to 2016. In comparison to Figure 13.5b, the significant in-phase ENSO relationship during 2006 to the end of the time series was found to coincide with the induction of the wet condition in the basin, which is forced by La Niña episodes. This result suggests that ENSO leads the variability of drought in the Congo basin. It is also noted that the anti-phase IOD relationship during 2013 to 2016 was consistent with the decreasing drought magnitude as well as increasing wet tendency.

In case of the Niger basin (Figure 13.7e, f), in-phase relationships with statistically significant coherence were observed between ENSO and WWSDI at 0.25–1-year band over 2003–2005 and 2015–2016, and at 1–4-year band over 2003–2005, over 2012–2013, and over 2014–2016, respectively. Additionally, significant anti-phase relationship between ENSO and WWSDI was observed over 2005–2007 at 1–4-year band. The WTC between IOD and WWSDI shows significant anti-phase relationship at 2–4-year band over 2013–2016. In comparison to Figure 13.5c, it is found that the in-phase ENSO relationship during 2003–2005 coincided partially with the decrease of drought; further, the in-phase ENSO relationship during 2012–2013 and 2014–2016 corresponded to the wet event in the WWSDI time series (Figure 13.5c), which is mainly controlled by induction of the La Niña phase. More so, the anti-phase IOD relationship appearing during 2013–2016 was found to agree with the induction of the wet condition, as shown in Figure 13.5c. However, the IOD influences were relatively lower than the impact of the ENSO in the basin.

Regarding the Zambezi basin (Figure 13.7g, h), the ENSO and WWSDI show significantly in-phase relationships at 1–2-year band over 2003–2005, and anti-phase relationship at 2–4-year band over 2011–2017. WTC plot between IOD and WWSDI shows that IOD and WWSDI were in in-phase relationships at 0–2-year band over 2004–2007, over 2010–2012, and over 2013–2017, and at 2–4-year band over 2013 to 2016. In comparison to Figure 13.5d, it found that the in-phase ENSO and IOD relationship agreed partially with the drought condition during the 2003–2005 period; more so, the anti-phase of ENSO and in-phase of IOD corresponded to the recovery period from 2011–2017. Overall, the acquired results indicate that both ENSO and IOD clearly impacted the drought fluctuation in the basin.

Concerning the Orange basin (Figure 13.7 i, j), the WTC between ENSO and WWSDI shows statistically in-phase relationships at 1–2-year band over 2010–2012. Additionally, at the 2–4-year band the ENSO was significantly dominant across all years. The WTC between IOD and WWSDI shows in-phase relationships with significant coherence at 2–4-year band over the entire time series. Moreover, at 3.5–4.5-year band, an in-phase relationship was observed over 2003–2009. The obtained results for IOD and ENSO over the Orange basin (Figure 13.7i, j) revealed that both indices have significant effect on drought variation over the basin.

Overall, the findings from wavelet coherence analysis revealed that the climate factors had a substantial effect on WWSDI, and the impact of El Niño Southern Oscillation Index (ENSO) was significantly high over most basins. More so, the Indian Ocean Dipole (IOD) was highly affected in the southern river basins.

13.4 SUMMARY

GRACE observations are an essential tool in hydro-climatological investigations. In this study, we generated the WWSDI based on combined TWS from GRACE and WGHM utilizing the CCR of each component as their weight to assess the occurrences of drought throughout the major African basins from January 2003 to December 2016.

Regarding CCR, SMS and SWS rank the second and third, while GWS change ranks the first and accounts for 56%, 61%, 47%, 64%, and 78% of TWS change in the Nile, Congo, Niger, Zambezi, and Orange basins, respectively. According to water component storage deficit (WCSD), different water components play different roles in response to drought events in the basins. The WSDI, SPI, SPEI, and scPDSI are correlated significantly against WWSDI over the Nile, Congo, Zambezi, and Orange basins. In the Niger basin, SPI is significantly correlated with WWSDI. Overall, our findings indicate that the WWSDI can successfully detect drought events over major basins in Africa. Based on WWSDI, the most severe droughts occurred in 2006, 2012, 2006, 2006, and 2003 in the Nile, Congo, Niger, Zambezi, and Orange basins, respectively. A significant wetting tendency was detected over the Nile, Niger, Zambezi, and Orange basins, while a mild trend was observed in the Congo basin. The rate of drought recovery for five basins was between 2.06 and 4.1 mm month^{-1}, and the average drought recovery time was between 5.1 and 12.3 months, respectively. The wavelet coherence analysis effectively demonstrated the teleconnections between climate indices and WWSDI. The influence of ENSO on drought was significantly high in most of the basins. IOD has strong impact on the southern river basins.

REFERENCES

Abd-Elbaky, M., Jin, S., 2019. Hydrological mass variations in the Nile River Basin from GRACE and hydrological models. Geod. Geodyn. 10, 430–438. https://doi.org/10.1016/j.geog.2019.07.004

AghaKouchak, A., Feldman, D., Hoerling, M., Huxman, T., Lund, J., 2015. Water and climate: Recognize anthropogenic drought. Nature. 524, 409–411. https://doi.org/10.1038/524409a

Ahmadalipour, A., Moradkhani, H., 2018. Multi-dimensional assessment of drought vulnerability in Africa: 1960–2100. Sci. Total Environ. 644, 520–535. https://doi.org/10.1016/j.scitotenv.2018.07.023

Alriah, M. A. A., Bi, S., Shahid, S., Nkunzimana, A., Ayugi, B., Ali, A., Bilal, M., Teshome, A., Sarfo, I., Elameen, A. M., 2021. Summer monsoon rainfall variations and its association with atmospheric circulations over Sudan. J. Atmos. Solar-Terrestrial Phys. 225, 105751. https://doi.org/10.1016/j.jastp.2021.105751

Bahaga, T. K., Fink, A. H., Knippertz, P., 2019. Revisiting interannual to decadal teleconnections influencing seasonal rainfall in the Greater Horn of Africa during the 20th century. Int. J. Climatol. 39, 2765–2785. https://doi.org/10.1002/joc.5986

Chen, J., Cazenave, A., Dahle, C., Llovel, W., Panet, I., Pfeffer, J., Moreira, L., 2022. Applications and challenges of GRACE and GRACE follow-on satellite gravimetry. Surv. Geophys. 43, 305–345. https://doi.org/10.1007/s10712-021-09685-x

Cui, A., Li, J., Zhou, Q., Zhu, R., Liu, H., Wu, G., Li, Q., 2021. Use of a multiscalar GRACE-based standardized terrestrial water storage index for assessing global hydrological droughts. J. Hydrol. 603, 126871. https://doi.org/10.1016/j.jhydrol.2021.126871

Dai, A., 2011. Drought under global warming: A review. Wiley Interdiscip. Rev. Clim. Chang. 2, 45–65. https://doi.org/10.1002/wcc.81

Diatta, S., Diedhiou, C. W., Dione, D. M., Sambou, S., 2020. Spatial variation and trend of extreme precipitation in West Africa and teleconnections with remote indices. Atmosphere (Basel). 11, 999. https://doi.org/10.3390/atmos11090999

Dobardzic, S., Dengel, C. G., Gomes, A. M., Hansen, J., Bernardi, M., Fujisawa, M., Intsiful, J., 2019.2019 State of climate services: agriculture and food security. World Meteorol. Organ. 43.

Ferreira, V., Asiah, Z., Xu, J., Gong, Z., Andam-Akorful, S., 2018. Land water-storage variability over West Africa: inferences from space-borne sensors. Water 10, 380. https://doi.org/10.3390/w10040380

Forootan, E., Khaki, M., Schumacher, M., Wulfmeyer, V., Mehrnegar, N., van Dijk, A. I. J. M., Brocca, L., Farzaneh, S., Akinluyi, F., Ramillien, G., Shum, C. K., Awange, J., Mostafaie, A., 2019. Understanding the global hydrological droughts of 2003–2016 and their relationships with teleconnections. Sci. Total Environ. 650, 2587–2604. https://doi.org/10.1016/j.scitotenv.2018.09.231

Grinsted, A., Moore, J. C., Jevrejeva, S., 2004. Application of the cross wavelet transform and wavelet coherence to geophysical time series. Nonlinear Process. Geophys. 11, 561–566. https://doi.org/10.5194/npg-11-561-2004

Hasan, E., Tarhule, A., Kirstetter, P.-E., 2021. Twentieth and twenty-first century water storage changes in the Nile River basin from GRACE/GRACE-FO and modeling. Remote Sens. 13, 953. https://doi.org/10.3390/rs13050953

Hassan, A., Jin, S., 2016. Water storage changes and balances in Africa observed by GRACE and hydrologic models. Geod. Geodyn. 7, 39–49. https://doi.org/10.1016/j.geog.2016.03.002

Huang, S., Huang, Q., Chang, J., Leng, G., 2016. Linkages between hydrological drought, climate indices and human activities: A case study in the Columbia River basin. Int. J. Climatol. 36, 280–290. https://doi.org/10.1002/joc.4344

Huang, Z., Yeh, P. J. F., Pan, Y., Jiao, J. J., Gong, H., Li, X., Güntner, A., Zhu, Y., Zhang, C., Zheng, L., 2019. Detection of large-scale groundwater storage variability over the karstic regions in Southwest China. J. Hydrol. 569, 409–422. https://doi.org/10.1016/j.jhydrol.2018.11.071

Huffman, G. J., Adler, R. F., Bolvin, D. T., Gu, G., Nelkin, E. J., Bowman, K. P., Hong, Y., Stocker, E. F., Wolff, D. B., 2007. The TRMM Multisatellite Precipitation Analysis (TMPA): Quasi-global, multi-year, combined-sensor precipitation estimates at fine scales. J. Hydrometeorol. 8, 38–55. https://doi.org/10.1175/JHM560.1

Hulsman, P., Savenije, H. H. G., Hrachowitz, M., 2021. Satellite-based drought analysis in the Zambezi River Basin: Was the 2019 drought the most extreme in several decades as locally perceived? J. Hydrol. Reg. Stud. 34, 100789. https://doi.org/10.1016/j.ejrh.2021.100789

IPCC, 2018. An IPCC Special Report on the impacts of global warming of 1.5°C above pre-industrial levels and related global greenhouse gas emission pathways, in the context of strengthening the global response to the threat of climate change, sustainable development, IPCC - Sr15 2, 17–20.

IPCC. 2022. IPCC 2022: Climate Change 2022: Impacts, Adaptation, and Vulnerability. Contribution of Working Group II to the Sixth Assessment Report of the Intergovernmental Panel on Climate Change. Cambridge University Press. In Press. 1–225.

Jiao, W., Wang, L., McCabe, M. F., 2021. Multi-sensor remote sensing for drought characterization: current status, opportunities and a roadmap for the future. Remote Sens. Environ. 256, 112313. https://doi.org/10.1016/j.rse.2021.112313

Jin, S., Zhang, T., 2016. Terrestrial water storage anomalies associated with drought in Southwestern USA from GPS observations. Surv. Geophys. 37, 1139–1156. https://doi.org/10.1007/s10712-016-9385-z

Khaki, M., Forootan, E., Kuhn, M., Awange, J., Longuevergne, L., Wada, Y., 2018. Efficient basin scale filtering of GRACE satellite products. Remote Sens. Environ. 204, 76–93. https://doi.org/10.1016/j.rse.2017.10.040

Khorrami, B., Gunduz, O., 2021. An enhanced water storage deficit index (EWSDI) for drought detection using GRACE gravity estimates. J. Hydrol. 603, 126812. https://doi.org/10.1016/j.jhydrol.2021.126812

Leblanc, M. J., Tregoning, P., Ramillien, G., Tweed, S. O., Fakes, A., 2009. Basin-scale, integrated observations of the early 21st century multiyear drought in southeast Australia. Water Resour. Res. 45. https://doi.org/10.1029/2008WR007333

Liu, X., Feng, X., Ciais, P., Fu, B., 2020a. Widespread decline in terrestrial water storage and its link to teleconnections across Asia and eastern Europe. Hydrol. Earth Syst. Sci. 24, 3663–3676. https://doi.org/10.5194/hess-24-3663-2020

Liu, X., Feng, X., Ciais, P., Fu, B., Hu, B., Sun, Z., 2020b. GRACE satellite-based drought index indicating increased impact of drought over major basins in China during 2002–2017. Agric. For. Meteorol. 291, 108057. https://doi.org/10.1016/j.agrformet.2020.108057

Lopez, T., Al Bitar, A., Biancamaria, S., Güntner, A., Jäggi, A., 2020. On the use of satellite remote sensing to detect floods and droughts at large scales. Surv. Geophys. 41, 1461–1487. https://doi.org/10.1007/s10712-020-09618-0

Masih, I., Maskey, S., Mussá, F. E. F., Trambauer, P., 2014. A review of droughts on the African continent: A geospatial and long-term perspective. Hydrol. Earth Syst. Sci. 18, 3635–3649. https://doi.org/10.5194/hess-18-3635-2014

Mgquba, S. K., Majozi, S., 2020. Climate change and its impacts on hydro-politics in transboundary basins: A case study of the orange-senqu river basin. J. Water Clim. Chang. 11, 150–165. https://doi.org/10.2166/wcc.2018.166

Mishra, A. K., Singh, V. P., 2010. A review of drought concepts. J. Hydrol. 391, 202–216. https://doi.org/10.1016/j.jhydrol.2010.07.012

Modanesi, S., Massari, C., Camici, S., Brocca, L., Amarnath, G., 2020. Do satellite surface soil moisture observations better retain information about crop-yield variability in drought conditions? Water Resour. Res. 56, e2019WR025855. https://doi.org/10.1029/2019WR025855

Müller Schmied, H., Caceres, D., Eisner, S., Flörke, M., Herbert, C., Niemann, C., Asali Peiris, T., Popat, E., Theodor Portmann, F., Reinecke, R., Schumacher, M., Shadkam, S., Telteu, C. E., Trautmann, T., Döll, P., 2021. The global water resources and use model WaterGAP v2.2d: Model description and evaluation. Geosci. Model Dev. 14, 1037–1079. https://doi.org/10.5194/gmd-14-1037-2021

Munday, C., Washington, R., 2019. Controls on the diversity in climate model projections of early summer drying over southern Africa. J. Clim. 32, 3707–3725. https://doi.org/10.1175/JCLI-D-18-0463.1

Ndehedehe, C. E., Awange, J. L., Agutu, N. O., Okwuashi, O., 2018. Changes in hydro-meteorological conditions over tropical West Africa (1980–2015) and links to global climate. Glob. Planet. Change 162, 321–341. https://doi.org/10.1016/j.gloplacha.2018.01.020

Nigatu, Z. M., Fan, D., You, W., Melesse, A. M., 2021. Hydroclimatic extremes evaluation using GRACE/GRACE-FO and multidecadal climatic variables over the Nile River Basin. Remote Sens. 13, 651. https://doi.org/10.3390/rs13040651

Nkunzimana, A., Shuoben, B., Guojie, W., Alriah, M. A. A., Sarfo, I., Zhihui, X., Vuguziga, F., Ayugi, B. O., 2021. Assessment of drought events, their trend and teleconnection factors over Burundi, East Africa. Theor. Appl. Climatol. 145, 1293–1316. https://doi.org/10.1007/s00704-021-03680-3

Omer, A., Zhuguo, M., Zheng, Z., Saleem, F., 2020. Natural and anthropogenic influences on the recent droughts in Yellow River Basin, China. Sci. Total Environ. 704, 135428. https://doi.org/10.1016/j.scitotenv.2019.135428

Papa, F., Crétaux, J.-F., Grippa, M., Robert, E., Trigg, M., Tshimanga, R. M., Kitambo, B., Paris, A., Carr, A., Fleischmann, A. S., de Fleury, M., Gbetkom, P. G., Calmettes, B., Calmant, S., 2022. Water Resources in Africa under Global Change: Monitoring Surface Waters from Space. Surv. Geophys. 1–51. https://doi.org/10.1007/s10712-022-09700-9

Running, S., Mu, Q., Zhao, M., Moreno, A., 2017. MOD16A2 MODIS/Terra Net Evapotranspiration 8-Day L4 Global 500m SIN Grid V006 [Data set]. NASA EOSDIS L. Process. DAAC 1.5, 34.

Sarfo, I., Shuoben, B., Beibei, L., Amankwah, S. O. Y., Yeboah, E., Koku, J. E., Nunoo, E. K., Kwang, C., 2022. Spatiotemporal development of land use systems, influences and climate variability in Southwestern Ghana (1970–2020). Environ. Dev. Sustain. 24, 9851–9883. https://doi.org/10.1007/s10668-021-01848-5

Satish Kumar, K., Venkata Rathnam, E., Sridhar, V., 2021. Tracking seasonal and monthly drought with GRACE-based terrestrial water storage assessments over major river basins in South India. Sci. Total Environ. 763, 142994. https://doi.org/10.1016/j.scitotenv.2020.142994

Schlosser, C. A., Strzepek, K., Gao, X., Fant, C., Blanc, É., Paltsev, S., Jacoby, H., Reilly, J., Gueneau, A., 2014. The future of global water stress: An integrated assessment. Earth's Futur. 2, 341–361. https://doi.org/10.1002/2014ef000238

Sun, Z., Zhu, X., Pan, Y., Zhang, J., Liu, X., 2018. Drought evaluation using the GRACE terrestrial water storage deficit over the Yangtze River Basin, China. Sci. Total Environ. 634, 727–738. https://doi.org/10.1016/j.scitotenv.2018.03.292

Swenson, S., Wahr, J., 2002. Methods for inferring regional surface-mass anomalies from Gravity Recovery and Climate Experiment (GRACE) measurements of time-variable gravity. J. Geophys. Res. Solid Earth 107, ETG 3–1-ETG 3–13. https://doi.org/10.1029/2001jb000576

Thomas, A. C., Reager, J. T., Famiglietti, J. S., Rodell, M., 2014. A GRACE-based water storage deficit approach for hydrological drought characterization. Geophys. Res. Lett. 41, 1537–1545. https://doi.org/10.1002/2014GL059323

United Nations Environment Programme, 2010. Africa water atlas 1, 1–336.

Vicente-Serrano, S. M., Beguería, S., Gimeno, L., Eklundh, L., Giuliani, G., Weston, D., El Kenawy, A., López-Moreno, J. I., Nieto, R., Ayenew, T., Konte, D., Ardö, J., Pegram, G. G. S., 2012. Challenges for drought mitigation in Africa: The potential use of geospatial data and drought information systems. Appl. Geogr. 34, 471–486. https://doi.org/10.1016/j.apgeog.2012.02.001

Vissa, N. K., Anandh, P. C., Behera, M. M., Mishra, S., 2019. ENSO-induced groundwater changes in India derived from GRACE and GLDAS. J. Earth Syst. Sci. 128, 1–9. https://doi.org/10.1007/s12040-019-1148-z

Wang, H., Chen, Y., Pan, Y., Li, W., 2015. Spatial and temporal variability of drought in the arid region of China and its relationships to teleconnection indices. J. Hydrol. 523, 283–296. https://doi.org/10.1016/j.jhydrol.2015.01.055

Wang, J., Chen, Y., Wang, Z., Shang, P., 2020. Drought evaluation over Yangtze River basin based on weighted water storage deficit. J. Hydrol. 591, 125283. https://doi.org/10.1016/j.jhydrol.2020.125283

Wells, N., Goddard, S., Hayes, M. J., 2004. A self-calibrating Palmer Drought Severity Index. J. Clim. 17, 2335–2351. https://doi.org/10.1175/1520-0442(2004)017<2335:ASPDSI>2.0.CO;2

West, H., Quinn, N., Horswell, M., 2019. Remote sensing for drought monitoring & impact assessment: Progress, past challenges and future opportunities. Remote Sens. Environ. 232, 111291. https://doi.org/10.1016/j.rse.2019.111291

Wu, T., Zheng, W., Yin, W., Zhang, H., 2021. Spatiotemporal characteristics of drought and driving factors based on the GRACE-derived total storage deficit index: A case study in Southwest China. Remote Sens. 13, 79.

Yang, P., Xia, J., Zhan, C., Qiao, Y., Wang, Y., 2017. Monitoring the spatio-temporal changes of terrestrial water storage using GRACE data in the Tarim River basin between 2002 and 2015. Sci. Total Environ. 595, 218–228. https://doi.org/10.1016/j.scitotenv.2017.03.268

Zhang, Y., He, B. I. N., Guo, L., Liu, D., 2019. Differences in response of terrestrial water storage components to precipitation over 168 global river basins. J. Hydrometeorol. 20, 1981–1999. https://doi.org/10.1175/JHM-D-18-0253.1

14 Flood Disaster Monitoring and Extraction Based on SAR Data and Optical Data

Minmin Huang and Shuanggen Jin

Flood is one of the most frequent natural disasters and normally causes large property damage and life losses. Therefore, it is important to monitor the flood disaster by remote sensing in time. SAR images have very high accuracy in water extraction. The vegetation index difference method (MSAVI method) and CDAT method can be used to extract vegetation inundated areas, but these two methods have their emphasis and limitations. In this chapter, flood disaster is extracted and mapped based on SAR data and optical data. The flood disaster monitoring method based on image classification has higher recognition accuracy than other methods in the completely inundated area and vegetation inundated area and also identifies a large number of other land types, which is greatly improved when compared with the three commonly used methods.

14.1 INTRODUCTION

Flood is one of the most frequent natural disasters and normally causes large property damage and life losses. Therefore, it is important to monitor the flood disaster by remote sensing in time. From the perspective of remote sensing data sources, flood remote sensing monitoring can be divided into flood extraction based on optical images, flood extraction based on SAR images, and flood extraction based on optical images and SAR images. There are two methods for flood extraction based on optical and SAR images. One is used after data combination, and the other is used after data fusion. The combination uses the bands and products of SAR and optical satellites directly to form a new layer group without other fusion processing[1]. After fusion, a set of new information or synthetic images are generated by a certain algorithm, and the original data and the fused data are used for water or inundated area extraction[2–3].

From a methodological point of view, the commonly used methods of flood remote sensing monitoring are mainly divided into supervised and unsupervised methods. The supervised method mainly uses image classification and image recognition technology to extract water or inundated area[4–6]. Neural network[7], support vector machine[8], and other methods have been applied to flood disaster extraction. There is no essential difference in image recognition based on the three types of data sources. This method relies on the quantity and quality of sample data and is rarely used in current research. Moreover, because there are fewer samples in the flooded area, the recognition targets of this kind of method are mostly water.

The commonly used threshold determination methods include the empirical method, the histogram bimodal[9] method, and OTSU[10–11]. The threshold method is most widely used because of its fast speed and simple principle. Its basic principle is to mark the image into two categories of water and non-water by setting a reasonable threshold according to the radiation characteristics of water to form a binary image[12]. Different satellites choose different data products when using the threshold method to extract water or inundated area according to their different band characteristics. When using optical data and threshold judgment methods to extract water or

DOI: 10.1201/9781003363118-14

waterlogging area, the commonly used data products are a single-band or multi-band calculation method of the original optical image (spectral index method). Water has high absorptivity and low reflectivity in the near infrared-short wave infrared band and low brightness in the image. According to its radiation characteristics, the near-infrared band is usually selected as the data source of the single-band threshold method based on optical data, and the segmentation thresholds of water and non-water are determined by appropriate threshold methods[13]. For optical images, the multi-band algorithm is used in more methods. The multi-band algorithm mainly constructs spectral index by band operation to extract water. The water index method is used more. Water information is highlighted by the band relationship to extract water. In addition to the water index, the vegetation index is also applied to the extraction of flood disasters. Based on the study of the long-term sequence NDWI (Normalized Difference Water Index) index[14], the vegetation index time series model is established, and the disaster judgment threshold is determined based on the variation law of the vegetation index time series. The mutation detection of the vegetation growth period is carried out to determine the flood range, which is suitable for large-scale and long-term agricultural disaster monitoring. In Reference[15], the vegetation area inundation extraction model based on vegetation index difference and empirical threshold method was proposed.

Different from the abundant band information of optical satellites, SAR data bands and products are relatively single. Since the water has a very low scattering coefficient on the SAR image and presents a dark tone[16], this feature is an important basis for distinguishing water and other ground objects. It has experienced the process from single threshold to multi-thresholds and realizes the image recognition from a single water to some specific types of inundated areas. The inundated detection of the single-threshold method is based on mathematical theory, and appropriate thresholds are selected to divide inundated and non-inundated areas. Most studies use appropriate methods to extract the open water bodies before and during disasters[17–18], and then change detection is carried out on the water bodies before and during the disaster. Reference[19] proposed a two-step automatic change detection method based on Sentinel-1 data, which can effectively mention the change of open water in the suburbs. This method only considers the areas completely inundated in as open water bodies and cannot detect areas with slight waterlogging; in Reference[20], an adaptive threshold method was proposed to improve the accuracy of real-time detection of water in SAR images and was applied in Huainan area. The results show that this method can quickly, effectively, and accurately extract water in large-scale SAR images. However, due to the complexity of the surface, there are different changes after inundation, and a single threshold cannot fully cover the inundation and non-inundation areas of various types of ground objects. To overcome the defects of a single threshold, a multi-threshold extraction model of complex inundated area was born: with the threshold method as the core, based on a single open water extraction, the extraction model of other typical categories was added, or the water extraction was carried out at multi-level scales[21–22]. In Reference[23], based on Sentinel-1 data, two types of inundated area extraction models for open water and vegetation area were constructed and applied to flood monitoring in the lower reaches of the Ilowadi River in Myanmar, and good results were obtained in the extraction of the cultivated land affected area. This method is often used to increase the corresponding land cover inundation extraction model for certain areas, which has a good effect on specific problems, but it is still not comprehensive. In Reference[24], the difference value was calculated by using the bands before and during the disaster, and then the difference value was determined by using the threshold method and the idea of the decision tree to extract the complete flooded area and farmland flooded area. In addition, the RGB synthesis method is also one of the commonly used methods. Based on the principle of RGB color synthesis, the radar images in pre-disaster are used as RGB bands to extract color anomaly areas[25]. The core of this method is the RGB color model, so the data processing makes it suitable for the RGB model with high requirements, and the data processing is subjective.

At present, there are two ways to extract the inundated area based on remote sensing: image recognition first and then change detection, andchange detection first and then image recognition. The supervisory method usually first carries on the image recognition, then carries on the change detection to the classified image. The threshold method is widely used in both ideas. In References[17–20], the threshold method is widely used in both ideas. In the literature, the image before and during the disaster is segmented by a threshold, and then the segmented image is changed. In References[14][24–26], the change information is obtained first, and then the appropriate method is selected to recognize the change detection results.

When the flood disaster extraction target based on SAR data is water, it only pays attention to the change of water area caused by flood and does not pay attention to the land-related information transformed into water. It only needs to know the scattering characteristics of water. When the extraction target is not limited to a single water extraction, the change of backscattering coefficient caused by inundation is important information to guide the extraction.

14.2 SATELLITE REMOTE SENSING DATA AND PROCESSING

14.2.1 SENTINEL-1 DATA

Sentinel-1, the first satellite in the European Space Agency (ESA)'s Copernicus program for environmental monitoring, is composed of A and B satellites. Sentinel-1A was launched on April 3, 2014 and Sentinel-1B was launched on April 25, 2016. The single-satellite revisit period is 12 days, and the double-satellite revisit period is 6 days. It is equipped with a C-band SAR sensor to monitor changes in the global ocean environment, surface deformation, and dynamic changes in terrestrial forests, surface water, and soil. Its orbit features are divided into ascending and descending orbits, distributed through the results in Level 0, Level 1, and Level 2. Level 0 data refers to unprocessed compressed data and its related information, which is the basic data for producing other high-level data. Level 1 data is a product of Level 0, which can be directly used by most users after internal calibration and Doppler centroid estimation. According to the processing method, it is divided into Single Look Complex (SLC)[27] and Ground Range Detected (GRD)[28]. Level 2 data is a geo-positioning geophysical product derived from Level 1 data. Sentinel-1 products in four imaging modes can provide radar images of different polarization modes, such as single polarization (HH, VV, HV, VH), and dual polarization (HH&HV, VV&VH)[29].

There are four data acquisition modes for Sentinel-1 data (Table 14.1): stripmap mode (SM), interferometric wide swath (IW), extra wide swath (EW), and wave mode (WM)[30].

(1) Interferometric wide swath

IW mode is the default mode for Earth observation, using a 250-km width and 5 m × 20 m combination of resolutions. The IW mode uses a progressive terrain scan method (TOPSAR, Terrain

TABLE 14.1

Four Imaging Modes and Parameters of Sentinel-1

Parameter Operating Mode	Incidence Angle	Resolution	Width	Polarization Mode
SM	20~45°	5×5 m	80 km	H H+HV, VH+VV, HH, VV
IW	29~46°	5×20 m	250 km	H H+HV, VH+VV, HH, VV
EW	19~47°	20×40 m	400 km	H H+HV, VH+VV, HH, VV
WM	22~38°	5×5 m	400 km	HH, VV

Observation with Progressive Scans SAR) to acquire three sub-strips. TOPSAR technology can ensure uniform image quality across the entire frame[31].

(2) Strip mode

SM mode is mainly used for emergency management in special situations. A strip imaging mode is available, which can be used for the continuity of ERS and ENVISAT tasks. The SM mode provides a resolution of 5 m × 5 m on an 80-km swath. The radar antenna can be adjusted to change the beam incident angle and elevation beam width. The main features of this mode are selectable incident angles and high resolution.

(3) Extra wide swath

EW mode is suitable for areas such as oceans and polar regions, where the coverage is wide and the revisit period can be low. The technology used in the EW mode is similar to the TOPSAR technology used in the IW mode. It uses five strips instead of three, so the resolution is lower, 20 m × 40 m. EW mode can also be used for interferometry like IW mode.

(4) Wave mode

The Wave mode, combined with a global ocean wave model, can help determine the direction, wavelength, and height of waves in the ocean. The waveform pattern consists of 20 km × 20k m strip images obtained alternately at two different angles of incidence. The strip patch switches the angle of incidence every 100 km.

The example data used in this book are GRD ground distance products of Level 1 in the IW mode from Sentinel-1A and B satellites. The IW mode has a width of 250 km and a spatial resolution of 5 m × 20 m. After processing, it can generate 10 m × 10 m products. The data is subjected to focus processing and slant distance-to-ground distance conversion, and its pixel value represents the amplitude information of the target object while the phase information is lost.

14.2.2 Sentinel-2 Data

The Sentinel-2 satellite is a satellite project jointly managed by the European Commission (EC) and ESA and will directly make an important contribution to global land monitoring, emergency response, and security services. Sentinel-2 satellite consists of two satellites A and B: Sentinel-2A and Sentinel-2B were launched on June 23, 2015, and March 7, 2017, respectively[32]. Sentinel-2 operates on a sun-synchronous orbit with an on-orbit altitude of 786 km and an orbital inclination of 98.5°. The local time for satellite imaging is 10:30. The satellite adopts three-axis attitude control, which can achieve superior accuracy and stability[33].

The revisit period of the Sentinel-2 satellite is 10 days for a single star and 5 days for a double star[34]. Sentinel-2 carries a multi-spectral imager instrument, which can provide data in 13 bands (Table 14.2), including visible light, near-infrared, and short-wave infrared, with ground resolutions of 10 m, 20 m, and 60 m, respectively[35], and the spatial resolution of different bands is different[36–37]. The Sentinel-2 is the only satellite with three bands in the red edge range, which can be used for vegetation health detection[38] and can also be used for spatial planning, agricultural environment monitoring, water monitoring, forest vegetation monitoring, etc. Sentinel-2 provides a total of four types of products: raw data Level 0, geometric rough correction products containing meta-information Level 1A, embedded GCP-optimized geometric models without corresponding geometric corrections radiance products Level 1B, and Level 1C of the atmospheric apparent reflectance product after orthorectification and sub-pixel geometric fine-tuning.

TABLE 14.2

Band Parameters of Sentinel-2 Satellite

Band	Center Wavelength (µm)	Resolution (m)
Band1 - Coastal aerosol	0.443	60
Band2 - Blue	0.490	10
Band3 - Green	0.560	10
Band4 - Red	0.665	10
Band5 - Vegetation red edge	0.705	20
Band6 - Vegetation red edge	0.740	20
Band7 - Vegetation red edge	0.783	20
Band8 - NIR	0.842	10
Band8A - Vegetation red edge	0.865	20
Band9 - Water vapor	0.945	60
Band10 - SWIR-Cirrus	1.375	60
Band11 - SWIR	1.610	20
Band12 - SWIR	2.190	20

14.2.3 Data Preprocessing

14.2.3.1 Sentinel-1 Data Preprocessing

There are two main methods for data preprocessing of Sentinel-1 GRD products. One is to use the online data set and online data processing platform of Google Earth Engine platform for processing. This method does not need to download the original product, which is suitable for large-scale research. The other is to download Sentinel-1 products to the local area and use professional desktop software for processing. In this paper, the second method is used for data processing. The data processing software is Sentinel Application Platform (SNAP) provided by ESA, which can be used to process Sentinel-1, Sentinel-2, Sentinel-3, and other software.

The main steps of Sentinel-1 GRD product pretreatment using SNAP software (Figure 14.1) are orbit correction, radiometric calibration, speckle filtering, multi-view processing, orthophoto correction (or terrain correction)[39], and logarithmic processing[40].

(1) Orbit correction. Sentinel-1 orbit correction is generally divided into coarse orbit correction and fine orbit correction. The coarse orbit information is included in the metadata of SAR products, and the accuracy is within 10 cm. The precision orbit file is generally released on the official orbit network after 21 days of product generation. The precision orbit file provides accurate satellite position and velocity information, and the position accuracy is within 5 cm after the correction of the precision orbit. When processing the data, the software will automatically retrieve whether there is a required track file under

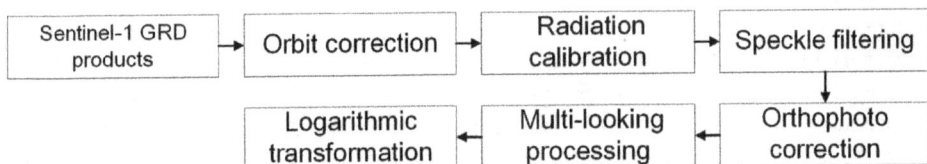

FIGURE 14.1   Sentinel-1 data preprocessing process.

the specified track file directory. When there is no track file corresponding to the processing product, SNAP software will automatically download the track file according to the SAR product metadata in the state of networking. It can also download the file of the corresponding date from its official track website and put it in. When there is a track file in the corresponding directory, it does not need to download many times.

(2) Radiation calibration. The radiometric calibration corrects the parameters of the satellite sensor and establishes the accurate relationship between the SAR image and the backscattering of the ground objects. Thus, the pixel value of the image represents the backscattering calibration vector of the reflector radar, and the digital pixel value is converted to the radiometric calibration backscattering, which allows the image intensity value to be converted to Sigma or Gamma zero value simply[41].

(3) Speckle filtering. Speckle noise is a major feature of SAR imaging, which is manifested as small spots on the image, reducing the quality of the image and affecting image segmentation and information extraction. In order to reduce the impact of speckle noise, a filtering algorithm is needed to reduce the impact of speckle noise. The commonly used filtering algorithms include the median filtering algorithm, Boxcar filtering algorithm, Lee filtering algorithm, refined Lee filtering algorithm, Frost filtering algorithm, and Gamma MAP filtering, among which the refined Lee filtering algorithm is used more.

(4) Orthophoto correction (or terrain correction). Since the imaging mode of SAR is side-looking imaging, large terrain fluctuation will lead to large geometric distortion of SAR image, leading to perspective contraction, overlapping, shadow, and other phenomena. It is necessary to carry out orthorectification of radar images to compensate for these distortions so that the geometric expression of the image is more likely to be close to the real situation. Orthophoto correction uses digital elevation model (DEM) to correct each pixel caused by the terrain geometry. SRTM DEM is selected for general DEM data.

(5) Multi-looking processing. The purpose of averaging the resolutions in the range and azimuth directions of the image is to obtain the same ground resolution (ground distance resolution) in the range and azimuth directions and improve the quality of the image. The determination of the number of views is an important part of multi-view. With the increase in the number of views, the speckle noise decreases and the image resolution decreases. The number of views is calculated according to the range resolution, azimuth resolution, and central incidence angle of oblique distance. Multi-view processing improves radiation resolution by reducing spatial resolution.

(6) Logarithmic transformation. The numerical range of the scattering coefficient is very different. To make the range controllable, researchers have established a series of data processing models. Sonic shell conversion is one of the commonly used methods. In this paper, the backward scattering coefficient is converted by the formula (14.1)[42].

$$\sigma^0 = 10 \times \log_{10} \sigma \tag{14.1}$$

Where σ is the original backscattering coefficient, and σ^0 is the transformed backscattering coefficient.

14.2.3.2 Sentinel-2 Data Preprocessing

The ESA's official website only provides the download of L1C data, which is the atmospheric apparent reflectance product after orthophoto correction and geometric precision correction, without atmospheric correction. The ESA defines the atmospheric bottom reflectance data after atmospheric correction as L2A, and the L2A data is produced by users themselves. The ESA provides the Sentinel-2 Sen2Cor process plug-in for L1C product preprocessing [43], which can be downloaded from the ESA's official website at present. The batch processing of Sentinel-2 can be carried out by using the Windows command line, and the obtained L2A data can be processed in SNAP for subsequent processing. The data are imported into SNAP software. In this paper, only bands 2, 3, 4, and 8 with 10-m resolution are used to export the preprocessed data to ENVI format data.

14.2.4 REMOTE SENSING DATA PROCESSING

14.2.4.1 Image Fusion

Different remote sensing sensors usually use different band intervals to record pixel information, and there are differences in spectral, spatial, and temporal resolutions[44]. Due to its wavelength characteristics, optical remote sensing is easily affected by cloudy and rainy weather, and the image quality is poor during floods, but in good weather conditions, it contains a large amount of spectral information, which enables it to classify objects with high accuracy. SAR images have the characteristics of all-day and all-weather, but due to their imaging characteristics, they are less effective in classifying ground objects. The flood disaster monitoring model in this study takes image classification and change detection as the core and SAR as the main data source. To provide the classification accuracy of SAR images, a certain algorithm is used to fuse Sentinel-1 and Sentinel-2 data.

Image fusion refers to the use of appropriate algorithms to combine the data of different sensors and different bands and combine the features of the original image and the information of the ground objects to generate a higher-performance remote sensing image[45–46]. Image fusion is divided into three categories: pixel-level fusion, feature-level fusion, and decision-level fusion[47–48]. Pixel-level fusion is the fusion of the basic level, which directly acts on the pixel information of the image to obtain new image pixels. Feature-level fusion uses the features extracted from the original image to perform fusion processing without changing the original image, which loses metaphysical meaning, usually for specific scenes and purposes; decision-level fusion is the highest-level fusion, based on the original image. Interpreting the results and synthesizing new results using certain decision-making methods can optimize the detection results of a single image to a certain extent. According to the results of change detection, the trend of the change of the scattering coefficient of the pixel is judged to infer whether there is water accumulation. Therefore, this study uses pixel-level fusion. Pixel-level fusion methods are mainly based on component substitution, multi-scale variation, model base, and their combinations.

14.2.4.2 Image Classification

Image classification is to divide each pixel in the image into different categories according to certain rules or algorithms according to its spectrum, spatial structure characteristics, or other information in different bands. Image classification can be divided into supervised classification and unsupervised classification according to whether there are samples in the classification process. It involves the type, distribution, and amount of change that is necessary to determine the ground type, boundary, and change trend before and after the change and then analyze the characteristics and causes of these dynamic changes.

Supervised classification is a commonly used statistical judgment classification with high accuracy. Various training samples are extracted from training venues of known categories, and each pixel point in the image is classified by selecting characteristic variables and determining discriminant functions or discriminant rules. A classification method is assigned to each given class[49]. Commonly used supervised classification methods are K-nearest neighbor, decision tree classifier, and Bayesian classifier. The main steps include 1) selecting the characteristic band; 2) selecting the training area; 3) selecting or constructing the training classifier; and 4) evaluating the classification accuracy. There are many deficiencies in the supervised classification algorithm in solving the classification problem; either it can only solve the linear problem, or it can solve the nonlinear problem, but the computational complexity is high, and the efficiency cannot meet the requirements. For this reason, the research hotspots in recent years—the three basic theories of support vector machines

(SVM) (structural risk theory, quadratic optimization theory, and nuclear space theory) are used to solve nonlinear problems[50].

Unsupervised classification is a classification process that divides the category of objects according to the statistical characteristics of the image itself and the distribution of natural point groups without prior category knowledge. Unsupervised classification methods are based on the statistical features of images and do not require known knowledge of specific objects. Using unsupervised classification can also better obtain the inherent distribution law of target data. Unsupervised classification methods include Bayesian learning, maximum likelihood classification, and clustering. The unsupervised Bayesian method and the maximum likelihood method are basically the same as the supervised Bayesian learning and the maximum likelihood method; the only difference is that the previous methods do not require training samples, while the subsequent methods require training samples.

14.2.4.3 Change Detection

Remote sensing change detection uses multi-temporal remote sensing data, uses a variety of image processing and pattern recognition methods to extract change information, and quantitatively analyzes and determines the characteristics and processes of surface changes. For remote sensing change detection, data selection is performed first, and then data preprocessing is performed, including radiometric correction, geometric correction, and image matching. The key task is to select appropriate information extraction and separation methods to detect the changing information and finally, evaluate the accuracy of the results[51]. Change information extraction is the core of remote sensing image change detection. The extraction results play a feedback role on the results of data preprocessing, and also play a decisive role in the accuracy of change detection results and subsequent processing. It can be divided into pixel-based change information extraction, object-based change information extraction, and fusion change information extraction. The pixel-based change detection algorithm uses pixels as the processing unit. This type of algorithm generally directly compares the input image with pixel-level spectral features, texture features, and other specific features (water, vegetation index, etc.) The difference image is obtained by processing the ratio and so on, and then the change information is extracted by the method of threshold segmentation. Some algorithms use each pixel in the image to compare one by one and then use a classifier or machine learning to detect changes. Among them, judging whether the pixel has changed is the key step, and its core is threshold segmentation and classification. Object-oriented change detection is based on image segmentation and classification. It integrates the spatial and spectral characteristics around pixels, combines homogeneous pixels to form objects, and then uses the object as a unit to perform spectral features, shape features, and texture features, spatial context neighborhood relationship and features with practical significance (such as vegetation coefficient, etc.) are compared to detect changes. This kind of method is the mainstream method of change detection at present and has a wide range of applications in building change detection, urban land management, and so on. Object-oriented change detection and pixel-based change detection are two different methods, but there are some similarities and intersections in the detection link, segmentation, and classification extraction.

14.3 MAPPING FLOOD DISASTERS FROM REMOTE SENSING

14.3.1 THRESHOLD DETERMINATION METHOD

14.3.1.1 Histogram Bimodal Method

The basic idea of the bimodal histogram method is that when the histogram has typical bimodal characteristics (Figure 14.2), the pixel value corresponding to the lowest point between the bimodal peaks is the segmentation threshold[52]. This method is simple and convenient, but there are higher requirements on the distribution ratio of target pixels and background pixels on the image.

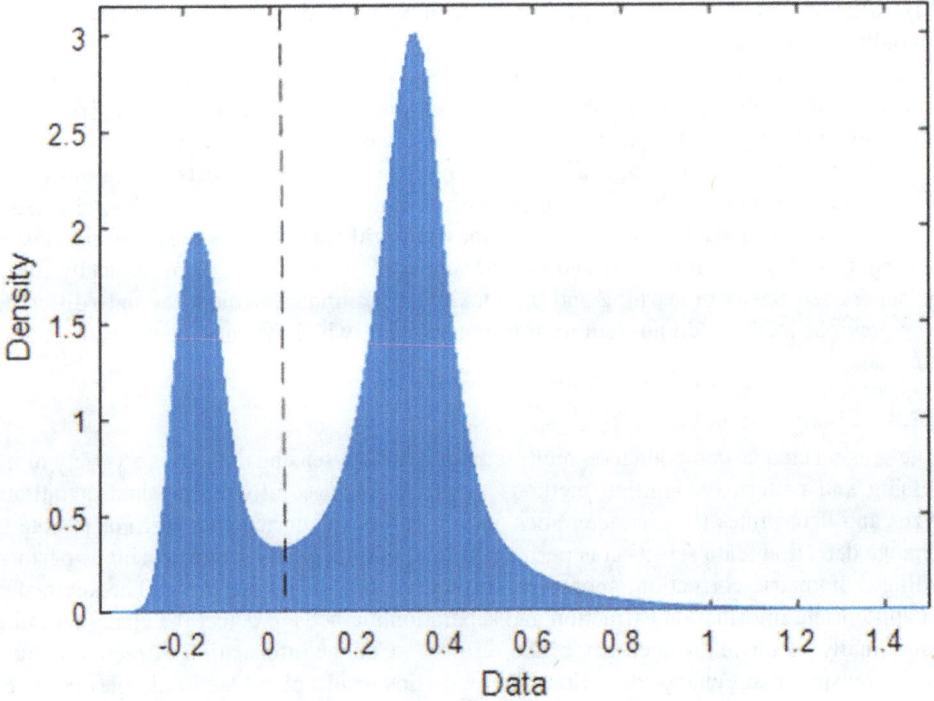

FIGURE 14.2   Pixel bimodal distribution histogram.

14.3.1.2   Maximum Inter-Class Difference Method

The OTSU method is one of the most commonly used threshold methods, also known as the maximum inter-class difference method. The OTSU threshold method divides the image into the target area and the background area with a single threshold and uses the inter-class variance as the standard to measure the difference between the two. When the variance is the largest, the difference between the two is considered to be the largest, and the threshold at this time is selected as the best threshold[53]. The specific algorithm is:

Assuming that the grayscale range of the image is [0, T] and the number of pixels $f(T_i)$ corresponding to the gray level is T_i, then the total pixel N of the image is

$$N = f(0) + f(1) + \ldots + f(T) = \sum_{i=0}^{T} f(T_i) \tag{14.2}$$

Suppose P_i is the probability of gray-level pixels in the image; that is, $P_i = {f(T_i)}/{N}$. The pixels in the image are divided into two parts, A and B, according to the gray level t. Then the probability ω_A and ω_B of the two parts appearing in the whole image are

$$\omega_A = \sum_{i=0}^{t} P_i \tag{14.3}$$

$$\omega_B = \sum_{i=t+1}^{T} P_i = 1 - \omega_A \tag{14.4}$$

The average gray value μ_A and μ_B of A and B are

$$\mu_A = \sum_{i=0}^{t} i \left(P_i / \omega_A \right) \tag{14.5}$$

$$\mu_B = \sum_{i=t+1}^{T} i \left(P_i / \omega_B \right) \tag{14.6}$$

The average gray value μ of the whole image is

$$\mu = \sum_{i=0}^{T} T_i \times P_i = \omega_A \times \mu_A + \omega_B \times \mu_B \tag{14.7}$$

The between-class variance σ^2 is

$$\sigma^2 = \omega_A \left(\mu_A - \mu \right)^2 + \omega_B \left(\mu_B - \mu \right)^2 = \omega_A \omega_B \left(\mu_A - \mu_B \right)^2 \tag{14.8}$$

In the range of [0,T], the step length of 1 is used to increase the threshold t in turn, which is the best threshold when σ^2 reaches the maximum.

14.3.2 Multi-Band Calculation Method

The multi-band calculation method is also called the spectral index method. The spectral index is mainly constructed through the mathematical calculation of the bands.

14.3.2.1 Water Index Method

The water index method uses the high absorption rate of the water in the near-infrared band, the mid-infrared band, and the high reflectivity in the green band to highlight the water information through band calculation[54–55]. Commonly used water indices include the Normalized Difference Water Index (NDWI)[56] and the Modified Normalized Water Index (MNDWI)[57]. McFeeters first proposed NDWI in 1996[58], using green wave and near-infrared wavebands to suppress the influence of vegetation and other backgrounds on water bodies, highlight water information, and calculate the difference between pre-disaster and post-disaster NDWI with a reasonable threshold. The improved normalized water index comprehensively considers the influence of soil, buildings, and other ground objects on water extraction, which can better reveal the fine characteristics of water, and is more suitable for water extraction in urban areas.

$$NDWI = \frac{Green - NIR}{Green + NIR} \tag{14.9}$$

$$MNDWI = \frac{Green - MIR}{Green + MIR} \tag{14.10}$$

14.3.2.2 Vegetation Index Method

The vegetation index can reflect vegetation cover and growth status[59]. Flood inundation or erosion will affect vegetation status. Many scholars also use the vegetation index to study flood extraction[14–15]. The normalized vegetation index, enhanced vegetation index, green normalized vegetation index, ratio vegetation index, and improved soil adjustment vegetation index are commonly used. Detailed information

TABLE 14.3

Detailed Information of Each Vegetation Index

Name	English and Abbreviations	Calculation Formula	Serial Number
Normalized Vegetation Index[60–61]	NDVI (normalized difference vegetation index)	$\dfrac{\text{NIR-Red}}{\text{NIR+Red}}$	(14.11)
Enhanced vegetation index[62]	EVI (enhanced vegetation index)	$\dfrac{2.5 \times\ \text{NIR-Red}}{\text{NIR} + 6 \times \text{Red} - 7.5 \times \text{Blue} + 1}$	(14.12)
Green Normalized Vegetation Index[63–64]	G NDVI (green normalized difference vegetation index)	$\dfrac{\text{NIR-Green}}{\text{NIR+Green}}$	(14.13)
Ratio Vegetation Index[65]	RVI (ratio vegetation index)	NIR / Red	(14.14)
Improvement of Soil to Adjust Vegetation Index[66–67]	MSAVI (modified soil adjusted vegetation index)	$\sqrt{(2 \times \text{NIR}+1)^2 - 8 \times \dfrac{\text{NIR-Red}}{2}}$	(14.15)

of each index is shown in Table 14.3. RVI, NDVI, and MSAVI, three vegetation index differences, were used to determine the scope of flood inundation[15]; the greater the difference in vegetation index shows that the greater the change of surface coverage or vegetation growth before and after the flood, the more likely to experience a flood. The difference results are extracted by an empirical method.

14.3.3 CHANGE DETECTION AND THRESHOLDING METHOD

The change detection and thresholding (CDAT) method first obtains the change information of the SAR images before and during the disaster and then analyzes the change results to determine the inundation range. References[23][68] use this method to extract the inundated areas in vegetation. The detailed processes are as follows: firstly, the absolute values of the SAR image products before and during the disaster are calculated, and then the difference between the absolute value products is calculated. Then, the idea of a decision tree and histogram method are used to determine several segmentation thresholds of difference products. Using the terrain data, according to the slope value, to eliminate part of the error caused by the terrain, when the terrain in the study area is flat, this part of the error can be ignored. Remove the 0 value; this part of the area did not flood.

Then, according to formula (14.16) and formula (14.17), the segmentation thresholds P_{D1} and P_{D2} for common inundated area and vegetation inundated area are determined respectively:

$$P_{D1} < (mean_{CD} - k_{f1} * dev_{CD}) \tag{14.16}$$

$$P_{D2} > (mean_{CD} + k_{f2} * dev_{CD}) \tag{14.17}$$

In formulas (14.16) and (14.17), $mean_{CD}$ is the mean value of difference change results, dev_{CD} is the standard deviation of the difference change result, k_{f1} and k_{f2} are the coefficients for determining the two thresholds, respectively. The experience points in reference[11] are 1.5 and 2.5, respectively.

Finally, the image is segmented according to the determined threshold, the image is divided into inundated area and vegetation inundated area, the inundated area is expanded according to the adjacent cluster analysis, and the inundated area is expanded according to the number of adjacent pixels (4 or 8) and the minimum pixel of each group. The number of pixels is divided, and the recognition rate of the inundated area after the adjacent cluster analysis is improved.

14.3.4 SUPERVISED CLASSIFICATION OF FLOOD EXTRACTION

Radar images reflect the reflection characteristics of ground objects, and optical images reflect the optical characteristics of ground objects in different wavelength bands. Radar images can sensitively reveal changes in the reflection characteristics of ground objects, and optical images can be used to classify ground objects with high precision. The supervised classification flood extraction and analysis method[69] uses radar images as the main data and optical images as auxiliary data to quickly assess the degree of flood inundation. Select the high-quality optical images closest to the disaster event, and use the supervised classification method to perform optical classification of the ground objects in the study area. Based on optical classification, the radar images before and during the disaster were added to supervise the classification again, and the classification images of the pre-disaster and during the disaster after adjustment according to the scattering characteristics of the ground objects were obtained. Information on changes in the classification of ground objects caused by floods can be obtained. Taking the backscattering coefficients change rules of various objects as prior knowledge and the average backscattering coefficients of objects as a reference, all objects' backscattering coefficients without inundation are regarded as different inundated degrees, and the water is completely inundated. The inundated area and degree can be determined by analyzing the change process of the classification of ground objects.

The flood extraction and analysis method for supervised classification include the following steps: (1) analyze and determine the basic types of ground objects in the study area; (2) perform optical classification of pre-disaster ground objects; (3) process the pre-disaster image group and the disaster image group; (4) supervise and classify the pre-disaster image group and the image group during the disaster; (5) count the average scattering rate of various ground objects and arrange them in ascending order; (6) determine the variation law of scattering character-istics of various ground objects with different inundated degrees; (7) number the objects in sequence; (8) detect changes in the classification of the objects before and during the disaster; (9) determine the inundation range and level according to the categories of the objects before and after the change.

This method transforms the change detection of the reflection feature caused by inundation into the change detection of the classification feature caused by the change of reflection feature, which avoids the complex threshold determination process of extracting the inundation area. The inunda-tion extraction model based on the classification change of ground object reflection characteristics is established; it provides a rapid evaluation method for flood inundation.

14.4   FLOOD EXTRACTION IN SHOUGUANG CITY

Taking Kouzi Village, Shouguang City, Shandong Province as an example, multi-source data are used to extract floods, and the advantages and disadvantages of different methods are compared.

In August 2018, the study area was affected by typhoon 'Rumbia'. The total rainfall observed at Shouguang Station from August 18 to August 20 was 182.6 mm in 3 days, and the maximum rainfall in 20 days was 120.1 mm. In addition, the flood discharge of the upstream reservoirs in the study area intensified the disaster, and the farmland villages were severely affected. After 25 days, the flood gradually subsided. Before and after this scene, there were six dates, July 27, August 2, August 8, August 14, August 20, and August 26, 2018, in the Sentinel-1 image erosion cover study area. August 20 and August 26 were during the disaster, and the other four dates were after the disaster. Rainfall occurred on August 14 on July 27, and no rainfall occurred during the past 72 hours before August 2. There was 13.4 mm of rainfall during the past 24 hours before August 8, and no runoff was generated, according to the SCS-CN model. The Sentinel-2 image cloud cover was large, and the image quality was poor on August 15 and 20, 2018, and the image quality was high on August 10, 2018.

14.4.1   WATER EXTRACTION BASED ON THRESHOLD METHOD

Using the S_Gamma data of Sentinel-1 GRD product VH polarization on August 20, 2018, the water area was extracted by the OTSU method and bimodal method. The NDWI products based on the satellite data of August 21, 2018, and the empirical method are used to determine the threshold to extract water. The pixel distribution histogram of the two products is shown in Figure 14.3 (a-b). When using SAR data for water extraction, the segmentation threshold of the OTSU method is 0.8756, and the segmentation threshold of the bimodal method is 1.4343. The area extracted by the bimodal method is larger than that of the OTSU water body, and the effect of the bimodal method is better in this case. When the NDWI index method based on the Planet satellite is used to extract water bodies, no research has shown that the NDWI index presents a clear feature similar to the bimodal distribution, which can guide the extraction of the water body. At first, many studies

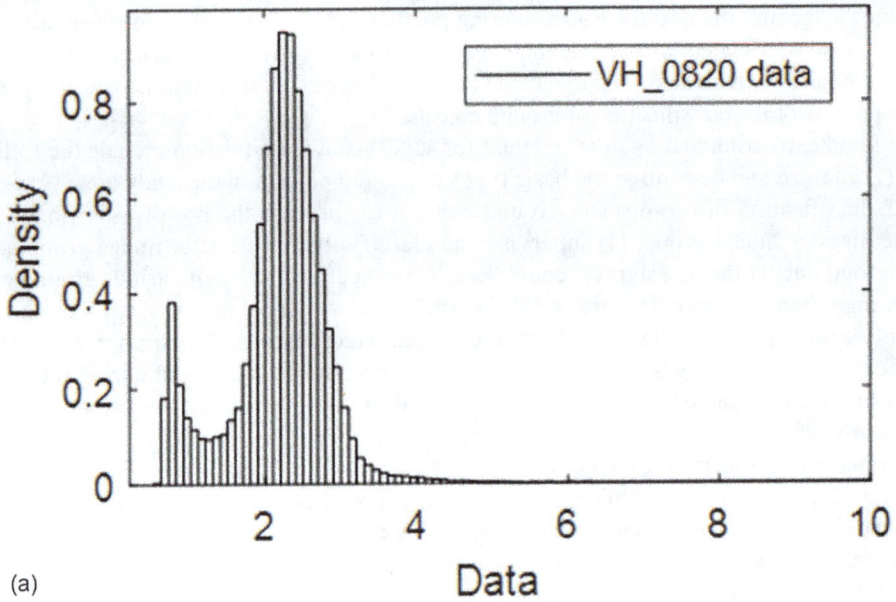

(a)

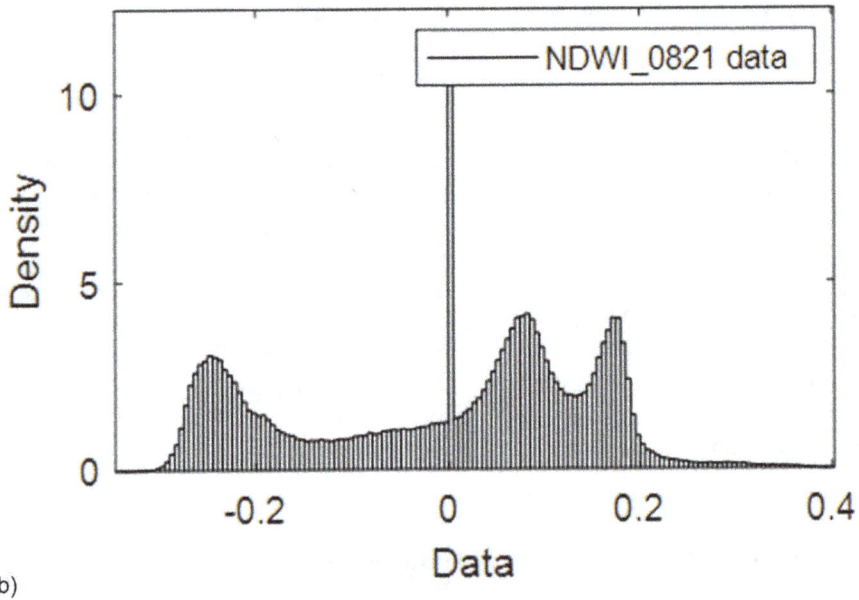

(b)

FIGURE 14.3   Multi-method water extraction results. (*Continued*)

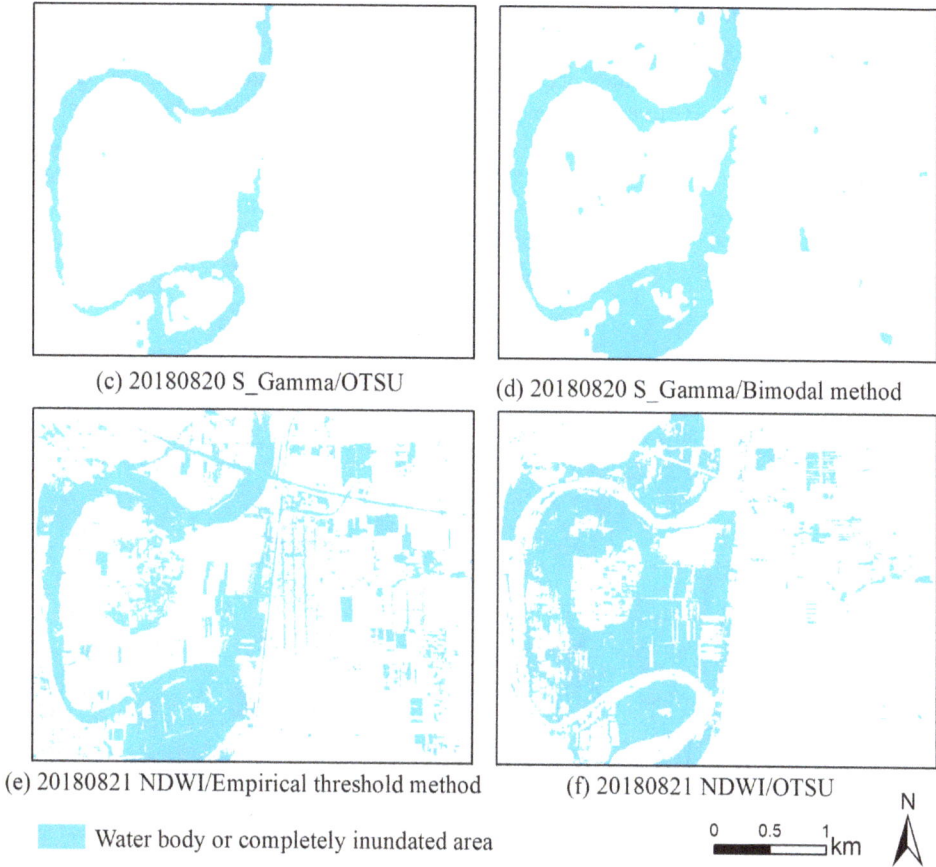

(c) 20180820 S_Gamma/OTSU   (d) 20180820 S_Gamma/Bimodal method

(e) 20180821 NDWI/Empirical threshold method   (f) 20180821 NDWI/OTSU

■ Water body or completely inundated area

0   0.5   1 km   N

FIGURE 14.3 (*Continued*)

directly used 0 as the segmentation value according to the band characteristics. However, with the deepening of the research, it is found that this error is large. In this paper, the optimal segmentation threshold is -0.1869 after many experiments using the empirical method, and the segmentation threshold determined by the OTSU method is 0.0784. From the extraction results, the empirical method is better than the OTSU method for the extraction of water by the optical index method, but the results are still greatly affected by roads, buildings, and other backgrounds. By comparing the results of extracting open water from SAR images and optical images based on the threshold method, it is found that the overall effect of SAR products is better than that of optical images. Therefore, the water results based on VH polarized S_Gamma products on August 20, 2018 and the bimodal method are used as the final results of open water.

14.4.2 EXTRACTION OF INUNDATED AREA BASED ON THE DIFFERENCE BETWEEN VEGETATION INDEXES

The vegetation inundated area was extracted using the vegetation index difference method to obtain the absolute value of the MSAVI vegetation index difference in pre-disaster and post-disaster study areas, and the inundated area was extracted by threshold segmentation of the difference results. This method is the same as water extraction based on the NDWI index; no research shows that the histogram distribution has the significance of guiding change extraction, and the threshold is determined by an empirical method. MSAVI products obtained from Planet images on August 10, 2018 were used as pre-disaster reference. Planet images on August 23 and 27, 2018 were used as post-disaster images, respectively. The image covers the whole area on August 10 and 27, and 68% of the left area of the image covers the study area on August 23. The absolute value of vegetation difference between the two periods after the disaster and that before the disaster is obtained, and the pixel distribution histogram of the results is shown in Figure 14.4 (a-b). Although there are obvious peaks and troughs in the histogram of these two results, it is found that the first trough of the two results indicates the segmentation value of vegetation area and other land types rather than the segmentation value of the inundated area and non-inundated area. The segmentation thresholds of these two products are calculated by the OTSU method, which are 0.3490 and 0.3451, respectively, which are

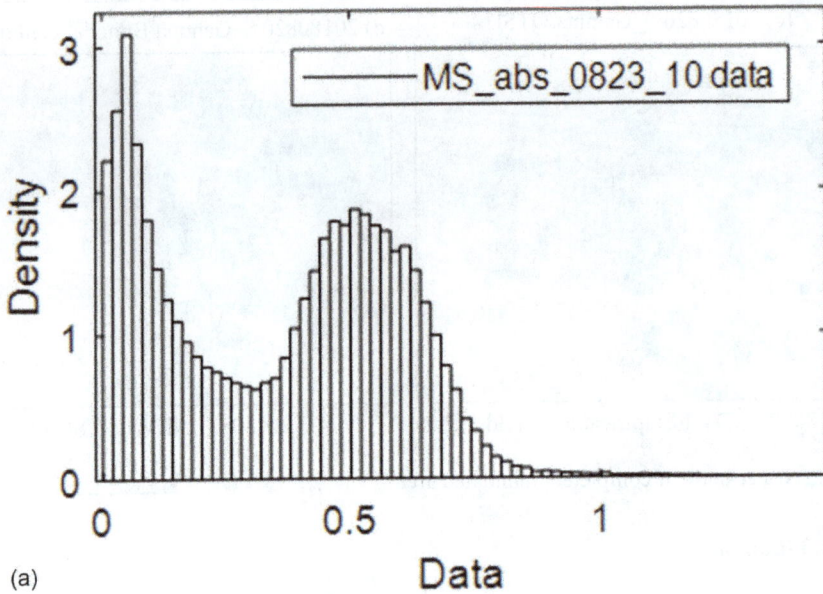

(a)

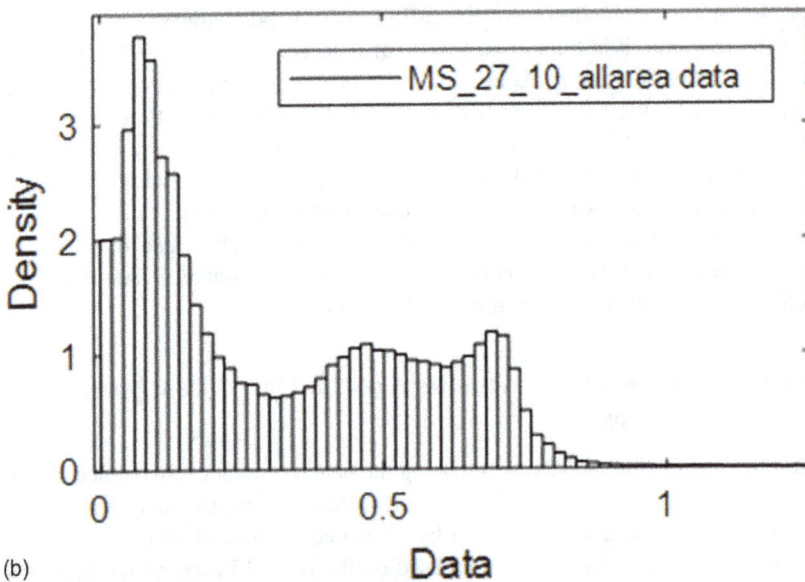

(b)

FIGURE 14.4   Extraction results of vegetation inundated area. (*Continued*)

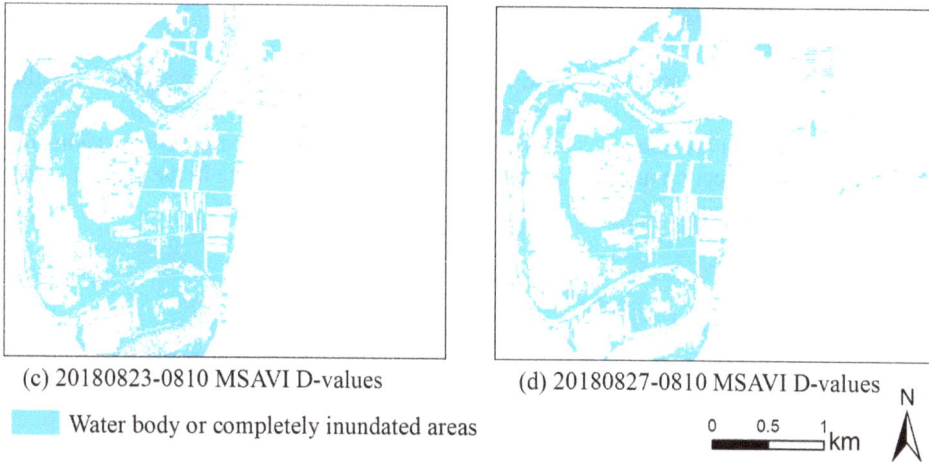

(c) 20180823-0810 MSAVI D-values

(d) 20180827-0810 MSAVI D-values

Water body or completely inundated areas

0   0.5   1 km

N

FIGURE 14.4 (Continued)

close to the first trough, rather than the threshold of dividing the inundated area and non-inundated area. The thresholds of the inundated area and non-inundated area were determined by multiple experiments with the empirical method, and the thresholds of two dates were 0.5174 and 0.6029. The difference between the results extracted from the two dates is small, and the recognition rate is high in the vegetation area. It is impossible to identify in other areas. According to the coverage range of the original image, the results of August 27 are selected as the vegetation-flooded area.

14.4.3 CHANGE DETECTION AND THRESHOLD EXTRACTION

The Sentinel-1 image of August 8, 2018 was used as the reference image before the disaster, and the Sentinel-1 image of August 20, 2018 was used as the disaster image. The change detection and threshold method (CDAT) was used to extract the inundated area. The S_Gamma products of the images before and during the disaster under VH and VV polarization were compared. The histogram and spatial distribution of the difference results are shown in Figure 14.5. The results of CDAT were segmented by the threshold method. The threshold values of the general inundated area and vegetation inundated area under VH polarization were 0.8460 and 0.9706, respectively. The threshold values of the general inundated area and vegetation inundated area under VV polarization were -1.0844 and 2.2604, respectively. According to Figures 14.5 (e) and (f), the recognition rate of vegetation under VH polarization is higher, and the recognition rate of the water body and building area under VV polarization is higher.

14.4.4 SUPERVISED CLASSIFICATION FLOOD EXTRACTION

There are five types of land in the study area: water, shedless farmland, shed farmland, road, and construction area. The Sentinel-2 image on August 10, 2018 was used for random forest classification to obtain the optical classification results of pre-disaster ground objects.

The radar images under different polarization modes in the study area before and during the disaster were pre-processed. That is, all the products were processed by radiation correction, geocoding, spatial filtering, and normalization. In order to increase the data difference, the backscattering coefficient is changed to power according to formula (14.18):

$$S\_Gamma = \sqrt{100 \times Gamma} \tag{14.18}$$

Gamma is the normalized backscattering coefficient, and S_Gamma is the changed backscattering coefficient product.

The radar images before and during the disaster are combined with the optical classification results of ground objects to form two-layer groups, and the scattering characteristics of ground objects before the disaster are taken as samples for random forest classification supervision and classification. The sample selection needs to ensure that the ground objects are in the state of no rainfall and that the results of Figure 14.6 are obtained respectively.

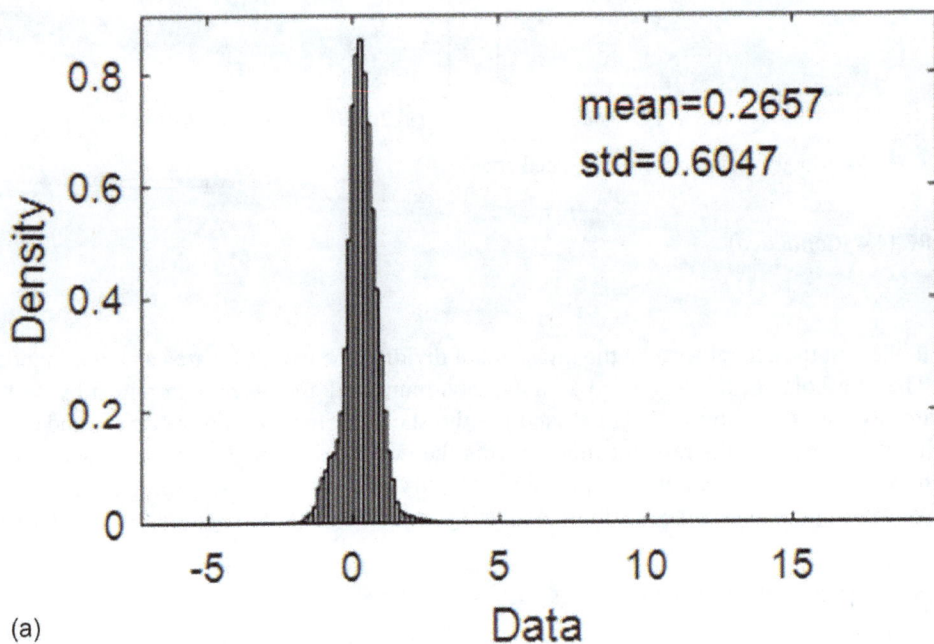

(a)

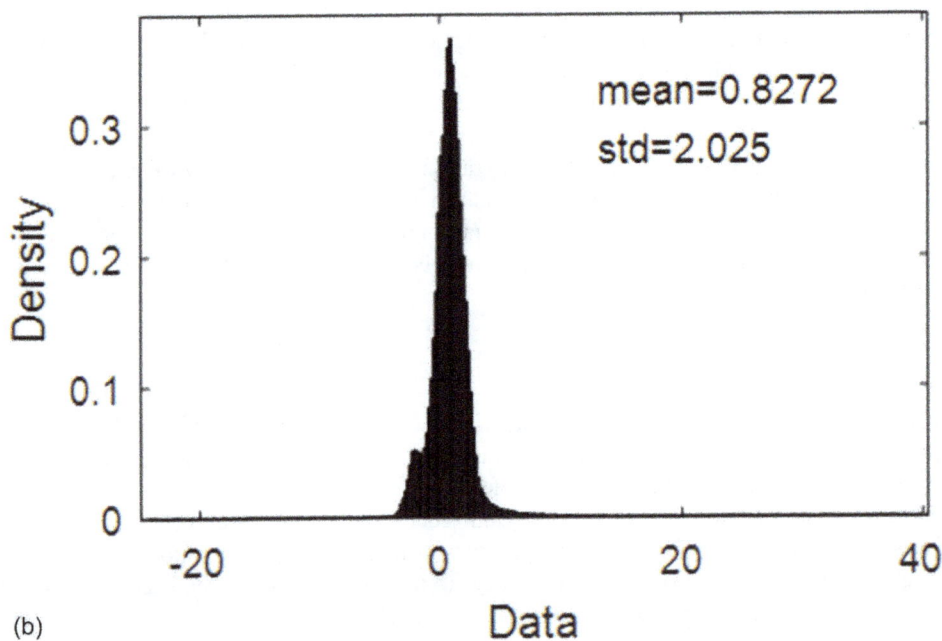

(b)

FIGURE 14.5   Extraction result of the CDAT method. (*Continued*)

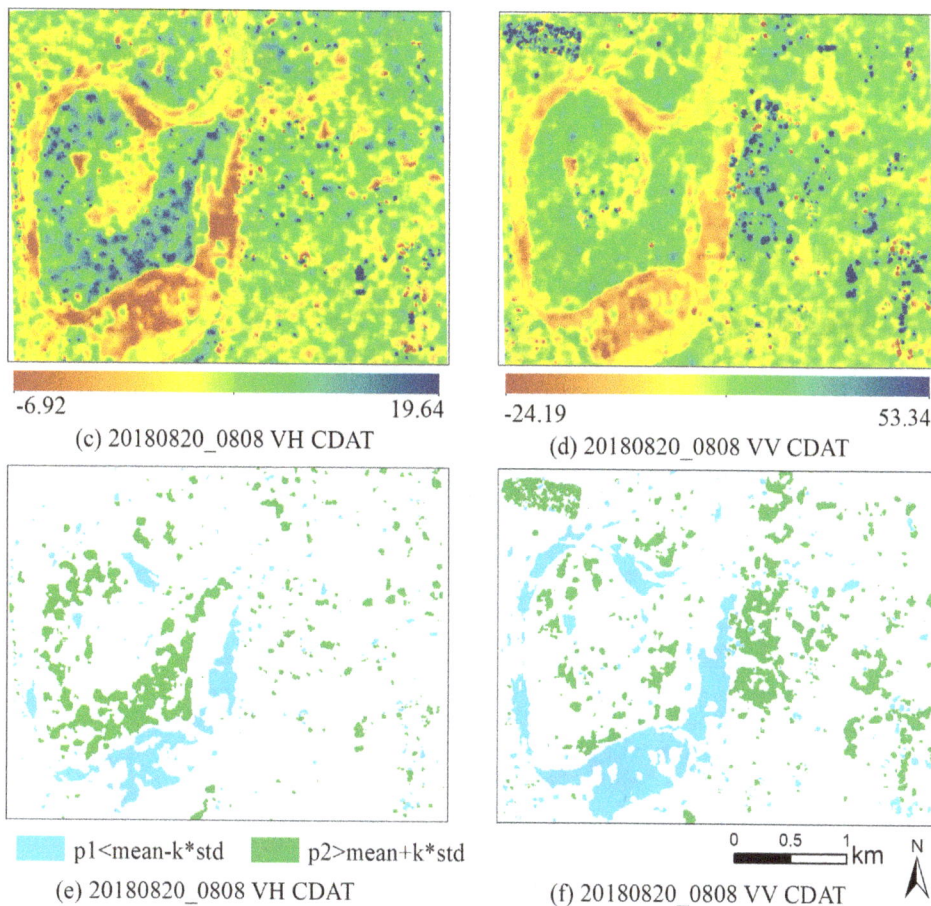

(c) 20180820_0808 VH CDAT

-6.92   19.64

(d) 20180820_0808 VV CDAT

-24.19   53.34

p1<mean-k*std   p2>mean+k*std

(e) 20180820_0808 VH CDAT

(f) 20180820_0808 VV CDAT

0   0.5   1 km   N

FIGURE 14.5 (*Continued*)

After completing the classification of ground objects, the average reflectance of each ground object is counted. Based on the Sentinel-1 image on August 2, 2018, the basic backscattering coefficients of five land types are calculated. The scattering coefficients and mean values of each land type are shown in Table 14.4.

In addition to vegetation, when other types of ground objects are flooded, the reflectivity decreases, and when vegetation is flooded, the reflectivity increases; the degree of inundation is divided into five grades: complete inundation, severe inundation, moderate inundation, mild inundation, and non-flood. When the local level of change is 1, it is considered to be slightly inundated; when the local level of change is 2, it is considered to be moderate inundation; when the local level changes to 3, it is considered to be severe inundation; when the local level changes to 4, it is completely inundated. Except for no-shed farmland, all other types change to the category with smaller reflectivity, while no-shed farmland changes to the category with larger reflectivity. When the category in the disaster is water, it is always considered to be completely inundated, and other situations are considered not to be inundated. The specific judgment rules are shown in Table 14.5.

The pixel detection method is used to detect the change in ground object classification before and during the disaster, and the results are shown in Figure 14.7.

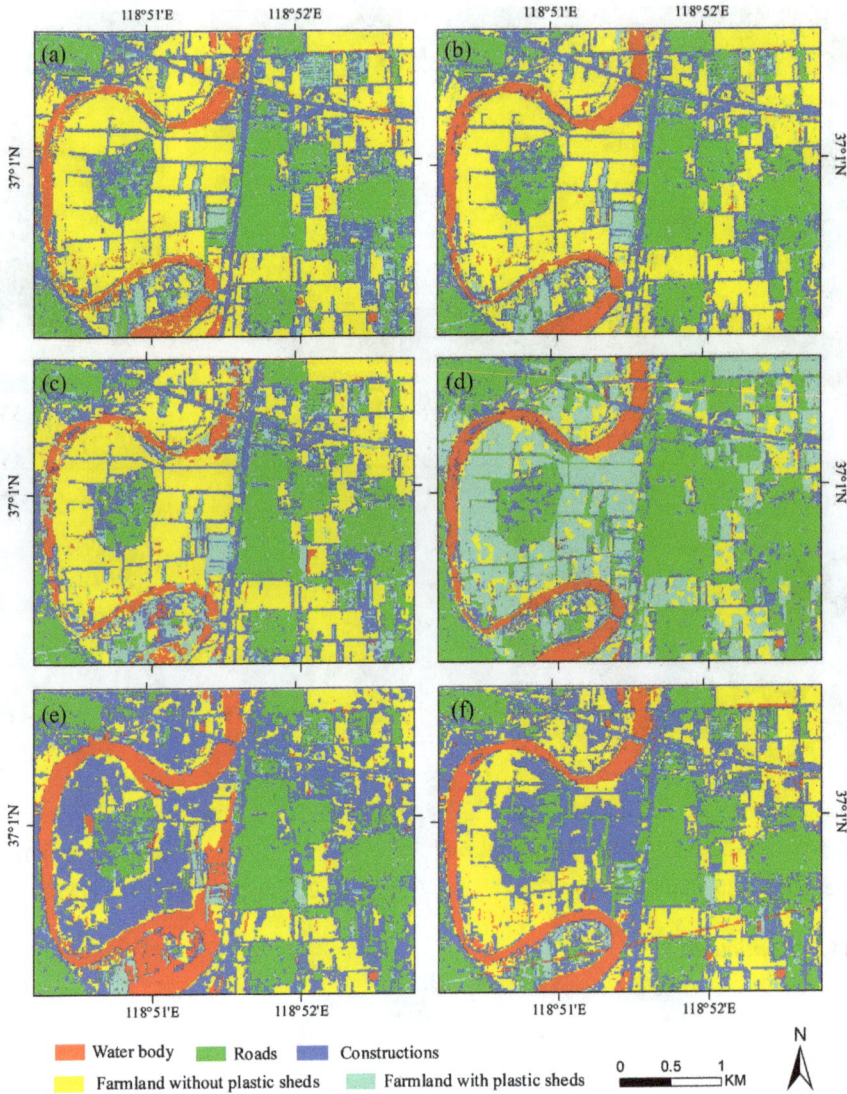

FIGURE 14.6 The results of the classifications of ground objects: (a) 2018–08–10 based on Sentinel-2 RGB ground object classification results; (b) 2018–08–02 ground object scattering feature classification; (c) 2018–07–21 ground object scattering feature classification; (d) 2018–07–27 ground object scattering feature classification; (e)2018–08–20 object scattering feature classification; (f) 2018–08–26 object scattering feature classification.

TABLE 14.4
Mean and Standard Deviation of Objects' Backscattering Coefficients (S_Gamma)

Objects	Serial Number	VV		VH	
		Mean	Standard Deviation	Mean	Standard Deviation
water	1	1.982416	0.863452	1.143954	0.35071
shedless farmland	2	3.46729	0.683308	1.876859	0.217285
shed farmland	4	3.768043	0.740086	1.909334	0.319949
road	8	3.863828	0.663861	2.014632	0.295261
construction area	16	5.151355	2.348478	2.20923	0.623642

TABLE 14.5

Judgment Rules for Flood Degree

Flood Degree	Difference Value	Pre-Disaster Object Number	Disaster Object Number
complete inundation	−15	16	1
	−7	8	1
	−3	4	1
	−1	2	1
severe inundation	−14	16	2
	14	2	16
moderate inundation	−12	16	4
	−6	8	2
	6	2	8
mild inundation	−8	16	8
	−4	8	4
	−2	4	2
	2	2	4
non-flood	other values	−	−

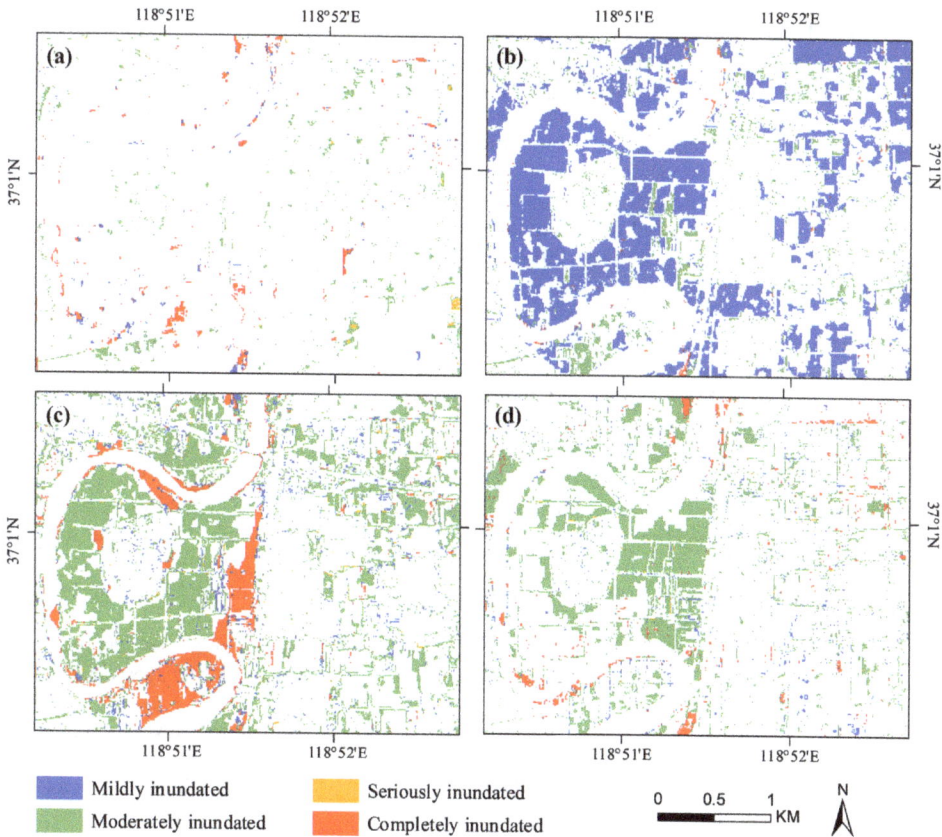

FIGURE 14.7 Flood extraction results:

(1) 2018–07–21; (2) 2018–07–27; (3) 2018–08–20; (4) 2018–08–26.

14.5 SUMMARY

SAR images have very high accuracy in water extraction. The vegetation index difference method (MSAVI method) and CDAT method can be used to extract vegetation inundated areas, but these two methods have their emphasis and limitations. The MSAVI method has a high recognition rate of vegetation inundated areas, but it can only identify some completely inundated areas and cannot identify the inundated information of other land types. The histogram trough of MSAVI is indicative of identifying vegetation and non-vegetation areas but not for identifying flood areas.

The recognition rate of the complete inundated area of the CDAT method is due to the MSAVI difference method, but the effect of the two polarization methods is quite different. The recognition area of the VH polarization vegetation inundated area is large, and the recognition area of VV polarization vegetation inundated area is small, but it can identify the inundated area of the building area, indicating that VH polarization is more suitable for this method to identify vegetation inundated area.

The flood disaster monitoring method based on image classification has higher recognition accuracy than other methods in the completely inundated area and vegetation inundated area and also identifies a large number of other land types, which is greatly improved when compared with the three commonly used methods.

REFERENCES

[1]   Li Jianfeng. Research on the rapid extraction of Sri Lanka's surface water and flood information based on multi-source remote sensing images and Google Earth Engine. Chang'an University, 2020.

[2]   Pan Zuyan, Gong Adu, Zhou Tinggang, et al. Research on extraction of flood submerged range based on fusion of TM and SAR images. International Conference on Remote Sensing. 2010.

[3]   Xie Qiuxia, Zhang Jiahui, Lu Kun, et al. Research and application of accurate flood information extraction based on typical remote sensing image fusion method. Disaster Science, 2017, 32(1): 183–186 + 204.

[4]   Wang Y. Seasonal change in the extent of inundation on floodplains detected by JERS-1 Synthetic Aperture Radar data. International Journal of Remote Sensing, 2004, 25(13): 2497–2508.

[5]   Sharifi A. Flood mapping using relevance vector machine and SAR data: A case study from Aqqala, Iran. Journal of the Indian Society of Remote Sensing, 2020, 48(9): 1289–1296.

[6]   Benoudjit A, Guida R. A novel fully automated mapping of the flood extent on SAR images using a supervised classifier. Remote Sensing, 2019, 11(7).

[7]   Skakun S. A Neural network approach to flood mapping using satellite imagery. Computing and Informatics, 2010, 29: 1013–1024.

[8]   Insom P, Cao C, Boonsrimuang P, et al. A support vector machine-based particle filter method for improved flooding classification. IEEE Geoscience and Remote Sensing Letters, 2015, 12(9): 1943–1947.

[9]   Jia Yilin, Zhang Wen, Meng Lingkui. Research on NDWI segmentation threshold selection method for GF-1 images. Remote Sensing of Land and Resources, 2019, 31(1): 95–100.

[10] Wang Jingming, Wang Shixin, Wang Futao, et al. Research on flood inundation extraction method based on Sentinel-1 SAR data. Disaster Science, 2021, 36(4): 214–220.

[11] An Chengjin, Niu Zhaodong, Li Zhijun. et al. Threshold comparison of typical Otsu algorithm and analysis of water segmentation performance of SAR images. Journal of Electronics and Information, 2010, 32(9): 2215–2219.

[12] Wang Hang, Qin Fen. A review of water extraction from remote sensing images. Surveying and Mapping Science, 2018, 43(5): 23–32.

[13] McMinn WR, Yang Q, Scholz M. Classification and assessment of water bodies as adaptive structural measures for flood risk management planning. Journal of Environmental Management, 2010, 91(9): 1855–1863.

[14] Zhang Zhanqian. Remote sensing monitoring model for flood disasters based on vegetation index time series. Northeast Agricultural University, 2019.

[15] Bi Jingjia, Huang Haijun, Liu Yanxia. Flood trace extraction and extent estimation using remote sensing and GIS technology. Remote Sensing Information, 2016, 31(6): 147–152.

[16] Suzanne JS, Stephen KH, John MM. Inundation area and morphometry of lakes on the Amazon river flood-plain, Brazil. Archiv fur Hydrobiologie, 1992, 123(4): 385–400.

[17] Martinis S, Kersten J, Twele A. A fully automated TerraSAR-X based flood service. ISPRS Journal of Photogrammetry and Remote Sensing, 2015, 104: 203–212.

[18] Otsu N. A Threshold Selection Method from Gray-Level Histograms. IEEE Transactions on Systems, Man, and Cybernetics, 1979, 9(1): 62–66.

[19] Li Y, Martinis S, Plank S, et al. An automatic change detection approach for rapid flood mapping in Sentinel-1 SAR data. International Journal of Applied Earth Observation and Geoinformation, 2018, 73: 123–135.

[20] Chen Lingyan, Liu Zhi, Zhang Hong. SAR image water detection based on water scattering characteristics. Remote Sensing Technology and Application, 2014(6): 963–969.

[21] Liang J, Liu D. A local thresholding approach to flood water delineation using Sentinel-1 SAR imagery. ISPRS Journal of Photogrammetry and Remote Sensing, 2020, 159: 53–62.

[22] Cao H, Zhang H, Wang C, et al. Operational flood detection using Sentinel-1 SAR data over large areas. Water, 2019, 11(4): 786.

[23] Sun Yayong, Huang Shifeng, Li Jiren, et al. Application of Sentinel-1A SAR data in flood monitoring in the lower reaches of the Ayeyarwady River in Myanmar. Remote Sensing Technology and Application, 2017, 32(2): 282–288.

[24] Long S, Fatoyinbo TE, Policelli F. Flood extent mapping for Namibia using change detection and thresholding with SAR. Environmental Research Letters, 2014, 9(3): 035002.

[25] Francisco CC, María DMM. Flood monitoring based on the study of Sentinel-1 SAR images: The Ebro River Case Study. Water, 2019, 11(12).

[26] Borah SB, Sivasankar T, Ramya MNS, et al. Flood inundation mapping and monitoring in Kaziranga National Park, Assam using Sentinel-1 SAR data. Environmental Monitoring and Assessment, 2018, 190(9): 520.

[27] Cheng Xia, Zhang Yonghong, Deng Min, et al. Recent surface deformation analysis of the Yellow River Delta by Sentinel-1A satellite. Surveying and Mapping Science, 2020, 45(2): 43–51.

[28] Zheng Shaolan, Liu Longwei. Rice information extraction from single-phase Sentinel-1A satellite SAR data. Geospatial Information, 2020, 18(4): 61–64.

[29] Chen Sainan, Jiang Mi. Application of Sentinel-1 SAR in flood range extraction and polarization analysis. Journal of Geo-Information Science, 2021, 23(6): 1063–1070.

[30] Li Xinyao. Remote sensing retrieval of surface soil moisture based on Sentinel-1 SAR data. Northwestern University, 2019.

[31] Liang Hanyue. Research on information extraction of early rice in Nanchang County, Jiangxi Province based on Sentinel-1 data. Chengdu University of Technology, 2017.

[32] Wang Zhenghui, Xin Cunlin, Sun Zhe, et al. Automatic extraction of aquatic vegetation groups in small lakes from Sentinel-2 data: Taking Cuiping Lake as an example. Remote Sensing Information, 2019, 34(5): 132–141.

[33] Tian Haifeng. Remote sensing identification of winter wheat in China's main producing areas based on Sentinel-1&2 satellite images. University of Chinese Academy of Sciences (Institute of Remote Sensing and Digital Earth, Chinese Academy of Sciences), 2019.

[34] Yi Qiuxiang. Estimation of cotton leaf area index based on Sentinel-2 multispectral data. Chinese Journal of Agricultural Engineering, 2019, 35(16): 189–197.

[35] Zhang Lei, Gong Zhaoning, Wang Qiwei, et al. The yellow river delta wetland information extraction based on Sentinel-2 image multi-feature optimization. Journal of Remote Sensing, 2019, 23(2): 313–326.

[36] Yang Lin, Ding Feng, Zhu Lingli, Zheng Zicheng, et al. Improvement and application of NDISI index based on Sentinel-2 and Landsat 8 data passed on the same day. Journal of Fujian Normal University (Natural Science Edition), 2020, 36(1): 70–78.

[37] Li Yuchen, Zhang Jun, Liu Chenli. Extraction method of small glacial lakes in Qianhu Mountain, Yunnan from Sentinel-2 image. Science of Surveying and Mapping, 2021, 46(4): 114–120.

[38] Liu Hao. Extraction of crop planting structure in Hetao irrigation area based on Sentinel-2 images. Resources and Environment in Arid Areas, 2021, 35(2): 88–95.

[39] Zhou Han, Ye Huping, Wei Xianhu, et al. Comparative study of water extraction methods based on Sentinel-1/2: A small water in Sri Lanka as an example. Journal of the University of Chinese Academy of Sciences, 2019, 36(6): 794–802.

[40] Luan Yujie, Guo Jinyun, Gao Yonggang, et al. Remote sensing monitoring and disaster analysis of Shouguang flood in 2018 based on Sentinel-1B SAR data. Journal of Natural Disasters, 2021, 30(2): 168–175.

[41] Chabai River Basin Based on Sentinel-1A and Landsat 8 Images. Guizhou University, 2019.

[42] Jiaozuo Guangli. Irrigation area based on Sentinel-1A data. Zhengzhou University, 2018.

[43] Wang Dejun, Jiang Qigang, Li Yuanhua, et al. Land use classification in farming areas based on Sentinel-2A/B time series data and random forest algorithm. Remote Sensing of Land and Resources, 2020, 32(4): 236–243.

[44] Zhao Yajie, Wang Lihui, Kong Xiangbing, et al. Research on land use classification based on Sentinel-2 and Landsat8 OLI data fusion. Journal of Fujian Agriculture and Forestry University (Natural Science Edition), 2020, 49(2): 248–255.

[45] Li Shutao, Li Congyu, Kang Xudong. Development status and future prospects of multi-source remote sensing image fusion. National Remote Sensing Bulletin, 2021, 25(01): 148–166.

[46] Luo Dong, Luo Hongxia, Liu Guangpeng, et al. Research on classification and fusion scale of sentinel-1A and sentinel-2A with different cloud cover. Hubei Agricultural Sciences, 2020, 59(05): 28–36+43.

[47] Ji Yongjie, Yue Cairong, Zhang Wangfei. Use fusion of SAR and optical images for land cover classification. Journal of Southwest Forestry University (Natural Sciences), 2016, 36(03): 158–162.

[48] Guo Lei, Yang Jihong, Shi Liangshu, et al. Comparative Study on SPOT6 Remote Sensing Image Fusion Methods. Remote Sensing of Land and Resources, 2014, 26(4): 71–77.

[49] Zhang Hao, Zhao Yunsheng, Chen Guanyu, Zhang Chunyuan. Research on Building Recognition and Classification in Remote Sensing Image Based on Support Vector Machine. Geological Science and Technology Information, 2016, 35(6): 194–199.

[50] Zhang Yu, Yang Haitao, Yuan Chunhui. A review of remote sensing image classification methods. Chinese Journal of Weaponry and Equipment Engineering, 2018, 39(8): 108–112.

[51] Tong Guofeng, Li Yong, Ding Weili, Yue Xiaoyang. A review of remote sensing image change detection algorithm. Chinese Journal of Image Graphics, 2015, 20(12): 1561–1571.

[52] Yang Xiuguo. Research and analysis of image threshold segmentation method. East China Normal University, 2009.

[53] Adaptive threshold calculation method for open-set voiceprint recognition based on Dajin algorithm and deep learning. Journal of Jilin University (Science Edition), 2021, 59(4): 909–914.

[54] Chen Wenqian, Ding Jianli, Li Yanhua, et al. Water extraction method based on domestic GF-1 remote sensing images. Resources Science, 2015, 37(6): 1166–1172.

[55] Duan Qiuya, Meng Lingkui, Fan Zhiwei, et al. Research on the applicability of water information extraction method from GF-1 satellite images. Remote Sensing of Land and Resources, 2015, 27(4): 79–84.

[56] Duan Jiwei, Zhong Jiusheng, Jiang Li, et al. Extraction method of ultra-green water index in post-rain flood area based on GF-2 image. Geography and Geographic Information Science, 2021, 37(3): 35–41.

[57] Li Zhihong, Li Wangping, Wang Yu, et al. Comparison and verification of two novel water extraction methods based on discrete particle swarm optimization. Journal of Earth Information Science, 2021, 23(6): 1106–1117.

[58] McFeeters SK. The use of Normalized Difference Water Index (NDWI) in the delineation of open water features. International Journal of Remote Sensing, 1996, 17: 1425–1432.

[59] Georgios K, Ioannis M, Aragonés D, et al. Fast and automatic data-driven thresholding for inundation mapping with Sentinel-2 data. Remote Sensing, 2018, 10(6): 910.

[60] Wang Changqing, Xing Yanqiu, Wang Xianyi, et al. Inversion study of leaf water content in coniferous forest combining Sentinel-1B and Landsat8 data. Forest Engineering, 2018, 34(4): 27–35+69.

[61] Guo Wenting, Zhang Xiaoli. Vegetation classification based on Sentinel-2 time series and multiple features. Journal of Zhejiang A&F University, 2019, 36(5): 849–856.

[62] Dai Shengpei, Luo Hongxia, Zheng Qian, et al. Comparative study on remote sensing estimation models of rubber forest leaf area index in Hainan Island. Smart Agriculture (Chinese and English), 2021, 3(10): 45–54.

[63] Testa S, Soudani K, Boschetti L, et al. MODIS-derived EVI, NDVI and WDRVI time series to estimate phenological metrics in French deciduous forests. International Journal of Applied Earth Observation and Geoinformation, 2018, 64: 132–144.

[64] Kong d, Zhang Y, Gu X, et al. A robust method for reconstructing global MODIS EVI time series on the Google Earth Engine. ISPRS Journal of Photogrammetry and Remote Sensing, 2019, 155: 13–24.

[65] Taddeo S, Dronova I, Depsky N. Spectral vegetation indices of wetland greenness: Responses to vegetation structure, composition, and spatial distribution. Remote Sensing of Environment, 2019, 234: 111467.

[66] Sankaran S, Zhou J, Khot L, et al. High-throughput field phenotyping in dry bean using small unmanned aerial vehicle based multispectral imagery. Computers and Electronics in Agriculture, 2018, 151: 84–92.

[67] Wang Ze, Zhao Liangjun, Niu Kai, et al. A review on the extraction method of vegetation coverage based on remote sensing images. Agriculture and Technology, 2021, 41(14): 25–29.

[68] Xu D, Wang C, Chen J, et al. The superiority of the normalized difference phenology index (NDPI) for estimating grassland aboveground fresh biomass. Remote Sensing of Environment, 2021, 264: 112578.

[69] Huang M, Jin S. Rapid Flood Mapping and Evaluation with a Supervised Classifier and Change Detection in Shouguang Using Sentinel-1 SAR and Sentinel-2 Optical Data. Remote Sensing, 2020, 12(13): 2073.

15 Flood Hydrological Simulation and Risk Assessment Based on GIS

Minmin Huang and Shuanggen Jin

15.1 INTRODUCTION

The establishment of an urban waterlogging simulation model based on the principles of hydrology and hydrodynamics can make up for the shortcomings of traditional methods such as precipitation estimation of waterlogging disasters, realize the accurate simulation of rainfall production and concentration process, effectively predict the depth and spatial distribution of urban surface ponding, and thus achieve the prediction of direct disaster causing factors. Different production practices have different requirements for models, and with the development of disciplines and social technology, different hydrological models have emerged. The hydrological model has gone through the black-box model, conceptual model stage, and distributed hydrological model research stage. The conceptual hydrological model refers to the use of abstract and generalized equations to describe the water cycle process of a basin. It usually has a certain physical basis and a certain experience. The model structure is simple and practical. The advantage of distributed hydrological model that the model parameters have clear physical meaning, can be solved by continuous equation and dynamic equation, can more accurately describe the water cycle process, and has strong adaptability. Surface inundation analysis of urban rainstorm refers to the possible water depth and inundation of urban surface obtained by reasonably distributing the water into the regional space according to the urban terrain after obtaining the total amount of urban rainstorm water. There are two main types of flooding algorithms based on DEM: passive flooding and active flooding. Passive inundation means that all grids whose elevation value is lower than a given water level will enter the submerged area without considering connectivity, while active inundation considers regional connectivity; that is, regional floods can only inundate the area that it can flow through.

For the special terrain like a crater, passive inundation may generate inundation areas inside and outside the crater. In the active inundation scenario, if the external flood does not reach, the inundation area will only be outside the crater in the end. These two cases are of practical significance. The first case is equivalent to a large area of uniform precipitation in the whole area, and ponding may occur in low-lying areas. The second kind is equivalent to the high-risk flood flowing to the nearby area, which is similar to the flood bursting or the surging flood caused by local rainstorm spreading around. Neither active nor passive diffusion algorithms have solved the problem of catchment area boundary, which often leads to unreasonable phenomena at the catchment area boundary. In actual diffusion, there is no real limiting boundary, and water may spread across the catchment area, which will lead to water redistribution. It shows that water distribution is a dynamic process and needs to be constantly adjusted according to the actual diffusion situation.

Scientific flood disaster risk assessment is the premise and demand of flood disaster risk management. Using scientific risk assessment methods can make wise decisions on the prevention and control of rainstorm and flood disasters under changing environments, and timely, effectively, and continuously increase the comprehensive management of flood disasters. As the basis and technical

DOI: 10.1201/9781003363118-15

support of disaster prevention and mitigation decision-making, flood disaster risk assessment has important practical significance.

15.2   HYDROLOGICAL SIMULATION BASED ON GIS TECHNOLOGY

15.2.1   FLOOD INUNDATION ANALYSIS BASED ON GIS TECHNOLOGY

The flood inundation analysis based on GIS technology has three parts: rainfall runoff, surface rainwater conflux, and numerical simulation of surface inundation depth (Figure 15.1). Rainfall runoff refers to the forming net rain after deducting various losses in rainfall, i.e., mainly through evaporation, plant interception, surface depression filling, and soil infiltration. There is an obvious relationship between surface runoff and soil moisture[1]. Light rain can increase soil moisture at 0~10 cm soil depths, and moderate rain can increase it at 10–20 cm depths[2]. Scholars found that

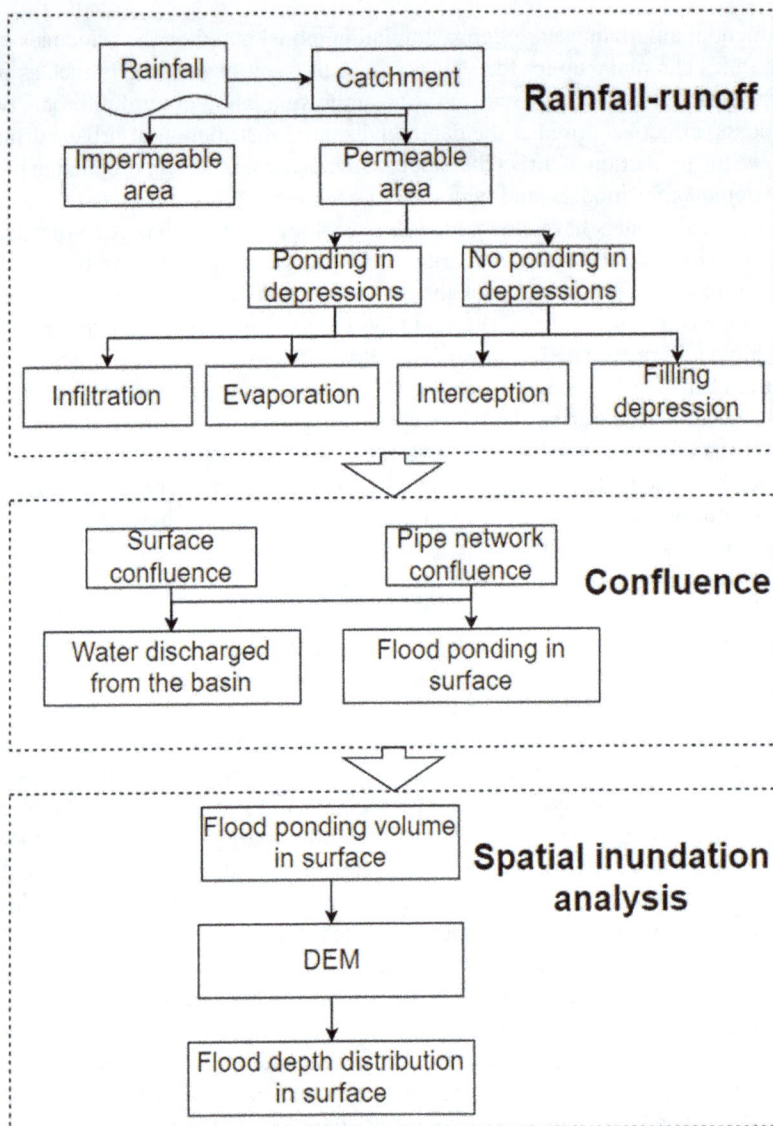

FIGURE 15.1   Process diagram of rainfall-producing and confluence process.

when the rainfall is greater than 200 mm in three days, flooding disasters generally occur, while when the rainfall is less than 30 mm, the probability of flood is small[3].

Surface rainwater conflux refers to the process in which runoff or floodwater flows from the surface into a pipeline network or river. In our urban waterlogging simulation, the confluence of the pipe network is the link between surface water and drainage, which is the most critical part of the inundation analysis. On the other hand, rivers and ditches are the main drainage channels for rural or suburban areas without pipelines.

The numerical simulation of surface inundation depth refers to the possible surface flood depth according to the terrain after determining the total rainstorm water. The flooding algorithms based on DEM mainly include "inundation without a source" and "inundation with a source". Inundation without a source means that the grids with lower elevation values than the given water level are inundated areas. When regional connectivity is considered, the flood can only inundate areas with flowing through. As such, in this study, connectivity is not considered.

15.2.2 Rainfall-Runoff

Rainfall runoff refers to the process of forming net rain after deducting various losses from rainfall, among which losses mainly include evaporation, plant interception, surface filling, soil infiltration, etc. Commonly used runoff calculation methods include the runoff coefficient method[4], the SCS-CN method[5], the full storage runoff method[6], and the infiltration curve method.

Rainfall is a complex process, and its intensity and duration have a great impact on ponding. Scholars have studied the critical rainfall of surface runoff and flood.

According to the research on the soil moisture of different soil layers by rainfall, the soil moisture changes under different rainfall intensities show a "rising period" – "plateau period" – "water withdrawal period". There is an obvious threshold relationship between surface runoff and soil moisture. Light rain can supplement soil moisture of 10 cm, and moderate rain can supplement soil moisture of 10–20 cm. It shows that there is almost no runoff on the surface under moderate and light rainfall, and there will be almost no accumulation of water on the surface after the rainfall.

By analyzing historical disaster data, some scholars found that when the rainfall in 3 days is more than 200 mm, flood disasters will occur, while when it is less than 30 mm, the possibility of flood disasters is small. This study gives the critical value of flood and non-flood under the 72-hour duration, but no more studies reveal the impact of the previous rainfall duration and intensity on the surface ponding.

15.2.3 Surface Rainwater Conflux

Surface rainwater conflux refers to the process in which runoff or floodwaters flow into the pipeline network or river channel from the surface[7–8]. In the analysis of urban waterlogging, the confluence of the pipe network is the link for water exchange and drainage and is the most critical part of the inundation analysis. For rural or suburban areas without pipes, rivers and ditches were considered as the main drainage channel.

15.2.4 Numerical Simulation of Surface Inundation Depth

The numerical simulation of surface inundation depth refers to the possible surface flood depth according to the terrain after determining the total rainstorm water. The flooding algorithms based on DEM mainly include "inundation without a source" and "inundation with a source". Inundation without a source means that the grids with lower elevation values than the given water level are inundated areas. When regional connectivity is considered, the flood can only inundate areas by flowing through them[9]. According to the basic principle of inundation analysis, only when the study area is similar to a regular cuboid pool, the water depth of any area is the same. However, such an ideal surface usually does not exist in reality. In reality, surface water occurs in local depressions, starting from the lowest terrain, and with the increase of rainfall, the area of water accumulation increases, forming small ponds. However, the depth of ponding in the same ponding area is inconsistent. For the same place, the greater the rainfall, the deeper the water. During the same rainfall, the depth of water accumulation depends on the local topography.

15.3 HYDROLOGICAL MODEL

15.3.1 SCS-CN MODEL

The critical aspect of constructing the relationship between rainfall and surface runoff is to establish a convenient and effective rainfall-runoff model. The Soil Conservation Service Curve Number (SCS-CN) model was developed based on climatic characteristics and multi-year hydrological runoff data from the United States. It has been widely used due to its simple results, few parameters, and high accuracy. The SCS-CN model includes a water balance equation and two basic assumptions[10]. The balanced equation is as follows (1):

$$P = I_a + F + Q \tag{15.1}$$

where P is rainfall (mm), Q is runoff depth (mm), F is cumulative infiltration (mm), and I_a is initial loss (mm).

Equation (15.2) is an assumption of equal proportions:

$$\frac{Q}{P - I_a} = \frac{F}{S} \tag{15.2}$$

The assumption of a proportional relationship between the initial rainfall loss and the potential stagnant storage is described in Equation (15.3):

$$I_a = \lambda S \tag{15.3}$$

where S is the maximum infiltration of the watershed (mm), I_a is the initial loss (mm), and λ is the initial loss rate.

The runoff depth can be calculated by combining Equations (15.1) and (15.3)[11]:

$$Q = \begin{cases} \dfrac{(P - \lambda S)^2}{(P + (1 - \lambda)S)}, & P > \lambda S \\ 0, & P \leq \lambda S \end{cases} \tag{15.4}$$

where S is calculated by the CN (Curve Number) coefficient in Equation (15.5) with the following statistical relationship:

$$S = \frac{25400}{CN} - 254 \tag{15.5}$$

Runoff CN is the model's main parameter, which reflects the comprehensive characteristics of the soil moisture degree (antecedent moisture condition [AMC]), slope, soil type, and land use status in the early stage of the basin. Table 15.1 shows the CN value of each category under the average soil humidity (AMC II)[12]. The greater the CN value is, the worse the maximum water storage capacity of the basin is.

TABLE 15.1

CN Value Lookup Table

Category	Water	Grassland	Cultivated Land	Construction	Road
CN	100	69	71	98	98

15.3.2 SWMM MODEL

SWMM is a stormwater management model researched and developed by the Environmental Protection Agency of the United States. The model has a high degree of commercialization and can be used well even if you do not understand the model principle. The SWMM model is widely used in urban drainage design and management, flood prevention project planning, and so on. The SWMM model has strong applicability and can simulate the generation, accumulation, scouring, and transport processes of non-point source pollution loads in the watershed, as well as the runoff and confluence processes on the surface and the transport process in the pipeline network under rainfall conditions. The SWMM model has lower data requirements and the input data is easy to obtain. The model can customize any input time interval, and can also output the results of any integer step size, and there is no limit to the area of the input model area.

The SWMM model contains many modules, which can be divided into calculation modules and service modules according to their properties. The calculation models mainly include runoff module, transportation module, storage/treatment module, etc. The service modules mainly include the rainfall module, statistics module, joint module, etc.

15.3.2.1 SWMM Model Parameters

There are two basic parameters needed to simulate the runoff-producing and flow concentration of the urban rainfall by using the SWMM model: precipitation data and hydrological data.

15.3.2.1.1 Precipitation data

Precipitation data is the specific rainfall time series of the rainfall events to be simulated. The SWMM model can read rainfall time series with various time intervals. As long as the rainfall time series used in a simulation process have equal intervals, the corresponding time and rainfall values can be directly and manually entered through the rainfall module of the SWMM model. The rainfall time series data can also be processed into a. dat file that can be directly recognized by the SWMM model according to the template, and the file can be directly read.

15.3.2.1.2 Hydrological data

The catchment area is the unit for hydrological calculation using the SWMM model, so determining the catchment area division of the study area is the primary task of applying the SWMM model[13]. The catchment area is an important input parameter of the SWMM model, and its attributes can be divided into two parts: deterministic attributes and uncertain attributes. The area and feature width can be directly obtained by GIS tools. The area ratio of impervious area, the Manning coefficient of impervious area, and the Manning coefficient of the permeable area can be calculated by GIS under land classification. The attribute of the rain gauge in the catchment area specifies the rain gauge used in each catchment area. Only one rain gauge can be specified in each catchment area. The rainfall time series corresponding to the rain gauge is the source of runoff in each catchment area. There is only one drainage outlet in each catchment area. After rainfall-runoff and surface confluence, the remaining water flows directly into the drainage outlet to exchange water between drainage networks. The remaining uncertain attributes of the catchment area include the storage capacity of the surface depressions in the impervious area, the storage capacity of the surface depressions in the permeable area, the maximum infiltration rate, the minimum infiltration rate, and the attenuation constant, which need to be determined by the method of parameter calibration.

The inspection well is the transportation hub for surface confluence and pipe network confluence. The water after surface confluence enters the pipeline through the inspection well for water exchange between catchment areas. The attributes of the inspection well include geographical location, upper bottom elevation, lower bottom elevation, and depth. Nodes and pipes located upstream of the drainage outlet and whose bottom elevation is greater than the drainage outlet do not necessarily participate in the calculation of water exchange. Therefore, generalizing the drainage network with a reasonable standard is beneficial to remove redundancy, reduce complexity, and reduce errors.

Pipes are passages through which water volumes in different catchments are exchanged. The main attributes of the pipeline are the name of the starting and ending nodes, the diameter of the pipe, the shape of the pipe, the roughness of the pipe wall, and the slope of the pipeline. The slope of the pipeline is obtained by dividing the bottom elevation difference between the starting and ending nodes by the length of the pipe. The degree of water interaction is determined by these properties of the pipe together.

The water outlet is the outlet that discharges the water volume of the model to the outside, and its properties include the bottom elevation and the inflow time series. The inflow time series can set the speed of the external water flowing into the model, which is suitable for the drainage outlet located on the edge of the river; when the water level does not exceed the drainage outlet, the water in the river will pour into the model.

15.3.2.2 SWMM Model Functional Modules

The main modules used when using the SWMM model to simulate urban waterlogging are the rainfall module, runoff module, and transport module.

The SWMM model sets the rainfall time series of each rain gauge through the rainfall module. Each rain gauge has a unique time series, but there can be multiple rain gauges and rainfall time series in a simulation. Each rain gauge and rainfall time series are independent of each other.

The runoff module of the SWMM model is the hydrological model. After the rainfall in the catchment area is obtained, the nonlinear reservoir slope catchment after the interception of rainfall by surface plants and infiltration of rainfall by unsaturated soil can be calculated, and various micro impacts that reduce and delay rainfall and runoff can be simulated.

The transport module of the SWMM model is the hydraulic model, which can simulate the flow of surface runoff and external water flow in pipes and channels. The simulation results can be used to analyze information such as pipe flow, depth, and water accumulation at nodes.

The specific processes involved in each module are shown in Figure 15.2.

15.3.2.3 SWMM Model Production and Convergence Theory

The core calculation module of the SWMM model is the calculation of runoff and confluence. The water accumulation at the nodes is finally determined through the calculation of runoff and confluence. The SWMM model uses the Horton infiltration model and the nonlinear reservoir model to calculate the runoff[14]. At the initial stage of rainfall, the surface water content is low, the infiltration rate is high, and there is no surface runoff. At present, it is the interception stage of surface plants. With the progress of rainfall, the rainfall intensity increases continuously, the rainfall intensity starts to be greater than the infiltration rate, and the ground starts to have ponding, which starts to enter the flow generation stage. As the rainfall continues, the rainfall intensity continues to increase, the depressions are full, and the surface ponding in places with low permeability becomes more. The whole region is in the process of runoff generation and collection, and the ponding is concentrated at the water outlet of the basin. At present, it is the confluence stage.

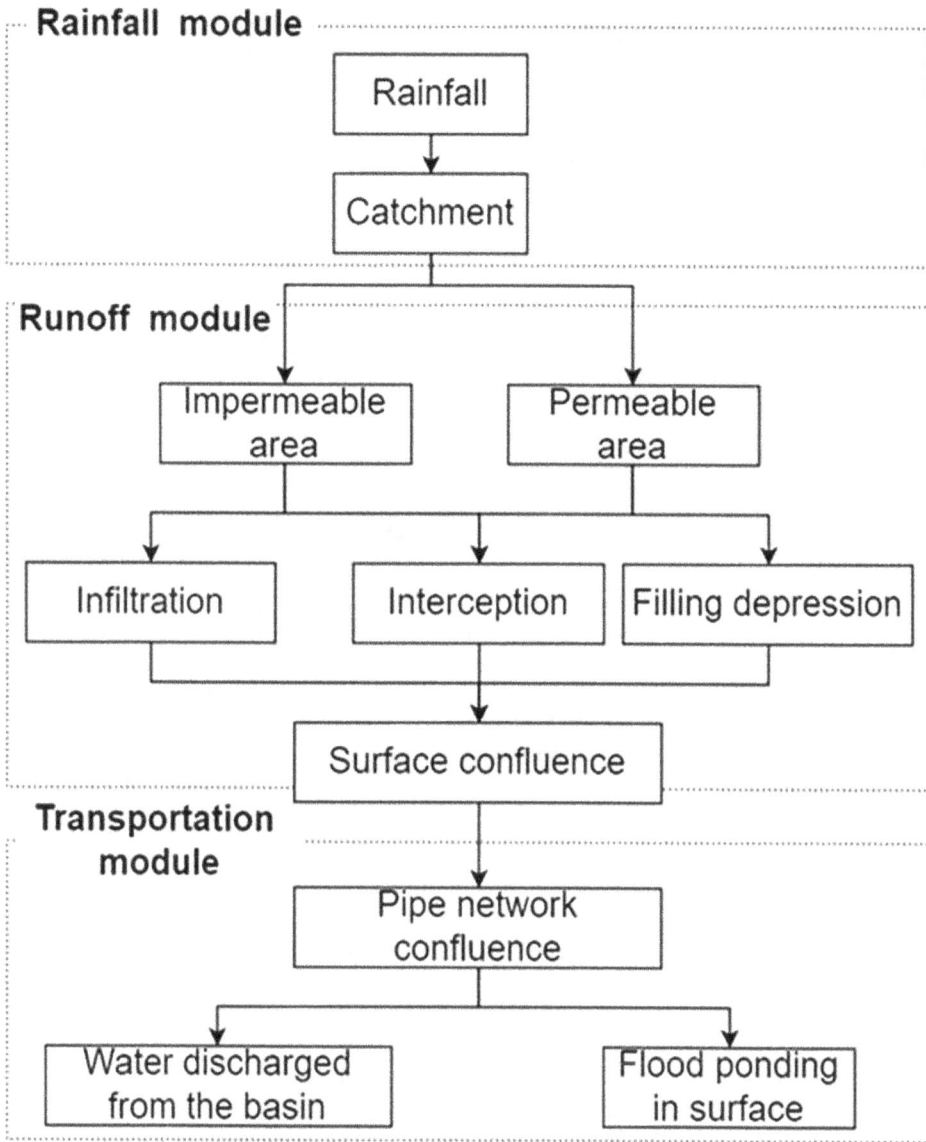

FIGURE 15.2 SWMM model simulation flow chart.

15.3.2.3.1 Rainfall-Runoff Calculation

Rainfall-runoff refers to the forming net rain after deducting various losses in rainfall, i.e., mainly through evaporation, plant interception, surface depression filling, and soil infiltration[15]. It is obtained by subtracting rainfall loss from total rainfall:

$$W_{runoff} = Q_{sum} - q_{in} - q_{ex} \qquad (15.6)$$

Q_{sum} is the total amount of precipitation, q_{in} is the amount of infiltration rainfall, and q_{ex} is the amount of rainwater evaporation. In general, the model does not consider evaporation.

Q is the cumulative precipitation during the rainfall period, and its unit is m³. The rainfall depth is obtained according to the cumulative value of the measured precipitation of the rain gauges associated with each catchment area, and its unit is mm, and the product is multiplied by the area of the catchment area (unit: m²). That is, the total amount of precipitation in each catchment area is obtained. The rainfall data is the time series data of the measured precipitation.

q_{in} is the amount of water infiltrated into the surface soil. In the SWMM model, the general rainfall loss only considers the infiltration loss of the soil. The Horton infiltration model is used in the infiltration part, which describes the various characteristics of the infiltration rate with time during the rainfall process. The calculation equation of soil infiltration reduction is as follows:

$$q_{in} = I_t * \Delta t * S, I_t = f_0 + (f_i - f_0)e^{-\alpha t} \tag{15.7}$$

In Equation (15.7), I_t is the average infiltration rate within the time of Δt (time interval, in s), in mm/s; S is the area of the permeable zone, in m²; f_o is the final infiltration rate, in mm/s; f_i is the initial infiltration rate, in mm/s; α is the infiltration lapse rate, in mm/s.

15.3.2.3.2 Surface Rainwater Conflux Calculation

Urban rainfall confluence refers to the process in which the runoff produced by rainfall converges to the outlet of the basin[16]. The confluence process is greatly affected by the underlying surface factors. Different underlying surfaces have different confluence and convergence effects. The urban underlying surface is complex, and the gap between regions is large. Most of them are impermeable ground. The SWMM model generalizes the study area into several sub-basins (catchment areas), describes the surface characteristics through attributes such as water permeability and Manning coefficient, and constructs a generalized urban pipe network model without changing the drainage capacity of the pipe network, which can effectively simulate the urban surface confluence and pipe network confluence.

Surface confluence is the process of confluence from the surface to the outlet of the watershed after runoff in each catchment area[17]. The SWMM model uses a nonlinear reservoir model to simulate the surface confluence. The basic idea is that each catchment area is generalized as a nonlinear reservoir, the water depth of the reservoir is very shallow, the rainfall is the input, the surface confluence and infiltration are the output, and the outflow at the outlet of the basin is a nonlinear function of the water depth. The calculation principle of the nonlinear reservoir model is as follows (taking the calculation of runoff in the permeable area as an example):

$$\text{Continuity equation} \quad \frac{dV}{dt} = A\frac{dd}{dt} = A*i - Q \tag{15.8}$$

In Equation (15.8), V is the amount of accumulated water in the catchment area, $V = A \times d$; d is the water depth; A is the area of the catchment area; i is the net rain; Q is the outflow:

$$Q = W \cdot \frac{1.49}{n} \cdot (d - d_p)^{5/3} \cdot S^{1/2} \tag{15.9}$$

In Equation (15.9), W is the overflow width of the catchment area; n is the surface Manning roughness coefficient; dp is the surface stagnant water depth; S is the width of the catchment area. Combining Equations (15.8) and (15.9) and simplifying the nonlinear differential equation for solving the water depth d produces the following:

$$\frac{dd}{dt} = i - \frac{1.49W}{A \cdot n}(d - d_p)^{5/3} \cdot S^{1/2} = i + WCON \cdot (d - d_p)^{5/3} \quad WCON = -\frac{1.49WS^{1/2}}{An} \tag{15.10}$$

The overflow width W, slope S, and roughness n of the catchment area are calculated to obtain the parameter $WCON$. The equations are solved using the finite difference method, and the values of net inflow and net outflow are the average values over the time step. Equation (15.10) is processed and the result is as follows:

$$\frac{d_2-d_1}{\Delta t}=i+WCON\left[d_1+\frac{1}{2}(d_2-d_1)-d_p\right]^{5/3} \tag{15.11}$$

In Equation (15.11), d_1 and d_2 represent the initial water depth and the water depth of the final state, respectively. The New-Raphson iteration method is used to solve the problem, and we get:

$$F=\Delta d-\Delta t\left(WCON\beta^{5/3}+i\right) \tag{15.12}$$

In Equation (15.12), F is the Newton function:

$$\beta=d_1+\frac{1}{2}(d_2-d_1)-d_p \tag{15.13}$$

Differentiate the Newton function:

$$\frac{dF}{d(\Delta d)}=1-\Delta t\cdot\frac{5}{6}K\beta^{2/3} \tag{15.14}$$

Finally, the recursive function of AA is obtained:

$$(\Delta d)_{n+1}=(\Delta d)_n-\frac{F_n}{dF_n/d(\Delta d)} \tag{15.15}$$

The method just described can calculate d_2, and substitute it into Equation (15.9) to obtain the outflow at the end of the time step.

Pipe network confluence refers to the process of dredging, collecting, and draining the runoff in the catchment area through the urban drainage pipe network system, and finally draining the drainage outlet in the basin[18].

The calculation method of pipe network drainage is generally based on the Manning equation. The pipe network drainage Q_p is determined by the drainage capacity and drainage duration of the pipe network. The calculation form is as follows:

$$Q_p=\frac{1}{n}\frac{\pi d^2}{4}\left(\frac{d}{4}\right)^{\frac{2}{3}}S^{\frac{1}{2}}\Delta t \tag{15.16}$$

In Equation (15.16), n is the roughness of the pipe wall; d is the pipe diameter, in m; S is the slope of the bottom of the pipe (the ratio of the height difference of the pipe end to the projected length of the pipe in the horizontal direction); Δt is the drainage duration.

15.4 FLOOD DIFFUSION ALGORITHMS

15.4.1 SURFACE FLOOD DIFFUSION ALGORITHM WITH NON-SOURCE

In actual scenarios, floods always occur in areas with low local topographies. Floods gradually diffuse according to the digital elevation model (DEM) and terrain connectivity. After calculating the basin runoff Q, the flood depth and total flood water exist in the following relationship in each catchment according to the water balance theory:

$$\left(\sum_{i=0}^{n} Q_i\right) * Area = \left(\sum_{j=0}^{m} H - H_j\right) * Area \qquad (15.17)$$

where $Area$ is the area of one grid (m²), Q_i is the runoff of the grid i (m), n is the number of grids in the current catchment area, H is the flood elevation in the catchment (m), H_j is the elevation of the grid j (m), and m is the number of flooded grids in the catchment.

A diffusion algorithm with non-source is used to calculate flood depth. Literature[19] assumes that the flood starts from the grid with the lowest elevation in the catchment and gradually diffuses to grids with higher elevations. Flood diffusion ceases when flood elevation H is lower than all non-flood grids.

15.4.2 An Algorithm with Source Inundation and Dynamic Water Distribution

15.4.2.1 Inundation Algorithm Principle

The basic idea of the algorithm with source inundation and dynamic water distribution is as follows. The diffusion sources are nodes. The water spreads around based on terrain with trial method and water balance principle. The result of rationality is judged after each round of diffusion. If the current flood elevation is higher than the elevations of grids neighboring, the diffusion area is expanded to a circle bigger than before; otherwise it will diffuse again in the new area, until the flood elevation is lower than elevations of grids neighboring all flooded grids. After all nodes diffusions are completed, it judges whether the grids are involved in two or more nodes diffusion process. If there are some nodes whose diffusion areas are intersected, the node is made as a new one and the nodes' volume is merged as the new node's volume, and then it diffuses again with the basic idea. It realizes water dynamic allocation and flood dynamic diffusion by judging the rationality of flood depth and area constantly during the diffusion process. The flow direction depends on the difference in grids' elevations without any human intervention, and the flood can flow into any grid around it. There is no boundary in the algorithm, and the inundation end depends on terrain and flood volume, which more matches the actual circumstances.

Specific steps of the algorithm are given in the following with Figure 15.3 and Figure 15.4:

(1) Read the DEM to get each grid's elevations.
(2) Read all nodes' information, including nodes' number, volume, and location.
(3) Diffuse the flood from the first node, assuming that the flood volume of the current node is V, the flood elevation is H, and A is the area of each grid.
(4) Diffuse the flood from the center (source) grid to outside, write the diffusion area of round n as Rn. R1 is the diffusion area in the first round, just shown as the area masked with r1 in Figure 15.3; R2 is not area masked with r2 in Figure 15.3 but r1+r2, so Rn should be r1+r2+. . .+rn, while the current H is that:

$$H = \frac{V}{A} \qquad (15.18)$$

where the unit of H is m, the unit of V is m³, and the unit of A is m².

The H should be compared with the elevation of grids outside around the diffusion area, which is written as compare area. If H is smaller than the minimum of the elevation of compared area, the node's diffusion comes to the end, or else comes to step (5). After all nodes diffusion has been done, it comes to step (6).

(5) Make the diffusion area a circle larger than the former, and diffuse the flood among the new area. Assume that n grids are flooded and their elevations are H1, H2, . . ., Hn. H with trial method and the principle of water balance can be calculated as:

$$\left(i = \sum_{i=0}^{n} H - H_i\right) * A = V \qquad (15.19)$$

where the unit of H and H_i are m, the unit of V is m^3, the unit of A is m^2.

If H is smaller than the minimum of the elevations of all grids surrounding the diffusion area, the node's diffusion comes to the end; otherwise repeat the step. After all nodes diffusion has been done it comes to step (5).

(6) Judge whether there is any grid involved in two or more nodes' diffusion process. If there is not, the flood inundation is finished and the result is returned. If there is, it makes the nodes into a new one and merges the nodes' volumes as the new node's volume with repeating step (4).

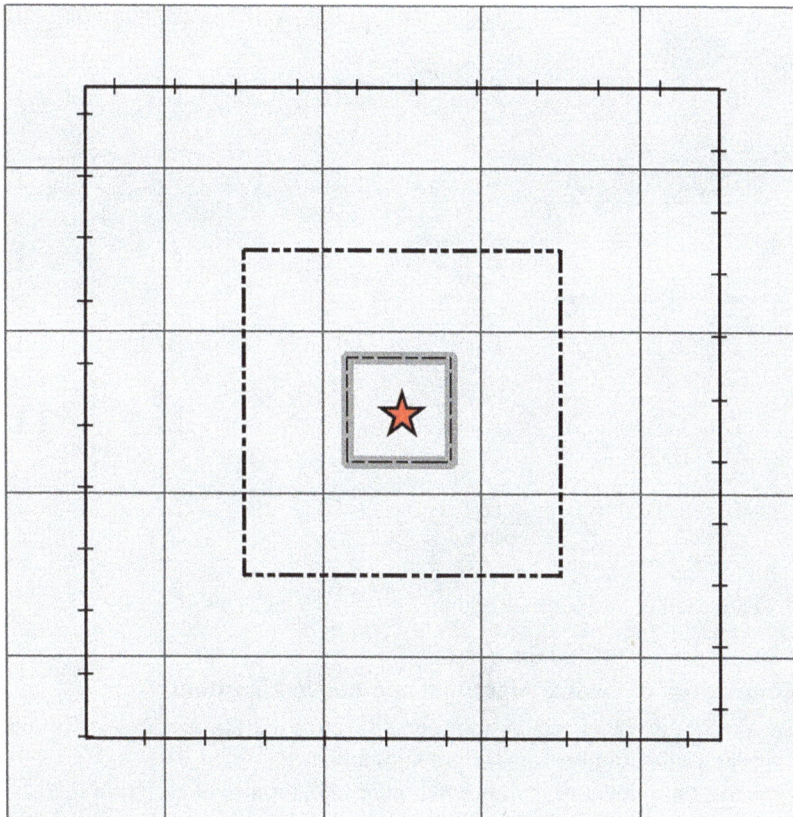

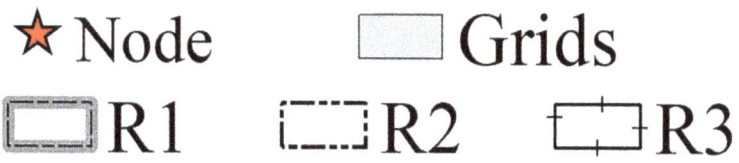

FIGURE 15.3 Auxiliary graph of the source algorithm.

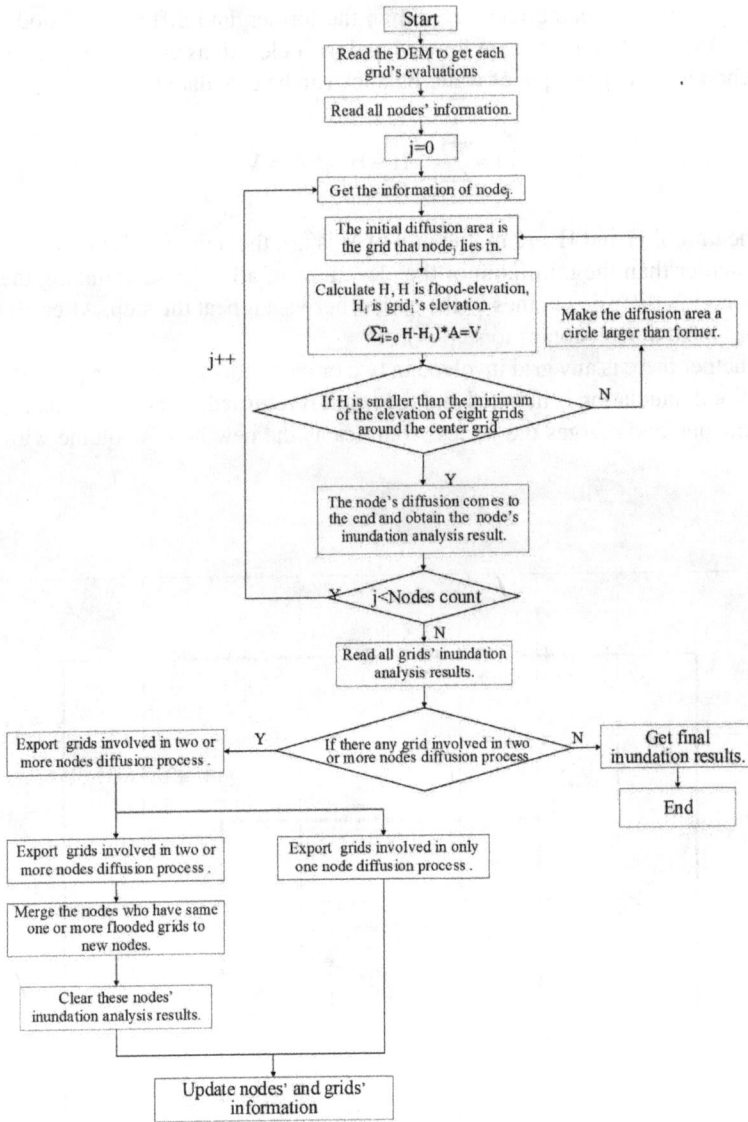

FIGURE 15.4 Flow chart of the source algorithm.

15.4.2.2 Comparison of Passive Algorithm and Active Algorithm

A comparative analysis of the effects of non-source (Section 4.2.1) and source diffusion algorithms (Section 4.2.2) is carried out in the form of graphics and text.

Now suppose that the study area is shown in Figure 15.5, the size of each grid is 10 m × 10 m, and the number shown on the figure is the elevation value of each grid unit (unit is m). The five-pointed star represents the location of the node; that is, the grid where the diffusion source is located. The DEM of the study area is used to analyze the surface catchment area with hydrological tools, and finally two catchment areas as shown in the figure are obtained, and the nodes in the figure exist in the catchment area 1.

(1) It is assumed that after a rainfall, the accumulation of water in the diffusion source (node) is 430 m³ and the diffusion is carried out.

(a)

8.9	9	9.5	9	9.3
9.1	8.4	6.3	9.1	8.8
8.2	8.1 ★	8.3	9.5	9.4
8.6	9.2	9.5	8.9	8.6
9.3	9.5	8.9	9.1	8.8

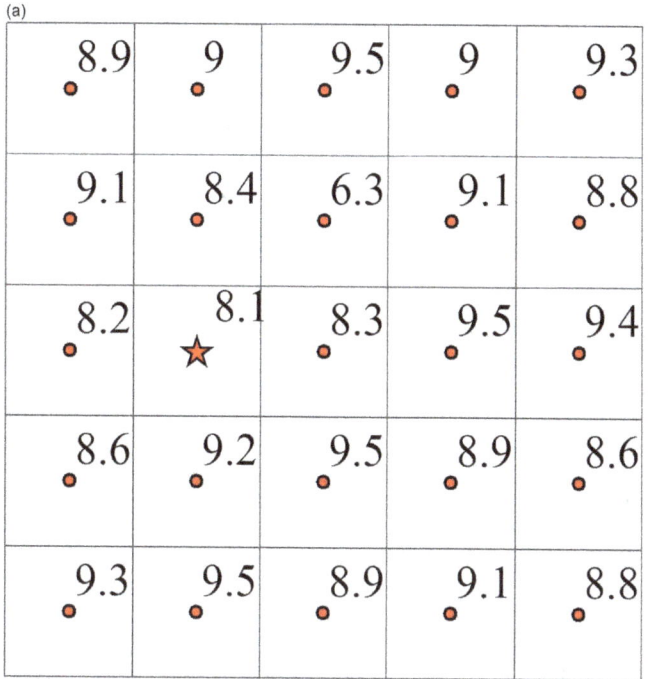

★ Node • Grid center point ▭ Grid

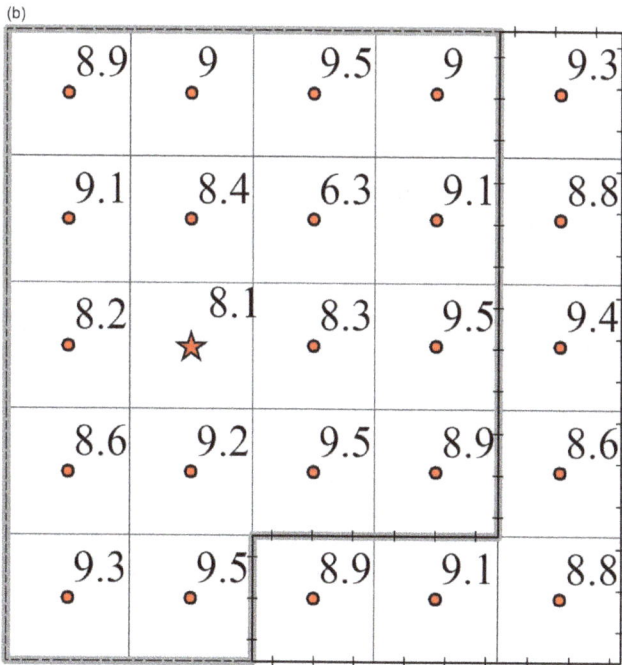

(b)

8.9	9	9.5	9	9.3
9.1	8.4	6.3	9.1	8.8
8.2	8.1 ★	8.3	9.5	9.4
8.6	9.2	9.5	8.9	8.6
9.3	9.5	8.9	9.1	8.8

★ Node • Grid center point ▭ Grid

▭ Catchment 1   ▭ Catchment 2

FIGURE 15.5 DEM (a) and catchment area information (b) of the example area.

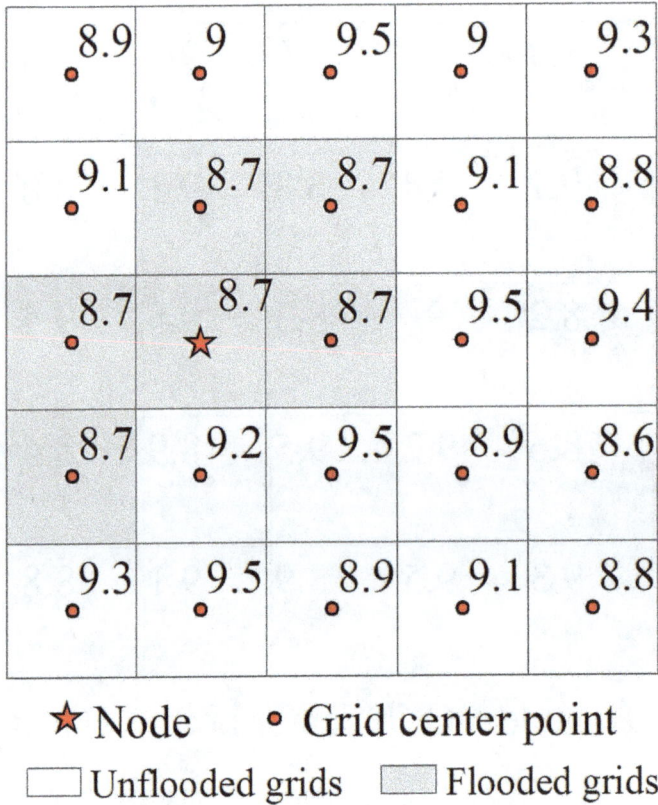

FIGURE 15.6   The result of waterlogging in active diffusion scenario 1.

  a. Active Diffusion Algorithm: Diffusion is first performed only on the diffusion source grid. At this time, the water level of the waterlogging is higher than the water level of the outer circle, and the diffusion range needs to be expanded to the outside, and the situation shown in Figure 15.6 is obtained. The waterlogging water level is 8.7 meters, which is less than the elevation value of all the grids in its outer circle, then the diffusion of the diffusion source ends here, and Figure 15.4 shows the final diffusion of the diffusion source.
  b. Passive Diffusion Algorithm: Diffusion starts from the lowest part of the catchment area, and then searches for the grid with the lowest elevation with the most flooding in turn, and obtains the final distribution Figure 15.7.
     Comparison of results: the waterlogging level and the waterlogging range are the same. Reason analysis: The catchment area presents a depression effect centered on the grid where the node is located. In the current situation, the water volume is concentrated in the same grid around the node.
 (2) It is assumed that after a rainfall, the accumulated water volume of the diffusion source (node) is 730 m^3.
  a. Active Diffusion Algorithm: Take the node as the center, and the result after two circles of diffusion is performed is shown in Figure 15.6a. At this time, if the waterlogged water level is greater than its adjacent grid, the node needs to be re-diffused, and the diffusion range will be increased by one circle. Finally, the situation shown in Figure 15.8b is obtained. The final waterlogged water level is 9.00714, which is approximated to 9.01.

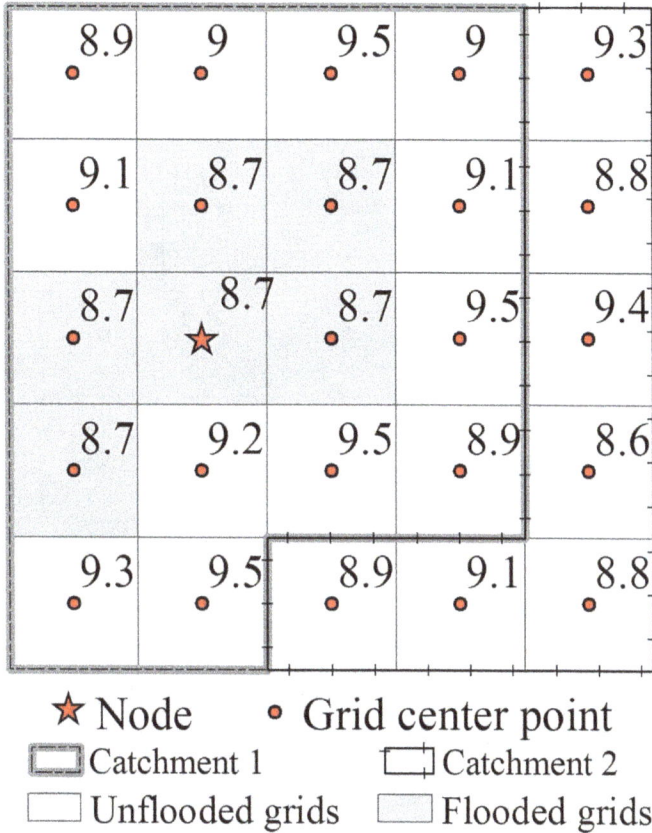

8.9	9	9.5	9	9.3
9.1	8.7	8.7	9.1	8.8
8.7	8.7 ★	8.7	9.5	9.4
8.7	9.2	9.5	8.9	8.6
9.3	9.5	8.9	9.1	8.8

★ Node   • Grid center point
☐ Catchment 1   ☐ Catchment 2
☐ Unflooded grids   ☐ Flooded grids

FIGURE 15.7   Schematic diagram of waterlogging in passive diffusion scenario 1.

    b. Passive Diffusion Algorithm: Diffusion starts from the lowest point of the catchment area, and sequentially finds the grid with the lowest elevation with the most flooding, and obtains the final distribution in Figure 15.9.

Comparison of results: The accumulated water level and the accumulated water range are inconsistent. The accumulated water range obtained by active diffusion is wider and the water level is lower. The active diffusion algorithm forms a continuous distribution in the whole world. The passive diffusion algorithm has the accumulated water level greater than the elevation value of the adjacent grid at the boundary, which is inconsistent with the actual situation and unreasonable.

Reason analysis: Due to the existence of the boundary of the catchment area, the passive algorithm stops the diffusion when it encounters the boundary, while the active diffusion algorithm proposed in this paper will take the node as the center and diffuse in the global area, and there is no boundary limit.

(3) Assuming that there are two nodes with similar positions in this area, after a rainfall, the water accumulation of these two nodes is 730 m³ and 110 m³ respectively.

    a. Active Diffusion Algorithm: The waterlogging accumulation diffusion is performed on these two nodes respectively to obtain the waterlogging accumulation situation as shown in Figure 15.10a. The waterlogging accumulation area formed by these two nodes intersects, so it is necessary to combine the waterlogging accumulation of these two nodes: namely, add the water volume of these two nodes to form a new node. The diffusion range of the new node is the outer polygon of the waterlogging accumulation

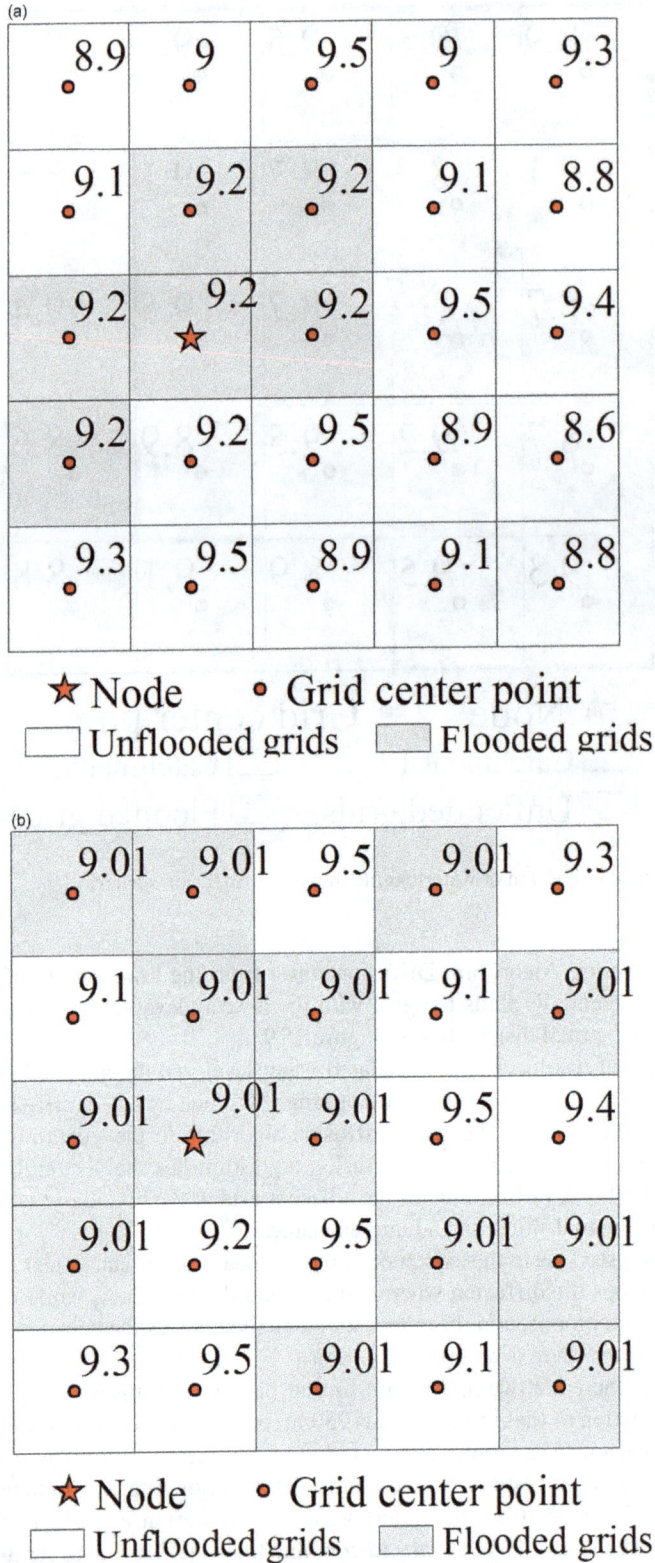

FIGURE 15.8   Simulation of waterlogging in active diffusion scenario 2.

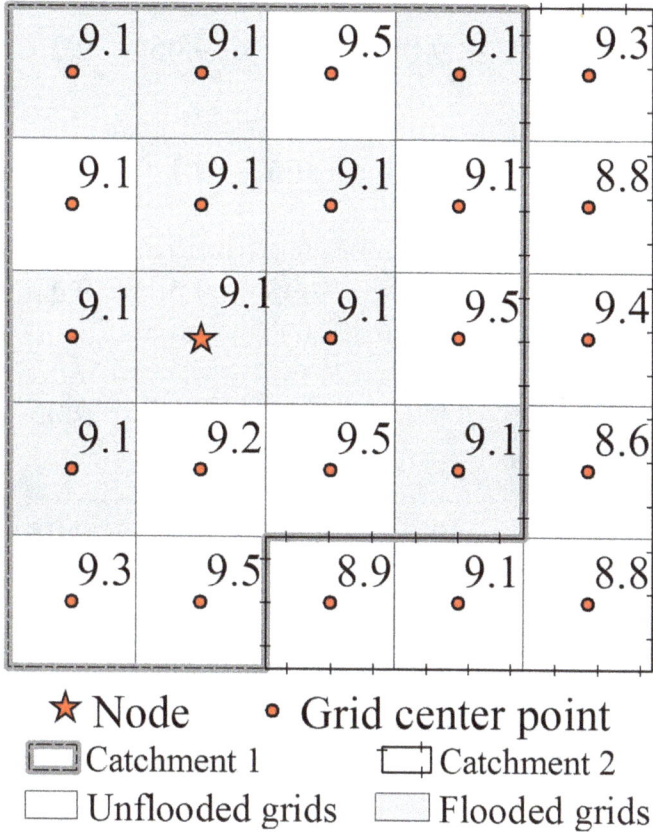

FIGURE 15.9   The result of waterlogging in passive diffusion scenario 2.

area of these two nodes, and finally obtain the waterlogging accumulation result as shown in Figure 15.10b, The final accumulated water level is 9.0533, and the approximate value is 9.05.

b. Passive Diffusion Algorithm: Combine the accumulated water volume of the two nodes as the final accumulated water volume of the catchment area, and diffuse in the catchment area.

Comparison of results: the waterlogging level and the waterlogging range are inconsistent, the active diffusion algorithm obtains a wider waterlogging range, the water level is lower, and the passive diffusion algorithm is unreasonable at the boundary (Figure 15.11).

Reason analysis: As in scenario 2, the boundary of the catchment area limits the diffusion range of accumulated water, but the active diffusion algorithm proposed in this paper does not have boundary restrictions.

Conclusion: Combining these three simulation scenarios, the results of the two algorithms are not much different when the amount of water accumulation is small. In the case of large water volume, the active ponding diffusion algorithm proposed in this paper is more reasonable than the passive diffusion algorithm, which can fully reflect the water flow path according to the terrain height. The passive diffusion algorithm is limited by the boundary. There will be unreasonable situations at the boundary of the catchment area, and passive diffusion can be used for simple surface inundation analysis when rainfall is not large.

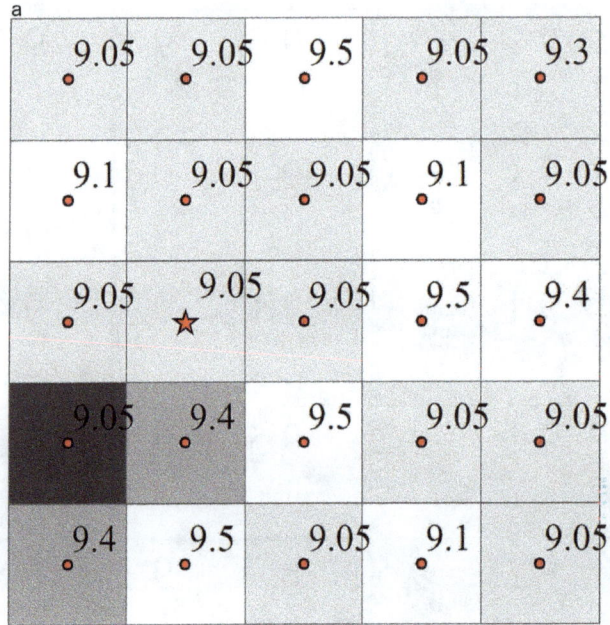

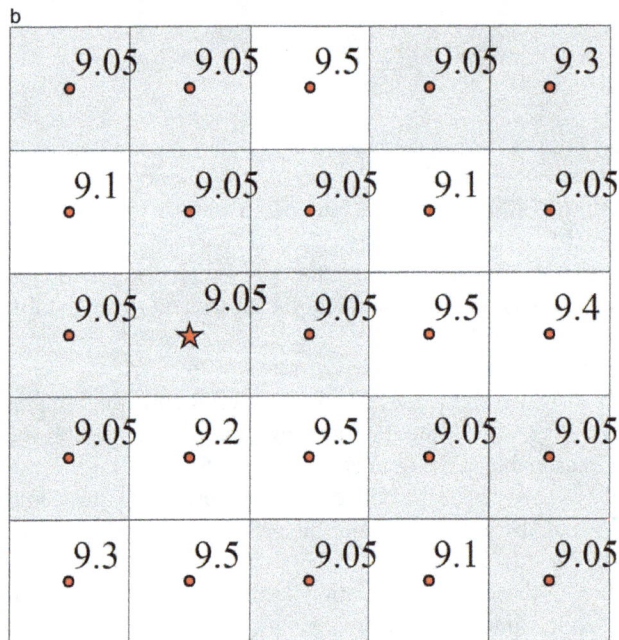

FIGURE 15.10   Simulation of flooding in active diffusion scenario 3.

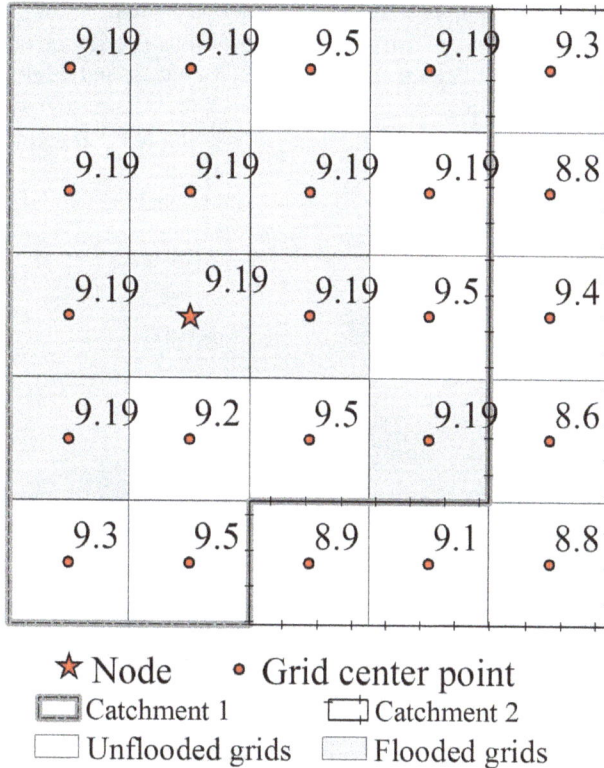

FIGURE 15.11 The result of waterlogging in passive diffusion scenario 3.

15.4.3 ANALYSIS OF SURFACE WATER BY COMBINING DEM AND REMOTE SENSING IMAGES

For the surface ponding analysis combined with DEM and remote sensing images, the inundation range is usually obtained by using remote sensing images, and then the elevation-based hydrological analysis is carried out in combination with the inundation range and DEM data to determine the ponding depth. Reference[20] used Sentinel-1 images before and after the disaster and the RGB composite change detection method to obtain the surface inundation range. For each submerged patch, the waterlogging height DEM_{max} was determined according to equation (20). For all the grids whose elevations are less than the waterlogging height within the same submerged patch, the waterlogging depth H_i is determined according to Equation (15.21):

$$DEM_{max} = \overline{DEM} + 2\sigma_{DEM} \tag{15.20}$$

$$H_i = \max(0, DEM_{max} - DEM_i) \tag{15.21}$$

In Equations (15.19) and (15.20), σ_{DEM} is the elevation standard deviation of all grids in the same waterlogging patch, and DEM_i is the height of the ith grid.

15.4.4 NUMERICAL SIMULATION EXAMPLE OF RAINSTORM WATER DEPTH

15.4.4.1 Numerical Simulation of Standing Water Based on SWMM Model and Active Diffusion Algorithm

Taking Longwen District of Zhangzhou City as an example, the area surrounded by Shuixian Avenue in the north, Zhanghua East Road in the south, Jiulong Avenue in the East, and Longwen

North Road in the west is selected, with an area of about 6.61 square kilometers. This area is the economic activity center of Longwen District, with concentrated population and buildings. In addition, there are green land, forest land, farmland, water body, roads, and other land use types in the demonstration area of the study area. The drainage network in this area is mainly distributed along the road, with 4,424 inspection wells, 34 drainage outlets, and 4,331 pipelines. The specific geographic location of the sample area is shown in Figure 15.12.

Using GIS and remote sensing technology to process and analyze the data, various parameters such as the catchment area and the generalized pipe network data required for building the model are obtained (Figure 15.13).

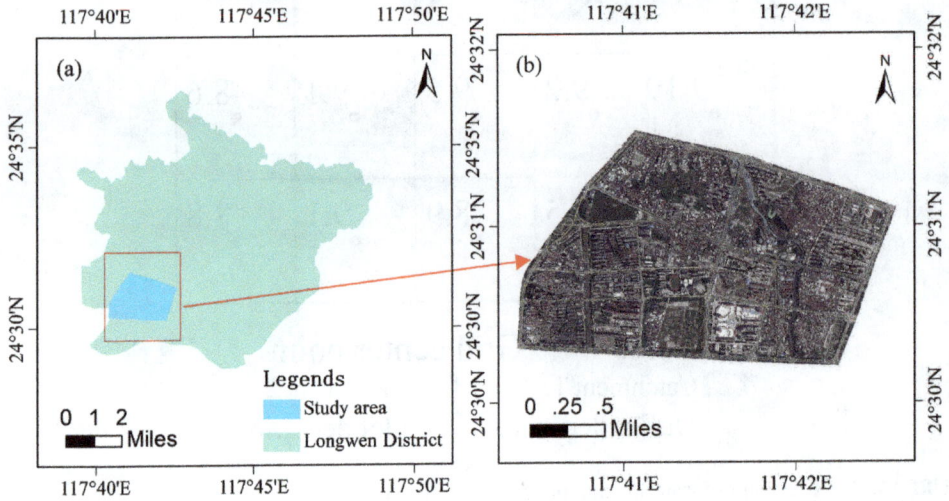

FIGURE 15.12 Basic information of the sample area.

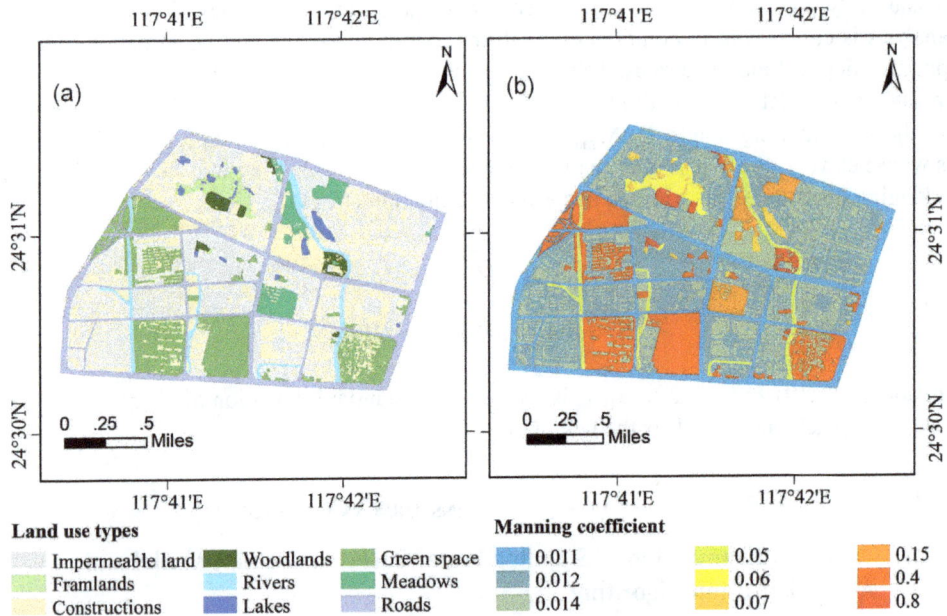

FIGURE 15.13 Modeling parameters of the SWMM model in the example area. (*Continued*)

(c)

Outlets
Nodes
Conduits
Study area

1:20,000

(d)

Nodes
Conduits
Study area

1:20,000

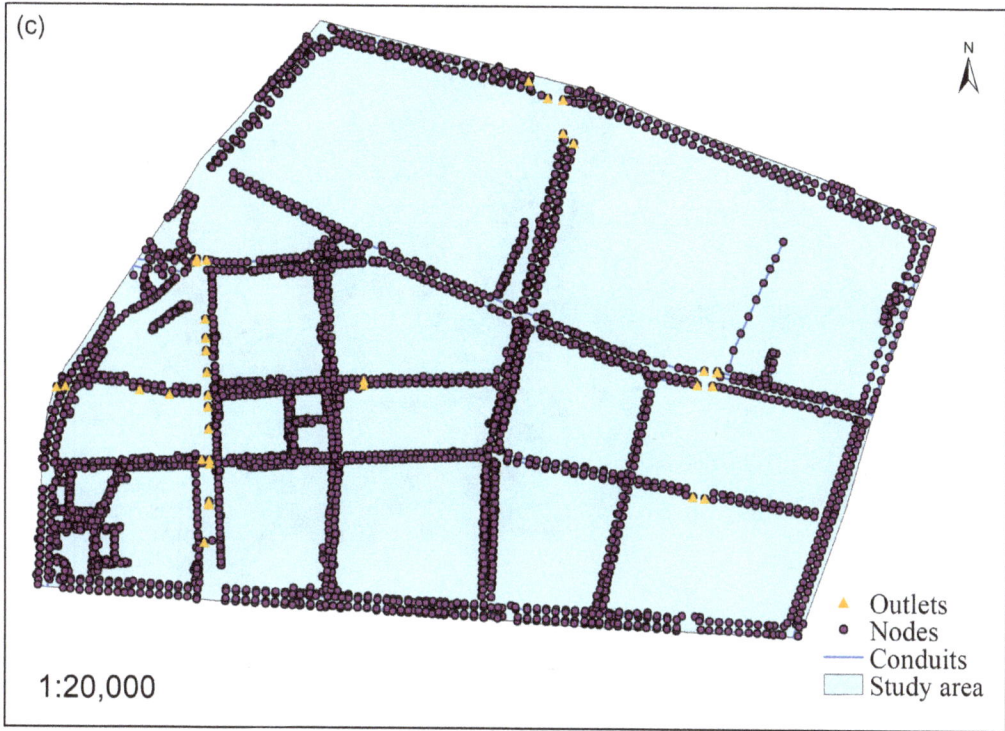

FIGURE 15.13 (*Continued*)

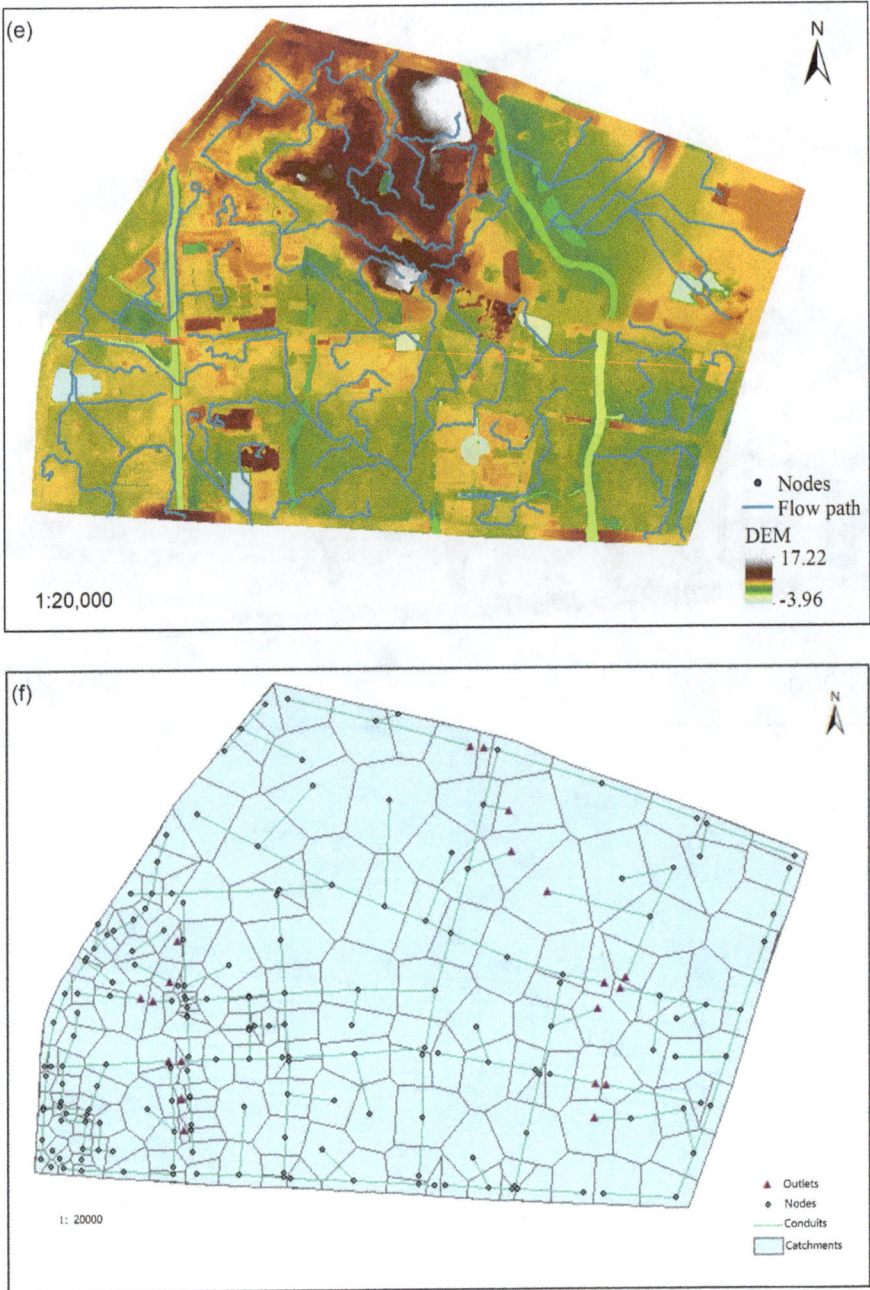

FIGURE 15.13 (*Continued*)

The rainfall on June 18, 2016 is simulated, and the rainfall data adopt the data of the Changfu Village automatic station in the study area. The specific rainfall time series is shown in Table 15.2. Taking 30 minutes as the time interval, the SWMM model is used to simulate the urban rainfall process, and the time series of water accumulation at the nodes is obtained. The active water diffusion algorithm is used to distribute the water volume to the surface grid cells, and the water depth of each grid is obtained. The rainstorm inundation analysis is carried out at each moment, and the simulation results of water accumulation at each moment are output as Figure 15.14.

TABLE 15.2

Rain Gauge Precipitation Time Series on June 18, 2016

Date (YYMMDD)	Time (hh:mm)	Precipitation (mm)
20160618	13:00	13.3
20160618	13:30	33.9
20160618	14:00	1.2
20160618	14:30	0.4
20160618	15:00	0.2
20160618	15:30	0.1

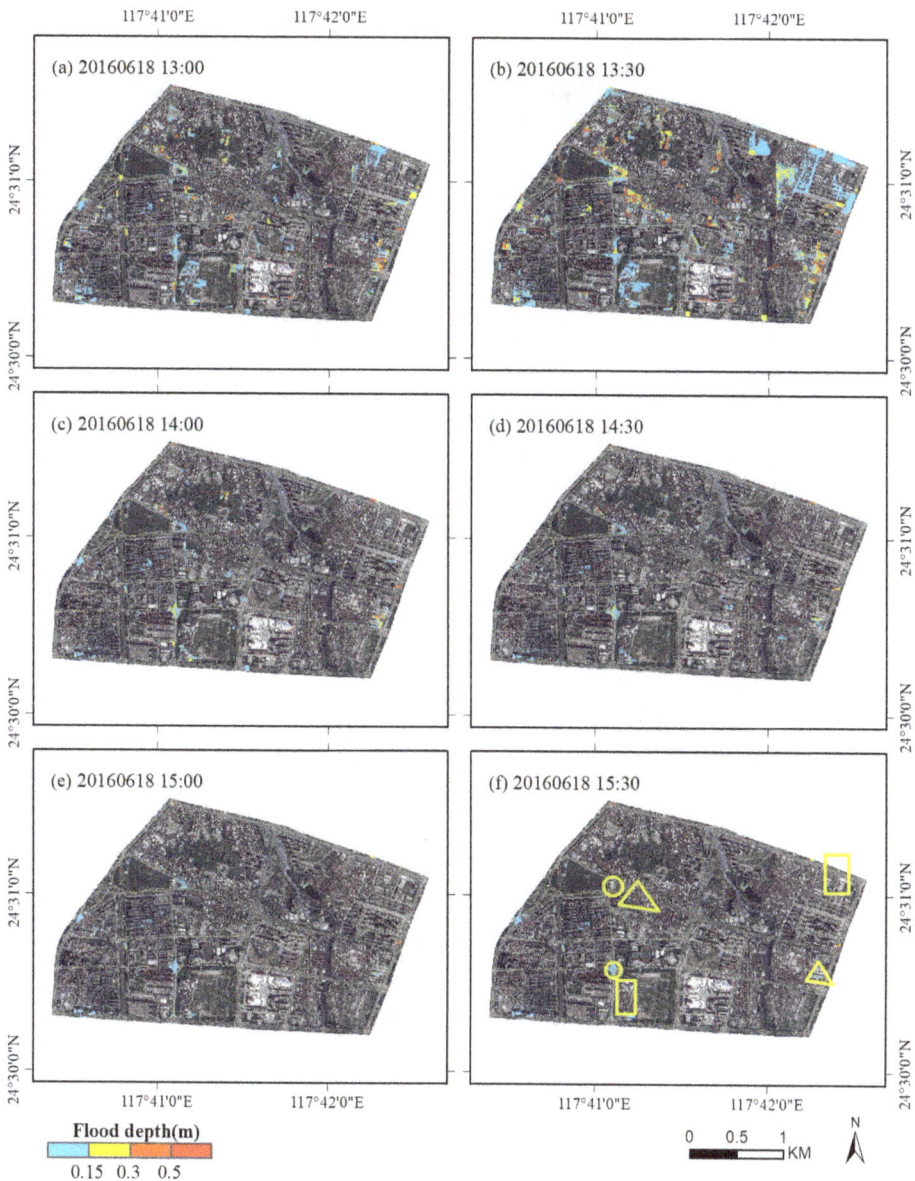

FIGURE 15.14 Simulation results of source algorithm on June 18, 2016.

15.4.4.2 Numerical Simulation of Stagnant Water Based on SCS-CN Model and Passive Algorithm

Taking a part of Kouzizi Village, Shouguang City, Shandong Province as an example, the SCS-CN model and the passive stagnant water diffusion algorithm (section 4.1) are used. The water accumulation depth simulation was performed for two rainfall processes on November 18, 2020 and May 16, 2018. The 24-hour rainfall on November 18, 2020 was 37.7 mm, and the 24-hour rainfall on May 16, 2018 was 68.9 mm. The example area is divided into five categories: water body, vegetation, greenhouse, building area, and road. According to the distribution of land types, the average CN value of the example area is calculated to be about 78.92, and S is 67.84. According to the calculation of the SCS-CN model, when P > 13.57 mm, the sample area begins to produce surface runoff. Figure 15.15 shows basic information such as remote sensing and DEM in the sample area.

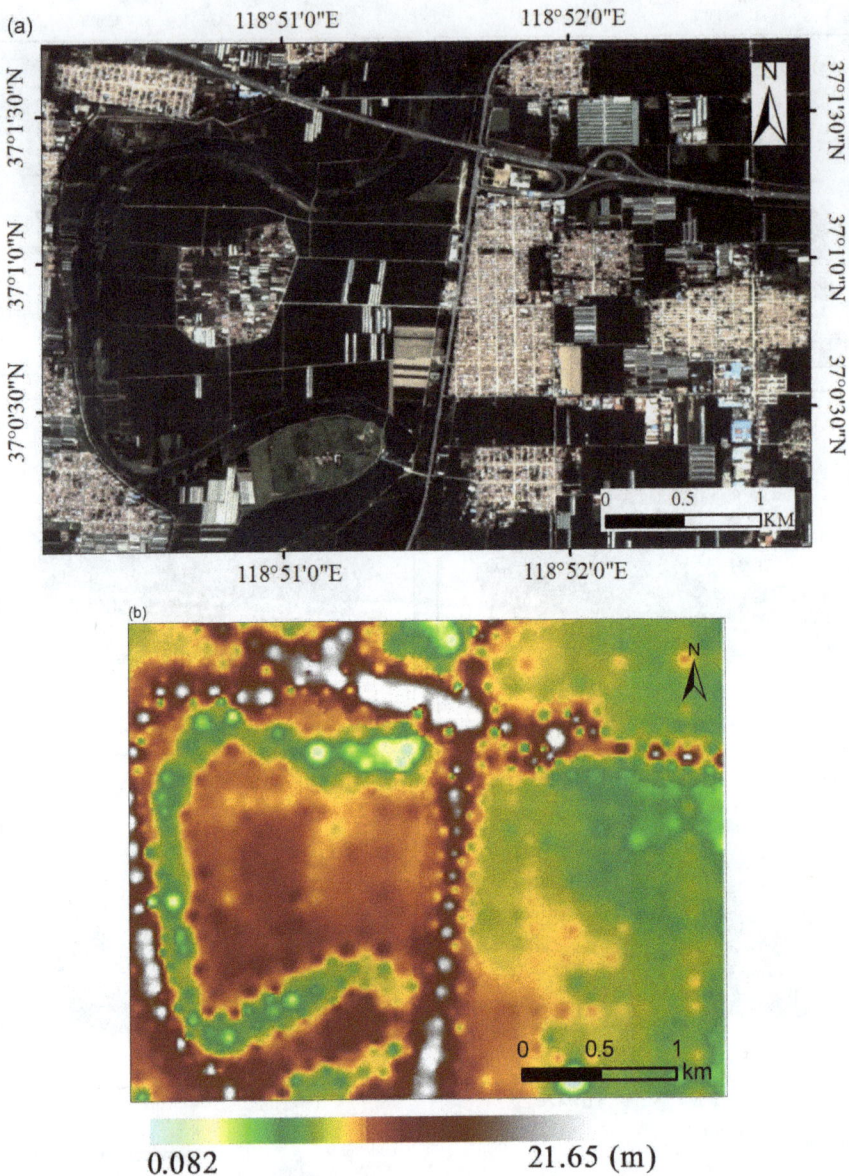

FIGURE 15.15　Basic information of the sample area. (*Continued*)

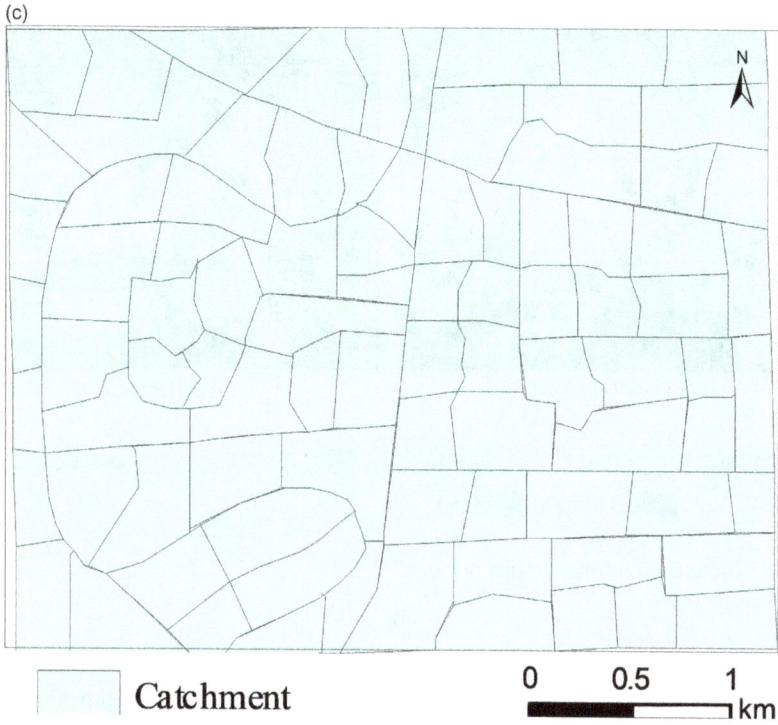

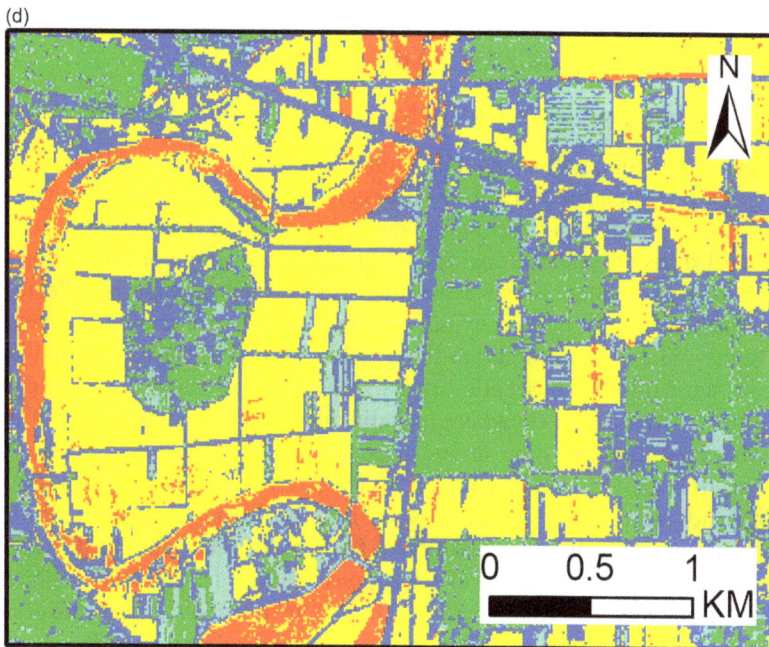

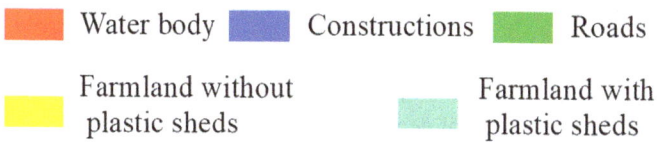

FIGURE 15.15 (*Continued*)

(a) 20201118   (b) 20180516

Flood depth(m)   0~0.1   0.1 ~ 0.3   0.3 ~ 0.5   0   0.5   1 km   N

0.5 ~ 1   > 1

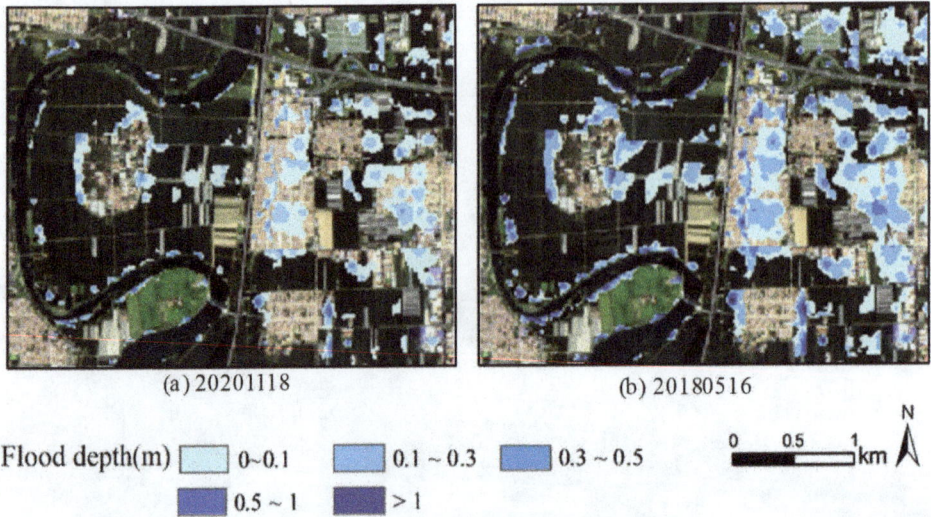

FIGURE 15.16   Simulation results of water depth.

Use the SCS-CN model to calculate the runoff of each grid in each rainfall, count the total runoff of each catchment, and use the passive diffusion algorithm to calculate the depth of water accumulation in each grid. The result is Figure 15.16.

15.5   FLOOD DISASTER RISK ASSESSMENT

15.5.1   THE CONNOTATION OF DISASTER SYSTEM

From the perspective of connotation and definition of flood risk, current scholars have not unified understanding of the elements of flood disaster system. Some scholars believe that flood disaster risk is composed of hazard, exposure, and vulnerability. Hazard refers to the intensity and spatial distribution of disasters, such as the depth, duration, and scope of inundation[21]. Exposure generally refers to the population and economic distribution that may be affected during the flood process, and vulnerability generally refers to the damage degree of the bearing body under the disaster. Some scholars think that the flood risk is composed of hazard and susceptibility. Although the vulnerability and susceptibility show the same meaning in some degree, the connotation of vulnerability is more abundant, including not only the vulnerability of the carrier[22], but also the distribution information of social economy. In recent years, many scholars have proposed a flood disaster vulnerability or sensitivity assessment different from these two theories[23]. Vulnerability refers to the degree or possibility of disaster occurrence. Rainfall index is the direct disaster-causing factor, the condition of flood occurrence, and the disaster result is the final performance of rainfall after a series of actions on the surface. The rainfall index in the traditional risk factors is distinguished from the disaster result, and the method does not consider the disaster-bearing body factor.

15.5.2   RISK EXPRESSION

The United Nations Department of Humanitarian Affairs published the definition of natural disaster risk in 1992: risk is the expected loss of people's lives, property, and economic activities caused by specific natural disasters in a certain region and a given period of time. Scholars use "risk degree" to express the quantified disaster risk.

TABLE 15.3

Judgment Scale of Relative Importance of Evaluation Indicators

Scale	Implication
1	i is as important as j
3	i is slightly more important than j
5	i is significantly more important than j
7	i is strongly important compared to j
9	i is extremely important compared to j
2, 4, 6, 8	The importance of I compared to j is between adjacent odd numbers

The connotation of natural disaster system is different, and the corresponding expression of disaster risk degree is also different. At present, there are mainly four viewpoints on the expression of disaster risk degree: risk = hazard, risk = hazard + vulnerability, risk = probability loss, risk = hazard result. With the deepening of research, the first expression given by domestic and foreign scholars for the United Nations disaster reduction strategy has the highest recognition[24].

15.5.3 ANALYTICAL HIERARCHY PROCESS METHOD

Judgment matrix is the basis of analytical hierarchy process (AHP), and it is a numerical expression for judging the relative importance of various evaluation indicators. The comparative scale a_{ij} is used to express the relative importance of the i index and the j index in the object layer, and the relative importance measure between the two factors is called the judgment scale. For the weight analysis of multiple factors, the pairwise comparison method can be used, and only two factors are compared with each other in the importance of n factors at a time (Table 15.3).

The comparison scale a_{ij} has the characteristics of $a_{ij}>0$, $a_{ij}=1/a_{ji}$, and $a_{ii}=1$ ($i, j=1, 2, 3, \ldots, n$). The formed comparison judgment matrix is a positive and inverse matrix, and the judgment matrix P has the following form:

$$P=\begin{bmatrix} 1 & a_{12} & \cdots & a_{1n} \\ 1/a_{12} & 1 & & a_{2n} \\ \vdots & & \ddots & \vdots \\ 1/a_{1n} & 1/a_{2n} & \cdots & 1 \end{bmatrix}$$

(15.21)

According to the comparison and judgment matrix P, the eigenvectors corresponding to the largest eigenroot and the largest eigenvalue are obtained.

The importance ranking process of AHP is a subjective process. In order to reduce errors caused by one-sidedness, a consistency check is required. A consistency test mainly means that when X1 is more important than X2 and X2 is more important than X3, then X1 must be more important than X3. When the judgments are completely consistent, there should be $\lambda_{max} = n$. Define the consistency index CI as

$$CI = \frac{\lambda max - n}{n - 1}$$

(15.22)

λ_{max} is the largest characteristic root of the comparative judgment matrix P, and n is the number of indicators.

TABLE 15.4

Consistency Check RI Lookup Table

Order	1	2	3	4	5	6	7	8	9	10	11	12	13
RI	0	0	0.52	0.89	1.12	1.26	1.36	1.41	1.46	1.49	1.52	1.54	1.56

CR is the consistency ratio:

$$CR = \frac{CI}{RI} \tag{15.23}$$

When CR < 0.1, the consistency is considered acceptable and the index assignment is reasonable. The average random consistency index RI lookup table of the judgment matrix is shown in Table 15.4.

15.5.4 ENTROPY METHOD

Both the entropy method and the information gain theory are based on the information entropy theory[25]. The information gain theory judges the importance or contribution of features to classification by analyzing the information gain of features. The entropy method can determine the index weight. The entropy method can be used to evaluate the weight results of the analytic hierarchy process numerically, so as to find less risky solutions that tend to be consistent.

As a measure of the uncertainty of the system, the larger the information, the smaller the uncertainty, the smaller the entropy, and the lower the risk degree. Conversely, the smaller the information, the larger the uncertainty, the larger the entropy, and the higher the risk degree.

The concept of information quantity is based on the pan-system observation and control theory. By composing the observation and control object array and the risk observation and control index, the full information content (I_T) measured by the observation and control index is obtained[26–28], namely:

$$I_T = \log_2 n \tag{15-24}$$

In Equation (15.24), I_T is the total amount of information, and n is the total number of risk observation and control indicators.

According to the concept of information amount, after a series of control measures, the reduction of uncertainty before control can be known, that is, the amount of free information (I_T). The calculation equation is

$$I_F = -\sum_{i=1}^{n} p_i \log_2 p_i \tag{15.25}$$

In Equation (15.25), p_i is the probability value (i = 1, 2, . . ., n) corresponding to the i-th observation and control object, and $\sum_{i=1}^{n} p_i = 1$.

Risk is the amount of unknown information that is unavoidable under certain constraints before the implementation of control measures or risk management, that is, the amount of constraint information I_B:

$$I_B = I_T - I_F \tag{15.26}$$

which is

$$I_B = \log_2 n + \sum\nolimits_{i=1}^{n} p_i \log_2 p_i, \ (i = 1, 2, \cdots, n), \quad i = 1 \ 2 \ \ldots \ n \qquad (15.27)$$

I_B is a decrease in the uncertainty of regulation money or pre-management caused by some intrinsic distribution. Its essence is to display and map the intrinsic measure of risk[26]. Under the premise of a certain I_T, the larger the value of I_B, the smaller the value of I_F, so that the effect of risk control and management measures on the risk itself is smaller, and the risk is larger.

15.5.5 Examples of Disaster Risk Assessments

The flood disaster risk assessment framework of "risk" + "vulnerability" is selected, and the risk is composed of disaster-causing factors and disaster-pregnant environment. Through disaster monitoring and simulation, the rainfall is directly transformed into the distribution of ponding degree to realize the assessment of direct disaster-causing factors. The elevation and river distance that have a greater impact on the flood results are selected as the indicators for disaster-pregnant environment assessment. In flood and waterlogging disaster risk assessment, land use and population density are generally selected as assessment indicators of disaster-bearing bodies, and land use is selected as assessment indicators of disaster-bearing bodies. As an important channel for post-disaster rescue, the distance between each pixel and the road is used as an evaluation index of disaster prevention and mitigation capacity. The flood and waterlogging risk assessment system and data sources of various indicators in this study are shown in Table 15.5.

When grading and assigning points to each risk assessment indicator, the higher the inundation, the greater the risk. The degree of submergence is graded, and four grades of 1, 2, 3, and 4 are assigned to no water accumulation, mild water accumulation, moderate water accumulation, and complete submersion.

The lower the elevation, the more prone to flooding. The elevation of the study area is divided into 4 grades (unit is meter) using the natural breakpoint method, and each grade is assigned to 4, 3, 2, and 1 in turn.

The closer you are to the river, the more prone to flooding. Use ArcGIS to generate the river buffer to judge the distance from the pixel to the river, divide the river distance into 4 levels, and assign the distance to 4, 3, 2, and 1 from near to far.

TABLE 15.5
Flood Risk Assessment System

Target Layer	Criterion Layer	Indicator Type	Number	Indicator Layer	Data Sources
Flood disaster risk	Dangerous	Direct hazard	F1	Submerged level	FMIFSC remote sensing monitoring or disaster simulation model
		Disaster-pregnant environment	F2	Ground elevation	1:5000 National Basic Confidential Surveying and Mapping Data
			F3	River distance	Remote Sensing Image Digitization
	Vulnerability	Disaster-bearing body	F4	Land use	
		Disaster Prevention and Mitigation Capability	F5	Road distance	

TABLE 15.6

Weights and Constraints of Each Scheme

Program	Submerged Level	Land Use	Road Distance	River Distance	Ground Elevation	CI	CR	I_B
1	0.468392	0.268058	0.143553	0.075858	0.044138	0.016241	0.014501	0.417328
2	0.473565	0.282092	0.128735	0.079582	0.036027	0.030954	0.027638	0.452154
3	0.502474	0.256983	0.12681	0.078294	0.035439	0.043424	0.038771	0.482991
4	0.449418	0.292022	0.137571	0.084536	0.036453	0.025641	0.022894	0.415601

When land use is used as a disaster-bearing body indicator in risk assessment, the social attributes of land types are mainly considered, and the degree of influence of land types on disaster formation is not considered. The relative importance of each category is listed as follows: construction area, road, greenhouse, farmland, water body. The higher the importance, the greater the risk. According to its importance, the categories are assigned 4, 3, 2, 1, and 0 in turn.

As an evaluation index for disaster prevention and mitigation, roads are the main channel for escape and post-disaster rescue, so the closer you are to the road, the greater the risk. According to the distance from each pixel to the road, it is divided into 1, 2, 3, and 4 in order from near to far.

The relative importance of flood disaster risk assessment indicators is in order: inundation degree, land use, road distance, river distance, and ground elevation. According to the relative importance, AHP is used to construct several weight schemes, and finally four schemes satisfying the consistency test (CR < 0.1) are determined. The index weight result is taken as the risk observation and control index, and the entropy method is used to calculate the constraint information of each scheme. The detailed indexes, CI, Cr, and values of each scheme are shown in Table 15.6. The CI and CR values of scheme 1 are the smallest among the eight schemes, and the value of scheme 4 is the smallest. According to the pan-system control theory, the smaller the I_B, the smaller the risk. Finally, scheme 4 is selected as the weight scheme of each indicator, and the weights of each indicator are 0.449418, 0.292022, 0.137571, 0.084536, and 0.036453.

According to the risk assessment framework, choose the expression "Risk (Risk) = Risk (H) Vulnerability (V)" to quantify the risk value.

The risk score of a single pixel is R, and the total risk score of the on-site rainfall process is R_sum, which is calculated as

$$R = f(H,V) = w_1 * f_1 = w_2 * f_2 + w_3 * f_3 + w_4 * f_4 + w_5 * f_5 \qquad (15.28)$$

In Equation (15.28), w_i is the weight corresponding to the i evaluation index, and S is the score corresponding to the i evaluation index.

Based on the flood disaster risk assessment framework, the disaster risk under four design rainfall events (Table 15.7) was assessed. Based on the risk assessment results of Design Scenario 3, the natural breakpoint method is used to divide the risk results into four levels, which are divided into low-risk areas, medium-risk areas, high-risk areas, and high-risk areas. The spatial distribution of risk zoning under the four design rainfall scenarios is shown in Figure 15.17, and the proportion of risk levels of each rainfall scenario is calculated. The results are shown in Table 15.7.

Combined with the analysis in Figure 15.17 and Table 15.8, it can be seen that under each rainfall scenario, the medium-risk area is the largest and most widely distributed among all risk levels. It is mainly distributed in the two main roads, the construction area and most of the farmland on the west side of the main road in the study area. The second lowest proportion is that the low-risk areas are mainly distributed in the farmland area on the east side of the main road. The high-risk areas accounted for the lowest proportion, and the spatial distribution of higher-risk areas and high-risk areas were similar, mainly distributed in farmland along the river and a small part of roads.

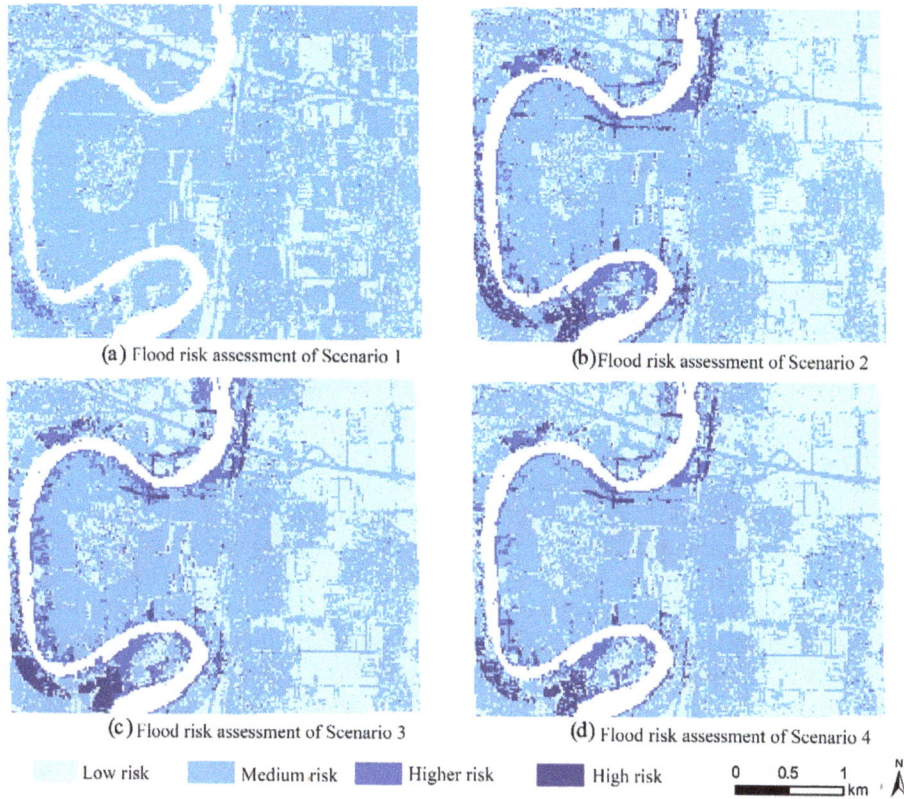

(a) Flood risk assessment of Scenario 1   (b) Flood risk assessment of Scenario 2

(c) Flood risk assessment of Scenario 3   (d) Flood risk assessment of Scenario 4

Low risk   Medium risk   Higher risk   High risk   0 0.5 1 km

FIGURE 15.17 Risk zoning for the design rainfall scenario.

TABLE 15.7
Design Storm Rainfall Process Table

Serial Number	Rainfall process (mm)			Rain Days	Total Rainfall (mm)
	Day 1	Day 2	Day 3		
1	0	0	140	1	140
2	0	110	115	2	220
3	0	85	140	2	
4	60	85	115	3	260

TABLE 15.8
Disaster Risk Assessment Results of the Designed Rainstorm Scenario

Rain Scenario	Rain Days	Total Rainfall (mm)	Low Risk	Medium Risk	Higher Risk	High Risk
Rain Scenario 1	1	140	24.92%	74.05%	1.04%	0.00%
Rain Scenario 2	2	220	36.91%	53.33%	6.65%	3.11%
Rain Scenario 3	3	220	37.92%	51.59%	7.43%	3.05%
Rain Scenario 4	4	260	38.92%	53.34%	5.45%	2.29%

15.6   CONCLUSION

The hydrological analysis based on GIS technology, with rainfall and terrain as the basic data, and a hydrological model as the core, combined with the surface inundation analysis algorithm, can realize the numerical simulation of surface ponding depth and the conversion from rainfall factor to the spatial distribution of ponding, but the data requirements are complex and the simulation accuracy is greatly affected by the model and algorithm. The SCS-CN model is widely used because of its simple results, few parameters and high accuracy. The SWMM model is one of the most commonly used stormwater models. Most of its parameters have physical meanings. Because it is open source and free, the secondary development of the SWMM model based on GIS technology is a common method to study the numerical simulation of urban waterlogging. It can realize the "fine and dynamic" simulation of urban rainfall and waterlogging process. Both passive diffusion algorithm and active diffusion algorithm have their applicability and can solve specific problems, but there are also some limitations, which need to be optimized according to the research objectives in practical applications.

Flood and waterlogging disaster risk assessment uses numerical values to quantitatively assess the probability and distribution of risk occurrence, which can provide the most intuitive decision-making reference for disaster prevention and reduction. However, the process of disaster risk assessment is usually subjective, and the assessment content is closely related to the research objectives. By establishing a flood disaster simulation model in the study area, the disaster simulation results of specific scenarios are obtained, and the risk assessment of specific rainfall scenarios is carried out, so as to realize the dynamic assessment of flood disaster risk and the direct assessment of disaster causing factors. Through the superposition of disaster data and socio-economic information, a dynamic assessment of flood risk is achieved[23].

REFERENCES

[1]   Wang S., Bao X., Rong Y., et al. Study on soil moisture variation and runoff characteristics on typical karst slope under different rainfall intensities. Research of Agricultural Modernization, 2020, 41(5): 889–898.

[2]   Yang K., Lin R., Li Y. Effects of rainfall on soil moisture in different soil layers. Agriculture and Technology, 2015, 35(8): 200–201.

[3]   Zhang X., Luo J., Chen L., et al. Zoning of Chinese flood hazard risk. Journal of Hydraulic Engineering, 2000, (3): 3–9.

[4]   Deng P. Study on the flood-peak runoff coefficient method of urban storm water pipe design and mathematic modell. Water & Wastewater Engineering, 2014, 40(5): 108–112.

[5]   Wang Y., Wang T., Xiang B. Study on improvement and application of urban AMC in SCS model. Journal of China Hydrology, 2011, 31(4): 23–26,57.

[6]   Ren B., Zhou S., Deng R. Analysis of property and calculating method of urban surface runoff yield. Journal of University of South China (Science and Technology), 2006, 20(1): 8–12.

[7]   Ren B., Deng R. Analyses of properties and calculation methods of urban surface rainwater conflux. China Water & Wastewater, 2006, 22(14): 39–42.

[8]   Zhang X., Feng J., Liu F. Development and application of storm flood computation model for urban rain pipe network. Water Resources and Power, 2008, 26(5): 40–42, 103.

[9]   Zhang D., Liu R., Zhang Y., et al. The design and implement of a new algorithm to calculate source flood submerge area based on DEM. Journal of East China University of Technology (Natural Science), 2009, (2): 181–184.

[10] Lu Y., Deng Y., Deng H., et al. Improved rainstorm flood simulation algorithm and its application in Huaihe river basin. Advances in Meteorological Science and Technology, 2020, 10(5): 121–129.

[11] Cheng J. Spatial division and feature analysis of the disaster risk of urban water logging in Fuzhou based on the SCS-CN. Geospatial Information, 2020, 18(4): 92–95+124+7.

[12] Tian T., Mou F., Wang J., et al. Impact of land use change on surface runoff in the main urban area of Chongqing. Research of Soil and Water Conservation, 2021, 28(4): 128–135.

[13] Qin P., Lei K., Qiao F., et al. Impact of sub-catchment size delineation on urban non-point source pollution simulation using SWMM. Environmental Science & Technology, 2016, (6): 179–186.

[14] Yang B. Research and application of automatic calibration of parameters of hydrological forecasting model in flood watershed. Journal of Dalian University of Technology, 2008, 15–16.

[15] Hua Y., Wang Z., Han Z., et al. Simulation experiment study on runoff generation and confluence characteristics of rainfall on typical urban underlying surface. Journal of Tianjin University of Technology, 2016, (6): 48–53.

[16] Zou X., Liu J., Zhang Y., et al. Calculation of kinematic wave for urban surface rainfall runoff. Journal of China Hydrology, 2015, 35(2): 12–16.

[17] Xu Z., Wu L., Zou H., et al. The forecasting model of urban pipe network convergence node pump station. Proceedings of the 2011 International Conference on Software Engineering and Multimedia Communication (SEMC 2011 V1), 2011: 169–172.

[18] Song J., Zhu Y. Application of GIS technology to urban earthquake prevention and disaster reduction. China Earthquake Engineering Journal, 2002, 24(1): 85–91.

[19] Wang W. Research on urban Waterlogging simulation based on GIS and SWMM model. Journal of Nanjing University of Information Science & Technology (Natural Science Edition), 2017.

[20] Jo M.-J., Osmanoglu B., Zhang B., Wdowinski S. Flood extent mapping using dual-polarimetric Sentinel-1 synthetic aperture radar imagery. Proceedings of the ISPRS Technical Commission III Midterm Symposium on "Developments, Technologies and Applications in Remote Sensing". Proceedings of the ISPRS Technical Commission III Midterm Symposium on "Developments", 2018: 936–938.

[21] Huang M., Yang S., Qi W., et al. Numerical simulation of urban waterlogging reduction effect in Guyuan Sponge City. Water Resources Protection, 2019, 35(5): 13–18 + 39.

[22] Huang G., Luo H., Lu X., et al. Study on risk analysis and zoning method of urban flood disaster. Water Resources Protection, 2020, 36(6): 1–6 + 17.

[23] Chen W, Li Y, Xue W, et al. Modeling flood susceptibility using data-driven approaches of naïve Bayes tree, alternating decision tree, and random forest methods. Science of the Total Environment, 2020, 701: 134979.

[24] Liao Y., Nie C., Yang L., et al. An overview of the risk assessment of flood disaster. Progress in Geography, 2012, 31(3): 361–367.

[25] Yu H., Wang Y. A viewing-controlling analysis to risk. Science & Technology Progress and Policy, 1997, (3): 71–74.

[26] Li X., Huang Q., Leon F., et al. Pansystems observation-control model of periphery and its application to water resources. Journal of Lanzhou University (Natural Sciences), 2005, (5): 19–24.

[27] Yu H., Yu K., Wu X. New technique to estimate information quantitated. Science & Technology Progress and Policy, 2001, (7): 105–106.

[28] Li X., Wei X. Evolution risk analysis of regional soil erosion: A case study of Jinghe River Basin. Geographical Research, 2011, 30(8): 1361–1369.

16 Numerical Weather Forecast and Multi-Meteorological Data Fusion Based on Artificial Intelligence

Yuanjian Yang, Shuai Wang, Wenjian Zheng,
Shaohui Zhou, Zexia Duan, and Mengya Wang

16.1 INTRODUCTION

Artificial intelligence (AI) can obtain beneficial insights and improve insufficient understanding of the physical mechanisms in the weather and climate system, and machine learning models can naturally handle a variety of different weather and climate observation and simulation data sources (Li et al., 2021; Haupt et al., 2021; Wang et al., 2019, 2020; Vannitsem et al., 2021; Zeng et al., 2020), such as numerical weather and climate prediction results, radar, satellite, site observations, etc., and even decision data (natural language), which is almost impossible with existing weather climate models. The machine learning is to find an algorithm to fit the data with a lot of parameters. Traditional machine learning algorithms can be summarized as follows: linear regression, logistic regression, SVM (support vector machine), decision tree and random forests, boosting, clustering, dimension reduction, Bayesian learning, graphic models and sparse learning. Deep learning algorithms can be summarized as follows: neural network, CNN (convolutional neural network), RNN (recurrent neural network), and LSTM (long short-term memory networks). This chapter will introduce two aspects of artificial intelligence in numerical weather forecast correction and multi-source Met data fusion and reconstruction in details.

16.2 ARTIFICIAL INTELLIGENCE NUMERICAL WEATHER PREDICTION

A revolutionary change in weather forecasting occurred during the 1950s. Technological advances have allowed us to use models to simulate atmospheric motion, a method that is fast and accurate in forecasting operations. The numerical model remains at the heart of weather forecasting even now. Using basic physics, the numerical model can predict storms before their formation. In recent years, previous studies showed that AI technology can provide new opportunities for the development of objective weather forecasting methods by using direct physical simulations with AI big data-driven approaches (Haupt et al., 2021; Yang et al., 2022; Vannitsem et al., 2021; Zhang et al., 2022). In particular, more and more researches apply AI technology to all aspects of forecasting business, such as 1) short-term nowcasting technology such as identification and extrapolation of strong convection; 2) based on numerical weather forecast results; and 3) short-term climate prediction technology such as seasonal forecasting, etc. (Geer, 2021; Haupt et al., 2021; Kashinath et al., 2021; Yang et al., 2022; Zhang et al., 2022). Taking deep learning techniques as an example, the growing studies on deep learning techniques have been applied widely to weather forecasting, including statistical postprocessing, ensemble forecasting, analog ensemble, statistical downscaling, data-driven forecasting models and extreme weather forecasting.

DOI: 10.1201/9781003363118-16

The terrain of southern China is high in the west and low in the east, where plains, hills, basins and plateaus intertwine in a complex manner. The complex micro-topography forms the micro-meteorology. Micro-meteorology refers to the changes of small-scale meteorological elements in the near-surface atmosphere and upper soil caused by certain tectonic features (such as micro-topography, frost, rain, etc.). However, the change of micro-meteorology elements will not change the characteristics of weather and climate determined by large-scale processes (advection, frontal) greatly. There is a lot of inaccuracy in the temperature forecast under micro-topographic conditions, so the forecast results need to be corrected for accuracy. This chapter focuses on the correction and accuracy analysis of air temperature forecasts in southern China with the latitude range of 17.3°N–32.1°N and the longitude range of 89.6°E–119.4°E.

16.2.1 Introduction to WRF Model

The Weather Research and Forecasting Model (WRF model) is one of the effective means for forecasting meteorological elements such as temperature, humidity and wind pressure, and can effectively improve the accuracy of ice forecasting under complex terrain. Although the WRF model has updated the high-altitude data information, for the study area of the five southern China provinces under complex terrain conditions, the resolution of the terrain data provided by the WRF model still cannot meet the simulation needs of low-level meteorological elements, resulting in a relatively large error in the forecast results of meteorological elements, resulting in a large icing prediction error. Therefore, it is necessary to introduce the SRTM3DEM terrain data into the WRF model for the simulation of meteorological elements, so as to improve the simulation accuracy of icing. In addition, the atmospheric boundary layer is greatly affected by the underlying surface, and the turbulence characteristics are more obvious. In the simulation study of meteorological elements using the WRF model, special attention should be paid to the selection of the parameterization scheme of the boundary layer. In recent years, based on numerical models at home and abroad, it has become a hotspot to carry out the meteorological element simulation of different boundary layer parameterization schemes and to analyze the difference of the simulation effect of each meteorological element.

The WRF model is a new generation of mesoscale numerical model. Here we select version 4.2, which is a mesoscale weather forecast model developed by scientific research institutions such as the Center for Environmental Prediction (NCEP) in the United States. It is widely used in research and forecasting of weather systems, climate change and environmental pollution. It is written in F90 language and has the characteristics of portability, scalability and high efficiency. ArakawaC grid points are used in the horizontal direction, and terrain-following mass coordinates are used in the vertical direction. It is mainly divided into ARW (the Advanced Research WRF) and NMM (the Nonhydrostatic Mesoscale Model), which are managed and maintained by NCEP and National Center for Atmospheric Research (NCAR), respectively. ARW is mainly used in meteorological scientific research, and NMM is mainly used in meteorological business. As the core of the system, WRF-NMM is flexible, which is a perfect and efficient atmospheric simulation system, and can be operated in parallel. It has the characteristics of multiple nesting and rich parameterization scheme design. The parameterization scheme mainly includes the parameterization scheme of microphysical processes, long-wave radiation, short-wave radiation, boundary layer, near-surface layer, land surface process and cumulus convection.

The calculation process of the WRF model mainly includes four parts: input data, preprocessing system WPS of WRF model, main part of model system and model post-processing system. The preprocessing system (WPS) of the WRF model is used for real-time data processing. The functions include defining the simulation area, interpolating terrain data (such as terrain, soil type, etc.) into the simulation area, interpolating other model data into the simulation area and model coordinates. The main part of WRF system is a key component of the whole system. For different physical processes, appropriate parameterization schemes are selected for forecasting and simulation research. The post-processing system re-analyzes, extracts and visualizes the output of the WRF model, including conversion programs such as NCL and Grads, interpolating the forecast grid to the normal grid, calculating the diagnostic output, and processing and analyzing the output of the model.

16.2.2   WRF Model Data Source

When forecasting meteorological elements, the WRF model uses the global forecast system (GFS) developed by NCEP. The GFS includes data related to the atmosphere and land such as temperature, precipitation and wind data. The system is updated every 6 hours, at 0:00, 6:00, 12:00 and 18:00, respectively. The data at each time can be used for forecasting the next 8 days. The time scale of the measured data in the study area is 1 hour, so the time interval of the forecast data of the WRF model is also set to 1 hour.

16.2.3   Mode Scheme Setting

Using the WRF model, combined with the daily data resolution of $0.25° \times 0.25°$, the starting time of forecast is 18:00 UTC, the time resolution of forecast is 3 hours, and the NCEP/GFS forecast field data are used as the initial field and the forecast field data and lateral boundary conditions of the WRF model; obtaining surface static data with a resolution of 500 m provided by MODIS satellites, such as topography, soil data, and vegetation coverage; combined with the two-layer grid nesting layer, the forecast area is shown in Figure 16.1.

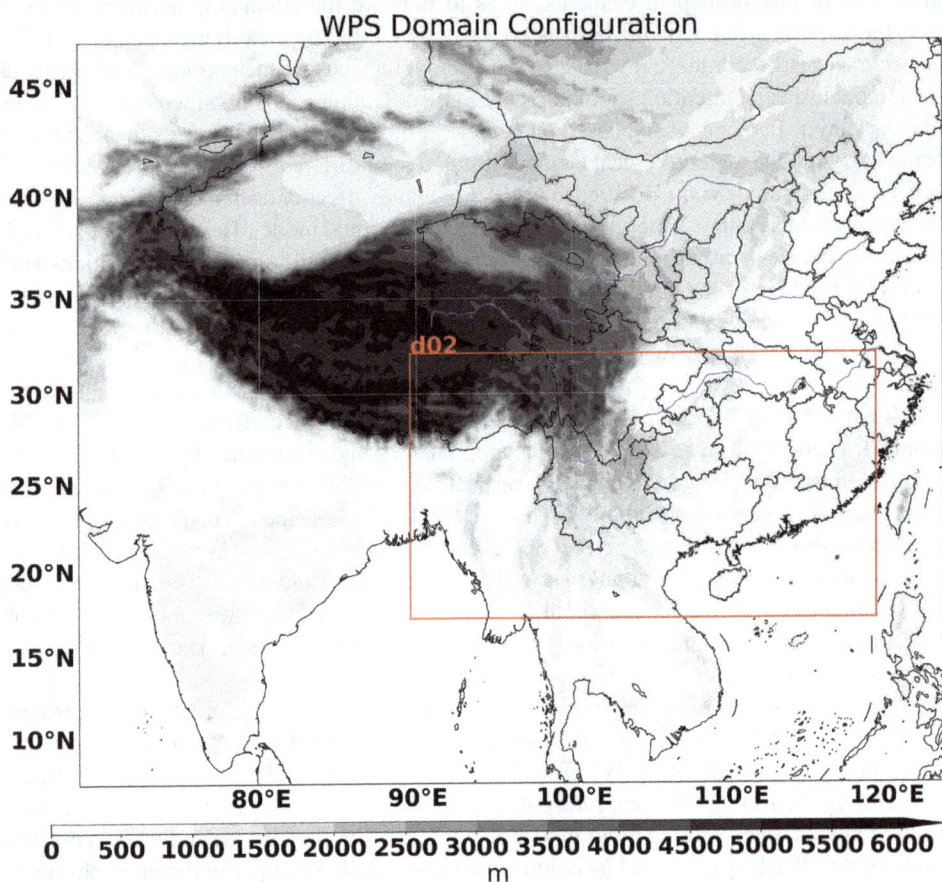

FIGURE 16.1   Schematic diagram of WRF model simulation and domains 2 (d02) area.

The grid numbers are 600×500, 967×535; the horizontal grid resolutions are 9 km, 3 km; and the grid center points are set at 29°N and 96°E. Combined with the parameter named "CONUS" parameterization scheme: the microphysics scheme is the Thompson scheme; the cumulus parameterization scheme is the Tiedtke scheme; the long and short wave radiation schemes are the RRTMG scheme; the boundary layer and near-ground parameterization schemes are both the MYJ scheme; the pavement process scheme is generated by the Noah pavement process scheme. WRFOUT numerical weather forecast file includes temperature, humidity, precipitation and other meteorological elements. The WPS configuration process is shown in Figure 16.2.

16.2.4 SRTM3DEM Data

The SRTM3 data (with a resolution of about 90 m) used in this paper belongs to the fourth version. The measurement is jointly completed by NASA and other institutions with high precision, covering more than 80% of the land surface, which can be obtained directly and free of charge. It is realistic, but its data format cannot be directly applied by the WRF model, which brings inconvenience to analysis and research.

The three main steps for WRF forecasting flow and artificial intelligence (random forest) correction flow chart are shown in Figure 16.2, including WRF model forecasting, random forest model training and testing, and forecast correction and verification.

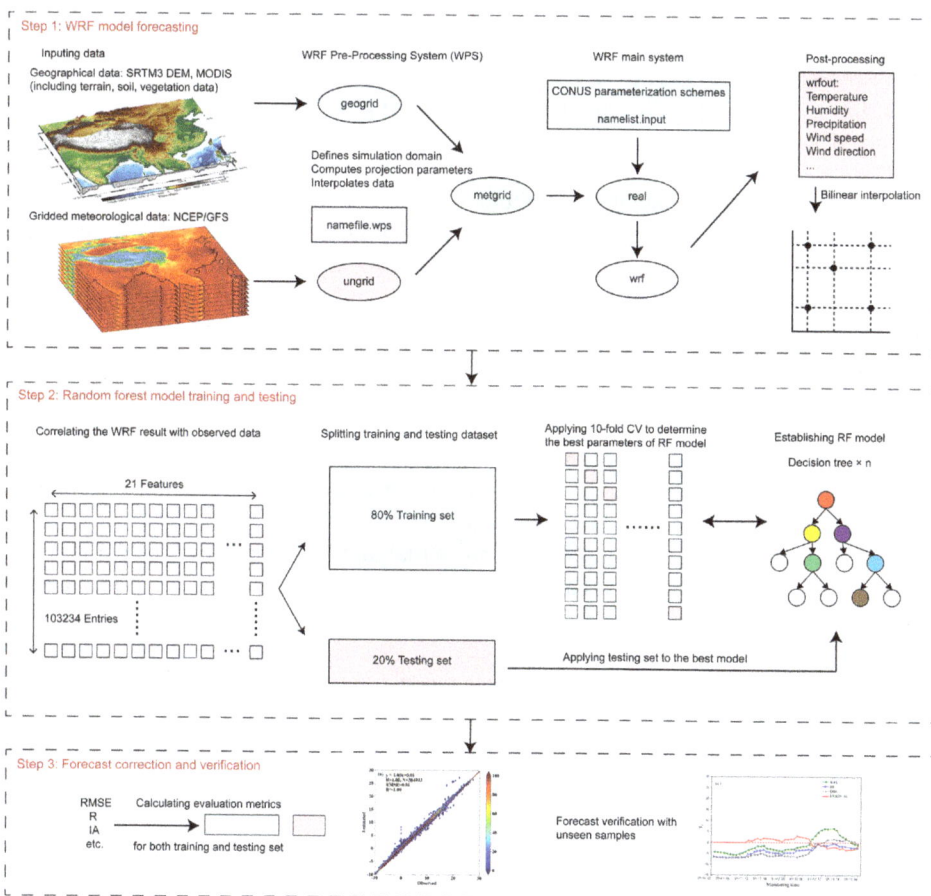

FIGURE 16.2 WRF forecasting flow and artificial intelligence (random forest) correction flow chart.

16.2.5 Temperature Prediction in South China

During the period from December 13, 2020 to December 19, 2020, the 2 m temperature, 2 m wet bulb temperature, relative humidity, precipitation, altitude, 10 m wind speed and 10 m wind direction predicted by the WRF model with a grid resolution of 3 km were selected. By using bilinear interpolation method, these forecast meteorological elements were interpolated to 364 towers with known longitude and latitude, combined with comprehensive monitoring data of icing on high-voltage transmission lines of 364 base towers at the same time, including temperature, humidity and ice thickness. In addition, combining with the known information of these towers, including the operating unit, line name, tower number, terminal number, longitude and latitude, altitude, slope, pass orientation, phase and geographic information, it will be interpolated to the tower according to the time dimension. The WRF forecast data and the actual monitoring data of the tower are combined to form the training set and test set of the random forest algorithm, excluding the data whose temperature value is greater than 40°C or less than -30°C, the matrix size of the total data set is 10,323,421; namely 103,234 rows of data, nearly 21 sets of feature columns. The ratio of training set and test set is 4:1, in which the test set does not participate in model training. The training set is subjected to 10-fold cross validation, and the Bayesian parameter optimization method is used to find the parameter settings of the random forest algorithm with the highest accuracy in the training set, the specific parameters including n_estimators, max_features, max_depth and min_samples_split. The detailed steps are given as follows.

> Step A: Using the bilinear interpolation method, the 2 m temperature, 2 m wet bulb temperature, relative humidity, precipitation, altitude, 10 m wind speed and 10 m wind direction, which are forecasted by the WRF model with a grid resolution of 3 km, are interpolated to 364 towers with known latitude and longitude.
>
> Step B: Comprehensive monitoring data of icing on high-voltage transmission lines of 364 towers at the same time, including temperature, humidity and thickness of ice coating; in addition, combined with the known information of these towers: operating unit, line name, tower number, terminal number, longitude and latitude, altitude, slope, pass orientation, phase and geographic information. According to the time dimension, the WRF forecast data are interpolated to the tower and the actual monitoring data of the tower, which are combined to form the training set and test set of the random forest algorithm (see Figure 16.3).
>
> Step C: Using the Bayesian parameter optimization method, it is found that the parameter settings of the random forest algorithm with the highest accuracy in the training set. The specific parameters are n_estimators = 197, max_features = 20, max_depth = 30 and min_samples_split = 2. The R^2 coefficients of determination of the temperature training and test sets on the prediction tower are 0.999 and 0.997, respectively, as shown in Figure 16.3. The residuals of the actual value of the tower temperature and the predicted value of the random forest are also shown in Figure 16.4.
>
> Step D: Selecting the random forest correction, actual and WRF predicted temperature on a transmission line in the test set that did not participate in the modeling, and comparing the correction and actual residuals, as shown in Figure 16.4.

The value of the testing set that did not participate in model training is about 0.997. Referring to similar work at home and abroad, the error statistic can be used; that is, 1 minus the predicted value or the coincidence index between the predicted value and the observed value. The specific formula is as follows:

$$IA = 1 - \frac{\sum_{i=1}^{N} (\phi_i)^2}{\sum_{i=1}^{N} (|p_i - \bar{O}| + |O_i - \bar{O}|)^2}$$

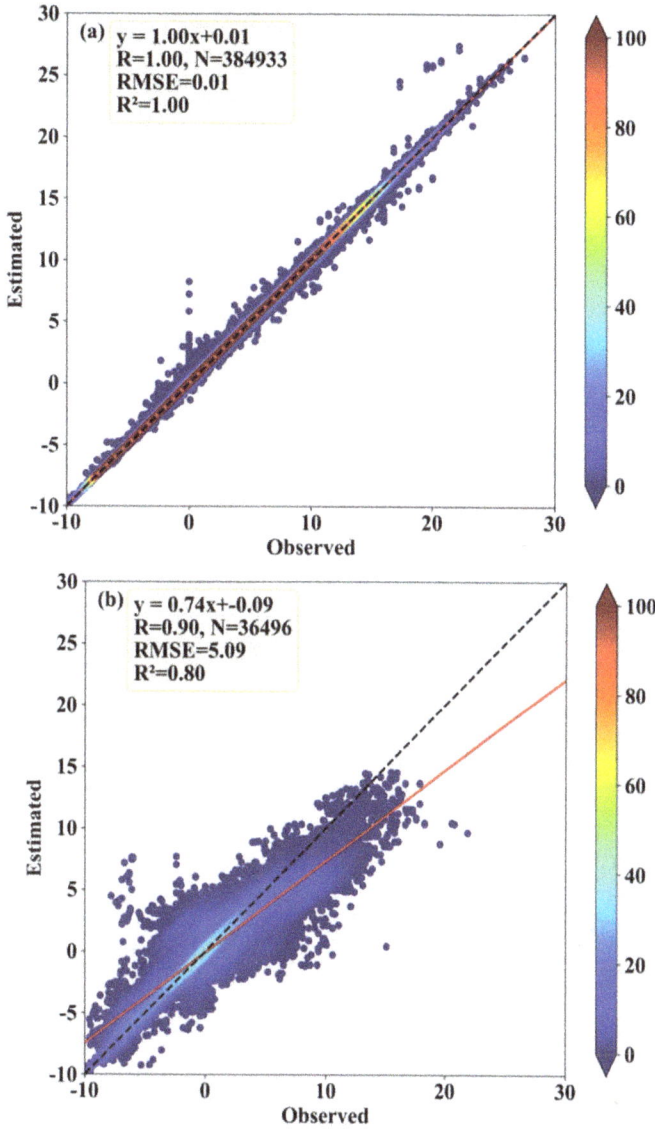

FIGURE 16.3 Density chart of the actual value (x-axis) and predicted value (y-axis) of the temperature model training set (a) and test set (b) (the larger the color scale, the smaller the distance between the points).

where ϕ_i is difference between the predicted value p_i and the observed value O_i for sample NO. i, N is total samples, $\overline{O}$ is mean value of observations. When it is equal to 1, it indicates that the spatiotemporal change trend between the predicted value and the observed value is in good agreement.

In the previous section, the value in the test set that did not participate in model training in the temperature correction was calculated to be about 0.075%, and the value in the test set that did not participate in model training in the relative humidity correction was calculated to be about 0.178%. Note that in the test set, the predicted temperature is about 0°C, and there is a large deviation, about ±4°C.

Through the machine learning approach, the forecast results of meteorological elements under micro-topographic conditions can be effectively corrected. In this way, a more accurate

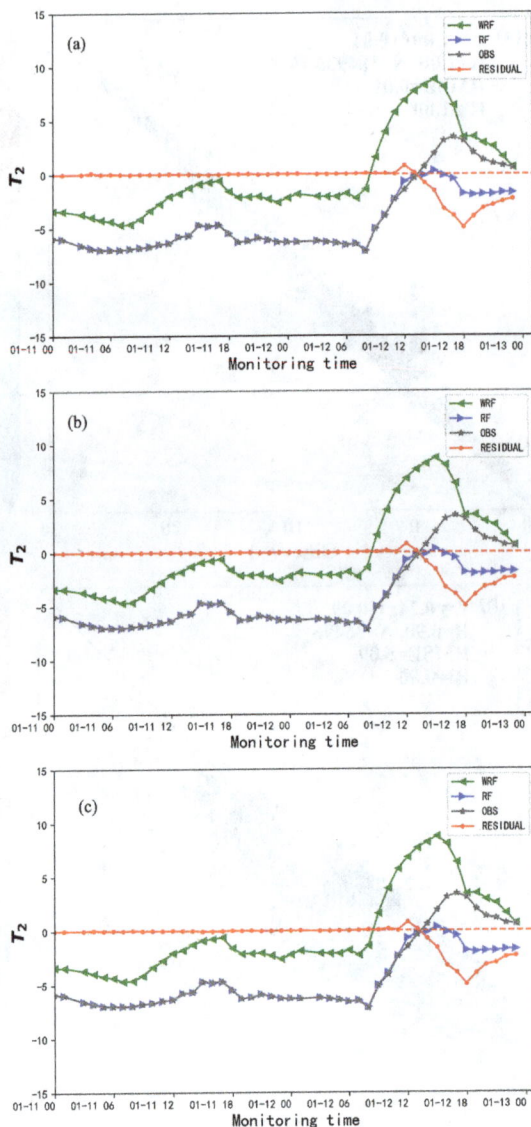

FIGURE 16.4 The WRF 2 m temperature, random forest correction temperature, actual tower temperature and correction residual error in three phases (positive, negative, ground) from January 11 to 13, 2021.

meteorological numerical prediction model can be obtained, and the short-term fixed-point forecast of meteorological elements in the southern China region can be realized.

16.3 MULTI-SOURCE METEOROLOGICAL DATA FUSION AND RECONSTRUCTION

AI can be effectively applied to the development of high-resolution meteorological data sets, such as data quality control, homogenization, data interpolation/interpolation, data fusion and inversion, and the construction of reanalysis data (Li et al., 2021; Zeng et al., 2020; Wang et al., 2019, 2020). This section will take gross primary productivity (GPP) estimating as an example to introduce the application of machine learning in GPP inversion, fusion and reconstruction.

Accurately quantifying crop GPP can provide valuable information on the ecosystem's carbon cycle, agricultural applications and climate change (Malmström et al., 1997; Zhu et al., 2018; Wang et al., 2017; Xie et al., 2020). For assessing GPP in crops, eddy covariance (EC) system, satellite-driven methods and process-based models are frequently employed. Among them, the EC technique allows direct and continuous monitors of land-atmosphere net ecosystem exchange (NEE) (Baldocchi, 2003). The gathered NEE data are routinely partitioned into GPP and ecosystem respiration (Lasslop et al., 2010). However, these EC measurements only represent the fluxes at the scale of the tower footprint, with an along-wind extent ranging between hundreds of meters and several kilometers (John et al., 2013). Although over 600 EC flux towers are operating in the world, their point-based measurements are insufficient to cover continuous regions in space (Lee et al., 2020).

To deal with the problem of spatial discontinuity in EC technique, satellite remote sensing, light-use efficiency models and process-oriented land surface models are adopted (Baldocchi, 2003). However, remote sensing-based GPP may not fully guarantee the accuracy of data. For instance, Wang et al. (2017) assessed the latest Moderate Resolution Imaging Spectroradiometer (MODIS) GPP product (MOD17A2H) at different biome types against global EC flux-estimated GPP and found that MOD17A2H GPP performed poorly at both annual (coefficient of determination: 0.62) and 8-day scales (coefficient of determination: 0.52). Thus, a more advanced calibration model is required for large-scale applications (Reeves et al., 2005). Process-based land surface models (e.g., Community Land Model [Post et al., 2018] and Simple Biosphere Model 2 [Wang et al., 2007]) have been designed for cropland GPP, but are subject to complicated scientific assumptions and model parameters (Ueyama et al., 2013).

GPP in crops is a complicated non-linear function due to the spatial heterogeneity of vegetation and soil properties, and the temporal heterogeneity of the environmental factors (meteorological conditions and agricultural managements) (Lee et al., 2020; Post et al., 2018). Currently, data-driven machine learning algorithms are another popular method for predicting GPP because they can elucidate precisely nonlinear processes of CO_2 exchanges in agroecosystems (Cutler et al., 2007). Although in principle they are black-box models, machine learning methods, (e.g., model tree ensembles [Jung et al., 2011], support vector machines [Ueyama et al., 2013], neural network models [Dou et al., 2018; Tramontana et al., 2020] and random forest models [Zeng et al., 2020; Reitz et al., 2021; Cai et al., 2020; Chen et al., 2019]) have good performance for multi-ecosystem GPP estimations. For instance, Tramontana et al. (2015) quantified the 8-day GPP and the mean European annual carbon budget across ecosystems (e.g., forest, grassland, cropland and wetland) by using random forest (RF) algorithm, remote sensing and EC data. Recently, the RF model was also adopted to upscale the EC-based GPP to regional scales in an arid and semi-arid area in northwestern China (Yu et al., 2021). Previous studies have evaluated the latest MOD17A2H GPP product across various ecosystems (e.g., forests and grasslands) with global EC data (Zhu et al., 2018; Wang et al., 2017). However, the validation has rarely been performed for double-cropping agriculture, especially in rice–wheat rotation cropland, which is the most extensive land cover type in the northern Yangtze River Delta (NYRD) region, China (Timsina et al., 2001). Furthermore, the existing studies on the GPP changes in the NYRD mainly focused on the temporal characteristics of carbon exchanges (Chen et al., 2015; Duan & Yang et al., 2021a; Ge et al., 2018), leaving a knowledge gap with respect to the upscaling GPP and its calibration to the MOD17A2H GPP product.

Therefore, a random forest (RF) machine learning algorithm for GPP (GPP_{RF}) was developed for rice–wheat double-cropping fields by integrating multi-source satellite remote sensing images as well as ground measurements. Based on the foregoing data, the main objectives were to (1) assess the performance of the MODIS GPP product (GPP_{MOD}) through comparison with EC-estimated GPP (GPP_{EC}) and determine the driving factors of GPP; (2) extrapolate the GPP from the single-site scale to multi-site scales; and (3) calibrate the GPP_{MOD} over the rice–wheat rotation cropland in the NYRD.

16.3.1 STUDY AREA AND DATA

The NYRD is composed of northern Anhui and Jiangsu provinces and Shanghai, ranging between 114°–122°E and 29°–36°N (Figure 16.5). The NYRD covers an area of 176,960 km², consisting of 73% cropland, 16% grassland, 5% built-up land, 4% water bodies and 2% forest (Figure 16.5). Three EC flux sites were representative of typical rice–wheat rotation cropland landscapes found over this cropland (Figure 16.5, inset map) (Timsina et al., 2001; Chen et al., 2015). The soil pH value (H_2O), soil organic carbon and soil total nitrogen in topsoil (0–0.3 cm) for our study area mainly ranged between 5.5–7.2, 1.2–2% and 0.1–0.15 %, respectively, according to the results of Shangguan et al. (2013). Here, the winter wheat grows from November to late May. At the beginning of June, the rice paddies were flooded, plowed and harrowed to incorporate the wheat straw residue from the last wheat growing season (Duan and Grimmond et al., 2021). Then, one-month-old rice seedlings were transplanted to the leveled field in middle June and harvested in early November (Figure 16.6a), which can be indicated by the seasonal dynamics of 8-day leaf area index (LAI) averaged from 23 weather stations and 3 EC stations during 2014–2018 (Figure 16.6b). The rice/wheat canopy height can reach about 1–1.2 m at the peak LAI growing seasons. The local climate is sub-tropical monsoon-type, with a mean annual (2014–2018, calculated from the 23 surface meteorological stations in Figure 16.5) air temperature of 16°C and rainfall of 1,100 mm.

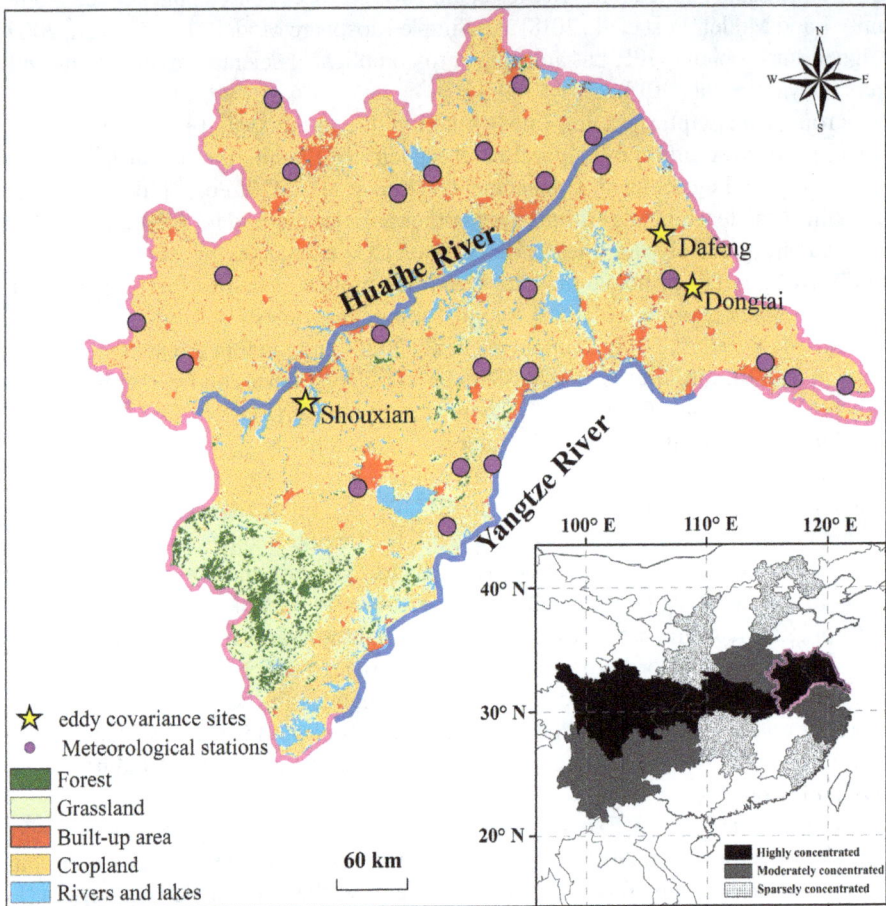

FIGURE 16.5 MODIS landcover maps (resolution: 500 m) in 2016 and the meteorological stations in the North Yangtze River Delta region. The inset map indicates the distribution of rice–wheat rotation cropland areas in China.

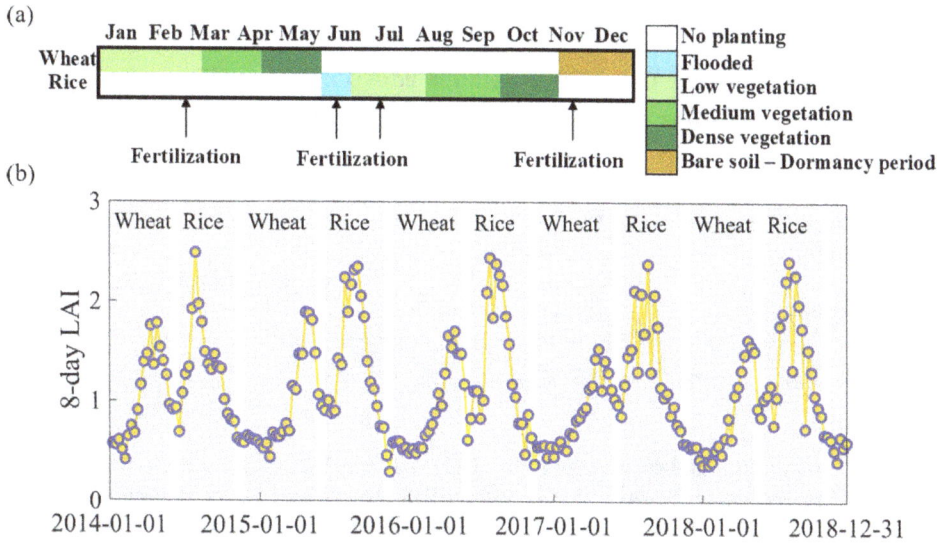

FIGURE 16.6 (a) Crop calendars for the rice and wheat in the North Yangtze River Delta region. (b) Time series of 8-day leaf area index (LAI) for the rice–wheat rotation croplands averaged from 23 weather stations and 3 EC stations during 2014–2018 of the North Yangtze River Delta.

Flux data from three rice–wheat rotation cropland EC stations within the study area—Shouxian site in Anhui, Dongtai, and Dafeng sites in Jiangsu—were selected for model training and prediction (Figure 16.5). At Shouxian site, the EC sensors were mounted 2.5 m above the ground, consisted of a three-dimensional sonic anemometer (CSAT3, Campbell Scientific Incorporation, USA) along with a CO_2/H_2O open-path infrared gas analyzer (EC 150, Campbell Scientific Incorporation, USA). At Dongtai and Dafeng sites, virtual temperature and wind velocity components were monitored using a three-dimensional sonic anemometer (CSAT3, Campbell Scientific, Inc., USA). To measure H_2O and CO_2 density, a fast-response open-path gas analyzer (LI-7500, LI-COR Biosciences, Inc., USA) was used. The installation height of the sensors for the Dongtai site was 10 m, whereas for the Dafeng site it was 6.3 m above the ground. As mentioned in the previous studies (Duan & Yang et al., 2021b; Ge et al., 2018; Duan and Grimmond et al., 2021), three EC sites are relatively flat, with more than 90% of the flux primarily contributed by the cropland. EddyPro 5.2.1 (LI-COR Inc., 2015) software was applied to calculate hourly CO_2 fluxes and to correct for CO_2 canopy storage to gain NEE values. Data pre-processing in the EddyPro software mainly included averaging and statistical tests (Lee et al., 2005), time lag compensation, double coordinate rotation, spectral correction (Moncrieff et al., 2005) and the Webb-Pearman-Leuning density correction (Wutzler et al., 2018). The poor quality fluxes (EddyPro quality check flag value = 2) were further discarded. The REddyProc R package (https://www.bgc-jena.mpg.de/bgi/index.php/Services/REddyProcWebRPackage) inputted pre-processed half-hourly EC data and supported further processing (Wutzler et al., 2018). Firstly, a quality-check and filtering were performed based on the relationship between observed flux and friction velocity to discard biased data (Papale et al., 2006). Then, the flux data were gap-filled using the marginal distribution sampling approach (Reichstein et al., 2005). NEE was separated into GPP and ecosystem respiration based on the nighttime partitioning algorithms (Reichstein et al., 2005). The gap-filled hourly GPP data were summed to compute cumulative GPP for daily, 8 day, seasonal and annual time resolution for further analysis (Wagle et al., 2021). Data from these three sites processed using the same methods. Details of the agricultural practices and processing methods at these three sites can be obtained from the references in Table 16.1 (Duan et al., 2021; Ge et al., 2018; Duan and Grimmond et al., 2021).

TABLE 16.1

Characteristics of the Three Rice–Wheat Rotation Eddy Covariance Sites

Station	Location	Altitude (m)	Period	T_{ave} (°C)	P_{ave} (mm)	Reference
Shouxian	(32.44°N, 116.79°E)	27	July 15, 2015–April 24, 2019	16	1115	Duan and Yang et al. (2021a)
Dongtai	(32.76°N, 120.47°E)	2	December 1, 2014–November 30, 2017	13	1484	Duan and Grimmond et al. (2021)
Dafeng	(33.21°N, 120.28°E)	1	November 16, 2015–November 29, 2016	15	1060	Ge et al. (2018)

Note: T_{ave}, annual mean air temperature; P_{ave}, annual cumulative precipitation.

Hourly air temperature and relative humidity (RH) at 23 automatic stations were obtained from the China Meteorological Administration in 2014–2018. The hourly vapor pressure deficit (VPD) was estimated with relative humidity and air temperature data following the World Meteorological Organization Commission for Instruments and Methods of Observation Guide conversion equation (Yang et al., 2021). The hourly surface downward solar radiation (DSR) ERA5 reanalysis data were provided by the European Center for Medium-Range Weather Forecasts at a 0.25° spatial resolution.

Land cover maps were available in a 500-m spatial resolution of the MODIS MCD12Q1 product for the year 2016 (Figure 16.5, [Friedl et al., 2019]). The 16-day Normalized Difference Vegetation Index (NDVI) data during 2014–2018 were obtained from the MODIS MOD13Q1 product with a 250-m resolution (Didan, 2015). The 8-day Fraction of Photosynthetically Active Radiation (FPAR) and LAI data were derived from the 500-m spatial resolution of MODIS MOD15A2H (Myneni et al., 2015). The MODIS GPP product MOD17A2H (version 6) had an 8-day temporal resolution and 500-m spatial resolution (Running et al., 2015). All of these data sets were downloaded from https://ladsweb.modaps.eosdis.nasa.gov/search/. These MODIS products were quality-controlled to exclude anomalous pixel interference.

16.3.2 RF AND UPSCALING METHODS

RF is a fast and flexible machine learning algorithm, which is often used for analyzing the classification and regression tasks (Breiman, 2001). This model can successfully process high dimensional and multicolinear data, being insensitive to overfitting (Belgiu and Dragut, 2016). The RF model provides a feature-selection tool to identify the importance of the predictor. Feature importance is defined as the contribution of each variable to the model, with important variables showing a greater impact on the model evaluation results (Liu et al., 2021). In this section, a GPP prediction model based on RF framework was proposed. The flowchart of estimating, upscaling GPP and calibrating MOD17A2H GPP product with the RF model was shown in Figure 16.7, including four steps as follows:

(1) Variable selection and data matching. Crop photosynthesis is a complicated process affected by shortwave radiation, air temperature, vapor pressure deficit, soil edaphoclimatic conditions and fertilization at the canopy scale, etc. At the ecosystem level, GPP is closely related to light, water and canopy phenology (Yu et al., 2021; Xiao et al., 2008). Based on the previous literatures as well as our current available data, nine input explanatory variables: NDVI, LAI, FPAR, DSR, daily maximum air temperature (T_{max}), daily minimum air temperature (T_{min}), daily mean air temperature (T_{mean}), VPD, and RH were chosen for predicting the GPP dynamics in the NYRD region. Because RF model training requires a large number of samples, MODIS data were linearly interpolated from 8-day/16-day to daily values to match the input parameters, following a previous study by Reitz et al. (2021).

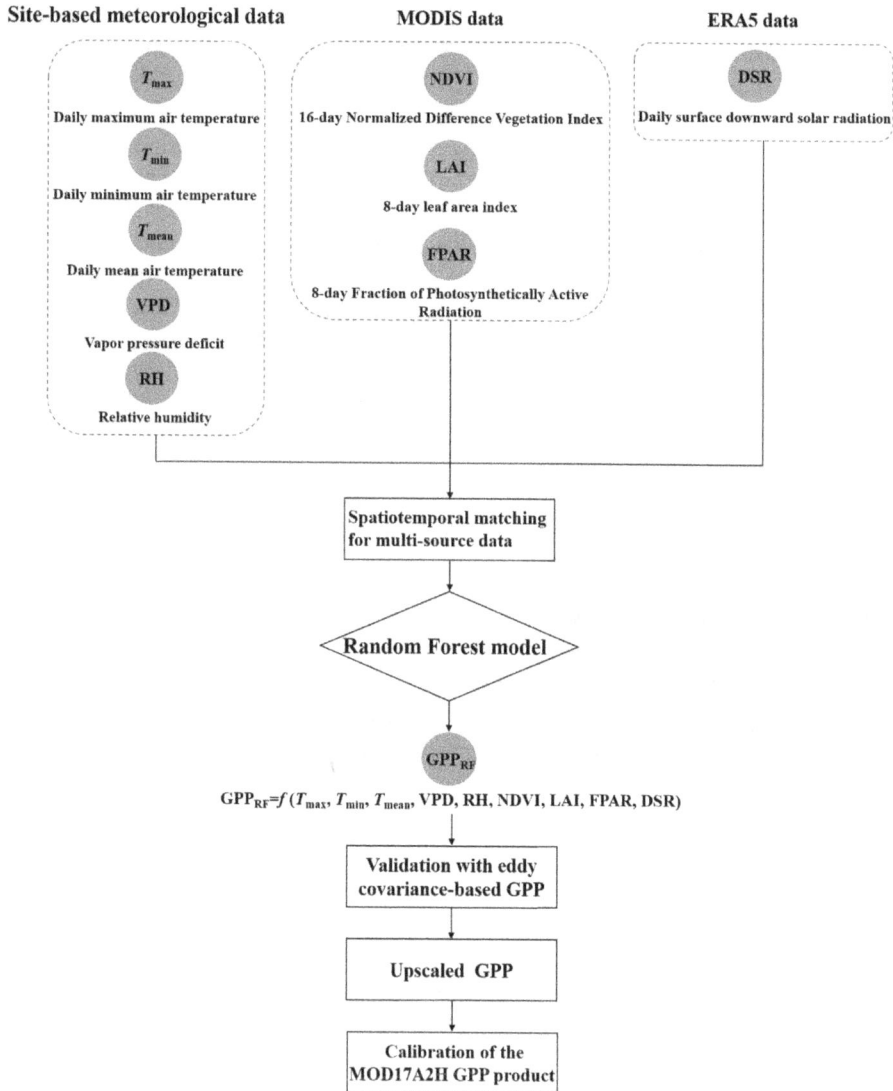

FIGURE 16.7   Flowchart of the random forest model for estimating, validating, upscaling gross primary production (GPP), and calibrating MOD17A2H GPP product.

(2) RF model constructing, training and testing. In this paper, 90% (the rest 10%) of the EC data at Shouxian and Dongtai during the entire observation period were employed to train (validate) the RF model, and 100% of the EC data at Dafeng were applied to validate. The Shouxian and the Dafeng site were independent of each other with a negligible autocorrelation between them, since these sites are about 300–400 kilometers far away from each other (Figure 16.5). Here, 10-fold cross-validation (CV) algorithm was applied to weaken the overfitting (Cai et al., 2020). In 10-fold CV experiments, all the training data at the Shouxian and Dongtai sites during the entire observation period were randomly partitioned into ten equal sized subsamples. Of the ten subsamples, nine subsamples were used as the training data and the remaining one was the testing data. This CV process should repeat ten times, with all ten subsamples used exactly once as the testing data. The ten results from the folds were averaged to produce a single estimation.

To select the best model, we adjusted the four hyperparameters of the RF model based on Bayesian optimization (Baareh et al., 2021; Frazier, 2018): the number of trees to grow (n_estimators), minimum sample number placed in a node prior to the node being split (msplit), maximum number of features considers to split a node (Mfeatures), and maximum levels' number in each decision tree (Mdepth). Three statistical metrics—the index of agreement (IA) (Willmott, 1982), the coefficient of determination (R^2), and the root mean square error (RMSE)—were used to examine the simulated performance of the 10-fold CV results. The range of IA is 0–1, and a better correspondence between the observed and modeled results often occurs when it approaches 1 (Zhang et al., 2008). Therefore, n_estimators = 219, msplit = 2, Mfeatures = 9 and Mdepth = 32 were set for the final RF model.

(3) GPP upscaling. The general relationships between GPP$_{RF}$ and explanatory data were first trained at site level, and then applied regionally by using regional surface meteorological stations of explanatory variables as follows: GPP$_{RF}$ = f (T_{max}, T_{min}, T_{mean}, VPD, RH, NDVI, LAI, FPAR, DSR).

(4) MOD17A2H GPP product calibrating. Based on the upscaled results of GPP$_{RF}$ and GPP$_{MOD}$ at the station scale, a relationship between GPP$_{RF}$ and GPP$_{MOD}$ was built. The calibration function was then applied from the site scale to the regional scale.

16.4 RESULTS AND ANALYSIS

16.4.1 INTRASEASONAL VARIATIONS OF GPP

MOD17A2H GPP has been extensively employed to evaluate the terrestrial carbon balance (Zhu et al., 2018). However, to have confidence in GPP$_{MOD}$, it is critical to validate it against *in situ*

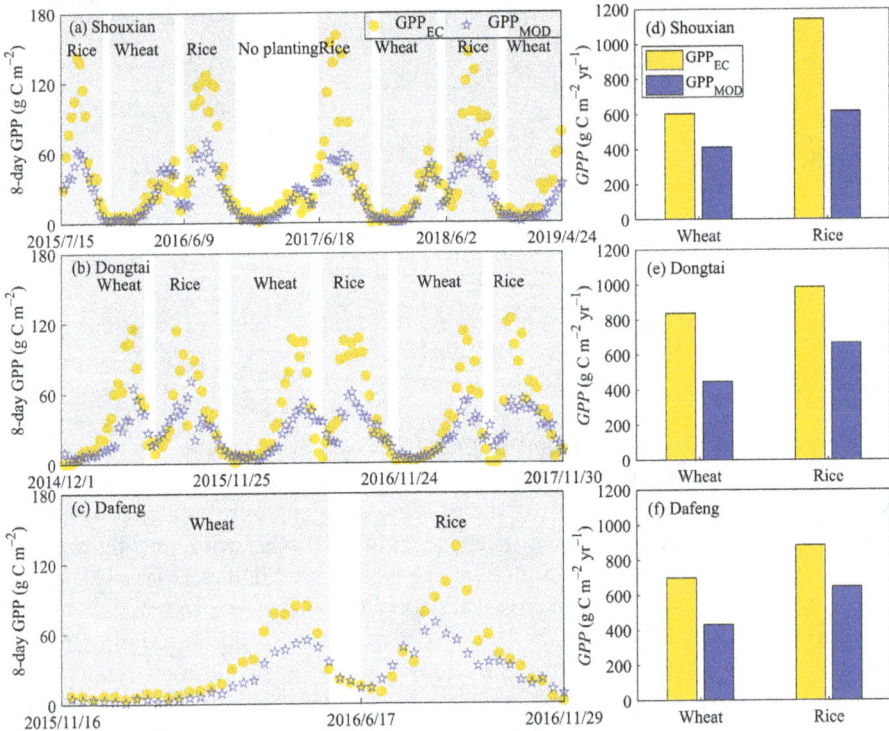

FIGURE 16.8 (a)–(c) 8-day averaged gross primary production (GPP) measured by EC (GPP$_{EC}$) and MOD17A2H (GPP$_{MOD}$), and (d)–(f) seasonal cumulative GPP for rice and wheat growth seasons.

measurements (Zhang et al., 2008). As shown in Figures 16.5a–c, the 8-day GPP$_{MOD}$ and GPP$_{EC}$ exhibit close agreement in their seasonal patterns, with peaks in July (May) for the rice (wheat) growing seasons across the three rice–wheat rotation cropland sites. The GPP increases during the no-planting period in 2017 (Figure 16.8a), mainly due to the weed photosynthesis. GPP$_{MOD}$ underestimated GPP$_{EC}$ during the rice (wheat) active growing periods from July to September (from March to May), with IA = 0.56 and RMSE = 47 g C m^{-2} (IA = 0.61 and RMSE = 29 g C m^{-2}) across the three sites. However, GPP$_{MOD}$ performed well during the intercropping periods from late May to early June (or late November), with IA = 0.77 and RMSE = 8 g C m^{-2} across the three sites.

The seasonal cumulative GPP$_{EC}$ at the three cropland sites was larger for the summer rice growing seasons (1170, 1066 and 889 g C m^{-2} for Shouxian, Dongtai and Dafeng, respectively) than for wheat (609, 848 and 701 g C m^{-2}, respectively) (Figures 16.8d–f). The seasonal cumulative GPP$_{MOD}$ was significantly lower than GPP$_{EC}$ during the wheat growth seasons, with a 32%–47% underestimation of the seasonal cumulative GPP$_{EC}$ at the three sites; the seasonal average GPP$_{MOD}$ was 27%–47% lower than the seasonal cumulative GPP during the summer rice growth seasons (Figures 16.8d–f).

16.4.2 Driving Factors of GPP on a Seasonal Scale

The possible drivers related to the GPP variations in the NYRD were investigated by the RF model in Figure 16.9 to assess their relative contributions. NDVI was the most important factor in modulating GPP, accounting for 56% of the overall variable importance. As illustrated in Figure 16.10, GPP showed the strongest positive correlation with NDVI, with the highest Pearson correlation coefficient (r) of 0.74, which was consistent with the variable importance value in Figure 16.9. In addition to NDVI, there were another three dominant variables; namely, LAI, DSR, T_{max} and FPAR, with importance values of 13%, 10%, 8% and 3%, respectively. NDVI and LAI were important indicators of the phase of terrestrial photosynthesis, which tracked well the crop phenological dynamics

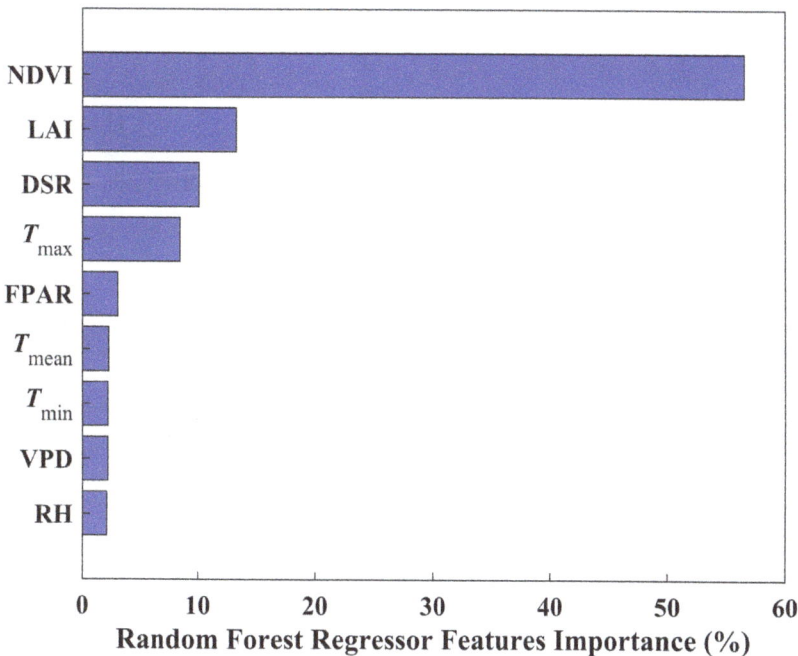

FIGURE 16.9 Feature importance for the random forest model in the North Yangtze River Delta region.

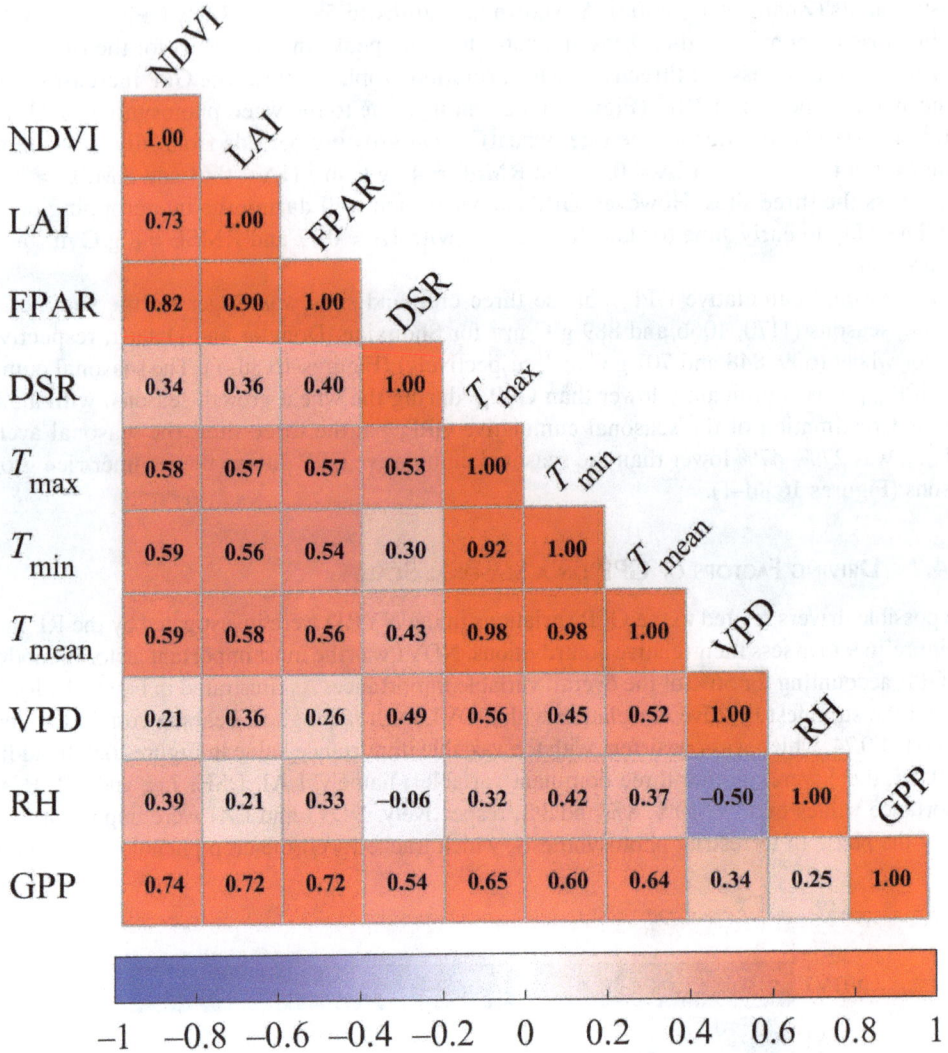

FIGURE 16.10 Correlations among the GPP (gross primary productivity) and input variables. The correlations were calculated by all training data from the Dongtai and Shouxian sites.

over time (Tramontana et al., 2015; Willmott, 1982; Rahman et al., 2005). DSR and FPAR covaried with light to a large degree—the source of energy for photosynthesis in vegetation (Alberto et al., 2009). T_{max} played a critical role in the chemical reactions of biological processes (Tramontana et al., 2020). In contrast, the impact of T_{mean}, T_{min}, VPD and RH on GPP was not obvious, exhibiting the lowest relative importance, with values of 2%, 2%, 2% and 2%, respectively. Although FPAR was highly correlated with NDVI (Figure 16.10), the importance of FPAR was so low in the RF model. This was because NDVI was directly derived from the satellite spectrum, while FPAR was indirectly calculated based on LAI and the physical models. The uncertainties in the MODIS LAI product can be attributed to the input data (surface reflectance or radiation data), model imperfections, and the inversion process (Fang et al., 2019). The Pearson correlation coefficient for T_{mean} ($r = 0.64$) lay in the range of those for T_{max} ($r = 0.65$) and T_{min} ($r = 0.60$), because T_{mean} incorporated both day- and nighttime conditions (Figure 16.10). Generally, all these predictors were involved to different degrees in CO_2 exchange processes. Vegetation indices (i.e., NDVI and LAI), related to the phenological properties of the plants, had the greatest influence on GPP variations. In terms of meteorological factors, DSR, T_{max} and FPAR carried the information of the light-dependent reactions of photosynthesis, which had a moderate effect on the GPP changes. In particularly, T_{mean}, T_{min}, VPD and RH showed weak influences on the GPP cycles.

16.4.3 RANDOM FOREST MODEL EVALUATION

The RF model performed well for both the training ($R^2 = 0.99$, RMSE = 0.42 g C m^{-2} d^{-1}) and testing ($R^2 = 0.89$, RMSE = 2.8 g C m^{-2} d^{-1}) data set (Figures 16.11a and 11b). This indicated that the input variables in the RF model were representative and can well capture the temporal characteristics of GPP. RF model also proved to be good at the validation site (i.e., Dafeng site), in which the seasonal distributions of GPP$_{RF}$ showed high correlation and coherence with GPP$_{EC}$ (IA = 0.94, Figure 3.8c). All sites exhibited double peaks, with the peaks during the rice growth season being higher than those during the wheat growth season (Figure 16.12), which is a common pattern in this double-cropping field (Figure 16.11). The R^2 and RMSE at the validation site (i.e., Dafeng site) were 0.80 and 4.39 g C m^{-2} d^{-1} (Figure 16.11d)—a result that was similar to that across global FLUXNET sites conducted by Tramontana et al. (2016) in which the R^2 ranged from 0.61–0.81. Hence, the RF model was deemed suitable for GPP prediction at unknown stations as well as regional GPP upscaling.

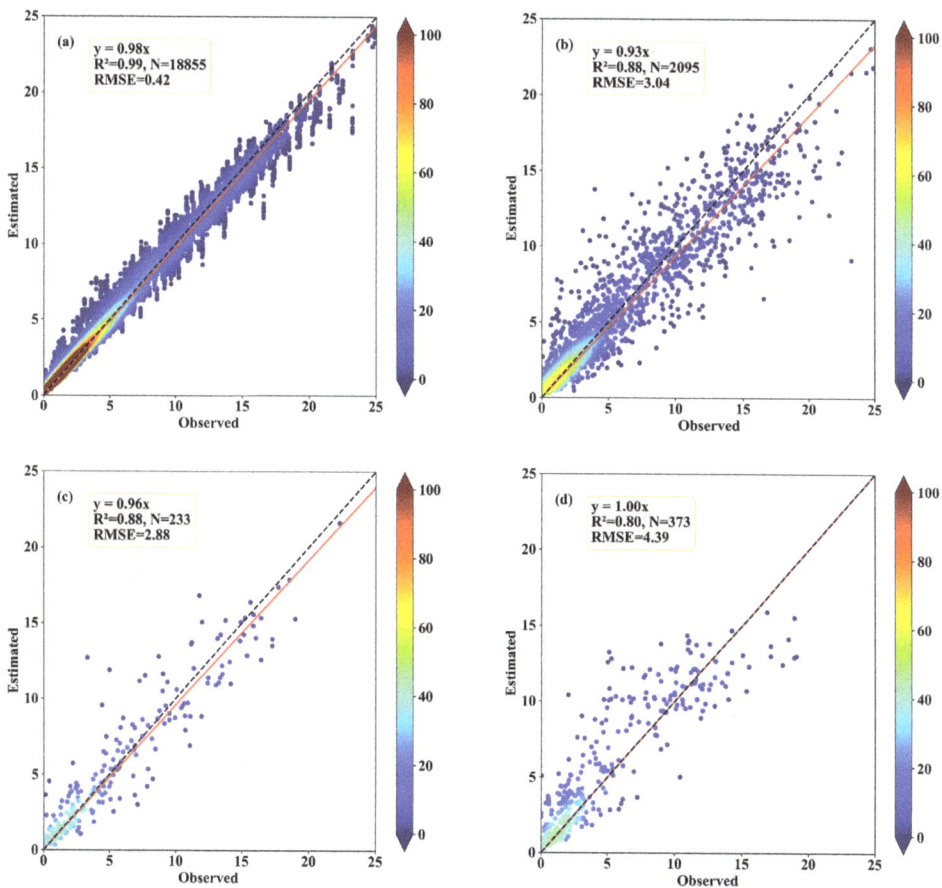

FIGURE 16.11 Scatter density plots results for the random forest model in predicting gross primary productivity in the (a) 10-fold cross-validation training set, (b) 10-fold cross-validation testing set, (c) validation by the rest samples at Shouxian and Dongtai sites, and (c) validation by all samples at Dafeng site.

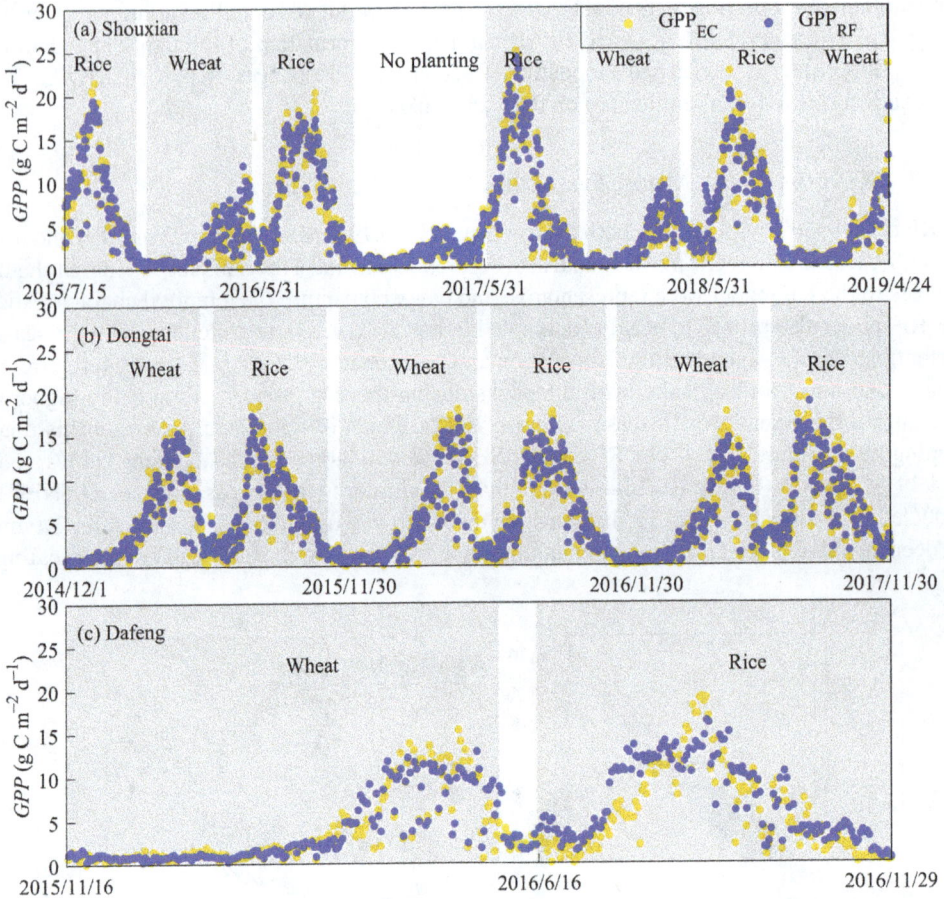

FIGURE 16.12 Daily gross primary productivity (GPP) measured by EC (GPPEC) and predicted by the random forest (RF) model (GPPRF) at (a) Shouxian, (b) Dongtai, and (c) Dafeng during the rice and wheat growth seasons.

According to previous studies, random forests are much less likely to overfit than other models because they consists of many weak classifiers that are trained independently on completely different subsets of the training data. In addition, we also employed other machine learning methods to build GPP prediction models. Generally, the simulations of the RF-based GPP models show a better performance with respect to other machine learning methods (e.g., decision tree regression, support vector machine, artificial neural network, and deep belief network.

16.4.4 Upscaled GPP

Figures 16.13a–c show that the regional RF-modeled cumulative seasonal GPP (GPP_{RF}) averaged from 23 weather stations and 3 EC stations during 2014–2018 was much higher for the rice growth seasons (924 g C m^{-2}) than that for the wheat growth seasons (532 g C m^{-2}). This relationship (cumulative seasonal GPP in the rice growth season > cumulative seasonal GPP in the wheat growth season) was also be confirmed by the GPP_{MOD} in Figures 16.13d–f. For our study sites, the annual mean for GPP_{MOD} and GPP_{RF} averaged from 23 weather stations and 3 EC stations were 966 g C m^{-2} and 1548 g C m^{-2}, respectively. Figures 16.13g–i show the difference between the two GPP products across all sites (relative error at each site as computed by [$GPP_{MOD} - GPP_{RF}$] ×100/GPP_{RF}) during

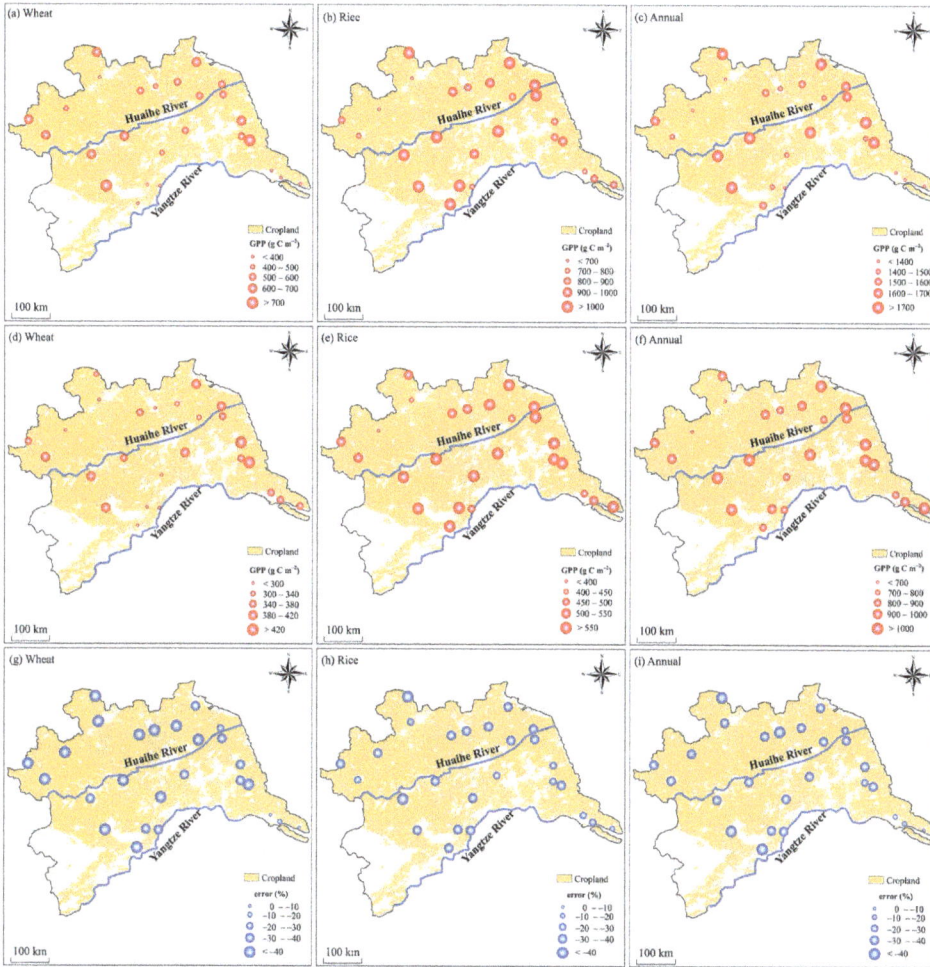

FIGURE 16.13 Spatial distributions of gross primary productivity (GPP) (a–c) predicted by the random forest (RF) model (GPP$_{RF}$), (d–f) measured by the MODIS product (MOD17A2H) (GPP$_{MOD}$), and (g–i) their difference ([GPP$_{MOD}$ – GPP$_{RF}$] ×100/GPP$_{RF}$) at the station scale for wheat growth seasons, rice growth seasons, and the whole year during 2014–2018. The dark yellow background represents cropland.

2014–2018. Relative errors exhibited negative values across all sites, with -18%–-46% for rice growing seasons, -1%–-50% for wheat growing seasons and -5%–-47% for the whole year, respectively. In general, relative errors during the wheat growing seasons were relatively larger than those during the rice growing seasons/the whole year at most sites (Figures 16.13g–i).

To examine the spatial consistency of the GPP$_{RF}$ dynamics among the upscaled sites, Figure 16.14 shows the seasonal variations in GPP$_{RF}$ among 23 weather stations and 3 EC stations during 2014–2018 over the rice–wheat rotation cropland in the North Yangtze River Delta region. The daily mean GPP$_{RF}$ averaged from 23 weather stations and 3 EC stations during 2014–2018 for wheat was lower than 2 g C m^{-2} d^{-1} during the winter extensive bare soil period (December–February). It started to increase in the active tillering stage (March) and reached a maximum of about 8–10 g C m^{-2} d^{-1} during the heading stage (late April), and next decreased to around 4 g C m^{-2} d^{-1} at harvest. The largest daily GPP$_{RF}$ for rice paddies occurred in late July, with a peak value of about 11 g C m^{-2} d^{-1}, suggesting that the rice biological activities (e.g., photosynthetic rates) were quite strong at this stage. After then, daily GPP$_{RF}$ decreased to approximately 1 g C m^{-2} d^{-1} at rice harvest (Figure 16.14). Generally,

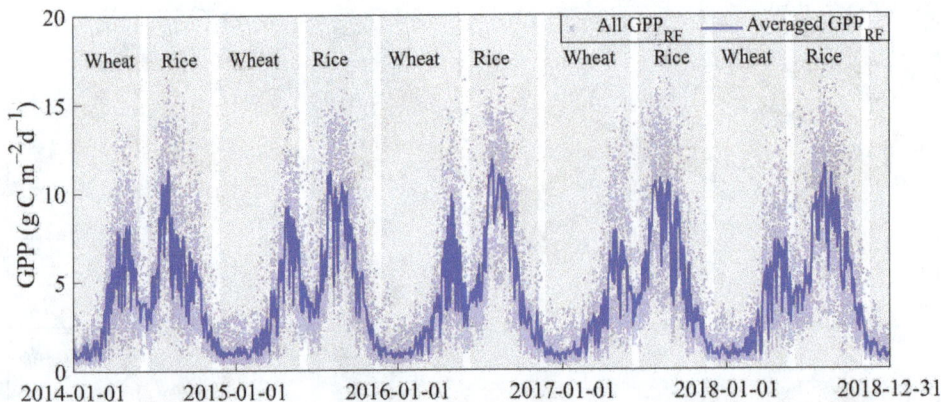

FIGURE 16.14 Seasonal variations in gross primary productivity (GPP) predicted by the random forest (RF) model (GPP$_{RF}$). The blue line represents the daily GPP$_{RF}$ averaged from 23 weather stations and 3 EC stations during 2014–2018 over the rice–wheat rotation cropland in the North Yangtze River Delta region.

a good consistency was found in the seasonal variation among all upscaled sites, exhibiting similarly temporal characteristics of the real GPP over the rice-wheat rotation system (Figure 16.12).

16.4.5 Calibration of the MOD17A2H GPP Product

Based on the upscaled results of GPP$_{RF}$ and GPP$_{MOD}$ at the station scale in the previous section, the relationship between GPP$_{RF}$ and GPP$_{MOD}$ is shown in Figure 16.15. Here, the daily GPP$_{RF}$ was aggregated to 8-day sums to match the 8-day GPP$_{MOD}$ product.

Then, the linear relationship between GPP$_{MOD}$ (Figures 16.16a–c) and the calibrated GPP$_{MOD}$ (GPP$_{CMOD}$) (Figures 16.16d–f) at the grid scale was established as follows:

$$GPP_{CMOD} = \begin{cases} 1.5 \times GPP_{MOD}, \textit{ for wheat} \\ 1.7 \times GPP_{MOD}, \textit{ for rice} \\ 1.6 \times GPP_{MOD}, \textit{ for annual} \end{cases} \quad (16.1)$$

Both GPP$_{MOD}$ and GPP$_{CMOD}$ exhibited a higher value during the rice growth seasons than that during the wheat growth seasons. The annual mean GPP in most parts of the NYRD varied from 2–4 g C

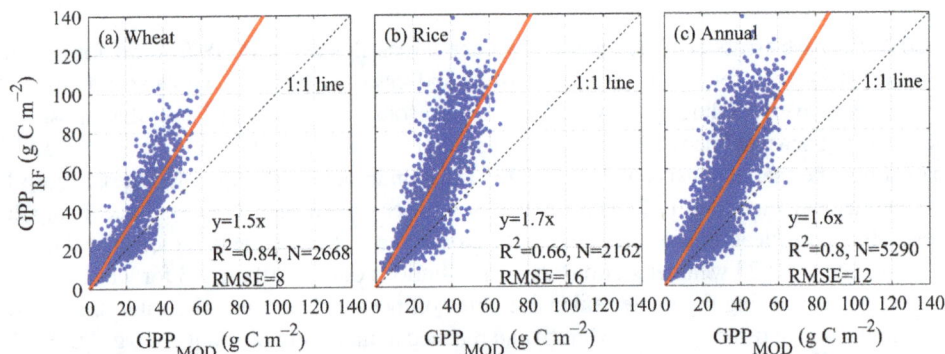

FIGURE 16.15 Relations between the gross primary productivity (GPP) predicted by the random forest (RF) model (GPP$_{RF}$) and that measured by MODIS (GPP$_{MOD}$) for the (a) wheat growth seasons, (b) rice growth seasons, and (c) the annual mean during 2014–2018.

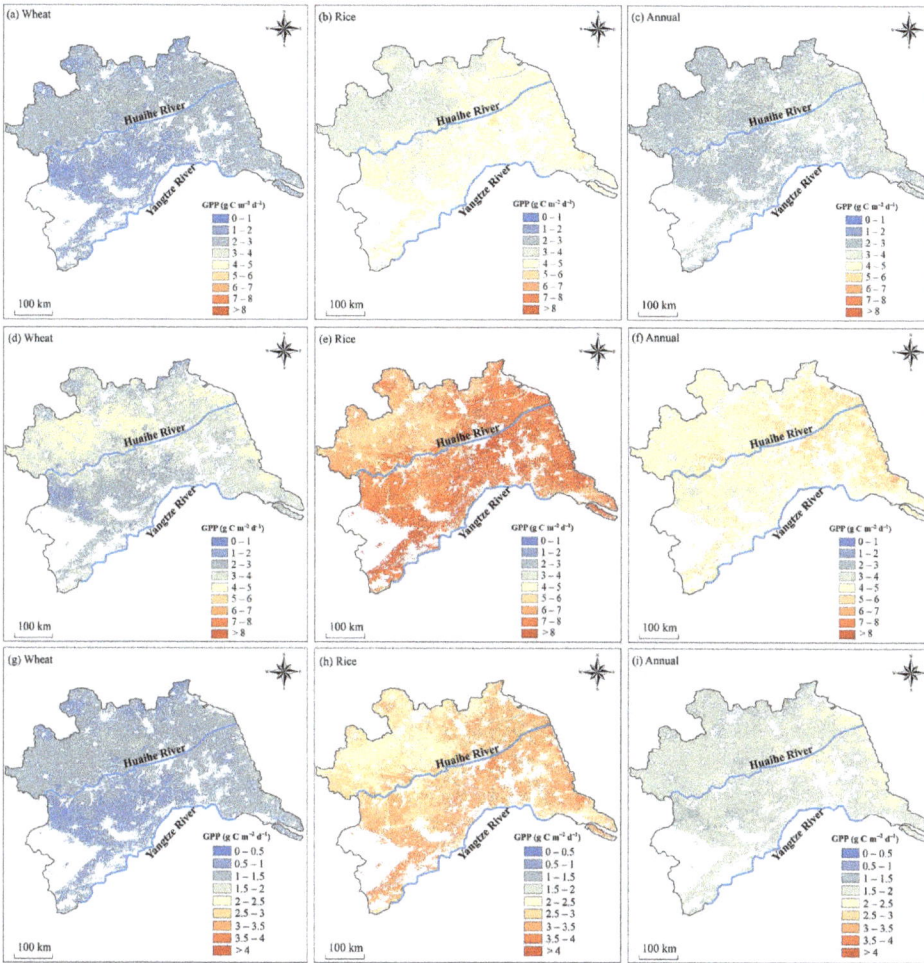

FIGURE 16.16 Spatial patterns of the gross primary productivity (GPP) (a–c) measured by MODIS (GPP$_{MOD}$), (d–f) measured by MODIS and then calibrated (GPP$_{CMOD}$), and (g–i) their difference (GPP$_{CMOD}$ minus GPP$_{MOD}$) (ΔGPP) at the grid scale for wheat growth seasons, rice growth seasons, and the annual mean during 2014–2018.

m^{-2} d^{-1} for GPP$_{MOD}$ and 4–6 g C m^{-2} d^{-1} for GPP$_{CMOD}$, with the higher values in the eastern coastal areas of the NYRD (Figures 16.16c and 16.16f). Here, sea–land breezes prevail, carrying rainfall and water resources sufficient to favor crop growth. Figures 16.16g–i show the seasonal mean of the error ranges of daily GPP (ΔGPP, as computed by subtracting GPP$_{MOD}$ [Figures 16.16a–c] from GPP$_{CMOD}$ [Figures 16.16d–f]) during 2014–2018. ΔGPP in most parts of the NYRD ranged between 2 g C m^{-2} d^{-1} and 4 g C m^{-2} d^{-1} during the rice growing seasons, while they were smaller (i.e., 0–1.5 g C m^{-2} d^{-1}) for wheat. The probability density function (PDF) of ΔGPP in the NYRD is shown in Figure 16.17. The PDF of ΔGPP varies seasonally, which play a pivotal role in regulating the carbon dynamics in the NYRD. Rice paddies had a broader distribution in the peak PDF of the mean ΔGPP than that for wheat fields, i.e., between 0.75 and 1.25 g C m^{-2} d^{-1} during the wheat growth seasons, between 2.5 and 3.25 g C m^{-2} d^{-1} during the rice growth seasons, and around 1.75 g C m^{-2} d^{-1} for the annual mean.

16.4.6 POTENTIAL DISCREPANCY BETWEEN GPP$_{EC}$ AND GPP$_{MOD}$

Figure 16.8 has shown the inconsistency between GPP$_{MOD}$ and GPP$_{EC}$ at the three sites, which can be attributed to three aspects: (a) input parameters such as FPAR data and meteorological conditions

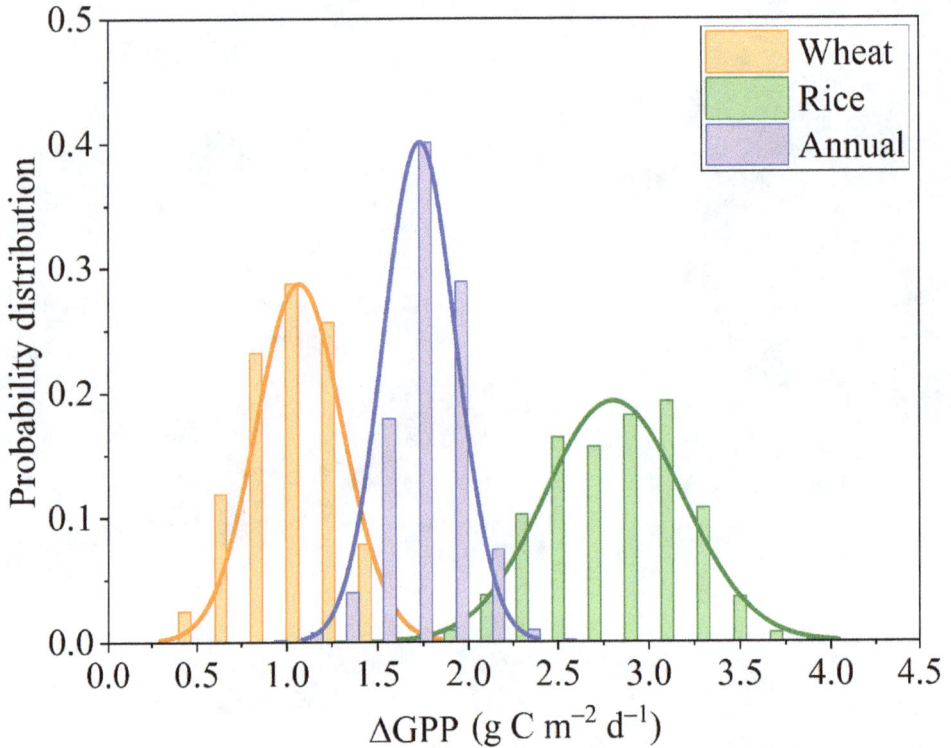

FIGURE 16.17   Probability distribution functions of the error ranges of daily GPP at the grid scale (ΔGPP) in the NYRD during 2014–2018.

(Zhang et al., 2008); (b) the uncertainties in the MOD17A2H GPP algorithm (Wang et al., 2017); and (c) the spatial mismatch between remotely sensed pixels and EC footprints (Wagle et al., 2020; Gelybó et al., 2013). In the past few decades, a wide network of sites has been established across various ecosystems and climate regions; for example, AmeriFlux, Integrated Carbon Observation System, National Ecological Observatory Network, and FluxNet (Franssen et al., 2010; McGloin et al., 2018). These EC sites provide potential opportunities to annually update the cropland sites in the land cover maps, redefine the MOD17 cropland parameters and greatly improve GPP_{MOD} at regional or even global scales. To our best knowledge, China has the largest area of the rice–wheat rotation croplands in the world. In China, they are practiced widely along the Yangtze River Basin (Figure 16.5 inset map), covering around 13 Mha in total (Timsina et al., 2001). This rotation cropland system is a non-negligible part of the agroecosystem. The MOD17 algorithm defines only 11 land cover classes, i.e., one type of cropland, one type of woodland, two types of grasslands, two types of shrubland and five types of forests (Zhang et al., 2008). Therefore, the large discrepancy between GPP_{MOD} and GPP_{EC} over the rice–wheat rotation indicates that the parameters in the MOD17 product should be modified and more types of cropland (e.g., double-cropping or mixed-cropping systems) should be defined.

Meanwhile, MOD17 GPP product has fine resolution with 500 m, so high-quality MOD17 GPP product can be employed to accurately assess the ecosystem's carbon cycle and agricultural productions. Particularly, out-of-academy precision agriculture or commercialized precision agriculture put forward higher requirements for the accuracy evaluation of GPP product. Nowadays, EC-based calibration to MODIS products is a common practice. In the present work, due to limited three EC sites over rotation cropland areas in eastern China, they cannot be used to stand for the whole area. Therefore, we proposed the machine learning-based GPP prediction model for 23 meteorological sites by using multi-source data to derive more virtual EC sites (23 sites) over the whole area, which can offer more ground-based GPP samples for calibrating MOD17 GPP. Generally, the simulations of the RF-based GPP models show a better performance with respect to other machine learning methods (e.g., decision tree regression, support vector machine, artificial neural network, and deep belief network), which is consistent with the results in Yu et al. (2021). In our present work, thus, RF-based upscaling and calibrating methods are more suitable over large-scale agroecosystem areas if EC measurements, meteorological observations and MODIS data were available.

The GPP estimated from EC flux measurements over rice–wheat rotation cropland can represent the amount of carbon uptake by the main land cover type in the NYRD area. To obtain multiple samples for calibration of the MOD17A2H GPP product, an RF model for estimating GPP was designed by integrating multi-source satellite retrievals and *in situ* ground observations during 2014–2018 over the rice–wheat double-cropping fields of eastern China. The RF model showed that multiple co-acting factors (NDVI, LAI, DSR, T_{max}, and FPAR) modulate GPP dynamics. GPP_{RF} performed well when compared with GPP_{EC}, with a R^2 of 0.99 and RMSE of 0.42 g C m^{-2} d^{-1}, indicating these explanatory variables are reasonably representative and reliable for regional GPP upscaling. The regional upscaled cumulative seasonal GPP_{RF} in rice paddies (924 g C m^{-2}) was roughly 2 times higher than that in a wheat field (532 g C m^{-2}) at the station scale, probably because of the much longer growing season and lower LAI of wheat. Compared with GPP_{EC}, this indicates that GPP_{MOD} underestimates GPP during the active crop growth stages but performs well during the crop rotation periods. Based on the upscaled results of GPP_{RF} at the station scale, the functional relationship between GPP_{MOD} and GPP_{RF} at the grid scale was established to calibrate the GPP_{MOD}. The error range of ΔGPP (GPP_{RF} minus GPP_{MOD}) was higher for rice paddies than for wheat fields, i.e., between 0.75 and 1.25 g C m^{-2} d^{-1} during the wheat growth seasons, between 2.5 and 3.25 g C m^{-2} d^{-1} during the rice growth seasons, along with an annual mean of 1.75–2 g C m^{-2} d^{-1}.

To sum up, the GPP in rice–wheat rotation agroecosystem is considerably diverse and varies with the seasons. Our findings are potentially applicable in terms of the climate response of greenhouse gases over wide-scale cropland areas. Our research demonstrates that RF machine learning is a powerful and expedient modeling tool for estimating and even calibrating the MODIS GPP product. In future, it would be worthwhile using global FLUXNET data, multi-source satellite observations and machine learning methods to simulate the GPP in more ecosystem types (e.g., grassland and forests) and climate zones at large scales to fully understand the nature of global carbon dynamics.

16.5   CONCLUSIONS

In general, artificial intelligence (AI) techniques could bring about great opportunities for the development of weather forecasting and meteorological big data. Based on AI techniques, a weather forecasting model driven by meteorological big data could be established which will have great potential to realize model optimization. Furthermore, beneficial understandings and accurate predictions could be achieved based on AI techniques even if we lack in sufficient understanding and prior knowledge of the physical mechanism of weather and climate system. Therefore, the problem of "huge cost" in numerical weather forecasting models could be solved theoretically by using AI techniques; meanwhile the accuracy of model prediction could also be improved. Currently, the application of AI techniques, particularly deep learning technique, in the field of weather forecasting is still at an initial stage. Hence, challenges will exist when applying AI techniques into the intelligent grid forecasting business system, including algorithm selection, data foundation, multi-source data fusion, model interpretability, credibility, usability and engineering. For now, hybrid forecasting models based on the fusion of numerical models (physical simulations) and machine learning (data-driven) are still mainstream. However, the development of multi-source data fusion methods based on machine learning and the creation of high-resolution meteorological data sets will also have great advantages, including quality control and homogenization; data imputation or interpolation; data fusion; inversion; construction of reanalysis data sets.

Despite the integration of weather forecasting and meteorological big data, AI techniques have also been widely and deeply applied in the other fields of atmospheric science. For example, AI techniques have been applied into short-term climate predictions, such as seasonal forecasting, etc., based on machine learning in terms of climate prediction. Besides, they have also been applied in physical process diagnosis and prediction methods of multi-scale climate events (e.g., monsoon, El Niño, North Pacific Oscillation, etc.), fusion technology and data assimilation by utilizing numerical models (physical simulation) and machine learning (data-driven) in the field of assimilation. In terms of mode simulation, machine learning is used to accelerate some components of numerical models, optimization of physical process parameterization, etc. Accounting for weather-climate change impacts and auxiliary decision-making models, the advantages of machine learning are to characterize the impacts of weather and climate change on nature, society, economy, etc., as well as to build impact assessment and auxiliary decision-making models in order to provide valuable decision-making support for national and professional meteorological users.

16.5.1 Acknowledgments

This study was supported by the Natural Science Foundation of China (42222503).

REFERENCES

Alberto, M.C.R.; Wassmann, R.; Hirano, T.; Miyata, A.; Kumar, A.; Padre, A.; Amante, M. CO2/heat fluxes in rice fields: Comparative assessment of flooded and non-flooded fields in the Philippines. Agricultural and Forest Meteorology 2009, 149, 1737–1750, https://doi.org/10.1016/j.agrformet.2009.06.003.

Baareh, A.K.; Elsayad, A.; Al-Dhaifallah, M. Recognition of splice-junction genetic sequences using random forest and Bayesian optimization. Multimedia Tools and Applications 2021, https://doi.org/10.1007/s11042-021-10944-7.

Baldocchi, D.D. Assessing the eddy covariance technique for evaluating carbon dioxide exchange rates of ecosystems: past, present and future. Global Change Biology 2003, 9, 479–492, https://doi.org/10.1046/j.1365-2486.2003.00629.x.

Belgiu, M.; Dragut, L. Random forest in remote sensing: A review of applications and future directions. Isprs Journal of Photogrammetry and Remote Sensing. 2016, 114, 24–31, https://doi.org/10.1016/j.isprsjprs.2016.01.011.

Breiman, L. Random forests. Machine Learning 2001, 45, 5–32, https://doi.org/10.1023/A:1010933404324.

Cai, J.C.; Xu, K.; Zhu, Y.H.; Hu, F.; Li, L.H. Prediction and analysis of net ecosystem carbon exchange based on gradient boosting regression and random forest. Applied Energy 2020, 262, https://doi.org/10.1016/j.apenergy.2020.114566.

Chen, C.; Li, D.; Gao, Z.; Tang, J.; Guo, X.; Wang, L.; Wan, B. Seasonal and interannual variations of carbon exchange over a rice-wheat rotation system on the North China Plain. Advances in Atmospheric Sciences. 2015, 32, 1365–1380, https://doi.org/10.1007/s00376-015-4253-1.

Chen, Y.; Shen, W.; Gao, S.; Zhang, K.; Wang, J.; Huang, N. Estimating deciduous broadleaf forest gross primary productivity by remote sensing data using a random forest regression model. Journal of Applied Remote Sensing 2019, 13, https://doi.org/10.1117/1.jrs.13.038502.

Cutler, D.R.; Edwards Jr., T.C.; Beard, K.H.; Cutler, A.; Hess, K.T.; Gibson, J.; Lawler, J.J. Random Forests for classification in ecology. Ecology 2007, 88, 2783–2792, https://doi.org/10.1890/07-0539.1.

Didan, K. MOD13Q1 MODIS/Terra Vegetation indices 16-day L3 global 250m SIN Grid V006. distributed by NASA EOSDIS Land Processes DAAC. 2015, https://doi.org/10.5067/MODIS/MOD13Q1.006.

Dou, X.; Yang, Y.; Luo, J. Estimating forest carbon fluxes using machine learning techniques based on eddy covariance measurements. Sustainability 2018, 10, https://doi.org/10.3390/su10010203.

Duan, Z.; Grimmond, C.; Gao, C.Y.; Sun, T.; Liu, C.; Wang, L.; Li, Y.; Gao, Z. Seasonal and interannual variations in the surface energy fluxes of a rice–wheat rotation in Eastern China. Journal of Applied Meteorology and Climatology 2021, 60, 877–891, https://doi.org/10.1175/JAMC-D-20-0233.1.

Duan, Z.; Yang, Y.; Wang, L.; Liu, C.; Fan, S.; Chen, C.; Tong, Y.; Lin, X.; Gao, Z. Temporal characteristics of carbon dioxide and ozone over a rural-cropland area in the Yangtze River Delta of eastern China. Science of the Total Environment 2021a, 757, https://doi.org/10.1016/j.scitotenv.2020.143750.

Duan, Z.; Yang, Y; Zhou, S.; Gao, Z.; Zong, L.; Fan, S.; Yin, J. Estimating Gross Primary Productivity (GPP) over Rice–Wheat-Rotation Croplands by Using the Random Forest Model and Eddy Covariance Measurements: Upscaling and Comparison with the MODIS Product. Remote Sensing 2021b, 13, https://doi.org/10.3390/rs13214229

Fang, H.; Zhang, Y.; Wei, S.; Li, W.; Ye, Y.; Sun, T.; Liu, W. Validation of global moderate resolution leaf area index (LAI) products over croplands in northeastern China. Remote Sensing of Environment 2019, 233, 111377, https://doi.org/10.1016/j.rse.2019.111377.

Franssen, H.J.H.; Stöckli, R.; Lehner, I.; Rotenberg, E.; Seneviratne, S.I. Energy balance closure of eddy-covariance data: A multisite analysis for European FLUXNET stations. Agricultural and Forest Meteorology 2010, 150, 1553–1567, https://doi.org/10.1016/j.agrformet.2010.08.005.

Frazier, P.I. A tutorial on Bayesian optimization. 2018.

Friedl, M., D. Sulla-Menashe. MCD12Q1 MODIS/Terra+Aqua Land Cover Type Yearly L3 Global 500m SIN Grid V006. distributed by NASA EOSDIS Land Processes DAAC. 2019, https://doi.org/10.5067/MODIS/MCD12Q1.006.

Ge, H.; Zhang, H.; Zhang, H.; Cai, X.; Song, Y.; Kang, L. The characteristics of methane flux from an irrigated rice farm in East China measured using using the eddy covariance method. Agricultural and Forest Meteorology 2018, 249, 228–238, https://doi.org/10.1016/j.agrformet.2017.11.010.

Geer, A. J. Learning earth system models from observations: Machine learning or data assimilation?. Philosophical Transactions of the Royal Society A 2021, 379(2194), 20200089

Gelybó, G.; Barcza, Z.; Kern, A.; Kljun, N. Effect of spatial heterogeneity on the validation of remote sensing based GPP estimations. Agricultural and Forest Meteorology 2013, 174–175, 43–53, https://doi.org/10.1016/j.agrformet.2013.02.003.

Haupt, S. E; Chapman, W; Adams S V; et al. 2021. Towards implementing artificial intelligence post-processing in weather and climate: Proposed actions from the Oxford 2019 workshop. Philosophical Transactions of the Royal Society A, 379(2194): 20200091

John, R.; Chen, J.; Noormets, A.; Xiao, X.; Xu, J.; Lu, N.; Chen, S. Modelling gross primary production in semi-arid Inner Mongolia using MODIS imagery and eddy covariance data. International Journal of Remote Sensing 2013, 34, 2829–2857, https://doi.org/10.1080/01431161.2012.746483.

Jung, M.; Reichstein, M.; Margolis, H.A.; Cescatti, A.; Richardson, A.D.; Arain, M.A.; Arneth, A.; Bernhofer, C.; Bonal, D.; Chen, J.; et al. Global patterns of land-atmosphere fluxes of carbon dioxide, latent heat, and sensible heat derived from eddy covariance, satellite, and meteorological observations. Journal of Geophysical Research: Biogeosciences 2011, 116, https://doi.org/10.1029/2010JG001566.

Kashinath K; Mustafa M; Albert A; et al. 2021. Physics-informed machine learning: Case studies for weather and climate modelling. Philosophical Transactions of the Royal Society A, 379(2194): 20200093

Lasslop G; Reichstein M; Papale D; Richardson AD; Arneth A. Barr A; Stoy P; Wohlfahrt G. Separation of net ecosystem exchange into assimilation and respiration using a light response curve approach: critical issues and global evaluation. Global Change Biology 2010, 16, 187–208, https://doi.org/10.1111/j.1365-2486.2009.02041.x.

Lee, B.; Kim, N.; Kim, E.-S.; Jang, K.; Kang, M.; Lim, J.-H.; Cho, J.; Lee, Y. An artificial intelligence approach to predict gross primary productivity in the forests of South Korea using satellite remote sensing data. Forests 2020, 11, 1000.

Lee, X.; Massman, W.J.; Law, B.E. Handbook of Micrometeorology: A Guide for Surface Flux Measurement and Analysis; Springer: 2005.

Li, X., Yang, Y., Mi, J., Bi, X., Zhao, Y., Huang, Z., Liu, C., Zong, L., and Li, W. Leveraging machine learning for quantitative precipitation estimation from Fengyun-4 geostationary observations and ground meteorological measurements, Atmospheric Measurement Techniques 2021, 14, 7007–7023, https://doi.org/10.5194/amt-14-7007-2021.

LI-COR, Inc. EddyPro® version 5.2.1 Help and User's Guide. LI-COR, Inc.: Lincoln, NE, 2015.

Liu, J.; Zuo, Y.; Wang, N.; Yuan, F.; Zhu, X.; Zhang, L.; Zhang, J.; Sun, Y.; Guo, Z.; Guo, Y.; et al. Comparative analysis of two machine learning algorithms in predicting site-level net ecosystem exchange in major biomes. Remote Sensing 2021, 13, 2242.

Malmström, C.M.; Thompson, M.V.; Juday, G.P.; Los, S.O.; Randerson, J.T.; Field, C.B. Interannual variation in global-scale net primary production: Testing model estimates. Global Biogeochemical Cycles 1997, 11, 367–392, https://doi.org/10.1029/97GB01419.

McGloin, R.; Šigut, L.; Havránková, K.; Dušek, J.; Pavelka, M.; Sedlák, P. Energy balance closure at a variety of ecosystems in Central Europe with contrasting topographies. Agricultural and Forest Meteorology 2018, 248, 418–431, https://doi.org/10.1016/j.agrformet.2017.10.003.

Moncrieff, J.; Clement, R.; Finnigan, J.; Meyers, T. Averaging, Detrending, and Filtering of Eddy Covariance Time Series. In Handbook of Micrometeorology: A Guide for Surface Flux Measurement and Analysis, Lee, X., Massman, W., Law, B., Eds.; Springer Netherlands: Dordrecht, 2005; pp. 7–31.

Myneni, R.; Y. Knyazikhin; T. Park. MOD15A2H MODIS/Terra Leaf Area Index/FPAR 8-Day L4 Global 500m SIN Grid V006. distributed by NASA EOSDIS Land Processes DAAC. 2015, https://doi.org/10.5067/MODIS/MOD15A2H.006.

Papale, D.; Reichstein, M.; Aubinet, M.; Canfora, E.; Bernhofer, C.; Kutsch, W.; Longdoz, B.; Rambal, S.; Valentini, R.; Vesala, T.; et al. Towards a standardized processing of Net Ecosystem Exchange measured with eddy covariance technique: algorithms and uncertainty estimation. Biogeosciences 2006, 3, 571–583, https://doi.org/10.5194/bg-3-571-2006.

Post, H.; Hendricks Franssen, H.J.; Han, X.; Baatz, R.; Montzka, C.; Schmidt, M.; Vereecken, H. Evaluation and uncertainty analysis of regional-scale CLM4.5 net carbon flux estimates. Biogeosciences 2018, 15, 187–208, https://doi.org/10.5194/bg-15-187-2018.

Rahman, A.F.; Sims, D.A.; Cordova, V.D.; El-Masri, B.Z. Potential of MODIS EVI and surface temperature for directly estimating per-pixel ecosystem C fluxes. Geophysical Research Letters 2005, 32, https://doi.org/10.1029/2005GL024127.

Reeves, M.C.; Zhao, M.; Running, S.W. Usefulness and limits of MODIS GPP for estimating wheat yield. International Journal of Remote Sensing 2005, 26, 1403–1421, https://doi.org/10.1080/01431160512331326567.

Reichstein, M.; Falge, E.; Baldocchi, D.; Papale, D.; Aubinet, M.; Berbigier, P.; Bernhofer, C.; Buchmann, N.; Gilmanov, T.; Granier, A.; et al. On the separation of net ecosystem exchange into assimilation and ecosystem respiration: review and improved algorithm. Global Change Biology 2005, 11, 1424–1439, https://doi.org/10.1111/j.1365-2486.2005.001002.x.

Reitz, O.; Graf, A.; Schmidt, M.; Ketzler, G.; Leuchner, M. Upscaling net ecosystem exchange over heterogeneous landscapes with machine learning. Journal of Geophysical Research: Biogeosciences 2021, 126, e2020JG005814, https://doi.org/10.1029/2020JG005814.

Running, S.; Q. Mu; M. Zhao. MOD17A2H MODIS/Terra Gross Primary Productivity 8-Day L4 Global 500m SIN Grid V006. distributed by NASA EOSDIS Land Processes DAAC. 2015, https://doi.org/10.5067/MODIS/MOD17A2H.006.

Shangguan, W.; Dai, Y.; Liu, B.; Zhu, A.; Duan, Q.; Wu, L.; Ji, D.; Ye, A.; Yuan, H.; Zhang, Q.; et al. A China data set of soil properties for land surface modeling. Journal of Advances in Modeling Earth Systems 2013, 5, 212–224, https://doi.org/10.1002/jame.20026.

Timsina, J.; Connor, D.J. Productivity and management of rice–wheat cropping systems: issues and challenges. Field Crops Research 2001, 69, 93–132, https://doi.org/10.1016/S0378-4290(00)00143-X.

Tramontana, G.; Ichii, K.; Camps-Valls, G.; Tomelleri, E.; Papale, D. Uncertainty analysis of gross primary production upscaling using Random Forests, remote sensing and eddy covariance data. Remote Sensing of Environment 2015, 168, 360–373, https://doi.org/10.1016/j.rse.2015.07.015.

Tramontana, G.; Jung, M.; Schwalm, C.R.; Ichii, K.; Camps-Valls, G.; Raduly, B.; Reichstein, M.; Arain, M.A.; Cescatti, A.; Kiely, G.; et al. Predicting carbon dioxide and energy fluxes across global FLUXNET sites with regression algorithms. Biogeosciences 2016, 13, 4291–4313, https://doi.org/10.5194/bg-13-4291-2016.

Tramontana, G.; Migliavacca, M.; Jung, M.; Reichstein, M.; Keenan, T.F.; Camps-Valls, G.; Ogee, J.; Verrelst, J.; Papale, D. Partitioning net carbon dioxide fluxes into photosynthesis and respiration using neural networks. Global Change Biology 2020, 26, 5235–5253, https://doi.org/10.1111/gcb.15203.

Ueyama, M.; Ichii, K.; Iwata, H.; Euskirchen, E.S.; Zona, D.; Rocha, A.V.; Harazono, Y.; Iwama, C.; Nakai, T.; Oechel, W.C. Upscaling terrestrial carbon dioxide fluxes in Alaska with satellite remote sensing and support vector regression. Journal of Geophysical Research: Biogeosciences 2013, 118, 1266–1281, https://doi.org/10.1002/jgrg.20095.

Vannitsem, S.; Bremnes, J. B.; Demaeyer. J.; et al. 2021. Statistical postprocessing for weather forecasts: Review, challenges, and avenues in a big data world. Bulletin of the American Meteorological Society 102(3): E681–E699. https://doi.org/10.1175/BAMS-D-19-0308.1

Wagle, P.; Gowda, P.H.; Neel, J.P.S.; Northup, B.K.; Zhou, Y. Integrating eddy fluxes and remote sensing products in a rotational grazing native tallgrass prairie pasture. Science of The Total Environment 2020, 712, 136407, https://doi.org/10.1016/j.scitotenv.2019.136407.

Wagle, P.; Gowda, P.H.; Northup, B.K.; Neel, J.P.S.; Starks, P.J.; Turner, K.E.; Moriasi, D.N.; Xiao, X.; Steiner, J.L. Carbon dioxide and water vapor fluxes of multi-purpose winter wheat production systems in the U.S. Southern Great Plains. Agricultural and Forest Meteorology 2021, 310, 108631, https://doi.org/10.1016/j.agrformet.2021.108631.

Wang, J.W.; Denning, A.S.; Lu, L.X.; Baker, I.T.; Corbin, K.D.; Davis, K.J. Observations and simulations of synoptic, regional, and local variations in atmospheric CO2. Journal of Geophysical Research-Atmospheres 2007, 112, https://doi.org/10.1029/2006jd007410.

Wang, H.; Li, J.; Gao, M.; Chan, T.-C.; Gao, Z.; Zhang, M.; Li, Y.; Gu, Y.; Chen, A.; Ho, H. C.; Yang, Y. Spatiotemporal variability in long-term population exposure and lung cancer mortality attributable to PM2.5 across the Yangtze River Delta (YRD) Region over 2010–2016: A multistage approach, Chemosphere, 2020, https://doi.org/10.1016/j.chemosphere.2020.127153

Wang, H.; Li, J.; Gao, Z.; Yim, S. H.; Shen, H.; Ho, H. C.; Li, Z.; Zeng, Z.; Liu, C.; Li, Y.; Ning, G.; Yang, Y. High-spatial-resolution population exposure to PM2.5 pollution based on multi-satellite retrievals: A case study of seasonal variation in the Yangtze River Delta, China in 2013. Remote Sensing, 2019, 11, https://doi.org/10.3390/rs11232724

Wang, L.; Zhu, H.; Lin, A.; Zou, L.; Qin, W.; Du, Q. Evaluation of the Latest MODIS GPP Products across Multiple Biomes Using Global Eddy Covariance Flux Data. Remote Sens. 2017, 9, https://doi.org/10.3390/rs9050418.

Willmott, C.J. Some comments on the evaluation of model performance. Bulletin of the American Meteorological Society 1982, 63, 1309–1313, https://doi.org/10.1175/1520-0477(1982)063 < 1309:scoteo>2.0.co;2.

Wutzler, T.; Lucas-Moffat, A.; Migliavacca, M.; Knauer, J.; Sickel, K.; Sigut, L.; Menzer, O.; Reichstein, M. Basic and extensible post-processing of eddy covariance flux data with REddyProc. Biogeosciences 2018, 15, 5015–5030, https://doi.org/10.5194/bg-15-5015-2018.

Xiao, J.; Zhuang, Q.; Baldocchi, D.D.; Law, B.E.; Richardson, A.D.; Chen, J.; Oren, R.; Starr, G.; Noormets, A.; Ma, S.; et al. Estimation of net ecosystem carbon exchange for the conterminous United States by combining MODIS and AmeriFlux data. Agricultural and Forest Meteorology 2008, 148, 1827–1847, https://doi.org/10.1016/j.agrformet.2008.06.015.

Xie, X.Y.; Li, A.N.; Tan, J.B.; Jin, H.A.; Nan, X.; Zhang, Z.J.; Bian, J.H.; Lei, G.B. Assessments of gross primary productivity estimations with satellite data-driven models using eddy covariance observation sites over the northern hemisphere. Agricultural and Forest Meteorology 2020, 280, 14, https://doi.org/10.1016/j.agrformet.2019.107771.

Yang, D.; Xu, X.; Xiao, F.; Xu, C.; Luo, W.; Tao, L. Improving modeling of ecosystem gross primary productivity through re-optimizing temperature restrictions on photosynthesis. Science of The Total Environment 2021, 788, 147805, https://doi.org/10.1016/j.scitotenv.2021.147805.

Yang, X.; Kan, D.; Yuejian, Z. 2022. Progress and challenges of deep learning techniques in intelligent grid weather forecasting. Acta Meteorologica Sinica, 80(5): 1–19 https://doi.org/10.11676/qxxb2022.051

Yu, T.; Zhang, Q.; Sun, R. Comparison of machine learning methods to up-scale gross primary production. Remote Sensing 2021, 13, https://doi.org/10.3390/rs13132448.

Zeng, J.; Matsunaga, T.; Tan, Z.-H.; Saigusa, N.; Shirai, T.; Tang, Y.; Peng, S.; Fukuda, Y. Global terrestrial carbon fluxes of 1999–2019 estimated by upscaling eddy covariance data with a random forest. Scientific Data 2020, 7: 313, https://doi.org/10.1038/s41597-020-00653-5.

Zhang, Y.; Mingxuan, C.; Lei, H.; Linye, S.; Lu, Y. Multi-element deep learning fusion correction method for numerical weather prediction. Acta Meteorologica Sinica, 2022, 80(1): 153–167 https://doi.org/10.11676/qxxb2021.066

Zhang, Y.; Yu, Q.; Jiang, J.; Tang, Y. Calibration of Terra/MODIS gross primary production over an irrigated cropland on the North China Plain and an alpine meadow on the Tibetan Plateau. Global Change Biology 2008, 14: 757–767, https://doi.org/10.1111/j.1365-2486.2008.01538.x.

Zhu, X.Y.; Pei, Y.Y.; Zheng, Z.P.; Dong, J.W.; Zhang, Y.; Wang, J.B.; Chen, L.J.; Doughty, R.B.; Zhang, G.L.; Xiao, X.M. Underestimates of grassland gross primary production in MODIS standard products. Remote Sensing 2018, 10: 16, https://doi.org/10.3390/rs10111771.**Cloud Phase**

Index

Note: Page numbers in *italic* indicate a figure and page numbers in **bold** indicate a table on the corresponding page.

For Product Safety Concerns and Information please contact our EU
representative GPSR@taylorandfrancis.com
Taylor & Francis Verlag GmbH, Kaufingerstraße 24, 80331 München, Germany